Methods in Enzymology

Volume 111
STEROIDS AND ISOPRENOIDS
Part B

METHODS IN ENZYMOLOGY

EDITORS-IN-CHIEF

Sidney P. Colowick   Nathan O. Kaplan

Methods in Enzymology

Volume 111

Steroids and Isoprenoids

Part B

EDITED BY

John H. Law

DEPARTMENT OF BIOCHEMISTRY
UNIVERSITY OF ARIZONA
TUCSON, ARIZONA

Hans C. Rilling

DEPARTMENT OF BIOCHEMISTRY
UNIVERSITY OF UTAH SCHOOL OF MEDICINE
SALT LAKE CITY, UTAH

1985

ACADEMIC PRESS, INC.

(Harcourt Brace Jovanovich, Publishers)

Orlando   San Diego   New York   London
Toronto   Montreal   Sydney   Tokyo

COPYRIGHT © 1985, BY ACADEMIC PRESS, INC.
ALL RIGHTS RESERVED.
NO PART OF THIS PUBLICATION MAY BE REPRODUCED OR
TRANSMITTED IN ANY FORM OR BY ANY MEANS, ELECTRONIC
OR MECHANICAL, INCLUDING PHOTOCOPY, RECORDING, OR
ANY INFORMATION STORAGE AND RETRIEVAL SYSTEM, WITHOUT
PERMISSION IN WRITING FROM THE PUBLISHER.

ACADEMIC PRESS, INC.
Orlando, Florida 32887

United Kingdom Edition published by
ACADEMIC PRESS INC. (LONDON) LTD.
24–28 Oval Road, London NW1 7DX

LIBRARY OF CONGRESS CATALOG CARD NUMBER: 54-9110
ISBN 0-12-182011-4

PRINTED IN THE UNITED STATES OF AMERICA

85 86 87 88   9 8 7 6 5 4 3 2 1

Table of Contents

CONTRIBUTORS TO VOLUME 111 ix
PREFACE xiii
VOLUMES IN SERIES xv

Section I. Methodology

1. A Comparison of Methods for the Identification of Sterols	WILLIAM R. NES	3
2. High-Performance Liquid Chromatography of Sterols: Yeast Sterols	RUSSELL J. RODRIGUEZ AND LEO W. PARKS	37
3. High-Performance Liquid Chromatography of Bile Acids	WILLIAM H. ELLIOTT AND ROGER SHAW	51
4. Mass Spectrometry of Bile Acids	JAN SJÖVALL, ALEXANDER M. LAWSON, AND KENNETH D. R. SETCHELL	63
5. General Carotenoid Methods	GEORGE BRITTON	113
6. Gas Chromatography of Isoprenoids	WALTER JENNINGS AND GARY TAKEOKA	149
7. High-Performance Liquid Chromatography of Carotenoids	MANFRED RUDDAT AND OSCAR H. WILL III	189
8. Isolation and Assay of Dolichol and Dolichyl Phosphate	W. LEE ADAIR AND R. KENNEDY KELLER	201
9. Invertebrate Carotenoproteins	P. F. ZAGALSKY	216
10. Separation of Mevalonate Phosphates and Isopentenyl Pyrophosphate by Thin-Layer Chromatography and of Short-Chain Prenyl Phosphates by Ion-Pair Chromatography on a High-Performance Liquid Chromatography Column	PETER BEYER, KLAUS KREUZ, AND H. KLEINIG	248
11. Reversed-Phase High-Performance Liquid Chromatography of C_5 to C_{20} Isoprenoid Benzoates and Naphthoates	TSAN-HSI YANG, T. MARK ZABRISKIE, AND C. DALE POULTER	252

Section II. Sterol Metabolism

12. Lecithin–Cholesterol Acyltransferase and Cholesterol Transport	CHRISTOPHER J. FIELDING	267

13. Assay of Cholesteryl Ester Transfer Activity and Purification of a Cholesteryl Ester Transfer Protein	YASUSHI OGAWA AND CHRISTOPHER J. FIELDING	274
14. Cholesterol Acyltransferase	JEFFREY T. BILLHEIMER	286
15. Sterol Carrier Protein	MARY E. DEMPSEY	293
16. Epoxide Hydrolases in the Catabolism of Sterols and Isoprenoids	BRUCE D. HAMMOCK, DAVID E. MOODY, AND ALEX SEVANIAN	303
17. Biosynthesis and Interconversion of Sterols in Plants and Marine Invertebrates	L. J. GOAD	311
18. Yeast Sterols: Yeast Mutants as Tools for the Study of Sterol Metabolism	LEO W. PARKS, CYNTHIA D. K. BOTTEMA, RUSSELL J. RODRIGUEZ, AND THOMAS A. LEWIS	333
19. Enzymatic Dealkylation of Phytosterols in Insects	YOSHINORI FUJIMOTO, MASUO MORISAKI, AND NOBUO IKEKAWA	346
20. Side-Chain Cleavage of Cholesterol by Gas Chromatography–Mass Spectrometry in a Selected Ion Monitoring Mode	MASUO MORISAKI, MIKIO SHIKITA, AND NOBUO IKEKAWA	352
21. Purification from Rabbit and Rat Liver of Cytochromes P-450 Involved in Bile Acid Biosynthesis	STEFAN ANDERSSON, HANS BOSTRÖM, HENRY DANIELSSON, AND KJELL WIKVALL	364
22. Biosynthesis and Metabolism of Ecdysteroids and Methods of Isolation and Identification of the Free and Conjugated Compounds	H. H. REES AND R. E. ISAAC	377
23. Ecdysone Conjugates: Isolation and Identification	C. HETRU, B. LUU, AND J. A. HOFFMANN	411
24. Ecdysone Oxidase	J. KOOLMAN	419
25. Ecdysteroid Carrier Proteins	J. KOOLMAN	429
26. Ecdysone 3-Epimerase	MALCOLM J. THOMPSON, GUNTER F. WEIRICH, AND JAMES A. SVOBODA	437
27. 20-Hydroxyecdysone Binding Protein from Locust Hemolymph	R. FEYEREISEN	442
28. Ecdysone 20-Monooxygenase	GUNTER F. WEIRICH	454
29. Measurement and Characterization of Ecdysteroid Receptors	BECKY A. SAGE AND JOHN D. O'CONNOR	458

Section III. Metabolism of Other Isoprenoids

30. Dolichol Kinase, Phosphatase, and Esterase Activity in Calf Brain	Carlota Sumbilla and Charles J. Waechter	471
31. Purification and Properties of the Juvenile Hormone Carrier Protein from the Hemolymph of *Manduca sexta*	Ronald C. Peterson	482
32. Analysis of Juvenile Hormone Esterase Activity	Bruce D. Hammock and Richard M. Roe	487
33. Cellular Juvenile Hormone Binding Proteins	Ernest S. Chang	494
34. Experimental Techniques for Photoaffinity Labeling of Juvenile Hormone Binding Proteins of Insects with Epoxyfarnesyl Diazoacetate	Glenn D. Prestwich, John K. Koeppe, Gae E. Kovalick, John J. Brown, Ernest S. Chang, and Ambarish K. Singh	509
35. Radiochemical Assay for Juvenile Hormone III Biosynthesis *in Vitro*	R. Feyereisen	530
36. Epoxyfarnesoic Acid Methyltransferase	Gunter F. Weirich	540
37. S-Adenosylmethionine : γ-Tocopherol Methyltransferase (*Capsicum* Chromoplasts)	Bilal Camara	544
38. Dolichyl Pyrophosphate Phosphatase in Brain	Malka G. Scher and Charles J. Waechter	547

Author Index 555
Subject Index 581

Contributors to Volume 111

Article numbers are in parentheses following the names of contributors.
Affiliations listed are current.

W. LEE ADAIR (8), *Department of Biochemistry, University of South Florida College of Medicine, Tampa, Florida 33612*

STEFAN ANDERSSON (21), *Department of Pharmaceutical Biochemistry, University of Uppsala, S-751 23 Uppsala, Sweden*

PETER BEYER (10), *Institut für Biologie II, Universität Freiburg, D-7800 Freiburg, Federal Republic of Germany*

JEFFREY T. BILLHEIMER (14), *Department of Biological Sciences, Drexel University, Philadelphia, Pennsylvania 19104*

HANS BOSTRÖM (21), *Department of Pharmaceutical Biochemistry, University of Uppsala, S-751 23 Uppsala, Sweden*

CYNTHIA D. K. BOTTEMA (18), *Endocrinology Division, Stanford University School of Medicine, Stanford, California 94305*

GEORGE BRITTON (5), *Department of Biochemistry, University of Liverpool, Liverpool L69 3BX, England*

JOHN J. BROWN (34), *Department of Entomology, Washington State University, Pullman, Washington 99164*

BILAL CAMARA (37), *Laboratoire de Régulations Métaboliques et Différenciation des Plastes, Université Pierre et Marie Curie, 75230 Paris Cedex 05, France*

ERNEST S. CHANG (33, 34), *Bodega Marine Laboratory, Bodega Bay, California 94923, and Department of Animal Science, University of California, Davis, Davis, California 95616*

HENRY DANIELSSON (21), *Department of Pharmaceutical Biochemistry, University of Uppsala, S-751 23 Uppsala, Sweden*

MARY E. DEMPSEY (15), *Department of Biochemistry, University of Minnesota Medical School, Minneapolis, Minnesota 55455*

WILLIAM H. ELLIOTT (3), *Edward A. Doisy Department of Biochemistry, St. Louis University School of Medicine, St. Louis, Missouri 63104*

R. FEYEREISEN (27, 35), *Departments of Entomology and Agricultural Chemistry, Oregon State University, Corvallis, Oregon 97331*

CHRISTOPHER J. FIELDING (12, 13), *Cardiovascular Research Institute, and Department of Physiology, University of California Medical Center, San Francisco, California 94143*

YOSHINORI FUJIMOTO (19), *Department of Chemistry, Tokyo Institute of Technology, Tokyo 152, Japan*

L. J. GOAD (17), *Department of Biochemistry, University of Liverpool, Liverpool L69 3BX, England*

BRUCE D. HAMMOCK (16, 32), *Departments of Entomology and Environmental Toxicology, University of California, Davis, Davis, California 95616*

C. HETRU (23), *Laboratoire de Biologie Générale (associé au Centre National de la Recherche Scientifique), Université Louis Pasteur, 67000 Strasbourg, France*

J. A. HOFFMANN (23), *Laboratoire de Biologie Générale (associé au Centre National de la Recherche Scientifique), Université Louis Pasteur, 67000 Strasbourg, France*

NOBUO IKEKAWA (19, 20), *Department of Chemistry, Tokyo Institute of Technology, Tokyo 152, Japan*

R. E. ISAAC (22), *Department of Biochemistry, University of Liverpool, Liverpool L69 3BX, England*

WALTER JENNINGS (6), *Department of Food Science and Technology, University of California, Davis, Davis, California 95616*

R. KENNEDY KELLER (8), *Department of Biochemistry, University of South Florida College of Medicine, Tampa, Florida 33612*

H. KLEINIG (10), *Institut für Biologie II, Universität Freiburg, D-7800 Freiburg, Federal Republic of Germany*

JOHN K. KOEPPE (34), *Department of Biology, University of North Carolina, Chapel Hill, North Carolina 27514*

J. KOOLMAN (24, 25), *Physiologisch-Chemisches Institut, Philipps-Universität Marburg, D-3550 Marburg, Federal Republic of Germany*

GAE E. KOVALICK (34), *Department of Biology, University of North Carolina, Chapel Hill, North Carolina 27514*

KLAUS KREUZ (10), *Botanisches Institut, Universität Kiel, D-2300 Kiel, Federal Republic of Germany*

ALEXANDER M. LAWSON (4), *Clinical Mass Spectrometry Section, Clinical Research Centre, Harrow HA1 3UJ, Middlesex, England*

THOMAS A. LEWIS (18), *Department of Microbiology, Oregon State University, Corvallis, Oregon 97331*

B. LUU (23), *Laboratoire de Chimie Organique des Substances Naturelles (associé au Centre National de la Recherche Scientifique), Université Louis Pasteur, 67000 Strasbourg, France*

DAVID E. MOODY (16), *Departments of Entomology and Environmental Toxicology, University of California, Davis, Davis, California 95616*

MASUO MORISAKI (19, 20), *Kyoritsu College of Pharmacy, Tokyo 105, Japan*

WILLIAM R. NES (1), *Department of Biological Sciences, Drexel University, Philadelphia, Pennsylvania 19104*

JOHN D. O'CONNOR (29), *Department of Biology, University of California, Los Angeles, Los Angeles, California 90024*

YASUSHI OGAWA (13), *Collagen Corporation, 2500 Faber Place, Palo Alto, California 94303*

LEO W. PARKS (2, 18), *Department of Microbiology, Oregon State University, Corvallis, Oregon 97331*

RONALD C. PETERSON (31), *Department of Biochemistry, Uniformed Services University of the Health Sciences, Bethesda, Maryland 20814*

C. DALE POULTER (11), *Department of Chemistry, University of Utah, Salt Lake City, Utah 84112*

GLENN D. PRESTWICH (34), *Department of Chemistry, State University of New York, Stony Brook, New York 11794*

H. H. REES (22), *Department of Biochemistry, University of Liverpool, Liverpool L69 3BX, England*

RUSSELL J. RODRIGUEZ (2, 18), *Department of Microbiology, Oregon State University, Corvallis, Oregon 97331*

RICHARD M. ROE (32), *Department of Entomology, North Carolina State University, Raleigh, North Carolina 27695*

MANFRED RUDDAT (7), *Department of Molecular Genetics and Cell Biology, and Department of Biology, University of Chicago, Chicago, Illinois 60637*

BECKY A. SAGE (29), *Department of Biology, University of California, Los Angeles, Los Angeles, California 90024*

MALKA G. SCHER (38), *Department of Biological Chemistry, University of Maryland School of Medicine, Baltimore, Maryland 21201*

KENNETH D. R. SETCHELL (4), *Department of Pediatrics, Childrens Hospital Medical Center, Cincinnati, Ohio 45229*

ALEX SEVANIAN (16), *Institute for Toxicology, School of Pharmacy, and Department of Pathology, University of Southern California School of Medicine, Los Angeles, California 90033*

ROGER SHAW (3), *LBI-Basic Research Program, NCI-Frederick Cancer Research Facility, Frederick, Maryland 21701*

MIKIO SHIKITA (20), *National Institute of Radiological Sciences, Anagawa, Chiba 260, Japan*

AMBARISH K. SINGH (34), *Department of Chemistry, State University of New York, Stony Brook, New York 11794*

JAN SJÖVALL (4), *Department of Physiological Chemistry, Karolinska Institutet, S-104 01 Stockholm, Sweden*

CARLOTA SUMBILLA (30), *Department of Biological Chemistry, University of Maryland School of Medicine, Baltimore, Maryland 21201*

JAMES A. SVOBODA (26), *Insect Physiology Laboratory, Agricultural Research Service, United States Department of Agriculture, Beltsville, Maryland 20705*

GARY TAKEOKA (6), *Department of Food Science and Technology, University of California, Davis, Davis, California 95616*

MALCOLM J. THOMPSON (26), *Insect Physiology Laboratory, Agricultural Research Service, United States Department of Agriculture, Beltsville, Maryland 20705*

CHARLES J. WAECHTER (30, 38), *Department of Biological Chemistry, University of Maryland School of Medicine, Baltimore, Maryland 21201*

GUNTER F. WEIRICH (26, 28, 36), *Insect Physiology Laboratory, Agricultural Research Service, United States Department of Agriculture, Beltsville, Maryland 20705*

KJELL WIKVALL (21), *Department of Pharmaceutical Biochemistry, University of Uppsala, S-751 23 Uppsala, Sweden*

OSCAR H. WILL III (7), *Department of Biology, Augustana College, Sioux Falls, South Dakota 57192*

TSAN-HSI YANG (11), *1004 South Oakley Boulevard, Chicago, Illinois 60612*

T. MARK ZABRISKIE (11), *Department of Chemistry, University of Utah, Salt Lake City, Utah 84112*

P. F. ZAGALSKY (9), *Department of Biochemistry, Bedford College, University of London, London NW1 4NS, England*

Preface

Steroids and isoprenoids were last reviewed in *Methods in Enzymology* in 1969 in Volume XV which was edited by R. B. Clayton. Significant articles are also scattered throughout the first six volumes of this series.

Since then remarkable changes have taken place in the field of steroids and isoprenoids. High-performance liquid chromatography (HPLC) has revolutionized the preparation and analysis of both substrates and products. Analytical procedures now include high-resolution mass spectrometry and gas chromatography–mass spectrometry. Representative enzymes for nearly all of the reactions leading from acetyl-CoA to farnesyl pyrophosphate have been purified to homogeneity, and many other enzymes of polyprenol synthesis and metabolism have been extensively purified from a variety of sources. It is to be noted that several of the enzymes necessary for the conversion of farnesyl pyrophosphate to cholesterol have been solubilized from microsomes, and very significant purifications have been achieved.

An enzyme that has been relatively neglected in Volumes 110 and 111 is 3-hydroxy-3-methylglutaryl-CoA reductase. There are two reasons for this. First, this enzyme was the subject of several articles in Volume 71 of this series. In addition, this area of research is in a period of flux since it is now apparent that the reductase is an intrinsic rather than an extrinsic microsomal protein and much of the earlier work was carried out with a proteolytically degraded molecule.

Volumes 110 and 111 are each divided into three sections. Section I of Volume 110 deals with the enzymes of the early stages of terpenogenesis, and the reactions are common for the biosynthesis of all isoprenoids, i.e., the synthesis of isopentenyl pyrophosphate from acetyl-CoA. Included is a consideration of alternate metabolism of mevalonate. Section II covers the linear head-to-tail (1'-4) condensations as well as the head-to-head (1'-1-4) condensations of terpene biosynthesis. The products of the 1'-4 condensation vary from 3 to 9 isoprene units in length, depending on the system, while squalene and carotenoids are produced in the head-to-head reactions. Reactions between terpenes and nonterpenoid molecules are also described. The final section of this volume covers the cyclization of terpenes in plants as well as the isolation of terpenes from insects, opening the area of biosynthesis of these unusual molecules.

The first section of Volume 111 presents methodology for the isolation and characterization of substrates and products. Several of the articles deal with HPLC and others with mass spectrometry as useful techniques.

Section II on sterol metabolism deals with the transformation of sterols and steroids in a variety of species from vertebrates to insects to plants. Chapters on proteins necessary for the transport and for the acylation of sterols are also included. The final section considers the metabolism of nonsteroidal polyterpenes in vertebrate, plant, and insect systems.

The size and breadth of the area of research represented by the title of these two volumes is far too great to have allowed comprehensive coverage. However, we hope that the material presented is representative. It should provide a reasonable entry into the literature even if it does not present the reader with an "instant" solution to the problem at hand.

We would especially like to thank the authors who contributed to these volumes. The enthusiasm and promptness that they exhibited were most gratifying. Special thanks are due to the staff of Academic Press, both for their encouragement and for knowing what was going on when we did not. Finally, we are indebted to our secretaries, Ellie Moreland and Anne Kidd, for their role in bringing these volumes to fruition.

JOHN H. LAW
HANS C. RILLING

METHODS IN ENZYMOLOGY

EDITED BY

Sidney P. Colowick and Nathan O. Kaplan

VANDERBILT UNIVERSITY
SCHOOL OF MEDICINE
NASHVILLE, TENNESSEE

DEPARTMENT OF CHEMISTRY
UNIVERSITY OF CALIFORNIA
AT SAN DIEGO
LA JOLLA, CALIFORNIA

I. Preparation and Assay of Enzymes
II. Preparation and Assay of Enzymes
III. Preparation and Assay of Substrates
IV. Special Techniques for the Enzymologist
V. Preparation and Assay of Enzymes
VI. Preparation and Assay of Enzymes (*Continued*)
Preparation and Assay of Substrates
Special Techniques
VII. Cumulative Subject Index

METHODS IN ENZYMOLOGY

EDITORS-IN-CHIEF

Sidney P. Colowick and Nathan O. Kaplan

VOLUME VIII. Complex Carbohydrates
Edited by ELIZABETH F. NEUFELD AND VICTOR GINSBURG

VOLUME IX. Carbohydrate Metabolism
Edited by WILLIS A. WOOD

VOLUME X. Oxidation and Phosphorylation
Edited by RONALD W. ESTABROOK AND MAYNARD E. PULLMAN

VOLUME XI. Enzyme Structure
Edited by C. H. W. HIRS

VOLUME XII. Nucleic Acids (Parts A and B)
Edited by LAWRENCE GROSSMAN AND KIVIE MOLDAVE

VOLUME XIII. Citric Acid Cycle
Edited by J. M. LOWENSTEIN

VOLUME XIV. Lipids
Edited by J. M. LOWENSTEIN

VOLUME XV. Steroids and Terpenoids
Edited by RAYMOND B. CLAYTON

VOLUME XVI. Fast Reactions
Edited by KENNETH KUSTIN

VOLUME XVII. Metabolism of Amino Acids and Amines (Parts A and B)
Edited by HERBERT TABOR AND CELIA WHITE TABOR

VOLUME XVIII. Vitamins and Coenzymes (Parts A, B, and C)
Edited by DONALD B. MCCORMICK AND LEMUEL D. WRIGHT

VOLUME XIX. Proteolytic Enzymes
Edited by GERTRUDE E. PERLMANN AND LASZLO LORAND

VOLUME XX. Nucleic Acids and Protein Synthesis (Part C)
Edited by KIVIE MOLDAVE AND LAWRENCE GROSSMAN

VOLUME XXI. Nucleic Acids (Part D)
Edited by LAWRENCE GROSSMAN AND KIVIE MOLDAVE

VOLUME XXII. Enzyme Purification and Related Techniques
Edited by WILLIAM B. JAKOBY

VOLUME XXIII. Photosynthesis (Part A)
Edited by ANTHONY SAN PIETRO

VOLUME XXIV. Photosynthesis and Nitrogen Fixation (Part B)
Edited by ANTHONY SAN PIETRO

VOLUME XXV. Enzyme Structure (Part B)
Edited by C. H. W. HIRS AND SERGE N. TIMASHEFF

VOLUME XXVI. Enzyme Structure (Part C)
Edited by C. H. W. HIRS AND SERGE N. TIMASHEFF

VOLUME XXVII. Enzyme Structure (Part D)
Edited by C. H. W. HIRS AND SERGE N. TIMASHEFF

VOLUME XXVIII. Complex Carbohydrates (Part B)
Edited by VICTOR GINSBURG

VOLUME XXIX. Nucleic Acids and Protein Synthesis (Part E)
Edited by LAWRENCE GROSSMAN AND KIVIE MOLDAVE

VOLUME XXX. Nucleic Acids and Protein Synthesis (Part F)
Edited by KIVIE MOLDAVE AND LAWRENCE GROSSMAN

VOLUME XXXI. Biomembranes (Part A)
Edited by SIDNEY FLEISCHER AND LESTER PACKER

VOLUME XXXII. Biomembranes (Part B)
Edited by SIDNEY FLEISCHER AND LESTER PACKER

VOLUME XXXIII. Cumulative Subject Index Volumes I–XXX
Edited by MARTHA G. DENNIS AND EDWARD A. DENNIS

VOLUME XXXIV. Affinity Techniques (Enzyme Purification: Part B)
Edited by WILLIAM B. JAKOBY AND MEIR WILCHEK

VOLUME XXXV. Lipids (Part B)
Edited by JOHN M. LOWENSTEIN

VOLUME XXXVI. Hormone Action (Part A: Steroid Hormones)
Edited by BERT W. O'MALLEY AND JOEL G. HARDMAN

VOLUME XXXVII. Hormone Action (Part B: Peptide Hormones)
Edited by BERT W. O'MALLEY AND JOEL G. HARDMAN

VOLUME XXXVIII. Hormone Action (Part C: Cyclic Nucleotides)
Edited by JOEL G. HARDMAN AND BERT W. O'MALLEY

VOLUME XXXIX. Hormone Action (Part D: Isolated Cells, Tissues, and Organ Systems)
Edited by JOEL G. HARDMAN AND BERT W. O'MALLEY

VOLUME XL. Hormone Action (Part E: Nuclear Structure and Function)
Edited by BERT W. O'MALLEY AND JOEL G. HARDMAN

VOLUME XLI. Carbohydrate Metabolism (Part B)
Edited by W. A. WOOD

VOLUME XLII. Carbohydrate Metabolism (Part C)
Edited by W. A. WOOD

VOLUME XLIII. Antibiotics
Edited by JOHN H. HASH

VOLUME XLIV. Immobilized Enzymes
Edited by KLAUS MOSBACH

VOLUME XLV. Proteolytic Enzymes (Part B)
Edited by LASZLO LORAND

VOLUME XLVI. Affinity Labeling
Edited by WILLIAM B. JAKOBY AND MEIR WILCHEK

VOLUME XLVII. Enzyme Structure (Part E)
Edited by C. H. W. HIRS AND SERGE N. TIMASHEFF

VOLUME XLVIII. Enzyme Structure (Part F)
Edited by C. H. W. HIRS AND SERGE N. TIMASHEFF

VOLUME XLIX. Enzyme Structure (Part G)
Edited by C. H. W. HIRS AND SERGE N. TIMASHEFF

VOLUME L. Complex Carbohydrates (Part C)
Edited by VICTOR GINSBURG

VOLUME LI. Purine and Pyrimidine Nucleotide Metabolism
Edited by PATRICIA A. HOFFEE AND MARY ELLEN JONES

VOLUME LII. Biomembranes (Part C: Biological Oxidations)
Edited by SIDNEY FLEISCHER AND LESTER PACKER

VOLUME LIII. Biomembranes (Part D: Biological Oxidations)
Edited by SIDNEY FLEISCHER AND LESTER PACKER

VOLUME LIV. Biomembranes (Part E: Biological Oxidations)
Edited by SIDNEY FLEISCHER AND LESTER PACKER

VOLUME LV. Biomembranes (Part F: Bioenergetics)
Edited by SIDNEY FLEISCHER AND LESTER PACKER

VOLUME LVI. Biomembranes (Part G: Bioenergetics)
Edited by SIDNEY FLEISCHER AND LESTER PACKER

VOLUME LVII. Bioluminescence and Chemiluminescence
Edited by MARLENE A. DELUCA

VOLUME LVIII. Cell Culture
Edited by WILLIAM B. JAKOBY AND IRA PASTAN

VOLUME LIX. Nucleic Acids and Protein Synthesis (Part G)
Edited by KIVIE MOLDAVE AND LAWRENCE GROSSMAN

VOLUME LX. Nucleic Acids and Protein Synthesis (Part H)
Edited by KIVIE MOLDAVE AND LAWRENCE GROSSMAN

VOLUME 61. Enzyme Structure (Part H)
Edited by C. H. W. HIRS AND SERGE N. TIMASHEFF

VOLUME 62. Vitamins and Coenzymes (Part D)
Edited by DONALD B. MCCORMICK AND LEMUEL D. WRIGHT

VOLUME 63. Enzyme Kinetics and Mechanism (Part A: Initial Rate and Inhibitor Methods)
Edited by DANIEL L. PURICH

VOLUME 64. Enzyme Kinetics and Mechanism (Part B: Isotopic Probes and Complex Enzyme Systems)
Edited by DANIEL L. PURICH

VOLUME 65. Nucleic Acids (Part I)
Edited by LAWRENCE GROSSMAN AND KIVIE MOLDAVE

VOLUME 66. Vitamins and Coenzymes (Part E)
Edited by DONALD B. MCCORMICK AND LEMUEL D. WRIGHT

VOLUME 67. Vitamins and Coenzymes (Part F)
Edited by DONALD B. MCCORMICK AND LEMUEL D. WRIGHT

VOLUME 68. Recombinant DNA
Edited by RAY WU

VOLUME 69. Photosynthesis and Nitrogen Fixation (Part C)
Edited by ANTHONY SAN PIETRO

VOLUME 70. Immunochemical Techniques (Part A)
Edited by HELEN VAN VUNAKIS AND JOHN J. LANGONE

VOLUME 71. Lipids (Part C)
Edited by JOHN M. LOWENSTEIN

VOLUME 72. Lipids (Part D)
Edited by JOHN M. LOWENSTEIN

VOLUME 73. Immunochemical Techniques (Part B)
Edited by JOHN J. LANGONE AND HELEN VAN VUNAKIS

VOLUME 74. Immunochemical Techniques (Part C)
Edited by JOHN J. LANGONE AND HELEN VAN VUNAKIS

VOLUME 75. Cumulative Subject Index Volumes XXXI, XXXII, and XXXIV-LX
Edited by EDWARD A. DENNIS AND MARTHA G. DENNIS

VOLUME 76. Hemoglobins
Edited by ERALDO ANTONINI, LUIGI ROSSI-BERNARDI, AND EMILIA CHIANCONE

VOLUME 77. Detoxication and Drug Metabolism
Edited by WILLIAM B. JAKOBY

VOLUME 78. Interferons (Part A)
Edited by SIDNEY PESTKA

VOLUME 79. Interferons (Part B)
Edited by SIDNEY PESTKA

VOLUME 80. Proteolytic Enzymes (Part C)
Edited by LASZLO LORAND

VOLUME 81. Biomembranes (Part H: Visual Pigments and Purple Membranes, I)
Edited by LESTER PACKER

VOLUME 82. Structural and Contractile Proteins (Part A: Extracellular Matrix)
Edited by LEON W. CUNNINGHAM AND DIXIE W. FREDERIKSEN

VOLUME 83. Complex Carbohydrates (Part D)
Edited by VICTOR GINSBURG

VOLUME 84. Immunochemical Techniques (Part D: Selected Immunoassays)
Edited by JOHN J. LANGONE AND HELEN VAN VUNAKIS

VOLUME 85. Structural and Contractile Proteins (Part B: The Contractile Apparatus and the Cytoskeleton)
Edited by DIXIE W. FREDERIKSEN AND LEON W. CUNNINGHAM

VOLUME 86. Prostaglandins and Arachidonate Metabolites
Edited by WILLIAM E. M. LANDS AND WILLIAM L. SMITH

VOLUME 87. Enzyme Kinetics and Mechanism (Part C: Intermediates, Stereochemistry, and Rate Studies)
Edited by DANIEL L. PURICH

VOLUME 88. Biomembranes (Part I: Visual Pigments and Purple Membranes, II)
Edited by LESTER PACKER

VOLUME 89. Carbohydrate Metabolism (Part D)
Edited by WILLIS A. WOOD

VOLUME 90. Carbohydrate Metabolism (Part E)
Edited by Willis A. Wood

VOLUME 91. Enzyme Structure (Part I)
Edited by C. H. W. HIRS AND SERGE N. TIMASHEFF

VOLUME 92. Immunochemical Techniques (Part E: Monoclonal Antibodies and General Immunoassay Methods)
Edited by JOHN J. LANGONE AND HELEN VAN VUNAKIS

VOLUME 93. Immunochemical Techniques (Part F: Conventional Antibodies, Fc Receptors, and Cytotoxicity)
Edited by JOHN J. LANGONE AND HELEN VAN VUNAKIS

VOLUME 94. Polyamines
Edited by HERBERT TABOR AND CELIA WHITE TABOR

VOLUME 95. Cumulative Subject Index Volumes 61–74 and 76–80
Edited by EDWARD A. DENNIS AND MARTHA G. DENNIS

VOLUME 96. Biomembranes [Part J: Membrane Biogenesis: Assembly and Targeting (General Methods; Eukaryotes)]
Edited by SIDNEY FLEISCHER AND BECCA FLEISCHER

VOLUME 97. Biomembranes [Part K: Membrane Biogenesis: Assembly and Targeting (Prokaryotes, Mitochondria, and Chloroplasts)]
Edited by SIDNEY FLEISCHER AND BECCA FLEISCHER

VOLUME 98. Biomembranes [Part L: Membrane Biogenesis (Processing and Recycling)]
Edited by SIDNEY FLEISCHER AND BECCA FLEISCHER

VOLUME 99. Hormone Action (Part F: Protein Kinases)
Edited by JACKIE D. CORBIN AND JOEL G. HARDMAN

VOLUME 100. Recombinant DNA (Part B)
Edited by RAY WU, LAWRENCE GROSSMAN, AND KIVIE MOLDAVE

VOLUME 101. Recombinant DNA (Part C)
Edited by RAY WU, LAWRENCE GROSSMAN, AND KIVIE MOLDAVE

VOLUME 102. Hormone Action (Part G: Calmodulin and Calcium-Binding Proteins)
Edited by ANTHONY R. MEANS AND BERT W. O'MALLEY

VOLUME 103. Hormone Action (Part H: Neuroendocrine Peptides)
Edited by P. MICHAEL CONN

VOLUME 104. Enzyme Purification and Related Techniques (Part C)
Edited by WILLIAM B. JAKOBY

VOLUME 105. Oxygen Radicals in Biological Systems
Edited by LESTER PACKER

VOLUME 106. Posttranslational Modifications (Part A)
Edited by FINN WOLD AND KIVIE MOLDAVE

VOLUME 107. Posttranslational Modifications (Part B)
Edited by FINN WOLD AND KIVIE MOLDAVE

VOLUME 108. Immunochemical Techniques (Part G: Separation and Characterization of Lymphoid Cells)
Edited by GIOVANNI DI SABATO, JOHN J. LANGONE, AND HELEN VAN VUNAKIS

VOLUME 109. Hormone Action (Part I: Peptide Hormones)
Edited by LUTZ BIRNBAUMER AND BERT W. O'MALLEY

VOLUME 110. Steroids and Isoprenoids (Part A)
Edited by JOHN H. LAW AND HANS C. RILLING

VOLUME 111. Steroids and Isoprenoids (Part B)
Edited by JOHN H. LAW AND HANS C. RILLING

VOLUME 112. Drug and Enzyme Targeting (Part A) (in preparation)
Edited by KENNETH J. WIDDER AND RALPH GREEN

VOLUME 113. Glutamate, Glutamine, Glutathione, and Related Compounds (in preparation)
Edited by ALTON MEISTER

Section I

Methodology

[1] A Comparison of Methods for the Identification of Sterols

By WILLIAM R. NES

Introduction

There is a wide variety of methods available for the identification and quantitation of sterols. This chapter summarizes the usefulness of some of the more important chromatographic and spectral techniques. Several general points are worth emphasizing at the outset.

1. Identification cannot be achieved by the use of a single method, although NMR, especially ^{13}C NMR, comes close to providing a pattern of information unique enough to represent a "fingerprint." Thus the proper identification of a sterol usually rests not on a single measurement but rather on a pattern of characteristics which are unique for the sterol.

2. Chromatographic techniques separate sterols from each other, and the degree of separation (based on rates of movement) tells us something about structure. Knowing the way structure influences chromatographic mobilities also permits us to make rational choices of chromatographic conditions for the separation of a given set of sterols.

3. Stereochemistry is an important element in the determination of chromatographic mobilities. Thus, as a first approximation, in adsorption, gas–liquid, as well as reversed-phase chromatography the flatter a sterol is the slower it moves. This is shown in Table I, and discussed at greater length for each of the methods in the next section. It will be remembered that in 5α-cholestan-3β-ol (which has a 5α-H atom) all of the carbon atoms in the tetracyclic nucleus lie approximately in two planes which are parallel to each other,[1] and at least in the solid state[1-3] the side chain is extended to the right (C-22 trans to C-13) in the usual view of the molecule. In this view, C-18 and C-19 are pointing toward the observer (front or β face) and C-3 is to the left. Sterols with a 5β-H, 14β-H, 17β-H (with a 17α-

---

[1] W. R. Nes and M. L. McKean, "Biochemistry of Steroids and Other Isopentenoids." University Park Press, Baltimore, Maryland, 1977.

[2] R. B. Turner, *in* L. F. Fieser and M. Fieser, "Natural Products Related to Phenanthrene," p. 620. Van Nostrand, Reinhold, Princeton, New Jersey, 1949.

[3] C. Romers, C. Altona, H. J. C. Lacops, and R. A. G. DeGraff, *Terpenoids Steroids* **4**, 531 (1974).

TABLE I
COMPARISON OF THE EFFECTS OF STRUCTURAL FEATURES OF STEROLS ON MOBILITY IN DIFFERENT CHROMATOGRAPHIC SYSTEMS[a]

Relative mobility of sterols with feature	Comments		
	Adsorption method	Gas–liquid method	Reversed-phase method
5β-H > 5α-H	Yes	Yes	Yes
14β-H > 14α-H	?	Yes	?
17β-H > 17α-H	?	Yes	?
20S > 20R	Yes	Yes	?
Δ^8 > 9β,19-cyclo	No separation except with AgNO$_3$	Yes	Yes
4α-CH$_3$ $\simeq$ 4α-H	4α-CH$_3$-sterol faster, but slower as sterane	No, 4α-CH$_3$-sterol is slower	Yes
4β-H > 4β-CH$_3$	?	Yes	Almost no effect
14α-CH$_3$ $\simeq$ 14α-H	Little or no effect	Almost no effect	Almost no effect
24-H > 24-CH$_3$	?	Yes	Yes
24-CH$_3$ > 24-C$_2$H$_5$	?	Yes	Yes
C$_{19}$ > C$_{21}$ > C$_{25}$ > C$_{27}$	Yes but weak	Yes	Yes
3α-OH > 3β-OH	Yes	Yes	Yes
6β-OH > 6α-OH	Yes	?	?
Z-Δ^{22} > E-Δ^{22}	?	Yes	Yes
Z-$\Delta^{17(20)}$ > E-$\Delta^{17(20)}$	Yes	Yes	Yes
E-$\Delta^{24(28)}$ > Z-$\Delta^{24(28)}$	No separation except with AgNO$_3$	Slight	No

[a] Yes indicates effect operates as shown in first column; ? implies knowledge is uncertain or unavailable.

side chain), or 9β,19-cyclo grouping are bent causing deviation from the flatness of 5α-cholestanol. Also, the 3α-, 4β-, 6β-, 10β-, 14α-, and 13β-positions are axial. Axial groups protrude from the general plane of the molecule and thicken it. The common natural sterols such as 5α-cholestanol are all thickened on the β-face by the presence of C-18 and C-19 (the so-called angular methyls). However, in 5α-cholestanol, cholesterol, and the other common sterols, the α-face has no protruding carbon atoms.

4. A stereochemical generality related to the previous point is that in gas–liquid and reversed-phase chromatography the larger and more extended the molecule, the slower it moves as summarized in Table I. Sterols with lower molecular weights, or sterols with a more compact shape as a result either of having side chains to the left [20S or Z-$\Delta^{17(20)}$] or of having side chains with a cis-Δ^{22}-double bond move faster than those with a larger molecular weight, with side chains to the right [20R or E-$\Delta^{17(20)}$], or with side chains having a trans-Δ^{22}-double bond, respectively.

A more extended discussion which includes adsorption chromatography is to be found in the next section.

5. The presence of polar groups (HO, C=O, C=C, etc.) has a strong impact on chromatographic mobility, but the direction and magnitude of the kind of effect depend strongly on the method. Polar groups usually retard the movement of sterols on adsorption and gas–liquid chromatography but increase it in reversed-phase systems.

6. The mechanism by which a sterol binds to a stationary (or mobile) phase tends to be independent of the chromatographic mode (column, layer, gas, liquid, etc.). For instance, adsorption phenomena can exist in a classic column procedure or in modern high-pressure liquid chromatography.

7. Only when coupled with some other method (spectrometry, flame ionization, densitometry, etc.) can chromatography be used for quantitative determination.

8. Some methods, e.g., flame ionization and mass spectroscopy, which are used for quantitative analysis destroy the sterol, while others, e.g., ultraviolet spectrometry, do not.

9. Quantitation depends on the measurement of some particular property, and the response of a given sterol will therefore usually depend on both the structure of the sterol and the kind of property (method). Two sterols may respond similarly in one assay but differently in another.

Methods

Ordinary Phase (Adsorption) Chromatography

Without Silver Ion. Adsorption chromatography involves binding of a substrate to the surface of a polar solid (the stationary phase) which may be of natural origin as in the case of alumina or of silica gel, or may be man made. The mobile phase usually should have only a moderate "polarity." The polarity, i.e., eluting power, of common solvents is n-hexane < cyclohexane < CCl_4 < C_6H_6 < ether < $CHCl_3$ < EtOAc < CH_2Cl_2 < acetone < acetonitrile < primary alcohols < HOAc. For alumina column chromatography in the author's laboratory we often use a gradient of ether into hexane in which the former represents 5, 10, 20, 40, 60, and 100% for 10 fractions each. The fraction volume is frequently 1.0 ml for each 10 g of Al_2O_3. The Al_2O_3 (dried at 120° overnight) is usually deactivated with 3 to 5% of water (w/w). The more water (which can be as high as 10%) the less adsorptive (but often the more selective) the Al_2O_3 and the less should be the eluting power of the mobile phase. For moderately deactivated alumina we commonly use 0.5–1 g of sterol for each 100 g of alumina. For thin or thick layer chromatography (TLC) we usually use a stationary phase of silica gel G and a mobile phase of ether–benzene (1:9, v/v). Either in the column (LC) or the thin or thick layer (TLC) modes this method of chromatography is characterized by hydrogen-bonding and other electronic attractions (e.g., dipole–dipole interactions) between the sterol and the stationary phase. Hydrogen bonding is stronger with alumina and silica gel when the sterol has groups which can act as proton donors (e.g., OH groups) than it is when groups (e.g., C=O groups) are present which can act only as proton acceptors or as dipoles. Adsorption chromatography can be performed either in the usual way with gravity flow or, with more modern microspheres and bonded phases, at elevated pressure.

As a result of hydrogen-bonding the influence of groupings on the rates of movement of a sterol will follow the order: alkyl < C=O < hindered OH < unhindered OH, where the progression from left to right in this series is equivalent to decreases in the rates of movement of the sterol. A frequently hindered hydroxyl group is at C-3, and the hindrance usually comes from C-4 in the form of a 4α-methyl group. In the case of the 4,4-dimethyl grouping still more steric hindrance is created by the 4β-methyl group. It follows that we would observe relative rates of movement as indicated by 7α-hydroxycholesterol < cholesterol ≃ sitosterol < 4α-methylcholesterol < 4,4-dimethylcholesterol ≃ lanosterol < cholestanone ≃ cholesteryl esters < cholestane. Adsorption chromatography is, therefore, a good method to separate sterols according to the kind and number of their oxygen functions. For instance, as a first approximation, all 3β-OH-sterols lacking steric hindrance will move at a similar rate, while 4,4-dimethylpolycycles, regardless of whether they are sterols or triterpenoids (e.g., lanosterol or β-amyrin, respectively) will move more or less together. This permits us to obtain a slower moving "4-desmethyl" and a faster moving "4,4-dimethyl" fraction each of which can then be used for subsequent analysis. Careful use of adsorption systems will permit separation of very closely related sterols.

Double bonds (which are weakly polar groups) usually have a small but definite influence and retard mobility.[4-6] On modern microspheres, e.g., μPorasil with hexane–benzene (9/1 by volume), separation can be achieved as follows in terms of mobility: $\Delta^0 > \Delta^5 > \Delta^{8(14)} > \Delta^{8(9)} > \Delta^7 > \Delta^{5,7} > \Delta^{8,14} > \Delta^{7,14}$.[5] Δ^5- and Δ^7-sterols, the former moving faster, can be separated (with base line resolution) by ordinary thin-layer chromatogra-

---

[4] W. R. Nes, J. M. Joseph, J. R. Landrey, and R. L. Conner, *J. Biol. Chem.* **253**, 2361 (1978).
[5] J. R. Thowsen and G. J. Schroepfer, Jr., *J. Lipid Res.* **20**, 681 (1979).
[6] I. R. Hunter, M. K. Walden, G. F. Bailey, and E. Heftman, *Lipids* **14**, 687 (1979).

phy.[7] Experimentally this separation is currently being done in the author's laboratory by Dr. Vipin Garg with thin or thick layers of silica gel G developed 3–6 times using 15% ether in benzene as solvent. Although the underlying reason for separation of Δ^5- from Δ^7-sterols has not been clearly demonstrated, it may have to do with the fact that the Δ^5-bond does and the Δ^7-bond does not have an interaction with the 3β-OH group. It is well known in sterol chemistry that the Δ^5-bond (but not the Δ^7-bond) can contribute electron density to C-3 and, for example, its 3β-oxygenation results in the 3,5-cyclosterol rearrangement.[8,9] This effect, which should strengthen the H—O bond, decreasing the ability of the sterol to act as an H-donor in H-bonding to the adsorbent, would therefore increase the observed chromatographic mobility of the sterol. Of course, there may also be some contribution to the relative mobilities by stereochemical effects derived from the presence of the Δ^5- and Δ^7-bonds at ring junctions.

Axial hydroxyls have less of a retarding effect than do their equatorial analogs presumably as a consequence of the importance of molecular dimensions. 3-Episterols, for instance, with an α and axial OH-group usually move faster (merging into the hindered OH region) than do the corresponding sterols with a β and equatorial OH-group.[10] Similarly, 6β-hydroxycholesterol (axial OH-group at C-6) moves faster than the 6α-epimer with an equatorial OH-group at C-6.[10] Neither proximity to C-19 nor a particular side of the molecule is crucial. The faster moving epimer in both cases is the one in which there is an OH-group situated 90° to the general plane of the molecule regardless of the side it is on. Similarly, the flatter A/B-trans- or 5α-steroids move slower[10] than do their A/B-cis- or 5β-analogs, as in the case of 3β-hydroxy-5α-cholestane with an equatorial OH-group which moves slower than either 3α- or 3β-hydroxy-5β-cholestane in which ring A bends to the α-side. For the four diastereoisomeric 3-hydroxycholestanes, the actual order of elution[10] is 3α,5α > 3α,5β > 3β,5β > 3β,5α. 5α-sapogenins also move slower[10] than their 5β-analogs as do 5α-bile acids compared to their 5β-analogs. Inversion of C-5, not the secondary effect on the HO-group, appears to be responsible for the change in mobility because 5β-cholestane lacking an HO-group moves nearly twice as fast as its isomer, 5α-cholestane.[11] The importance of flatness would suggest that cycloartenol might move faster than lanosterol, but the difference in rates of movement is negligible.

[7] W. R. Nes, K. Krevitz, J. Joseph, W. D. Nes, B. Harris, G. F. Gibbons, and G. W. Patterson, *Lipids* **12**, 511 (1977).
[8] W. R. Nes and J. A. Steele, *J. Org. Chem.* **22**, 1457 (1957).
[9] W. R. Nes and C. W. Shoppee, *J. Chem. Soc.* p. 93 (1956).
[10] B. P. Lisboa, this series, Vol. 15, p. 56.
[11] G. Ryback, *J. Chromatogr.* **116**, 207 (1976).

The ability of a sterol to have an extended as opposed to a compressed or compact shape is also an important element in determining its rate of migration. Sterols with the natural (20R)-configuration in which the preferred rotational isomer about the 17(20)-bond is the one with C-22 to the right in the usual view of the molecule are longer and they move significantly slower than the more compact (20S)-epimers in which C-22 should be preferentially on the left.[11-16] In the euphoids (inverted at C-13, C-14, and C-17), X-ray crystal structures[17] show the (20S)-epimer has the preferred conformation with C-22 to the right. J. R. Landrey in the author's laboratory has, however, found slight separation of the epimers (euphol/tirucallol) the S-epimer (tirucallol) being slower. Similarly, (E)-17(20)-dehydrosterols with C-22 fixed rigidly to the right by virtue of the double bond at C-17(20) move slower than the isomeric (Z)-analogs with C-22 to the left.[12,15]

While adsorption chromatography is usually not thought of as a way to separate molecules based on size, it has been shown by Ryback[11] that size is a contributing factor. When steranes were eluted from a column of alumina with pentane, the following retention volumes relative to 5α-cholestane were found: 5α-androstane (C_{19}), 0.56; 5α-pregnane (C_{21}), 0.69; halostane (C_{26}) (24,24-dimethyl-5α-cholane), 0.74; 5α-cholestane, 1.00; and 4-α-methyl-5α-cholestane, 1.05. Similarly, Hunger et al.[6] using 0.5% isopropanol in n-hexane ran a C_{18}-Porasil-B column in the adsorbent mode and found sitosterol emerged before stigmasterol which moved faster than campesterol which in turn had a greater mobility than cholesterol which was followed by ergosterol. The order of elution is reversed in terms of size compared to Ryback's experiments,[11] because with the 3β-OH-group present H-bonding is the dominant mechanism of interaction with the stationary phase, and as the size of the molecule increases the relative contribution of the OH-group decreases. In the absence of the 3β-OH-group 24β-methyl-5α-cholestane moves slower than 5α-cholestane,[11] but the 24α-methyl epimer moves faster[11] (perhaps due to a conformation which is more compact).

With Silver Ion (Argentation Chromatography). The influence of double bonds is often enhanced by the incorporation of 5–30% of silver nitrate into silica gel or alumina. The more a double bond can interact with silver ion the slower the compound moves, and this is the principal

[12] J. M. Joseph, Ph.D. Dissertation, Drexel University (1979).
[13] J. M. Joseph and W. R. Nes, *J. Chem. Soc., Chem. Commun.* p. 367 (1981).
[14] W. R. Nes, T. E. Varkey, and K. Krevitz, *J. Am. Chem. Soc.* **99**, 260 (1977).
[15] W. R. Nes, T. E. Varkey, D. R. Crump, and M. Gut, *J. Org. Chem.* **41**, 3429 (1976).
[16] W. R. Nes, B. C. Sekula, W. D. Nes, and J. H. Adler, *J. Biol. Chem.* **253**, 7218 (1978).
[17] W. D. Nes, R. Y. Wong, M. Benson, J. R. Landrey, and W. R. Nes, *Proc. Natl Acad. Sci. USA.* **81**, 5896 (1984).

basis for separation by argentation chromatography. As a first approximation, increasing ability to form a complex is associated with decreasing substitution on or near the double bond or decreasing steric hindrance. That is, the easier the silver ion can add to the double bond, the higher the concentration of the polar (positively charged) complex at any given mo-

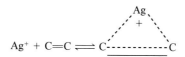

ment, and this leads to greater adsorption and a slower rate of movement. Consequently, more polar solvent systems, e.g., hexane–chloroform mixtures with small amounts of acetic acid or acetone[7,18] have been used. The polarity of the sterol is often reduced by acetylation. Sterols with an unhindered methylene (as in 24-methylenecholesterol) easily form complexes and move especially slowly. Δ^5-Sterols also readily add silver ion from the α-side and experimentally move more slowly than do Δ^0-sterols. Conversely, Δ^8-sterols with high (tetra) substitution on the double bond, as well as with proximity of the double bond to C-19 and a preference not to form a cis-B/C-ring junction (required of a silver ion complex on Δ^8), have little proclivity to form complexes and therefore move relatively fast. Another example of steric hindrance is in the mobility of Δ^{22}-sterols which increases with increasing size of the group on C-24 (relative rates: 24-C_2H_5 > 24-CH_3 > 24-H). The 24-ethylsterol, e.g., stigmasterol, moves so fast that it is difficult to separate it from the analog, e.g., sitosterol, lacking the Δ^{22}-bond.

Argentation chromatography has, for instance, been used to separate the monoenic Δ^5- from Δ^7-,[19,20] Δ^7- from Δ^8-,[21–23] and Δ^{23}-24-methyl- from Δ^{24}-24-methylsterols.[18] The conjugated dienic sterols ($\Delta^{7,14}$, $\Delta^{8,14}$, and $\Delta^{5,7}$) have also been separated by this method[24–26] as have the isomeric [cis–

---

[18] W. R. Nes, J. M. Joseph, J. R. Landrey, S. Behzadan, and R. L. Conner, *J. Lipid Res.* **22**, 770 (1981).
[19] R. Kammereck, W.-H. Lee, A. Paliokas, and G. J. Schroepfer, Jr., *J. Lipid Res.* **8**, 282 (1967).
[20] A. M. Paliokas and G. J. Schroepfer, Jr., *J. Biol. Chem.* **243**, 453 (1968).
[21] W.-H. Lee, R. Kammereck, B. N. Lutsky, J. A. McCloskey, and G. J. Schroepfer, Jr., *J. Biol. Chem.* **244**, 2033 (1969).
[22] A. M. Paliokas, W.-H. Lee, and G. J. Schroepfer, Jr., *J. Lipid Res.* **9**, 143 (1968).
[23] W.-H. Lee, B. N. Lutsky, and G. J. Schroepfer, Jr., *J. Biol. Chem.* **244**, 5440 (1969).
[24] B. N. Lutsky, J. A. Martin, and G. J. Schroepfer, Jr., *J. Biol. Chem.* **246**, 6737 (1971).
[25] B. N. Lutsky, H. M. Hsiung, and G. J. Schroepfer, Jr., *Lipids* **10**, 9 (1975).
[26] S. Huntoon, B. Fourcans, B. N. Lutsky, E. J. Parish, H. Emery, F. F. Knapp, Jr., and G. J. Schroepfer, Jr., *J. Biol. Chem.* **253**, 775 (1978).

trans-$\Delta^{24(28)}$] 24-ethylidenesterols.[27,28] The *E*-isomer, e.g., fucosterol, moves the faster. As expected, it is also possible to separate sterols based on the number of double bonds as in the case of Δ^5 (fast), $\Delta^{5,7}$ (intermediate), and $\Delta^{5,7,22(E)}$ (slow)[18,30] and Δ^5 (fast) and $\Delta^{5,24(28)}$ (slow).[27] An example of the shielding of the Δ^{22}-bond by the 24α-ethyl group is found in the resolution of 24α-ethylcholesta-5,7,22(*E*)-trien-3β-yl acetate from its slower moving 24-dealkyl analog [cholesta-5,7,22(*E*)-trien-3β-yl acetate].[29] Some quantitative examples of relative rates of movement (in parentheses) on silica gel G with 5% $AgNO_3$ and a mobile phase of hexane–chloroform with a trace of acetic acid are sitosterol (1.00) > stigmasterol (0.98) > fucosterol (0.78) > 28-isofucosterol (0.70) > 7-dehydrocholesterol (0.39) = 24-methylenecholesterol (0.39) > ergosterol (0.35).[27] Separation of nonconjugated dienols from each other can also be achieved if one of the double bonds is in an appropriate position. This is illustrated by the faster rate of movement of $\Delta^{8,24}$ (faster) from $\Delta^{7,24}$ (slower).[30]

The euphoids (euphol and tirucallol) which are epimeric at C-20 do not separate well from each other,[31] but in the author's laboratory J. R. Landrey finds a very slight difference with euphol moving the faster.

Reversed-Phase Chromatography

Several kinds of reversed-phase (RP) chromatography (i.e., chromatography in which the stationary phase is nonpolar) have been developed over the years. In earlier systems the polar groups of, e.g., paper, were masked by methylation or acetylation (as in the case of the OH-groups of cellulose) or by polymerization of some less polar chemical on the surface. In order to make a nonpolar stationary phase more directly, hydrocarbons have been impregnated into a solid carrier in a TLC system giving in a theoretical sense the simplest and most direct of the reversed phase systems.[32,33] As would be expected, the more polar the sterol is by virtue of HO-groups, double bonds, or size, the faster it moves,[33] because it binds less well with the nonpolar stationary phase, hence the term reversed phase. The RP systems are in may ways more sensitive to structural changes than are normal phase systems. This is particularly true with respect to the presence of double bonds and to the size of the sterol.

[27] W. R. Nes, P. A. G. Malya, F. B. Mallory, K. A. Ferguson, J. R. Landrey, and R. L. Conner, *J. Biol. Chem.* **246**, 561 (1971).
[28] T. Itoh, S. Sakurai, T. Tamura, and T. Matsumoto, *Lipids* **15**, 22 (1980).
[29] W. R. Nes, A. Alcaide, F. B. Mallory, J. R. Landrey, and R. L. Conner, *Lipids* **10**, 140 (1975).
[30] U. F. Taylor, A. Kisic, R. A. Pascal, Jr., A. Izumi, M. Tsuda, and G. J. Schroepfer, Jr., *J. Lipid Res.* **22**, 171 (1981).
[31] T. Itoh, T. Uetsuki, T. Tamura, and T. Matsumoto, *Lipids* **15**, 407 (1980).
[32] J. W. Copius-Peereboom and H. W. Beekes, *J. Chromatogr.* **9**, 316 (1962).
[33] N. J. de Souza and W. R. Nes, *J. Lipid Res.* **10**, 240 (1969).

As the number of double bonds increases or the size of the sterol decreases, the less binding there is to the stationary phase and the faster the molecule moves. This has been demonstrated clearly with a thin layer RP system (paraffin hydrocarbon/aq. acetone) devised in the author's laboratory.[33] Not only, for instance, can 5α-cholestanol, cholesterol, and desmosterol with zero, one, and two double bonds be separated with baseline resolution, but so can the homologs, cholesterol, 24α-methylcholesterol, and 24α-ethylcholesterol.

A great improvement in RP-systems was achieved by the covalent bonding of alkyloxy groups to polysaccharides such as Sephadex giving a lipophilic dextran.[7,34–40] Columns of this material, e.g., Lipidex 5000 with 5% hexane in methanol as solvent, have permitted preparative quantities (up to 100 mg/run) of sterols to be separated[7,36–40] with the same efficiency obtained in the older, analytical RPTLC system.[33] However, the technology for chemically uniting alkyl groups to silica instead of dextran has advanced so rapidly that the so-called "chemically bonded" or usually now just "bonded" silica phases resulting from this technology have for the most part superceded the lipophilic dextrans especially with the commercial appearance of large, preparative columns. These new bonded silica phases whether used analytically or preparatively have very small diameters (3–5 μm) and a high porosity which give the particles a high surface area permitting very efficient contact between sterol and alkyl group. This in turn allows equilibria to be easily established between the stationary and mobile phases yielding columns with as much as 10^4 theoretical plates which still retain sharp elution peaks. The methodology is called reversed-phase high-performance liquid chromatography or RPHPLC. Separation of sterols with very small differences in their structures can be achieved with these new RPHPLC columns.[5,41–47] In a typical

[34] J. Ellingboe, E. Nyström, and J. Sjövall, *Biochim. Biophys. Acta* **152**, 803 (1968).
[35] J. Ellingboe, E. Nyström, and J. Sjövall, *J. Lipid Res.* **11**, 266 (1970).
[36] P. M. Hyde and W. H. Elliott, *J. Chromatogr.* **67**, 170 (1972).
[37] R. A. Anderson, C. J. W. Brooks, and B. A. Knights, *J. Chromatogr.* **75**, 247 (1973).
[38] G. W. Patterson, M. W. Khalil, and D. R. Idler, *J. Chromatogr.* **115**, 153 (1975).
[39] W. R. Nes, J. M. Joseph, J. R. Landrey, and R. L. Conner, *J. Biol. Chem.* **255**, 11815 (1980).
[40] M. L. McKean and W. R. Nes, *Phytochemistry* **16**, 683 (1977).
[41] H. H. Rees, P. L. Donnahey, and T. W. Goodwin, *J. Chromatogr.* **116**, 281 (1976).
[42] E. Hansbury and T. J. Scallen, *J. Lipid Res.* **19**, 742 (1978).
[43] N. Kikuchi and T. Miki, *Mikrochim. Acta* **1**, 89 (1978).
[44] J. M. DiBussolo and W. R. Nes, *J. Chromatogr. Sci.* **20**, 193 (1982).
[45] K. S. Ritter, B. A. Weiss, A. L. Norrbom, and W. R. Nes, *Comp. Biochem. Physiol.* **71B**, 345 (1982).
[46] W. J. Pinto and W. R. Nes, *J. Biol. Chem.* **258**, 4472 (1983).
[47] W. J. Pinto, R. Lozano, B. C. Sekula, and W. R. Nes, *Biochem. Biophys. Res. Commun.* **112**, 47 (1983).

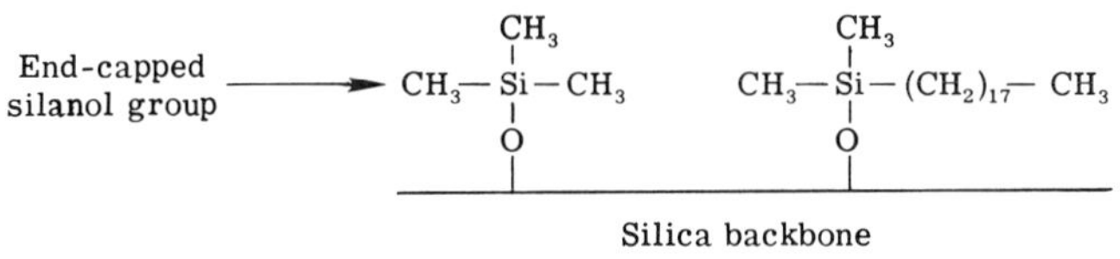

FIG. 1. Structure of C_{18} stationary phase end-capped with TMS.

stationary phase an —O—Si(CH$_3$)$_2$—(CH$_2$)$_{17}$CH$_3$ group is attached to one of the Si atoms of the silica (Fig. 1). This octadecyldimethylsilyloxy group is the heart of the so-called ODS or C_{18} RPHPLC packings which are now widely used both in columns and on plates. A C_8-packing is one in which C_{18} is replaced by C_8, and so forth. Such nonpolar stationary phases have been very successfully used for sterols with 100% acetonitrile as well as with acetonitrile and 10–20% methanol, isopropanol, or water as the mobile phase.

As noted above, reversed-phase chromatography is more sensitive to the presence of double bonds and to molecular size than is adsorption chromatography. Straight lines are obtained by the RP techniques when the size of the molecule (carbon number) is plotted against a suitable function of the chromatographic mobility either in thin layer[33] or column[44] modes. For instance, when log α_c [where α_c is the retention time (or volume) relative to that of cholesterol] is plotted against the number of carbon atoms in the side chain, a straight line is produced (Fig. 2). This linear relationship[44] shows that each CH$_2$-group added without branching increases the retention time (or volume) by the same percentage which from the slope of the line can be calculated to be 25% for a C_{18}-column and acetonitrile in the HPLC mode. Branching reduces the contribution. For instance, it is 16% at C-24 in the absence of nearby double bonds and still lower at C-4 and C-14 (Table II). Coupled with other information regularities of this sort become a powerful tool for the elucidation of structure or for a decision as to whether RPHPLC should be used to separate a particular mixture.[5,44,48,49]

The relationship between rate of movement and structure can be manipulated in an exacting mathematical sense for structural elucidation. If one divides the rate of movement of a sterol with functional group-X by the rate of movement of the sterol without this group, a contribution for X is obtained which has been termed its σ-value.[44] The greater the σ-value, the slower the sterol moves. A long list of σ-values has been compiled.[44] The σ-value when multiplied by the rate of movement of the sterol with-

---

[48] E. Heftmann and I. R. Hunter, *J. Chromatogr.* **165**, 283 (1979).
[49] M. Gassiot, E. Fernandez, G. Firpo, R. Carbó, and M. Martin, *J. Chromatogr.* **108**, 337 (1975).

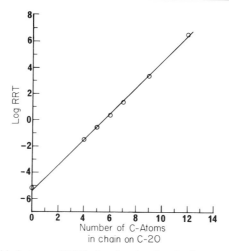

FIG. 2. Relationship between RRT (α_c) and molecular size in reversed-phase HPLC on a C_{18} column with acetonitrile. The sterols chromatographed were pregn-5-en-3β-ol and its (20R)-n-alkyl derivatives with varying numbers of carbon atoms (4, 5, 6, 7, 9, and 12) as plotted on the x axis where 0 represents the parent C_{21}-sterol. The standard for RRT (α_c) is cholesterol. Data are from J. M. DiBussolo and W. R. Nes, *J. Chromatogr. Sci.* **20**, 193 (1982).

out the structural feature gives the rate of movement of the sterol with the feature. Thus, the 25% contribution of a CH_2-group mentioned in the previous paragraph is equivalent to a σ-value of 1.25 which is the multiplicand to obtain the retention time of C_{n+1} from C_n for unbranched additions to the side chain.

Rate of movement in terms of k' is given relative to cholesterol by α_c in Eq. (1). The meaning of k' is given by Eq. (2). The V_0 is the void volume

$$\alpha_c = k' \text{ of test sterol}/k' \text{ of cholesterol} \tag{1}$$
$$k' = (V_s - V_0)/V_0 = (t_s - t_0)/t_0 \tag{2}$$

(expressed in time as t_0) and is the volume required to elute a very polar substance, e.g., uracil, which moves with the solvent front; V_s (t_s being the time analog) is the total elution volume for the sterol in question. That is, k' (the capacity factor) is the ratio of the corrected (for void volume) volume (or time) required to elute the sterol divided by the void volume (or time). The σ-factor is then given by Eq. (3) which after division of

$$\sigma = k' \text{ of sterol with feature}/k' \text{ of sterol without feature} \tag{3}$$

numerator and denominator by k' for cholesterol giving Eq. (4) simplifies by substitutions from Eq. (1) to give Eq. (5) in which the contribution (σ)

TABLE II
Contribution of Selected Molecular Features to Retention Time in RPHPLC with Acetonitrile on a C_{18} Column[a]

Feature	σ-Value[b]
Δ^4 (with 3β-OH)	0.55
Δ^5 (no 3-OH)	0.64
Δ^5 (with 3β-OH)	0.61
Δ^7	0.65
$\Delta^{8(14)}$	0.65
E-$\Delta^{17(20)}$	0.67
Z-$\Delta^{17(20)}$	0.63
E-Δ^{22} (no 24-C)	0.78
Z-Δ^{22} (no 24-C)	0.70
E-Δ^{22} (24-C)	0.86
$\Delta^{24(25)}$ (no 24-C)	0.66
$\Delta^{24(28)}$ (24-C_1)	0.64
$\Delta^{24(28)}$ (24-C_2)	0.71
$\Delta^{5,7}$	0.45
$\Delta^{25(27)}$ (24-C_2)	0.67
3β-OH (without Δ^5)	0.55
3β-OH (with Δ^5)	0.53
3α-OH (with Δ^5)	0.37
20-Keto	0.27
7β-OH (with Δ^5)	0.37
20α-OH	0.31
25-OH	0.41
5β-H (vs 5α-H)	0.87
4α-CH_3 (with Δ^7)	0.98
4β-CH_3 (with Δ^7)	1.02
14α-CH_3 (with Δ^7)	1.04
Unbranched, sat. CH_2 extenting side chain	1.26
20-CH_3 (added to cholesterol)	1.00
24α- or 24β-CH_3 (without E-Δ^{22})	1.16
24β-CH_3 (with E-Δ^{22})	1.24
24α- or 24β-C_2H_5 (without E-Δ^{22})	1.37
24α- or 24β-C_2H_5 (with E-Δ^{22})	1.51

[a] J. M. DiBussolo and W. R. Nes, *J. Chromatogr. Sci.* **20**, 193 (1982).

[b] σ was calculated by dividing the retention time (at constant flow rate) of the sterol with the feature by that of the sterol without the feature.

$$\sigma = \frac{k' \text{ of sterol with feature}}{k' \text{ of cholesterol}} \bigg/ \frac{k' \text{ of sterol without feature}}{k' \text{ of cholesterol}} \quad (4)$$

$$\sigma = \alpha_c \text{ of sterol with feature}/\alpha_c \text{ of sterol without feature} \quad (5)$$

of a group is readily treated in terms of α_c. If we define the retention time (RT_s) of the test sterol as $t_s - t_0$ in a system with constant rate of solvent movement, relative retention time (RRT) in column chromatography becomes equal to α_c [Eq. (6)], because t_0 will cancel [cf. Eq. (2)], and σ becomes a function of RRT [see Eq. (7)] as well as of RT itself, because

$$\alpha_c = \frac{(t_s - t_0)}{t_0} \bigg/ \frac{(t_{\text{chol.}} - t_0)}{t_0} = \frac{t_s - t_0}{t_{\text{chol.}} - t_0} = \frac{RT_s}{RT_{\text{chol.}}} = RRT \quad (6)$$

$$\sigma = \frac{RRT_s \text{ with feature}}{RRT_s \text{ without feature}} = \frac{RT_s \text{ with feature}}{RT_s \text{ without feature}} \quad (7)$$

the values of $RT_{\text{chol.}}$ cancel. Now, if we multiply the α_c, RT, or RRT of a parent by the σ-values of various groups added to it, we will get the α_c, RT, or RRT, respectively, of the final sterol [Eq. (8)].[44] For instance, the

$$\alpha_c \text{ (or } RT \text{ or } RRT) = \text{parental } \alpha_c(RT, RRT) \times \text{product of } \sigma\text{-values} \quad (8)$$

α_c for 5α-cholestane is 2.95 on a particular C_{18}-column developed with acetonitrile.[44] The σ-values for the Δ^5-bond and 3β-OH group are 0.64 and 0.53, respectively (Table II). The calculated α_c for cholesterol is then 2.95 × 0.64 × 0.53 or 1.00. The actual value is also 1.00. Such a calculation allows one to estimate where a given sterol would appear in a chromatogram and whether it would separate from some other sterol. Differences of about 3% in α_c-values will yield separate peaks. It is perhaps worth emphasizing that when $\sigma < 1$, as it usually is for double bonds, carbonyls and hydroxyl groups, an increasing effect of the group correlates with a decreasing σ-value, while increased effect correlates with a greater σ-value when $\sigma > 1$. For more detail, the original literature[44,45] should be consulted, but it is perhaps worth noting that there is a marked affect of the molecular environment on the σ-value of a group. A nearby (vicinal) double bond or steric hindrance, or departure from flatness have important detailed consequences which permit subtle changes in the α_c-values and therefore separation of closely related sterols. For instance, inversion of the configuration of C-3 of cholesterol reduces the α_c-value by a substantial 33% as shown by the σ-values for 3α- and 3β-hydroxyl groups (Table II). Bending the carbon skeleton at the A/B-ring (junction) (inversion at C-5) also reduces the α_c-value (by 13%), but despite a stereochemical alteration, the conversion of lanosterol to cycloartenol increases the

α_c-value (by 17%) because one of the polar groups (Δ^8) has been eliminated. Steric hindrance of methyl groups whether in the side chain or nucleus reduces their effect. Addition of CH_3 to C-20, C-4, or C-14 actually has almost no influence yielding σ-values as small as 1.00 to 1.04.

Gas–Liquid Chromatography

The movement of sterols in gas–liquid systems depends on the polarity and molecular weight of the molecule with each of these structural features making a major contribution to mobility. The larger or more polar the molecule, the slower is the movement. For instance, in the author's laboratory Joseph[12] found the logarithm of the retention time increased linearly with the number of carbon atoms in an homologous series of sterols in which the size of the side chain on C-20 was varied from 0 to 12 carbon atoms. Furthermore, by an extension of Hückel's molecular orbital method Gassiot et al.[49] were able to calculate the retention time of many sterols a priori from consideration of their molecular volume (a derivative of size), total electronic energy, and effective charge at a given position. The correlation found by Gassiot et al. between their theoretical values and the experimental ones obtained by Patterson[50] on SE-30 was remarkably good as seen in Table III.

The influence of sterol structure on mobility is a manifestation of the effect of structure on the ability of the molecule repetitively to vaporize from and to dissolve in the stationary liquid phase. Thus, the contribution which molecular weight (frequently correlating with size and volume) makes to mobility will depend mainly on how vaporization is altered, while the way changes in polarity influence mobility will also depend importantly on the nature of the liquid phase. In the homologous series, for instance, comprised of cholesterol, 24α-methylcholesterol (campesterol), and 24α-ethylcholesterol (sitosterol), where the molecular weight increases with no change in the number and kind of polar groups, each additional carbon atom and its associated two hydrogen atoms successively increases the retention time by about 29%, and this effect is influenced only slightly (a few percent) by the polarity of the liquid phase (cf. later for more detail).[50,51] In contrast, if instead of a CH_2-group a polar one is added, there is a much larger increase in the retention time (e.g., 83% on SE-30 for the 3β-OH group), and the increase is very strongly dependent on the polar character of the liquid phase.

[50] G. W. Patterson, *Anal. Chem.* **43**, 1165 (1971).
[51] G. A. Blondin, B. D. Kulkarni, J. P. John, R. T. van Aller, P. T. Russell, and W. R. Nes, *Anal. Chem.* **39**, 36 (1967).

TABLE III
COMPARISON OF GLC RETENTION TIMES CALCULATED[a] BY QUANTUM
CHEMISTRY WITH THOSE OBSERVED[b]

Sterol	Retention time (SE-30) of acetate relative to RT of cholesteryl acetate	
	Calculated[a]	Observed[b]
5,22(Z)-Cholestadienol	0.86	0.87
5,22,24-Cholestatrienol	0.93	0.94
5,7,22(E)-Cholestatrienol	0.94	0.99
5-Cholestenol	1.01	1.00
5α-Cholestanol	1.08	1.03
8,14-Cholestadienol	1.00	1.03
14α-Methyl-8-cholestadienol	1.06	1.04
8-Cholestenol	1.07	1.05
5,25-Cholestadienol	1.00	1.07
5,24-Cholestadienol	1.08	1.09
5,7-Cholestadienol	1.09	1.09
24β-Methyl-5,22-cholestadienol	1.14	1.12
7-Cholestenol	1.16	1.12
8,24-Cholestadienol	1.14	1.13
14α-Methyl-7-cholestenol	1.15	1.17
24β-Methyl-5,7,22(E)-cholestatrienol	1.28	1.22
24-Methyl-5,24(28)-cholestadienol	1.28	1.26
24α-Methyl-5-cholestenol	1.35	1.30
14α-Methyl-24-methyl-8,24(28)-cholestadienol	1.33	1.30
24β-Methyl-8,14-cholestadienol	1.34	1.33
14α,24β-Dimethyl-8-cholestadienol	1.40	1.34
24β-Methyl-8-cholestenol	1.36	1.36
4,4,14-Trimethyl-8-cholestenol	1.43	1.41
24β-Methyl-5,7-cholestadienol	1.43	1.42
24α-Ethyl-5,22(E)-cholestadienol	1.44	1.42
24-Methyl-7,24(28)-cholestadienol	1.42	1.42
24β-Methyl-7-cholestenol	1.49	1.46
24ξ-Ethyl-7,22(E)-cholestadienol	1.58	1.58
24α-Ethyl-5-cholestenol	1.59	1.63
24α-Ethyl-8-cholestenol	1.65	1.70
24α-Ethyl-7,25-cholestadienol	1.78	1.73
14α-Methyl-24α-ethyl-8-cholestenol	1.83	1.92

[a] Each value was calculated theoretically from the molecular volumes and electronic status by M. Gassiot, E. Fernandez, G. Firpo, R. Carbó, and M. Martin, *J. Chromatogr.* **108,** 337 (1975).

[b] Values were measured experimentally by G. W. Patterson, *Anal. Chem.* **43,** 1165 (1971).

TABLE IV
LIQUID PHASES FOR GAS–LIQUID CHROMATOGRAPHY

Designation	Structural type	McReynold's constant
SE-30 (ultraphase)	Methylsilicone	15
OV-1	Dimethylsilicone	16
UCCW-982	Vinylmethylsilicone	17
SE-30	Methylsilicone	17
Dexsil-300	Polycarboranesiloxane	47
OV-17	Phenylmethylsilicone	119
OV-25	—	178
XE-60	Methylcyanosilicone	—
SP-2340	Cyanosilicone	—
QF-1	Trifluoropropylsiloxane	148
Hi-EFF-8BP	Cyclohexanedimethanol succinate	271
Silar-5CP	Cyanopropylphenylsilicone	319
SP-1000	Carbowax 20M plus substituted terephthalic acid	332
STAP	—	345

Some of the liquid phases in common use are shown in Table IV. They are of two basic types: the nonpolar silicones, e.g., SE-30 in which the silicon atoms of the polymer bear alkyl groups as in $\cdots$ Si(CH$_3$)$_2$—O—Si(CH$_3$)$_2$—O $\cdots$, and polar phases both of the silicone (actually silicon ethers as shown, not ketones as originally thought) and other types. In the polar silicones one or more of the methyl groups in the structure just given are replaced by a polar group. An example is XE-60 which has both methyl and nitrile (C≡N) groups. Still more polar phases are found in the usual carbon chemistry as with polyesters such as HI-EFF-8BP. The overall polarity is measured by McReynold's constants[52] which increase as the polarity increases.

By relating the retention time to the McReynold's constant we can arrive at a quantitative assessment of the interaction between the sterol and the liquid phase. That binding of polar groups of the sterol occurs to polar groups of the liquid phase can be seen from Fig. 3 where the data of Homberg and Bielefeld[53] have been plotted. It will be seen that as the McReynold's constant increases, there is a very much larger decrease in the mobility, i.e., an increase in the retention time, with steroids containing an oxygen function than there is with the corresponding hydrocarbon. For instance, the retention time of 5α-cholestan-3-one rises 3-fold from 1.93 relative to the desoxy analog (5α-cholestane) to 5.76 as the McRey-

---

[52] W. O. McReynolds, *J. Chromatogr. Sci.* **8**, 685 (1970).
[53] E. Homberg and B. Bielefeld, *J. Chromatogr.* **180**, 83 (1979).

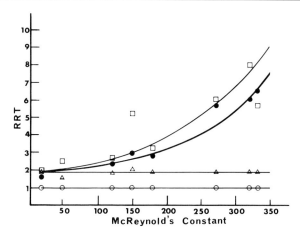

FIG. 3. Relationship between relative retention time (*RRT*; relative to 5α-cholestane) in GLC and the McReynold's constant for the stationary phase. 5α-Cholestane, ○; cholesteryl TMS derivative, △; 5α-cholestan-3β-ol, ●; and 5α-cholestan-3-one, □. Data are taken from E. Homberg and B. Bielefeld, *J. Chromatogr.* **180**, 83 (1979).

nold's constant increases from 17 to 332. On the other hand, when the oxygen function is protected from interactions, as it is in the trimethylsilyl (TMS) ethers, the relative retention time is not significantly influenced by the McReynold's constant.

Interaction between double bonds and the liquid phase is variable but small with low McReynold's constants[53] in the case of Δ^1, Δ^2, Δ^4, Δ^5, Δ^7, $\Delta^{3,5}$, $\Delta^{4,6}$, $\Delta^{2,6}$, and $\Delta^{5,7}$ (retention times varying from 0.96 to 1.10 relative to 5α-cholestane for McReynold's constants less than 50). When the McReynold's constant becomes large, there is but a small increase in the relative retention times of the sterenes (to 1.07–1.26) and only slightly more for the conjugated steradienes (to 1.22–1.57).[53] Similar (small) increases occurring in relative retention time as the polarity of the liquid phase is raised have been observed for the Δ^{22}-bond as well as for the $\Delta^{24(28)}$-bond in both the methylene and ethylidene groups.[50,54]

The foregoing discussion shows that an increase in the polarity of the liquid phase can greatly increase the separation of some steroids. A good example, derived from the interaction of the hydroxyl group with the liquid phase, is the separation of cholesterol from its corresponding hydrocarbon (Δ^5-cholestene). The ratio (alcohol to hydrocarbon) of retention times is 1.77 at McReynold's constant 17 and 6.12 at McReynold's constant 332.[53] A more important though less dramatic case is the separa-

[54] J. A. Ballantine, J. C. Roberts, and R. J. Morris, *J. Chromatogr.* **103**, 289 (1975).

tion of cholesterol from its dihydro derivative, 5α-cholestan-3β-ol. The Δ^5-bond (in hydrocarbons, 3-alcohols, and 3-trimethylsilyl ethers) actually has a slightly negative effect on retention time when the McReynold's constant is low (< 50) but with increasing McReynold's constant there develops an interaction which can be sufficiently strong for the double bond to have a small (5%) positive effect on retention time.[53] Thus, the separation of the 5α-stanol from the Δ^5-stenol both increases and reverses for the free alcohols as the McReynold's constant is raised.[53] The crossover point is near McReynold's constant 100.

Conversely, a decrease in the polarity of the liquid phase tends to increase the influence of the nonpolar 24-ethyl group and to do so more than for the 24-methyl group. Thus, on HI-EFF-8BP the TMS-ethers of campesterol and stigmasterol have exactly the same retention times (1.32 relative to cholesteryl trimethylsilyl ether). However, on SE-30 the retention time of the campesteryl TMS ether is essentially unchanged (1.31) but that of the stigmasteryl TMS ether has actually increased 7% to 1.40. Although this is a small change, it permits resolution.[54] Here the methyl groups of the SE-30 liquid phase (cf. Table IV) are interacting with the 24-alkylsterol and retarding movement more than are the ester groups of HI-EFF-8BP (cf. Table IV). The observed increase in retention time, incidentally, would not be expected to be due to the presence of the Δ^{22}-bond, and experimentally the relative retention time of cholesta-5,22(E)-dien-3β-yl TMS ether actually falls 5% as the McReynold's constant is reduced from 345 to 17.[54]

Addition of a CH_2-group to the side chain as already mentioned, increases the retention time by about 29%. Similarly, a substantial but smaller increment is found with additions to C-4 to give the 4α-methylsterols[50,51] or 4α-methylsteranes.[11] The equatorial 4α-addition gives an increment of about 15% on SE-30,[11,50] but when the CH_2-group is added β and axially the retention time is increased by the usual 30% or so[11,55] on SE-30[11] or SP2250.[55] Steric hindrance of the oxygen function at C-3 probably does not account for why the 4α-substituent gives a smaller increment to the retention time than does a similar addition to the side chain, because it is independent of the presence[50,55] or absence[11] of the 3β-OH group. It is noteworthy that a 4β-methyl group causes no major change in the thickness or flatness of the molecule, although there is a slight thickening in the 4α-methyl case. However, a 14α-methyl group cause a substantial thickening because there are no other substituents on the α-face. The decrease in flatness brought about by protrusion of the methyl group out from the otherwise nearly planar α-surface seems almost exactly to

[55] T. M. Buttke and K. Bloch, *Biochemistry* **20**, 3267 (1981).

counteract the effect of the increase in molecular weight, because the increment in retention time is experimentally a negative 1–2% on SE-30.[50] A departure from flatness caused by inversion of C-14 from 14α-H to 14β-H also is associated with a negative effect on the retention time both with (euphoids)[56] and without[11] inversion of C-17. This "α-face effect" is also shown at the other end of the molecule. Inversion of C-3 from the usual arrangement as a 3β-alcohol to give 3α-hydroxysterols strongly reduces (by 10–25%) the retention time even though there is no change in the molecular weight.[50,56,57] The negative effect of inversion at C-3 is observed both with and without (in the 5α-series) a Δ^5-bond as well as on both polar and nonpolar phases. A bend at the A/B-ring junction caused by inversion of C-5 from 5α-H to 5β-H which throws much of ring A onto the α-face also is accompanied by a reduction in the retention time.[11,50,56] Perhaps sterol–sterol packing in the liquid phase occurs on the α-face in which case groups strongly protruding onto this side of the molecule would reduce condensation from the gas phase. This explanation would lead to the expectation that 9β,19-cyclosterols, e.g., cycloartenol, which are bent (at the B/C-ring junction) toward the α-face might move more slowly than their Δ^8-analogs, e.g., lanosterol. Actually, however, cycloartenol moves about 18% slower than lanosterol (even though the Δ^8-bond has a small positive effect[50]) for reasons which are not yet clear. However, the compact (20S)-sterols move faster than their extended (20R)-analogs,[12–14,56] the compact (20R)-euphoids move faster than the extended (20S)-euphoids,[56] the compact (Z)-17(20)-dehydrocholesterol moves faster than its more extended (E)-isomer,[15] and the bent (Z)-22-dehydrocholesterol moves faster than the extended (E)-isomer.[50]

From a knowledge of the contribution (σ^G where G stands for GLC to distinguish this σ from σ in HPLC) which each group makes to the retention time, it is possible to make a calculation of the retention time of any given sterol. The σ^G-values are obtained by division of the retention time of the sterol with the feature by that of the sterol lacking the feature (cf. the section on reversed-phase chromatography). In practice the relative retention times can be used for this division, because the retention time of the standard will cancel. The retention time (or relative retention time) can be calculated from the value for a parent, such as 5α-cholestane, times the product of the σ^G-values for the groups (or stereochemical changes) added to the parent. These relationships are shown by the equations which follow.

[56] T. Itoh, H. Tani, K. Fukushima, T. Tamura, and T. Matsumoto, *J. Chromatogr.* **234**, 65 (1982).
[57] M. Tohma, Y. Nakata, and T. Kurosawa, *J. Chromatogr.* **171**, 469 (1979).

$$\sigma^G = RRT \text{ with feature}/RRT \text{ without feature} \qquad (9)$$

$$RT_u = RT_p \times \text{product of } \sigma^G\text{-values} \qquad (10)$$

$$RRT_u = RRT_p \times \text{product of } \sigma^G\text{-values} \qquad (11)$$

where RT = retention time, RRT = relative retention time, u = unknown, p = parent, and σ^G = contribution of molecular feature

The σ^G-values for a number of groups and stereochemical alterations are given in Table V. They can be used not only to calculate the rate of movement of a given sterol which is useful for structural elucidation as well as in the decision as to whether the sterol will separate from some other one in a given system, but the σ-values also can be used directly to calculate separation factors for two or more sterols.

Much more data are available in the original literature than is shown in Table V. Resolutions for many pairs of sterols have been recorded in the recent literature.

TABLE V
CONTRIBUTION[a] OF SELECTED MOLECULAR FEATURES TO GLC RETENTION TIMES[b]

Feature[c]	σ^G		
	SE-30	OV-17	Hi-EFF-8P
Δ^1	0.99	1.02	1.12
Δ^2	0.99	1.01	1.13
Δ^4	0.99	1.01	1.05
Δ^5	1.01	1.03	1.09
Δ^7	1.11	1.19	1.24
Δ^8	1.02	1.06	1.01
$\Delta^{5,7}$	1.08	1.21	1.51
$\Delta^{8,14}$	0.99	—	1.11
E-Δ^{22}	0.91	0.93	0.91
Z-Δ^{22}	0.87	0.90	0.88
E-Δ^{23}	—	1.03	—
$\Delta^{24,25}$	1.09	1.21	1.29
$\Delta^{24(28)}$ (with 24-C_1)	0.94	1.03	1.08
E-$\Delta^{24(28)}$ (with 24-C_2)	0.98	—	1.08
Z-$\Delta^{24(28)}$ (with 24-C_2)	1.01	—	1.16
$\Delta^{25(27)}$	—	1.01	—
3α-OAc	2.24	2.65	4.56
3α-OTMS	1.64	—	—
3β-OAc	2.54	3.05	5.18
3β-OTMS	2.04	1.87	1.91

(*Continued*)

TABLE V (Continued)

Feature[c]	σ^G		
	SE-30	OV-17	Hi-EFF-8P
3β-OAc (with Δ^5)	2.47	2.99	5.18
3β-OTMS (with Δ^5)	1.99	1.83	1.90
3β-OH	1.83	2.32	5.68
3β-OH (with Δ^5)	1.78	2.36	6.01
6α-OH	1.60	1.96	—
3-Keto	1.93	2.71	6.09
3-Keto (with Δ^4)	2.30	3.42	9.21
3-Keto (with Δ^7)	1.92	2.64	5.73
7-Keto	1.66	2.26	4.26
4α-CH$_3$	1.15	1.12	1.09
4β-CH$_3$	1.22	1.18	
24α-CH$_3$	1.30	1.312	1.32
24β-CH$_3$	1.30	1.307	1.32
24α-C$_2$H$_5$	1.64	1.634	1.59
24β-C$_2$H$_5$	1.64	1.626	—
24α-C$_2$H$_5$ (with Δ^{22})	—	1.54	—
24β-C$_2$H$_5$ (with Δ^{22})	—	1.55	—
Extension of side chain by an unbranched CH$_2$-group	—	1.29	—
9β,19-Cyclo	—	1.23(?)	—
14α-CH$_3$ (with Δ^8)	0.99	—	0.93
14α-CH$_3$ (with Δ^7)	1.04	—	0.97
5β (vs 5α) (no 3-ol)	0.92	—	—
5β (vs 5α) (with 3β-ol)	0.83	0.80	0.76
14β (vs 14α)	0.87	—	—
20S (vs 20R)	0.91	0.89	—

[a] The contribution σ^G is derived by dividing the retention time or relative retention time of the sterol with the feature by that of the sterol without the feature.

[b] The data used to calculate σ^G are from E. Homberg and B. Bielefeld [*J. Chromatogr.* **180**, 83 (1979)], G. W. Patterson [*Anal. Chem.* **43**, 1165 (1971)], G. Ryback [*J. Chromatogr.* **116**, 207 (1976)], M. Tohma, Y. Nakata, and T. Kurosawa [*J. Chromatogr.* **171**, 469 (1979)], J. A. Ballantine and J. C. Roberts [*J. Chromatogr.* **103**, 289 (1975)], T. Itoh, H. Tani, K. Fukushima, T. Tamura, and T. Matsumoto [*J. Chromatogr.* **234**, 65 (1982)], and J. M. Joseph [Ph.D. Dissertation, Drexel University (1980)].

[c] Unless otherwise noted, there is a 5α-H-atom in cases where the configuration is important.

The development of capillary columns has been particularly valuable.[56,58-63] Instead of filling a column fully with the stationary phase ("packed columns"), various investigators and companies have bound either the stationary phase to the inside surface of a column of capillary dimensions giving a "wall coated open tubular column" (WCOT) or have bound a support (covered with a liquid phase) to the inside wall giving "support coated open tubular columns" (SCOT). The WCOT and SCOT methods have yielded a much more efficient transfer of sterol (so-called "mass transfer") to the liquid phase than has been possible previously resulting in truer equilibria between gas and liquid phases within a given segment of the column. By sharpening the peaks the better mass transfer has allowed longer columns to be used with an increased number of theoretical plates. Some important separations not achievable with packed columns have therefore been reported with capillary columns. These include the separation of 24α- and 24β-methyl and ethyl sterols.[56,58] The β-epimers move faster than the ones, respectively, with an α-configuration. However, a vicinal effect exists with the (E)-Δ^{22}-bond, at least in the 24-ethyl series, resulting in a reversal of the relative rates of movement. For the pairs chondrillasterol/spinasterol, poriferasterol/stigmasterol, and 25(27)-dehydrochondrillasterol/25(27)-dehydrospinasterol the Δ^{22}-24β-epimer moves slower than the isomer with the α-configuration. Equivocal results have been reported in the same paper for the Δ^{22}-24-methylsterols.[58] Labeling of the chromatographic tracing is reversed relative to the table of retention times.

Combination (GC-MS) of mass spectrometry (MS) with gas–liquid chromatography (GLC or GC) has been extensively employed, but the technique has been hampered to some extent by the bleeding of the liquid phase from the column, which causes an elevated MS base line. Several laboratories[54,64-67] have worked with some success with more thermostable liquid phases, but phases such as SE-30 and OV-17 have been used without much difficulty. An example of a new thermostable phase used by

[58] R. H. Thompson, Jr., G. W. Patterson, M. J. Thompson, and H. T. Slover, *Lipids* **16,** 694 (1981).
[59] C. G. Edmonds, A. G. Smith, and C. J. W. Brooks, *J. Chromatogr.* **133,** 372 (1977).
[60] E. Tomori, E. Orban, and A. Maderspack, *J. Chromatogr.* **191,** 261 (1980).
[61] S.-N. Lin and E. C. Horning, *J. Chromatogr.* **112,** 483 (1975).
[62] J. M. Halket and B. P. Lisboa, *J. Chromatogr.* **189,** 267 (1980).
[63] R. S. Deedler, J. J. M. Ramaekers, J. H. M. Van Den Berg, and M. L. Wetzels, *J. Chromatogr.* **119,** 99 (1976).
[64] J. A. Ballantine, K. Williams, and R. J. Morris, *J. Chromatogr.* **198,** 111 (1980).
[65] J. A. Ballantine, K. Williams, and R. J. Morris, *J. Chromatogr.* **166,** 491 (1978).
[66] J. A. Ballantine and K. Williams, *J. Chromatogr.* **148,** 504 (1978).
[67] S. G. Wakeham and N. M. Frew, *Lipids* **17,** 831 (1982).

several laboratories is a polyphenyl ether sulfone known as POLY-S-176 (formerly PZ-176). Dexsil 300GC and SE-30 Ultraphase are also considered especially thermostable.[64]

Liquid phases are primarily characterized by their McReynold's constants (Table IV and Fig. 3), but in some cases (as has been reported for QF1 and ketosteroids[53]) special affinities exist which exaggerate a particular separation. The affinity of the liquid phase, whether special or not, can also be coupled with advantage to derivatization of the sterol. Thus, while SP-2401 as the liquid phase permits an excellent separation of cholesterol (faster) and 5α-cholestanol (slower) by as much as 5%, conversion to their trifluoroacetates allowed an increase in the separation by as much as 14% of the retention time.[68] This is the largest separation of a Δ^5/Δ^0 pair which has been reported.

Gas–liquid chromatography is usually carried out without artifact formation, but in some cases reactions occur. These include the dehydration of the allylic Δ^4-3β-hydroxysterols[53] as well as of the 20-hydroxysterols found in the author's laboratory (unpublished). More complicated pyrolysis with the production of many peaks occurs with 3β-hydroxy-Δ^6-5α,8α-epidioxides[51,69] and with 5α,6α-epoxy-3β,7α-diols.[69] The peaks obtained constitute a fingerprint pattern which is quite useful for identification. Vitamin D also reacts during gas chromatography with reclosure of the 9(10)-bond yielding pyro- and isopyrocalciferol (isomers of the natural $\Delta^{5,7}$-sterols).[70]

Ultraviolet Spectrometry

Ultraviolet (UV) spectra are most useful for identification and quantitation when conjugated ethylenic or carbonyl double bonds, e.g., in the $\Delta^{5,7}$- or 3-keto-Δ^4-systems, are present. In such cases the absorption band is intense and is shifted fully into the usable portion (205–400 nm) of the UV region. The observed λ_{max} is sensitive to the type, position, and extent of conjugation of the unsaturated system. Isolated ethylenic double bonds have $\lambda_{max} < 205$ nm, and isolated carbonyl groups, while having λ_{max} in the vicinity of 290 nm, have very low ε-values (< 100). Conjugated systems by contrast have $\epsilon > 5000$ in most of the cases commonly encountered. Dorfman[71] has compiled the most extensive listing of UV absorption of sterols which is available. Table VI gives some selected data from

---

[68] K. A. Klause and M. T. Subbiak, *J. Chromatogr.* **103**, 170 (1975).
[69] P. A. G. Malya, J. R. Wright, and W. R. Nes, *J. Chromatogr. Sci.* **9**, 700 (1971).
[70] H. Ziffer, W. J. A. Vanden Heuvel, E. O. Haahti, and F. C. Horning, *J. Am. Chem. Soc.* **82**, 6411 (1960).
[71] L. Dorfman, *Chem. Rev.* **53**, 47 (1953).

TABLE VI
ULTRAVIOLET ABSORPTION OF STEROIDS WITH SELECTED CONJUGATED SYSTEMS

Conjugated system	λ_{max} (nm)	ε	Reference
$\Delta^{3,5}$ (3-enol acetate)	235	19,000	a
$\Delta^{4,6}$	232	21,500	a
	239	23,500	
	248	16,000	
$\Delta^{5,7}$	271	11,400	a
	282	11,900	
	293	6,900	
$\Delta^{6,8(9)}$	275	5,000	a
$\Delta^{6,8(14)}$	253	17,000	a
$\Delta^{6,8(14)}$ (3,5-cyclo)	262	24,200	b
$\Delta^{7,9(11)}$	236	12,800	a
	243	14,500	
	251	10,100	
$\Delta^{7,14}$	242	10,000	a
$\Delta^{8,14}$	248	18,500	a
$\Delta^{4,6,8(14)}$	283	33,000	a
$\Delta^{5,7,9(11)}$	312	10,400	a
	324	11,800	
	339	7,400	
$\Delta^{5,7,14}$	319	16,200	a
$\Delta^{6,8,14}$	280	5,700	c
$\Delta^{6,8(14),9(11)}$ (3,5-cyclo)	244	16,100	b
	295	12,100	
$\Delta^{5,7,9(11),14}$	250	9,860	d
	259	14,400	
	269	15,000	
	304	21,800	
$\Delta^{22,24(28)}$ (24-C_1)	232	19,000	e
Δ^4-3-Keto	241	16,600	a
Δ^4-6-Keto	244	6,300	a
Δ^5-7-Keto	239	12,000	a

[a] L. F. Dorfman, *Chem. Rev.* **53,** 47 (1953). Solvent was ethanol.
[b] W. R. Nes and J. A. Steele, *J. Org. Chem.* **22,** 1457 (1957). Solvent was isooctane.
[c] E. J. Parish, M. Tsuda, and G. J. Schroepfer, Jr., *Chem. Phys. Lipids* **25,** 111 (1979).
[d] W. R. Nes, *J. Am. Chem. Soc.* **78,** 193 (1956). Solvent was isooctane.
[e] W. R. Nes, P. A. G. Malya, F. B. Mallory, K. A. Ferguson, J. R. Landrey, and R. L. Conner, *J. Biol. Chem.* **246,** 561 (1971).

Dorfman's compilation, together with that from the more recent literature. Some of the absorption bands show fine structure, i.e., peaks (or shoulders not listed in Table VI) on one or both sides of the main peak, and these can be diagnostically useful. When the ends of a conjugated system are close to each other, as for a diene system ($\Delta^{5,7}$, etc.) in a single ring (homoannular), the ϵ-value is small compared to the value for conjugated systems in which the ends are far from each other, as in the heteroannular $\Delta^{4,6}$ case. An empirical calculation (Woodward's rules) can be made[71] for the value of λ_{max} which agrees well with the experimental values. The carbonyl-containing conjugated systems exhibit marked solvent effects not observed with the polyenes. Thus, λ_{max} for the Δ^4-3-ketone system shifts from 241 to 230 nm when alcohol is replaced by hexane.[71]

Mass Spectrometry

Mass spectrometry (MS) has two main uses. The first and simplest is to determine the molecular weight (MW) of the sample (and the empirical formula under special conditions of high resolution). This immediately discriminates between various alternatives. In MS carried out by the "field desorption" technique the molecule (M) is ionized giving M^+, which from its mass gives MW, but in ordinary MS, M^+, after being formed, then undergoes fragmentation. The resulting fragmentogram leads to the second use to which MS can be put, structural elucidation, because different groupings in the molecule influence fragmentation in different ways and to different extents. The technique of producing and analyzing fragmentograms has been highly developed for sterols (especially in Djerassi's laboratory, see refs. 72–75 as keys to the extensive literature). Although the rules for interpreting mass spectra can be very complicated, some of the more important aspects include the fact that cleavage of a carbon–carbon bond will be enhanced by increased substitution on either of the carbon atoms or by the presence of an allylic, i.e., γ,δ, double bond (where the carbons on the single bond are taken as α and β). Thus, loss of C-19 will occur and there can be more cleavage of this sort with Δ^5- and Δ^8-sterols than with sterols having a Δ^7-bond as seen in Table VII for the sum of M^+-CH_3 and $M^+-CH_3-H_2O$. When the γ,δ-double bond is conjugated with another one, as in $\Delta^{5,7}$, the loss of C-19 proceeds to a greater extent. Similarly, a Δ^7- or Δ^8-14α-methylsterol will

[72] L. N. Li, U. Sjöstrand, and C. Djerassi, *J. Org. Chem.* **46**, 3867 (1981).
[73] I. J. Massey and C. Djerassi, *J. Org. Chem.* **44**, 2448 (1979).
[74] H. M. Fales, G. W. Milne, H. U. Winkler, H. D. Beckey, J. N. Damico, and R. Barron, *Anal. Chem.* **47**, 207 (1975).
[75] F. F. Knapp and G. J. Schroepfer, Jr., *Chem. Phys. Lipids* **17**, 466 (1976).

TABLE VII
MASS FRAGMENTOGRAM OF SELECTED STEROLS[a]

Fragment	Relative intensity[b] (%) of fragment								
	Structural features[b] added to 5α-cholestan-3β-ol								
	Δ^5	Δ^7	$\Delta^{5,22}$	$\Delta^{7,22}$	$\Delta^{5,7,22}$	$\Delta^{8,24}$	$\Delta^{5,25(27)}$	$\Delta^{5,22,25(27)}$	4,4,14α-CH$_3$-$\Delta^{8,24}$
M$^+$	100	100	66	28	92	100	100	19	43
M$^+$-CH$_3$	30	28	9	10	3	55	25	5	100
M$^+$-C$_2$H$_5$								14	
M$^+$-H$_2$O	45	2	11	1	5		17	4	
M$^+$-CH$_3$-H$_2$O	30	5	13	1	100	11	28	4	50
M$^+$-C$_2$H$_5$-H$_2$O								11	
M$^+$-C$_3$H$_5$-H$_2$O					43				
M$^+$-C$_3$H$_7$				13					
M$^+$-84 or 85	40		22				18	7	
M$^+$-C$_7$H$_{14}$			5				45		
M$^+$-C$_7$H$_{10-12}$O	50		42	31	26			41	
M$^+$-SC[c]	25	21	23	100		18	75		
M$^+$-SC-2H	7		59					100	
M$^+$-SC-H$_2$O	30	70	100	40	55		33	58	
M$^+$-SC-C$_3$H$_6$	27	21	18	18			38		
M$^+$-SC-C$_3$H$_8$	14	21	21	20			33		
M$^+$-SC-C$_3$H$_6$-H$_2$O	40	20	45	3	35		67	26	

[a] Data are taken from W. R. Nes, K. Krevitz, J. Joseph, W. D. Nes, B. Harris, G. F. Gibbons, and G. W. Patterson [*Lipids* **12**, 511 (1977)], U. F. Taylor, A. Kisic, R. A. Pascal, A. Izumi, M. Tsuda, and G. J. Schroepfer, Jr. [*J. Lipid Res.* **22**, 171 (1981)], and B. C. Sekula and W. R. Nes [*Phytochemistry* **19**, 1509 (1980)].

[b] The Δ^5 column represents a composite derived from cholesterol and its 24α- and 24β-methyl and 24α- and 24β-ethyl derivatives. The Δ^7 column is a composite of lathosterol and its 24α-ethyl derivative, and the $\Delta^{5,22}$ and $\Delta^{7,22}$ columns are for the 24α- and 24β-ethylsterols. No clear distinction has been found between the fragmentograms of 24α- and 24β-epimers. The $\Delta^{5,7,22}$ column is for ergosterol which has a 24β-methyl group, and the columns for $\Delta^{5,25(27)}$ and $\Delta^{5,22,25(27)}$ represent 24β-ethylsterols. The $\Delta^{8,24}$ and 4,4,14α-$\Delta^{8,24}$ columns are for 24-H-sterols (zymosterol and lanosterol, respectively). All fragmentations are not shown. For instance, lanosterol yields a series of three peaks with combined intensities of 50 for M$^+$−CH$_3$−SC- a C$_3$-unit with varying amounts of H-atoms. These are variants of the fragmentation in the table. Also in some cases a fragmentation shown in the left-hand column may be present but the assignment is ambiguous. A blank space means no fragment observed, one of only very low intensity was found, or information was not available.

lose the 14α-methyl group very readily; Δ^{22}-sterols will lose the side chain [by cleavage at the C-17(20)-bond] more readily than the 22,23-dihydro analogs; Δ^{25}- and especially $\Delta^{22,25}$-sterols lose a 24-ethyl group if present; and 24-ethylidene-sterols [$\Delta^{24(28)}$] cleave at the bond between C-22 and C-23. Loss of the side chain, which occurs quite commonly, at least to some extent, yields a valuable diagnostic fragment for locating extra carbon atoms in the side chain (SC). Thus, 24-methylcholesterol gives a very strong M^+ for 28 carbon atoms and an M^+-SC for only 19 carbon atoms. Had the methyl group been at C-14 with Δ^5, M^+-SC would have had 14 more mass units. If it were Δ^8-14α-CH_3, M^+-CH_3 would have had a much greater intensity than M^+.

The importance of degree of substitution in promoting cleavage is especially evident in breaks within the nucleus which occur at the ring junctions such as cleavage through ring B of 9,19-cyclosterols, removing 112 mass units comprised of the C-3 oxygen, 7 carbons, and the appropriate hydrogens. At the other end of the molecule both in sterols with and without the 9,19-cyclo group ring D (C_3 and an appropriate number of H-atoms) together with the side chain is often lost. A loss of C_3 with or without loss of H_2O from ring A can also occur from the side chain itself by cleavage at C-24/C-25 (loss of the isopropyl group). This is especially strong in Δ^{22}-24-alkylsterols. Loss of water from ring A also occurs giving M^+-H_2O and M^+-H_2O-CH_3. This fragmentation is enhanced by Δ^5- or especially a $\Delta^{5,7}$-unsaturation. The dehydration and effect of unsaturation on it is exaggerated by ester formation. Thus, while ergosterol gives a strong M^+, its acetate under the same conditions gives little or none. Examination of Table VII will illustrate some of these principles. As already alluded to, the various fragmentations can occur in combinations to varying degrees resulting in rather distinctive spectra in many cases. One of the important features, however, not easily distinguished, is the configuration at C-20 or at C-24 when the latter bears an additional carbon.

Nuclear Magnetic Resonance Spectra

The spectra derived from nuclear magnetic resonance (NMR) permit hydrogen and carbon atoms from a single position to be detected. An internal standard (tetramethylsilane, TMS) is used, and the frequency of absorption is measured in cycles per second (cps or Hz) displacement, called the chemical shift, from the absorption peak for TMS. This displacement is then divided by the applied RF in MHz and given the symbol δ (in parts per million, ppm). An increase in δ is described as a downfield shift and a decrease as an upfield shift. Natural hydrogen and carbon are isotopic mixtures mostly of ^1H and ^2H and of ^{11}C, ^{12}C, ^{13}C, and ^{14}C, respectively. Even though ^{13}C is a minor component, it, along with ^1H, is selected for study owing to spin characteristics. These nuclides of course then act as tags or labels for all the H- and C-atoms in the sterol. The NMR spectra so obtained show resolved signals (absorption peaks) for hydrogen at many of the positions in a sterol and for carbon, remarkably, at nearly all of the positions. As in other physical properties, ethylenic and carbonyl double bonds, hydroxyl groups and stereochemistry strongly influence the detail. This can then be used diagnostically. It is also possible to determine how many H-atoms are on a given C-atom from the multiplicity of the absorption band. This arises from a coupling of the C and H nuclei yielding several possible excited states with similar though not identical energies. For ^{13}C spectra the coupling applies to the H-atoms on the carbon in question, and a singlet is obtained for quaternary carbon (no H-atoms), a doublet for CH, a triplet for CH_2, and a quartet for CH_3. For ^1H spectra the coupling occurs with the H-atoms on the adjacent carbons. The multiplicity is $n + 1$ where n is the number of H-atoms on the carbon atom(s) attached to the carbon atom holding the H-atom(s) in question. Thus, the three H-atoms on C-19 of cholesterol give a singlet (because C-10 is quaternary), while the three H-atoms on C-26 give a doublet (because C-25 has one H-atom).

^1H NMR Spectra. It has been possible with different spectra derived by the application of decoupling and nuclear Overhauser enhancement as well as by other techniques to assign chemical shifts to all of the protons in 11-hydroxyprogesterone (Table VIII). Assignments have also been made to many of the protons in sterols such as cholesterol. Of special significance are the protons (all three of which have the same chemical shift) on a methyl group, because the intensity of the absorption band will be three times as much as for a single proton. The band therefore rises well above the bands for protons in other positions. This is especially true for the singlets from C-18 and C-19 protons and doublets for protons arising from, e.g., C-21, C-26, and C-27. When the multiplicity is greater than 1 the separation of the peaks (within the doublet, triplet, etc.) is given by the coupling constant (J) and the δ-value is taken from the center of the band. Also of interest are protons on double bonds and on carbon atoms bearing an oxygen atom (OH, OCH_3, $OCOCH_3$, etc.), because in both cases the downfield shift caused by the unsaturation or oxygen function puts the absorption in a region of the spectrum not usually occupied permitting the presence of a band to be used diagnostically. Thus, in cholesterol the protons on C-3α, C-6, C-18, C-19, C-21, C-26, and C-27 are readily found experimentally and appear at 3.50, 5.36, 0.68, 1.01, 0.91, 0.87, and 0.87 ppm. In the terminal methyl (C-26 and C-27) doublet at 0.87

TABLE VIII
CHEMICAL SHIFTS IN ^{1}H NMR FOR ALL PROTONS OF
11β-HYDROXYPROGESTERONE

Position	Chemical shift (δ)		
	Hall and Sanders[a]	Arnold et al.[b]	Nes and McKean[c]
1α	1.81	—	—
1β	2.18	—	—
2α	2.35	—	—
2β	2.47	—	—
4	—	—	5.75
6α	2.23	—	—
6β	2.48	—	—
7α	1.06	—	—
7β	2.00	—	—
8	2.00	—	—
9	1.00	—	—
11α	4.40	—	—
12α	1.65	—	—
12β	2.19	—	—
14	1.11	—	—
15α	1.75	—	—
15β	1.33	—	—
16α	1.69	—	—
16β	2.19	—	—
17	2.43	—	—
18	—	0.92	—
19	—	1.45	—
21	—	—	2.12

[a] Spectra were determined in CDCl$_3$ with an uncertainty of 0.01 ppm by L. D. Hall and J. K. M. Sanders, *J. Chem. Soc., Chem. Commun.* p. 368 (1980).

[b] Based on increments for the Δ^4-3-keto, 11β-hydroxy, and 17β-acetyl moieties obtained with spectra of several hundred steroids in CDCl$_3$ obtained by W. Arnold, W. Meister, and G. Englert, *Helv. Chim. Acta* **57,** 1559 (1974).

[c] Values for progesterone from the compilation of W. R. Nes and M. L. McKean, "Biochemistry of Steriods and Other Isopentenoids," p. 111. University Park Press, Baltimore, Maryland.

one can actually observe a shoulder at 220 MHz or greater indicating that the two methyl groups do not have exactly the same signal. When a methyl or ethyl group is present at C-24 the terminal methyl absorptions (C-26 and C-27) are well resolved into two doublets. This effect of one group on another allows structural deductions to be made as illustrated by

TABLE IX
INFLUENCE OF SELECTED GROUPINGS ON THE ^{1}H NMR SIGNALS FROM C-18 AND C-19

Grouping	Increment[a]	
	C-18	C-19
5α-Androstane	0.688	0.787
Δ^2	0.018	−0.023
Δ^5	0.032	0.223
Δ^6	0.052	−0.085
Δ^7	0.117	−0.011
$\Delta^{8(14)}$	0.178	−0.132
$\Delta^{9(11)}$	−0.074	0.115
Δ^{14}	0.242	0.031
3α-OH	0.001	−0.011
3β-OH	0.004	0.036
3β-OAc	0.005	0.054
3β-OCH$_3$	0.002	0.020
3α-OH + Δ^5	0.041	0.225
3β-OH + Δ^5	0.034	0.232
3β-OH + $\Delta^{5,7}$	−0.026	0.166
4β-OH	0.003	0.260
6α-OH	−0.006	−0.007
6β-OH	0.033	0.202
7α-OH	0.000	−0.007
7β-OH	0.022	0.024
3-Keto	0.039	0.237
3-Keto-Δ^4	0.068	0.406
17α-OH	−0.041	0.007
17β-OH	0.042	0.006
17β-C$_2$H$_5$	−0.136	−0.005
17β-COCH$_3$	−0.080	0.000
17β-C$_8$H$_{17}$[b]	−0.040	−0.008
17β-C$_9$H$_{19}$[c]	−0.030	−0.004
17β-C$_{10}$H$_{21}$[d]	−0.039	−0.002
17β-C$_9$H$_{17}$ (trans-Δ^{22})	−0.025	−0.007
17β-C$_{10}$H$_{19}$ (trans-Δ^{22})	−0.018	−0.001

[a] From data in CDCl$_3$ of W. Arnold, W. Meister, and G. Englert, *Helv. Chim. Acta* **57**, 1559 (1974).
[b] Cholesterol side chain.
[c] 24ξ-Methylcholesterol side chain.
[d] 24ξ-Ethylcholesterol side chain.

shifts in the signals for the C-18 and C-19 methyl groups of 5α-androstane given in Table IX taken from the extensive compilation (292 different nuclear substituents) of Arnold et al.[76] The increment listed is added to the value of 5α-androstane to give the value for the steroid in question. Table X shows the effect of structural changes on the signals not only for C-18 and C-19 but also for methyl groups in the side chain. Shifts associated with changes in unsaturation, stereochemistry, and presence of methyl groups at C-4, C-14, and C-24 are evident. The chirality at C-20[13,14] as well as at C-24[77–80] of sterols with [79,80] and without[77,78] a Δ^{22}-bond and the cis–trans isomerism about the $\Delta^{24(28)}$-bond of 24-ethylidenesterols[78,81,82] has been especially well demonstrated with ^{1}H NMR.

^{13}C *NMR Spectra.* The spectra arising from ^{13}C excitation have been the subject of a great deal of recent examination, because so much information is obtainable. For instance, Eggert and Djerassi[83] have considered the chemical shifts for a variety of double bonds. Blunt and Stothers[84] have listed the spectra of no less than 413 steroids and have discussed them at length. Smith[85] has also listed and discussed many spectra. Lanosterol and related compounds, e.g., euphol, have been studied by S. A. Knight.[86] The discrimination between 24α- and 24β-alkylsterols has been reported from several laboratories.[87–90] The ^{13}C spectra of a variety of biosynthetic precursors to cholesterol have been published[91] as have the spectra of 9,19-cyclosterols.[92] The spectra of many sterols, e.g., 20-

[76] W. Arnold, W. Meister, and G. Englert, *Helv. Chim. Acta* **57**, 1559 (1974).
[77] W. R. Nes, K. Krevitz, and S. Behzadan, *Lipids* **11**, 118 (1976).
[78] I. Rubinstein, L. J. Goad, A. D. H. Clague, and L. J. Mulheirn, *Phytochemistry* **15**, 195 (1976).
[79] T. Matsumoto, N. Shimizu, T. Itoh, T. Iida, and A. Nishioka, *Phytochemistry* **22**, 789 (1983).
[80] L. J. Goad, G. G. Holz, Jr., and D. H. Beach, *Phytochemistry* **22**, 425 (1983).
[81] D. J. Frost and J. P. Ward, *Tetrahedron Lett.* **34**, 3779 (1968).
[82] J. Artand, M. C. Iatrides, C. Tisse, J. P. Zahra, and J. Estienne, *Analusis* **8**, 277 (1980).
[83] H. Eggert and C. Djerassi, *J. Org. Chem.* **46**, 5399 (1981).
[84] J. W. Blunt and J. B. Stothers, *Org. Magn. Reson.* **9**, 439 (1977).
[85] W. B. Smith, in "Annual Reports of NMR Spectroscopy" (G. A. Webb, ed.), p. 199. Academic Press, New York, 1978.
[86] S. A. Knight, *Org. Magn. Reson.* **6**, 603 (1974).
[87] Y. Fujimoto and N. Ikekawa, *J. Org. Chem.* **44**, 1011 (1979).
[88] N. Koizumi, Y. Fujimoto, T. Takeshita, and N. Ikewawa, *Chem. Pharm. Bull.* **27**, 38 (1979).
[89] J. L. C. Wright, A. G. McInnes, S. Shimizu, D. G. Smith, and J. A. Walter, *Can. J. Chem.* **56**, 1898 (1978).
[90] J. E. Zarembo and W. R. Nes, *J. Nat. Prod.* **44**, 7 (1981).
[91] M. Tsuda and G. J. Schroepfer, Jr., *J. Org. Chem.* **44**, 1290 (1979).
[92] F. Khuong-Huu, M. Sangare, V. M. Chari, A. Bekaert, M. Devys, M. Barbier, and G. Lukacs, *Tetrahedron Lett.* p. 1787 (1975).

TABLE X
SELECTED METHYL GROUP SIGNALS IN THE ^{1}H NMR SPECTRA OF VARIOUS STEROLS[a]

δ-Values for protons on indicated carbon

Sterol	C-18	C-19	C-21	C-26/C-27	C-28	C-29
5α-Cholestan-3β-ol	0.65	0.81	0.89	0.87	—	—
Cholesterol	0.68	1.01	0.91	0.87	—	—
Cholesta-8(14)-en-3β-ol	0.86	0.70	0.95	0.89	—	—
Cholesta-7,14-dien-3β-ol	0.85	0.92	0.97	0.88	—	—
Lathosterol	0.54	0.80	0.92	0.87/0.86	—	—
(E)-22-Dehydrocholesterol	0.69	1.02	1.01	0.86	—	—
(Z)-22-Dehydrocholesterol	0.72	1.02	0.96	0.89/0.90	—	—
24-Dehydrocholesterol	0.69	1.02	0.95	1.59/1.68	—	—
7-Dehydrocholesterol	0.61	0.94	0.94	0.87	—	—
Zymosterol	0.59	0.93	0.93(?)[b]	1.58/1.66	—	—
4α-CH$_3$-Lathosterol	0.54	0.84	0.92	0.87	—	—
24α-CH$_3$-Cholesterol	0.68	1.01	0.91	0.80/0.85	0.77	—
24β-CH$_3$-Cholesterol	0.68	1.01	0.92	0.78/0.86	0.77	—
24α-CH$_3$-Lathosterol	0.54	0.80	0.92	0.80/0.85	0.78	—
24β-CH$_3$-Lathosterol	0.54	0.80	0.92	0.78/0.85	0.77	—
24α-CH$_3$-(E)-22-Dehydrocholesterol	0.69	1.02	0.00	0.82/0.83	0.91	—
24β-CH$_3$-(E)-22-Dehydrocholesterol	0.69	1.01	1.01	0.82/0.83	0.91	—
24β-CH$_3$-(E)-22-Dehydrolathosterol	0.54	0.80	1.01	0.82/0.83	0.91	—
24α-CH$_3$-7-Dehydrocholesterol	0.62	0.95	0.93	0.80/0.85	0.78	—
24α-CH$_3$-7,(E)-22-Bisdehydrocholesterol[c]	0.63	0.95	1.03	0.83/0.84	0.92	—
24β-CH$_3$-7,22-Bisdehydrocholesterol	0.63	0.95	1.04	0.83/0.84	0.92	—

Compound						
24α-C₂H₅-Cholesterol	0.68	1.01	0.92	0.81/0.84	—	0.85
24β-C₂H₅-Cholesterol	0.68	1.01	0.93	0.81/0.83	—	0.86
24α-C₂H₅-Lathosterol	0.54	0.80	0.93	0.82/0.84	—	0.85
24β-C₂H₅-Lathosterol	0.54	0.80	0.93	0.82/0.84	—	0.86
24α-C₂H₅-(E)-22-Dehydrocholesterol	0.70	1.01	1.02	0.80/0.85	—	0.80
24β-C₂H₅-(E)-22-Dehydrocholesterol	0.70	1.02	1.03	0.79/0.85	—	0.81
24α-C₂H₅-(E)-22-Dehydrolathosterol	0.54	0.80	1.02	0.80/0.85	—	0.80
24β-C₂H₅-(E)-22-Dehydrolathosterol	0.55	0.80	1.03	0.79/0.85	—	0.82
24-Methyl-24(28)-dehydrocholesterol	0.70	1.02	0.98	1.03	—	—
24-Ethyl-(E)-24(28)-dehydrocholesteryl acetate	0.69	1.02	0.99	0.98	—	1.58
24-Ethyl-(Z)-24(28)-dehydrocholesteryl acetate	0.68	1.02	0.95	0.98	—	1.59
24β-CH₃-25(27)-Dehydrocholesterol[d]	0.67	1.02	0.92	1.62	0.99	—
24β-C₂H₅-25(27)-Dehydrocholesterol	0.68	1.01	0.91	1.57	—	0.80
4α-Methylstigmasterol[e]	0.71	1.05	1.02	0.85/0.80	—	0.82
4β-Methylstigmasterol[f]	0.70	1.06	1.02	0.85/0.80	—	0.82
Lanosterol[g]	0.69	1.00	0.91	1.60/1.68	—	—
24-Dihydrolanosterol[g]	0.69	1.00	0.89	0.87	—	—

[a] All spectra were with sterol in CDCl₃. Except as noted the data are from I. Rubinstein, L. J. Goad, A. D. Clague, and L. J. Mulheim [*Phytochemistry* **15**, 195 (1976)], W. R. Nes, K. Krevitz, J. Joseph, W. D. Nes, B. Harris, G. F. Gibbons, and G. W. Patterson [*Lipids* **12**, 511 (1977)], and T. Iida, T. Tamura, and T. Matsumoto [*J. Lipid Res.* **21**, 326 (1980)].
[b] U. F. Taylor, A. Kisic, R. A. Pascal, A. Izumi, M. Tsuda, and G. J. Schroepfer, Jr., *J. Lipid Res.* **22**, 171 (1981). Doublet for C-21 not given, probably being under C-19 singlet.
[c] W. R. Nes, J. M. Joseph, J. R. Landrey, S. Behzadan, and R. L. Conner, *J. Lipid Res.* **22**, 770 (1981).
[d] G. Romeo and M. A. Toscano, *J. Nat. Prod.* **46**, 187 (1983); C-26 at 4.64 and 4.68.
[e] 4α-CH₃ at 1.11.
[f] 4β-CH₃ at 1.14.
[g] B. C. Sekula and W. R. Nes, *Phytochemistry* **19**, 1509 (1980); 4α-CH₃ at 0.98; 4β-CH₃ at 0.81; 14α-CH₃ at 0.88.

TABLE XI
^{13}C NMR Spectra of Selected Steroids[a]

	δ-Value (ppm)				
Position	Cholesterol	24β-Methyl-cholesterol	24α-Methyl-cholesterol	Lathosterol	Cholest-4-en-3-one
1	37.3	37.5	37.4	37.1	37.1
2	31.6	31.9	31.8	31.3	34.2
3	71.6	72.0	71.8	70.7	200.1
4	42.2	42.6	42.4	37.8	126.0
5	140.6	141.0	140.9	40.2	168.5
6	121.4	121.9	121.7	29.6	73.1
7	31.9	32.2	32.0	117.2	38.6
8	31.9	32.2	32.0	139.3	29.8
9	50.2	50.4	50.3	49.4	53.7
10	36.5	36.8	36.6	34.1	37.9
11	21.1	21.3	21.2	21.5	21.0
12	39.8	40.0	39.9	39.5	39.4
13	42.3	42.6	42.4	43.2	42.5
14	56.8	57.0	56.9	54.9	55.9
15	24.3	24.5	24.4	22.9	24.1
16	28.3	28.4	28.3	27.9	28.1
17	56.2	56.3	56.3	56.1	56.1
18	11.9	12.1	11.9	11.8	12.0
19	19.4	19.6	19.4	12.9	19.2
20	35.8	36.4	36.0	36.1	35.7
21	18.8	19.1	18.8	18.8	18.7
22	36.2	34.0	33.8	36.1	36.1
23	23.9	30.9	30.4	23.9	23.8
24	39.5	39.3	39.0	39.4	39.6
25	28.0	31.7	32.5	27.9	27.9
26/27	22.6/22.8	17.9/20.7	18.3/20.2	22.5/22.7	22.5/22.8
28	—	15.7	15.5	—	—

[a] From J. E. Zarembo and W. R. Nes, *J. Nat. Prod.* **44,** 7 (1981); J. W. Blunt and J. B. Stothers, *Org. Magn. Reson.* **9,** 439 (1977); M. Tsuda and G. J. Schroepfer, Jr., *J. Org. Chem.* **44,** 1290 (1979).

epicholesterol, have been measured in the author's laboratory.[93] Since each carbon atom yields a signal and since introduction of a grouping frequently influences the signals coming from adjacent carbons, very complicated, extensive, and unique spectra are obtained. The voluminous data are beyond the scope of this chapter to present in any detail. How-

[93] J. E. Zarembo, Ph.D. Dissertation, Drexel University (1980).

ever, the spectra for cholesterol and related steroids (Table XI) will reveal that strong downfield shifts are obtained with carbon-carbon double bonds (usually appearing at 110–170 ppm), carbonyl groups (usually appearing in the range 170–221 ppm), and carbinol groups (usually appearing in the 60–80 ppm region). Saturated carbons, on the other hand, are found at smaller δ-values (10–60). However, since the measurements are accurate to 0.2 ppm it will be seen from Table XI that very slight differences in the environment of saturated carbon atoms still can result in distinct changes in δ. This is illustrated by the differences in values for the ring carbons, e.g., 32.0 for C-7 and 21.1 for C-11 in Δ^5-sterols, as well as for the influence of configuration at C-24 on C-25 in the 24-methylcholesterols where δ decreases 0.8 ppm when the 24α-methyl proceeds to 24β-methyl.

[2] High-Performance Liquid Chromatography of Sterols: Yeast Sterols

By RUSSELL J. RODRIGUEZ and LEO W. PARKS

Introduction

Biological membranes comprise a very complex array of hydrophobic, hydrophilic, and amphipathic components. In order to study membranes in detail it is necessary to devise techniques for isolation and analysis of individual components. The roles of lipids in biological and nonbiological membranes have been an intense area of interest over the last 30 years. One group of lipids in particular, the sterols, have been analyzed by thin-layer and gas–liquid chromatography, mass spectroscopy, nuclear magnetic resonance, and infrared and ultraviolet spectroscopy. Although these techniques are invaluable analytical tools, they are unsatisfactory with regard to sample recovery, particularly on a preparative scale. Column chromatography facilitates sample recovery and provides desirable separations (e.g., isomer separation), but can be time consuming with some separations taking up to 3 days.[1,2] In the last decade development of high-pressure liquid chromatography (HPLC) has allowed for considera-

[1] I. D. Frantz, *J. Lipid Res.* **4,** 176 (1963).
[2] P. D. Klein and P. A. Szczepanik, *J. Lipid Res.* **3,** 460 (1962).

ble reduction in the time required for sterol analysis without sacrificing efficiency of sample recovery or resolution.

Design of the System

Development of an HPLC system for the separation of sterols involves the following factors: (1) the type of HPLC chromatography and hence the type of column to be used, (2) the method of detection, and (3) the solvent system. Sterols have been chromatographed on HPLC columns packed with either reverse phase,[3-5] normal phase,[6,7] silver nitrate impregnated normal phase,[6] or adsorption phase material[8]; or a combination of column types.[9,10] There are benefits and disadvantages to each of the various modes of HPLC. Detailed comparisons between the different modes of HPLC for sterols have been discussed in the literature.[6,9,11] Normal and adsorption phase chromatography can be more limiting than reverse phase with regard to the number of solvents that can be used. However, normal phase HPLC can excel in isomeric separations.[12] It is evident from the literature that the different packing materials available for the various modes of chromatography differ drastically with regard to elution and separation profiles for sterols. The mechanism by which these stationary phases influence separation of sterols is not certain. It has been indicated in the reverse phase system that sterol polarity,[11] molecular weight, and molecular structure[6] are involved.

There are two detection methods used for monitoring sterols in HPLC analysis: ultraviolet (UV) absorbance and refractive index (RI) changes. A comparison between these two detection methods[6] indicates that monitoring sterols by UV absorbance (254 nm) is more sensitive than by RI changes. However, only sterols with conjugated double bonds have strong absorbances at high UV wavelengths. Data published by Hunter *et al.* indicates that sterols which lack characteristic absorbance at high UV wavelengths have strong absorbance at low UV wavelengths (e.g., 205 nm).[7] When sterol elution is monitored by UV absorbance the solvents comprising the mobile phase must have UV absorbance cutoffs below the

[3] R. J. Rodriguez and L. W. Parks, *Anal. Biochem.* **119,** 200 (1982).
[4] E. Hansbury and T. J. Scallen, *J. Lipid Res.* **19,** 742 (1978).
[5] P. J. Trocha, S. J. Jasne, and D. B. Sprinson, *Biochemistry* **16,** 4721 (1977).
[6] H. Colin, G. Guiochon, and A. Siouffi, *Anal. Chem.* **51,** 1661 (1979).
[7] I. R. Hunter, M. K. Walden, and E. Heftmann, *J. Chromatogr.* **153,** 57 (1978).
[8] J. R. Thowsen and G. J. Schroepfer, Jr., *J. Lipid Res.* **20,** 681 (1979).
[9] E. Hansbury and T. J. Scallen, *J. Lipid Res.* **21,** 921 (1980).
[10] H. H. Rees, P. L. Donnahey, and T. W. Goodwin, *J. Chromatogr.* **116,** 281 (1976).
[11] E. Heftmann and I. R. Hunter, *J. Chromatogr.* **165,** 283 (1979).
[12] R. A. Pascal, Jr., C. L. Farris, and G. J. Schroepfer, Jr., *Anal. Biochem.* **101,** 15 (1980).

wavelength being monitored. For this reason detection at 205 nm limits the number of solvents which can be used.

Once the column type and detection method have been chosen it is necessary to determine which solvents are applicable. In general there are three factors to be considered: (1) reactivity of solvents with sterols, (2) miscibility of solvents if more than one is used, and (3) solvent viscosity. If sterols are detected by UV absorbance then the UV cutoff of the solvent(s) must also be considered.

Interaction of mobile phase solvents with sterols determines both elution volumes and resolutions. Due to the hydrophobic nature of sterols, nonpolar solvents such as hexane are used to decrease elution volumes. Polar solvents such as water can be incorporated to increase elution volumes. In reverse phase HPLC solvents must be used such that the stationary phase will be sufficiently hydrophobic for sterol interaction. For reverse phase HPLC this usually involves relatively polar solvents.

Solvent miscibility is an important consideration in determining optimum solvent ratios for sterol elution and separation. For example, although hexane and water are not ordinarily miscible, they can both be incorporated into a mobile phase by using the proper ratios.[13] Viscosity of the mobile phase solvents is important with regard to ultimate flow rates and hence time of analysis. High viscosity solvents must be pumped through columns at lower flow rates than low viscosity solvents and therefore increase analysis times.

The last consideration in HPLC analysis of sterols is the chemical form in which the sterols are chromatographed. Sterols have been analyzed either in free form[3-7,9] or in a form which has been derivatized at the C-3–OH. The derivatized forms include steryl acetates,[8-10] benzoates,[13] *p*-nitrobenzoates,[14] and long chain fatty acid esters.[15]

In this paper we describe a reverse phase HPLC system for analyzing free sterols. Chromatographic properties of sterols are compared to determine effects of specific functional moieties on elution volumes.

Materials and Methods

Apparatus

An Altex model 332 gradient HPLC is used for chromatographic analyses in our laboratory. Sterols are detected with a Hitachi model 155-40 variable wavelength spectrophotometer operated at 210 nm. Sterols pos-

---

[13] S. I. Schlager and H. Jordi, *Biochim. Biophys. Acta* **665**, 355 (1981).
[14] F. A. Fitzpatrick and S. Siggia, *Anal. Chem.* **45**, 2310 (1973).
[15] R. M. Carroll and L. L. Rudel, *J. Lipid Res.* **22**, 359 (1981).

sessing $\Delta^{5,7}$ unsaturations are also monitored at 282 nm. Sample injections of 25 μl for analytical and 250 μl for semipreparative analyses are made using an Altex model 210 rheodyne sample injection valve. Purified sterols are collected either manually or automatically with a Buchler model LC-100 fraction collector coupled to an ISCO model 2150 peak separator.

Columns and Chromatographic Conditions

For analytical studies sterols are separated on a 25-cm × 4.6-mm-i.d. stainless-steel column packed with 5 μm ODS-Ultrasphere particles (Altex). Semipreparative purification is accomplished on a 25-cm × 10-mm-i.d. column packed with the same material. To increase column life samples are first pumped through a 4.0-cm × 4.6-mm-i.d. precolumn [packed with 5 μm Nucleosil C_{18} material (Rainin)] which immediately precedes either Ultrasphere column. A trinary mobile phase of ethanol, methanol, and water (10:86:4, v/v/v) is pumped through the 25-cm × 4.6-mm-i.d. column at a rate of 1.5 ml/min for analytical separations and 6 ml/min through the 25-cm × 10-mm-i.d. column for semipreparative purifications. All three columns have been used continually for 2 years in our laboratory and require only intermittent cleansing with 100% methanol (20–30 column volumes) when ghost peaks or column desiccation are evident. Under both analytical and semipreparative conditions back pressures for our systems are less than 3000 psi (50% of maximum). Occasionally the back pressure may increase but can be easily corrected by replacing packing frits in the precolumn. Columns may be saturated with mobile phase solvents and sealed for storage at room temperature. If solvents which comprise a small percentage of the mobile phase, such as water, are delivered to the analytical column by one pump, very high back pressures may ensue resulting in low flow rates. To eliminate this problem solvents are premixed and delivered isocratically through a single pump. Under these conditions flow rates of 2.5–3 ml/min and at least 6 ml/min can be achieved for the analytical and semipreparative columns respectively.

Solvents

Methanol and 100% ethanol are of reagent grade and glass distilled prior to use. Water is double distilled. Distillation of all three solvents is necessary to eliminate any mobile phase UV interference at 210 nm. All solvents are filtered through 934AH grade Reeve Angel glass filters after distillation. Under these conditions degassing of solvents has no effect on sterol elution or separation. To ensure chromatographic reproducibility methanol and water (95.5:4.5 v/v) should be mixed before ethanol is added to 10%.

Equations

To determine the effect of specific structural moieties on sterol elution volumes, it is necessary to calculate sterol retention values. Resolution of sterols is determined by calculating retention values relative to a standard (separation factor). Retention of solutes is determined by calculating capacity factors (K) using the following equation:

$$K = (V_1 - V_0)/V_0$$

where V_1 is the retention volume of the solute and V_0 is the dead volume of the column. Dead volume for our system is determined by injecting 5 μl of acetone into the column and monitoring its elution at 300 nm. The dead volume is equal to the product of the elution time of acetone and the flow rate. For our analytical column this is 2.33 min × 1.5 ml/min = 3.49 ml. Separation of sterols relative to sitosterol is determined by calculating separation factors (α) with the following equation: $\alpha = K_2/K_1$ where K_2 is the capacity factor for sitosterol and K_1 is for the compound of interest. For a more detailed explanation of HPLC equations and terminology see review by Brown and Krstulovic.[16]

Strains and Growth Conditions

Organisms used in these studies are *Saccharomyces cerevisiae* strains S288c (a wild-type haploid from the Berkeley stock culture collection) and the sterol mutants 3701b-n3,[17] JR4,[18] and 8R1.[19] Cells are grown aerobically at 27° in medium containing 1% yeast extract, 0.5% tryptone, and 2% dextrose.

Cell Extraction and Sample Preparations

Sterols from cell samples are obtained from stationary phase yeast cultures (10–50 ml) by acid labilization[20] followed by alkaline saponification and hexane extraction.[21] Cell extracts were originally prepurified by

[16] P. R. Brown and A. M. Krstulovic, *Anal. Biochem.* **99**, 1 (1979).
[17] L. W. Parks, *CRC Crit. Rev. Microbiol.* **6**, 301 (1978).
[18] J. R. Ramp, Honor's Thesis, Oregon State University, Corvallis (1980).
[19] M. T. McCammon and L. W. Parks, *J. Bacteriol.* **145**, 106 (1981).
[20] R. B. Gonzales and L. W. Parks, *Biochim. Biophys. Acta* **489**, 507 (1977).
[21] R. B. Bailey and L. W. Parks, *J. Bacteriol.* **124**, 606 (1975).

thin-layer chromatography and clarified by passage over short sodium sulfate columns. However, we find that centrifugation in a table top centrifuge (International Clinical Model cl) at 2500 g for 1 min is ideal for clarification. Centrifugation allows for direct analysis of saponified cell extracts without thin layer chromatographic prepurification. Sterols obtained commercially or from cell extracts are dissolved in isopropanol (IPA) for analysis. Dissolution of lipids in IPA is greatly enhanced by brief sonication (in a Bransonic 200 water bath sonicator) and/or heating at 50°. In some semipreparative purifications desirable concentrations of sterols were not obtainable in IPA. Under these conditions a mixture of IPA-acetone (3:1, v/v) is very effective for solubilizing higher concentrations. The presence of acetone has no effect on sample elution, peak symmetry, or peak separations. If tetrahydrofuran is used to dissolve sterols there is considerable tailing of peaks.

Sterol Standards

Desmosterol was purchased from Organon, West Orange, NJ. Brassicasterol, Δ^4-cholestenol, lathosterol, and Δ^7-campestenol were purchased from Research Plus Inc., Bayonne, NJ. Stigmasterol was obtained from CalBiochem., campesterol from Supelco, and sitosterol from Applied Science. Cholesterol, ergosterol, 7-dehydrocholesterol, and lanosterol were purchased from Sigma. 24-Methylcholesterol, lophenol, 24-methylenecholesterol, fucosterol, and $\Delta^{8(14)}$-cholestenol were gifts from Dr. Henry Kircher. Ignosterol was extracted from S288c cells grown with 15-azasterol[22]; ergosta-7,22-dien-3β-ol was isolated from 3701b-n3[23]; zymosterol, cholestatrienol, and cholestatetraenol were obtained from 8R1[19]; and 14-methylergosta-8,24(28)-dien-3β-ol was extracted from JR4.[18]

Elution Properties of Sterols

HPLC elution properties of the sterols used in our recent studies are shown in the table.[3] By comparing data for structurally similar sterols it is possible to determine the effects of individual double bonds and alkyl groups on elution properties. In general double bonds decrease elution volumes and alkyl groups increase elution volumes.

Effect of Specific Structural Moieties

Unsaturations. Chromatographic properties of sterols are affected both by the number and molecular location(s) of unsaturations. The effect

[22] P. R. Hays, L. W. Parks, and A. C. Oelschlager, *Lipids* **12,** 666 (1977).
[23] L. W. Parks, F. T. Bond, E. D. Thompson, and P. R. Starr, *J. Lipid Res.* **13,** 311 (1972).

HPLC PROPERTIES OF STEROLS

Sterol	Elution volume (ml)	k'	α relative to sitosterol
Cholesta-5,7,22,24-tetraen-3β-ol (cholestatetraenol)	21.0	5.01	3.08
Cholesta-5,7,24-trien-3β-ol (cholesatrienol)	27.6	6.91	2.23
Cholesta-8,24-dien-3β-ol (zymosterol)	31.4	7.97	1.94
Cholesta-5,24-dien-3β-ol (desmosterol)	33.4	8.54	1.81
14-Methylergosta-8,24(28)-dien-3β-ol (C-29 sterol)	34.5	8.86	1.74
Ergosta-5,7,22-trien-3β-ol (ergosterol)	35.1	9.03	1.71
Cholesta-5,7-dien-3β-ol (7-dehydrocholesterol)	36.6	9.46	1.63
Ergosta-8,14-dien-3β-ol (ignosterol)	36.6	9.46	1.68
24-Methylenecholest-5-en-3β-ol (24-methylenecholesterol)	36.8	9.54	1.62
Cholest-8(14)-en-3β-ol [$\Delta^{8(14)}$-cholestenol]	38.2	9.96	1.55
Ergosta-5,22-dien-3β-ol (brassicasterol)	41.5	10.86	1.42
Ergosta-7,22-dien-3β-ol (3701b-n3 sterol)	41.0	10.71	1.44
Cholest-4-en-3β-ol (Δ^4-cholestenol)	41.8	10.94	1.41
Cholest-7-en-3β-ol (lathosterol)	42.5	11.14	1.39
Cholest-5-en-3β-ol (cholesterol)	44.0	11.57	1.33
24-Ethylcholesta-5,24(28)-dien-3β-ol (fucosterol)	44.7	11.81	1.31
4,4,14-Trimethylcholesta-8,24-dien-3β-ol (lanosterol)	46.0	12.14	1.27
24α-Ethylcholesta-5,22-dien-3β-ol (stigmasterol)	50.3	13.37	1.16
24α-Methylcholest-7-en-3β-ol (Δ^7-campestenol)	50.5	13.43	1.15
24α-Methylcholest-5-en-3β-ol (campesterol)	51.0	13.57	1.14
24β-Methylcholest-5-en-3β-ol (24-methylcholesterol)	51.0	13.57	1.14
4α-Methylcholest-7-en-3β-ol (lophenol)	53.0	14.19	1.09
24α-Ethylcholest-5-en-3β-ol (sitosterol)	57.6	15.46	1.00
4,4,14-Trimethylcholest-8-en-3β-ol (24-dihydrolanosterol)	60.0	16.14	0.96

of the number of double bonds is best illustrated by comparing elution profiles of cholesterol, $\Delta^{5,7}$-cholestadienol, $\Delta^{5,7,24}$-cholestatrienol, and $\Delta^{5,7,22,24}$-cholestatetraenol (Fig. 1, the table). As the number of unsaturations on the cholesterol molecule increases there is a concomitant decrease in elution volume. However, this decrease is not numerically proportional to the number of unsaturations. The effect of the molecular location of double bonds, is observed by comparing zymosterol ($\Delta^{8,24}$), desmosterol ($\Delta^{5,24}$), and 7-dehydrocholesterol ($\Delta^{5,7}$). Although these sterols all have two unsaturations, the different molecular locations result in different elution volumes (see the table).

Nuclear Unsaturations. In order to determine the effect of specific unsaturations it is necessary to analyze isomeric sterols. By comparing isomeric sterols from the table we are able to assess the effect on elution volumes due to unsaturations at the C-4(5), C-5(6), C-7(8), C-8(9), and C-

[24] A. M. Pierce, A. M. Unrau, and A. C. Oelschlager, *Can. J. Biochem.* **57**, 201 (1979).

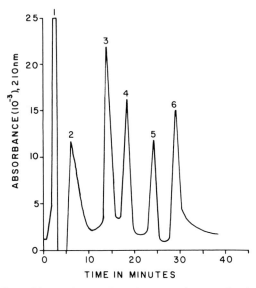

FIG. 1. The effect of increasing number of unsaturations on the elution of cholesterol derivatives monitored at 210 nm. Peaks: 1 and 2, solvent; 3, cholestatetraenol; 4, cholestatrienol; 5, cholestadienol; 6, cholesterol.

8(14) positions. To help illustrate these comparisons we have included Fig. 2 which depicts a C_{32} sterol molecule with ring letter and carbon number designations. HPLC analyses of Δ^4-cholestenol, cholesterol, Δ^7-cholestenol, and $\Delta^{8(14)}$-cholestenol (see the table) indicate that the effect these unsaturations have in decreasing elution volumes is $\Delta^{8(14)} > \Delta^4 > \Delta^7 > \Delta^5$. The effect of a $\Delta^{8(9)}$ unsaturation is determined by comparing zy-

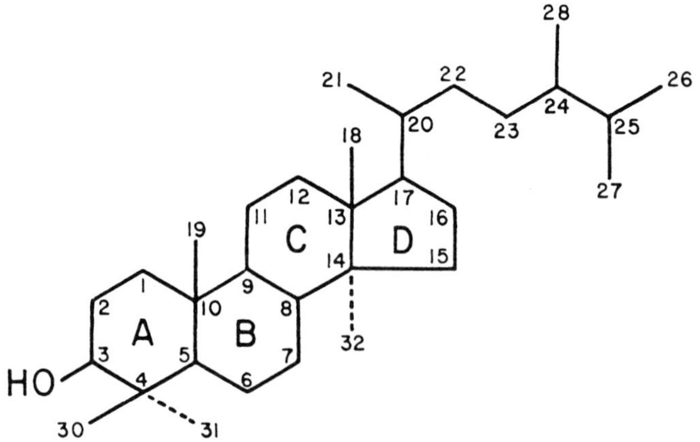

FIG. 2. Structure, carbon numbering, and ring letter designation for a C-32 sterol.

mosterol ($\Delta^{8,24}$) and desmosterol ($\Delta^{5,24}$). The elution volume difference between the $\Delta^{8,24}$ and the $\Delta^{5,24}$ is identical to the difference between the Δ^4 and Δ^5 isomers so the effect of a C-8(9) unsaturation is the same as a Δ^4 double bond. These comparisons indicate that the ability of nuclear unsaturations to decrease elution volume is $\Delta^{8(14)} > \Delta^{8(9)} = \Delta^{4(5)} > \Delta^{7(8)} > \Delta^{5(6)}$.

Side-Chain Unsaturations. Side-chain unsaturations are analyzed by comparing desmosterol and cholesterol for a $\Delta^{24(25)}$ unsaturation and 24-methylcholesterol with brassicasterol for a $\Delta^{22(23)}$ unsaturation. Desmosterol eluted in 10.6 ml less than cholesterol and brassicasterol eluted in 9.5 ml less than 24-methylcholesterol. From these data it appears that a $\Delta^{24(25)}$ double bond has a greater effect on sterol elution than a $\Delta^{22(23)}$ double bond. Originally[3] we indicated that a $\Delta^{24(28)}$ unsaturation may be similar to a $\Delta^{24(25)}$. However by more direct analyses we find that a $\Delta^{24(28)}$ unsaturation has a greater effect on elution properties than a $\Delta^{24(25)}$ double bond. This is observed by comparing fucosterol ($\Delta^{5,24(28)}$) with sitosterol (Δ^5) or 24-methylenecholesterol ($\Delta^{5,24(28)}$) with 24-methylcholesterol (Δ^5). In both cases the decrease in elution volume attributed to the $\Delta^{24(28)}$ double bond is approximately 4 to 5 ml greater than that seen with a $\Delta^{24(25)}$ unsaturation.

Mechanism of the Unsaturation Effect. These data indicate that double bonds alter the polarity of sterol molecules. With the exception of the 3β-OH, the remainder of a saturated sterol molecule is very nonpolar. The presence of double bonds in a sterol increases its relative polarity and decreases its reversed phase HPLC elution volume. The 3β-OH has some effect on the polarity of the A and B rings of the sterol due to electronic dispersion and hence unsaturations in these rings have less influence on elution volumes than unsaturations further from the 3β-OH. This explains the relative abilities of different double bonds to decrease elution volume as shown above to be $\Delta^{8(14)} > \Delta^{8(9)} = \Delta^4 > \Delta^7 > \Delta^5$. As shown in the table, Δ^4-cholestenol elutes in slightly less volume than Δ^7-cholestenol (lathosterol), yet its double bond is closer to the 3β-OH than that of lathosterol. This may be explained by the formation of a resonant hybrid of Δ^4-cholestenol. This would result if the oxygen bound at C-3 possessed sufficient double bond character to delocalize electron density between it and the Δ^4 unsaturation. The Δ^5 and Δ^7 unsaturations are not close enough to the 3β-OH to participate in this type of electron delocalization and therefore elute in slightly greater volumes. In support of this explanation Δ^4-cholestenol showed much stronger absorbance at 210 nm than either cholesterol or lathosterol.

Nuclear Alkyl Groups. The presence of alkyl groups on the sterol molecule increases hydrophobicity and overall bulk. The effect of one C-4 methyl group can be determined by comparing lophenol (4-methylcholest-7-en-3β-ol) and lathosterol (cholest-7-en-3β-ol) (see the table). The presence of a methyl at C-4 of lophenol causes it to elute in 10 ml greater than lathosterol. This large increase is anticipated since the nonpolar methyl is in close proximity to the polar 3β-OH and acts to decrease polarity. If such effects were additive then lanosterol [$\Delta^{8,24(25)}$], with two methyls at C-4 and a methyl at C-14, should elute in at least 20 ml more than zymosterol [$\Delta^{8,24(25)}$] which lacks these methyl groups. The difference of only 14.6 ml shown in the table indicates that although comparisons can be made and the effects of individual moieties assessed, the effect(s) are not additive for multiple functional groups. A direct assessment of the effect of a C-14 methyl cannot be made from the sterols in the table.

Side-Chain Alkyl Groups. The effect of alkyl groups on the sterol side chain was analyzed for methyl and ethyl moieties. The presence of a C-24 methyl group caused 24-methylcholesterol to elute in a volume 7 ml greater than that of cholesterol. Sitosterol has an ethyl group at the C-24 and elutes in 13.6 ml greater than cholesterol, a difference almost twice that attributed to a single methyl. In this system identical alkyl groups at C-24 in either α or β configuration are inseparable. This is demonstrated with campesterol and 24β-methylcholesterol. Data in the table indicate that the ability of alkyl groups to increase sterol elution volume is 4,4,14-trimethyl > C-24 ethyl > C-4 methyl > C-24 methyl.

Double Bonds versus Alkyl Groups on the Side Chain. When comparing alkyl groups and double bonds on the sterol side chain, the effect of double bonds on elution volumes is greater than that of methyl groups and approximately equal to that of ethyl groups. A comparison of 7-dehydrocholesterol and ergosterol illustrates the first of these two observations. These sterols are identical in ring structure and differ in that ergosterol has both a C-24 methyl and a Δ^{22} unsaturation, and 7-dehydrocholesterol has neither of these moieties. If the effect of the methyl group (which acts to increase elution volume) and the double bond (which acts to decrease elution volume) were equal these sterols would have coeluted. When the C-24 methyl group is replaced by an ethyl group the effect of a side chain unsaturation is virtually nullified as seen by comparing stigmasterol and campesterol.

Detection of Free Sterols

All sterols used in these studies with the exception of cholestanol and ergostanol (data not shown) had strong absorbance at 210 nm. Only those sterols with conjugated double bonds had UV absorbing properties above this wavelength. Figure 3 shows HPLC analyses of a sterol mixture monitored at 210 and 254 nm. All eight sterols tested are detectable at 210 nm, but only two (possessing $\Delta^{5,7}$ conjugated double bonds) show absorbance

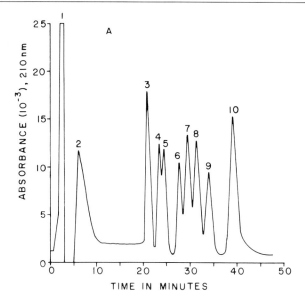

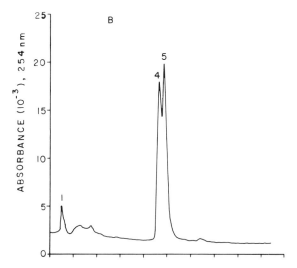

FIG. 3. (A) HPLC separation of sterol standards monitored at 210 nm. Peaks: 1 and 2, solvent; 3, zymosterol; 4, ergosterol; 5, 7-dehydrocholesterol; 6, brassicasterol; 7, cholesterol; 8, lanosterol; 9, stigmasterol; 10, sitosterol. (B) HPLC separation of the same standards monitored at 254 nm. Peaks: 1, solvent; 4, ergosterol; 5, 7-dehydrocholesterol. Each sterol peak in these graphs represents approximately 3 μg of material.

at 254 nm. The fact that cholestanol and ergostanol (saturated sterols) are not detectable by UV absorbance suggests that a free sterol must contain at least one unsaturation to show significant absorbance at 210 nm. Extinction coefficients of sterols at 210 nm varied significantly depending on the number and location of double bonds, making direct quantitation difficult. Variation in extinction coefficient is clearly demonstrated in Fig. 3 where sterols were added in approximately equal molar concentrations.

Application to the Yeast System

The major sterol of the yeast *Saccharomyces cerevisiae* is ergosterol. The biosynthetic pathway for ergosterol is very complex and involves many intermediates.[24] Approaches to studying membrane physiology of this organism with regard to sterol metabolism have involved (1) using sterol mutants which possess enzymatic defects in the late stages of ergosterol biosynthesis and accumulate sterols other than ergosterol,[25] (2) growth of cells anaerobically such that they require sterol,[26] and (3) use of heme mutants which are auxotrophic for sterol under aerobic conditions.[27]

In studying sterol metabolism of yeast we have taken two of these approaches: isolation and analysis of sterol mutants, and conversion and deposition of sterols by supplemented sterol auxotrophs. For auxotrophic feeding experiments and sterol mutant screening it was necessary to develop a system for sterol purification, separation, and identification. In order to decrease time of large scale analyses and to avoid sterol oxidation during saponification of derivatized forms, the system was developed for analysis of free sterols. To obtain maximum sensitivity of detection, sterols were monitored for UV absorbance at 210 nm.

The sterol biosynthetic pathway of yeast involves a large number of intermediates and therefore allows for numerous possible mutations. These mutations may lead to a variation in normal sterols involving single unsaturations or methyl groups. Since sterols differing in the number and location of double bonds and methyl groups can be separated by the HPLC procedure described, it is possible to use this technique to screen large numbers of mutants for altered sterol patterns. However, these mutations may lead to synthesis of sterols which elute close to or with the normal sterol of yeast (ergosterol) so an additional screening method is recommended. One such method is determining UV spectra on the HPLC purified sterol peaks to identify conjugated unsaturations. An example of

[25] C. A. McLean-Bowen and L. W. Parks, *J. Bacteriol.* **143,** 1325 (1981).
[26] W. R. Nes, B. C. Sekula, W. D. Nes, and J. H. Adler, *J. Biol. Chem.* **253,** 6218 (1978).
[27] R. J. Rodriguez and L. W. Parks, *Arch. Biochem. Biophys.* **225,** 861 (1983).

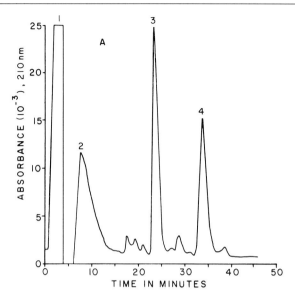
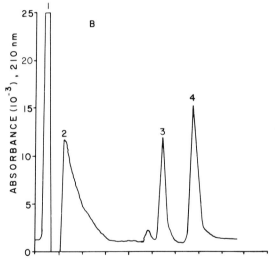

FIG. 4. (A) HPLC analysis of the sterol fraction from S288c. Peaks: 1 and 2, solvent; 3, ergosterol; 4, campesterol added as a standard. (B) HPLC analysis of the sterol fraction from 3701β-n3. Peaks: 1 and 2, solvent; 3, ergosta-7,22-dien-3β-ol sterol (3701b-n3); 4, campesterol added as a standard.

this coupled HPLC-UV technique is demonstrated in Fig. 4. Extracts from S288c and 3701b-n3 were analyzed by HPLC showing two major peaks that separated by 5.9 ml (the difference of one nuclear unsaturation). UV spectra revealed that the S288c sterol had the characteristic absorbance of ergosterol, while the 3701b-n3 sterol had no absorbance other than at 210 nm. In addition, this coupled technique has been used to identify yeast mutants possessing defects in sterol biosynthesis involving Δ^5-desaturation, $\Delta^{8\rightarrow7}$-isomerization, C-14 demethylation, C-24 methylation, and $\Delta^{24(28)}$-methylene reduction.

We have also found the HPLC to be invaluable for purifying large quantities of sterols and stanols for feeding experiments with yeast sterol auxotrophs.[27] Analysis by HPLC revealed that commercially available cholestanol contains cholesterol as a contaminant. Upon removal of contaminating cholesterol by HPLC purification of cholestanol, we were able to demonstrate that cholestanol alone is insufficient to satisfy the growth requirement of yeast sterol auxotrophs.[28]

Summary

It is evident that the high-pressure liquid chromatograph is an excellent tool for studying sterol metabolism. As noted in the text, the individual effects of unsaturations and alkyl groups on reverse phase elution volumes cannot be extrapolated to predict quantitative effects of multiple functional groups. The mechanism(s) of retention seems more complex than can be explained simply by polarity or hydrophobicity. Since the molecular location of these functional moieties seems critical, retention and separation of sterols may involve specific structural configurations and hence specific interactions with the stationary phase material. The association we have drawn between polarity and HPLC elution may indeed be a secondary effect of another phenomenon. Future studies may unveil the true mechanism(s) of HPLC retention and separation, and allow for the construction of HPLC systems which will separate all isomeric combinations of sterols at the analytical level.

The simplicity, rapidity, and reproducibility of these methods make the coupled technique very useful for investigating sterol metabolism. Application of this technique to analyzing putative sterol mutants, purifying sterols for auxotrophic feeding, and analyzing the metabolism of supplemented sterols in auxotrophs provides for significant advances in membrane physiology.

[28] R. J. Rodriguez, F. R. Taylor, and L. W. Parks, *Biochem. Biophys. Res. Commun.* **106,** 435 (1982).

Acknowledgments

We are grateful to Dr. Henry Kircher for supplying some of the sterols used in these studies. We also acknowledge Gael Kurath for editorial assistance and Carlene Pelroy for typing this manuscript. This work was supported by a grant (PCM-7827609) for the National Science Foundation and by a Public Health Service grant (AM-05190) from the National Institutes of Health.

[3] High-Performance Liquid Chromatography of Bile Acids

By WILLIAM H. ELLIOTT and ROGER SHAW

Bile acids are determined by a variety of analytical procedures, but several of these methods require hydrolysis or solvolysis of conjugates as part of an integrated procedure. Conjugates of the carboxyl group of the bile acid with amino acids (normally glycine or taurine), or derivatives of nuclear hydroxyl groups such as sulfates or glucuronides, or the free bile acids can now be analyzed by HPLC without removal of the conjugating moiety, and thus, provide identification and quantitation of these naturally occurring substances. Because of our interest in the 5α- or allo bile acids, free and conjugated 5β- and 5α-bile acids have been available in this laboratory for such studies. To avoid the formation of artifacts from deconjugation or solvolysis, initial studies were directed at separation and identification of the conjugated bile acids.[1]

Instrumentation

Equipment from Waters Associates used in these studies included Model 6000A pumps, Model U6K injectors, Model 401 differential refractometers (RI), a Schoeffel Model 770 (or a Waters Model 480) variable UV flow detector, a Model 660 gradient programmer, and radial compression modules (RCM). Columns include Waters μBondapak/C_{18}, μPorasil, and "fatty acid analysis" columns (each 30 cm × 3.9 mm i.d. with 10 μm particle size), Radial-PAK C_{18} cartridges (10 cm × 8 mm i.d., 10 or 5 μm particle size), and Radial-PAK silica cartridges (10 cm × 8 mm i.d., 10 μm particle size). Columns of other manufacturers have been used, but data

---

[1] Further details on this subject can be acquired from a more lengthy review; cf. W. H. Elliott and R. Shaw, in "Steroid Analysis by HPLC" (M. P. Kautsky, ed.,), p. 1. Dekker, New York, 1981.

obtained with them will not be included in this report because of less experience with them. Waters guard columns for the 30-cm columns or RCM cartridges were used with each column. The Schoeffel (now Kratos) UV detector functions at 190–200 nm when purged with a slow stream of nitrogen to remove atmospheric gases. The life of the lamp is shortened by use at this low wavelength, so usage under these conditions should be restrictive. Samples are injected into the coupled HPLC system via a 10-μl syringe.

Solvents

Commercial organic solvents were initially either glass distilled or purchased from Burdick and Jackson Laboratories. Currently, commercial HPLC grade solvents with a high grade of purity are used as received; solvents retained on the shelf or in storage for months should be purified before use. Aliphatic solvents containing halogen are corrosive to stainless steel, and should be washed from stainless-steel columns and pumps as quickly as feasible. Deionized distilled water was doubly distilled or purified with a commercial system (Millipore Water Purification System). When the UV detector was used at 193 nm, solvent mixtures were degassed, saturated with nitrogen, and degassed again prior to usage. The use of a magnetic stirrer is helpful as degassing is carried out with an aspirator. The use of nitrogen to displace dissolved gases was not necessary when the RI detector was used. Final degassing is essential so that the pump will compress only the solvent.

Studies on the Mobilities of Bile Acids under Specific Conditions

General

The mobility of a number of conjugated and free bile acids has been studied[2] with a single μBondapak/C_{18} column with 10 μm particles, a U6K loop injector, a Model 6000 pump, an R401 differential refractometer set at 8×, and a solvent system of 2-propanol/10 mM phosphate buffer, pH 7.0 (8:17) at a flow rate of 1 ml/min. At the end of the working day the column was washed with methanol–water (63:37). The use of an automatic timer for this operation is recommended. The mobility of each bile acid is represented by the capacity factor, k', which was calculated as:

$$k' = (t_R - t_0)/t_0 \tag{1}$$

[2] R. Shaw, M. Rivetna, and W. H. Elliott, *J. Chromatogr.* **202,** 347 (1980).

TABLE I
HPLC MOBILITIES OF SOME COMMON CONJUGATED
AND FREE BILE ACIDS[a]

Compound[b]	rk'	Compound[b]	rk'
T-3α-OH	1.77	GA-3α,7α,12α-(OH)$_3$	0.29
T-3α,7α-(OH)$_2$	0.68	3α-OH	2.15
T-3α,7β-(OH)$_2$	0.25	3β-OH	1.47
T-3α,12α-(OH)$_2$	0.83	3=O	1.74
T-3α,6α,7α-(OH)$_3$	0.15	3α,6α-(OH)$_2$	0.36
T-3α,6β,7β-(OH)$_3$	0.15	3α,6β-(OH)$_2$	0.24
T-3α,7α,12α-(OH)$_3$	0.31	3α,7α-(OH)$_2$	0.80
TA-3α,7α,12α-(OH)$_3$	0.32	3α,7β-(OH)$_2$	0.34
G-3α-OH	1.65	3-(OH)-7=O	0.36
G-3α,7α-(OH)$_2$	0.65	3α,12α-(OH)$_2$	1.00
G-3α,7β-(OH)$_2$	0.24	3α,6α,7α-(OH)$_3$	0.26
G-3α,12α-(OH)$_2$	0.83	3α,6β,7β-(OH)$_3$	0.17
G-3α,7α,12α-(OH)$_3$	0.30	3α,7α,12α-(OH)$_3$	0.35

[a] Chromatographic conditions: 2-propanol/10 mM potassium phosphate (pH 7.0) (8:17), μBondapak/C$_{18}$ column (30 cm × 0.4 cm i.d., 10 μm particle); flow rate 1 ml/min. Mobilities expressed as relative capacity factors, rk', relative to deoxycholic acid [3α,12α-(OH)$_2$]. From ref. 2.

[b] All compounds have 5β configuration except those designated A which stands for allo or 5α. T, Taurine conjugated; G, glycine conjugated; OH, hydroxyl; =O, oxo or keto.

where t_R is the retention time of a solute R, and t_O the retention time of the first peak appearing in the chromatogram. Since capacity factors were found to vary because of a number of factors (e.g., variability in solvent composition, pH, quantity of eluted material), it was more practical to include an internal standard S in the chromatographic run, so that the mobility of solute R is denoted as the relative capacity factor, rk', which is defined as

$$rk' = k'_R/k'_S = (t_R - t_0)/(t_S - t_0) \tag{2}$$

Deoxycholic acid was chosen as the internal standard in these studies. The value of rk' was almost constant despite slight variations in experimental conditions. The relative capacity factors of selected free and conjugated bile acids are given in Table I. The accuracy and reproducibility of reversed-phase HPLC permit one to assign a numerical value, the contribution function, for each substituent to the mobility of a compound. This value is related linearly to the logarithmic function of rk'. Knowledge of contribution functions for all the substituents allows prediction of the mobility of a compound.[2]

Monooxygenated Bile Acids

A study of the mobility of hydroxy and oxo compounds with substitutions at 3, 6, 7, or 12α positions under the conditions stated above has provided basic information on the structure–mobility relationship of bile acids, and generated contribution functions for individual substituents. Bile acids interact with the C_{18} stationary phase via their hydrophobic β-surface. If the hydrophobic side is disturbed by the presence of β-hydroxyl group, such materials are eluted sooner than the α-analogs, except for compounds with 6-hydroxyl substituents. Compounds with oxo substituents, whose orientation is intermediate between β- and α-hydroxyl groups, are not retained as long as the α-hydroxyl derivatives, even though the former are less polar than the latter. With α-hydroxyl substituents, retention time increases in the order 3α-OH ≈ 6α-OH < 7α-OH < 12α-OH, as steric hindrance around the hydroxyl group increases. Thus, the free spatial volume available to a hydroxyl group affects its retention time. However, no resolution was observed between 5β- and 5α-epimers of bile acids containing α-hydroxyl groups.[2]

Polyoxygenated Bile Acids

Chenodeoxycholic acid [(3α,7α-(OH)$_2$] and deoxycholic [3α,12α-(OH)$_2$] acids are readily separated under the above stated conditions. In the same system, the 5β-isomers of deoxycholic acid, hyodeoxycholic acid [3α,6α-(OH)$_2$], 6β-hyodeoxycholic [3α,6β-(OH)$_2$], and hyocholic acid [3α,6α,7α-(OH)$_3$] are eluted separately from their 5α-epimers. Resolution of 5-epimers of other common polyoxygenated bile acids is unsuccessful under these conditions.

The mobilities of most of the common polyoxygenated bile acids can be predicted by summing the contribution functions of the individual substituents. This is true for bile acids that contain only α-hydroxyl groups which are not adjacent to each other. For vicinal α-hydroxyl groups, the free spatial volume available to each group is diminished, resulting in a reduced solubility of the bile acid in the polar mobile phase and a longer tention time. For compounds containing both β- and α-hydroxyl groups, neither surface of the steroid nucleus can fully interact hydrophobically with the stationary phase. Thus, the retention time is shorter than calculated. It is clear from the above arguments that the stereochemistry and conformation of a molecule contribute significantly to its mobility in reversed-phase HPLC.

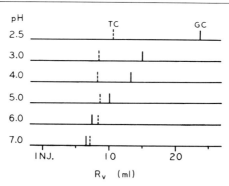

FIG. 1. Effect of pH on the retention volumes (R_v) of taurocholate (T; dashed line) and glycocholate (GC; solid line). Conditions: μBondapak/C_{18} column; 2-propanol/8.8 mM potassium phosphate (8:17) with the pH of the phosphate buffer adjusted as indicated, at a flow rate of 1 ml/min. Detection was with a differential refractometer. Reprinted with permission of the publishers and the authors.[1,3]

Conjugated Bile Acids

Under the above conditions taurine and glycine conjugated 5β- and 5α-bile acids substituted variously at 3α, 6, 7, and 12α positions with hydroxyl groups are eluted slightly ahead of the comparable free bile acid. This is not unexpected since the acidic groups are completely ionized at pH 7.0 (Fig. 1) while the number of methylene groups in the amino acids is small compared with those in the steroid entity. In fact, corresponding taurine and glycine conjugates overlap as a single peak. We have found also that the contribution functions for taurine and glycine are very similar for the 11 tauro- and 11 glycoconjugates analyzed.[2]

Among the various isomers, the conjugates of ursodeoxycholic acid [3α,7β-$(OH)_2$] are eluted far ahead of those of chenodeoxycholic acid [3α,7α-$(OH)_2$] (Table I) for reasons previously given. But the conjugates of 5β-isomers are not resolved from those of 5α-isomers. A particularly useful finding is the excellent separation achieved for the corresponding conjugates of chenodeoxycholic acid and deoxycholic acid (Table I). These positional isomers which are usually found in body fluids have been difficult to separate without prior chemical modifications.

Reversed-Phase Liquid Chromatography (RPLC)
of Conjugated Bile Acids

The following discussion, data, and illustrations emphasize the influence of the pH of the solvent system and of the oxygenated nuclear substituents on the order of elution of taurine and glycine conjugated bile

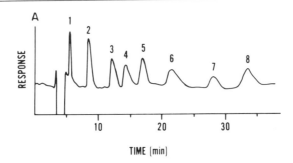

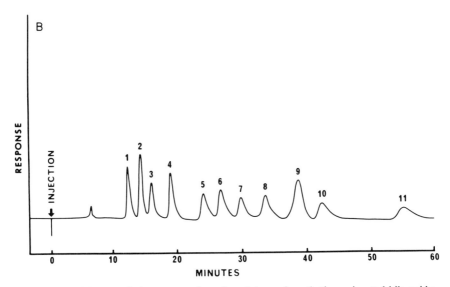

FIG. 2. (A) Reversed-phase separation of a mixture of synthetic conjugated bile acids: Peak (1) tauro-α-muricholate; (2) taurocholate; (3) taurochenodeoxycholate; (4) taurodeoxycholate; (5) glycocholate; (6) taurolithocholate; (7) glycochenodeoxycholate; (8) glycodeoxycholate. The system included a "fatty acid analysis" column with a differential refractometer, 2-propanol/8.8 mM potassium phosphate buffer (pH 2.5) (8 : 17) at a flow rate of 1 ml/min. Reprinted by permission of the publisher and the authors.[4] (B) Reversed-phase chromatography of some common conjugated bile acids: (1) tauroursodeoxycholate, (2) glycoursodeoxycholate, (3) taurocholate, (4) glycocholate, (5) taurochenodeoxycholate, (6) taurodeoxycholate, (7) glycochenodeoxycholate, (8) glycodeoxycholate, (9) testosterone acetate (internal standard), (10) taurolithocholate, (11) glycolithocholate. The system consisted of a μBondapak/C_{18} column eluted with a mixture of acetonitrile/methanol/0.03 M phosphate (pH 3.4) (1 : 6 : 3) and a Shimazu UV spectrometer Model SPD-1 set at 200 nm was used for detection. Reprinted from F. Nakayama and M. Nakagaki, *J. Chromatogr.* **183,** 287 (1981) by courtesy of the authors and of Marcel Dekker, Inc.

acids. With pK_a values of taurocholic (TC) and glycocholic (GC) acids of 1.4 and 4.4, respectively, a pH of RPLC solvents above 5.0 will ensure the presence of the ionic form of both acids, whereas at lower pH glycocholic acid exists as the undissociated acid. This is shown graphically[3] in Fig. 1, where TC was well separated on a μBondapak/C_{18} column from GC at pH 2.5, but at pH 6.0 or 7.0 the order of elution was interchanged. Figure 2A shows the separation of a synthetic mixture of taurine and glycine conjugates on a "fatty acid analysis" column using the RI detector and a solvent mixture of 2-propanol/8.8 mM potassium phosphate buffer (pH 2.5) (8:17) at a flow rate of 1 ml/min. Material collected from each peak was identical in TLC with the authentic standard. Figure 2B shows a similar separation of a synthetic mixture of 10 conjugated bile acids with testosterone as internal standard on a μBondapak/C_{18} column eluted with a mixture of acetonitrile/methanol/30 mM phosphate buffer (pH 3.4) (1:6:3) with a UV detector set at 200 nm.

Figure 3 illustrates the separation of major conjugated bile acids from human bile at pH 2.5 (A) and 4.7 (B). Table II provides capacity factors (k') and relative capacity factors (rk') of 10 conjugated bile acids in several solvent systems and at different pH; all relative capacity factors in Table II are referenced to taurodeoxycholate as 1.00. System E utilized the Radial Compression Module with a Radial-Pak C_{18} cartridge. Bile used for HPLC analysis can be deproteinized by dropwise addition into ethanol followed by centrifugation, or dropwise addition of bile into ethanol in an ultrasonic bath, followed by centrifugation. Analysis of rat bile on a "fatty acid analysis" column with a UV detector with a solvent system of 2-propanol/8.8 mM phosphate buffer, pH 2.5 (8:17) demonstrated that tauro-β- and α-muricholates have the same mobility; however, β-muricholate is present in only trace amounts in bile of the rat with an active enterohepatic circulation.[4]

To establish identification of components of numbered peaks of Fig. 3A, collections of material eluted from the detector corresponding to the numbered peaks provided samples whose identity was confirmed by TLC and mass spectrometry.[3] Quantitation of eluted materials (e.g., Fig. 2A) was established with the Schoeffel UV monitor (Fig. 4A). The limit of detection at 193 nm was 0.1 and 0.2 nmol for TC and GC, respectively.[1,4,5] With the RI detector set at 8× the lower limits were 0.04–0.06 μmol of each component.[1,6] Bloch and Watkins reported a lower detection limit of

[3] R. Shaw and W. H. Elliott, *Lipids* **13**, 971 (1978).
[4] R. Shaw, J. A. Smith, and W. H. Elliott, *Anal. Biochem.* **86**, 450 (1978).
[5] R. Shaw and W. H. Elliott, *Biomed. Mass Spectrom.* **5**, 433 (1978).
[6] R. Shaw and W. H. Elliott, *Anal. Biochem.* **74**, 273 (1976).

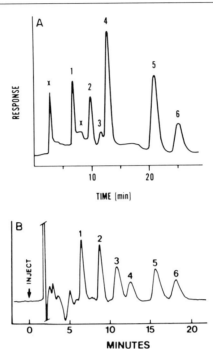

FIG. 3. (A) A chromatogram of a deproteinated human gallbladder bile at pH 2.5. (X) Unknown; (1) taurocholate; (2) taurochenodeoxycholate; (3) taurodeoxycholate; (4) glycocholate; (5) glycochenodeoxycholate; (6) glycodeoxycholate. The system included a "fatty acid analysis" column with exactly the same conditions given for Fig. 2A. Reprinted by permission of the publishers and the authors.[4] (B) Chromatogram of human gallbladder bile at pH 4.7. (1) Taurocholic acid, (2) glycocholic acid, (3) taurochenodeoxycholic acid, (4) taurodeoxycholic acid, (5) glycochenodeoxycholic acid, (6) glycodeoxycholic acid. The system included a μBondapak/C_{18} column, a solvent system of methanol/water/acetic acid + 10 N sodium hydroxide (pH 4.7) (65:35:3.27), and a flow rate of 2.0 ml/min. A differential refractometer was used. Reprinted by permission of the publisher and the authors.[7]

0.01 μmol.[7] Detection was linear to higher concentrations as shown in Fig. 4A and B. Total bile acid content determined by HPLC analyses of six samples of rat bile agreed within ±5% of the values obtained by use of preparations of 3α-hydroxysteroid dehydrogenase.

After numerous injections of bile samples, the guard column becomes contaminated with bile pigments, and must be replaced, or refilled with new column packing. These pigments may collect on the top 2–3 mm of the packing of the analytical column, and should be carefully replaced with new packing. Column resolution deteriorated after 5 months of usage

[7] C. Bloch and J. B. Watkins, *J. Lipid Res.* **19**, 510 (1978).

TABLE II
Capacity Factors (k') and Relative Capacity Factors (rk') of Some Conjugated Bile Acids[a]

	A[b] (2.5)[c]		B[b] (3.4)	C[b] (4.7)		D[b] (7.0)		E[b] (7.0)	
	k'	rk'	rk'	k'	rk'	k'	rk'	k'	rk'
TC	1.33	(0.43)	(0.60)	3.21	(0.43)	1.04	(0.57)	1.09	(0.50)
TUDC	1.27	(0.41)	(0.46)	1.95	(0.26)	0.83	(0.41)	1.00	(0.34)
TCDC	2.47	(0.79)	(0.90)	6.26	(0.85)	2.47	(0.86)	2.41	(0.82)
TDC	3.13	(1.00)	(1.00)	7.40	(1.00)	2.90	(1.00)	3.18	(1.00)
TLC	5.47	(1.74)	(1.59)	14.04	(1.90)	6.36	(1.60)	6.64	(1.83)
GC	3.37	(1.19)	(0.71)	4.87	(0.66)	1.06	(0.59)	1.05	(0.49)
GUDC	3.04	(0.97)	(0.53)	—	—	0.88	(0.40)	1.00	(0.34)
GCDC	6.87	(2.19)	(1.12)	9.65	(1.30)	2.13	(0.97)	1.35	(0.77)
GDC	8.20	(2.61)	(1.26)	11.38	(1.54)	2.88	(0.85)	3.05	(0.97)
GLC	15.5	(4.96)	(2.07)	—	—	5.50	(1.63)	6.05	(1.68)

[a] Explanation of relative capacity factors is given in the text. All rk' are referenced to taurodeoxycholate (TDC) as 1.00. C, cholate; UDC, ursodeoxycholate; CDC, chenodeoxycholate; LC, lithocholate; G, glyco; T, tauro.

[b] System A: "Fatty acid analysis" column; 2-propanol/8.8 mM potassium phosphate (pH 2.5) (8:17); flow rate 1 ml/min; differential refractometer; from Shaw et al.[4] System B: μBondapak/C$_{18}$ column; acetonitrile/methanol/30 mM phosphate buffer (pH 3.4) (10:60:30); flow 0.5 ml/min; Shimazu Model SPD-1 UV detector at 200 nm. Reproduced by permission from J. Chromatogr. **183**, 287 (1980) and Drs. F. Nakayama and M. Nakagaki. System C: μBondapak/C$_{18}$ column; methanol/water/acetic acid (65:35:3.27) and 10 N sodium hydroxide to pH 4.7; flow 2 ml/min; differential refractometer. Reproduced by permission of the publishers and the authors.[7] System D: μBondapak/C$_{18}$ column; 2-propanol/10 mM potassium phosphate buffer (pH 7.0) (8:17); flow 1 ml/min; differential refractometer; from Shaw et al.[2] System E: Radial-PAK C$_{18}$ cartridge (10 μm particle); 2-propanol/10 mM phosphate buffer (pH 7.0) (8:17); flow 1 ml/min; differential refractometer; reproduced by permission of Marcel Dekker, Inc., and the authors.[1]

[c] pH.

with solvent at pH 2.5, particularly with respect to resolution of TDC and GC which was not improved by a change in pH of the buffer. It has been suggested by a manufacturer that column efficiency may be restored to some extent by elution at the end of the day with a liter of 10 mM sodium citrate (pH 3.5) to remove bound cations.

RPLC of Free Bile Acids

Free bile acids are eluted from a "fatty acid analysis" column (pH 2.5) after their taurine conjugates and before their glycine conjugates. For

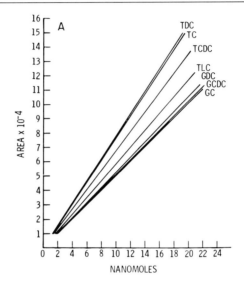
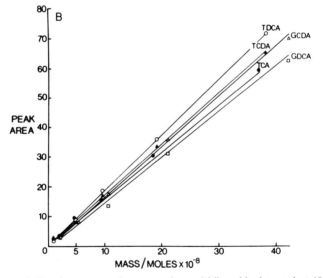

FIG. 4. (A) Calibration curves of some conjugated bile acids detected at 195 nm. TC, Taurocholate; TCDC, taurochenodeoxycholate; TDC, taurodeoxycholate; TLC, taurolithocholate; GC, glycocholate; GCDC, glycochenodeoxycholate; GDC, glycodeoxycholate. The system consisted of a "fatty acid analysis" column, 2-propanol/8.8 mM potassium phosphate, pH 2.5 (8:17), a flow rate of 1 ml/min, a Schoeffel UV monitor set at 0.1 AUF, and a Hewlett-Packard Integrator 3385A. Duplicate determinations of five concentrations ranging from 2 to 20 nmol were made for each compound. Data points are not shown to avoid ambiguity. Correlation coefficients of all lines fell within the range 0.9998–0.9983.

TABLE III
CAPACITY FACTORS (k') AND RELATIVE
CAPACITY FACTORS (rk') OF SEVERAL COMMON
BILE ACIDS[a]

	F^b (2.5)[c]		G^b (7.0)	
	k'	rk'	k'	rk'
C	2.21	(0.57)	1.27	(0.45)
HDC	2.29	(1.59)	1.45	(0.48)
UDC	2.14	(0.56)	1.32	(0.46)
CDC	3.57	(0.93)	3.17	(0.86)
DC	3.86	(1.00)	4.09	(1.00)
LC	6.36	(1.65)	9.23	(2.00)

[a] Relative capacity factors (rk') are related to deoxycholic acid (DC) as 1.00 and are given in parentheses.
[b] System F: "Fatty Acid Analysis" column; 2-propanol/8.8 mM potassium phosphate, pH 2.5 (2:3); flow 1 ml/min; differential refractometer. From Shaw et al.[4] System G: Radial-PAK C_{18} cartridge; 2-propanol/10 mM potassium phosphate (pH 7.0) (8:17); flow 1 ml/min; differential refractometer. Reproduced with permission from Marcel Dekker, Inc. and the authors.[1]
[c] pH.

those who prefer chromatography of the undissociated bile acids, capacity factors for six common bile acids eluted at pH 2.5 from a "fatty acid analysis" column are given in Table III, system F. For comparison of capacity factors of bile acids at pH 7.0 on a μBondapak column (Table I), system G (Table III) provides capacity factors for common bile acids on a Radial-PAK C_{18} cartridge at pH 7.0. Figure 5 illustrates the effect of the rate of elution of four bile acids from the same Radial-PAK C_{18} cartridge with a programmed flow of the solvent system. Relative capacity factors in Table III are related to deoxycholate.

Reprinted by permission of Marcel Dekker, Inc. and the authors.[1] (B) Calibration curve of some conjugated bile acids as determined with a Model R401 differential refractometer: TDCA, taurodeoxycholate; TCDA, taurochenodeoxycholate; GCDA, glycochenodeoxycholate; TCA, taurocholate; GDCA, glycodeoxycholate. The system consisted of a μBondapak/C_{18} column eluted with a mixture of methanol/water/acetic acid + 10 N sodium hydroxide (pH 4.7) (65:35:3.27). Reprinted by permission of the publisher and authors.[7]

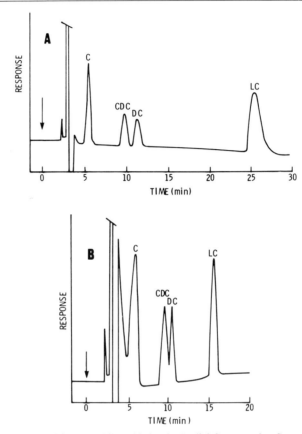

FIG. 5. Separation of four free bile acids by the Radial Compression Separation System: C, cholate; CDC, chenodeoxycholate; DC, deoxycholate; LC, lithocholate. The system included 2-propanol/10 mM potassium phosphate (pH 7.0) (8 : 17), and a Model R401 differential refractometer set at 8×. Chromatogram A was achieved at a flow rate of 1 ml/min and chromatogram B programmed for 1 to 4 ml/min over 14 min according to gradient curve 10 of Model 660 programmer. Reprinted by permission of Marcel Dekker, Inc. and the authors.[1]

RPLC of Sulfates

The 3-sulfates of cholate, deoxycholate, chenodeoxycholate, and lithocholate and of their taurine and glycine conjugates were separated with a C_{18} column (25 cm × 4.6 mm. i.d.). The UV detector was set at 210 nm, and the solvent mixture was 0.5% aqueous ammonium carbonate/ acetonitrile delivered at a flow rate of 2.1 ml/min.[8] To shorten the elution

---

[8] J. Goto, H. Kato, and T. Nambara, *Lipids* **13,** 908 (1978).

time the 3-sulfates of taurodeoxycholate, lithocholate, glycolithocholate, and taurolithocholate were eluted in 8 min with a solvent ratio of 5:2; all other 3-sulfates including that of taurodeoxycholate were eluted within 15 min with a solvent ratio of 13:4.

Normal-Phase Chromatography

Since silica columns were available commercially before reversed-phase columns, separations of bile acids and their derivatives were first studied in HPLC with such columns.[1] Free bile acids can be separated by the same type of normal-phase liquid–liquid partition chromatography as employed in TLC. The procedures were hampered by the solubilization of silica in aqueous polar phases. Esters of several types have been separated on columns of Corasil, μPorasil, Partisil, and Micropak-NH_2 or in larger quantities on preparative columns (12 × 2 in. i.d.) of silica of larger particle size. Conjugated bile acids have been similarly studies on the above analytical columns. Glycine and some taurine conjugates of 5β- and 5α-bile acids are generally better resolved in a system of five columns of Corasil II connected in series.[3,9]

Acknowledgments

Support from Grants HL-07878 and CA-16375 for some of the studies reported here is gratefully acknowledged. This communication constitutes Paper LXXIV from this laboratory.

[9] R. Shaw and W. Elliott, *Lipids* **15**, 805 (1980).

[4] Mass Spectrometry of Bile Acids

By JAN SJÖVALL, ALEXANDER M. LAWSON, and KENNETH D. R. SETCHELL

Introduction

Mass spectrometry (MS), particularly in combination with gas chromatography (GC-MS), has become an indispensable method for many studies of bile acids and their metabolism. Three major areas of application can be distinguished: (1) identification and structure determination, (2) quantitative analysis, and (3) kinetic and metabolic studies using stable isotopes. Several reviews have appeared, covering in particular the fragmentation patterns of trimethylsilyl (TMS) ethers, acetates, and trifluoroacetates of methyl hydroxy- and oxocholanoates.[1-4] Additional information can be obtained from reviews of the fragmentation of steroids[5-7] since mass spectra of sterols and sterol derivatives are often very similar to those of methyl cholanoates.

The aim of this review is to provide information on the selection of suitable conditions for qualitative and quantitative MS and GC-MS, and for the identification or partial characterization of bile acids by these methods. While it may become possible to obtain detailed information about complex bile acid mixtures by soft ionization of crude biological materials and tandem MS-MS analysis, the two most common methods today are GC-MS analysis of derivatized unconjugated bile acids, and direct MS of fractions obtained from low- or high-performance liquid chromatographic systems. It is assumed in the following discussion that some type of purification of the biological material has been carried out prior to the mass spectrometric analysis.

Conditions for Mass Spectrometry

The conditions for introducing bile acids and producing spectra are similar to those for other steroids and have been developed in a similar sequence. The chemical form and composition of a bile acid-containing sample dictate the most suitable mode of sample introduction, while the ionization method employed is largely dependent on the information required from the analysis. Table I lists several of the options available for the different bile acid classes.

Ionization Methods

Electron Impact (EI). The most commonly available and used method of ionization is electron impact, in which a beam of energetic electrons

[1] J. Sjövall, P. Eneroth, and R. Ryhage, in "The Bile Acids" (P. P. Nair and D. Kritchevsky, eds.), p. 209. Plenum, New York, 1971.
[2] W. H. Elliott, in "Biomedical Applications of Mass Spectrometry" (G. R. Waller, ed.), p. 291. Wiley, New York, 1972.
[3] P. A. Szczepanik, D. L. Hachey, and P. D. Klein, *J. Lipid Res.* **17,** 314 (1976).
[4] W. H. Elliott, in "Biomedical Applications of Mass Spectrometry. First Supplementary Volume" (G. R. Waller and O. C. Dermer, eds.), p. 229. Wiley, New York, 1980.
[5] H. Budzikiewicz, in "Biomedical Applications of Mass Spectrometry" (G. R. Waller, ed.), p. 251. Wiley, New York, 1972.
[6] Z. C. Zaretskii, "Mass Spectrometry of Steroids." Wiley, New York, 1976.
[7] H. Budzikiewicz, in "Biomedical Applications of Mass Spectrometry. First Supplementary Volume" (G. R. Waller and O. C. Dermer, eds.), p. 211. Wiley, New York, 1980.

TABLE I
SAMPLE INTRODUCTION AND IONIZATION METHODS FOR DIFFERENT
BILE ACID CLASSES[a]

Chemical form	Sample introduction method	Ionization
Free unconjugated bile acids	Direct probe	EI or CI
	FD emitter	FD
	FAB target	FAB
	LD target	LD
	LC interface	CI
Derivatized unconjugated bile acids	Direct probe	EI or CI
	GC/MS	EI or CI
Conjugated bile acids		
Amino acid	Direct probe	EI
	FAB target	FAB
	FD emitter	FD
	LD target	LD
	LC interface	CI
	GC-MS as derivatives	EI
Sulfates (with and without glycine or taurine)	FAB target	FAB
Glucuronides	Direct probe	CI
	GC-MS as derivatives	EI

[a] EI, Electron impact; CI, chemical ionization; FD, field desorption; FAB, fast atom bombardment; LD, laser desorption; LC, liquid chromatography.

(typically 20–70 eV) collide with the sample molecules in the ion chamber. The loss of an electron from a molecule during this process produces a positively charged radical ion with sufficient excess energy to lead to fragmentation of the molecule. The abundance of ions at each mass value depends on a combination of the structure of the compound and the energetics of the processes involved. The temperature of the ion source also influences the mass spectra of thermally labile molecules and, in the case of bile acids, free hydroxyl groups are particularly susceptible to dehydration. The ion source, however, has to be maintained at a temperature some 20–30° above the volatilization temperature of the sample to prevent adsorption.

An ionizing voltage of 70 eV ensures that the abundance of ions from fragmentation of chemical bonds is maximized and this voltage has been used in many bile acid studies. As the electron energy is decreased less fragmentation is usually observed until, at the ionization potential (approximately 6–12 eV), only the ionized intact molecule is produced. It has been shown, for example, that in spectra of methyl ester TMS ether derivatives of bile acids, the relative abundance of diagnostically useful ions is greater at 22.5 eV than at higher energies.[8] For these and other reasons 22.5 eV can be considered an optimal ionization voltage for qualitative and quantitative analyses but this does not exclude the use of higher energies.

Chemical Ionization (CI). Chemical ionization was developed to permit a much "softer" or low energy ionization which would minimize the energy transferred to the molecule and hence reduce the extent of fragmentation of chemical bonds. This is achieved by ionizing a reagent gas by electron impact, and bringing about ionization of the sample molecules by ion molecule interactions.

A wide variety of reagent gases has been investigated and the most commonly used are the Brønsted acid reagent systems which react primarily by proton transfer (see ref. 9) to produce a protonated form of the sample molecule. This is known as a quasimolecular ion. Typical reagent gases of this type are H_2, N_2O/H_2, CH_4, H_2O, $i\text{-}C_4H_{10}$, and NH_3 with their corresponding reactant ions being H_3^+, N_2OH^+, CH_5^+, $H^+(H_2O)_n$, $C_4H_9^+$, and $H^+(NH_3)_n$, respectively. Negative ion reagent gas systems which produce negatively charged quasimolecular ions of the type $[M - H]^-$ utilize reactive species such as H^-, NH_2^-, OH^-, O^-, and Cl^-.

In addition to the value of obtaining molecular weight information, chemical ionization offers advantages in quantitative studies. These accrue from the reduced fragmentation and hence usually greater sensitivity possible from confining the ionization to fewer ions. The yield of selected ions may in some cases by some 10–100 times greater than with electron impact. The isomeric nature of groups of bile acids, however, makes their differentiation very difficult using CI when samples are introduced on a direct insertion probe. As a result it is advisable to prepare volatile derivatives and, with GC-MS, take advantage of the additional chromatographic data when making identifications.

Chemical ionization mass spectra of bile acids are most often recorded as derivatives, e.g., methyl esters,[10] methyl ester acetates,[3,11–13] and

[8] B. Almé, A. Bremmelgaard, J. Sjövall, and P. Thomassen, *J. Lipid Res.* **18**, 339 (1977).
[9] A. G. Harrison, "Chemical Ionization Mass Spectrometry." CRC Press, Boca Raton, Florida, 1983.
[10] G. M. Muschik, L. H. Wright, and J. A. Schroer, *Biomed. Mass Spectrom.* **6**, 266 (1979).
[11] K. Kuriyama, Y. Ban, T. Nakashima, and T. Murata, *Steroids* **34**, 717 (1979).
[12] B. R. DeMark and P. D. Klein, *J. Lipid Res.* **22**, 166 (1981).
[13] T. Murata, S. Takahashi, S. Ohnishi, K. Hosoi, T. Nakashima, Y. Ban, and K. Kuriyama, *J. Chromatogr.* **239**, 571 (1982).

methyl ester TMS ethers.[14,15] Some free bile acids have also been studied.[16]

Quasimolecular ions are not always present and in the methane CI spectra of the methyl cholanoates such as lithocholate, deoxycholate, chenodeoxycholate, and cholate, the major ions represent facile loss of the hydroxyl groups as molecules of water.[10] Similar results are obtained with isobutane and the methyl ester acetate derivatives with loss of acetic acid,[3] and in TMS ethers of methyl cholanoates by loss of trimethylsilanol.[14] Jelus and co-workers demonstrated, however, that the molecular ion of the TMS ether of methyl chenodeoxycholate could be obtained by charge exchange using a N_2/NO reagent gas.[14,15]

The most effective mean of obtaining molecular weight information on bile acid methyl ester acetates[12] and of cholic acid[16] has been by using ammonia as a reagent gas. The study of 28 bile acids by DeMark and Klein[12] showed the $[M + NH_4]^+$ as the base peak in all spectra at an ion source temperature of 160°. The electrophilic attachment of NH_4^+ is a less exothermic process than proton addition from $C_4H_9^+$ leading to almost no fragmentation and only little loss of acetic acid. Hence greater sensitivity can be achieved, particularly in the ion selective mode (see Selected Ion Monitoring).

Direct Chemical Ionization (DCI). This ionization method involves coating the sample on a wire, or similar support, at the end of a probe then inserting it into, or very close to, the electron beam in the ion source. The mechanism by which ions characteristic of the molecular weight are produced has not yet been clarified but it would appear that the neutral sample molecules are desorbed from the inert surface of the wire, frequently assisted by rapid heating of the probe tip, and are ionized almost immediately by the ion plasma before they suffer collisions with other surfaces. In this way the thermal effects encountered in conventional direct probe introduction are reduced. The method has been applied to many compounds including ionic salts and thermally labile organic molecules[17] and may well be useful for analysis of bile acids.

Field Desorption (FD). In FD the sample is deposited from solution onto a fine tungsten wire activated by growth of a dense matrix of microneedles on the surface.[18] When the wire is placed in the region of a high

---

[14] B. L. Jelus, B. Munson, and C. Fenselau, *Biomed. Mass Spectrom.* **1**, 96 (1974).

[15] B. Jelus, B. Munson, and C. Fenselau, *Anal. Chem.* **46**, 729 (1974).

[16] A. K. Bose, H. Fujiwara, B. N. Pramanik, E. Lazaro, and C. R. Spillert, *Anal. Biochem.* **89**, 284 (1978).

[17] D. F. Hunt, J. Shabanowitz, F. K. Botz, and D. A. Brent, *Anal. Chem.* **49**, 1161 (1977).

[18] H. D. Beckey, "Principles of Field Ionization and Field Desorption Mass Spectrometry." Pergamon, Oxford, 1977.

electric field (approximately 10^8 V cm^{-1}) and heated, ions are produced by ionic desorption of sample molecules. While giving molecular weight information, the spectra are often only transient and compare unfavorably in ion sensitivity to electron impact ionization.

Ionization by FD has been used to obtain the spectra of thermally labile or involatile compounds and in most cases intense molecular ions are observed with little or no fragmentation. Sodium salts of a number of the common bile acids and their conjugates have been studied[19–21] and the quasimolecular ions in all cases were the [M + Na]$^+$ ions with intense double charged ions at [M + 2Na]$^{2+}$.

The value of field desorption to bile acid analysis lies in establishing the molecular weight of free or conjugated bile acids where this information is not available by other methods. Stereoisomers cannot be differentiated by FD and, as with direct insertion probe, the bile acid needs to be in pure form if any confidence is to be placed on the interpretation. Nevertheless, in combination with good chromatographic separation, field desorption does provide a means of establishing the nature of conjugated bile acids without hydrolysis.

Laser Desorption (LD). Quasimolecular ions have been obtained by laser desorption ionization of underivatized free and conjugated bile acids and also their sodium salts.[22,23] The negative ion spectra of the compounds so far studied contained [M − H]$^-$ ions in every case, and in the low mass region cluster ions of C_n^- and C_nH^- arising from pyrolysis. When sodium is present in the sample the sodium cationized species [M + Na]$^+$ are the predominant ions in the positive mode and fragmentation is not significant. In some instances the [M + H]$^+$ ion is observed.[23]

LD is not yet widely available and its utility for bile acids requires further evaluation.

Fast Atom Bombardment (FAB). The principle of ionizing materials from the solid state using high velocity particles has been known for some considerable time. The introduction of ion guns which could produce molecular beams of neutral molecules and the concept of dissolving the sample in a matrix to prevent premature destruction of the sample, forms the basis of the presently used technique for organic molecules.[24] It is

---

[19] D. E. Games, M. P. Games, A. H. Jackson, A. H. Olavesen, M. Rossiter, and P. J. Winterburn, *Tetrahedron Lett.* **27**, 2377 (1974).

[20] R. Shaw and W. H. Elliott, *Biomed. Mass Spectrom.* **5**, 433 (1978).

[21] J. O. Whitney, S. Lewis, K. M. Straub, M. M. Thaler, and A. L. Burlingame, *Koenshu-Iyo Masu Kenkyukai* **6**, 33 (1981).

[22] R. J. Cotter, *Anal. Chem.* **53**, 719 (1981).

[23] R. J. Day, J. Zimmerman, and D. M. Hercules, *Spectrosc. Lett.* **14**, 773 (1981).

[24] M. Barber, R. S. Bordoli, G. V. Garner, D. B. Gordon, R. D. Sedgwick, and L. W. Tetler, *Biochem. J.* **197**, 401 (1981).

important that the sample molecules dissolve in the matrix. A glycerol or similar liquid of low volatility is used as the matrix and coated thinly on a metal target (approximately 15 mm^2 area). The sample as a solid or in solution (1–2 μl) is mixed with the matrix and the target bombarded in the ion source by a beam of argon or xenon atoms (10 kV, 10 μA). Depending on the nature of the compound, the proportions of negative to positive ions produced by FAB will differ and complementary information is often obtained by studying both positive and negative ion mass spectra.

In a similar way to field desorption, FAB has extended the range of bile acids amenable to mass spectrometry and in particular to the less thermally stable and more involatile conjugated bile acids. EI and CI have proved effective for free and derivatized unconjugated bile acids but give poor molecular ions of conjugated acids in most cases. Bile acid sulfates are especially difficult and will degrade even under FD conditions. FAB can effectively ionize these conjugates and give spectra with quasimolecular ions of the intact molecule. Practical advantages of FAB over FD include faster and easier sample handling and more intense and stable ion beams.

The general pattern of FAB ionization of conjugated bile acids has been established.[21,25–29] Useful spectra can be obtained from small or submicrogram amounts of sample. In positive ion spectra of the sodium salts of sulfated, taurine and glycine conjugated bile acids the protonated molecular ion [M + H]$^-$ and the sodium addition ion [M + Na]$^+$ are the most commonly observed ions. In the case of the sodium salts of taurine conjugates and sulfates the [M + Na]$^+$ ion is favored, and in spectra of sulfates an attachment ion containing a further sodium is found while the [M + H]$^+$ ions are absent. The glycine conjugates[26,30] usually give both [M + H]$^+$ and [M + Na]$^+$, while the positive quasimolecular ions of the sodium salts of sulfated glycine or taurine conjugates are of low intensity.

A comparison of the molecular ion regions of positive and negative ion mass spectra from a selected group of conjugated bile acids is shown in Table II. The spectra of the sodium salts of glycine and taurine conjugates

---

[25] J. O. Whitney, S. Lewis, K. M. Straub, F. C. Walls, A. L. Burlingame, and M. M. Thaler, *Proc. 29th Annu. Conf. Mass Spectrom. Allied Top., 1981* p. 829 (1981).

[26] J. O. Whitney and A. L. Burlingame, *Koenshu-Iyo Masu Kenkyukai* **7**, 3 (1982).

[27] J. G. Liehr, C. F. Beckner, A. M. Ballatore, and R. M. Caprioli, *Proc. 30th Annu. Conf. Mass Spectrom. Allied Top., 1982* p. 234 (1982).

[28] Y. Itagaki, T. Higuchi, Y. Naito, J. Goto, and T. Nambara, *Proc. 30th Annu. Conf. Mass Spectrom. Allied Top., 1982* p. 236 (1982).

[29] J. O. Whitney, V. Ling, D. Grunberger, M. M. Thaler, and A. L. Burlingame, *Koenshu-Iyo Masu Kenkyukai* **8**, 47 (1983).

[30] A. M. Ballatore, C. F. Beckner, R. M. Caprioli, N. E. Hoffman, and J. G. Liehr, *Steroids* **41**, 197 (1983).

TABLE II
QUASIMOLECULAR IONS IN FAB SPECTRA OF BILE ACID CONJUGATES[a]

Bile acid conjugates	Molecular weight	Positive ion spectra[b]		Negative ion spectra[b]	
		[M + H]⁺	[M + Na]⁺	[M − H]⁻	[M − Na]⁻
Glycoconjugates					
Glycodeoxycholic acid	449	450(8)	472(1)	448(10)	—
Sodium glycochenodeoxycholate	471	472(55)	494(24)	470(1)	448(94)
Glycocholic acid	465	466(10)	488(3)	464(9)	—
Tauroconjugates					
Sodium taurodeoxycholate	521	522(18)	544(76)	520(6)	498(99)
Sodium taurochenodeoxycholate	521	522(5)	544(21)	520(10)	498(98)
Sodium taurocholate	537	538(36)	560(82)	536(15)	514(100)
Sulfate conjugates					
Sodium chol-5-enoate-3-*O*-sulfate	476	—	499(4)	475(14)	453(57)
Sodium lithocholate-3-*O*-sulfate	478	—	501(27)	477(32)	455(100)
Sodium chenodeoxycholate-3-*O*-sulfate	494	—	517(8)	493(38)	471(84)
Sulfated glycoconjugates					
Disodium glycolithocholate-3-*O*-sulfate	557	558(3)	580(1)	556(0.5)	534(16)
Sulfated tauroconjugates					
Disodium taurolithocholate-3-*O*-sulfate	607	608(0.4)	630(1.7)	606(1.5)	584(18)
Disodium taurochenodeoxycholate-3-*O*-sulfate	623	624(0.3)	646(2)	622(0.3)	600(3)

[a] Samples run on Varian (Finnigan) MAT-731 using FAB gun (M-Scan Ltd, Ascot, Berkshire, UK) with stainless-steel target and glycerol matrix.
[b] Numbers in parentheses are relative intensities.

show little fragmentation but those of the sulfate analogs are more complex and also show positive ions due to loss of $NaHSO_4$. The sulfates usually have ions at m/z 143 and 165 which have been assigned as $[Na_2HSO_4]^+$ and $[Na_3SO_4]^+$, respectively,[28] and in negative mode an ion for $[HSO_4]^-$ is observed at m/z 97.

The sodium salts of unconjugated bile acids also give FAB spectra with quasimolecular ions $[M + H]^+$ and $[M + Na]^+$ dominating in the positive mode and $[M - Na]^-$ and $[M - H]^-$ in the negative mode. In the positive ion spectra of free unconjugated acids, however, intense ions due to loss of water are characteristic and the base peak represents an ion due to complete water loss (i.e., m/z 359, 357, and 355 from the monohydroxy, dihydroxy, and trihydroxy bile acids, respectively).[26] Ions resulting from elimination of water are not present in the negative ion spectra.

Isomeric bile acids have almost identical FAB spectra and cannot be distinguished without prior chromatographic separation. Despite this disadvantage the analysis of raw unfractionated bile and duodenal juice has been demonstrated and can give rapid qualitative information on the major bile acids present and their conjugation states.[26,29]

Sample Introduction

Direct Insertion Probe. The earlier use of direct insertion probe, in which the sample is placed in a small heatable crucible and positioned close to the ionization region, had the advantage of utilizing lower volatilization temperatures and hence reducing thermal effects experienced with gas chromatographic inlets. However, with improvements in GC and the use of derivatives, this advantage has been outweighed in applications where complex mixtures of bile acids are present. In such situations direct probe analysis leads to mixed spectra in which the ions from major components swamp those of minor constituents. Nevertheless, direct insertion can be used effectively when samples of individual bile acids are available and for introducing the less volatile conjugated bile acids.

The temperature required for volatilization from direct probe depends largely on the degree of hydroxylation and conjugation. Almost all free unconjugated bile acids reported volatilize below about 120° while glycine conjugates require higher temperature (225–250°).[20]

Gas Chromatographic Inlet. GC-MS has added greatly to the ability to detect and characterize the wide range of bile acid structures which are excreted by man and animals. While the gas chromatographic column may only serve in some cases as a convenient inlet for MS, the retention characteristics of peaks from complex mixtures are essential to identifying the many stereoisomeric compounds which give closely similar mass spectra or for which the reference spectra may not be available to the analyst. Gas–liquid chromatography (GLC) is routinely employed on its own to quantify and characterize bile acids and there is a well-established literature on its use, particularly in the application of packed columns. The choice of liquid phases for packed column GLC is extremely varied but in GC-MS it is limited to phases which are thermally stable within the temperature range required for separation and will not bleed into the mass spectrometer to give a high background of ions. The GLC conditions for GC-MS have been reviewed in the past.[2,4]

The most appropriate combination of derivative and liquid phase is one which permits adequate separation of the components at temperatures at which the phase is thermally stable and no adsorption of the sample takes place. Numerous derivatives of bile acids have been investigated and the acetates, trifluoroacetates, and TMS ethers of the methyl cholanoates are best characterized and understood. They are readily and quantitatively prepared. Table III contains a list of most of the derivatives which have been used for the GC-MS analysis of bile acids together with some general notes on their utility.[31–50]

The mass spectrometric fragmentation of the various derivatives has an important influence on the choice of derivative for particular applications (see Mass Spectra of Unconjugated Bile Acids). The requirements for quantification, in addition to quantitative preparation and stability, are

---

[31] A. Kuksis, in "Lipid Chromatographic Analysis" (G. V. Marinetti, ed.), 2nd ed., Vol. 2, p. 479. Dekker, New York, 1976.
[32] R. Galeazzi, E. Kok, and N. Javitt, *J. Lipid Res.* **17,** 288 (1976).
[33] I. M. Yousef, M. M. Fisher, J. J. Myher, and A. Kuksis, *Anal. Biochem.* **75,** 538 (1976).
[34] P. A. Szczepanik, D. L. Hachey, and P. D. Klein, *J. Lipid Res.* **19,** 280 (1978).
[35] T. Nakashima, Y. Ban, K. Kuriyama, and T. Takino, *Jpn. J. Pharmacol.* **29,** 667 (1979).
[36] P. Child, A. Kuksis, and J. J. Myher, *Can. J. Biochem.* **57,** 639 (1979).
[37] H. Miyazaki, M. Ishibashi, M. Inoue, M. Itoh, and T. Kubodera, *J. Chromatogr.* **99,** 553 (1974).
[38] G. Karlaganis, R. P. Schwarzenbach, and G. Paumgartner, *J. Lipid Res.* **21,** 377 (1980).
[39] T. Iida, F. C. Chang, T. Matsumoto, and T. Tamura, *J. Lipid Res.* **24,** 211 (1983).
[40] T. Cronholm, I. Makino, and J. Sjövall, *Eur. J. Biochem.* **24,** 507 (1972).
[41] K. Carlström, D. N. Kirk, and J. Sjövall, *J. Lipid Res.* **22,** 1225 (1981).
[42] K. Imai, Z. Tamura, F. Mashige, and T. Osuga, *J. Chromatogr.* **120,** 181 (1976).
[43] R. Edenharder and J. Slemr, *J. Chromatogr.* **222,** 1 (1981).
[44] B. C. Musial and C. N. Williams, *J. Lipid Res.* **20,** 78 (1979).
[45] H. Miyazaki, M. Ishibashi, and K. Yamashita, *Biomed. Mass Spectrom.* **5,** 469 (1978).
[46] A. Fukunaga, Y. Hatta, M. Ishibashi, and H. Miyazaki, *J. Chromatogr.* **190,** 339 (1980).
[47] S. H. G. Andersson and J. Sjövall, *J. Chromatogr.* **289,** 195 (1984).
[48] S. Barnes, D. G. Pritchard, R. L. Settine, and M. Geckle, *J. Chromatogr.* **183,** 269 (1980).
[49] G. A. De Weerdt, R. Beke, H. Verdievel, and F. Barbier, *Biomed. Mass Spectrom.* **7,** 515 (1980).
[50] A. Bremmelgaard and J. Sjövall, *J. Lipid Res.* **21,** 1072 (1980).

TABLE III
DERIVATIVES FOR GLC AND GC-MS OF BILE ACIDS

Derivative	Comments	GLC liquid phases	References
Methyl ester	Less suitable for polyhydroxylated bile acids due to polarity. Prepare with fresh diazomethane	SE-30; OV-1; OV-17; QF-1	1,2,4,31
Methyl ester acetate	Good storage stability	SP-525; Poly S-179; PPE-20; OV-210; OV-225; AN-600	1,3,31–36
Methyl ester trifluoroacetate	High volatility—lower column temperatures than for acetates. Thermal loss of trifluoroacetic acid on GLC column	QF-1	31
Methyl ester trimethylsilyl ether	Good thermal stability and high volatility—good GLC and MS properties. Storage stability limited	SE-30; OV-1; OV-17; SP-2340; Hi-Eff 8 BP; PPE-20; Poly I-110; PEG 20,000	2,4,31,32,37–39
Methyl ester O-methyloxime	For oxo bile acids, syn and anti form. May not be quantitative on GLC. Helpful for MS interpretation. Hydroxyl groups can be silylated	SE-30; OV-1	40,41
Hexafluoroisopropyl ester trifluoroacetate	Avoids use of diazomethane. Simple preparation, good resolution. High molecular weights	QF-1	42,43
Perheptafluorobutyryl	Simultaneous deconjugation (partial) and derivatization of conjugates. Artifacts with oxo groups. No separation of 3,6- and 3,7-dihydroxy bile acids. High molecular weights	QF-1; OV-225	43,44
Methyl ester alkyldimethylsilyl ethers	Longer retention times than equivalent TMS ethers. High mass ions for selected ion monitoring	OV-101/Dexil 300-GC SE-30; OV-1	45–47
Permethyl	Single step preparation, very stable, good GLC-properties. Forms enol ethers of oxo groups	SE-53; SP-2250	48,49
Acetonide	Derivative of vicinal hydroxyl groups, cis in the rings	SE-30; OV-1	50

the presence of ions in the spectra of adequate intensity for detection, and which are sufficiently characteristic to allow specific monitoring of individual acids. In this regard the ammonia CI spectra of methyl ester acetates and the EI spectra of methyl ester alkyldimethyl ethers have useful ions in the higher mass range.

While packed glass columns have been essential to the development of GC-MS in the past and will continue to be used, they may be replaced by capillary columns for most applications. These columns (15–30 m long, 0.25–0.5 mm i.d.) give much higher separation efficiencies, inversely related to phase thickness and internal diameter. The most recent capillary columns with chemically bonded phases have high thermal stability (temperature limit 375° for OV-1) which reduces column bleed, and these columns can be cleaned by solvents. The columns made of fused silica are flexible and easy to handle. This avoids any difficulty in making connections with the injector and permits introduction of the column end into the ion source. The need for a carrier gas separator or other transfer system with potential thermal and adsorption effects is eliminated and the mass spectrometer can readily handle the gas flow from the capillary column (0.5–2 ml/min).

Capillary columns place increased demands on the performance of the mass spectrometer due to their low loading capacity and the narrow peak widths. As the compounds are eluted in very tight bands from the column, sensitivity of detection of a given amount of compound is higher with a capillary column than a packed column. However, the mass spectrometer has to be scanned much more rapidly to obtain a spectrum not biased by the rapidly changing sample concentration. When repetitive scanning is in operation (see Repetitive Scanning Techniques) 2–3 scans a second may be desirable. Since the number of ions collected decreases with increasing scan rate sensitivity can become a problem.

Several injection systems for capillary columns are available[51] and each has been applied to bile acid analysis. Due to discrimination effects, the split mode is not suitable for quantitative studies, and the large proportion of the sample passed to waste is a further limitation. The solid injector on the other hand, which employs a dropping glass needle,[52] allows for the concentration of a volume of sample solution prior to injection. The same principle is used in automatic injection systems where the samples are loaded in individual glass capillaries and mounted in a carousel. These methods reduce contamination of the column and minimize

---

[51] W. Jennings, "Gas Chromatography with Glass Capillary Columns." Academic Press, New York, 1980.

[52] P. M. J. Van den Berg and T. P. H. Cox, *Chromatographia* **5**, 301 (1972).

solvent front. In these respects on-column injection of solutions is troublesome.

The methyl ester TMS ether derivatives of bile acids have been the most extensively studied with regard to their separation on capillary columns.[38,39,53–56] The stationary phases include methyl- (SE-30, OV-1), phenylmethyl- (OV-17), and cyanopropyl- (SP-2340) silicones and polyethylene glycols (PEG 20,000, PEG-HT, Carbowax 20 M). In general, if a wide range of bile acids require analysis then nonselective phases such as OV-1 or SE-30 offer fewer limitations than the selective phases. On the selective polyethylene glycol phases, bile acid methyl ester TMS ethers separate in a similar order as on Hi-Eff 8 BP used in packed columns (increasing retention time with decreasing number of trimethylsiloxy groups, usually better separation of epimers than on SE-30). The oxo-containing acids are particularly retarded on PEG 20,000 and on PEG-HT this is accentuated[54] to an extent which undermines the utility of the phase for rapid analysis unless only hydroxylated bile acids are of interest.

No liquid phase can in itself adequately separate all the possible isomers of bile acids on capillary columns although the situation is greatly improved over packed columns for bile acid profile analysis. However, identification of incompletely resolved peaks can usually be achieved from consecutive mass spectra taken over the elution profile of an unresolved multiplet. Alternatively, unresolved components on one phase may well resolve on another. Iida and co-workers[39] investigated OV-1, OV-17, and SP-2340 for the separation of methyl ester TMS ether derivatives and came to the conclusion that sequential use of these phases could provide most separations. Monohydroxylated acids, for example, separate well on OV-1, but the 7α- and 12α-isomers failed to separate on OV-17 and SP-2340. The 3α,7β- and 3β,7β-diols are inseparable on OV-1 but all four 3,7-diol epimers are resolved on the other columns. While 3,12-diols prove difficult on OV-1 and SP-2340 they can be separated on OV-17. The 3,7,12-trihydroxy epimers are not fully separated by SP-2340, and OV-1 and OV-17 are required for resolving 3β,7β,12α- and 3β,7β,12β-trihydroxycholanoates. It should be pointed out that columns from different sources may give quite different retention indices for the same bile acid derivative. Judging from published data, the order of elution of isomers may even be different. However, the very high retention indices on OV-1 reported by Iida et al.[39] are difficult to explain on this basis.

[53] T. Laatikainen and A. Hesso, *Clin. Chim. Acta* **64**, 63 (1975).
[54] N. Tanida, Y. Hikasa, and T. Shimoyama, *J. Chromatogr.* **240**, 75 (1982).
[55] K. D. R. Setchell, in "Bile Acids in Gastroenterology" (L. Barbara, R. H. Dowling, A. F. Hofmann, and E. Roda, eds.), p. 1. MTP Press Ltd., Lancaster, 1982.
[56] K. D. R. Setchell, A. M. Lawson, N. Tanida, and J. Sjövall, *J. Lipid Res.* **24**, 1085 (1983).

An illustration of the resolution of a mixture of bile acids from rat feces on an OV-1 capillary column is shown in Fig. 1.[56] The complexity of the mixture of acids present underlines the need for mass spectrometric identification.

The methyl ester dimethylethylsilyl, dimethyl-*n*-propylsilyl, and dimethylisopropylsilyl ether derivatives of 12 bile acids have been compared with their corresponding TMS derivatives on SE-30.[46] Improved separation of the bile acids was observed with the dimethylethyl derivative being the most favorable. The compounds elute in sequence according to the number of hydroxyl groups in the parent molecule and the mass spectra tend to yield characteristic fragment ions with abundant intensity in the high mass region. The application of these derivatives in the quantitative analysis of bile acids at very low concentrations in human tissue is of considerable interest.[57] It remains to be shown, however, that the alkyldimethylsilyl derivatives are superior to the TMS ethers for resolving more complex isomeric mixtures of bile acids.

Liquid Chromatographic Inlet. High-pressure liquid chromatography (HPLC) has now become a routine method for the analysis of bile acids using different methods of detection. To establish HPLC systems for handling complex mixtures of bile acids it is necessary to determine the chromatographic mobilities and identities of individual components. Liquid chromatography–mass spectrometry (LC-MS) would be extremely useful for such studies.

The instrumentation for LC-MS continues to develop (see ref. 58) but application to bile acid analysis has been limited. Games and co-workers[59] demonstrated separation of lithocholic, cholic, chenodeoxycholic, taurocholic, and glycocholic acids on an octadecylsilane column and, under CI conditions with ammonia as reagent gas, they obtained the expected mass spectra. Taurocholic acid failed to yield molecular weight information (e.g., $[M + H]^+$ or $[M + NH_4]^+$), the highest mass ion at m/z 479 representing loss of two molecules of water. Any thermal effects experienced by direct probe introduction are likely to occur on the belt interface used in these studies, and it has been suggested that derivatization may help to overcome such volatility problems.[60] Combination with FAB ionization might become an alternative procedure.

[57] J. Yanagisawa, M. Itho, M. Ishibashi, H. Miyazaki, and F. Nakayama, *Anal. Biochem.* **107,** 75 (1980).

[58] C. G. Edmonds, J. A. McCloskey, and V. A. Edmonds, *Biomed. Mass Spectrom.* **10,** 237 (1983).

[59] D. E. Games, C. Eckers, J. L. Gower, P. Hirter, M. E. Knight, E. Lewis, K. R. N. Rao, and N. C. Weerasinghe, *in* "Current Developments in the Clinical Applications of HPLC, GC and MS" (A. M. Lawson, C. K. Lim, and W. Richmond, eds.), p. 97. Academic Press, New York, 1980.

[60] M. A. Quilliam and E. Y. Osei-Twum, *Proc. 28th Annu. Conf. Mass Spectrom. Allied Top., 1980* p. 612 (1980).

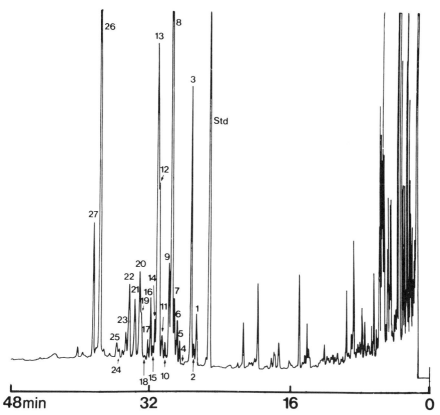

FIG. 1. Profile of methyl ester TMS ethers of unconjugated bile acids in rat feces analyzed on a 25 m × 0.25 mm OV-1 capillary column (220° isothermal for 5 min, then programmed to 285° at a rate of 2°/min). Helium was the carrier gas at a flow rate of about 2 ml/min and the sample was introduced with a dropping needle system. The following compounds are indicated: 1, 3α-Hydroxy-5α-cholanoic; 2, 3β-hydroxy-5β-cholanoic; 3, 3α-hydroxy-5β-cholanoic (lithocholic); 4, unknown monohydroxy bile acid; 5, 3α,12α-dihydroxy-5α-cholanoic; 6, 3α,12β-dihydroxy-5β-cholanoic; 7, 3α,12β-dihydroxy-5α-cholanoic; 8, 3α,12α-dihydroxy-5β-cholanoic (deoxycholic); 9, 3β-hydroxy-5α-cholanoic; 10, 3α,6β-dihydroxy-5β-cholanoic and 3-oxo-12α-hydroxy-5β-cholanoic; 11, unknown dihydroxy + trihydroxy bile acids; 12, 3α,6β,7α-trihydroxy-5β-cholanoic (α-muricholic); 13, 3α,6α-dihydroxy-5β-cholanoic (hyodeoxycholic); 14, 3β,6α-dihydroxy-5β-cholanoic; 15, unknown tetrahydroxy bile acid; 16, 3β,12α-dihydroxy-5α-cholanoic; 17, 12-oxo-3α-hydroxy-5α-cholanoic; 18, unknown trihydroxy bile acid; 19, 3β,12β-dihydroxy-5α-cholanoic; 20, 12-oxo-3α-hydroxy-5β-cholanoic; 21, unknown monooxo-monohydroxy bile acid; 22, 3α,6β,7β-trihydroxy-5β-cholanoic (β-muricholic); 23, unknown compound; 24, monooxo-dihydroxy and monooxo-monohydroxy bile acids; 25, 12-oxo-3β-hydroxy-5α-cholanoic; 26, 3α,6α,7β-trihydroxy-5β-cholanoic (ω-muricholic); 27, 3α,6α,7β-trihydroxy-5α-cholanoic. From ref. 56 with permission.

At present HPLC and MS are used in a discontinuous mode in which samples are separated by HPLC and collected fractions subjected to mass spectrometry.[61,62] The most appropriate MS introduction and ionization methods can be selected; direct introduction of the free compounds or derivatization and GC-MS with EI or CI, and FD or FAB to give molecular weight information on intact conjugates.

Mass Spectra of Unconjugated Bile Acids

General Fragmentation Patterns

Fragmentation can most easily be influenced in two ways: by the method of ionization, and by the choice of derivative. In studies of biological materials it may be necessary to influence fragmentation to fulfill the objectives: (1) determination of molecular weight, (2) production of diagnostically significant fragment ions for identification, and (3) production of specific ions for selected ion monitoring.

Electron impact ionization (see Electron Impact) is presently the most common method to obtain fragmentation patterns of diagnostic use. A number of derivatives have been used (see Gas Chromatographic Inlet) and we prefer TMS ethers because their fragmentation is usually more informative.

Irrespective of the derivative, bile acids fragment along several common pathways under electron impact. Conclusions regarding the nature of the fragments have usually been drawn from analogies between spectra of a large number of bile acids.[1-4] The origin and structures of most common ions have been confirmed from high resolution mass spectral data,[63] and mechanistic studies of the elimination of free and esterified hydroxyl groups have provided explanations for several previously observed characteristics of spectra of bile acid derivatives.[64-66]

While fragmentation mechanisms may be very complex, much structural information can be gained from the mass of fragment ions and losses by simple arithmetic considerations. An important step in the characterization of a bile acid from its mass spectrum is to establish the mass of a molecular ion. Table IV is provided as an aid in the calculation of molecu-

---

[61] R. Shaw and W. H. Elliott, *Lipids* **13,** 971 (1978).
[62] G. Mingrone, A. V. Greco, L. Boniforte, and S. Passi, *Lipids* **18,** 90 (1983).
[63] T. Iida, F. C. Chang, T. Matsumoto, and T. Tamura, *Biomed. Mass Spectrom.* **9,** 473 (1982).
[64] J. R. Dias and B. Nassim, *Org. Mass Spectrom.* **13,** 402 (1978).
[65] J. R. Dias and B. Nassim, *J. Org. Chem.* **45,** 335 (1980).
[66] J. R. Dias, *J. Org. Chem.* **44,** 4572 (1979).

TABLE IV
MASS VALUES FOR CALCULATION OF MOLECULAR WEIGHTS OF COMMONLY
USED DERIVATIVES OF BILE ACIDS

Group	Mass increase per substituent	Mass increase upon derivatization
Norcholane	316	—
Methyl(ene)	14	—
Carboxyl	44	—
Carbonyl	28	—
Double bond	−2	—
Hydroxyl	16	—
Methoxy	30	14
Acetoxy	58	42
Hexafluoroisopropoxy	166	150
Trifluoroacetoxy	112	96
Difluoromethylene	50	—
Trimethylsiloxy	88	72
Isopropylidenedioxy (acetonide)	72	40
Oxo	14	—
Oxime	29	15

lar weights. In many cases a molecular ion is not seen and its mass may be deduced indirectly from fragment ions. Some derivatives are better than others for this purpose: alkyldimethylsilyl ethers[45–47,67] give an intense ion at m/z [M − alkyl]$^+$. When a molecular weight cannot be deduced, alternative ionization methods may be used (see Ionization Methods).

The three most general types of fragmentation of bile acids are loss of a methyl group (angular or from a TMS group), loss of hydroxyl groups as water or its equivalent in derivatives, and loss of the side chain.[1–4,6] These losses are seen in different combinations and result in fragment ions of varying intensities depending on the positions and configuration of hydroxyl groups and the type of derivative.[1,2,6] TMS ethers of polyhydroxycholanoates or trihydroxycholanoates with vicinal hydroxyl groups often lose the last trimethylsiloxy group without a hydrogen (−89).[1,2] In almost all cases an ABCD-ring ion is seen at m/z 257, 255, 253, and 251 for mono-, di-, tri-, and tetrahydroxycholanoates, respectively, and 14 Da higher per oxo group in the ring system. A recent comparison of 26 isomeric methyl 3-, 7-, and 12-monohydroxy-, 3,7-, and 3,12-dihydroxy-, and 3,7,12-trihydroxy-5β-cholanoates clearly illustrates the influence of position and configuration of substituents.[63] The differences in spectra

[67] Y. Nishikawa, K. Yamashita, M. Ishibashi, and H. Miyazaki, *Chem. Pharm. Bull.* **26**, 2922 (1978).

between these isomeric bile acids are often seen in an analogous way in spectra of acetates and TMS ethers. This shows that stereoselective fragmentations, although influenced by temperature and other mass spectrometric variables, can be of help in the assignment of a bile acid structure. However, the retention time is usually of greater value (see Gas Chromatographic Inlet). Establishment of the configuration at C-5 is frequently a problem, which may often be solved by oxidation and comparison of the product with synthetic oxo-5α- and oxo-5β-cholanoates.[1] When appropriate reference compounds are not available nuclear magnetic resonance spectrometry can provide positive information about the stereochemistry of hydroxyl groups in the ring system.

Cleavage through the D-ring is also common and produces peaks at m/z [ABCD-27] (loss of C-16, C-17) and [ABCD-42] (loss of C-15–C-17), i.e., at m/z 230, 228, and 226 and at m/z 215, 213, and 211, respectively, for saturated cholanoates with one, two, and three hydroxyl groups, respectively, in the ABC-rings. Analogous ions may be seen still containing a hydroxyl (or equivalent derivative) group[65] but this is rarely the case for TMS ethers. Ions due to cleavage through the D-ring are also absent or of low intensity in spectra of cholanoates with neighboring hydroxyl groups in the A- or B-rings.

This general fragmentation pattern is seen also in spectra of neutral steroids with a side chain.[5–7,68] It may sometimes be difficult to decide whether a mass spectrum is due to a C_{27} steroid or a methyl cholanoate with the same number of hydroxy or oxo groups. The side chain is 2 Da less in the steroid (113 vs 115) which is equivalent to a double bond. Thus, it is advisable to separate neutral from acidic steroids on a lipophilic ion exchanger prior to GC-MS analysis.[56]

Diagnostic Fragmentation Patterns

In addition to the general fragmentation ions, the mass spectrum has ions which aid in the localization of the substituents. Different derivatives may be of different value in this respect, and the following discussion is focused on the use of TMS ethers. Irrespective of the derivative analyzed one should always be aware of the possible presence of derivatives formed as artifacts. Use of methanol, ethanol, ethyl acetate, or acetic acid in isolation procedures can produce methyl or ethyl esters or acetates depending on the solvent. Ethyl esters are commonly observed as artifacts,[69] but they have also been suggested to occur naturally.[70] Alkyl

[68] L. Aringer and L. Nordström, *Biomed. Mass Spectrom.* **8**, 183 (1981).
[69] W. S. Harris, L. Marai, J. J. Myher, and M. T. R. Subbiah, *J. Chromatogr.* **131**, 437 (1977).
[70] M. I. Kelsey and S. A. Sexton, *J. Steroid Biochem.* **7**, 641 (1976).

TABLE V
SETS OF IONS[a]

B-ol m/z	B-diol m/z	B-triol m/z	B-tetrol m/z	B-olone m/z	B-diolone m/z	B-one m/z	B-dione m/z	B-trione m/z
462	550	623	636	476	564	388	402	416
447	535	548	621	461	549	373	384	398
372	460	458	546	386	474	370	370	384
357	445	443	531	371	456	356	353	380
318	370	369	456	271	384	318	329	366
276	345	368	441	253	369	315	292	343
264	255	343	367	368	269	292	287	301
257	229	253	366	293	251	273	269	283
230	228	261	351	292	261	255	260	265
215	201	247	314	269	242	246	245	261
331	405	226	261	249	366	233	192	243
129	262	609	251	229	359	231		185
	249	444	246	226	351	150		
	243	403	225	177	341	121		
	213	285	224	313	279			
	208	195	217	316	409			
	331	217	209	231	319			
	129	329	197	154	243			
	231	327	195	121	229			
	386	158	182	208	121			

[a] Sets of ions for reconstruction of fragment ion current chromatograms useful in screening for common types of C_{24} bile acids analyzed by repetitive scanning GC/MS of the methyl ester TMS ether derivatives. A maximum of 20 chromatograms are stacked and the ions are listed in order from top to bottom, ions at the bottom being more selective for positional isomers.[8] B, methyl cholanoate.

ethers may also be artifacts or naturally occurring derivatives.[3] Diazomethane should always be prepared immediately before use to limit the methylation to the carboxyl group and to avoid side reactions (ring expansion) with oxo bile acids.

The most commonly encountered variations in the structure of bile acids from natural sources are position of hydroxyl groups, oxo groups, and double bonds in the ring structure, and structure of the side chain. The ions discussed below are not absolutely specific nor are they necessarily formed from all bile acids with a given substituent. Generalizations have been made to assist in the preliminary evaluation of a mass spectrum. Series of ions that are useful in screening for common bile acid structures in repetitive scanning GC-MS analyses are given in Table V.

Hydroxyl Groups in the A-Ring. With few exceptions, naturally occurring bile acids have an oxygen function at C-3. Hydroxyl groups in this position, whether derivatized or not, are lost after those at C-7 and C-12.[6,64-66] Differences are seen depending on configuration.[1,2] The loss of water, trimethylsilanol, acetic acid, or equivalent is often followed by a retro-Diels–Alder elimination of C-1–C-4.[1,2,5] The mass of the resulting ions depends on the number and nature of substituents in the BCD-rings and on the structure of the side chain. In the spectra of 3-hydroxy-, 3,7-, and 3,12-dihydroxy-, and 3,7,12-trihydroxycholanoates, ions appear at m/z 318, 316, and 314, respectively. Formation of these ions requires the presence of at least one hydroxy or oxo group in the A-ring.

Bile acids hydroxylated at C-1 have been found in urine of healthy adults[8] and infants,[71] in meconium[72] and in increased amounts in patients with liver disease.[73] In all cases the identification has been aided by the typical ion given by TMS ethers of mass 217 that contains C-1–C-3 and the two trimethylsiloxy groups. This ion corresponds to m/z 129 (see Hydroxy Bile Acids with Unsaturated Ring System) with an additional trimethylsiloxy group at C-1. It should be pointed out that only $1\beta,3\alpha$ isomers in the 5β-series have been studied[41] and the possibility exists that some isomers may not fragment to m/z 217. The 1,3-bistrimethylsiloxy structure also gives rise to ions at m/z 142 and 143, assumed to contain C-1–C-4 and one trimethylsiloxy group. These ions are prominent only in the absence of hydroxyl groups in the B-ring, and they are also found in TMS ethers of other 3-hydroxy bile acids with a double bond or an additional hydroxyl group in the A-ring. Bile acids with a 1,3,7-tristrimethylsilyloxy structure also yield ions at m/z 182 and 195[8,73] believed to contain the A-ring after cleavage of the C-9,10 and C-5,6 or C-6,7 bonds, respectively. Derivatives of 1-hydroxy-3-oxo bile acids have not been studied but spectra of TMS ethers of neutral steroids with a 1-hydroxy-3-oxo-Δ^4 structure give an intense ion due to loss of 116 daltons, possibly representing C-1 and C-2 with the trimethylsiloxy group.[74]

Bile acids with a hydroxyl group at C-2, previously known from *Arapaima gigas*[75] have recently been found in man.[71,76] In the only published spectrum of a TMS ether of an authentic bile acid (methyl $2\beta,3\alpha,7\alpha,12\alpha$-tetrahydroxy-$5\beta$-cholanoate) the ion at m/z 143 (together with m/z 142 diagnostically useful for neutral steroids with a 2,3-bistrimethylsiloxy structure[77,78]) has a low intensity, possibly because of fragmentation in-

---

[71] B. Strandvik and S.-Å. Wikström, *Eur. J. Clin. Invest.* **12**, 301 (1982).
[72] P. Back and K. Walter, *Gastroenterology* **78**, 671 (1980).
[73] A. Bremmelgaard and J. Sjövall, *Eur. J. Clin. Invest.* **9**, 341 (1979).
[74] B. P. Lisboa and J.-Å Gustafsson, *Eur. J. Biochem.* **6**, 419 (1968).
[75] G. A. D. Haslewood and L. Tökés, *Biochem. J.* **126**, 1161 (1972).
[76] P. T. Clayton, D. P. R. Muller, and A. M. Lawson, *Biochem. J.* **206**, 489 (1982).
[77] J.-Å. Gustafsson, B. P. Lisboa, and J. Sjövall, *Eur. J. Biochem.* **6**, 317 (1968).
[78] T. A. Baillie, H. Eriksson, J. E. Herz, and J. Sjövall, *Eur. J. Biochem.* **55**, 157 (1975).

duced by other substituents. However, the BCD-ring side chain fragment ion appears at m/z 314 which for a tetrahydroxycholanoate means that two hydroxyl groups are in the A-ring.[1,8] Cleavage between C-2 and C-3 is facilitated so that the ion of mass 243 formed from a 3,7-bistrimethylsiloxy structure (see below) becomes intense in the presence of a 2-trimethylsiloxy group. If a 2,3-dihydroxy bile acid structure is suspected, reaction with dimethoxypropane may be helpful[50] since only the cis-diols will react to form acetonides.[78] Reaction with methaneboronic acid, not often used in GC-MS analysis of bile acid derivatives, can also be very useful since both cis- and trans-diols may form cyclic boronate esters.[79] Derivatives of 2-hydroxy-3-oxo bile acids have not been studied but spectra of neutral steroids with this structure also show peaks at m/z 142 and 143.[80]

We have not found any studies of mass spectra of bile acids hydroxylated at C-4 or C-5. However, spectra of TMS ethers of three cholestane-3,5-diols have been reported.[68,81] In all cases intense peaks were present at m/z 143 and 243. The former ion is probably the same as that given by a 3-trimethylsiloxy-Δ^4 structure while the latter was suggested to contain C-3 to C-7 with substituents (i.e., an isomer of the ion formed from a 3,7-bistrimethylsiloxy structure).[81]

Hydroxyl Groups in the B-Ring. Mass spectra of 3,6- and 3,7-hydroxylated bile acids can be very similar, especially when a 12-hydroxy group is also present.[82] The mass spectrum of methyl 7-hydroxycholanoate and its trifluoroacetate shows peaks at m/z 249, 262, and 276 formed by cleavage through the B-ring and containing the CD-rings and side chain.[1,2,63,83] The two former ions are also formed from TMS ethers of 3,7-dihydroxycholanoates, their masses varying depending on the length of the side chain. When a 12-trimethylsiloxy group is introduced, m/z 247 (equivalent to m/z 249) is small or absent and a peak is usually seen at m/z 261. This is a major ion in the spectrum of the TMS ether of methyl allocholate showing that the configuration at C-5 influences fragmentation through the B-ring.[1] An analogous intense ion at m/z 303 is seen in the spectrum of the corresponding cholestanoic acid derivative.[84] Similar large differences between 3,7-dihydroxycholanoates isomeric at C-3 and C-5 are not seen, neither with free[85] nor with silylated[86] hydroxyl groups.

An ion of mass 243 is always seen in spectra of TMS ethers of 3,7-

---

[79] C. J. W. Brooks, W. J. Cole, H. B. McIntyre, and A. G. Smith, *Lipids* **15,** 745 (1980).
[80] J.-Å. Gustafsson, B. P. Lisboa, and J. Sjövall, *Eur. J. Biochem.* **5,** 437 (1968).
[81] C. J. W. Brooks, W. Henderson, and G. Steel, *Biochim. Biophys. Acta* **296,** 431 (1973).
[82] B. Almé and J. Sjövall, *J. Steroid Biochem.* **13,** 907 (1980).
[83] T. Harano, K. Harano, and K. Yamasaki, *Steroids* **32,** 73 (1978).
[84] K. Okuda, M. G. Horning, and E. C. Horning, *J. Biochem. (Tokyo)* **71,** 885 (1972).
[85] S. A. Ziller, E. A. Doisy, Jr., and W. H. Elliott, *J. Biol. Chem.* **243,** 5280 (1968).
[86] H. Eriksson, W. Taylor, and J. Sjövall, *J. Lipid Res.* **19,** 177 (1978).

dihydroxycholanoates. It consists of C-3–C-7 with the trimethylsiloxy groups.[1,2] The same ion is also seen in spectra of TMS ethers of isomeric cholates but it is not significant when additional hydroxyl groups are introduced (except at C-2, see above).

A trimethylsiloxy group at C-6 usually produces fragment ions similar to those discussed above. Thus, spectra of derivatives of 3,6-dihydroxy-5β-cholanoates show a peak at m/z 249 and sometimes at m/z 243. However, m/z 263 is more intense than m/z 262 and a peak is also seen at m/z 276. Both peaks are shifted by 14 Da in the spectrum of the ethyl ester, indicating their relationship to the ions formed from the 7-hydroxycholanoates. TMS ethers of 3,6,12-trihydroxycholanoates may also give peaks at m/z 243 and 247 and are difficult to differentiate from the TMS ethers of 3,7,12-trihydroxycholanoates.[82] The diagnostically most useful ion is formed by fragmentation through the A-ring with loss of C-1–C-4, the trimethylsiloxy group and one hydrogen (M-145 and/or M-90-145).[40] The resulting peaks at m/z 405 (dihydroxycholanoates) and m/z 493 and/or 403 (trihydroxycholanoates) are more intense in spectra of the 6β than the 6α isomers,[40] and may be very weak in the case of 3,6,12-substituted bile acids.[8,82]

Another ion useful for the differentiation of 3,6- from 3,7-bistrimethylsiloxycholanoates is found at m/z 323. An analogous ion is seen in spectra of TMS ethers of cholestane- and stigmastane-3,6-diols (m/z 321 and 349, respectively).[68,81] It is very intense for TMS ethers of cholestane-5,6-diols and -3,5,6-triols[81] and was suggested to arise by cleavage of the C-5,6, C-8,9, and C-12,13 bonds and thus to contain the side chain and D-ring with the attached C-18, C-8, C-7, and C-6 with the trimethylsiloxy group. An ion of this type is not seen when a 12-trimethylsiloxy group is present.[82]

Bile acids with hydroxyl groups at C-6 and C-7, previously known only from rats, mice (muricholic acids), and pigs (hyocholic and hyodeoxycholic acids), have also been found by GC-MS analysis of human urine.[8,71–73,87] Both 5β and 5α isomers of 3,6,7-trihydroxycholanoates have been prepared[88] and studied by GC-MS. The typical ions are m/z 285 and 195, formed by cleavage of the C-9,10 and C-6,7 bonds and consisting of the A-ring and C-6, with m/z 195 retaining only one trimethylsiloxy group.[1,2,4] The 7β epimers, whether of the 5α or 5β series give very intense ions of these masses (one of them being the base peak) while the 7α epimers have m/z 458 (M − 2 × 90) or m/z 369 (M − 2 × 90 − 89) as the base peak.[1,2,4] The TMS ether of 3α,6α,7α,12α-tetrahydroxy-5β-cholanoate gives an analogous spectrum with a base peak at m/z 546 (M − 2 ×

---

[87] J. A. Summerfield, B. H. Billing, and C. H. L. Shackleton, *Biochem. J.* **154**, 507 (1976).
[88] M. I. Kelsey, M. M. Mui, and W. H. Elliott, *Steroids* **18**, 261 (1971).

90), an intense peak at m/z 367 (M $-$ 3 $\times$ 90 $-$ 89), and a significant peak at m/z 195.[8,50,76]

The fragmentation may be complicated by simultaneous presence of additional hydroxyl groups in the A-ring. Thus, the spectrum of the TMS ether of a tentatively indentified 2,3,6,7-tetrahydroxycholanoate shows peaks which were interpreted to result from loss of part of the A-ring (M-219, cf. m/z 129 in Hydroxy Bile Acids) or to contain the entire A-ring (m/z 271 and 181 due to cleavage of the C-5,6 and C-9,10 bonds).[71] The validity of these interpretations will have to be established by studies of synthetic bile acids.

When vicinal hydroxyl groups are suspected, GC-MS identification may be aided by a reaction with dimethoxypropane. The cis-6,7-diols form acetonides which are easily separated from nonreacted positional and stereoisomers.[50] The retention times may shift,[50,89] and the mass spectra show typical peaks at m/z [M $-$ 15], [M $-$ 75] (loss of acetonide group) and/or [M $-$ 75 $-$ n $\times$ 90].[50] The production of [M $-$ 15] ions is of particular value for the establishment of a molecular weight. Preparation of cyclic boronate esters may also be useful.[79]

Hydroxyl Groups in the C-Ring. Derivatized 12-hydroxy groups undergo facile elimination.[1,2,63,65,68] This is followed by loss of the side chain, assumed to be activated by formation of an 11,12 double bond.[1,65,68] Thus, TMS ethers of isomeric 3,12-dihydroxycholanoates give a base peak either at m/z 255 (M $-$ 2 $\times$ 90 $-$ 115) or at m/z 345 (M $-$ 90 $-$ 115). The ratio between intensities of m/z 255 and 345 is higher for 5β than 5α isomers[1,2,86] presumably because of a difference in the rate of elimination of the 3-trimethylsiloxy group.

The other important feature is an ion at m/z 208, assumed to arise by cleavage of the C-11,12 and C-8,14 bonds with loss of trimethylsilanol.[1,68,86] The mass of this ion changes with the structure of the side chain and with substituents in the D-ring (e.g., to m/z 296 in the presence of a 15- or 16-trimethylsiloxy group). The ion is not seen in the spectrum of the TMS ether of methyl cholate and its 5α isomer, but it is intense in the spectrum of the 7β epimer. Thus, it appears possible that the more facile elimination of the 7α-trimethylsiloxy group will direct fragmentation away from the C-ring.

Hydroxyl Groups in the D-Ring. Bile acids with a hydroxyl group at C-14 have not been studied. This tertiary hydroxyl group in steroids is only slowly converted into a TMS ether,[90] which may aid in the identification of such possible bile acids.

---

[89] F. Kern, Jr., H. Eriksson, T. Curstedt, and J. Sjövall, *J. Lipid Res.* **18**, 623 (1977).
[90] L. Starka, J.-Å. Gustafsson, and J. Sjövall, *FEBS Lett.* **1**, 269 (1968).

TMS ethers of bile acids with a 15-hydroxy group give ions at m/z 243 and [M − 245].[41] This is in analogy with ions formed from TMS ethers of 15-hydroxylated steroids. Analogous ions are also found in spectra of C_{21}-, C_{27}-, and C_{29}-steroids with a 15-trimethylsiloxy group.[68,91] The ion m/z 243 can be assumed to consist of C-15–C-17, the side chain and the C-15 trimethylsiloxy group. The loss of 245 Da corresponds to the loss of side chain and D-ring generally observed for bile acid derivatives. In the presence of a 12-oxo group, the loss is 243 Da,[41] i.e., corresponding to [M − 155] in 12-oxo bile acids without substituents in the D-ring or side chain.

Bile acids with a hydroxyl group at C-16 are found in some snakes. The mass spectrum of the TMS ether of methyl $3\alpha,12\alpha,16\alpha$-trihydroxy-$5\beta$-cholanoate (pythocholate) shows peaks at m/z 245 and [M − 245] probably arising from the C-ring and side chain and their loss, respectively. A peak at m/z 191 [$(CH_3)_3\overset{+}{Si}O=CH-O-Si(CH_3)_3$ indicates migration of a trimethylsiloxy group.[92] Peaks at m/z 462 (M − 176) and m/z 481, 391, 301 (M − 157 followed by losses of trimethylsilanol), which have not been explained, are also indicative of rearrangements of trimethylsilyl groups. It should be pointed out that a lactone can be formed between the carboxyl and 16-hydroxy groups. The molecular ion of the lactone TMS ether at m/z 534 is an aid in the identification.

Hydroxy Bile Acids with Unsaturated Ring System. Unsaturated bile acids occur naturally, but they may also be artifacts formed by long term storage, e.g., in the bile of a 3200-year-old Egyptian mummy,[93] or under harsh hydrolytic or solvolytic conditions. They also arise from elimination of thermally labile derivatives in the GLC system.[1,94,95]

Mass spectra of unsaturated bile acids are in most cases very similar to those of bile acids carrying a hydroxyl group at the unsaturated carbon atom. For this reason it may be difficult to decide whether a peak represents the molecular ion of an unsaturated bile acid derivative or a fragment ion formed by loss of a hydroxyl function. Soft ionization methods or hydrogenation may have to be used to make this decision, or there may be fragment ions characteristic of the position of a double bond. Mass spectra of unsaturated steroids have been studied in great detail.[5-7] Kuksis and co-workers have made systematic studies of the liquid and gas chromatographic behavior[36] and mass spectra[94,95] of methyl ester acetate

---

[91] J.-Å. Gustafsson and J. Sjövall, *Eur. J. Biochem.* **6**, 236 (1968).
[92] J.-Å. Gustafsson, R. Ryhage, J. Sjövall, and R. M. Moriarty, *J. Am. Chem. Soc.* **91**, 1234 (1969).
[93] A. Kuksis, P. Child, J. J. Myher, L. Marai, I. M. Yousef, and P. K. Lewin, *Can. J. Biochem.* **56**, 1141 (1978).
[94] P. Child, A. Kuksis, and L. Marai, *Can. J. Biochem.* **57**, 216 (1979).
[95] A. Kuksis and P. Child, *Lipids* **15**, 770 (1980).

derivatives of unsaturated bile acids prepared by dehydration of the common bile acids.[94] The molecular ions are usually more intense than for the corresponding saturated bile acid.

Bile acids with a 2,3- or 3,4-double bond may be formed by bacterial action on 3-sulfates.[96] They are readily distinguished by the retro-Diels–Alder elimination of the A-ring (M − 54) from the 2,3- and not from the 3,4-unsaturated compounds,[5,6,94] and the metabolite from lithocholic acid sulfate was shown to have a 3,4-double bond.[96,97]

Hydroxy bile acids with a 4,5-double bond could conceivably be formed by reduction of a 3-oxo-Δ^4 structure. They have not been studied by mass spectrometry but there are many reports on mass spectra of TMS ethers of neutral 3-hydroxy-Δ^4 steroids. The ions typical of this structure are at m/z 142 and 143, containing C-1–C-4 with the trimethylsiloxy group.[68,81,98] They are found also in the presence of a 6-trimethylsiloxy group.[81] The mass spectrum of methyl 3α-methoxy-7α-acetoxychol-4-enoate showed the expected m/z 249 (see Hydroxyl Groups in the B-Ring) and an ion at m/z 138 that was suggested to consist of the A-ring with its substituents and C-6.[99]

The influence of a 5,6-double bond on the fragmentation has been extensively studied. Fragmentation through the B-ring occurs both with the methyl ester of 3-hydroxy-5-cholenoic acid and with the acetate and TMS ether. The methyl ester gives an intense peak at m/z 277 containing the CD-rings, side chain and two carbons from the B-ring (cf. m/z 276 from 7-hydroxycholanoate).[4,100,101] This ion is also seen with the acetate,[101] trifluoroacetate,[1] and TMS ether.[102] All derivatives of methyl 3-hydroxy-5-cholenoate give a peak at m/z 249 favored by allylic cleavage through the B-ring. This ion is also seen in the presence of a 7-methoxy group[83] whereas a 7-trimethylsiloxy group is eliminated with such ease that m/z [M − 90] becomes by far the predominant peak.[68,81] The ion m/z 247 is very weak when a 12-hydroxy group is introduced.[8]

The most characteristic ions are formed by cleavage in the A-ring resulting in m/z 129, [M − 129], and [M − 56].[102] The ion m/z 129[102–104] is

[96] M. I. Kelsey, J. E. Molina, S.-K. S. Huang, and K.-K. Hwang, *J. Lipid Res.* **21,** 751 (1980).
[97] R. H. Palmer, in "Bile Acids in Human Disease" (P. Back and W. Gerok, eds.), p. 65. Schattauer, Stuttgart, 1972.
[98] C. J. W. Brooks, E. C. Horning, and J. S. Young, *Lipids* **3,** 391 (1968).
[99] K. Harano, T. Harano, K. Yamasaki, and D. Yoshioka, *Proc. Jpn. Acad.* **52,** 453 (1976).
[100] S. H. M. Naqvi, *Steroids* **22,** 285 (1973).
[101] T. Harano and K. Harano, *Kawasaki Med. J.* **2,** 119 (1976).
[102] I. Makino, J. Sjövall, A. Norman, and B. Strandvik, *FEBS Lett.* **2,** 161 (1971).
[103] P. Eneroth, K. Hellström, and R. Ryhage, *J. Lipid Res.* **5,** 245 (1964).
[104] J. Diekman and C. Djerassi, *J. Am. Chem. Soc.* **32,** 1005 (1967).

not as specific as originally thought since it can arise from many sites of steroid TMS ethers.[105] However, the combined presence of the three ions makes the 3-trimethylsiloxy-Δ^5 structure highly probable.[8] The fragment of mass 56 has been shown to consist of C-1–C-3 which is lost with migration of the trimethylsilyl group to the charged fragment.[105,106]

Bile acids with a 6,7-double bond are formed in bacterial 7-dehydroxylation. The presence of this double bond induces cleavage through the B-ring, and m/z 249 (or 247 after loss of a 12-hydroxy function) is a significant peak.[1,88,94] In studies of methyl 3-acetoxycholenoates it was noted that m/z 262 was formed from the Δ^5 but not from the Δ^6 compounds.[94] This difference is not consistent for TMS ethers.

A bile acid with a 7,8-double bond has been found in the bile of a fish.[107] The methyl ester acetates and trifluoroacetates of 3,12-dihydroxy-7-cholenoic acid give no ion at m/z 247, thus distinguishing the Δ^7 from the Δ^6 isomer.[94] No ion diagnostically useful for a 7,8-double bond has been described but the double bond is readily isomerized to the 8(14) position, and derivatives of this isomer show a very pronounced loss of side chain, C-17 and C-16 (-142, see General Fragmentation Patterns) facilitated by allylic cleavage of the C-15,16 bond.[94,107]

Bile acids with a 9,11-double bond do not occur naturally. The acetates of methyl 3α,12α/β-dihydroxy-9(11)-cholenoate have been analyzed by GC-MS.[94] An ion of possible diagnostic value was present at m/z 386, possibly formed by loss of acetic acid (from C-3) and CH_2CO with retention of a hydroxyl group at C-12.

Bile acids with an 11,12-double bond show a pronounced loss of the side chain,[1,94,108] similar to that seen for derivatives of 12-hydroxy bile acids.

Bile acids with an 8,14-double bond were mentioned above and bile acids with double bonds in the D-ring have not been studied by mass spectrometry.

Oxo Bile Acids. The fragmentation of steroids with an oxo group in different positions has been studied for several decades and in great detail, particularly by Djerassi and co-workers, and has been reviewed.[5–7] Spectra of monooxosteroids often show ions characteristic of the position of the oxo group and can be used for the location of hydroxyl groups

---

[105] C. J. W. Brooks, D. J. Harvey, B. S. Middleditch, and P. Vouros, *Org. Mass Spectrom.* **7,** 925 (1973).

[106] I. Björkhem, J.-Å. Gustafsson, and J. Sjövall, *Org. Mass Spectrom.* **7,** 277 (1973).

[107] A. Kallner, *Acta Chem. Scand.* **22,** 2353 (1968).

[108] A. E. Cowen, A. F. Hofmann, D. L. Hachey, P. J. Thomas, D. T. E. Belobaba, P. D. Klein, and L. Tökés, *J. Lipid Res.* **17,** 231 (1976).

following oxidation, e.g., by selective enzymes.[79,109,110] However, when several oxo and hydroxyl groups are present, competing fragmentation pathways often obscure the features specific for the oxo groups.[1,2] Conversion to an oxime derivative is often of little help. Reduction with sodium borohydride and borodeuteride with analysis of the labeled and unlabeled hydroxy acids may be suggested as an additional means of determining the position. Conversion to enol ethers (or esters) may also be helpful.

Bile acids with a 3-oxo group may be formed both in the liver and by the intestinal microflora. When the A/B fusion is cis (5β) the mass spectra often show a peak due to retro-Diels–Alder elimination of the A-ring (−70).[1,2,5,6,111] Hydroxyl groups (or their derivatives) are usually lost before this fragmentation is seen. Enol-TMS ethers of 3-oxosteroids, best prepared with basic catalyst,[47,112] show a typical pair of ions at m/z 142, 143,[1,98] m/z 142 arising by a retro-Diels–Alder rearrangement. Conversion to the O-methyloxime can be achieved selectively[40] and the syn and anti isomers are formed in equal amounts producing a double peak upon GLC on a selective phase. The O-TMS-oximes are better for this purpose since their isomers separate also on a nonpolar phase.[113] In the presence of a 4,5 double bond, the two isomers are formed in proportions 1:2 (4-syn, 4-anti, in order of retention time). The oxime derivatives of simple 3-oxo-Δ^4 steroids give a typical peak at m/z 153 (methyloxime) or 211 (TMS-oxime), corresponding to m/z 124 from the underivatized structure [which also gives an intense peak due to loss of ketene (−42)].[5,6] These are seen with the derivatives of 3-oxo-4-cholenoic acid but not when a 7-hydroxy (TMS) group is present. However, m/z 125, 137, and 153 are all formed from the methyloxime-TMS ether derivative of 6α-hydroxy-3-oxo-5β-cholanoate.[82] In the presence of a 6-trimethylsiloxy group, 3-oxo-Δ^4-steroids give an intense ion at m/z M − 56.[114]

Microbial metabolism of bile acids may yield products with a 1,4-dien-3-one structure,[110,115,116] which give mass spectra having intense ions at m/z 121 and/or 122[5,6] (analogous to m/z 124 above) also in the presence of

[109] C. J. W. Brooks, W. J. Cole, T. D. V. Lawrie, J. MacLachlan, J. H. Borthwick, and G. M. Barrett, *J. Steroid Biochem.* **19**, 189 (1983).
[110] I. A. McDonald, V. D. Bokkenheuser, J. Winter, A. M. McLernon, and E. H. Mosbach, *J. Lipid Res.* **24**, 675 (1983).
[111] S. V. Hiremath and W. H. Elliott, *Steroids* **38**, 465 (1981).
[112] E. M. Chambaz, C. Madani, and A. Ros, *J. Steroid Biochem.* **3**, 741 (1972).
[113] R. Tandon, M. Axelson, and J. Sjövall, *J. Chromatogr.* **302**, 1 (1984).
[114] B. P. Lisboa, J.-Å. Gustafsson, and J. Sjovall, *Eur. J. Biochem.* **4**, 496 (1968).
[115] P. J. Barnes, J. D. Baty, R. F. Bilton, and A. N. Mason, *Tetrahedron* **32**, 89 (1976).
[116] M. E. Tenneson, J. D. Baty, R. F. Bilton, and A. N. Mason, *Biochem. J.* **184**, 613 (1979).

a 7-hydroxy group.[115,116] It is possible that analysis of otherwise underivatized methyl esters using the direct probe will yield the most informative spectra from these structures (see also ref. 117). Peaks at m/z 152 and 150 are indicative of 7-hydroxy-4-en-3-one and 7-hydroxyl-1,4-dien-3-one structures, respectively (cleavage of C-7,8 and C-9,10 bonds)[116] (see also 7-oxo).

Bile acids with a 1-oxo group could conceivably be formed from the 1-hydroxylated bile acids. Oxidation of the 1,3-dihydroxy structure of methyl 1β,3α,12α-trihydroxy-5β-cholanoate leads to 1,3-diones which, upon treatment with diazomethane, yield the 3-enol-1-oxo and 1-enol-3-oxo methyl ethers in proportions of 2 : 1.[41] Both give an intense peak at m/z 139 (and a smaller one at m/z 152) assumed to consist of the entire A-ring. One of them gives an ion at m/z 98 assumed to be formed by retro-Diels–Alder rearrangement of the A-ring of the 3-enol ether.[41]

Bile acids with a 6-oxo group occur naturally. The lability of the 5-hydrogen in these compounds should be remembered, since it leads to isomerization under GC-MS conditions. The O-methyloximes of methyl 3,6-dioxo-5α/β-cholanoates are very useful both for differentiation from 3,7-dioxocholanoate and for determination of the configuration at C-5.[118] The base peak is at m/z 138 (presumably the A-ring with substituent) for the 5β and at m/z 460 (M) for the 5α epimer. The same difference was not seen with the 3,6,12-trioxocholanoates epimeric at C-5, but relative intensity relationships between m/z 503 (M), 472 (M − 31) and 348 (M − 155) differentiated between 3,6,12- (base peak m/z 348) and 3,7,12- (base peak m/z 472) trioxocholanoates.[82] Loss of 29 Da is a diagnostically useful feature for TMS ethers of 3-hydroxy-6-oxosteroids.[119] The base peak in spectra of the TMS ethers of methyl 3α-hydroxy-6-oxo-5α-cholanoate and 3β-hydroxy-5α-cholestan-6-one[68] is m/z [M − 29], and presence of this ion distinguishes cholestane-3,6-diones from corresponding 3,7-, 3,12-, 3,14-, and 3,24-dioxosteroids.[68] Loss of the A-ring[5,6] results in a base peak at m/z 333 (M − 55) in the spectrum of methyl 6-oxo-5β-cholanoate[2] and m/z [M − 142] when a trimethylsiloxy group is present at C-3.

Bile acids with a 7-oxo group fragment through the B- and C-rings.[1,2,5–7] The latter process yields m/z 178 and 192 from methyl 7-oxo and 3,7-dioxocholanoates, respectively, and is more pronounced for the 5α-epimers.[1,2] Cleavage of the C-5,6 and C-9,10 bonds with loss of the entire A-ring yields an ion at m/z 292 not seen when a 12-hydroxy group is present. A peak at m/z 150 may be of diagnostic use (see ref. 1); together

---

[117] F. J. Brown and C. Djerassi, *J. Org. Chem.* **46**, 954 (1981).
[118] J. G. Allen, G. H. Thomas, C. J. W. Brooks, and B. A. Knights, *Steroids* **13**, 133 (1969).
[119] J. Sjövall and M. Axelson, *Vitam. Horm.* (*N.Y.*) **39**, 31 (1982).

with m/z 290 and 192 it distinguishes cholestane-3,7-dione from the other diones studied.[68] Loss of the oxo group as water often occurs to a greater extent with 7-oxo than with other oxo bile acids.

A 12-oxo group induces the most characteristic fragmentation of all carbonyl groups.[5,6] Cleavage of the C-13,17 and C-14,15 bonds with transfer of a hydrogen to the oxygen at C-12 yields a peak at m/z [M − 155]. This is seen in all 12-oxo bile acids studied[1–4] and is given also by the methyloxime derivative. A peak at m/z 121 is frequently seen in spectra of 12-oxocholanoates[1,3,94] (see also ref. 68). Its structure is not established; it probably consists of the C-ring in a 1-hydroxy-2-methyl tropylium ion.[120]

Few bile acids with a 15-oxo group have been studied. Cleavages of the C-13,17 and C-14,15 bonds with loss of the D-ring and side chain as fragments of mass 169 and 198 were seen with methyl 3,12,15-trioxo-5β-cholanoate and its O-methyloxime, respectively.[41,121] These are analogous to [M − 155] in 12-oxocholanoates not substituted in the D-ring. Another important ion is at m/z 211 in the spectrum of the 3,12,15-trioxo compound.[121] Analogous ions are found at m/z 209 and 237 in spectra of 15-oxosteroids without C-ring substituents and with a C_8 and C_{10} side chain, respectively,[68] and at m/z 155 in the spectrum of methyl 3,11,15-trioxo-5α-androstane-17β-carboxylate.[91] Thus, this fragmentation occurs both in the presence and absence of oxygen substituents in the C-ring. It is not seen with an O-methyloxime group at C-15. These ions are formed by cleavage of the C-8,14 and C-12,13 bonds and contain the D-ring, side chain, and an additional hydrogen on the C-15 oxygen.[5,6]

Bile acids with a 16-oxo group have not yet been found. Loss of the side chain with retention of a hydrogen followed by loss of the C-18 methyl group (M − 115 − 15) may be expected from methyl cholanoates carrying a 16-oxo group.[5,6]

Structure of the Side Chain. The *length* of the side chain can usually be deduced from the mass difference between an ABCD-ring fragment ion formed by loss of hydroxyl functions and side chain, and the ion formed by loss of only the hydroxyl functions.[1,2] Many studies have used this evidence for the characterization of primitive bile acids of the C_{27} series,[122–133] and steroidal acids with 1–3 carbon atoms in the side chain

---

[120] L. Tökés and C. Djerassi, *Steroids* **6**, 493 (1965).

[121] M. Kimura, M. Kawata, M. Tohma, A. Fujino, K. Yamazaki, and T. Sawaya, *Chem. Pharm. Bull.* **20**, 1883 (1972).

[122] G. A. D. Haslewood, "The Biological Importance of Bile Salts." North-Holland Publ., Amsterdam, 1978.

[123] S. Y. Karmat and W. H. Elliott, *Steroids* **20**, 279 (1972).

[124] I. G. Anderson, G. A. D. Haslewood, R. S. Oldham, B. Amos, and L. Tökés, *Biochem. J.* **141**, 485 (1974).

found in meconium[134,135] or formed by microbial action.[115,116] However, the side chain difference is not always seen in this way. Thus, derivatives of C_{20} (and C_{21}) acids do not give a useful odd-numbered ABCD-ring ion.[135] In these cases loss of side chain with part or all of the D-ring (see General Fragmentation Patterns) is diagnostically significant. Ions containing the side chain, e.g., as formed from bile acid derivatives with substituted B-ring, are also useful. This is exemplified by m/z 193 and 221, respectively, (CD-rings + side chain) in derivatives of C_{20} and C_{22} acids with a 3-hydroxy-Δ^5 structure.[134,135] As an aid in the determination of the length of the side chain one may compare spectra of the ethyl and methyl esters or prepare and analyze a 1 : 1 mixture of the protium and trideutero forms of the latter.[136] Derivatives of the alcohol formed upon LiAlH$_4$ reduction may also be analyzed.[134,135]

Depending on the derivative and probably on the electron voltage, a side chain fragment ion is often seen.[1] Since it is in the low mass region its diagnostic significance may be limited. However, it can be the base peak in spectra of bisnor bile acid derivatives (m/z 88). Losses of part of the side chain from the molecular ion (-73, -87) are of diagnostic value for oxo bile acids.[1,2] A fragment ion at m/z [ABCD + 28] indicates cleavage of the C-20,22 bond and loss of the terminal part of the side chain (see below).

Both methyl and ethyl homologues of C_{27} bile acids have been found and partially or fully characterized by GC-MS analysis of bile from primitive animals.[128,137] An unusual C_{29} bile acid has been found as a major

---

[125] R. F. Hanson, J. N. Isenberg, G. C. Williams, D. Hachey, P. Szczepanik, P. D. Klein, and H. L. Sharp, *J. Clin. Invest.* **56**, 577 (1975).
[126] G. A. D. Haslewood, S. Ikawa, L. Tökés, and D. Wong, *Biochem. J.* **171**, 409 (1978).
[127] I. G. Anderson, T. Briggs, G. A. D. Haslewood, R. S. Oldham, H. Schüren, and L. Tökés, *Biochem. J.* **183**, 507 (1979).
[128] M. Une, N. Matsumoto, K. Kihira, M. Yasuhara, T. Kuramoto, and T. Hoshita, *J. Lipid Res.* **21**, 269 (1980).
[129] Y. Noma, M. Une, K. Kihira, M. Yasuda, T. Kuramoto, and T. Hoshita, *J. Lipid Res.* **21**, 339 (1980).
[130] M. Une, T. Kuramoto, and T. Hoshita, *J. Lipid Res.* **24**, 1468 (1983).
[131] G. S. Tint, B. Dayal, A. K. Batta, S. Shefer, T. Joanen, L. McNease, and G. Salen, *J. Lipid Res.* **21**, 110 (1980).
[132] G. S. Tint, B. Dayal, A. K. Batta, S. Shefer, T. Joanen, L. McNease, and G. Salen, *Gastroenterology* **80**, 114 (1981).
[133] S. S. Ali, E. Stephenson, and W. H. Elliott, *J. Lipid Res.* **23**, 947 (1982).
[134] J. St. Pyrek, R. Lester, E. W. Adcock, and A. T. Sanghvi, *J. Steroid Biochem.* **18**, 341 (1983).
[135] J. St. Pyrek, R. Sterzycki, R. Lester, and E. Adcock, *Lipids* **17**, 241 (1982).
[136] M. Inoue, M. Ito, M. Ishibashi, and H. Miyazaki, *Chem. Pharm. Bull.* **22**, 1949 (1974).
[137] B. Amos, I. G. Anderson, G. A. D. Haslewood, and L. Tökés, *Biochem. J.* **161**, 201 (1977).

component among serum bile acids in patients with the Zellweger syndrome,[138] who lack peroxisomes in the hepatocytes and excrete C_{27} and side chain hydroxylated C_{27} bile acids in bile.[138,139] The C_{29} acid contained two carboxyl groups in the side chain, as indicated by an ABCD-ring ion at m/z 253 and an ion at m/z 482 formed by loss of the hydroxyl functions from the molecular ion.[138] Ions containing the entire side chain were 28 Da greater in the ethyl ester. Furthermore, a peak at m/z 281 (253 + 28) indicated that the additional substituents were beyond C-21. Detailed comparisons of mass spectra with respect to side chain fragment ions and NMR spectrometry revealed the structure $3\alpha,7\alpha,12\alpha$-trihydroxy-27a,27b-dihomo-5β-cholestane-26,27b-dioic acid.[140,141]

Bile acids with an unsaturated side chain have been found in many GC-MS studies of bile.[11,35,89,129,133,137] Depending on the methods of isolation, solvolysis or hydrolysis, their possible formation as artifacts should be considered.[137,138] Loss of an unsaturated side chain occurs both by the normal cleavage of the C-17,20 bond and with migration and loss of two nuclear hydrogens. Thus, ABCD-ring ions are also obtained which simulate the presence of an extra double bond in the nucleus.[111,123,133,142] However, these ions are present in much lower abundance with bile acid derivatives[133] than with unsubstituted sterenes.[5,6,142]

A 22,23-double bond in a C_{28} bile acid derivative has not been reported to give a diagnostically useful fragment ion or loss.[129] A 23,24-double bond in methyl ester TMS ether derivatives of di- and trihydroxycholestenoic acids results in allylic cleavage of the C-20,22 bond[5,6] and base peaks at m/z 283 and 281 (ABCD-rings + C-20,21) in spectra of di- and trihydroxycholestenoates, respectively.[129,133] These ions are also given by the corresponding 24,25-cholestenoic acid derivatives but in much lower abundance.[133] The ion m/z 281 is also seen in spectra of the TMS ether of cholestane-$3\alpha,7\alpha,12\alpha$-triol and in the corresponding Δ^{24} and Δ^{25} compounds where the migration of nuclear hydrogens results in a cluster at m/z 279–282.[143]

Both neutral steroids and bile acids may contain side chain hydroxyl groups. This frequently leads to characteristic fragment ions or losses of the side chain, particularly in the case of bile alcohols. Since the latter

[138] G. G. Parmentier, G. A. Janssen, E. A. Eggermont, and H. J. Eyssen, *Eur. J. Biochem.* **102**, 173 (1979).
[139] R. F. Hanson, P. Szczepanik-Van Leeuwen, G. C. Williams, G. Grabowski, and H. L. Sharp, *Science* **203**, 1107 (1979).
[140] G. Janssen and G. Parmentier, *Steroids* **37**, 81 (1981).
[141] G. Janssen, S. Toppet, and G. Parmentier, *J. Lipid Res.* **23**, 453 (1982).
[142] S. G. Wyllie and C. Djerassi, *J. Org. Chem.* **33**, 305 (1968).
[143] G. S. Tint, B. Dayal, A. K. Batta, S. Shefer, F. W. Cheng, G. Salen, and E. H. Mosbach, *J. Lipid Res.* **19**, 956 (1978).

occur in the same biological samples as bile acids, commonly encountered ions formed from their TMS ethers are listed in Table VI.[143-152] Most of these characteristic ions are formed by cleavage α to the trimethylsiloxy groups. The possibility of migration of trimethylsilyl from the side chain to the nuclear fragment should be kept in mind.[153]

The methyl ester TMS ether of a 22-hydroxylated 3α,7α,12α-trihydroxycholestanoic acid gives a very intense base peak at m/z 217 consisting of C-22–C-27 with the trimethylsiloxy group.[126] Smaller peaks at m/z 127 and 185 arise by loss of 90 and 32 Da, respectively, from this ion. Losses of trimethylsilanol and C-22–C-27 give ions of very low abundance (m/z 383, 293). A weak ABCD-ring ion is at m/z 253. A 22-hydroxycholestan-26-oic acid structure will also form a lactone which gives ions due to loss of side chain and to cleavage of the C-22,23 bond.[126] An ion m/z 113 represents the lactone ring.[126]

An intense ion at m/z 284 in the spectrum of the TMS ether of methyl 3α,7α,22-trihydroxy-5β-cholanoate corresponds to loss of trimethylsilanol and the C-22–C-24 part of the side chain with transfer of a hydrogen to the steroid nucleus.[152] ABCD-ring fragment ions were at m/z 253, 254, 255 and at m/z 344, indicating migration of one or two hydrogens from the nucleus to the lost side chain fragment (as with unsaturated side chains). Unfortunately the spectrum below m/z 200 was not reported and there were no comments on possible side chain fragment ions, e.g., at m/z 175.

The 23-hydroxylated bile acids present in some snakes and seals[122] have not been reported to give diagnostically useful ions as methyl ester TMS ether derivatives. The origin of an ion at m/z 143 in the spectrum of the 23-hydroxycholic acid derivative from human urine is unknown.[8] Cleavage of the C-22,23 bond and loss of the terminal side chain has been reported to occur with the methyl 23-hydroxycholanoates at low ionization energy (15 eV).[154]

[144] J.-Å. Gustafsson and J. Sjövall, *Eur. J. Biochem.* **8,** 467 (1969).
[145] T. Cronholm and G. Johansson, *Eur. J. Biochem.* **16,** 373 (1970).
[146] A. Kibe, S. Nakai, T. Kuramoto, and T. Hoshita, *J. Lipid Res.* **21,** 594 (1980).
[147] P. Eneroth and J.-Å. Gustafsson, *FEBS Lett.* **5,** 99 (1969).
[148] K. Kihira, T. Kuramoto, and T. Hoshita, *Steroids* **27,** 383 (1976).
[149] K. Kihira, M. Yasuhara, T. Kuramoto, and T. Hoshita, *Tetrahedron Lett.* **8,** 687 (1977).
[150] Y. Noma, Y. Noma, K. Kihira, M. Yasuhara, T. Kuramoto, and T. Hoshita, *Chem. Pharm. Bull.* **24,** 2686 (1976).
[151] G. Karlaganis, B. Almé, V. Karlaganis, and J. Sjövall, *J. Steroid Biochem.* **14,** 341 (1981).
[152] K. Kihira, Y. Morioka, and T. Hoshita, *J. Lipid Res.* **22,** 1181 (1981).
[153] S. J. Gaskell, A. G. Smith, and C. J. W. Brooks, *Biomed. Mass Spectrom.* **2,** 148 (1975).
[154] A. Kutner, R. Jaworska, W. Kutner, and A. Grzeszkiewicz, in "Advances in Steroid Analysis" (S. Görög, ed.), p. 333. Akadémia Kiadó, Budapest, 1982.

TABLE VI
SIDE CHAIN FRAGMENT IONS AND LOSSES GIVEN BY TMS ETHERS OF SIDE CHAIN
HYDROXYLATED STEROLS AND BILE ALCOHOLS

Side chain	Position(s) of -OTMS	Fragment(s)[a] Ion(s)	Loss(es)[b]	References
C-20–C-27	C-20	*201*	−85	81
	C-22	*173*,83	−71	144
	C-23	*159*,143	−57	143,145
	C-24	159,*145*,129	−43	68,98,143–145
	C-25	*131*	—	68,98,143,145
	C-26	(103)[c]	—	68,81,143–145
C-20–C-25[d]	C-24	*131*	−145	146
C-20–C-27	C-20,22	*173*	−173	147[e]
	C-22,25	261,*171*,131	−159	143,148
	C-23,25	247,157,*131*	−145	143
	C-24,25	233,143,*131*	−131	143
	C-24,26	233,143,*103*	−131	143
	C-25,26	*219*,129,103	−103	143,149
	C-26,27	Acetate analyzed		150
C-20–C-26	C-24,25	156,*129*,117	−117	151
	C-24,26	Acetate analyzed		150
C-20–C-25	C-24,25	*129*	−103	146
C-20–C-23	C-22,23	Not given	−103	152

[a] The most prominent ions are italicized.
[b] Also with additional loss of trimethylsilanol.
[c] m/z 103 only found in ref. 143, possibly dependent on mass spectrometric conditions.
[d] With a methyl group attached to C-24.
[e] Hydroxyl group at C-20 underivatized.

The 24-hydroxylated $3\alpha,7\alpha$- and $3\alpha,7\alpha,12\alpha$-trihydroxycholestanoic acids are of particular interest as intermediates in the biosynthesis of C_{24} bile acids. Spectra of the methyl ester of the tetrahydroxy acid (varanic acid) have been described[155] also with chemical ionization,[125] and the TMS ether has been used for GC-MS identification in many studies but without detailed discussion of the mass spectrum. Some peaks have been described which may aid in the localization of a 24-hydroxy group. The TMS ether of methyl varanate[130,133] and its 12-deoxy[133] and 24-methyl[130] analogs give an intense [M − 15] ion. A small but distinct peak at m/z 189[133] may correspond to the base peak at m/z 203 in the 24-methyl homolog,[130] representing C-24–C-27 with the trimethylsiloxy group. Loss of 87

[155] T. Kuramoto, H. Kikuchi, H. Sanemori, and T. Hoshita, *Chem. Pharm. Bull.* **21**, 952 (1973).

Da, probably C-25–C-27, together with trimethylsilanols gives a series of peaks of low but significant intensity ending at m/z 323 and 321 for the tri- and tetrahydroxy derivatives, respectively.[130,133] Analogous peaks are found in the CI spectrum.[125] Mass spectra of probable 5α isomers have indicated differences in relative intensities analogous to those described for C_{24} acids (see Hydroxyl Groups in the B-Ring).

A 25-hydroxylated derivative of 3α,7α,12α-trihydroxy-5β-cholestanoic acid has been found in the gastric contents from neonates with high intestinal obstruction.[76] Its methyl ester TMS ether gave a spectrum with small but distinct peaks at m/z 709, 619, 529, 439, and 349 corresponding to loss of the terminal carbomethoxy group (-59) and trimethylsilanols, and indicative of the position of the trimethylsiloxy group in the side chain. There was no side chain fragment ion but absence of an M $- 4 \times 90$ ion and the prominent loss of 245 Da (side chain) from m/z 498 to m/z 253 seem to indicate the presence of a tertiary trimethylsiloxy group.

One 26-hydroxylated bile acid has been studied.[76,128,129] The TMS ether of methyl 3α,7α,12α,26-tetrahydroxy-5β-cholestanoate gave a spectrum with no peaks indicative of the position of the hydroxyl group in the side chain.[128] Reduction of the ester with $LiAlH_4$ and analysis of the TMS ether of the resulting 26,27-diol aided in the identification,[76,128] confirmed by synthesis of the authentic acid.[128]

Mass Spectra of Conjugated Bile Acids

The nonvolatile and labile nature of these bile acids make them best suited for analysis with soft ionization methods (see Ionization Methods). However, a number of unusual conjugates of bile acids have been identified by direct probe electron impact mass spectrometry,[156–158] and Elliott and co-workers have performed detailed studies of the mass spectra of the glycine (as methyl ester) and taurine conjugates of common 5β and 5α bile acids.[4,20] Bile acids conjugated with glucuronic acid have been analyzed as methyl ester acetates by direct probe and chemical ionization[159,160] and as methyl ester TMS ethers by GC-MS and electron impact ionization.[82,161]

---

[156] J. J. Myher, L. Marai, A. Kuksis, I. M. Yousef, and M. M. Fisher, *Can. J. Biochem.* **53**, 583 (1975).

[157] I. M. Yousef and M. M. Fisher, *Can. J. Physiol. Pharmacol.* **53**, 880 (1975).

[158] P. P. Nair, A. I. Mendeloff, M. Vocci, J. Bankowski, M. Gorelik, G. Herman, and R. Plapinger, *Lipids* **12**, 922 (1977).

[159] P. Back, *Hoppe-Seyler's Z. Physiol. Chem.* **357**, 213 (1976).

[160] P. Back and D. V. Bowen, *Hoppe-Seyler's Z. Physiol. Chem.* **357**, 219 (1976).

[161] B. Almé, Å. Nordén, and J. Sjövall, *Clin. Chim. Acta* **86**, 251 (1978).

The spectra of conjugated bile acids have been discussed in detail by Elliott[4] and only a few points will be mentioned. Molecular ions were obtained with the methyl ester acetates of glycocholic acid and N^α-cholylornithine, -arginine, and -histidine.[156,157] The ornithine and arginine conjugates cyclize to the same lactam (M = 504) on the probe and give intense ions at m/z 156 and 169 containing the lactam ring and arising by cleavage of the C-22,23 and C-21,22 bonds, respectively. The same mode of fragmentation of methyl ester acetates gave ions at m/z 230 and 243 for the ornithine, m/z 314 and 327 for the arginine and m/z 253 and 266 for the histidine conjugates.[156] Common to all spectra were losses of acetic acid and the amino acid moiety. Similar types of spectra were obtained with the N^α-lysine conjugate while the N^ε-lysine conjugate did not form a lactam.[158]

The spectra of methyl esters of glycine conjugated bile acids showed an intense peak at m/z 131 due to rearrangement of the side chain.[20] Loss of 131 Da was also seen as were losses of glycine methyl ester (−89) and longer parts of the side chain together with water. Nuclear fragment ions as seen with unconjugated bile acids were also prominent. The free acids gave essentially analogous spectra.

Taurine conjugates were characterized by loss of water and 108 or 167 Da, the latter consisting respectively of the taurine moiety (without nitrogen) and the taurine moiety with C-22, C-23, and C-24. Important ions were formed by cleavage of the C-20,22 bond and loss of the side chain. Nuclear fragment ions were also seen and differences in relative intensities were observed for isomeric bile acids.

While bile acid sulfates have only been analyzed with FAB mass spectrometry, glucuronides were originally identified by using chemical ionization of the methyl ester acetates on a direct probe.[159,160] Few peaks, but typical of glucuronides were obtained: m/z 317 and 257 from the sugar moiety, and m/z 373, 371, and 369, representing the steroid nucleus with side chain of mono-, di-, and trihydroxycholanoates, respectively.

Electron impact spectra of methyl ester TMS ether derivatives of bile acid glucuronides, whether conjugated at C-3 or C-6, give essentially analogous spectra.[82] Thus, the number of hydroxyl groups can be deduced from an intense peak at one of m/z 373, 371, or 369 (459). A base peak at m/z 217 and peaks at m/z 204 and 317 arise from the derivatized glucuronyl moiety. These derivatives can also be analyzed by GC-MS and the ions mentioned are suitable for screening of repetitively scanned analyses (see Applications).[161] The position of conjugation can be determined by periodate oxidation which yields a formate ester at the site of conjugation, followed by oxidation of the free hydroxyl groups, hydrolysis of the formate, and GC-MS analysis of the methyloxime-TMS ether derivative.[82]

Quantitative Analysis

Almost all methods for quantitative analyses of bile acids by mass spectrometry are based on the use of a gas chromatographic inlet. This provides the necessary separation of the components of the mixture prior to their ionization, and the retention index is one of the parameters that determines the specificity. The type of derivative is chosen on the basis of the separation required and the MS characteristics. Ions are generated which are characteristic of the bile acid derivative and can be selected to add further to the specificity of the measurement.

Two principal modes of operation have been utilized for quantitative GC-MS analyses: (1) the repetitive scanning technique and (2) selected ion monitoring.

Repetitive Scanning Techniques

Principles. The repetitive scanning technique requires the use of a medium-sized computer to acquire and process data which are generated. Single scans during the emergence of a gas chromatographic component may be satisfactory for identification purposes; however, to enable quantitation it is essential to make multiple scans during the elution of the components of the mixture, thereby sampling over the whole GLC peak. This is achieved by computer control of initial dwell period, scan time, scan interval, and mass range so that repetitive acquisition of spectra occurs throughout the entire chromatographic run, regardless of the elution of components. Scan cycles of short duration (less than 1 sec for most modern mass spectrometers) are possible over the mass ranges required in bile acid analysis, which is essential to ensure that sufficient data points are obtained over narrow capillary GLC peaks.

During a GLC run of 30 min duration up to 1000 or more spectra may be acquired and stored by the computer. As a means of reducing the need for storage of spectra, a continuously increasing interval between scans can be incorporated so that an unnecessary number of scans will not be taken as peak widths increase under isothermal gas chromatographic conditions.[162] As an alternative, temperature programming will reduce the number of scans, but for quantitative analysis particularly with packed columns this is less desirable due to the changing intensities of background ions from column bleed, thereby making background correction more difficult.

Once acquired, the computer stores the data essentially as a three-dimensional picture, representing all mass (m/z) values over the scanning

---

[162] R. Reimendal and J. Sjövall, *Anal. Chem.* **44**, 21 (1972).

range, the intensity or abundance of each ion and the scan number which can be transformed to retention time or index.

When all ion intensities are summed for each scan and plotted, the result is a total ion current (TIC) chromatogram which will resemble closely the flame ionization detector GLC profile and can be used to correlate spectrum number with gas chromatographic peak, thereby enabling the user to withdraw and inspect the mass spectrum of any particular component if definitive identification is required. In addition to the TIC chromatogram, selected ion current chromatograms may be constructed of any m/z value and quantification is based upon the peak areas obtained when ions are chosen for specific bile acid structures (Fig. 2).[163] Pertinent ions for the quantitative determination of bile acids as their methyl ester TMS ethers by these techniques can be selected from those given in Table V and ref. 8.

Application. Examples of quantitation of bile acids by repetitive scan procedures are limited. Using a standard mixture of cholic, chenodeoxycholic, and deoxycholic acids and a sample of human bile, to which bile acids labeled with deuterium atoms were added as internal standards, Miyazaki *et al.*[37] made comparisons of the peak areas and amounts calculated from ion current chromatograms of three selected m/z values characteristic for the methyl ester TMS ethers of each of these bile acids. With the exception of the chenodeoxycholic acid derivative, no significant differences were found between the three different m/z values used for each bile acid, and the discrepancy with the derivative of chenodeoxycholic acid was due to an interference from cholesterol TMS ether. A coefficient, defined as the reciprocal of the relative intensity, was calculated for the fragment ions monitored for each bile acid derivative, and this could be used to check the purity of the peak being quantified and to distinguish and quantify isomers which give rise to the same ions but differing in relative abundance.

De Weerdt *et al.* quantitated bile acids as their permethyl derivatives using repetitive magnetic scanning over the mass range 100–550 Da.[49] Polydeuterated bile acids were used as internal standards and quantitation was made on the basis of the areas under the peaks of reconstructed ion current chromatograms. Although not indicated in their paper, it must be assumed that a single m/z value was chosen for each bile acid and that this was most probably the value of the base peak in the spectrum. Examples demonstrating a satisfactory linearity, reproducibility (coefficient of variation of approximately 5%), and sensitivity (2 ng of bile acid injected) of the technique were reported.

---

[163] P. Back, J. Sjövall, and K. Sjövall, *Med. Biol.* **52**, 31 (1974).

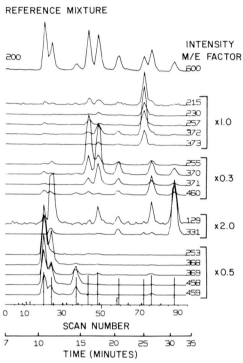

FIG. 2. Reconstructed fragment ion current chromatograms obtained from a repetitive scanning GC-MS analysis (packed column, Hi-Eff 8 BP) of the methyl ester TMS ether derivatives of a mixture of reference bile acids. The upper chromatogram represents total ionization between m/z 200 and 600. The intensities of the individual ion currents have been multiplied by a factor given to the right of the m/z (M/E) values. The four sets of ions have been selected as typical for the derivatives of saturated mono-, di-, and trihydroxycholanoates and 3β-hydroxy-5-cholenoate, respectively. The vertical lines on the x axis are indications of peaks located by the computer and represent the derivatives of (from the left): cholic acid, cholesterol, hyocholic, deoxycholic, chenodeoxycholic, hyodeoxycholic, lithocholic, ursodeoxycholic, and 3β-hydroxy-5-cholenoic acids. Adapted from ref. 163 with permission.

The repetitive scanning mode of operation has been extensively used by Sjövall and co-workers. With manual manipulation of the acquired data, the user selects appropriate point on either side of the peaks in the selected ion current chromatograms and following subtraction of the averaged background intensity, the areas of these peaks can be converted to mass by comparison with the areas obtained in a similar way for reference compounds.[162,164] To eliminate errors in injecting reproducible volumes of

[164] M. Axelson, T. Cronholm, T. Curstedt, R. Reimendal, and J. Sjövall, *Chromatographia* **7**, 502 (1974).

samples, a fixed amount of a suitable internal standard is added. m/z values are similarly selected for the internal standard and the areas of the peaks derived in these chromatograms serve to correct for differences in sample volumes injected and permit a comparison to be made between sample and reference compounds.

Automatic quantitative determination by computer obviates the need for the user to inspect ion current chromatograms. This type of program has been described by Axelson et al. and may operate using either external or internal standards.[164] Compounds are selected with their relative retention times and up to about 10 characteristic m/z values and the computer is programmed to search for peaks within specified retention "windows." Once a peak is identified, the retention time is noted and the summed intensities of the ions over the peak computed after baseline correction and background subtraction procedures. Peak height or peak area can be translated into TIC equivalents, thereby allowing comparisons to be made between different compounds and response factors determined. When reference compounds are unavailable it must be assumed that the total ion yield will be similar to that obtained for the same amount of a structurally similar compound and quantitation will consequently be an approximation.

Axelson et al.[164] observed no significant differences in the quantitative results when comparisons were made between values for peak height and peak area for a mixture of cholic, chenodeoxycholic and lithocholic acids as their methyl ester TMS ethers analyzed on a 2% Hi-Eff 8 BP column. Coefficients of variation of 5.3–8.6 and 6.1–14.0% were found for the synthetic mixture and the same bile acids identified in a sample of urine, respectively.

A refined and further automated version of this method has been used in several studies of urinary and plasma bile acids.[8,73,165,166] This program has the names and specific ions for about 30 of the bile acids which occur in urine, together with the names and ions of the external standards and coprostanol. The latter is used as an internal standard to check the injected volume. The program needs a request to search for peaks of the individual bile acid derivatives as well as the respective standards within a given range of relative retention times. When found by the computer, the peak areas in the fragment ion current chromatograms are converted to total ion current equivalents and compared to the peak area given by the external standard. The daily excretion is then calculated using given information on volumes and proportions of extracts.

Interferences due to chemical background and peak heterogeneity in

---

[165] P. Thomassen, *Eur. J. Clin. Invest.* **9**, 425 (1979).
[166] A. Bremmelgaard and B. Almé, *Scand. J. Gastroenterol.* **15**, 593 (1980).

biological samples will have pronounced effects upon accuracy and precision. If the value for the percentage contribution of the specific ions to the total ionization is calculated for each bile acid and compared with the corresponding value for the authentic reference compound, an indication of peak homogeneity and thus a measure of interference from coeluting compounds can be obtained. At present, manual inspection of the values produced by the program is necessary to eliminate errors due to interfering compounds. For bile acid derivatives having similar retention indices, careful selection of the m/z values used in the construction of ion current chromatograms may permit quantification of complex mixtures, otherwise differences in the relative abundance of the characteristic ions in the spectra can be utilized as was indicated by Almé et al.[8] for three coeluting bile acid derivatives. Alternatively, the utilization of capillary columns will go some way to overcoming the problem of overlapping or coeluting compounds.

In general, irrespective of whether quantitative determination of bile acids by repetitive scanning procedures is carried out manually or automatically by the computer, for most bile acids the lower limit for reliable quantitation is approximately 0.01–0.05 nmol (~4–20 ng) depending upon the mass spectrum, GLC characteristics and retention time, but the limit of detection for qualitative analysis may be slightly lower. Improvements in sensitivity are possible by limiting the mass range which is scanned, but structural information is lost if retrieval of the complete mass spectrum is later required for identification purposes.

While capillary columns afford increased resolution, the need for fast scan speeds places constraints upon sensitivity. Furthermore, their limited capacity (100–200 ng) is frequently a disadvantage when analyzing complex mixtures because to quantitate minor components in the profile, it is often necessary to overload the column with sample leading to deterioration in the chromatographic resolution and distortion of peak shapes.[167]

One area in which repetitive scanning GC-MS computer methods have proved particularly useful is in studies of the *in vivo* metabolic conversion of administered compounds which have been labeled with stable isotopes. The mass spectrometer can provide data both on the number and abundance of heavy atoms in a metabolite, using only nanogram amounts of material. This combined information is much more difficult to obtain with the aid of radioactive isotopes. In addition, mass spectrometry may yield information on the position of the isotopic atoms.

[167] J. J. Vrbanac, C. C. Sweeley, and J. D. Pinkson, *Biomed. Mass Spectrom.* **10,** 155 (1983).

The methods of GC-MS analysis are obviously of critical importance and there is a large literature on the subject of isotope analysis. A number of studies are relevant for bile acid analysis,[164,168–171] and the original literature should be consulted for information on sensitivity and precision, scanning versus switching methods and factors influencing the accuracy. The analysis of mixtures of molecules containing many different numbers of isotopic species may be achieved by repetitive accelerating voltage scanning over an appropriate narrow mass range as exemplified in Fig. 3.[172] Besides instrumental factors and isotope separation effects, interference by contaminating compounds giving ions of the same mass as the unlabeled or labeled bile acids is the major source of error. This is particularly true in experiments with multiple labeling where populations of bile acid molecules with 0–17 deuterium atoms have been measured.[164,173]

Several examples of the use of GC-MS in evaluating ^{13}C-, ^{2}H-, and ^{18}O-labeling patterns of bile acids in rats have been reported. In particular GC-MS has proved useful in investigating the rate and extent of incorporation of ^{13}C and ^{2}H after the administration of [^{13}C]-, [1,1-$^{2}H_2$]-, or [2,2,2-$^{2}H_3$]-ethanol[40,172–178] and of ^{18}O after inhalation of $^{18}O_2$.[179] The results have permitted conclusions to be drawn regarding compartmentation of acetate and coenzyme pools as well as pools and turnover of cholesterol and bile acids. Analyses at high resolution (100,000), using direct probe introduction allows separate determination of ^{2}H- and ^{13}C-containing ions of the same integer mass.[176,177]

[168] P. D. Klein, J. R. Haumann, and W. J. Eisler, *Clin. Chem. (Winston-Salem, N.C.)* **17**, 735 (1971).
[169] P. D. Klein, J. R. Haumann, and D. L. Hachey, *Clin. Chem. (Winston-Salem, N.C.)* **21**, 1253 (1975).
[170] D. A. Schoeller, *Biomed. Mass Spectrom.* **3**, 265 (1976).
[171] D. A. Schoeller, *Biomed. Mass Spectrom.* **7**, 457 (1980).
[172] Z. R. Vlahcevic, T. Cronholm, T. Curstedt, and J. Sjövall, *Biochim. Biophys. Acta* **618**, 369 (1980).
[173] T. Cronholm, A. L. Burlingame, and J. Sjövall, *Eur. J. Biochem.* **49**, 497 (1974).
[174] T. Cronholm, I. Makino, and J. Sjövall, *Eur. J. Biochem.* **26**, 251 (1972).
[175] T. Cronholm, H. Eriksson, S. Matern, and J. Sjövall, *Eur. J. Biochem.* **53**, 405 (1975).
[176] D. M. Wilson, A. L. Burlingame, S. Evans, T. Cronholm, and J. Sjövall, *in* "Stable Isotopes. Applications in Pharmacology, Toxicology and Clinical Research" (T. A. Baillie, ed.), p. 205. Macmillan, London, 1978.
[177] D. M. Wilson, A. L. Burlingame, D. Hazelby, S. Evans, T. Cronholm, and J. Sjövall, *Proc. 25th Annu. Conf. Mass Spectrom. Allied Top., 1977* p. 357 (1977).
[178] T. Cronholm, J. Sjövall, D. M. Wilson, and A. L. Burlingame, *Biochim. Biophys. Acta* **575**, 193 (1979).
[179] I. Björkhem and A. Lewenhaupt, *J. Biol. Chem.* **254**, 5252 (1979).

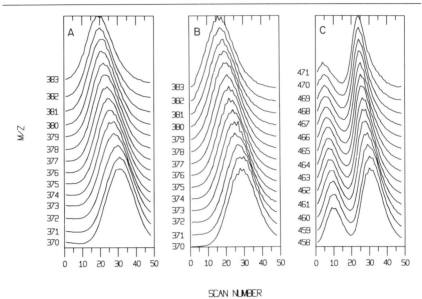

Fig. 3. Ion current chromatograms obtained in a GC-MS analysis of methyl ester TMS ethers of bile acids from a bile fistula rat given [2,2,2-^{2}H$_3$]ethanol for 24 hr. The cluster of ions at m/z [M − 2 × 90] was scanned repetitively (24 Da in 2 sec) during elution of the compounds. Intensity values acquired at 10 kHz were bunched to give 12 values per mass spectrometric peak. Following baseline subtraction, the central peak area (sum of 6 values) was used to construct fragment ion current chromatograms for molecules with 0–13 deuterium atoms. The chromatograms are shown normalized to their respective highest values. The isotope separation (retention times are seen to decrease with increasing number of ^{2}H atoms) is taken into account by using the central peak area (about 10 scans) of each individual chromatographic peak in the subsequent calculation of the isotopic composition. This calculation includes background subtraction and correction for natural isotope abundance and presence of [M − 1] ions, as determined by analysis of the derivative of the unlabeled bile acid.[164] Molecules containing 13 ^{2}H atoms (top chromatograms) constitute about 0.5% of the total population. (A) Allochenodeoxycholic acid from the sulfate fraction; (B) chenodeoxycholic acid from the taurine conjugate fraction; (C) α-muricholic and cholic acids from the taurine conjugate fraction. From ref. 172 with permission.

Selected Ion Monitoring

Principles. Selected ion monitoring (SIM) was first described by Henneberg[180] and has also been referred to as selected ion detection, single or multiple ion monitoring, and mass fragmentography.

In single ion monitoring the user selects an ion from the mass spectrum which is either specific for, or characteristic of, the compound of

---

[180] D. Henneberg, *Fresenius' Z. Anal. Chem.* **183**, 12 (1961).

interest and this is focused and continuously recorded. Usually, more than one ion is selected and recorded by maintaining a constant magnetic field and rapidly switching the accelerating voltage to focus ions of different mass in turn. A disadvantage of this method is the limited range which the accelerating voltage can be switched (corresponding to 10–30% of lowest mass value) without causing defocusing effects.

Modern magnetic instruments utilize computer feedback control, usually by locking on one of the masses arising from the column bleed or a reference compound introduced into the ion source, to maintain focusing during switching. Switching of the magnetic field while maintaining a constant accelerating voltage overcomes the limitations of mass range, but switching rates are then slower and consequently the method is less sensitive.

Quadrupole instruments have been widely employed for SIM because of rapid switching rates over wide mass ranges. However, most SIM methods for bile acids have used magnetic sector instruments (Table VII).

The most important advantage of using the SIM technique for quantitation compared with the repetitive scanning mode, is the greatly enhanced (10^2- to 10^3-fold) sensitivity attained as a result of the prolonged integration times of the signal. Comparisons between the two techniques have been reported.[162,181] The major drawback of the SIM technique for quantitation is that the selection of m/z values to be monitored must precede the analysis and therefore all other information is lost.

The main factors which will influence the sensitivity of the technique include (1) the GLC characteristics of the compound, (2) the number of ions monitored; sensitivity decreases with increasing number of ions, (3) the mass spectrometric resolution; sensitivity usually decreases as the resolution is increased, and (4) the degree of interference from column bleed, impurities or compounds coeluting with the compound of interest; this is often minimized by monitoring ions of high mass values.

Application. The SIM technique has been used for the measurement of bile acids in serum, interstitial fluid, and tissue where the relatively low concentrations generally preclude quantitation by repetitive scanning of the full mass range. A variety of GLC columns, derivatives, and internal standards have been employed in SIM methods for bile acids and these are detailed in Table VII.[182–195] Quantitation of the bile acids is carried out

[181] B. S. Middleditch and D. M. Desiderio, *Anal. Chem.* **45**, 806 (1973).
[182] K. D. R. Setchell, A. M. Lawson, E. J. Blackstock, and G. M. Murphy, *Gut* **23**, 637 (1982).
[183] T. C. Bartholomew, J. A. Summerfield, B. H. Billing, A. M. Lawson, and K. D. R. Setchell, *Clin. Sci.* **63**, 65 (1982).
[184] K. D. R. Setchell and A. Matsui, *Clin. Chim. Acta* **127**, 1 (1983).

by relating either the peak height or peak area response for any ion monitored to that of a specific ion given by an internal standard.

In studies of healthy subjects and patients with liver disease, bile acid profiles of serum and skin interstitial fluid were examined (Fig. 4).[182-184] Removal of cholesterol (and related sterols) which has been the subject of much attention in GC-MS analyses of bile acids[37,67] was achieved by lipophilic gel ion exchange chromatography.[8,184] Although alternative derivatives (see Gas Chromatographic Inlet) have been advocated as a solution to the problem of C_{27}-sterol interference,[45,67,185] it is preferable to remove sterols from the sample prior to GC-MS to prevent an overloading of the GLC column and loss of chromatographic resolution.

The most accurate results are obtained with analogs labeled with heavy isotopes as internal standards. When added to the sample they compensate for losses which may occur during the extraction, isolation, and purification steps, differences in yields during derivatization, and, if used in excess, will act as carriers to minimize losses due to adsorption in the GC-MS system.

The principal considerations for the selection of a labeled standard are (1) chemical stability of the incorporated isotope atoms in that part of the molecule which is to be selected for monitoring; (2) high isotopic enrichment to minimize the influence of naturally occurring isotopes; (3) sufficient number of isotope atoms to minimize interference from the analyte; and (4) nature of isotope: isotope effects with ^{13}C or ^{18}O are minimal compared to those with ^2H.

In some of the earliest examples of the use of isotope dilution techniques for bile acids, Björkhem and co-workers[186-188] employed a pentadeuterated cholic acid and trideuterated deoxycholic acid as the internal standards for the determination of the major bile acids in serum from the

---

[185] T. Beppu, Y. Seyama, T. Kasama, and T. Yamakawa, *J. Biochem.* (*Tokyo*) **89,** 1963 (1981).
[186] B. Angelin and I. Björkhem, *Gut* **18,** 606 (1977).
[187] B. Angelin, I. Björkhem, and K. Einarsson, *J. Lipid Res.* **19,** 527 (1978).
[188] J. Ahlberg, B. Angelin, I. Björkhem, and K. Einarsson, *Gastroenterology* **73,** 1377 (1977).
[189] I. Björkhem and O. Falk, *Scand. J. Clin. Lab. Invest* **43,** 163 (1983).
[190] Y. Akashi, H. Miyazaki, and F. Nakayama, *Clin. Chim. Acta* **133,** 125 (1983).
[191] H. Takikawa, H. Otsuka, T. Beppu, Y. Seyama, and T. Yamakawa, *J. Biochem.* (*Tokyo*) **92,** 985 (1982).
[192] M. Shino, Y. Nezu, T. Tateyama, K. Sakaguchi, K. Katayama, J. Tsutsumi, and K. Kawabe, *Yakugaku Zasshi* **99,** 421 (1979).
[193] I. Björkhem, B. Angelin, K. Einarsson, and S. Ewerth, *J. Lipid Res.* **23,** 1020 (1982).
[194] H. Takikawa, T. Beppu, Y. Seyama, and T. Wada, *Gastroenterol. Jpn.* **18,** 246 (1983).
[195] M. Tohma, Y. Nakata, H. Yamada, T. Kurosawa, I. Makino, and S. Nakagawa, *Chem. Pharm. Bull.* **29,** 137 (1981).

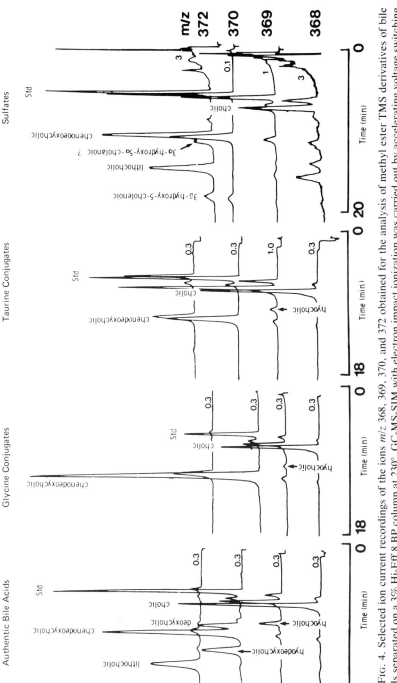

FIG. 4. Selected ion current recordings of the ions m/z 368, 369, 370, and 372 obtained for the analysis of methyl ester TMS derivatives of bile acids separated on a 3% Hi-Eff 8 BP column at 230°. GC-MS-SIM with electron impact ionization was carried out by accelerating voltage switching between the four ions. Chromatograms are shown for a mixture of authentic bile acids and for the glycine, taurine, and sulfate conjugated bile acid fractions isolated by lipophilic gel ion exchange chromatography of serum from a patient with primary biliary cirrhosis. Ions at m/z 368, 370, and 372 are characteristic of trihydroxy-, dihydroxy-, and monohydroxycholanoate derivatives. The internal standard, corprostanol is detected using the ion m/z 370. The ion m/z 369 was selected to detect the derivative of hyocholic acid. From ref. 184 with permission.

TABLE VII
SUMMARY OF SELECTED ION MONITORING TECHNIQUES FOR THE DETERMINATION OF BILE ACIDS IN BIOLOGICAL SAMPLES

Analyte	Internal standard[a]	GLC conditions and derivative[b]	Mass spectrometry conditions[c]	Ions monitored (m/z) for analyte/internal standard	References
Cholic	[2,2,3,4,4-^{2}H$_5$]CA	1.5% SE-30	EI (20 eV)	368/373	186–188
Chenodeoxycholic	[11,11,12-^{2}H$_3$]DCA	210–220° isothermal	Accelerating voltage switching	370/373	
Deoxycholic	[11,11,12-^{2}H$_3$]DCA	Me-TMS	Ion source temperature 260°	370/373	
[11,12-^{2}H$_3$]Chenodeoxycholic and many other bile acids	12-oxo-CDCA	3% OV-1 280° isothermal Me-propionate	EI	372/384 368,370,372/384	192
Lithocholic	[6,6,7,7-^{2}H$_4$]LCA	25 m SE-30 capillary	EI (22.5 eV)	461/465	57
Deoxycholic	[6,6,7,7,8-^{2}H$_5$]DCA	270° isothermal	Accelerating voltage switching	563/568	
Chenodeoxycholic	[11,11,12,12-^{2}H$_4$]CDCA	Solid injection	Ion source temperature 290°	459/463	
Ursodeoxycholic	[11,11,12,12-^{2}H$_4$]UDCA	Et-DMES		563/567	
Cholic	[11,11,12-^{2}H$_3$]CA			665/668	
Lithocholic	[11,11,12-^{2}H$_3$]DCA	3% QF-1	EI (70 eV)	622/623	185, 194
Deoxycholic	[11,11,12-^{2}H$_3$]DCA	215° isothermal	Accelerating voltage switching	620/623	
Chenodeoxycholic	[11,11,12-^{2}H$_3$]DCA	HFIP-TFA	Ion source temperature 270°	620/623	
Ursodeoxycholic	[11,11,12-^{2}H$_3$]DCA			620/623	
Cholic	[11,11,12-^{2}H$_3$]DCA			618/623	
Ursodeoxycholic	Glyco-5β-chol-3-enoic acid	1.5% Poly I-110 and 1.5% SE-30	EI (70 eV)	370/372	195
[11,12-^{2}H$_2$]Ursodeoxycholic		270 and 250°	Accelerating voltage switching	372/372	
Related bile acids		isothermal, Me-TMS	Ion source temperature 290°	368,370,386/372	

Glucuronides of deoxycholic, lithocholic, chenodeoxycholic, ursodeoxycholic, and cholic	[11,11,12-^{2}H$_3$]DCA 3-glucuronide [2,2,4,4,-^{2}H$_4$]LCA glucuronide	As for ref. 185	As for ref. 185	191
Deoxycholic 3-sulfate	[11,11,12-^{2}H$_3$]DCA 3-sulfate			
Unconjugated, glycine, taurine, and sulfate conjugates of mono-, di-, and trihydroxy bile acids	Coprostanol	3% Hi-Eff 8 BP 230° isothermal Me-TMS	EI (70 eV) Accelerating voltage switching Ion source temperature 285°	372,370,368/370	182–184
Monooxo bile acids after NaB^{2}H$_4$ reduction	[2,2,3,4,4-^{2}H$_5$]CA [2,2,3,4,4-^{2}H$_5$]CDCA [11,11,12-^{2}H$_3$]DCA	1.5% SE-30 260–290° Me-TMS	EI (20 eV) Accelerating voltage switching	624/623 and 623/628 371/370 and 370/373 256/255 and 255/258	193
Cholic Chenodeoxycholic Deoxycholic	[2,2,3,4,4-^{2}H$_5$]CA [2,2,3,4,4-^{2}H$_5$]CDCA [11,11,12-^{2}H$_3$]DCA	1.5% SE-30 210–250° Me-TMS	EI (20 eV) Accelerating voltage switching	623/628 370/373 255/258	189
Lithocholic Deoxycholic Chenodeoxycholic Ursodeoxycholic Cholic	Glycol[6,6,7,7-^{2}H$_4$]LCA Glycol[6,6,7,7,8-^{2}H$_5$]DCA Glycol[11,11,12,12-^{2}H$_4$]CDCA Glycol[11,11,12,12-^{2}H$_4$]UDCA Glycol[11,11,12-^{2}H$_3$]CA	25 m SE-30 capillary 270° isothermal Solid injection Et-DMES	EI (22.5 eV) Ion source temperature 290°	461/465 563/568 459/463 563/567 665/668	190

^{a} CA, Cholic acid; DCA, deoxycholic acid; LCA, lithocholic acid; CDCA, chenodeoxycholic acid; UDCA, ursodeoxycholic acid.
^{b} Me, Methyl ester; Et, ethyl ester; TMS, trimethylsilyl ether; DMES, dimethylethylsilyl ether; HFIP, hexafluoroisopropyl ester; TFA, trifluoroacetate.
^{c} EI, Electron impact.

portal vein and the peripheral circulation. Both interal standards were free of unlabeled molecules. In the absence of a deuterium-labeled chenodeoxycholic acid, trideuterated deoxycholic acid was used to quantitate both dihydroxy bile acids as their methyl ester TMS ethers, and standard curves were constructed for each individual bile acid. In later work [2,2,3,3,4-^{2}H$_5$]chenodeoxycholic acid was substituted.[189] Although these labeled standards compensate for procedural losses, the accuracy of the technique relies upon the assumption that the hydrolysis of bile acid conjugates is quantitative and that no degradation occurs during this stage. Internal standards conjugated with glycine and taurine would be preferable to assess this step. A solvolysis step was not included in the method, as was the case in the later work of Beppu et al.[185] and therefore the values quoted for bile acid concentrations are underestimated because measurable quantities of sulfated bile acids occur in the serum of healthy subjects[166] and may be elevated in cholestatic liver disease.[183]

Conjugated bile acids labeled with deuterium in the steroid nucleus have recently been introduced for analysis of bile acids. In one study, five glycine conjugates were added as internal standards to liver tissue and bile.[190] The samples were extracted, solvolyzed, and hydrolyzed with alkali, and the bile acids were analyzed by GC-MS of the ethyl ester dimethylethylsilyl ether derivatives. Ions of high mass were monitored to obtain the concentrations of mono-, di-, and trihydroxy bile acids.

Takikawa et al. described a SIM technique for the determination of bile acid glucuronides and sulfates in serum.[191] The internal standards [11,11,12-^{2}H$_3$]deoxycholic acid 3-glucuronide and 3-sulfate and [2,2,4,4-^{2}H$_4$]lithocholic acid 3-glucuronide were added to serum, and following enzymic hydrolysis of the glycine and taurine conjugates, glucuronide and sulfate conjugated bile acids were separated on the ion exchanger piperidinohydroxypropyl Sephadex LH-20.[196] The glucuronides were hydrolyzed with β-glucuronidase and sulfates solvolyzed to release free bile acids which were subsequently measured as their hexafluoroisopropyl trifluoroacetate derivatives by SIM of suitable ions of high mass. Bile acid glucuronides and sulfates were reported to comprise 8.7 and 11.2%, respectively, of the total serum bile acids.

In a novel attempt to determine the approximate concentration of monooxo bile acids in portal venous and peripheral blood using SIM, Björkhem et al. reduced the samples with sodium borodeuteride to yield monodeuterated hydroxy bile acids, and then calculated the ratio of these products to the endogenous hydroxy bile acids by measuring the

[196] J. Goto, M. Hasegawa, H. Kato, and T. Nambara, *Clin. Chim. Acta* **70,** 141 (1978).

(M + 1)/M ratio.[193] From knowledge of the concentration of each endogenous hydroxy bile acid, which was determined separately as described in their earlier reports, a semiquantitative assessment of the concentration of oxo bile acids was obtained. The structure of the oxo acids cannot be accurately determined, and if this information is required, capillary column GC-MS with monitoring of ions selective for oxo bile acid derivatives should be performed on nonreduced samples.[55]

All SIM methods reviewed above are based on GC-MS. Preliminary studies have also been reported on the use of FAB mass spectrometry for direct quantitative analysis of bile acids in bile.[26] [11,12-^{2}H$_2$]-Chenodeoxycholic acid was used as internal standard and response factors for cholic, glycocholic, and taurocholic acids were determined relative to this standard using the [M − 1]$^-$ ion and scan averaging. While the values were different for the three acids, reproducibility was reasonable (coefficient of variation 20–30%). However, further studies using more appropriate labeled standards have indicated pronounced interference between different types of bile acids and this method for direct quantitation requires further study (J. O. Whitney, personal communication).

Labeling with Stable Isotopes in Kinetic Studies

The availability of GC-MS has opened new possibilities to study metabolism with the aid of stable isotopes. The studies can be classified into two groups: those in which a labeled bile acid is administered and its metabolism and turnover are determined,[197–204] and those in which precursors of the carbon, hydrogen, and oxygen atoms are administered and the rates of labeling in different positions are determined.[172–179] The latter type was discussed earlier.

[197] J. B. Watkins, D. Ingall, P. Szczepanik, P. D. Klein, and R. Lester, *N. Engl. J. Med.* **288**, 431 (1973).
[198] W. F. Balistreri, A. E. Cowen, A. F. Hofmann, P. A. Szczepanik, and P. D. Klein, *Pediatr. Res.* **9**, 757 (1975).
[199] T. Tateyama, Y. Nezu, M. Shino, K. Sakaguchi, K. Katayama, and J. Tsutsumi, *Koenshu-Iyo Masu Kenkyukai* **3**, 207 (1978).
[200] F. Kern, Jr., G. T. Everson, B. DeMark, C. McKinley, R. Showalter, W. Erfling, D. Z. Braverman, P. Sczcepanik-Van Leeuwen, and P. D. Klein, *J. Clin. Invest.* **68**, 1229 (1981).
[201] G. Everson, B. DeMark, P. Klein, R. Showalter, C. McKinley, and F. Kern, Jr., *Gastroenterology* **80**, 1114 (1981).
[202] T. Nishida, H. Miwa, M. Yamamoto, T. Koga, and T. Yao, *Gut* **23**, 751 (1982).
[203] B. R. DeMark, G. T. Everson, P. D. Klein, R. B. Showalter, and F. Kern, Jr., *J. Lipid Res.* **23**, 204 (1982).
[204] F. Stellaard, R. Schubert, and G. Paumgartner, *Biomed. Mass Spectrom.* **10**, 187 (1983).

Bile acids labeled with deuterium or ^{13}C have been synthesized for turnover studies.[108,205-207] Labeling with deuterium vicinal to carbons carrying a hydroxyl group is not satisfactory, since these atoms may be labilized via oxidation and enolization. Thus, 11,12-$^2H_2$- (in the absence of a 12-hydroxy group) and 24-^{13}C-labeled bile acids have been most commonly used. The measurements of relative abundances of labeled and unlabeled molecules after injection of the former are usually performed with a SIM technique. Most methods have relied on collection of bile to get sufficient amounts of bile acids for analysis.[168,169,197,198,200,202] However, the improved methods of isolation and the development of instrumentation now permit the analysis of bile acids in serum.[199,201,203,204] DeMark et al.[203] prepared the methyl ester acetate derivative of chenodeoxycholic and cholic acids and used ammonia chemical ionization (ion source temperature 160°, pressure 1 Torr) to yield a predominant high mass ammonium adduct ion $[M + 18]^+$ for monitoring. The ratios of the ions m/z 509/508 and m/z 567/566 were determined and the mean natural abundance of molecules with one ^{13}C atom were 33.24 and 35.25%, respectively, for the reference bile acids. This compares with mean values of 33.18% for chenodeoxycholic and 35.22% for cholic acid isolated from serum. These determinations were achieved with precisions of 0.36–0.57% coefficient of variation. This means that a 1% excess of labeled molecules will be measured with a standard deviation of about 0.1–0.2 (i.e., precision 10–20%) which compares with the values obtained by repetitive scanning over a narrow mass range.[164] When 25 mg of [24-^{13}C]cholic acid was administered orally to an adult and fractional turnover rates, pool size, and synthesis rates were calculated, the agreement between values obtained from analyses of serum and bile was good.[201] Similarly, Tateyama et al. using oral [11,12-$^2H_2$]chenodeoxycholic acid, demonstrated satisfactory correlations between serum and biliary chenodeoxycholic acid kinetics.[199] Although relatively large volumes of serum were required for these studies, it was suggested that kinetic studies using volumes of serum as small as 0.1–0.5 ml would be possible with the use of capillary columns and enhanced instrument sensitivity, thus rendering the technique suitable for investigations in children and newborn infants. This was later supported by Stellard et al.[204] who determined [24-^{13}C]chenodeoxycholic acid kinetics using only 2% of the 1–2 ml volumes of serum analyzed and capillary column GC-MS with electron impact ionization. Bile acids were

---

[205] D. L. Hachey, P. A. Szczepanik, O. W. Berngruber, and P. D. Klein, *J. Labelled Cmpd.* **9,** 703 (1973).

[206] K.-Y. Tserng and P. D. Klein, *J. Lipid Res.* **18,** 400 (1977).

[207] T. A. Baillie, M. Karls, and J. Sjövall, *J. Labelled Cmpd. Radiopharm.* **14,** 849 (1978).

extracted using Amberlite XAD-2 resin and following solvolysis and enzymatic hydrolysis the methyl ester TMS ethers were prepared. GC-MS was performed on a quadrople instrument using computer controlled SIM of ions m/z 368–373 and m/z 460–461. The integration time at each mass was 0.2 sec with a total cycle time of 1.7 sec. The isotope ratios were determined from the average of 10 values over each GLC peak. No evidence for contamination was found in these chromatograms, and natural isotopic abundances for standard and serum bile acids were measured with a coefficient of variation of 0.2–0.9%. The metabolic conversion of chenodeoxycholic to lithocholic and ursodeoxycholic acids could be followed by monitoring of ions appropriate for the derivatives of these acids.

Conclusions

GC-MS is of great value for qualitative and quantitative analysis of bile acids. The improved instrumentation and methods of isolation have opened new possibilities to use stable isotopes in metabolic and kinetic studies. In the next few years we may expect advances in the use of soft ionization methods such as FAB, particularly in combination with high-performance liquid chromatography and tandem MS-MS. The methods for computer evaluation of data have to be improved so that the scientist can obtain the qualitative and quantitative data from GC-MS analyses in a more digested form after partial computer interpretation.

[5] General Carotenoid Methods

By GEORGE BRITTON

Introduction

Approximately 500 different naturally occurring carotenoids have now been characterized, if glycosides and enantiomers are included. In addition to the hydrocarbon carotenes, there is a wide range of xanthophylls containing oxygen functions, most commonly hydroxy, keto, epoxy, methoxy, and carboxy groups. Some acyclic carotenoids occur widely, e.g., lycopene, but more common are the monocyclic and bicyclic compounds. Cyclization in the carotenoid series is restricted to the formation of a six-membered (or occasionally five-membered) ring at one or both

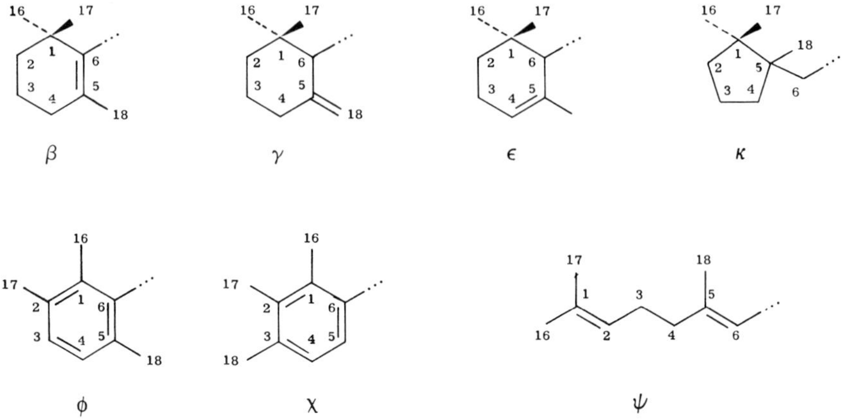

FIG. 1. Carotenoid end groups.

ends of the molecule. Extensive cyclization such as is seen in the triterpenoid series is not encountered.

Traditionally, naturally occurring carotenoids have been given trivial names, usually derived from the name of the biological source from which they were first isolated. In recent years, however, a semisystematic nomenclature has been devised for carotenoids, so that the name conveys structural information.[1] According to this scheme, the carotenoid molecule is considered in two halves, and the nature of the end group of each half is designated. Seven end groups have been recognized and are designated by Greek letter prefixes (Fig. 1). Each carotenoid is considered to be formally a derivative of a parent carotene, specified by the two Greek letters describing the end groups, and changes in hydrogenation level and the presence of substituent groups are indicated by use of conventional prefixes and suffixes. Thus, the acyclic lycopene with two ψ-end groups becomes ψ,ψ-carotene, and β-carotene should be called β,β-carotene. These structures and the carotene numbering scheme are illustrated in Fig. 2, along with the structures of *trans*-phytoene (7,8,11,12,7',8',11',12'-octahydro-ψ,ψ-carotene), lutein (β,ε-carotene-3,3'-diol), and astaxanthin (3,3'-dihydroxy-β,β-carotene-4,4'-dione), carotenoids which will be encountered later, and which provide good examples of the semisystematic naming of hydrogenated and oxygen-substituted carotenoids. In keeping with the common practice in biology and biochemistry, carotenoids will be referred to throughout this article by

---

[1] IUPAC Commission on the Nomenclature of Organic Chemistry and IUPAC-IUB Commission on Biochemical Nomenclature, *Biochemistry* **10**, 4827 (1971).

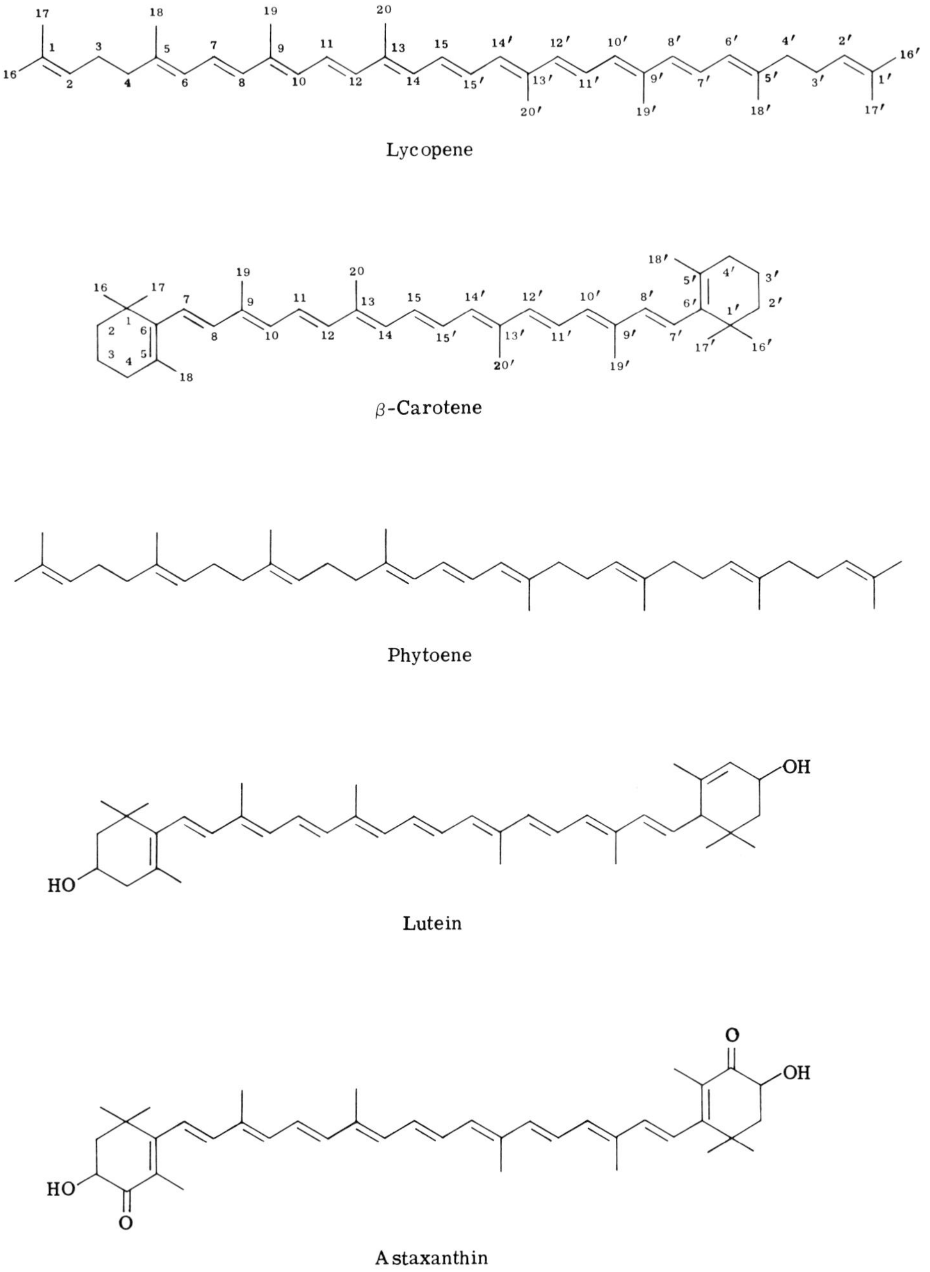

FIG. 2. Structures and numbering system of some carotenoids.

their trivial names, but the semisystematic names are also given in Table V.

Although the conventional carotenoids are C_{40} tetraterpenoids, some C_{30} "diapocarotenoids" (structurally and biosynthetically triterpenes) and C_{50} "homocarotenoids" (substituted C_{40} structures) occur in some bacteria. These have properties very similar to those of the C_{40} carotenoids, and are handled in the same way.

Many books and review articles are available dealing with all aspects of the chemistry and biochemistry of carotenoids. For a survey of the carotenoid field in general, especially chemistry, the monograph of Isler[2] is a monumental source of information. The proceedings of an International Symposium on Carotenoids series also contain an extensive range of useful articles on all aspects of carotenoid science.[3-8]

Complementary to this are the two volumes of "The Biochemistry of the Carotenoids" by Goodwin,[9] which include the most comprehensive tabulated data on the natural occurrence and distribution of carotenoids and a monograph compiled by Bauernfeind[10] which deals with technological aspects of carotenoids as food colorants and vitamin A precursors.

In their solubility properties, the carotenoids are like other groups of higher isoprenoids. The carotenes are typical nonpolar hydrocarbons, the xanthophylls more polar yet still insoluble in water. The special and characteristic properties of the carotenoids are a consequence of the large number of conjugated double bonds, constituting a long, light-absorbing chromophore. The absorption of visible light is, in many ways, an advantage, since the light absorption spectrum is a convenient first means of identification and provides a basis for accurate and sensitive quantitative analysis. The high degree of unsaturation, however, is also responsible for the instability of the carotenoids to oxygen, light, heat, acid, etc. It is because of this instability that precautions need to be taken when handling

[2] O. Isler, ed., "Carotenoids." Birkhaeuser, Basel, 1971.
[3] "Carotenoids other than Vitamin A—1," *Pure Appl. Chem.* **14,** 215 (1967).
[4] "Carotenoids other than Vitamin A—2," *Pure Appl. Chem.* **20,** 365 (1969).
[5] "Carotenoids other than Vitamin A—3," *Pure Appl. Chem.* **35,** 1 (1973).
[6] B. C. L. Weedon, ed., "Carotenoids—4 (Berne, 1975)." Pergamon, Oxford, 1977; also published in *Pure Appl. Chem.* **47,** 97 (1976).
[7] T. W. Goodwin, ed., "Carotenoids—5 (Madison, 1978)." Pergamon, Oxford, 1979; also published in *Pure Appl. Chem.* **51,** 435, 857 (1979).
[8] G. Britton and T. W. Goodwin, eds., "Carotenoid Chemistry and Biochemistry." Pergamon, Oxford, 1982.
[9] T. W. Goodwin, "The Biochemistry of the Carotenoids," 2nd ed., Vol. 1. Chapman & Hall, London, 1980; Vol. 2, (1984).
[10] J. C. Bauernfeind, ed., "Carotenoids as Colorants and Vitamin A Precursors." Academic Press, New York, 1981.

carotenoids, in order to minimize oxidation and isomerization. This chapter will concentrate on the special precautions and procedures that must be used in carotenoid work but are not necessarily applicable to work with other groups of isoprenoid compounds. It cannot be a fully comprehensive review of carotenoid methodology for workers already having considerable experience in the field. Rather, it presents a guide to the main special areas of carotenoid methods largely for biochemists and biologists with little or no previous experience with carotenoids. No complex carotenoid chemistry is included, nor is a detailed discussion of the physicochemical techniques such as mass spectrometry and nuclear magnetic resonance spectroscopy which are widely used by carotenoid chemists and which require a great deal of expertise for interpretation of the results. Extensive details of all relevant modern methods of carotenoid biochemistry are given in a monograph,[11] and consultation of other articles in the present series,[12,13] as well as a very thorough survey of methods applied to plant carotenoids[14] is also recommended.

General Procedures and Precautions

Because of the long conjugated double bond system the carotenoids generally are much less stable than most other isoprenoids, being sensitive to oxygen, light, heat, acids, and, in some cases, alkali, and especially to combinations of these factors, e.g., light and oxygen. A range of stringent precautions must be observed if losses of material are to be minimized. With the refinement of analytical techniques, small amounts of cis-isomers, oxidation products, etc. can readily be detected. Their production as artifacts must therefore be avoided in order to ensure that any mixture isolated or sample analyzed is unchanged from that which was originally present in the natural tissue under investigation. Speed of manipulation can be very important. All procedures that could unavoidably introduce risks of oxidation, isomerization etc. should be carried out as rapidly as possible.

Protection against Oxygen

Oxygen, especially in combination with light and/or heat, is the most destructive factor and can result in the formation of oxidized artifacts or

---
[11] G. Britton and B. H. Davies, "Methods of Carotenoid Biochemistry and Vitamin A Analysis." Academic Press, New York (in preparation).
[12] G. Britton and T. W. Goodwin, this series, Vol. 18, Part C, p. 654.
[13] S. Liaaen-Jensen and A. Jensen, this series, Vol. 23, p. 586.
[14] B. H. Davies, in "Chemistry and Biochemistry of Plant Pigments" (T. W. Goodwin, ed.), 2nd ed., Vol. 2, p. 38. Academic Press, New York, 1976.

the disappearance of material due to complete oxidative breakdown. The acyclic carotenoids, especially phytofluene and ζ-carotene, seem to be most susceptible to oxidative decomposition. The addition of an antioxidant such as butylated hydroxytoluene (BHT, 2,6-di-*t*-butyl-*p*-cresol) at the time of tissue extraction provides a useful way of minimizing oxidation, provided the antioxidant can be removed easily when pure materials are required for MS or NMR analysis. Carotenoid samples and carotenoid-containing extracts must always be stored *in vacuo* or in a completely inert atmosphere (N_2, Ar). Slight traces of residual oxygen are sufficient to cause decomposition even at freezer temperatures. Solutions should be flushed with nitrogen through a capillary tube for a few minutes before storage. Saponification and other reactions should normally be carried out under an inert atmosphere of nitrogen, and pyrogallol may be added to saponification mixtures as an antioxidant. Thin-layer chromatography of the more unstable carotenoids should also be carried out in an inert atmosphere, and exposure of carotenoids to air while they are adsorbed on thin layers before or after chromatography must be as short as possible. The final evaporation of solvent from carotenoid samples is done by blowing a gentle stream of nitrogen on to the solution.

Peroxides, such as may accumulate in diethyl ether and related solvents on storage, can also lead to destruction or to the formation of epoxycarotenoids. Freshly distilled, peroxide-free solvents must be used.

Protection against Light

Direct sunlight or UV light are particularly harmful in causing trans–cis isomerization and must be avoided. Low intensity diffuse daylight or subdued artificial light is acceptable for most carotenoid manipulations, but precautions must be taken to exclude light from carotenoid samples when these are adsorbed on chromatographic materials. Thus chromatography columns should be wrapped in aluminum foil or black paper (which can easily be moved to allow brief inspection of the column), and TLC tanks covered with black cloth or kept in a dark cupboard. Carotenoid solutions which may need to be left for prolonged periods, e.g., during evaporation, saponification etc. should also be protected from light.

Protection against Heat

Carotenoids in general, and xanthophylls in particular, may undergo isomerization and structural modification if heated, either as solids or in solution. They should therefore not be subjected to heat except when this is unavoidable. Solvents with low boiling points are preferred as these can subsequently be removed without the need for excessive heating. Solvents should be evaporated below 40° on a rotary evaporator, not by conventional distillation. When saponification is considered desirable, this should be done, if possible, in the cold (refrigerator or room temperature). Carotenoid samples should be stored in a freezer at $-20°$ or below, with oxygen totally excluded.

Avoidance of Acid or Alkali

Almost all carotenoids are at risk of oxidative decomposition, cis–trans isomerization or in some cases dehydration if subjected to acid conditions, especially in the presence of light and/or oxygen. The acid-catalyzed isomerization of 5,6-epoxycarotenoids is a well known diagnostic test (see below) but can also occur during inadvertent exposure to acid during extraction and purification. Acidic chromatographic adsorbents such as silicic acid and silica gel (even some HPLC materials) can bring about this isomerization unless neutralized, and acidic solvents, including chloroform which usually contains traces of HCl, are also dangerous. Acidic reagents and strong acids should not be used in rooms where carotenoids are being handled.

Although most carotenoids are stable to alkali and not destroyed by saponification, some, notably α-ketols (i.e., those containing the 3-hydroxy-4-oxo β-ring as in astaxanthin), fucoxanthin, peridinin, and related compounds, and, of course, carotenoid esters may be altered by treatment with even weak alkali. Saponification must be avoided if it is suspected that any such compounds may be present.

Purification of Solvents, Adsorbents, and Reagents

For all manipulations, especially chromatography, the purest possible solvents should be used. All solvents must be dried and redistilled before use, and peroxides must be removed from diethyl and other ethers by distillation from reduced iron powder, calcium hydride, or for small volumes, $LiAlH_4$. The solvents should be stored in dark bottles. Benzene should be avoided because of its toxicity; toluene can usually be substituted for benzene in, for example, chromatography. When carotenoid samples are to be subjected to MS, NMR analysis etc., solvent purification must be absolutely rigorous otherwise the evaporation of large volumes of solution may result in contamination of the carotenoid with substantial amounts of solvent-derived impurity. Passage of petroleum ether, diethyl ether etc. through a column of active alumina or silica, and distillation through a fractionating column is recommended. All other materials, e.g., adsorbents, drying agents, reagents, must also be of a satisfactory state of purity to preclude the introduction of extraneous materials into carotenoid samples. Plastic or polythene wash bottles, pipets, stoppers, etc. contain plasticizers which are readily leached out by organic solvents and should never be used in this kind of work.

Extraction and Saponification

Extraction

Carotenoids are widely distributed in a variety of natural tissues, both soft and hard. Extraction techniques must therefore be adapted to suit the tissue under investigation with due regard to the general precautions that must be taken in carotenoid work. Many detailed procedures have been published. The following general points and guidelines should be observed.

Carotenoids should be extracted from undamaged and preferably fresh material. Extractions should be performed as rapidly as possible after the material is obtained so as to avoid oxidative or enzymatic degradation; thus, for example, rapid enzymatic carotenoid destruction occurs immediately when leaves are cut. If tissues cannot be extracted immediately they should be stored in a freezer or freeze-dried. Air drying or dehydration by solvent is not recommended because of possible oxidation and loss of carotenoids.

Since most tissues contain much water, and carotenoids are not water soluble, they must be extracted with water-miscible polar organic solvents, usually acetone, methanol, or ethanol. Extraction with chloroform–methanol mixtures, widely used for other lipids, is not recommended for carotenoids because of the likely presence of HCl in the chloroform. Efficient extraction usually requires mechanical tissue disruption. Microbial cells, especially yeasts and some nonphotosynthetic bacteria, may be broken by, for example, passage through the French press or shaking with glass beads in a Braun MSK Homogenizer before extraction with solvent. Plant and animal tissues are usually homogenized directly in the organic solvent with a suitable electric blender. The "Ultra-Turrax" or "Polytron" homogenizer with motor mounted above the blades is most versatile and satisfactory, being suitable for large or small (0.5 g) samples and usable in a range of vessels (beaker, conical, or round-bottomed flask, tube etc.) For very hard materials, e.g., the shell (carapace) of crustaceans, the sample may need to be ground mechanically to a powder to permit efficient extraction. For small samples, grinding the tissue in acetone with clean sand (pestle and mortar) may be used but quantitative extraction by this means is not easy to achieve. Some workers have used long-term (overnight or longer) soaking with the extracting solvent in the cold as an alternative to mechanical blending. If this procedure is used, precautions must be taken to exclude air. Rapid and efficient extraction from microbial cells, including unicellular algae, can often be achieved without the need for mechanical cell disruption, by using hot solvent. The advantage of rapid extraction may outweigh the possible disadvantages of using hot solvent. When tissues contain a large amount of water, the first extraction with acetone or methanol may remove very little pigment but will effectively dry the tissue and thus allow efficient pigment removal during subsequent extractions. Samples of dried tissue can often be extracted with a water-immiscible solvent such as diethyl ether, but more efficient extraction is usually achieved by first moistening the sample with a little water and then using acetone or methanol in the usual way. Mechanical homogenization can introduce air; this effect may be reduced by addition of an antioxidant or small pieces of solid CO_2 (the latter has the added advantage of cooling the extract), or by directing a gentle stream of N_2 into the homogenizing vessel. The addition of $NaHCO_3$ to neutralize plant tissues containing acid is recommended.

In quantitative work the carotenoid content will eventually be correlated with the amount of tissue taken. The wet weight of tissue may obviously be obtained easily by weighing before extraction. With many tissues, however, especially from plants, the water content may be very variable, e.g., the water content of leaves varies substantially depending on environmental conditions, perhaps by as much as 50–100% during the course of a day. It is thus frequently more reliable to correlate the carotenoid content with some other parameter, cell or tissue dry weight generally being the easiest to determine experimentally.

Many of these points are illustrated in a general procedure for the extraction of leaf carotenoids, given in detail below. This procedure can easily be adapted to suit other tissues or other carotenoids, following the general guidelines summarized above.

Procedure. The clean tissue is rapidly weighed and cut into small pieces. Solid sodium bicarbonate ($\sim$1 g/10 g sample) or solid CO_2 (dry ice) or antioxidant may be added to neutralize acid in the tissues or minimize the risk of oxidation, respectively. The mixture is then homogenized for up to 2 min, in acetone (or methanol or ethanol, if preferred), $\sim$10 ml/g sample. The homogenate is filtered by suction either through a sintered glass funnel or two layers of filter paper, and the solid residue is recovered for further extraction. The extraction procedure is repeated with fresh solvent until no more pigment is removed. Usually three extractions suffice, but for some difficult or high water-content tissues five or six repetitions may be necessary to achieve quantitative pigment removal. The pigment-free solid debris may be used to determine the lipid-free dry weight of the tissue. After removal of sodium bicarbonate if this has been added, by thorough washing with water, the solid debris is dried to constant weight in an oven at 110°. The pigment-containing acetone extracts are combined. In large-scale work the combined extracts are concentrated (rotary evaporator) to one-half to one-third of the original volume; this is not necessary in small scale work. The acetone solution is then mixed with at least an equal volume of freshly redistilled ether in a separating funnel, and water, or preferably sodium chloride solution, is added until two layers separate. The separating funnel should not be shaken vigorously; efficient mixing can be achieved by swirling, with much less risk of forming an emulsion. The use of a sodium chloride solution instead of water also reduces this risk; if an emulsion does form this can usually be resolved easily by adding a little ethanol or solid sodium chloride. The lower aqueous phase is run off and reextracted with ether. The combined ether extracts containing the carotenoids are washed three times with water to remove the original extracting solvent (acetone etc). The ethereal solution may be dried by adding anhydrous sodium sulfate and allowing it to stand for at least 1 hr. The sodium sulfate is then filtered off by suction and washed with fresh dry ether to remove all adsorbed pigment and the combined ether solution is evaporated on a rotary evaporator. If required, the extract may be saponified at this point (see below). Final traces of solvent are removed under a stream of nitrogen. A little absolute ethanol may be added if required, to facilitate removal of any traces of water that may remain, but if this is done it is recommended that the residue be redissolved in hexane or light petroleum and reevaporated to make sure that any residual ethanol is removed before the sample is subjected to chromatography. In quantitative work, the total amount of lipid extracted may be determined by weighing, and the total carotenoid content obtained spectrophotometrically (see below).

Saponification

Alkaline hydrolysis or saponification used to be carried out on carotenoid-containing extracts almost as a matter of routine. It may still be useful both for destroying large amounts of either chlorophyll, which can mask the presence of minor carotenoids in an extract, or neutral lipid (fat, oil) which can make chromatography difficult, and also for hydrolysis of carotenoid esters. Care must always be taken, however, to ensure that saponification has no detrimental effect on the carotenoids present. Thus extracts suspected to contain fucoxanthin or astaxanthin and related compounds containing the 3-hydroxy-4-oxo β-ring must never be saponified in the normal way. If the slightest doubt exists about the advisability of saponification, a small scale trial experiment should be performed in which a small aliquot of the extract is saponified, and the extracts before and after saponification are compared chromatographically (e.g., by TLC or HPLC). Only if no changes are detected in the pigment pattern should saponification be undertaken.

Procedure. The carotenoid-containing extract is dissolved in ethanol and sufficient potassium hydroxide solution (60% w/v) added to bring the final overall KOH concentration to 6–10%. The mixture is then allowed to stand under nitrogen in the dark at or below room temperature for about 12 hr (overnight). An equal volume of diethyl ether is then added, followed by water or NaCl solution until two layers form. The aqueous phase is reextracted with ether and the combined ethereal extracts are washed with water until free from alkali (more ether can be added at any time, should this prove necessary). Care must be taken to avoid emulsions. The saponified extracts are dried and evaporated in the usual way.

Note. It is essential that no trace of acetone is present in the extract to be saponified. Carotenoid aldehydes, e.g., β-citraurin, readily undergo aldol condensation with acetone under alkaline conditions, giving rise to artifacts. Also polymerization of acetone in the presence of alkali produces oils which can become serious impurities and which can reduce the efficiency of subsequent chromatography.

Saponification of Astaxanthin Esters

Although astaxanthin and its esters are normally converted irreversibly into astacene by oxidation under the alkaline conditions of saponification, an anaerobic procedure has recently been described which allows the hydrolysis and minimizes destruction, though some racemization at C-3 and C-3' may occur.[15] The enzymatic hydrolysis of astaxanthin esters, fucoxanthin etc. with cholesterol esterase is also reported to proceed efficiently and to be useful in small scale work.[16]

Separation and Purification

General Features

Extracts usually contain a mixture of different carotenoids. For some purposes, especially quantitative determination of carotenoid composi-

---

[15] R. K. Muller, K. Bernhard, H. Mayer, A. Ruttimann, and M. Vecchi, *Helv. Chim. Acta* **63,** 1654 (1980).
[16] P. B. Jacobs, R. D. LeBoeuf, S. A. McCommas, and J. D. Tauber, *Comp. Biochem. Physiol. B* **B72,** 157 (1982).

tions, it may be sufficient only to separate the individual carotenoids from each other; the presence of colorless contaminants will not interfere with spectrophotometric analysis. In other work, especially when characterization by mass spectrometry, NMR or IR spectroscopy is required, or in studies involving isotopic labeling, purification of carotenoids must be rigorous.

Nonchromatographic Procedures

Some nonchromatographic procedures, formerly used more widely, are still sometimes employed to advantage. Most useful of these are phase separation and methods to precipitate sterols. Phase separation involves the partition of an extract between two immiscible solvent phases of different polarity, usually petroleum ether and aqueous (90%) methanol. Nonpolar compounds, e.g., carotenes, carotene epoxides, xanthophyll esters, are recovered from the petroleum ether epiphase, polar xanthophylls from the lower or hypophase. A refinement of the liquid–liquid partition procedure, countercurrent distribution, has been used to separate individual carotenoids.[17]

Two methods have been used frequently to remove large amounts of sterols from extracts to facilitate subsequent purification of carotenoids. A substantial amount of sterol may precipitate from petroleum ether solution left overnight at $-10°$, but a more efficient procedure involves precipitation of the sterols as their digitonides.

More detailed information about these nonchromatographic procedures can be obtained, if desired, from other articles,[11,14] but they have largely been superseded by the much more efficient chromatographic methods now available.

Carotenoid Chromatography

The technique of chromatography was originally developed to separate plant pigments, including carotenoids, on columns of sugar or starch. Column chromatography remains extremely valuable, especially for preliminary separations and large-scale work, while for smaller samples and delicate separations thin-layer chromatography (TLC) is widely used. Being so easy to detect by their absorption of visible light, carotenoids also lend themselves well to separation, purification and analysis by high-performance liquid chromatography (HPLC). Gas–liquid chromatography is not suitable for carotenoids because of the general instability of the compounds at high temperature, but GLC separation of perhydrogenated

[17] E. C. Grob, H. Pfander, U. Leuenberger, and R. Signer, *Chimia* **25**, 332 (1971).

carotenoids has been used successfully.[14,18] Chromatography on paper impregnated with aluminum oxide or kieselguhr is a valuable analytical technique and details are given elsewhere in this series.[13] Carotenoid HPLC is discussed in detail earlier in this volume,[19] so the present article will consider only the "conventional" procedures of column and thin-layer chromatography.

In general, column and thin-layer chromatography or a combination of the two can be used to separate and purify all carotenoids in an extract. Except in very large or very small scale work, the usual strategy is first to use column chromatography to achieve separation of the extract into fractions containing groups of carotenoids of approximately similar polarity and then TLC to separate the individual pigments in each fraction and bring these to the required state of purity for further study. For very large scale separation and purification it may be more practical to use only column chromatography, for very small scale work only TLC (or HPLC). Preliminary examination of a small sample of the extract by TLC or HPLC will allow a strategy to be devised for the purification (see below).

The choice of adsorbents for carotenoid chromatography is extensive; sucrose, cellulose, starch, $CaCO_3$, $Ca_3(PO_4)_2$, $Ca(OH)_2$, CaO, $MgCO_3$, MgO, $ZnCO_3$, Al_2O_3, silicic acid, silica gel, kieselguhr, microcell C, and various mixtures of these have all been used widely and successfully. The investigator should be aware, however, that some of these materials can cause problems. Thus alumina (Al_2O_3) cannot be used to purify carotenoids containing the 3-hydroxy-4-oxo β-ring (as in astaxanthin) or the 3-hydroxy-4-oxo-2-nor β-ring (of actinioerythrol), even if the 3-hydroxy groups are esterified. Irreversible oxidation to the diosphenol end group (e.g., astacene) or 3,4-dioxo-2-nor end group as in violerythrin occurs very rapidly and the products are virtually impossible to elute from the alumina. Indiscriminate use of alumina, especially combined with exposure to light, can also result in unwanted cis–trans isomerization, though this may perhaps only be detected by very sensitive methods such as HPLC or NMR spectroscopy. With acid-washed alumina and other acidic materials such as silicic acid or silica gel, it is likely that substantial isomerization of carotenoid 5,6-epoxides to the corresponding 5,8-furanoid oxides will occur. Neutral alumina or silica should normally be used. Microcell C is usually considered to be an inert material but has recently been shown[20] to oxidize β-carotene to isocryptoxanthin in good yield

[18] R. F. Taylor and M. Ikawa, this series, Vol. 67, p. 233.
[19] M. Ruddat and O. H. Will, this volume [7].
[20] D. B. Rodriguez, Y. Tanaka, T. Katayama, K. L. Simpson, T.-C. Lee, and C. O. Chichester, *J. Agric. Food Chem.* **24**, 819 (1976).

under certain conditions, and must therefore be used with caution. Finally if solvent mixtures containing acetone are used with basic adsorbents, especially MgO, which have not been diluted by mixing with an inert filter aid, e.g., Celite, polymerization of the acetone may occur, leading to contamination of the recovered carotenoid samples.

It is important to realize that different adsorbent materials achieve separation in different ways. Thus the commonly used alumina and silica will separate compounds of different polarity, i.e., compounds containing the most polar substituent groups are most strongly adsorbed. Separation on materials such as MgO, $Ca(OH)_2$, etc. is determined by the number and type of double bonds in the molecule. Carotenoids with the most extensive conjugated double bond system are most strongly adsorbed on MgO, e.g., lycopene > neurosporene > ζ-carotene. Acyclic carotenoids are much more strongly held than cyclic ones having the same number of double bonds, and the position of the ring double bonds is also important, β-ring compounds being more strongly adsorbed than the corresponding ε-ring isomers, (e.g., β,β-carotene > β,ε-carotene). Polarity, i.e., the presence of OH groups etc., is of less importance. Chromatography on MgO is very useful for removing colourless noncarotenoid impurities, which are only very weakly adsorbed, from carotenoid samples. $Ca(OH)_2$, $ZnCO_3$ will readily separate carotenoid cis–trans isomers that may be difficult or impossible to separate on other materials. It is recommended that the purification of any carotenoid should include chromatography at least once on each kind of adsorbent material, i.e., separation on the basis of polarity and double bond arrangement.

An unfamiliar extract should be subjected to preliminary investigation by TLC of a small sample on silica gel G. A single spot of the extract is chromatographed first in petroleum ether to determine if any carotene hydrocarbons are present. The same plate is then run successively in 5, 25, and 50% diethyl ether in petroleum ether and finally in ether; carotene epoxides and xanthophyll esters, monooxocarotenoids, dioxo- and monohydroxycarotenoids, and xanthophylls containing at least two hydroxy groups will migrate in these solvents, respectively. From these results a detailed strategy is devised for the separation and purification of all the carotenoids in the extract.

Column Chromatography

Early chromatographic separations of carotenoids employed columns of sugar or starch. These materials can still be useful for the separation of very polar substances, and have the advantage of being completely inert. Many different adsorbents have since been used for column chromatography of carotenoids and may be particularly valuable in certain specialized applications. Thus MgO columns are very good for separating complex mixtures of carotenes and biosynthetic intermediates present in tomatoes and other tissues, Ca(OH)$_2$ and ZnCO$_3$ for cis–trans isomers of xanthophylls.

Column Chromatography on Alumina, Silicic Acid. Column chromatography is now most widely used for preliminary separation of extracts into fractions containing compounds of similar polarity. Alumina is the adsorbent usually selected, but silica (silicic acid, neutralized if necessary) is used if the presence of astaxanthin or similar compounds is suspected. Fractions are eluted successively with solvents of increasing polarity as determined by preliminary TLC examination (see above). Neutral alumina deactivated to Brockmann Grade III is most suitable since there is good correlation between the solvents needed to elute substances from this adsorbent and for TLC on silica gel.

Procedure. The alumina is first deactivated to Brockmann Grade III.[21] This can be achieved simply and satisfactorily by weighing the appropriate amount of neutral alumina (10 g/100–200 mg lipid for small scale work, but larger lipid : adsorbent ratios are suitable for large scale separations) into a beaker or flask, covering with dry, redistilled petroleum ether, pipetting in water (0.6 ml/10 g alumina), and stirring gently for 3–4 min to achieve even distribution of water. The slurry of alumina in petrol is then used to pack the column. For best separations the length of the column of adsorbent should be 10–20 times its width.

It is frequently advantageous to add a 1 cm layer of clean sand to the top of any column. This helps to prevent the column from being blocked if the concentrated extract applied should precipitate or if any carotenoid crystallizes. The sand layer can be agitated if necessary to restore solvent flow without the actual column being disturbed.

A solution of the lipid extract in the minimal volume of petroleum ether is applied to the column and elution is continued in petroleum ether. Extracts containing a large amount of carotenes, especially lycopene, may need to be dissolved in a little toluene and petroleum ether then added to give an overall toluene concentration not greater than 20%. This solution is applied to the column and elution continued with 20% toluene in petroleum ether. When this first eluate becomes colorless, i.e., when all the carotenes have been collected, elution is continued with diethyl ether–petroleum ether mixtures of increasing diethyl ether content, and suitable fractions are collected. When a certain solvent mixture is used to elute fractions containing only colorless substances that might otherwise

---

[21] H. Brockmann and H. Schodder, *Ber. Dtsch. Chem. Ges.* **74,** 73 (1941).

TABLE I
SOLVENTS SUITABLE FOR ELUTING VARIOUS CAROTENOID GROUPS
FROM COLUMNS OF NEUTRAL ALUMINA (BROCKMANN GRADE III)

Solvent[a]	Compounds eluted
P, 1% E/P or 10% T/P	Carotene hydrocarbons
5% E/P	Carotene epoxides
10% E/P	Xanthophyll esters
20% E/P	Monomethoxy- and dimethoxycarotenoids
	Monooxocarotenoids
	[Sterols and triterpenoids]
50% E/P	Monohydroxycarotenoids
60% E/P	Dioxocarotenoids
E or 5% EtOH/E	Dihydroxy- and trihydroxycarotenoids
	Xanthophyll epoxides
20% EtOH/E	Tetrahydroxycarotenoids
	Carotenoid glycosides

[a] P, Petroleum ether; E, diethyl ether; T, toluene; EtOH, ethanol.

contaminate subsequent carotenoid fractions, 100 ml of eluate should be collected for each 10 g of alumina in the column. Otherwise elution is continued with each solvent until any moving colored band has been collected completely. During the chromatography, the column must be protected from light by enclosing it in a sheath of aluminum foil, black cloth, or black paper which can easily be opened or removed to allow brief inspection. Table I summarizes the series of solvent mixtures used in a typical experiment, and the classes of compound eluted from the column by each mixture.

Note. Polar xanthophylls do not dissolve readily in petroleum ether. If these are present it may be necessary to dissolve the extract in a little diethyl ether and then dilute the sample with petroleum ether to give an overall diethyl ether concentration not greater than 5%. The first fraction collected in this case is eluted with 5% diethyl ether in petroleum ether, and will contain the carotenes but also may have carotene epoxides and noncarotenoid impurities. This fraction can, if required, be evaporated and rechromatographed, in petrol, on a second alumina column, if it contains too much material for direct TLC to be practicable.

This kind of preliminary separation of extracts into fractions can also be achieved with columns of silicic acid, preferably neutralized to pH 7. The same solvent mixtures are applicable as for the activity grade III alumina columns.

Separation of Carotene cis–trans Isomers on Alumina. While grade III alumina is useful for the isolation of a total carotene fraction and,

TABLE II
SEPARATION OF CAROTENE cis–trans ISOMERS ON
ACTIVATED (GRADE I) ALUMINA COLUMNS

Compound	Eluting solvent
15-*cis*-Phytoene	2% E/P[a]
trans-Phytoene	4% E/P
15-*cis*-Phytofluene	10% E/P
cis-α- and β-Carotenes	15–20% E/P
15-*cis*-ζ-Carotene	25–30% E/P
trans-α- and β-Carotenes	30–40% E/P
trans-β-Zeacarotene	50% E/P
trans-ζ-Carotene	60–100% E/P
trans-Neurosporene	0–5% A/E
trans-Lycopene	10–20% A/E

[a] A, Acetone; E, diethyl ether; P, petroleum ether.

under controlled conditions, gives some separation of β-carotene and lycopene, fully activated (Grade I) alumina allows the efficient separation not only of a range of different carotenes but also of their cis and trans isomers. It is, for example, the best conventional chromatographic method available for separating the biosynthetically important 15-*cis*- and all-*trans*-phytoene (eluted with 2 and 4% ether in petroleum ether respectively). The power of the method is illustrated by the separation of carotenes from a mutant strain of the green alga *Scenedesmus obliquus*.[22] The carotene mixture is applied to the activated alumina column in petrol and the individual compounds are eluted with the solvent compositions listed in Table II. The separation can be adapted to a shallow ether–petroleum ether solvent gradient and made semiautomatic.

Column Chromatography on Other Adsorbents. With many other adsorbents, e.g., starch, sugar, cellulose, MgO, $CaCO_3$, the method of zone chromatography is frequently used. These materials may be packed dry or as a slurry (in petroleum ether), but their small particle size results in a tightly packed column with slow flow rate. A diatomaceous filter aid, e.g., Hyflo-Super-Cel or Celite 545 (30–60% by weight) is often added to improve the flow rate. The stepwise elution procedure, as used for alumina, can also be used with these columns, but it is common to use a longer column, to separate a mixture into zones on this, and then to extrude the column of adsorbent while still damp with solvent, cut out the individual zones and elute the pigment with suitable solvent of greater polarity. For

[22] R. Powls and G. Britton, *Arch. Microbiol.* **115**, 175 (1977).

further details see ref. 14 and ref. 12 which outlines the separation of the carotenes of tomatoes by this method on a column of MgO.

Thin-Layer Chromatography (TLC)

Since its introduction in the late 1950s, TLC has been the most versatile and effective method for purifying carotenoids, and remains extremely useful even in laboratories with good HPLC facilities. TLC is used both for separation and purification and for partial identification of carotenoids by comparison with authentic samples. The most extensively used adsorbent is silica gel (kieselgel), and mixtures based upon magnesium oxide and other basic oxides, hydroxides, and carbonates are also most valuable. A general strategy that will normally yield a pure carotenoid sample is to chromatograph a carotenoid-containing fraction on silica gel G in the same solvent as was used to elute that fraction from the preliminary alumina column. Each individual band obtained from silica gel is then rechromatographed on MgO–kieselguhr G or a related adsorbent mixture, and final purification is achieved by further TLC on silica gel or alumina-G in a different solvent.

TLC on Silica Gel. Silica gel is the most widely used adsorbent in all TLC, including that of carotenoids. The separations obtained depend on polarity, the most polar carotenoids being most strongly adsorbed. Diethyl ether (E)–petroleum ether (P) mixtures are very suitable for most carotenoid separations. Thus for carotenes petroleum ether alone, 1% E/P, or 10% toluene in petroleum ether gives satisfactory results, for carotene epoxides 5% E/P, for xanthophyll esters 10% E/P, for monooxocarotenoids 30% E/P, for monohydroxy and dioxocarotenoids 50–60% E/P, and for dihydroxy-tetrahydroxycarotenoids, xanthophyll epoxides, astaxanthin, etc., diethyl ether alone. The very polar carotenoid glycosides can be separated with 5–10% ethanol in ether, or better, ethyl acetate–2-propanol–water (60:40:2). Good carotenoid separations are also obtained with acetone-petroleum ether mixtures. Elution is achieved with ether or acetone, a little ethanol being added if necessary. TLC on silica gel is not successful in separating ε- and β-ring carotenoid isomers, e.g., α- and β-carotenes, lutein and zeaxanthin, or 5,6-epoxycarotenoids from the corresponding 5,8-epoxides, but these can be resolved on MgO (see below). After a second TLC, on MgO, each carotenoid is further purified on silica gel G, this time with a different solvent mixture. Methanol–toluene, 2-propanol–petroleum ether, and carbon tetrachloride–ethyl acetate mixtures are suitable. Details may be obtained from other articles.[11,12,14]

Workers should be aware of the apparently anomalous behavior of some carotenoids on silica gel. A commonly encountered example is that of astaxanthin which has two hydroxy and two oxo groups and yet is less strongly adsorbed than zeaxanthin which has only two hydroxy groups. The behavior of carotenoids containing a 5,6-dihydroxy-5,6-dihydro β-ring is particularly unusual. These compounds are similar in polarity to carotenes, and totally unlike any other diol structures.[23]

It may be very difficult to resolve mixtures of a conventional xanthophyll, e.g., zeaxanthin or astaxanthin, and its 7,8-dehydro (acetylenic) analog, adsorption via the hydroxy groups being so strong that the very weak effect of the acetylene group is negligible. In such a case blocking the hydroxy groups by acetylation renders the small effect of the difference in saturation significant and allows separation.

TLC on Other Adsorbents. Magnesium oxide and related materials [e.g., $Ca(OH)_2$, $ZnCO_3$, $CaCO_3$] do not adhere well, if at all, to glass plates, to form stable thin layers. It is usual to include a binder; calcium sulfate, silica gel, kieselguhr, and starch have all been used for this purpose. We find a 1:1 mixture of MgO and kieselguhr G especially useful. Adsorption on MgO seems to depend mainly on the number and nature of double bonds in the molecule. In general the strength of adsorption increases with the length of the linear conjugated polyene chromophore. Acyclic carotenoids are adsorbed more strongly than cyclic ones, and β-ring compounds more strongly than the corresponding ε-ring isomers. This affinity for double bonds far outweighs the effect of simple polarity, so that the acyclic hydrocarbon lycopene is very much more strongly bound than the bicyclic diol zeaxanthin. MgO is extremely valuable for resolving mixtures such as α- and β-carotenes or lutein and zeaxanthin which are inseparable on silica gel, and also for purifying 5,6-epoxycarotenoids which are much less strongly held than the isomeric 5,8-epoxides. TLC on MgO is also a good method for removing colorless noncarotenoid impurities. If a developing solvent is used such that the carotenoid leaves the origin but runs with a low R_f value, colorless impurities will usually run virtually with the solvent front and are thus easily removed. TLC on MgO–kieselguhr does have some disadvantages, notably that substances tend to run as rather broad, tailing bands, especially when the chromatogram is overloaded (which happens very easily). Carotenoids must also be eluted from MgO layers as rapidly as possible, preferably before the solvent has fully evaporated, otherwise some pigment may be bound irreversibly, and lost. The usual solvents for TLC on MgO–kieselguhr G are mixtures of petroleum ether with acetone and/or toluene. Examples of the solvent compositions suitable for some carotenoids are

[23] C. H. Eugster, *in* "Carotenoids—5 (Madison, 1978)" (T. W. Goodwin, ed.), p. 463. Pergamon, Oxford, 1979.

TABLE III
SOLVENTS SUITABLE FOR TLC OF CAROTENOIDS ON
MgO : KIESELGUHR G (1 : 1)

Carotenoids	Solvent[a]
Phytoene	P[b]
Phytofluene, α-carotene, β-carotene	2–6% A/P or 10% T/P
ζ-Carotene	6–8% A/P or 10–15% T/P
α- and β-Zeacarotene	10–15% A/P or 30% T/P
δ- and γ-Carotene	
Lycopene, spirilloxanthin	A/T/P (2 : 2 : 1)
Violaxanthin, lutein, lutein epoxide	20% A/P
Zeaxanthin	25% A/P
Neoxanthin	30% A/P
Spheroidene, spheroidenone, hydroxyspheroidene	A/T/P (1 : 1 : 1)

[a] Different batches of MgO may vary in activity. The solvent compositions listed are usually suitable, but any new batch of MgO should be checked with standard compounds and the solvent compositions adjusted accordingly by increasing or decreasing the percentage of A or T.

[b] A, Acetone; P, petroleum ether; T, toluene.

listed in Table III. Detailed examples of carotenoid purifications, including MgO TLC, are given elsewhere.[11,12] Elution from MgO is best achieved with ether or acetone, plus ethanol or toluene if required for strongly adsorbed hydroxy- and 5,8-epoxycarotenoids, or long chromophore acyclic carotenoids, respectively.

The more complex system of Hager[24] [CaCO$_3$–MgO–Ca(OH)$_2$, 30 : 6 : 4] is very versatile, gives good separations, and is also recommended. Partition chromatography of carotenoids by reversed-phase TLC on kieselguhr, silica gel G, or cellulose layers impregnated with liquid paraffin or vegetable oil has been used to separate polar xanthophylls and xanthophyll esters. More details of these procedures and of an interesting system achieving separation on polyamide are given in ref. 14.

Some Practical Points

Precautions against Oxidation. It is while they are adsorbed on thin layers of active adsorbents that carotenoids are most vulnerable to the destructive effects of light and oxygen so it is vital that all relevant precautions should be observed to keep pigment losses as low as possible. In particular, samples should be applied to the chromatogram as rapidly as

[24] A. Hager and H. Stransky, *Arch. Mikrobiol.* **72,** 68 (1970).

possible and removed and eluted immediately the thin-layer plate is removed from the solvent tank. In order to reduce to a minimum the risk of oxidizing especially labile compounds, a gentle stream of nitrogen may be directed onto the chromatogram as the sample is being applied, and the developing tank flushed with nitrogen and sealed to provide an inert atmosphere during development, which is always carried out in the dark.

Precautions against Acid and Base. Unless the manufacturer specifies that it has been neutralized to pH 7, silica gel is usually acidic enough to cause some isomerization of carotenoid 5,6-epoxides, and should therefore be neutralized during preparation of the thin-layer plates by using a slurry of the silica gel in dilute ($\sim$0.05 M) KOH, saturated $NaHCO_3$ or pH 7 buffer instead of simply water. For astaxanthin and related compounds better separation is obtained, with less risk of oxidation, if citric acid is included during preparation of the plates ($\sim$pH 4).

Location of Carotenoids. Obviously since most carotenoids are strongly colored they can easily be located on thin-layer chromatograms. The colorless phytoene and phytofluene, however, must be located by examination of the developed chromatogram under UV light, phytoene appearing as a weakly violet zone and phytofluene fluorescing intensely greenish white. Phytoene is more readily detected on silica gel G F_{254} plates, when it quenches the fluorescence of the incorporated inorganic dye. For other adsorbents the developed chromatogram is sprayed with a solution of Rhodamine 6G in acetone and viewed under UV light; phytoene again quenches the fluorescence. Exposure to UV light should be as short as possible.

Elution. The individual carotenoid zone is rapidly scraped from the plate and the carotenoid eluted with a solvent more polar than that used for development. The adsorbent is removed by filtration or centrifugation. Small sintered glass funnels or filter sticks may be used for rapid removal of adsorbent by filtration under suction. A simple but effective method is to place the adsorbent in a small glass funnel with a small plug of absorbent cotton wool in the stem. Solvent is added dropwise until the top surface of the adsorbent is colorless, and elution continued until all the carotenoid has been removed. A few drops of ethanol applied at an early stage in the elution will usually allow even polar carotenoids to be recovered in a very small volume. If the method of mixing with solvent and centrifuging is used to remove the adsorbent, the process must be repeated 2–3 times to ensure quantitative elution.

Cochromatography. Together with determination of the light absorption spectrum, cochromatography with an authentic sample is a first step in the identification of a carotenoid. Cochromatographic comparison should be sought in at least two different systems, e.g., silica gel and MgO.

Ready-made TLC Plates. A wide range of ready-prepared TLC plates is now available. The usual adsorbent is silica gel coated onto glass, plastic, or metal foil. The layers are extremely uniform and strong; they can be written on in pencil without risk of disturbing the adsorbent layer. Although expensive, these plates can be cut to any size and resolution and reproducibility are extremely good.

Purification of Samples for MS, NMR

The procedures described above will give a carotenoid sample free from other contaminating carotenoids and suitable for normal light absorption spectroscopic and spectrophotometric analysis. Substantial amounts of impurities originating from the chromatographic adsorbents and solvents, however, can easily be introduced into the sample during the purification unless extreme care is taken. Such contaminants must obviously be excluded or removed before the sample is submitted for MS or NMR analysis. Purification of solvents to be used for preparing these samples must be extremely rigorous (see earlier) and TLC plates should be prewashed by running in solvent at least as polar as that to be used in the chromatography. If possible, samples should be eluted from TLC adsorbents with ether. Acetone or ethanol frequently has to be used, however, but since they are most likely to extract contaminants, only the minimum amounts of these solvents should be used. Carotenoid samples from TLC should be given an additional purification immediately before being subjected to MS or NMR analysis. A satisfactory procedure is to filter the sample through a small column of neutral alumina (3–4 cm in a Pasteur pipet is usually sufficient), the activity of which has been adjusted by addition of water so that the carotenoid will be eluted with ether or ether–petroleum ether mixtures. The sample is applied to the column in petroleum ether, and elution is begun with a solvent not sufficiently polar to remove the carotenoid. This eluate is discarded, and the carotenoid is then eluted with a more polar solvent mixture, evaporated, and, if intended for NMR analysis, dried under vacuum to remove traces of solvent.

Note. For astaxanthin and related 3-hydroxy-4-oxo β-ring carotenoids, a small column of silicic acid is used, since alumina would cause oxidation and irreversible adsorption.

Purification of Radioactive Samples

It can be extremely difficult to purify radioactively labeled carotenoids, especially when tritium is the isotope present. Small amounts (mass) of colorless impurity may be undetectable in the strongly colored carotenoid sample, but such impurities can be highly radioactive and lead to erroneous results when the carotenoid is radioassayed. It is recommended that, whenever possible, a derivative be made, purified, and radioassayed before the radioactivity can confidently be assigned to the carotenoid. Simple procedures that are suitable include acetylation or silylation of hydroxy compounds, $NaBH_4$ reduction of ketones, acid-catalysed isomerization of 5,6-epoxides, and formation of cis isomers.[12] It is highly unlikely that any contaminant would give a derivative that cannot be separated from the carotenoid derivative.

Tritium-labeled carotenoids present another problem. If a small amount of a high specific activity ^{3}H-labeled carotenoid is mixed, for example in acetone solution, with another unlabeled carotenoid, and the unlabeled substance is then purified, this is likely to retain considerable tritium radioactivity tenaciously, even after three TLC purification steps—silica, MgO, and silica. With mixtures of carotenoids that are closely related structurally, e.g., [^{3}H]β-carotene and canthaxanthin, [^{3}H]lycopene and rhodopin, at least six different chromatographic steps may be necessary to remove all the contaminating radioactive carotenoid. Workers wishing to demonstrate the incorporation of a ^{3}H-labeled carotenoid into another carotenoid must therefore perform stringent control experiments, utilize extremely rigorous purification, and exercise caution in interpreting their results. Even when HPLC is used, repeated purification is necessary and different stationary and mobile phases should be employed.

Spectroscopic Analysis

Light Absorption Spectroscopy

The chromatographic purification of a carotenoid will give useful information (polarity etc.) to help in its identification. After this, the first step in the characterization of a carotenoid is the determination of its light absorption spectrum. All carotenoids absorb light strongly in the visible, or in some cases the UV region of the spectrum, and both the position of the absorption maximum and the shape or fine structure of the spectrum hold valuable information about the carotenoid structure.

Position of the Absorption Maxima. The absorption spectra of most carotenoids exhibit three maxima. The positions of these maxima are influenced by the solvent in which the spectrum is determined, but in any given solvent they are characteristic of the chromophore of the individual pigment. The wavelength of maximum absorption (λ_{max}) is mainly a function of the length of the conjugated double-bond chromophore of the compound, and this wavelength increases as the number of conjugated double bonds increases (Table IVa). Thus phytoene and phytofluene, with three and five conjugated double bonds respectively, absorb in the UV and are colorless whereas any carotenoid with a chromophore of at least seven conjugated double bonds will absorb light in the visible region and therefore be colored. Cyclization affects the absorption spectrum, and a cyclic carotene in which the conjugation extends into the ring (β-ring) will normally have its absorption maximum at shorter wavelengths than the acyclic pigment with the same number of conjugated double bonds (cf. lyocpene, γ-carotene, (β,ψ-carotene) and β,β-carotene, Table IVb). The nonconjugated double bonds of the γ- and ε-rings do not contribute to the chromophore; thus γ,γ-carotene and ε,ε-carotene are essentially linear conjugated nonaenes with absorption spectra identical to that of neurosporene.

Substituents such as hydroxy and methoxy groups and nonconjugated oxo groups do not affect the chromophore and therefore have little or no effect on λ_{max}. Thus for example β-carotene, β-cryptoxanthin (β,β-caroten-3-ol), and zeaxanthin (β,β-carotene-3,3'-diol) have virtually identical absorption spectra with λ_{max} at 428, 450, and 476–478 nm (in ethanol). A carbonyl group in conjugation with the double bond system (e.g., at C-4 of a β-ring) effectively extends the chromophore. Thus echinenone (β,β-caroten-4-one) and canthaxanthin (β,β-carotene-4,4'-dione) absorb at 461 and 474 nm, respectively (in EtOH). The presence of the conjugated carbonyl group also alters the shape of the spectrum (see below). A list of the wavelengths of maximum absorption of some of the most common carotenoids in different solvents is given in Table V and more extensive compilations are available elsewhere.[10,11,14] In order to be sure of the accuracy of λ_{max} values recorded, the spectrophotometer should be calibrated by superimposing on the spectrum of the sample the appropriate bands from a holmium oxide filter (279.4, 287.5, 333.7, 360.9, 418.4, 453.2, and 536.2 nm).

Spectra of xanthophylls are usually determined in ethanol, those of carotenes in petroleum ether or hexane. The wavelengths of maximal absorption are markedly solvent dependent; Table IVc gives the λ_{max} of lycopene in a range of solvents. Because of this solvent dependence care must be taken to remove all traces of acetone, benzene, chloroform, dichloromethane, pyridine, etc. from samples whose spectra are to be determined in petroleum ether or ethanol, otherwise abnormally high absorption maxima may be obtained.

Carotenoid spectra are greatly affected by the presence of water in water-miscible solvents. At high water concentrations (30–50% in ethanol or acetone) the normal absorption maxima may be largely replaced by a new peak in the near-UV region, e.g., at 370 nm for lutein, 380 nm for

TABLE IV
EFFECTS OF VARIOUS STRUCTURAL FEATURES AND SOLVENTS ON THE LIGHT ABSORPTION SPECTRAL MAXIMA OF CAROTENOIDS

a. Effect of increasing the length of the conjugated double bond chromophore

Compound	Chromophore length (conjugated double bonds)	λ_{max} (nm) in petroleum ether
Phytoene	3	276,286,297
Phytofluene	5	331,348,367
ζ-Carotene	7	378,400,425
Neurosporene	9	414,439,467
Lycopene	11	444,470,502
3,4-Didehydrolycopene	13	465,492,527
3,4,3',4'-Tetradehydrolycopene	15	480,510,540

b. Effect of rings

Compound	Number of rings	Number of conjugated double bonds		λ_{max} (nm) in petroleum ether
		Polyene chain	Ring	
Lycopene	0	11	0	444,470,502
γ-Carotene	1	10	1	437,462,494
δ-Carotene	1	10	0	431,456,489
β-Carotene	2	9	2	425,449,476
α-Carotene	2	9	1	422,444,473
ε-Carotene	2	9	0	416,440,470

c. Effect of solvent on the absorption maxima of lycopene

Solvent	λ_{max} (nm)
Petroleum ether	444,470,502
Hexane	447,472,504
Diethyl ether	446,471,503
Ethanol	446,472,503
Acetone	448,474,505
Chloroform	458,484,518
Toluene	457,484,519
Benzene	455,487,522
Pyridine	462,490,525
Carbon disulfide	477,508,548

TABLE V
Light Absorption Maxima and Specific Absorbance Coefficients ($A_{1\,cm}^{1\%}$) of Some Carotenoids

Carotenoid [semisystematic name]	λ_{max} (nm)			Solvent	$A_{1\,cm}^{1\%}$
Actinioerythrin	470	496	529	P[a]	
[3,3'-dihydroxy-2,2'-dinor-β,β-carotene-4,4'-dione 3,3'-diacylate]	480	508	538	A	
		518		C	
Adonirubin		466		H	
[3-hydroxy-β,β-carotene-4,4'dione]		478		E	
		479		C	
Adonixanthin		460		P	
[3,3'-dihydroxy-β,β-caroten-4-one]		465		E	
Aleuriaxanthin	434	*460*	491	P	2600[b]
[1',16'-didehydro-1',2'-dihydro-β,ψ-caroten-2'-ol]	440	*463*	493	A	2440
Alloxanthin	(423)	450	479	P	
[7,8,7',8'-tetradehydro-β,β-carotene-3,3'-diol]	(427)	450	478	E	
	436	460	489	C	
Anhydrorhodovibrin	454	*482*	515	P	2700
[1-methoxy-3,4-didehydro-1,2-dihydro-ψ,ψ-carotene]	459	488	522	A	
	471	499	533	C	
Antheraxanthin	422	445	472	P	
[5,6-epoxy-5,6-dihydro-β,β-carotene-3,3'-diol]	422	444	472	E	
	430	456	484	C	
8'-Apo-β-caroten-8'-al		*457*		P	2640
		463		E	
		477		C	
10'-Apo-β-caroten-10'-al		*435*		P	2190
		451		E	
		462		C	
12'-Apo-β-caroten-12'-al		*414*		P	2160
8'-Apo-β-caroten-8'-oic acid ethyl or methyl ester		*445*	470	P	2500
Astacene		473		P	
[3,3'-dihydroxy-2,3,2',3'-tetrahydro-β,β-carotene-4,4'-dione]		478		E	
		482		A	
		494		C	
		498		Py	1690
Astaxanthin		468		P	
[3,3'-dihydroxy-β,β-carotene-4,4'-dione]		478		E	
		480		A	
		485		C	
Auroxanthin	380	400	425	P	
[5,8,5',8'-diepoxy-5,8,5',8'-tetrahydro-β,β-carotene-3,3'-diol]	379	*400*	425	E	1850
	385	413	438	C	
Azafrin		*409*		P	2200
[5,6-dihydroxy-5,6-dihydro-10'-apo-β-caroten-10'-oic acid]		426	447	C	

TABLE V (*Continued*)

Carotenoid [semisystematic name]	λ_{max} (nm)			Solvent	$A_{1\,cm}^{1\%}$
Bacterioruberin	464	*494*	528	P	2350
[2,2'-bis(3-hydroxy-3-methylbutyl)-3,4,3',4'-tetrade-	468	*498*	532	A	2540
hydro-1,2,1',2'-tetrahydro-ψ,ψ-carotene-1,1'-diol]	475	506	544	C	
Bisanhydrobacterioruberin	465	493	527	P	
[2,2'-bis(3-methylbut-2-enyl)-3,4,3',4'-tetradehydro-1,2,1',2'-tetrahydro-ψ,ψ-carotene-1,1'-diol]	471	498	533	A	
3,4,3',4'-Bisdehydro-β-carotene		*471*		P	2400
[3,4,3',4'-tetradehydro-β,β-carotene]					
Bisdehydrolycopene	480	510	540	P	
[3,4,3',4'-tetradehydro-ψ,ψ-carotene]	493	528	567	C	
Bixin	432	*456*	490	P	4200
[methyl hydrogen 9'-*cis*-6,6'-diapocarotene 6,6'-dioate]	443	470	502	C	
Caloxanthin	426	449	475	E	
[β,β-carotene-2,3,3'-triol]	432	458	484	C	
Canthaxanthin		*466*		P	2200
[β,β-carotene-4,4'-dione]		474		E	
		482		C	
Capsanthin	450	475	505	P	
[3,3'-dihydroxy-κ,κ-caroten-6'-one]	460	*483*	518	B	2072
Capsorubin	445	479	510	P	
[3,3'-dihydroxy-κ,κ-carotene-6,6'-dione]	460	*489*	523	B	2200
α-Carotene	422	*444*	473	P	2800
[β,ε-carotene]	423	444	473	E	
	424	448	476	A	
	433	457	484	C	
β-Carotene	425	*449*	476	P	2592
[β,β-carotene]		450	476	E	2620
	(429)	452	478	A	
	435	*461*	485	C	2396
β-Carotene 5,6-epoxide		447	478	P	
[5,6-epoxy-5,6-dihydro-β,β-carotene]	423	*444*	473	H	2590
		459	492	C	
β-Carotene 5,6,5',6'-diepoxide	417	*440*	468	H	2690
[5,6,5',6'-diepoxy-5,6,5',6'-dihydro-β,β-carotene]	418	442	471	E	
	422	448	477	C	
β,γ-Carotene	421	444	472	P	
γ,γ-Carotene	414	438	468	P	
γ-Carotene	437	*462*	494	P	3100
[β,ψ-carotene]	440	460	489	E	
	439	461	491	A	
	446	475	509	C	
δ-Carotene	431	*456*	489	P	3290
[ε,ψ-carotene]	440	470	503	C	

(*continued*)

TABLE V (*Continued*)

Carotenoid [semisystematic name]	λ_{max} (nm)			Solvent	$A_{1\,cm}^{1\%}$
ε-Carotene	416	*440*	470	P	3120
[ε,ε-carotene]	417	440	470	E	
ζ-Carotene	378	400	425	P	
[7,8,7′,8′-tetrahydro-ψ,ψ-carotene]	380	*400*	425	H	2555
	377	399	425	E	
Unsymmetrical ζ-carotene	374	395	419	P	
[7,8,11,12-tetrahydro-ψ,ψ-carotene]	374	395	420	H	
Chlorobactene	435	461	491	P	
[φ,ψ-carotene]	450	476	508	B	
Chloroxanthin	417	440	470	H	
[1,2,7′,8′-tetrahydro-ψ,ψ-caroten-1-ol]					
Chrysanthemaxanthin, flavoxanthin		421	450	P	
[5,8-epoxy-5,8-dihydro-β,β-carotene-3,3′-diol]	400	*421*	448	E	2100
		430	459	C	
β-Citraurin		457	487	P	
[3-hydroxy-8′-apo-β-caroten-8′-al]		450	476	H	
C.p. 450	425	450	478	P	
[2,2′-bis(4-hydroxy-3-methylbut-2-enyl)-β,β-carotene]	427	454	481	A	
	422	450	476	E	
Crocetin	400	422	*450*	P	4320
[8,8′-diapocarotene-8,8′-dioic acid]	401	423	447	E	
	413	435	462	C	
Crustaxanthin	423	448	475	H	
[β,β-carotene-3,4,3′,4′-tetrol]	425	450	477	E	
		457	482	C	
α-Cryptoxanthin	421	*445*	475	H	2636
[β,ε-caroten-3-ol]	434	456	485	C	
β-Cryptoxanthin	425	*449*	476	P	2386
[β,β-caroten-3-ol]	428	450	478	E	
Decaprenoxanthin	414	439	469	P	
[2,2′-bis(4-hydroxy-3-methylbut-2-enyl)-ε,ε-carotene]	416	440	470	E	
3,4-Dehydro-β-carotene		*461*		P	2330
[3,4-didehydro-β,β-carotene]					
3,4-Dehydrolycopene	465	*492*	527	P	3000
[3,4-didehydro-ψ,ψ-carotene]	465	492	525	E	
	481	514	551	B	
Diadinoxanthin	421	445	475	H	
[5,6-epoxy-7′,8′-didehydro-5,6-dihydro-β,β-carotene-3,3′-diol]	424	445	474	E	
	432	455	482	C	
Diatoxanthin	425	449	475	E	
[7,8-didehydro-β,β-carotene-3,3′-diol]	433	458	486	C	
7,8-Didehydroastaxanthin		474		E	
[3,3′-dihydroxy-7,8-didehydro-β,β-carotene-4,4′-dione]	478	495	522	Py	

TABLE V (Continued)

Carotenoid [semisystematic name]	λ_{max} (nm)			Solvent	$A_{1\,cm}^{1\%}$
2,2'-Dihydroxy-β-carotene	425	*452*	480	A	2060
[β,β-carotene-2,2'-diol]					
α-Doradecin		455	471	P	
[3,3'-dihydroxy-2,3-didehydro-β,ε-caroten-4-one]		457	473	E	
β-Doradecin		461	475	P	
[3,3'-dihydroxy-2,3-didehydro-β,β-caroten-4-one]		465		E	
α-Doradexanthin		455	470	P	
[3,3'-dihydroxy-β,ε-caroten-4-one]		458	478	E	
β-Doradexanthin (=adonixanthin)					
Echinenone		*458*	482	P	2158
[β,β-caroten-4-one]		461		E	
		460		A	
		471		C	
Eschscholtzxanthin	444	*472*	502	H	3269
[4',5'-didehydro-4,5'-*retro*-β,β-carotene-3,3'-diol]	446	472	503	E	
	448	474	505	A	
	457	481	511	C	
Fucoxanthin	425	446	473	P	
[5,6-epoxy-3,3',5'-trihydroxy-6',7'-didehydro-	426	449	475	E	
5,6,7,8,5',6'-hexahydro-β,β-caroten-8-one 3'-acetate]		460	478	C	
2-Hydroxy-β-carotene	430	*452*	479	A	2290
[β,β-caroten-2-ol]					
3-Hydroxyechinenone		457		P	
[3-hydroxy-β,β-caroten-4-one]		460		E	
		472		C	
Hydroxyspheroidene	429	454	486	H	
[1'-methoxy-3',4'-didehydro-1,2,7,8,1',2'-hexahydro-	440	466	500	C	
ψ,ψ-caroten-1-ol]					
Hydroxyspheroidenone	460	483	516	H	
[1'-hydroxy-1-methoxy-3,4-didehydro-1,2,1',2',7',8'-		487		E	
hexahydro-ψ,ψ-caroten-2-one]		501		C	
Isocryptoxanthin	425	448	475	P	
[β,β-caroten-4-ol]	425	450	477	E	
	438	464	489	C	
Isorenieratene, leprotene	426	*448*	475	P	2080
[φ,φ-carotene]	428	460	495	C	
Isozeaxanthin	427	*450*	475	P	2400
[β,β-carotene-3,3'-diol]	430	451	478	E	
	435	463	489	C	
Loroxanthin	425	446	473	E	
[β,ε-carotene-3,19,3'-triol]	431	455	482	C	
Lutein	421	445	474	P	
[β,ε-carotene-3,3'-diol]	422	*445*	474	E	2550
	435	458	485	C	

(*continued*)

TABLE V (*Continued*)

Carotenoid [semisystematic name]	λ_{max} (nm)			Solvent	$A_{1\,cm}^{1\%}$
Lutein epoxide, taraxanthin	420	443	472	P	
[5,6-epoxy-5,6-dihydro-β,ε-carotene-3,3'-diol]	420	442	471	E	
	433	453	483	C	
Lycopene (all-*trans*)	444	*470*	502	P	3450
[ψ,ψ-carotene]	446	472	503	E	
	448	474	505	A	
	458	484	518	C	
Lycophyll	444	470	501	P	
[ψ,ψ-carotene-16,16'-diol]	446	472	504	E	
Lycoxanthin	443	469	500	P	
[ψ,ψ-caroten-16-ol]	444	471	502	E	
Mutatochrome	407	428	452	P	
[5,8-epoxy-5,8-dihydro-β,β-carotene]	405	427	454	E	
		435	469	C	
Mutatoxanthin		426	456	P	
[5,8-epoxy-5,8-dihydro-β,β-carotene-3,3'-diol]	409	427	457	E	
		437	468	C	
Mytiloxanthin		470		E	
[3,3',8'-trihydroxy-7,8-didehydro-β,κ-caroten-6'-one]		480		B	
Myxoxanthophyll	448	473	503	E	
[2'-(β-L-rhamnopyranosyloxy)-3',4'-didehydro-1',2'-dihydro-β,ψ-carotene-3,1'-diol]	450	*478*	510	A	2160
Neochrome	401	*424*	451	E	2270
[5',8'-epoxy-6,7-didehydro-5,6,5',8'-tetrahydro-β,β-carotene-3,5,3'-triol]	408	434	460	B	
Neoxanthin	416	438	467	P	
[5',6'-epoxy-6,7-didehydro-5,6,5',6'-tetrahydro-β,β-carotene-3,5,3'-triol]	415	*439*	467	E	2243
	423	448	476	C	
Neurosporene	414	439	467	P	
[7,8-dihydro-ψ,ψ-carotene]	416	*440*	470	H	2918
	416	440	469	E	
	424	451	480	C	
Nostoxanthin	428	450	478	E	
[β,β-carotene-2,3,2',3'-tetrol]	432	457	485	C	
Okenone	460	*484*	516	P	2320
[1'-methoxy-1',2'-dihydro-χ,ψ-caroten-4'-one]	465	487	518	A	
Oscillaxanthin	468	492	526	E	
[2,2'-bis(β-L-rhamnopyranosyloxy)-3,4,3',4'-tetradehydro-1,2,1',2'-tetrahydro-ψ,ψ-carotene-1,1'-diol]	470	499	534	A	
	480	510	548	C	
Peridinin		455	485	P	
[5',6'-epoxy-3,5,3'-trihydroxy-6,7-didehydro-5,6,5',6'-tetrahydro-10,11,20-trinor-β,β-caroten-19',11'-olide 3-acetate]		475		E	
		465	502	B	

TABLE V (*Continued*)

Carotenoid [semisystematic name]	λ_{max} (nm)			Solvent	$A_{1\,cm}^{1\%}$
Phytoene	276	*286*	297	P	1250
[7,8,11,12,7',8',11',12'-octahydro-ψ,ψ-carotene]	276	*286*	297	H	915
Phytofluene	331	*348*	367	P	1350
[7,8,11,12,7',8'-hexahydro-ψ,ψ-carotene]	331	*347*	366	H	1577
Plectaniaxanthin	445	471	502	P	
[3',4'-didehydro-1',2'-dihydro-β,ψ-carotene-1',2'-diol]	454	*478*	509	A	2505
Prolycopene	414	*436*	463	P	1920
[7,9,7',9'-tetra-*cis*-ψ,ψ-carotene]		454	484	C	
Renieratene	457	476	507	B	
[φ,χ-carotene]					
Rhodopin	443	470	503	P	
[1,2-dihydro-ψ,ψ-caroten-1-ol]	445	470	501	E	
	455	482	516	C	
Rhodovibrin	455	483	516	P	
[1'-methoxy-3',4'-didehydro-1,2,1',2'-tetrahydro-ψ,ψ-	460	488	522	A	
caroten-1-ol]	469	498	532	C	
Rhodoxanthin	456	487	521	P	
[4',5'-didehydro-4,5'-*retro*-β,β-carotene-3,3'-dione]	458	*490*	524	H	2500
		496	538	E	
	482	510	546	C	
Rubixanthin	434	*460*	490	P	2750
[β,ψ-caroten-3-ol]	433	463	496	E	
	439	474	509	C	
Sarcinaxanthin	415	441	470	E	
[2,2'-bis(4-hydroxy-3-methylbut-2-enyl)-γ,γ-carotene]	423	451	480	C	
Siphonaxanthin		446	468	P	
[3,19,3'-trihydroxy-7,8-dihydro-β,ε-caroten-8-one]	448–455			E	
		466		C	
Spheroidene	427	454	486	P	
[1-methoxy-3,4-didehydro-1,2,7',8'-tetrahydro-ψ,ψ-	429	454	486	E	
carotene]	440	466	500	C	
	441	*468*	502	B	2785
Spheroidenone	460	483	515	P	
[1-methoxy-3,4-didehydro-1,2,7',8'-tetrahydro-ψ,ψ-		488		E	
caroten-2-one]	455	484	505	A	
		499		C	
Spirilloxanthin	465	492	525	P	
[1,1'-dimethoxy-3,4,3',4'-tetradehydro-1,2,1',2'-	465	491	526	E	
tetrahydro-ψ,ψ-carotene]	475	505	543	C	
	479	*510*	546	B	2470
3,4,3',4'-Tetrahydrolycopene	480	510	540	P	
[3,4,3',4'-tetradehydro-ψ,ψ-carotene]					

(*continued*)

TABLE V (Continued)

Carotenoid [semisystematic name]	λ$_{max}$ (nm)			Solvent	$A^{1\%}_{1\,cm}$
Torularhodin	467	*501*	537	P	2040
[3′,4′-didehydro-β,ψ-caroten-16′-oic acid]	463	*495*	532	E	
	483	*515*	554	C	1932
Torulene	454	*480*	514	P	3240
[3′,4′-didehydro-β,ψ-carotene]	456	*486*	520	E	
	467	501	537	C	
Tunaxanthin		438	468	P	
[ε,ε-carotene-3,3′-diol]	419	*440*	470	E	
	427	*450*	479	C	
Violaxanthin	416	440	465	P	
[5,6,5′,6′-diepoxy-5,6,5′,6′-tetrahydro-β,β-carotene-	419	*440*	470	E	2550
3,3′-diol]	426	*449*	478	C	
Violerythrin		566		A	
[2,2′-dinor-β,β-carotene-3,4,3′,4′-tetrone]					
α-Zeacarotene	398	*421*	449	H	1850
[7′,8′-dihydro-ε,ψ-carotene]					
β-Zeacarotene	406	*428*	454	P	2520
[7′,8′-dihydro-β,ψ-carotene]	406	*427*	454	H	1940
	405	428	455	E	
Zeaxanthin	424	*449*	476	P	2348
[β,β-carotene-3,3′-diol]	428	*450*	478	E	2540
	430	*452*	479	A	2340
	433	*462*	493	C	

[a] Solvents: P, petroleum ether (bp 40–60°); A, acetone; C, chloroform; E, ethanol; H, hexane; Py, pyridine; B, benzene.

[b] The $A^{1\%}_{1\,cm}$ values given are for the wavelength printed in italics.

zeaxanthin, 400 nm for violaxanthin, 370 nm for neoxanthin, 354 nm for lycopene. These intense peaks are believed to result from molecular aggregation. Carotenoids dispersed in an aqueous environment by the use of lipids, lipoproteins, or detergents normally exhibit absorption maxima at some 10–20 nm longer wavelength than those of an ethanolic solution.

Shape of the Absorption Spectra. The overall shape or fine structure of the absorption spectrum in most cases reflects the extent of planarity of the chromophore. Thus the spectra of acyclic compounds, e.g., lycopene, in which the conjugated double bond system can adopt an almost planar conformation, are characterized by sharp maxima and minima (persistence). This is true also of cyclic carotenoids such as ε,ε-carotene, and carotenoid epoxides (violaxanthin) in which conjugation is restricted to the linear and approximately planar central polyene chain. The presence of ring double bonds conjugated with the main polyene system results in

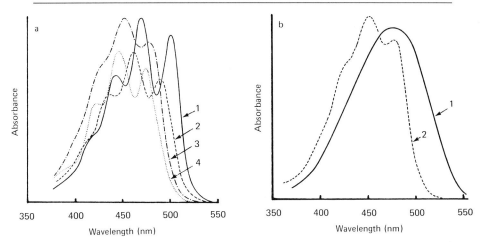

FIG. 3. (a) Light absorption spectra of some carotenes: (1) lycopene; (2) γ-carotene; (3) β-carotene; (4) α-carotene. (b) Light absorption spectra of canthaxanthin (1) and its NaBH$_4$ reduction production, isozeaxanthin (2). Absorbance scale is arbitrary.

the overall chromophore being twisted to some extent to relieve steric strain, giving a corresponding reduction in the "sharpness" of the absorption maxima and minima. Thus, in order of the extent of fine structure, lycopene (ψ,ψ-acyclic) > γ-carotene (β,ψ, monocyclic, one ring double bond conjugated) ≃ α-carotene (β,ε, bicyclic, only one ring double bond conjugated) > β-carotene (β,β, bicyclic, two ring double bonds in conjugation) (Fig. 3a). The loss of fine structure in the absorption spectrum is even more pronounced with cyclic carotenoids such as canthaxanthin that have oxo groups in conjugation with the polyene chain. The absorption spectrum of canthaxanthin in ethanol has only a single, rounded, almost symmetrical absorption peak in contrast to the usual three-peaked spectrum (Fig. 3b). A slight degree of fine structure remains if the spectra of canthaxanthin and similar ketones are determined in petroleum ether or hexane. A carotenoid sample cannot be considered identical to an authentic sample unless the shapes of the absorption spectra of the two samples are identical when determined in at least two different solvents.

Microscale Spectroscopic Tests. Several simple and rapid tests are available which can be performed on a microscale while the carotenoid sample is in the spectrophotometer cuvette, and which give diagnostic information about the presence of certain structural features. The conclusions reached as a result of these tests may be confirmed by recovery and chromatographic examination of the products.

Test for Carotenoid 5,6-Epoxides. The spectrum of a carotenoid is determined, in ethanol, and 1 drop of dilute acid (e.g., 0.1 M HCl) is then added to the solution in the cuvette, and the spectrum redetermined after 30 sec. If the absorption maxima shift to lower wavelength by 18–25 nm without significant change in spectral fine structure, this indicates that the original carotenoid contained one 5,6-epoxy β-ring (with or without a hydroxy group at C-3), which has undergone isomerization to the 5,8-epoxy or furanoid oxide structure with consequent shortening of the chromophore by one double bond. If the shift observed is about 40 nm, this indicates that the original compound possessed two 5,6-epoxy β-ring end groups, e.g., violaxanthin.

Test for Oxocarotenoids by Reduction with NaBH₄. Any compound whose absorption spectrum in ethanol exhibits little or no fine structure is likely to be an oxocarotenoid. To test for this, sodium borohydride (1–2 mg) is added to the ethanolic solution, the mixture is shaken, and the spectrum is redetermined after intervals of 30 sec, 5 min, and 30 min. A change in the shape of the spectrum from the single rounded peak to the normal three-peaked spectrum confirms that the original compound possessed one or more oxo groups conjugated with the double-bond system. The absorption maxima also shift to lower wavelengths as the conjugated oxo group is reduced to an alcohol group which does not contribute to the chromophore (e.g., canthaxanthin, λ_{max} 474 nm → isozeaxanthin, λ_{max} 430, *451,* 478 nm) (Fig. 3b). The time taken for the carbonyl groups to be completely reduced can also be diagnostic; aldehyde groups are reduced very rapidly (usually within 30 sec) whereas keto groups take much longer (up to 30 min). Chromatography will show that the reduced product is much more polar than the original oxo compound.

Iodine-Catalysed Cis–Trans Isomerization. The absorption spectrum of the carotenoid is determined in petroleum ether or hexane. The carotenoid solution is then treated with a few drops of a saturated solution of iodine in the same solvent, and the solution is illuminated, with the exclusion of oxygen, for 5–10 min. The spectrum is then redetermined. Iodine catalyses the photoisomerization of carotenoids, to produce a characteristic pseudo-equilibrium mixture of geometrical (cis–trans) isomers. The spectra of such mixtures are characteristic. If the original sample was a pure all-trans isomer, the wavelengths of maximum absorption are usually slightly decreased, and a reduction in spectral fine structure is seen, but the most obvious change is the appearance of a "cis-peak" at a characteristic wavelength in the UV, 142 ± 2 nm below the longest wavelength peak in the visible absorption spectrum of the all-trans isomer. Production of the pseudo-equilibrium mixture from an original cis-carotenoid should have the opposite effect, slightly increasing the λ_{max} and spectral fine structure and perhaps decreasing the intensity of the cis peak. In the extreme case, iodine-catalysed isomerization of poly-cis-carotenoids such as prolycopene (7,9,7',9'-tetra-*cis*-lycopene) a large increase in λ_{max} is observed (prolycopene λ_{max} 414, 436, 463 nm → 442, 468, 500 nm).

Other Spectroscopic Methods

Chromatographic behavior (including HPLC) and light absorption spectroscopic data allow tentative identification of a carotenoid. For full characterization and establishment of chirality, especially of a previously unknown compound, the sample should also be examined by mass spectrometry, CD, and NMR spectroscopy. These techniques are now used routinely by chemists but reliable interpretation of the results requires considerable experience and expertise, and is beyond the scope of this chapter. For information about the use of these techniques in the carotenoid field and the interpretation of the results the reader is referred to a series of comprehensive review articles, dealing specifically with mass spectrometry,[25,26] NMR spectroscopy[27] and circular dichroism,[28,29] or with spectroscopic methods in general.[30,31]

As a rough guideline a good mass spectrum can usually be obtained from about 10 to 20 µg of carotenoid, and modern high-resolution PFT NMR instruments require about 100–200 µg of sample for a ^{1}H spectrum, and about 5 mg or more for a natural abundance ^{13}C spectrum. It is now possible to obtain a CD spectrum from the amount of material needed to determine the light absorption spectrum.

It is essential that any carotenoid sample prepared for MS, NMR, or CD study must be rigorously purified (see above).

Quantitative Determination

It is standard practice to use a spectrophotometric method for the quantitative determination of carotenoids. Petroleum ether or hexane are the solvents normally used for carotenes, ethanol for xanthophylls. Oth-

---

[25] H. Budzikiewicz, in "Carotenoid Chemistry and Biochemistry" (G. Britton and T. W. Goodwin, eds.), p. 155. Pergamon, Oxford, 1982.

[26] M. E. Rose, in "Carotenoid Chemistry and Biochemistry" (G. Britton and T. W. Goodwin, eds.), p. 167. Pergamon, Oxford, 1982.

[27] G. Englert, in "Carotenoid Chemistry and Biochemistry" (G. Britton and T. W. Goodwin, eds.), p. 107. Pergamon, Oxford, 1982.

[28] K. Noack, in "Carotenoid Chemistry and Biochemistry" (G. Britton and T. W. Goodwin, eds.), p. 135. Pergamon, Oxford, 1982.

[29] V. Sturzenegger, R. Buchecker, and G. Wagniere, *Helv. Chim. Acta* **63**, 1074 (1980).

[30] W. Vetter, G. Englert, N. Rigassi, and U. Schwieter, in "Carotenoids" (O. Isler, ed.), p. 189. Birkhaeuser, Basel, 1971.

[31] G. P. Moss and B. C. L. Weedon, in "Chemistry and Biochemistry of Plant Pigments" (T. W. Goodwin, ed.), 2nd ed., Vol. 1, p. 149. Academic Press, New York, 1976.

ers may need to be used for crystalline samples that do not easily dissolve. The solvent used should not be too volatile (e.g., diethyl ether) or evaporation may alter the concentration even while the determinations are being made. The carotenoid is dissolved in an accurately known volume of the appropriate solvent and a 1 cm light path cuvette is filled with the solution; the matching reference cuvette is filled with the pure solvent. (For quantitative determination by measuring absorbance in the UV region, matched quartz cuvettes must be used, and the solvent must be free from UV-absorbing materials.) The absorbance ($\equiv$ extinction, optical density) of the solution is then determined at the appropriate wavelength (usually λ_{max}) and the amount of carotenoid calculated as described below. For the most accurate determinations, the absorbance recorded should be around 70–80% of the full scale value. If the solution originally prepared is too concentrated and therefore gives too high an absorbance reading, it must be diluted with solvent by an accurately known factor so that the overall volume of solution remains accurately known.

For most carotenoid determinations the specific absorbance coefficient $A_{1\,cm}^{1\%}$ ($\equiv$ specific extinction coefficient $E_{1\,cm}^{1\%}$) is used. This is the absorbance of a 1% (w/v) solution (1 g/100 ml) in a 1 cm path cuvette at the appropriate wavelength. Tables of $A_{1\,cm}^{1\%}$ (or $E_{1\,cm}^{1\%}$) values have been published,[10,11,14] and Table V gives values for a selection of carotenoids most likely to be encountered. Reported $A_{1\,cm}^{1\%}$ values may vary considerably for the same compound, especially if different solvents are used. The molar absorbance (extinction) coefficient, ε_{mol}, sometimes reported, is the absorbance of a 1 M solution. An arbitrary $A_{1\,cm}^{1\%}$ value of 2500 is usually taken when no experimentally determined value has been reported, for an unknown compound, or to give an estimate of total carotenoids in an extract.

Calculation. If x g of carotenoid in y ml solution gives an absorbance A at the given wavelength, then assuming a 1-cm path cuvette

$$x = Ay/(A_{1\,cm}^{1\%} \times 100)$$

Thus if 150 ml of a solution of an unknown amount x of a carotenoid (with $A_{1\,cm}^{1\%} = 2100$) is prepared and this solution gives an absorbance A of 0.70, then

$$\begin{aligned} x &= (0.70 \times 150)/(2100 \times 100) \text{ g} \\ &= (0.70 \times 150 \times 10^6)/(2100 \times 100) \text{ }\mu\text{g} \\ &= 500 \text{ }\mu\text{g} \end{aligned}$$

i.e., the solution contained 500 μg in 150 ml.

Quantitative Determination by HPLC. HPLC, employing detection by visible light absorption, is of course the most sensitive and accurate means of quantitative analysis of carotenoids. The worst problems of carotenoid losses encountered with conventional column chromatography and especially TLC do not arise, and the sensitivity of modern detectors allows accurate and reproducible quantitative analysis of very small samples. Details are given in another article in this volume[19] and elsewhere.[11]

[6] Gas Chromatography of Isoprenoids

By WALTER JENNINGS and GARY TAKEOKA

Introduction

The isoprenoids comprise an important class of naturally occurring compounds with a carbon skeleton made up of isoprene units (isoprene, 2-methyl-1,3-butadiene). The terpenes, which constitute the major portion of that class, have been extensively studied and their chromatographic separations have been the focus of a number of recent reviews.[1-4] Due to the tremendous amount of literature available on this subject, no attempt will be made to cover all areas comprehensively. Instead, we will emphasize the pioneering accomplishments, and newer trends and applications in this ever changing field.

The isoprenoids are a structurally diverse group of products whose members exhibit tremendous variability in both their chromatographic properties and in their stabilities. For example, some terpenes are very sensitive to acid-catalyzed rearrangements and thermal degradations, while the isoprenoid hydrocarbons are stable enough to have survived millions of years in geological samples. In attempting to organize this subject, it seemed logical to devote the first section to the area of sample

[1] R. Croteau and R. C. Ronald, *J. Chromatogr. Libr.* **22B**, 147 (1983).
[2] G. Zweig and J. Sherma, eds., "CRC Handbook Series in Chromatography, Terpenoids." CRC Press, Boca Raton, Florida, 1984.
[3] E. von Rudloff, *Adv. Chromatogr.* **10**, 173 (1974).
[4] C. J. Coscia, *in* "Chromatography—A Laboratory Handbook of Chromatographic and Electrophoretic Methods" (E. Heftmann, ed.), 3rd ed., p. 571. Van Nostrand-Reinhold, Princeton, New Jersey, 1975.

preparation. Due to the sensitive nature of many of these compounds, the prospects of introducing changes in sample composition during this step rarely receive sufficient emphasis. The related area of separation, including a critical evaluation of the entire gas chromatographic system, is also important and will be examined in the second section. Recent advances in fused silica capillary column technology, as well as in sample introduction (especially in on-column injection) have increased our analytical capabilities. Due to their superior resolution and more inert character, capillary columns are gradually replacing packed columns in isoprenoid analysis. The task of compound identification has been greatly aided by the use of combined gas chromatography–mass spectrometry (GC-MS). Isoprenoid analysis is complicated by the fact that certain structurally different terpenes have similar ion fragmentation patterns, making their identification by mass spectra alone impossible. The combination of retention indices and mass spectral data is sufficient, in most cases, to provide positive identification. Other spectroscopic techniques such as NMR, IR, and UV may be used to confirm identity, but they usually require much larger samples. A potentially powerful technique in conjunction with GC-MS is gas chromatography–Fourier transform infrared spectroscopy (GC-FTIR). The higher cost and lower sensitivity as compared to GC-MS will probably inhibit widespread usage of GC-FTIR, although its application to terpene analysis has been demonstrated.[5]

The third section will be devoted to providing specific chromatographic information about the different classes of isoprenoids ranging from the monoterpenes to the tetraterpenes, and finally to the acyclic isoprenoid hydrocarbons. Again, it should be emphasized that due to the labile nature of certain terpenes great care must be taken to avoid the formation of artifacts during sample storage, sample preparation, and subsequent chromatographic steps.

Sample Preparation

Steam Distillation

Steam distillation is a commonly used procedure for the isolation of essential oils from plant materials. Because the volatiles are usually recovered as a dilute solution, this step is often followed by extraction with an organic solvent such as pentane, hexane, dichloromethane, or ether.

Steam distillation at atmospheric pressure does subject the sample to high temperatures, and there is the danger of artifact formation. Pickett *et*

---

[5] V. F. Kalasinsky and J. T. McDonald, Jr., *J. Chromatogr. Sci.* **21**, 193 (1983).

al.[6] found that, depending on the pH conditions used, from 0 (geranyl propionate) to 94.3% (linalyl acetate) of some terpene alcohol esters was lost during steam distillation, and a variety of other terpenes was formed. In a comparison between cold-pressed and distilled lime oils, Azzouz et al.[7] reported that distillation caused the almost complete disappearance of α-thujene, neral, geranial, decanal, geranyl acetate, neryl acetate, α-elemene, and β-elemene, while apparently forming β-phellandrene, 1,4-cineole, p-cymene, α-fenchyl alcohol, α-terpineol, and β-terpineol. The isomerization of cis-dihydrocarvone to trans-dihydrocarvone during distillation of dill seed was reported by Koedam et al.[8] The latter isomerization was catalyzed by metals present in the seed and varied with the pH of the distillation water. In their isolation of volatiles from Abies × arnoldiana Nitz (Pinaceae), Koedam et al.[9] reported the occurrence of santene and 5,6-dimethyl-5-norbornen-exo-2-ol, apparently as artifacts formed during the distillation. The conversion of trans-sabinene hydrate into terpinen-4-ol and monoterpene hydrocarbons during a hydrodistillation procedure has also been observed.[10,11]

The use of vacuum steam distillation may significantly reduce the production of artifacts due to thermal degradation, but may also reduce the yield of some components and can introduce new sources of artifacts. Ferretti and Flanagan[12] postulated that the presence of benzothiazole in a vacuum steam distillate of stale nonfat dry milk[13] was an artifact originating from short sections of rubber tubing used in connecting parts of the vacuum system. The use of Teflon sleeves at the joints of distillation apparatus that contact solvent has been suggested to avoid sample contamination due to high vacuum grease.[14]

Kerven et al.[15] recently described an apparatus which allows simultaneous steam distillation of 10 individual plants for essential oil determination.

[6] J. A. Pickett, J. Coates, and F. R. Sharpe, *Chem. Ind. (London)* **5**, 571 (1975).
[7] M. A. Azzouz, G. A. Reineccius, and M. G. Moshonas, *J. Food Sci.* **41**, 324 (1976).
[8] A. Koedam, J. J. C. Scheffer, and A. Baerheim Svendsen, *Chem. Mikrobiol., Technol. Lebensm.* **6**, 1 (1979).
[9] A. Koedam, J. J. C. Scheffer, and A. Baerheim Svendsen, *J. Agric. Food Chem.* **28**, 862 (1980).
[10] A. Koedam and A. Looman, *Planta Med., Suppl.* 22 (1980).
[11] A. Koedam, J. J. C. Scheffer, and A. Baerheim Svendsen, *Perfum. Flavor.* **5**, 56 (1980).
[12] A. Ferretti and V. P. Flanagan, *J. Agric. Food Chem.* **21**, 35 (1973).
[13] A. Ferretti and V. P. Flanagan, *J. Agric. Food Chem.* **20**, 695 (1972).
[14] I. J. Jeon, G. A. Reineccius, and E. L. Thomas, *J. Agric. Food Chem.* **24**, 433 (1976).
[15] G. L. Kerven, W. Dwyer, S. Duriyaprapan, and E. J. Britten, *J. Agric. Food Chem.* **28**, 162 (1980).

Extraction

Extraction is used to isolate volatiles directly from the sample material or to concentrate volatiles from aqueous distillates obtained via steam distillation. Weurman[16] felt that the aqueous distillates should be concentrated prior to extraction to avoid large volumes of dilute extractant which must be subsequently concentrated, resulting in higher levels of solvent impurities. Direct extraction of the sample material may be preferred in cases where distillation causes changes in sample composition.[17]

The extraction solvent is normally chosen on the basis of its selectivity for the compounds of interest. Solvents commonly used for extraction of flavor volatiles are listed by Weurman.[16] Saturating the aqueous solutions with salts can raise the extraction efficiency though possibly at the expense of selectivity of the extraction solvent. To avoid losing volatiles to the atmosphere, Johnson et al.[18] recommended adding the extraction solvent before saturating the aqueous solution. Sodium chloride and sodium sulfate are commonly used salts.

In their study of apple essence volatiles, Schultz et al.[19] compared the extraction efficiency of various solvents. Ether gave a higher recovery of volatiles, particularly the low-molecular-weight alcohols, than did isopentane or 1,2-dichloro-1,1,2,2-tetrafluoroethane, whose selectivities were similar. Extraction with liquid carbon dioxide gave results similar to ether. Isopentane and fluorocarbon extracts contained the highest concentration of esters and aldehydes. Hardy[20] studied the recovery of 1-alkanols, 2-alkanones, and ethyl alkanoates by continuous extraction from dilute aqueous and aqueous alcoholic solutions using trichlorofluoromethane (Freon 11) as the solvent. Extraction efficiencies and recoveries for compounds larger than C_4 improved markedly with increasing chain length. Since the lower alcohols contribute little to the flavor, trichlorofluoromethane is useful for the extraction of volatiles from samples containing ethanol as a major constituent, i.e., alcoholic beverages.

Extractions can be performed in a batch or continuous mode (Fig. 1). To minimize the formation of thermally induced artifacts, low boiling solvents are preferred, since continuous extractions can be performed at or near room temperature. Also, since the volatiles are often concentrated

---

[16] C. Weurman, *J. Agric. Food Chem.* **17,** 370 (1969).

[17] A. Koedam, J. J. C. Scheffer, and A. Baerheim Svendsen, *Chem. Mikrobiol., Technol. Lebensm.* **6,** 1 (1979).

[18] J. H. Johnson, W. Gould, A. A. F. Badenhop, and R. M. Johnson, Jr., *J. Agric. Food Chem.* **16,** 255 (1968).

[19] T. H. Schultz, R. A. Flath, D. R. Black, D. G. Guadagni, W. G. Schultz, and R. Teranishi, *J. Food Sci.* **32,** 279 (1967).

[20] P. J. Hardy, *J. Agric. Food Chem.* **17,** 656 (1969).

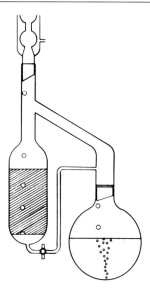

FIG. 1. Continuous liquid–liquid extractor for solvents denser than water. Reprinted from "Applications of Glass Capillary Gas Chromatography," Chromatographic Science Series, Vol. 15 (W. G. Jennings, ed.), p. 455 (1981), by courtesy of Marcel Dekker, Inc.

by distilling off the solvent, it may be better to use low-boiling solvents when analyzing for low-boiling volatiles. Weurman[16] argued that even with low-boiling solvents, the volatiles of interest are often lost during solvent removal.

For the analysis of solid samples, Soxhlet extraction is widely used. Typically, the acyclic isoprenoid hydrocarbons are extracted from shale or sediment samples with a benzene : methanol mixture in a Soxhlet apparatus. After removal of the solvent, the residue is fractionated by silica gel columns or by thin-layer plates. Jennings[21] placed a glass Soxhlet extraction in a pressurized container fitted with a cold finger condensor and utilized liquid carbon dioxide as the extracting solvent. The use of this technique to extract coal fly ash and banana headspace volatiles entrained on Porapak Q was demonstrated. Schultz and Randall[22] studied the extraction selectivity of carbon dioxide for various flavor constituents.

The use of supercritical fluids as extraction solvents has been applied to a number of materials. Vitzthum *et al.*[23] isolated black tea volatiles by

[21] W. G. Jennings, *J. High Res. Chromatogr.* **2**, 221 (1979).
[22] W. G. Schultz and J. M. Randall, *Food Technol.* **24**, 1283 (1970).
[23] O. G. Vitzthum, P. Werkhoff, and P. Hubert, *J. Agric. Food Chem.* **23**, 999 (1975).

extraction with supercritical carbon dioxide followed by steam distillation and entrainment of steam volatiles on Porapak Q. Hubert and Vitzthum[24] studied the extraction of hops, spices, and tobacco with supercritical fluids. They found that by varying the pressure and temperature conditions, the solvent power of the supercritical fluid could be varied without changing the solvent composition. The solubility behavior of a large number of natural products under pressure gradients up to 400 bar was investigated through a combination of supercritical fluid extraction and thin-layer chromatography.[25] Lipophilic compounds were readily extracted with carbon dioxide with the extraction becoming more difficult with the introduction of polar functional groups. Strongly polar substances such as sugars and amino acids could not be extracted in the range up to 400 bar. The steep change in the density and dielectric constant of carbon dioxide at 40° between 70 and 200 bar change its dissolving power and hence make fractionating separations possible.

Simultaneous Steam Distillation Extraction

A particularly valuable technique for the isolation of volatiles is simultaneous steam distillation extraction with a Likens and Nickerson apparatus[26,27] (Fig. 2). This technique is advantageous in that it requires only small amounts of solvent to extract large amounts of sample material. Maarse and Kepner[28] modified this unit through addition of a vacuum jacket to minimize premature condensation and a dry ice condenser to reduce the loss of solvent and volatiles. An external steam generator was added by Römer and Renner.[29] An interesting variation of the Likens and Nickerson apparatus was reported by Schultz et al.[30] (Fig. 3). Their apparatus had a mixing chamber for the vapors at the top of the condenser which was designed for thorough vapor mixing before condensation. The apparatus was tested with a 12 component model mixture (consisting of various esters, alcohols, and terpenes) at varying concentrations and experimental conditions. Most of the components showed nearly quantitative recovery after 1 hr of operation, although ethyl 3-hydroxyhexanoate required 4 hr to reach 90% recovery. After the simultaneous steam distil-

[24] P. Hubert and O. G. Vitzthum, *Angew. Chem., Int. Ed. Engl.* **17**, 710 (1978).
[25] E. Stahl, W. Schilz, E. Schütz, and E. Willing, *Angew. Chem., Int. Ed. Engl.* **17**, 731 (1978).
[26] S. T. Likens and G. B. Nickerson, *Proc. Am. Soc. Brew. Chem.* 5 (1964).
[27] G. B. Nickerson and S. T. Likens, *J. Chromatogr.* **21**, 1 (1966).
[28] H. Maarse and R. E. Kepner, *J. Agric. Food Chem.* **18**, 1095 (1970).
[29] G. Römer and E. Renner, *Z. Lebensm.-Unters. -Forsch.* **156**, 329 (1974).
[30] T. H. Schultz, R. A. Flath, T. R. Mon, S. B. Eggling, and R. Teranishi, *J. Agric. Food Chem.* **25**, 446 (1977).

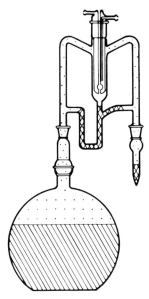

FIG. 2. Simultaneous distillation–extraction with a Likens and Nickerson apparatus. Reprinted with permission from S. T. Likens and G. B. Nickerson, *J. Chromatogr.* **21**, 1 (1966).

lation–extraction, the volatiles are further concentrated by solvent removal. This may induce changes in the extract due to loss of the low-boiling volatiles. To circumvent the problem, Godefroot et al.[31] designed a microversion of the Likens and Nickerson apparatus, which did not require further concentration by solvent evaporation (Fig. 4). Unlike the systems described above which used lighter-than-water solvents, the latter device used heavier-than-water solvents, such as dichloromethane. These researchers examined the essential oils of pepper, hops, and *Rhododandron simsii* (*Azalea indica*), using sample amounts ranging from 1 to 15 g. It was noted that for the hop and pepper samples, the maximum yield of essential oil was reached after 15 min of distillation–extraction time, while the flower sample required 1 hr to increase the recovery of the alcohols and high-boiling polar compounds.

A modification of the apparatus proposed by Godefroot et al.[31] was reported by Rijks et al.[32] Simultaneous steam distillation extraction produces a high extraction efficiency due to the continual extraction of con-

[31] M. Godefroot, P. Sandra, and M. Verzele, *J. Chromatogr.* **203**, 325 (1981).
[32] J. Rijks, J. Curvers, T. Noy, and C. Cramers, *in* "Proceedings of the 5th International Symposium on Capillary Chromatography" (J. Rijks, ed.), p. 520. Elsevier, Amsterdam, 1983.

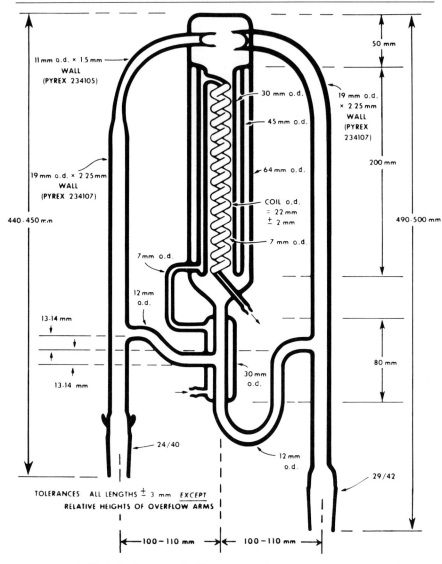

FIG. 3. Modified simultaneous distillation–extraction apparatus. From T. H. Schultz, R. A. Flath, T. R. Mon, S. B. Eggling, and R. Teranishi, *J. Agric. Food Chem.* **25,** 446 (1977). Copyright 1977 American Chemical Society.

densing sample vapor by condensing solvent vapor. However this technique has been reported to produce artifacts,[33,34] and operation at reduced pressure may be desirable.

[33] R. G. Clark and H. E. Nursten, *J. Sci. Food Agric.* **27,** 713 (1976).
[34] A. S. McGill and R. Hardy, *J. Sci. Food Agric.* **28,** 89 (1977).

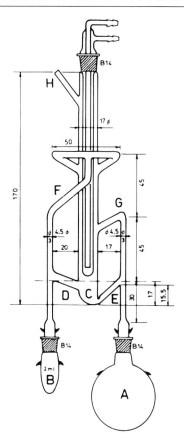

FIG. 4. Simultaneous distillation–extraction microapparatus. A, sample flask; B, solvent flask; C, demixing section, D and E, demixing return arms; F and G, vapor channels; H, inlet/vent. From M. Godefroot, P. Sandra, and M. Verzele, *J. Chromatogr.* **203,** 325 (1981).

Prefractionation

Because essential oils are complex mixtures, they are often separated into different fractions prior to gas chromatographic analysis. Using polar stationary phases, the oxygenated monoterpenes overlap with the sesquiterpenes, complicating the identification and quantitation of components. The separation of the oxygenated fraction from the terpene hydrocarbons may greatly aid the identification of trace components. The hydrocarbons can be separated from the total oil by selective elution with pentane or hexane from a silica gel column.[35] Tressl *et al.*[36] used adsorp-

---

[35] J. C. Kirchner and J. M. Miller, *Ind. Eng. Chem.* **44,** 318 (1952).
[36] R. Tressl, L. Friese, F. Fendesack, and H. Koppler, *J. Agric. Food Chem.* **26,** 1422 (1978).

tion chromatography with a silica gel–aluminum oxide (2:1) column to separate a beer extract into six fractions. They identified more than 110 volatile compounds in a German beer, 45 of which were characterized for the first time in beer. Their identification of a number of hop oil constituents was significant since these components tend to be masked by other beer components. The use of silica gel may, however, cause compositional changes in the essential oil due to isomerization reactions.

Hunter and Brogden[37] studied the isomerization of limonene with silica gel at 100 and 150°, and reported that it was initially isomerized to α-terpinene, γ-terpinene, terpinolene, and isoterpinolene, which subsequently underwent disproportionation and polymerization reactions. In studying black pepper oil constituents, Wrolstad and Jennings[38] reported the isomerization of sabinene into α-thujene, α-terpinene, γ-terpinene, limonene, β-phellandrene, and terpinolene during thin-layer chromatography. Evidence supported an acid-catalyzed isomerization of sabinene, since silica gel chromatostrips prepared by using 0.1 N NaOH in place of distilled water resulted in much lower isomerization.

To prevent the isomerization of sensitive terpenes during column chromatography, researchers have used stationary phases such as Emulphor-O[39] and Carbowax[40] to deactivate the silica gel. The rearrangement of citronellal has been reported despite the deactivation of silica gel with Carbowax.[41] In addition, since the stationary phases are not bonded to the silica, small amounts may elute from the column and contaminate the various terpene fractions.

The use of water is apparently an effective and simple method for deactivation of silica gel. Kubeczka[42] found no changes in a monoterpene hydrocarbon mixture after column chromatography on silica gel with a water content of 7%. Using acidic and basic washings followed by deactivation of the dried silica gel through the addition of water (5%), Scheffer *et al.*[43] were able to prevent the isomerization of various monoterpene hydrocarbons. They also reported that the water content of the silica gel should be kept low for efficient separation of the terpenes.

Severson *et al.*[44] reported a gas chromatographic method for the anal-

[37] G. L. K. Hunter and W. B. Brogden, Jr., *J. Org. Chem.* **28,** 1679 (1963).
[38] R. E. Wrolstad and W. G. Jennings, *J. Chromatogr.* **18,** 318 (1965).
[39] E. sz. Kováts and E. Kugler, *Helv. Chim. Acta* **46,** 1480 (1963).
[40] E. von Rudloff and F. W. Hefendehl, *Can. J. Chem.* **44,** 2015 (1966).
[41] F. W. Hefendehl, *Arch. Pharm. (Weinheim, Ger.)* **303,** 345 (1970).
[42] K. H. Kubeczka, *Chromatographia* **6,** 106 (1973).
[43] J. J. C. Scheffer, A. Koedam, and A. Baerheim Svendsen, *Chromatographia* **9,** 425 (1976).
[44] R. F. Severson, J. J. Ellington, R. F. Arrendale, and M. E. Snook, *J. Chromatogr.* **160,** 155 (1978).

ysis of hydrocarbons, terpenes, fatty alcohols, fatty acids, and sterols in flue-cured tobacco. The tobacco sample was hydrolyzed with potassium hydroxide, and following acidification, extracted with hexane. The hexane extract was separated into three fractions on a silicic acid column. The first fraction, eluted with hexane, consisted of the nonpolar lipids (hydrocarbons, neophytadiene). The second fraction, eluted with 25% benzene in hexane, was the terpene fraction (di- and triterpenes, phytol, paraffinic alcohols). The last fraction included the polar lipids (fatty acids, sterols, solanesol) and was eluted with a benzene–ethyl ether (3:1) mixture.

Although other chromatographic adsorbents such as Florisil[45] and alumina[46] have been used in prefractionations, silica is preferred due to its greater linear capacity and higher column efficiencies.[47]

The use of gel permeation chromatography to separate the pigments, flavonoids, and flavor fractions has been reported by Wilson and Shaw.[48] Due to its greater speed and resolution, high-performance liquid chromatography (HPLC) has advantages over conventional column chromatography in sample prefractionation. Because the sample material is in contact with the adsorbent for a much shorter time, there is less danger of rearrangements. Also, more samples can be run in an equivalent time and the improved separation may be helpful in the subsequent GC analysis. Teitelbaum[49] proposed a strategy using various separation techniques to analyze complex flavor mixtures. The first step involved a preliminary clean-up with a silica gel column. The next two steps involved the use of HPLC, first in the adsorption mode and then in the partition mode (normal or reverse phase). The last step used gas chromatography to separate the desired fractions. Jones et al.[50] used HPLC to separate a model mixture of monoterpenes. In their study of a model browning system, Yamaguchi et al.[51] used HPLC to fractionate a complex mixture of heterocyclic compounds (furans, thiophenes, pyrroles, thiazoles, oxazoles, pyrazines, and imidazoles), prior to gas chromatographic analysis. Kubeczka[52] used

[45] D. H. Miles, N. V. Mody, J. P. Minyard, and P. A. Hedin, *Phytochemistry* **14**, 599 (1975).
[46] H. Maarse and F. H. L. van Os, *Flavour Ind.* **4**, 477 (1973).
[47] L. R. Synder, *in* "Chromatography" (E. Heftmann, ed.), 3rd ed., p. 54. Van Nostrand-Reinhold, Princeton, New Jersey, 1975.
[48] C. W. Wilson, III and P. E. Shaw, *J. Agric. Food Chem.* **25**, 221 (1977).
[49] C. L. Teitelbaum, *J. Agric. Food Chem.* **25**, 466 (1977).
[50] B. B. Jones, B. C. Clark, Jr., and G. A. Iacobucci, *J. Chromatogr.* **178**, 575 (1979).
[51] K. Yamaguchi, S. Mihara, A. Aitoku, and T. Shibamoto, *in* "Liquid Chromatographic Analysis of Food and Beverages" (G. Charalambous, ed.), Vol. 2, p. 303. Academic Press, New York, 1979.
[52] K. H. Kubeczka, *in* "Flavor '81" (P. Schreier, ed.), 3rd Weurman Symp., p. 345. de Gruyter, Berlin, 1981.

reverse-phase HPLC (column material—LiChroprep RP18, 40 μm particles) to fractionate a mixture of oxygenated terpenoids, monoterpene hydrocarbons, and sesquiterpene hydrocarbons in 12 min.

The Direct Injection Technique

To avoid the preliminary isolation of sample volatiles, the direct injection technique can be used. In its simplest mode, the tissue or plant material is placed directly into the heated injector to release the volatiles for analysis into the gas chromatographic column. The method is advantageous when analyzing low-boiling components which may be obscured by a solvent peak and when analyzing large numbers of samples. Since only milligram amounts of material are required, the method may be valuable where only limited amounts of sample material are available. A drawback of the small sample size is that it may not be representative of a larger sample. In examining the leaves of several single branchlets of white and black spruce, von Rudloff[53] found the quantitative variation of volatiles from one leaf to another was far greater than the variation obtained from one tree to another when larger samples and steam-distillation were used. Hefendehl et al.[54] reported similar results with the leaves of the same mint shoot.

Von Rudloff[55] observed that for thin plant materials such as mint leaves and flat conifer needles, there was an almost quantitative release of volatiles, while for thicker plant materials such as juniper leaves, wood, or bark, the release was incomplete. In the above studies, von Rudloff used a separate preheating apparatus (Aerograph model 695 inductor) to release sample volatiles which were subsequently swept through the injection port onto the column. Karlsen[56] similarly designed an apparatus to sample biological material which was separate from the GC and hence required no modification of the injection port. The sampling of 5 mg of foliage material of *Rosmarinus officinalis* and the subsequent analysis on a 100 m open tubular column was demonstrated. The use of a heated injection port to release volatiles from plant material was reported by Henderson et al.,[57] who placed the sample in a glass liner tube that was then placed in the injection port. Senanayake et al.[58] simply used a stainless-steel basket to hold the sample in the injection port. Comparing the qualitative and quantitative recovery of volatiles from cinnamon stem

[53] E. von Rudloff, *Can. J. Bot.* **45**, 891 (1967).
[54] F. W. Hefendehl, E. W. Underhill, and E. von Rudloff, *Phytochemistry* **6**, 823 (1967).
[55] E. von Rudloff, *Recent Adv. Phytochem.* **2**, 127 (1969).
[56] J. Karlsen, *J. Chromatogr. Sci.* **10**, 642 (1972).
[57] W. Henderson, J. W. Hart, P. How, and J. Judge, *Phytochemistry* **9**, 1219 (1970).
[58] U. M. Senanayake, R. A. Edwards, and T. H. Lee, *J. Chromatogr.* **116**, 468 (1976).

bark, the latter workers found that the direct injection technique compared favorably with solvent extractions using carbon disulfide.

A superior method of simple introduction involves the encapsulation of the sample in glass or metal. Through use of a plunger assembly the volatiles are released by crushing or piercing the capsule, or by heating as in the case of capsules made from low melting metals, i.e., indium (mp 156°). Morgan and Wadhams[59] examined the volatiles from ant glands by sealing the samples in glass tubes. The tube was placed into the sample injector, heated to 210° for 5 min, then crushed to release the volatiles onto the column. Miller and Bertsch[60] recently checked the performance of a prototype capsule injector. The application of this automated all glass sampler was demonstrated by the analysis of volatiles from anthracite coal, black pepper, and perfumed soap.

GC System Considerations

The susceptibility of terpenes to rearrangements and degradations make necessary the use of clean and inert analytical systems. Due to the catalytic effects of metals, the use of injectors with glass liners, and glass columns, particularly fused silica capillary columns is strongly recommended. Vaporizing injectors subject the sample to a thermal shock which may cause decomposition of the more labile compounds. The accumulation of nonvolatile residues (particularly acidic residues) in the injector, caused by repeated injections of complex samples, can catalyze the degradation of samples subsequently injected.[61] A dirty insert can also cause adsorption effects which result in tailing peaks and reduced system performance.[62] The frequency of insert cleaning will be influenced by the type of samples injected.

The emergence of on-column injection with capillary columns allows us to utilize the powerful resolving power of capillary columns with an injection mode that offers the advantages of (1) minimizing sample decomposition due to thermal effects, since flash vaporization is avoided, (2) eliminating sample discrimination due to syringe effects, and (3) high reproducibility. Since the sample is deposited directly on the column, the accumulation of nonvolatile residues in the column inlet is a more severe problem. The use of immobilized phase capillaries permits backflushing

---

[59] E. D. Morgan and L. J. Wadhams, *J. Chromatogr. Sci.* **10**, 528 (1972).
[60] R. Miller and W. Bertsch, in "Proceedings of the Fourth International Symposium on Capillary Chromatography" (R. E. Kaiser, ed.), p. 267. Huethig Verlag, Heidelberg, West Germany, 1981.
[61] B. Mitzner, *Anal. Chem.* **36**, 242 (1964).
[62] K. Grob, Jr. and G. Grob, *J. High Res. Chromatogr.* **2**, 109 (1979).

with solvent to remove soluble residues from the column. Alternatively the inlet and outlet ends of the column can be inverted or short sections of column can be removed from the inlet end to restore column efficiency.[63,64] Perhaps the best solution is to use a "retention gap" as suggested by Grob.[65] Here a short section (0.6–3 m) of deactivated, uncoated column tubing is attached to the front of the analytical column. When samples containing compounds deleterious to stationary phase are injected, they do not contact the stationary phase and the resultant contributions to bleeding and peak tailing are avoided. The length of the analytical column is unchanged, so its efficiency is not affected. Additionally the "retention gap" may reduce or eliminate the problem of "band broadening in space" associated with the injection of samples at oven temperatures below the boiling point of the sample solvent.[65]

The column, being the heart of the chromatographic system, is of utmost importance. The use of gas chromatography was a major advance in the analysis of essential oils. Though much useful information has been derived with packed columns, capillary columns, with their superior efficiencies and inertness, are gradually replacing their use. Packed columns still have their place in preparative applications and in the analysis of some gases. One problem with packed columns is the activity of the support material. With decreased stationary phase loadings on the support, the absorptivity of the support becomes increasingly important. Acidic sites on the support can cause the isomerization and decomposition of sensitive terpenes.[61] Porous polymers such as Porapak Q and Tenax are of little value as support material for terpene analysis, but may be useful for concentrating materials from vapor sampling.

Capillary columns can be constructed from a variety of materials including nickel, stainless steel, glass, and fused silica. The preparation of nickel columns has been described by Bertsch et al.[66] Stainless-steel columns have been far more popular than nickel in spite of their activity toward polar compounds. Flath and Forrey[67] found that four reported pineapple components (ethyl lactate, 2,5-dimethyl-4-hydroxy-2,3-dihydro-3-furanone, 5-hydroxymethylfurfural, and p-allylphenol) failed to negotiate their stainless-steel open tubular column. The active metal surface may encourage isomerization or breakdown reactions of terpenes. In their

---

[63] K. Grob, Jr., *J. High Res. Chromatogr.* **1,** 307 (1978).
[64] F. Berthou and Y. Dreano, *J. High Res. Chromatogr.* **2,** 251 (1979).
[65] K. Grob, Jr., *J. Chromatogr.* **237,** 15 (1982).
[66] W. Bertsch, F. Shunbo, R. C. Chang, and A. Zlatkis, *Chromatographia* **7,** 128 (1974).
[67] R. A. Flath and R. R. Forrey, *J. Agric. Food Chem.* **18,** 306 (1970).

analysis of a lemon oil sample on a stainless-steel capillary column, Averill and March[68] noted a baseline rise between the neral and geranial peaks. No baseline change was observed when the sample was run on a glass capillary column indicating that the rise was probably due to a partial isomerization or decomposition caused by the active metal surface.

Glass, though not completely inert, is a better choice for column material than stainless steel. Glass contains a high level of metal oxides which must be removed by acid leaching. Fused silica, on the other hand, has a much lower metal oxide content (0.05 to 1 ppm) and a high tensile strength which make it the column material of choice at present.[69] Because metal oxides catalyze the breakdown of stationary phase, fused silica capillary columns can be operated at higher temperatures than can conventional glass capillary columns. It has also been observed that silicone phases coated on fused silica bleed at a lower rate than when coated on carefully leached soda-glass.[70] The flexibility and inherent straightness allows the optimum placement of the fused silica capillary column in the chromatographic system leading to maximum performance and inertness.[71]

Stationary Phase Considerations

Though the use of capillary columns has greatly enhanced the analysis of complex mixtures, separation problems remain which cannot be solved merely by a large number of theoretical plates. Due to the tremendous structural diversity of the isoprenoids, there is no single ideal liquid phase. In the analysis of essential oils and other complex mixtures, it is often desirable to employ several liquid phases with different polarity. With the large number of liquid phases available, many different types have been evaluated for their suitability in isoprenoid analysis.

The wide variety of available liquid phases often causes confusion among users trying to select a phase for a particular separation problem. Over 200 different liquid phases are used with packed columns; it is hoped that this trend does not spread to capillary columns, and that a more logical and selective approach is taken. Since many of the available liquid

---

[68] W. Averill and E. W. March, *Chromatogr. Newsl.* **4**, 20 (1976).
[69] S. R. Lipsky, *J. High Res. Chromatogr.* **6**, 359 (1983).
[70] L. Blomberg, J. Buijten, K. Markides, and T. Wännmam, in "Proceedings of the Fifth International Symposium on Capillary Chromatography" (J. Rijks, ed.), p. 99. Elsevier, Amsterdam, 1983.
[71] W. G. Jennings, "Comparisons of Fused Silica and Other Glass Columns in Gas Chromatography." Heuthig Verlag, Heidelberg, West Germany, 1981.

phases were not designed for use in GC, they often possess undesirable chromatographic properties, such as large batch to batch variation, high bleed rates, and low thermal stabilities. Efforts to reduce the number of liquid phases by proposing the use of standard or preferred stationary phases has been reported.[72,73]

Of the polar liquid phases, polyglycol type phases, particularly Carbowax 20M, are frequently used. They have relatively low maximum operating temperatures and high minimum operating temperatures, and are susceptible to trace levels of oxygen in the carrier gas, especially at higher temperatures (Carbowax 20M decomposes into trace amounts of acetaldehyde and acetic acid).[74] The thermal decomposition of polyethylene glycol 20M was recently investigated by Conder et al.,[75] who reported that trace levels of oxygen had a critical effect on phase decomposition. By incorporating an oxygen scrubber to reduce the oxygen concentration in the carrier gas from 10 ppm to 10 ppb the rate of oxidative degradation was reduced by almost a factor of five, and the temperature at which decomposition beings was raised from 160 to 200°. To attain the maximum thermal stability, Verzele et al.[76] stressed the importance of proper synthesis and purification of the polyethylene glycols. Metal ion contamination of the phase can catalyze its decomposition. The Ghent group has studied the properties of the Superoxes, polyethylene glycol polymers treated to contain only low levels of residual catalyst.[77,78] The gumlike character of the higher Superoxes permitted the researchers to prepare columns of high coating efficiencies (>95%). The polarity of the Superoxes is very similar to that of Carbowax 20M and they possess higher temperature stability than Carbowax 20M. Figure 5 shows the minimal bleed rate of a Superox-4 (mean molecular weight of 4×10^6) capillary column during a temperature programmed analysis. A disadvantage of the high-molecular-weight Superoxes is the low solubility and high viscosity of their solutions necessitating the use of a high-pressure reciprocating pump for the static coating procedure. Polar solutes in aqueous solutions

---

[72] J. J. Leary, J. B. Justice, S. T. Suge, S. R. Lowry, and T. L. Isenhour, *J. Chromatogr. Sci.* **11**, 201 (1973).

[73] S. Hawkes, D. Grossman, A. Hartkopf, T. Isenhour, J. Leary, and J. Parcher, *J. Chromatogr. Sci.* **13**, 115 (1975).

[74] H. E. Persinger and J. T. Shank, *J. Chromatogr. Sci.* **11**, 190 (1973).

[75] J. R. Conder, N. A. Fruitwala, and M. K. Shingari, *J. Chromatogr.* **269**, 171 (1983).

[76] M. Verzele, G. R. Redant, M. Van Roelenbosch, M. Godefroot, M. Verstappe, and P. Sandra, in "Proceedings of the Fourth International Symposium on Capillary Chromatography" (R. E. Kaiser, ed.), p. 239. Huethig Verlag, Heidelberg, West Germany, 1981.

[77] M. Verzele and P. Sandra, *J. Chromatogr.* **158**, 111 (1978).

[78] P. Sandra, M. Verzele, M. Verstappe, and J. Verzele, *J. High Res. Chromatogr.* **2**, 288 (1979).

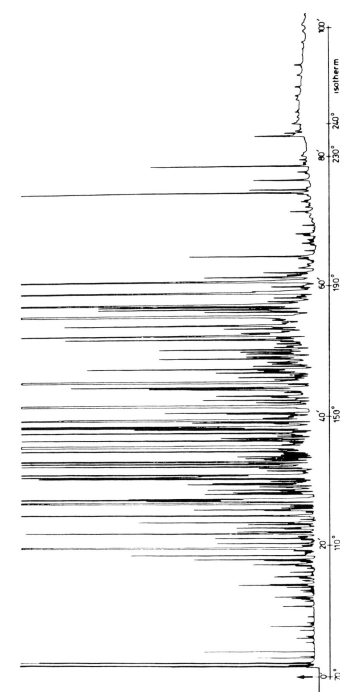

FIG. 5. Chromatogram of the oxygenated fraction of black pepper essential oil. Borosilicate capillary column, 40 m × 0.26 mm i.d., deactivated with N-cyclohexylazetindinol (CHAZ), and coated with Superox-4. Injection, 0.2 µl, split ~1:50 in an all-glass splitter. Temperature programmed from 70 to 240° at 2°/min; carrier gas (H$_2$) flow rate, 1.5 ml/min. Sensitivity setting: attenuation 1, range 10^{-11} (Varian 3700). From M. Verzele and P. Sandra, *J. Chromatogr.* **158**, 111 (1978).

have been successfully analyzed by direct injection onto fused silica capillary columns coated with Carbowax 20M.[79] With polyethylene glycol type phases, the molecular weight of the polymer influences its polarity (larger polymers such as Carbowax 20M are less polar than smaller polymers such as Carbowax 6000, 4000, 2000, and 600). Carbowax 20M has a wide molecular weight distribution which may lead to column bleed and polarity changes during its use.[80]

A major problem with Carbowax 20M lies in its low maximum-temperature limit, and its high minimum-temperature limit; at 55–65°, its chromatographic performance suffers catastrophically as it undergoes phase transition and solidifies; Superoxes suffer this same defect. The shorter-chain polyethylene glycols (e.g., Carbowax 400, 1500, 4000) possess lower minimum-temperature limits, but unfortunately their maximum-temperature limits are considerably lower. Another complication is that all of these hydrophilic phases are repelled by the hydrophobicity of the fused silica surface, leading to columns that have relatively low efficiencies and short lifetimes. This incompatibility has been largely overcome by bonding the stationary phase to the support surface, and several bonded polyethylene glycol-type phases are now available. Some of these exhibit somewhat higher maximum-temperature limits, but in most cases the minimum-temperature limit remains high. The most recent of these, Durabond Wax (DB-WAX), has both higher maximum-temperature, and lower minimum-temperature limits; good chromatography has been reported at 0° (column in ice water).[81] In addition, McReynolds constants on Carbowax 20M and Durabond Wax are almost identical, making this an especially useful phase for those with large amounts of retention data collected on Carbowax 20M. Figure 6 shows typical behavior.

Emulphor ON 870 (polyglycol monooctadecyl ether) and Ucon HB 5100 are polyglycol phases with moderately wide molecular weight distributions. The constant moderate bleed rate of the Emulphor phase make it a less desirable choice. The Ucons are polymers made of mixed ethylene and propylene oxides similar to the Pluronics. Pluronics were reported to have a narrower molecular weight distribution and hence a more precisely defined character.[82] Hlavay *et al.*[80] used Pluronic F68 capillary columns

---

[79] E. Bastian, H. Behlau, H. Husmann, F. Weeke, and G. Schomburg, in "Proceedings of the Fourth International Symposium on Capillary Chromatography" (R. E. Kaiser, ed.), p. 465. Huethig Verlag, Heidelberg, West Germany, 1981.

[80] J. Hlavay, A. Bartha, Gy. Vigh, M. M. Gazdag, and G. Szepesi, *J. Chromatogr.* **204**, 59 (1981).

[81] G. Takeoka and W. Jennings, *J. Chromatogr. Sci.* **22**, 177 (1984).

[82] Gy. Vigh, A. Bartha, and J. Hlavay, *J. High Res. Chromatogr.* **4**, 3 (1981).

for the quality control of essential oil samples and observed no changes in the retention characteristics of the column over a period of over 18 months. The suitability of Pluronics as liquid phases for capillary columns has been shown by Grob and Grob.[83] Due to their lower bleed rate and narrower molecular weight distribution, Pluronics are favored over other polyglycol phases. When using polyglycol phases, the high sensitivity of aldehydes to acid–base effects should be noted, as they may tail or fail to elute on neutral or basic columns.[83]

Esterification of Carbowax 20M with 2-nitroterephthalic acid produces the "free fatty acid phase," FFAP. This phase has been successfully coated on fused silica capillary columns though it suffers from oxygen sensitivity, similar to Carbowax 20M. It must be used with caution since it has been reported to adsorb aldehydes irreversibly.[84,85] The removal of aldehydes appears to be time and temperature dependent and seems to diminish with column use and/or exposure to air during storage. Withers[85] suggested that the removal may be due to an acid-catalyzed aldol condensation. FFAP has also been reported to dehydrate alcohols.[86] Despite these problems, FFAP has been widely used in terpene analysis. Hiltunen and Raisanen[86] studied the selectivity of nine different liquid phases for monoterpene hydrocarbons and found FFAP gave the best separation of Scots pine needle essential oil. Problems with other liquid phases have been observed by Klouwen and ter Heide[87] who found the decomposition of sabinene on silicone oil, and Gillen and Scanlon[88] who reported the enolization of isomenthone on alcohol-amine phases such as Theed and Quadrol. The role of the solid support in catalyzing solute reactions should not be overlooked, i.e., the enolization of isomenthone increased with decreasing liquid phase loading.[88]

Polyphenyl ether phases have been used to resolve conifer leaf oils.[89,90] Tyson[91] used a support coated open tubular (SCOT) column coated with polyphenyl ether (6 rings) for the gas chromatographic–mass spectrometric (GC-MS) analysis of various monoterpene hydrocarbons. Grob[92] has analyzed the polyphenyl ether phases (5 and 6 rings) by capillary gas chromatography and reported that they were nearly pure sub-

[83] K. Grob, Jr. and K. Grob, *J. Chromatogr.* **140**, 257 (1977).
[84] R. R. Allen, *Anal. Chem.* **38**, 1287 (1966).
[85] M. K. Withers, *J. Chromatogr.* **66**, 249 (1972).
[86] R. Hiltunen and S. Raisanen, *Planta Med.* **41**, 174 (1981).
[87] M. H. Klouwen and R. ter Heide, *J. Chromatogr.* **7**, 297 (1962).
[88] D. G. Gillen and J. T. Scanlon, *J. Chromatogr. Sci.* **10**, 729 (1972).
[89] E. von Rudloff, *Can. J. Chem.* **46**, 679 (1968).
[90] A. R. Vinutha and E. von Rudloff, *Can. J. Chem.* **46**, 3743 (1968).
[91] B. J. Tyson, *J. Chromatogr.* **111**, 419 (1975).
[92] K. Grob, *J. Chromatogr.* **198**, 176 (1980).

stances. A disadvantage is their high volatility which severely limits the maximum operating temperature, and which probably restricts their use to the relatively low boiling monoterpene hydrocarbons.

The silicones, available in a wide range of polarity, are the most widely used group of stationary phases in gas chromatography. Haken[93] compiled extensive information regarding the structure and chemical composition of silicones used in gas chromatography. Like the polyglycols, the thermal stability of the silicones is influenced by the level of residual catalyst. Proper selection of the catalyst and its subsequent removal are important in minimizing the thermal degradation of the phase. In his recent review of stationary phases in gas chromatography, Blomberg[94] discussed the stability, chemical properties, and cross-linking of silicones. The advent of cross-linked silicone stationary phases represents an important advance in capillary gas chromatography. Early attempts to coat polar phases on fused silica were often disappointing, and this was often attributed to improper wettability of the phase onto the column surface. However, using contact angle measurements, Bartle et al.[95] found that clean fused silica was a high-energy surface, completely wettable by most liquid phases. Compared to the nonpolar methyl silicones, the polar silicones experience a greater change in viscosity with temperature. This stationary phase viscosity is an important factor influencing the stationary phase film stability.[96] Phases whose viscosities become low at elevated temperatures will be more susceptible to rearrangements on the column surface than phases whose viscosities change only slightly with temperature. This leads to decreased column efficiency due to uneven film distribution. In this respect, gum bases are preferred over fluid phases. Peaden et al.[97] recently prepared 50 and 70% phenylmethylphenylpolysiloxane gum phases that were more viscous than any of the commercially available phenylpolysiloxane phases and produced thermally stable and efficient fused silica columns. The stability was further enhanced by a final cross-linking step. The cross-linking of silicone phases improves the film stability and is a particularly important step in the preparation of efficient polar capillary columns. The lowest polarity silicones are the methyl silicones. The methyl silicones experience a relatively small change in viscosity with temperature, which has been related

---

[93] J. K. Haken, *J. Chromatogr.* **141**, 247 (1977).
[94] L. Blomberg, *J. High Res. Chromatogr.* **5**, 520 (1982).
[95] K. D. Bartle, B. W. Wright, and M. L. Lee, *Chromatographia* **14**, 387 (1981).
[96] B. W. Wright, P. A. Peaden, and M. L. Lee, *J. High Res. Chromatogr.* **5**, 413 (1982).
[97] P. A. Peaden, B. W. Wright, and M. L. Lee, *Chromatographia* **15**, 335 (1982).

to their chemical structure.[96] The methyl silicone molecule has a helical structure with the methyl groups pointing outward. A rise in temperature increases the mean intermolecular distance, causing a decrease in viscosity; this decrease in viscosity is partly offset by expansion of the helices, thus decreasing the mean intermolecular distance. The result of these two opposing actions is that the net mean intermolecular distance is only slightly changed with temperature, and hence there is little change in the viscosity. The substitution of larger groups, such as cyanopropyl or phenyl, disturbs the helical structure, resulting in a greater change in viscosity with temperature. Cross-linking the stationary phase not only improves the thermal stability, but extends the lower temperature limit as well. Stark and Larson[98] studied the separation of C_1–C_5 hydrocarbons on fused silica capillary columns coated with cross-linked and non-cross-linked methyl silicone gums. The non-cross-linked columns suffered a sizable loss of efficiency at $-60°$ and lower, while the cross-linked columns showed a more gradual loss of efficiency. The addition of cross-links apparently suppresses the glass transition mechanism. Additional advantages of cross-linked stationary phases have been presented by Grob and Grob.[99,100] These include (1) elimination of phase stripping due to large volume on-column or splitless injections, (2) the ability to solvent wash the column to remove soluble nonvolatile residues, and (3) the ability to resilylate the phase, which end caps terminal silanol groups, resulting in lower bleed rates. By rendering the phase immobile, on-column and splitless injections can be performed without the danger of phase stripping. In addition, the ability to rinse the column to remove contaminants can sometimes restore column performance in columns damaged by intractable samples. The ability of cross-linked capillary columns to withstand aqueous injections has been demonstrated by Blomberg et al.[101] The introduction of water by direct injection or headspace sampling should be done with caution since water has been reported to cause retention time shifts on fused silica capillary columns.[102] The injection of acids or bases together with water can lead to acid or base catalyzed hydrolysis of the stationary phase. This generates terminal silanol groups, which catalyze the more rapid decomposition of the polymer chain. Through the release of polysiloxane rings of different sizes the

---

[98] T. J. Stark and P. A. Larson, *J. Chromatogr. Sci.* **20**, 341 (1982).
[99] K. Grob and G. Grob, *J. Chromatogr.* **213**, 211 (1981).
[100] K. Grob and G. Grob, *J. High Res. Chromatogr.* **5**, 349 (1982).
[101] L. Blomberg, J. Buijten, K. Markides, and T. Wannman, *J. Chromatogr.* **239**, 51 (1982).
[102] W. F. Burns, D. T. Tingey, and R. C. Evans, *J. High Res. Chromatogr.* **5**, 504 (1982).

columns show higher bleed rates which may limit their application in high temperature analyses, and particularly in GC-MS.[100] Prior to their resilylation step, Grob and Grob[100] recommended solvent washing the column, which necessitates the use of immobilized stationary phases. The washing step was followed by resilylation with diphenyltetramethyldisilazane (DPTMDS), which reduces the bleed rate on both new and used columns. An additional benefit of resilylation with used columns may be an increased column inertness, due to a cleaning effect (large-molecular-weight polar contaminants may be transformed into more soluble species). The efficacy of the resilylation procedure has been confirmed by Jennings.[103]

To aid in the tentative identification of unknown compounds, the analysis on a polar phase such as Carbowax 20M is combined with the analysis on a non-polar phase such as Apiezon L, SE-30, OV-101, or OV-1. The determination of retention indices on two different liquid phases can greatly aid in the identification of unknowns, particularly with certain terpenes which produce very similar mass spectra. The nonpolar phases generally elute solutes according to their boiling point. Among the nonpolar phases, the methyl silicones are generally the phase of choice due to their high thermal stability. The saturated hydrocarbon phase, squalane is of little practical value due to its low maximum operating temperature (100°).

Identification Techniques

The combination of mass spectral data along with retention indices determined on both a polar and nonpolar stationary phase is usually sufficient to identify unknown isoprenoids (the analysis of known standards should be carried out in conjunction with the unknowns). In some cases other spectroscopic techniques such as UV, IR, or NMR may be necessary to confirm the identity of the unknown. Since these techniques require larger sample amounts than does GC-MS, the unknowns are frequently isolated by preparative GC. Debrauwere and Verzele[104] used preparative GC (glass column 12 m × 11 mm i.d. 10% Carbowax 20M on Chromosorb G 30-40 mesh) to isolate various monoterpene hydrocarbons present in black pepper oil. They subsequently obtained structural confirmation of the collected terpenes by IR and NMR spectroscopy. Sample decomposition in the thermal conductivity detector cell (TCD) may be a

[103] W. Jennings, *J. Chromatogr. Sci.* **21**, 337 (1983).
[104] J. Debrauwere and M. Verzele, *J. Chromatogr. Sci.* **14**, 296 (1976).

problem in preparative GC.[105,106] Verzele[106] reported that an acid-contaminated TCD can cause sample isomerization. To minimize sample decomposition, it may be preferable to use a glass effluent splitter at the column outlet and direct 2–5% of the flow to a flame ionization detector (FID).[1]

The use of different column temperatures can assist in the qualitative analysis of complex mixtures. Because liquid phase polarity is temperature dependent,[107] changes in solute elution order can occur at different temperatures. The magnitude of retention index shift with temperature ($\Delta I/10°$) may provide additional information regarding the compound. The identification of terpenes such as myrcene, *cis-β-ocimene* and *trans-β-ocimene* in essential oils may be aided by this technique.[108]

The retention index system proposed by Kováts[109,110] is the most widely used retention method. The index expresses the retention behavior of a compound as equivalent to that of a hypothetical *n*-paraffin hydrocarbon. The accuracy of the retention index is influenced by a number of factors.[111] Using glass capillary columns coated with Carbowax 20M, Shibamoto *et al.*[112] reported that the film thickness also influenced the Kováts retention index (the index increased as the film thickness increased). When determining the retention indices of compounds such as alcohols, aldehydes, and esters on a Carbowax 20M column, it may be desirable to use ethyl esters as standards rather than normal hydrocarbons.[112] Saeed *et al.*[113] reported the retention indices of 20 monoterpene hydrocarbons on glass capillary columns coated with SE-30 and Carbowax 20M. Under their chromatographic conditions, these researchers found only a slight influence of the column material (soda-lime glass vs borosilicate glass) on the retention indices of the monoterpenes. Jennings and Shibamoto[114] have tabulated the retention indices of about 1200 flavor and fragrance compounds determined on glass capillaries coated with

---

[105] E. von Rudloff, *Can. J. Chem.* **38**, 631 (1960).
[106] M. Verzele, in "Preparative Gas Chromatography" (A. Zlatkis and V. Pretorius, eds.), p. 230. Wiley (Interscience), New York, 1971.
[107] K. Grob and G. Grob, *Chromatographia* **17**, 481 (1983).
[108] J. Karlsen and H. Siwon, *J. Chromatogr.* **110**, 187 (1975).
[109] E. sz. Kováts, *Adv. Chromatogr. (N.Y.)* **1**, 229 (1965).
[110] E. sz. Kováts, *Helv. Chim. Acta* **41**, 1915 (1958).
[111] F. J. van Lenten, J. E. Conaway, and L. B. Rogers, *Sep. Sci.* **1**, 12 (1977).
[112] T. Shibamoto, K. Harada, K. Yamaguchi, and A. Aitoku, *J. Chromatogr.* **194**, 277 (1980).
[113] T. Saeed, G. Redant, and P. Sandra, *J. High Res. Chromatogr.* **2**, 75 (1979).
[114] W. Jennings and T. Shibamoto, "Qualitative Analysis of Flavor and Fragrance Volatiles by Glass Capillary Gas Chromatography." Academic Press, New York, 1980.

methyl silicone OV-101 and polyethylene glycol Carbowax 20M liquid phases. Their book also includes the mass spectra of many of these compounds. The flexibility of fused silica capillary columns allows the placement of two columns in a common injection port, permitting the simultaneous determination of retention indices on two different liquid phases.[115] With complex samples, however, peak correlations may become difficult.

The combined technique of GC-MS represents the most powerful method in general use today for separating and identifying isoprenoids in complex mixtures. Masada[116] used GC-MS to determine the composition of various essential oils. The addition of computer capability can increase both the speed and the information gained in interpreting unknown mass spectra. Adams et al.[117] have developed a computer method utilizing mass spectra and retention data to aid in the identification of terpenes. Using their weighted similarities method they were able to distinguish spectra that appeared to be visually identical. Details on the fragmentation pathways and mass spectral data of the terpenes and terpenoids are also available.[114,118,119] The mass spectral data of various irregular monoterpenes have also been reported.[120] Some compounds, such as terpene alcohols and their esters, often fail to produce a measurable molecular ion with conventional electron impact or even the less vigorous chemical ionization MS. Hendriks and Bruins[121] used negative ion chemical ionization MS to obtain molecular weight data of various oxygenated terpenoids. The use of OH^- negative ion MS provides a novel method for identifying the acid moiety of an ester through the production of $(RCOO)^-$ ions.

Microreaction procedures can be useful in determining the correct chemical structure. The technique of carbon skeleton analysis by hydrogenation and hydrogenolysis reactions was developed and studied by Beroza et al.[122–124] Kepner and Maarse[125] used this technique to study

[115] R. J. Phillips, R. J. Wolstromer, and R. R. Freeman, "Application Note," AN 228-16. Hewlett-Packard Co., Avondale, Pennsylvania, 1981.
[116] Y. Masada, "Analysis of Essential Oils by Gas Chromatography and Mass Spectrometry." Wiley, New York, 1976.
[117] R. P. Adams, M. Granat, L. R. Hogge, and E. von Rudloff, J. Chromatogr. Sci. 17, 75 (1979).
[118] H. Budzikiewicz, C. Djerassi, and D. H. Williams, "Structure Elucidation of Natural Products by Mass Spectrometry," Vol. II. Holden-Day, San Francisco, California, 1964.
[119] C. R. Enzell and I. Wahlberg, in "Biochemical Applications of Mass Spectrometry" (G. R. Waller and O. C. Dermer, eds.), Suppl. Vol., p. 311. Wiley, New York, 1980.
[120] W. W. Epstein, L. R. McGee, C. Dale Poulter, and L. L. Marsh, J. Chem. Eng. Data 21, 500 (1976).
[121] H. Hendriks and A. P. Bruins, J. Chromatogr. 190, 321 (1980).
[122] M. Beroza, in "Column Chromatography" (E. Kováts, ed.), p. 92. Sauerlaender, Aarua, Switzerland, 1970.

hydrogenolysis of various monoterpenes. Maarse[126] extended this work in his study of selected sesquiterpenes. The analysis of terpenes is complicated by the fact that the hydrogenation and hydrogenolysis reactions do not produce single products. The vapor phase hydrogenation used in the above procedures can produce isomeric hydrocarbons which do not always reflect the original carbon skeleton. In addition, when the original compound contains branches and/or aliphatic rings, GC-MS may be necessary to identify the reaction products. The carbon skeleton technique is most valuable when only limited amounts of sample are available. Bierl-Leonhardt and DeVilbiss[127] recently applied a carbon skeleton technique for the examination of terpene-type alcohols. They subjected microgram amounts of the parent compounds to Pt catalyst and lithium aluminum hydride (LiAlH) at 250° in a sealed tube. The reaction products were subsequently dissolved in solvent and identified by GC-MS.

A complementary technique to GC-MS that is beginning to see wider usage is GC-FTIR, which can provide functional group information and a "fingerprint" of an unknown. Since every organic compound possesses its own unique "fingerprint," the IR spectra can be an important identification tool. Kalasinsky and McDonald[5] used GC/FTIR to examine terpenes present in plant extracts. Using a 6-ft packed column [10% Carbowax 20M on Chromosorb G (80/100 mesh)], they were able to identify nine components in a pine extract. The nondestructive nature of this technique allows its coupling to other detection modes such as MS. Wilkins et al.[128] described a GC-FTIR-MS configuration in which the IR spectrometer lightpipe was connected in series to the MS. In the same paper they also detail a parallel configuration in which the GC effluent is split, with one portion of the sample going to the FTIR, while the remainder goes to the MS. Using the latter configuration they were able to identify 12 out of 28 components detected in peppermint oil.

Recent advances in lightpipe interface design and improved lightpipes have increased the sensitivity so that useful spectra can be obtained with as little as 10–100 ng per component.[128] The sensitivity is strongly influenced by the nature of the compound being examined. The limited sample capacity of capillary columns of normal dimensions (0.25 mm i.d., d_f = 0.25 μm) may inhibit their use in GC-FTIR when the identification of

---

[123] M. Beroza and R. Sarmiento, *Anal. Chem.* **35**, 1353 (1963).
[124] M. Beroza and R. Sarmiento, *Anal. Chem.* **36**, 1744 (1964).
[125] R. E. Kepner and H. Maarse, *J. Chromatogr.* **66**, 229 (1972).
[126] H. Maarse, *J. Chromatogr.* **106**, 369 (1975).
[127] B. A. Bierl-Leonhardt and E. D. DeVilbiss, *Anal. Chem.* **53**, 936 (1981).
[128] C. L. Wilkins, G. N. Giss, R. L. White, G. M. Brissey, and E. C. Onyiruika, *Anal. Chem.* **54**, 2260 (1982).

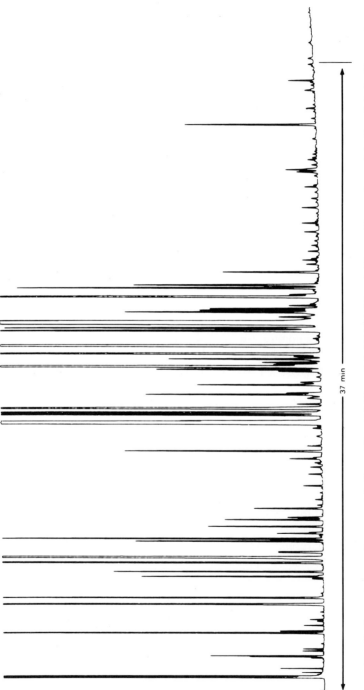

FIG. 6. Chromatogram of peppermint oil on a 30 m × 0.25 mm i.d. DB-WAX fused silica capillary column. Injection 0.1 μl with a pentane flush, split 1 : 100; temperature held at 35° for 0.5 min then programmed to 200° at 4°/min. (Reproduced courtesy of J & W Scientific.)

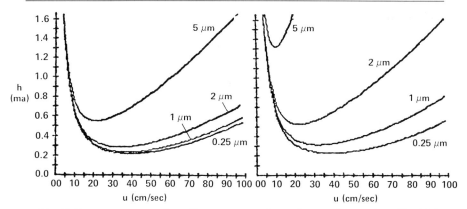

FIG. 7. Computer-generated van Deemter curves for various stationary phase film thicknesses. Left, SE-30, D_s taken as 1.67×10^{-5} cm^2/sec[136]; right, OV-225, D_s taken as 3.4×10^{-6} cm^2/sec[136]; Columns 30 m × 0.25 mm i.d.; hydrogen carrier with D_m taken as 0.34 cm^2/sec[137]; partition ratio $k = 5.0$. From W. Jennings, *Am. Lab.* **16**(1) 14 (1984). Copyright 1984 by International Scientific Communications, Inc.

minor components in a complex mixture is of interest. One approach for increasing the sample capacity is to use a wider bore column. Herres *et al.*[129] were able to identify 26 volatile components present in cherimoya (*Annona cherimolia*, Mill.) fruit using a wide bore capillary column (0.32 mm i.d., $d_f = 0.2$ μm) interfaced with an FTIR. An alternative approach to increase the sample capacity is to increase the stationary phase film thickness. Advances in phase immobilization have permitted the production of stable films with thicknesses up to 8 μm.[130] Ettre and co-workers[131,132] recently evaluated the theoretical aspects and performance of thick film columns. They used thick film columns with their GC-FTIR system and were able to identify 18 out of 24 components in a paint thinner. Sandra *et al.*[133] evaluated the efficiency of thick film capillary columns ($d_f = 5$ μm) and confirmed that nitrogen is the preferred carrier gas over hydrogen. Jennings[134,135] cautioned about the use of thick film columns with certain polar liquid phases because of their lower solute

[129] W. Herres, H. Idstein, and P. Schreier, *J. High Res. Chromatogr.* **6**, 590 (1983).
[130] K. Grob and G. Grob, *J. High Res. Chromatogr.* **6**, 133 (1983).
[131] L. S. Ettre, *Chromatographia* **17**, 553 (1983).
[132] L. S. Ettre, G. L. McClure, and J. D. Walters, *Chromatographia* **17**, 560 (1983).
[133] P. Sandra, I. Temmerman, and M. Verstappe, *J. High Res. Chromatogr.* **6**, 501 (1983).
[134] W. Jennings, *Rocky Mt. Conf., 25th, 1983* Paper 33 (1983).
[135] W. Jennings, *Am. Lab.* **16**, 14 (1984).

diffusivities (D_L). This results in sharply increasing h_{min} values and steeper van Deemter curves as the film thickness is increased (Fig. 7).[136,137] Capillary GC-FTIR was the subject of a recent review by Smith.[138] There is presently a lack of extensive vapor-phase infrared libraries which limits the use of GC-FTIR. This limitation can be expected to be resolved in the future. A more serious drawback is the relatively high price of the GC-FTIR system (comparable in price to a GC-MS system). The question of whether or not to invest in such a system must be judged on an individual basis and the analyst must decide whether the information gained justifies the purchase and operational costs to attain the spectra.

Terpenoids

Essential oils which commonly contain mono- and sesquiterpenes as constituents have also been extensively studied.[139] Largely due to advances in analytical methods, a great deal is known about their composition. Using MS, IR, and Kovats retention data, 35 mono- and sesquiterpenes were identified as components of coast redwood (*Sequoia sempervirens*) needle oil, 20 of which had not been previously reported.[140] Koyasako and Bernhard[141] used simultaneous distillation–extraction techniques to isolate the volatiles of kumquat peel. The extract was subjected to analysis using both Carbowax 20M (Fig. 8) and SE-30 WCOT fused silica capillary columns. Table I lists the elution order on the Carbowax 20M column of compounds identified in the essential oil. Using a deterpenation procedure, they were able to identify a number of minor oxygenated constituents.

Figure 9 shows a chromatogram of the essential oil of Saaz hops obtained with an SE-30 capillary column.[142] The essential oil, isolated by vacuum steam distillation, was analyzed by GC-MS (electron impact and chemical ionization); the identified components are listed in Table II.

[136] J. M. Kong and S. J. Hawkes, *J. Chromatogr. Sci.* **14**, 279 (1976).
[137] D. F. Ingraham, C. G. Shoemaker, and W. Jennings, *J. High Res. Chromatogr.* **5**, 227 (1982).
[138] S. L. Smith, *in* "Modern Aspects of Capillary Separation Methods" (M. Novotny, ed.). Wiley Interscience, in press.
[139] T. Shibamoto, *in* "Applications of Glass Capillary Gas Chromatography" (W. G. Jennings, eds.), Chromatogr. Sci. Ser., Vol. 15, p. 455. Dekker, New York, 1981.
[140] R. A. Okamoto, B. O. Ellison, and R. E. Kepner, *J. Agric. Food Chem.* **29**, 324 (1981).
[141] A. Koyasako and R. A. Bernhard, *J. Food Sci.* **48**, 1807 (1983).
[142] M. Verzele and P. Sandra, *in* "Applications of Glass Capillary Gas Chromatography" (W. G. Jennings, ed.), Chromatogr. Sci. Ser., Vol. 15, p. 535. Dekker, New York, 1981.

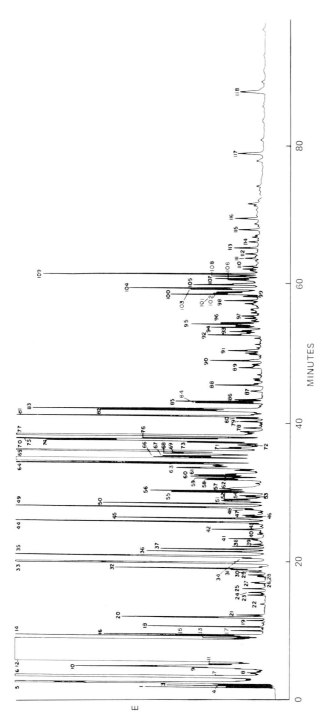

FIG. 8. Chromatogram of whole oil of kumquat split 1:100, on a 30 m × 0.25 mm i.d. WCOT fused silica capillary column coated with Carbowax 20M. Column temperature was held for 8 min at 70° then programmed to 180° at 2°/min. Injector and detector temperatures were 250°. For peak identification, see Table I. Reprinted from Journal of Food Science 1983 **45**(6), 1807. Copyright by Institute of Food Technologists.

TABLE I
COMPOUNDS IDENTIFIED IN THE ESSENTIAL OIL OF KUMQUAT[a]

Peak number	Compound	Peak number	Compound
1	Unknown	44	trans-2-Nonenal
2	Unknown	45	α-Bergamotene
3	1-Octene	46	β-Elemene
4	t-Butyl acetate	47	Isopulegol
5	Methyl isobutyrate	48	Sesquiterpene, MW = 204[b]
6	Isopropyl n-propionate	49	Sesquiterpene, MW = 204[b]
7	α-Pinene	50	Myrcenol
8	Hexanal	51	n-Undecanal
9	Sabinene	52	Linalyl n-propionate
10	β-Pinene	53	Citronellyl formate
11	Myrcene	54	γ-Elemene
12	Limonene	55	β-Terpineol
13	γ-Terpinene	56	Caryophyllene
14	trans-Ocimene	57	Sesquiterpene, MW = 204[b]
15	Isoamyl n-butyrate	58	Sesquiterpene, MW = 204[b]
16	p-Cymene	59	Citronellyl acetate
17	n-Octanol	60	Δ-Terpineol
18	Terpinolene	61	γ-Muurolene
19	2-Methyl heptanol	62	α-Terepineol
20	n-Hexanol	63	Neral
21	n-Heptyl formate	64	α-Humulene
22	cis-3-Hexanol	65	Terpinyl acetate
23	Cyclohexanol	66	Sesquiterpene, MW = 204[b]
24	n-Nonanol	67	Geraniol
25	2-Octanol	68	Dodecanal
26	Tetradecane	69	Unknown
27	Isobutyl 2-ethyl n-hexanoate	70	Carvone
28	Sesquiterpene, MW = 204[b]	71	Sesquiterpene, MW = 204[b]
29	1-Octen-3-ol	72	Unknown
30	cis-Linalool oxide	73	Sesquiterpene, MW = 204[b]
31	Sesquiterpene, MW = 204[b]	74	Sesquiterpene, MW = 204[b]
32	Dihydromyrcenol	75	Sesquiterpene, MW = 204[b]
33	trans-Linalool oxide	76	Sesquiterpene, MW = 204[b]
34	Sesquiterpene, MW = 204[b]	77	Nerol
35	Citronellal	78	γ-Cardinene
36	Norbornyl acetate	79	Carveol
37	2-Nonanol	80	Geranyl n-propionate
38	n-Decanal	81	Geraniol
39	Cyclohexyl n-butyrate	82	Unknown
40	Pentadecane	83	n-Undecanol
41	Linalool	84	Neryl n-butyrate
42	α-Copaene	85	2-Dodecanal
43	Unknown	86	Isoamyl decanoate

TABLE I (*Continued*)

Peak number	Compound	Peak number	Compound
87	Geranyl *n*-butyrate	103	Sesquiterpenoid, MW = 222[b]
88	Sesquiterpene, MW = 204[b]	104	Sesquiterpenoid, MW = 222[b]
89	Sesquiterpene, MW = 222[b]	105	Sesquiterpenoid, MW = 220[b]
90	Sesquiterpene, MW = 204[b]	106	Unknown
91	Geranyl tiglate	107	Unknown
92	Nerolidol	108	Unknown
93	Sesquiterpenoid, MW = 222[b]	109	Unknown
94	Sesquiterpenoid, MW = 220[b]	110	Unknown
95	Sesquiterpenoid, MW = 222[b]	111	Unknown
96	Sesquiterpenoid, MW = 220[b]	112	Unknown
97	Sesquiterpene, MW = 204[b]	113	Unknown
98	Sesquiterpenoid, MW = 222[b]	114	Unknown
99	Sesquiterpenoid, MW = 222[b]	115	Unknown
100	Sesquiterpenoid, MW = 222[b]	116	Unknown
101	Sesquiterpenoid, MW = 222[b]	117	Unknown
102	Sesquiterpenoid, MW = 222[b]	118	Unknown

[a] See Fig. 8.
[b] Compound tentatively identified.

A new class of hop constituents (variety Hersbrucker Spät), the tricyclic sesquiterpenes, was characterized by Tressl *et al.*[143] They identified more than 15 components (sesquiterpene hydrocarbons, epoxides, and alcohols) for the first time as hop constituents. Following distillation extraction with pentane–ether (1 : 1), the hop volatiles were fractionated by adsorption chromatography with a silica gel column. The fractions were concentrated and then subjected to capillary GC (WCOT column, 50 m × 0.32 mm i.d. coated with Carbowax 20M) and capillary GC-MS (WCOT column, 50 m × 0.32 mm i.d. coated with UCON). Figure 10 shows gas chromatograms of the sesquiterpene hydrocarbon fraction. It is evident that the variety Hersbrucker Spät contains more sesquiterpene hydrocarbons than the variety Northern Brewer (Table III).

Another possible method for concentrating terpenes is vapor phase sampling. Simon *et al.*[144] utilized a porous polymer trapping method (Tenax GC) to analyze the terpenoids from raw carrots, followed by elution of the trapped volatiles with diethyl ether. The composition of volatile organic compounds in the atmosphere has been investigated using

[143] R. Tressl, K.-H. Engel, M. Kossa, and H. Koppler, *J. Agric. Food Chem.* **31,** 892 (1983).
[144] P. W. Simon, R. C. Lindsay, and C. E. Peterson, *J. Agric. Food Chem.* **28,** 549 (1980).

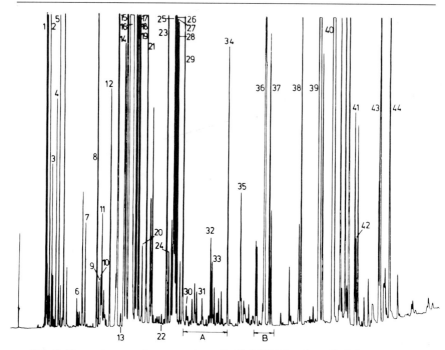

Fig. 9. Chromatogram of vacuum steam-distilled 1974 Saaz hop oil. Column: 90 m × 0.5 mm i.d. coated statically with SE-30. Column temperature-programmed from 80 to 200° at 2°/min; flow rate, 6 ml/min (H_2); sample size, 0.2 μl on column. For peak identification, see Table II. The two regions A and B contain compounds very important to hop aroma. Reprinted from "Applications of Glass Capillary Gas Chromatography," Chromatographic Science Series, Vol. 15 (W. G. Jennings, ed.), p. 535 (1981), by courtesy of Marcel Dekker, Inc.

vapor phase sampling with the sorbents polysorbimide and carbochrome.[145] Among the many hydrocarbons present, the terpenes limonene, α-pinene, and camphene were identified. Burns and Tingey[146] used GC-MS to examine the emission of hydrocarbon and oxygenated compounds (including monoterpenes) from live eucalyptus and sugar maple plants. Using a vapor sampling technique followed by direct injection, they analyzed hundreds of plant samples, each containing water, CO_2, and oxygen, on a bonded phase fused silica capillary column (WCOT column, 60 m × 0.25 mm i.d. DB-1, a bonded dimethylpolysiloxane, d_f = 1.0 μm). Despite these conditions, the column suffered no apparent degradation as evidenced by its retention time stability.

[145] B. V. Ioffe, V. A. Isidorov, and I. G. Zenkevich, *J. Chromatogr.* **142**, 787 (1977).
[146] W. F. Burns and D. T. Tingey, *J. Chromatogr. Sci.* **21**, 341 (1983).

TABLE II
COMPOSITION OF THE VACUUM-DISTILLED ESSENTIAL OIL OF SAAZ HOPS[a]

Peak number	Compound	Peak number	Compound
1	Acetone	22	cis-Linalool oxide[b]
2	Dimethylvinylcarbinol[a]	23	Methyl octanoate (branched)
3	3-Methylbutan-2-one	24	trans-Linalool oxide[b]
4	3-n-Propoxybut-1-ene	25	Linalool[b]
5	4-Methylpentan-2-one	26	Perilene
6	4-Mercapto-3-methylbutan-2-one (tentative?)	27	Aliphatic esters
		28	Aliphatic esters
7	Isoamyl acetate	29	Methyl octanoate
8	2-Methylbutyl-2-methylpropanoate	30	γ-Decalactone[b]
		31	Borneol
9	Methyl hexanoate	32	Terpineol-4
10	4-Hydroxy-4-methylpent-2-enoic acid lactone[b]	33	Methyl nonanoate (branched)
		34	Methyl nonanoate
11	Monoterpene	35	Geraniol
12	α-Pinene	36	Methyl-4,8-decadienoate[b]
13	6-Me-5-hepten-2-one	37	Methyl-4-decenoate[b]
14	Methyl heptanoate (branched)	38	α-Copaene
15	β-Pinene	39	β-Caryophyllene[b]
16	Myrcene	40	Humulene[b]
17	Isoamyl-2-methylpropanoate	41	Germacrene D
18	Pentyl-2-methylpropanoate	42	Calamenene
19	Methyl heptanoate	43	Oxygenated sesquiterpenoid
20	p-Cymene	44	Humuladienone[b]
21	Limonene		

[a] See Fig. 9.
[b] Typical for hop oil.

As mentioned earlier, retention data can be very useful in qualitative analysis. Retention data of various monoterpene alcohols on three different liquid phases have been reported.[147] Andersen and Falcone[148] have tabulated modified Kovats indices of selected sesquiterpenes on a variety of liquid phases.

The separation of terpene isomers by GC has generated considerable interest. Herling et al.[149] reported the complete separation of six positional isomers of p-menthene on a 30% $AgNO_3$–glycol column. The sepa-

[147] V. V. Bazyl'chik, N. P. Polyakova, A. I. Sedel'nikov, and E. A. Ionova, J. Chromatogr. **248,** 321 (1982).
[148] N. H. Andersen and M. S. Falcone, J. Chromatogr. **44,** 52 (1969).
[149] J. Herling, J. Shabtai, and E. Gil-Av, J. Chromatogr. **8,** 349 (1962).

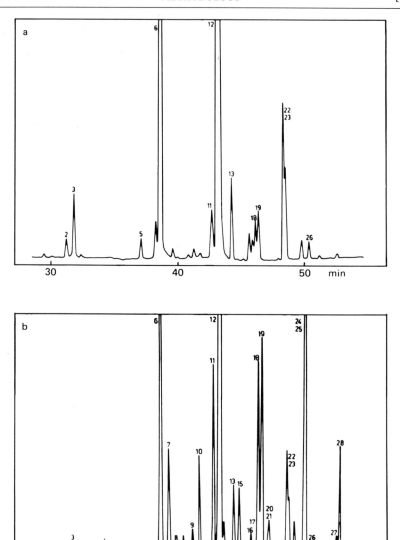

FIG. 10. Chromatograms of the sesquiterpene hydrocarbon fraction of hop varieties: (a) Northern Brewer; (b) Hersbrucker Spät. Column: 50 m × 0.32 mm i.d. coated with Carbowax 20M. Temperature programmed from 70 to 180° at 2°/min. Peak numbers correspond to the component numbers in Table III. Reprinted with permission from R. Tressl, K.-H. Engel, M. Kossa, and H. Koppler, *J. Agric. Food Chem.* **31**, 892 (1983). Copyright 1983 American Chemical Society.

TABLE III
SESQUITERPENES OF HOP (VARIETY HERSBRUCKER SPÄT)[a]

Peak number	Compound	Peak number	Compound
1	α-Cubebene	16	Germacrene D
2	α-Ylangene	17	δ-Guaiene
3	α-Copaene	18	β-Selinene
4	α-Gurjunene	19	α-Selinene
5	β-Cubebene	20	α-Muurolene
6	β-Caryophyllene	21	Bicyclogermacrene
7	Aromadendrene	22	γ-Cadinene
8	α-Guaiene	23	δ-Cadinene
9	γ-Element	24	Selina-4,7-diene
10	Alloaromadendrene	25	Selina-3,7-diene
11	β-Farnesene	26	α-Cadinene
12	Humulene	27	Calamenene
13	γ-Muurolene	28	Germacrene B
14	δ-Selinene	29	Cadalene
15	Viridiflorene		

[a] See Fig. 10.

ration of thujyl alcohol isomers has been reported.[150] Of several liquid phases tested, Carbowax 400 provided the best separation of menthol and menthone stereoisomers.[88] Enantiomers are analyzed by GC using two approaches. One way is to convert the enantiomers into diastereomers through the use of chiral reagents, followed by analysis on an achiral stationary phase. The second approach involves the direct separation of enantiomers on a chiral stationary phase.[151] However, to increase the substrate volatility and to improve the chromatographic selectivity, prior derivatization is generally preferred, even though enantiomers are separated by chiral stationary phases. The resolution of (±)-menthol and (±)-borneol as their tetra-O-acetyl-D-glucosides has been attained on a neopentyl glycol succinate column.[152] Konig et al.[153] achieved the enantiomer separation of various monoterpene alcohols as their isopropylurethanes on a capillary column coated with XE-60-S-valine-S-α-phenylethylamide (Fig. 11). The resolution of enantiomers on an achiral stationary phase has been attained through coinjection of racemates and a

[150] K. L. McDonald and D. M. Cartlidge, J. Chromatogr. Sci. **9**, 440 (1971).
[151] R. H. Liu and W. W. Ku, J. Chromatogr. **271**, 309 (1983).
[152] I. Sakata and K. Koshimizu, Agric. Biol. Chem. **43**, 411 (1979).
[153] W. A. Konig, W. Francke, and I. Benecke, J. Chromatogr. **239**, 227 (1982).

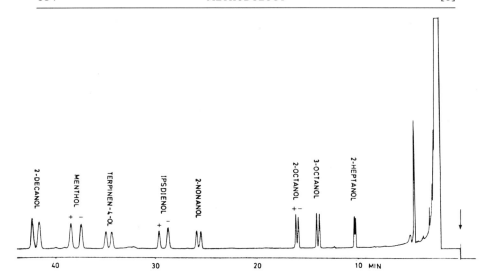

FIG. 11. Enantiomer separation of the isopropyl urethanes of chiral alcohols on a 40 m Pyrex glass capillary column coated with XE-60-S-valine-S-α-phenylethylamide. Column temperature, 120° (isothermal). From W. A. Konig, W. Francke, and I. Benecke, *J. Chromatogr.* **239**, 227 (1982).

volatile optically active resolving agent.[154] Various bicyclic ketone and alcohol enantiomers were separated through coinjection with a resolving agent such as *d*-camphor.

The presence of the isoprenoid hydrocarbons, *trans*-phyt-2-ene, phyt-1-ene, and phytane in bovine rumen liquor was confirmed through argenation TLC, GC-MS, and IR.[155] These isoprenoid compounds are apparently derived from dietary phytol (*trans*-3,7,11,15-tetramethyl-2-hexadecen-1-ol).

The separation of the diterpene resin acids (as methyl or TMS esters) is most effectively achieved through the use of at least two different capillary columns (one polar phase, one nonpolar phase). Samples should be treated with care, since diterpene resin acids have been observed to undergo isomerization during distillation.[156] The resin acid methyl esters should be dissolved in aromatic solvents, as they isomerize easily in chlorinated solvents such as CH_2Cl_2 and $CHCl_3$.[157] Retention data of 49 methyl esters of fatty and resin acids on glass capillary columns coated with butanediol–succinate polyester (BDS) and SE-30 have been re-

[154] P. D. Maestas and C. J. Morrow, *Tetrahedron Lett.* **14**, 1047 (1976).
[155] D. R. Body, *Lipids* **12**, 204 (1977).
[156] B. Holmbom, E. Avela, and S. Pekkala, *J. Am. Oil. Chem. Soc.* **51**, 397 (1974).
[157] M. Mayr, E. Lorbeer, and K. Kratzl, *J. Am. Oil. Chem. Soc.* **59**, 52 (1982).

ported.[158] With the BDS column there were no overlapping peaks while the SE-30 column could not resolve all fatty and resin acid esters in tall oil products. Mayr et al.[157] reported the Kovats indices of various diterpene resin acid methyl esters on four different glass capillary columns (SE-30, FFAP, DEGS, and BDS). Under the temperature programming conditions employed, the FFAP column separated the normally troublesome pair methyl levopimarate and methyl palustrate, but methyl 8,15-isopirmaradien-18-oate and methyl 8(14),15-pimaradien-18-oate coeluted. The retention characteristics of 77 diterpene resin acid methyl esters on glass capillary columns have been tabulated.[159] Six liquid phases (Silar 10C, BDS, SP-2330, SP-1000, SE-54, and SE-30), spanning a range of polarity, were investigated. Separation of methyl levopimarate and methyl palustrate was attained on the SE-30 column by lowering the oven temperature to 170°.

Hooper et al.[160] used GC-MS to analyze a mixture of pentacyclic triterpene acetates from a sow thistle extract. Though the peaks were incompletely resolved on the OV-1 capillary column, separation and mass spectra were obtained by mass chromatography. Retention data for various triterpenes have been reported by Ikekawa.[161] Derivatization of polar triterpenes can be used to increase their volatility and stability, hence minimizing or preventing degradation during analysis. Triterpene alcohols from various seed oils were characterized by Itoh et al.[162] Acetylation of the alcohol fraction was followed by separation by argenation TLC. Detection and identification of a number of minor components was facilitated through the combined use of two capillary columns (SCOT, Poly I-110, and OV-17) and GC-MS. The relative retention times and the methylene unit values of 42 tetracyclic triterpene alcohols and 22 pentacyclic triterpene alcohols, both as the acetate derivatives, were reported on OV-1 and OV-17 glass capillary columns (SCOT).[163] Triterpene alcohols were chromatographed as the trimethylsilyl (Me_3Si) ether derivatives by Gaydou et al.,[164] who examined the composition of neutral lipids in five Malagasy legume seed oils.

The analysis of carotenoids and related terpenoids by GC is limited by the thermal instability induced by the conjugated systems. While noncon-

[158] B. Holmbom, *J. Am. Oil Chem. Soc.* **54**, 289 (1977).
[159] D. O. Foster and D. F. Zinkel, *J. Chromatogr.* **248**, 89 (1982).
[160] S. N. Hooper, R. F. Chandler, E. Lewis, and W. D. Jamieson, *Lipids* **17**, 60 (1982).
[161] N. Ikekawa, this series, Vol. 15, p. 200.
[162] T. Itoh, T. Uetsuki, T. Tamura, and T. Matsumoto, *Lipids* **15**, 407 (1980).
[163] T. Itoh, H. Tani, K. Fukushima, T. Tamura, and T. Matsumoto, *J. Chromatogr.* **234**, 65 (1982).
[164] E. M. Gaydou, J.-P. Bianchini, and J. V. Ratovohery, *J. Agric. Food Chem.* **31**, 833 (1983).

jugated terpenoids are sufficiently stable to survive analysis, conjugated terpenoids usually require hydrogenation to prevent thermal destruction. Chloroform is reported to be the preferred solvent for carotenoid hydrogenation.[165] Derivatization of hydroxyl groups (as acetates or trimethylsilyl ethers) is recommended and should be performed after hydrogenation.[166] The retention data of an extensive number of hydrogenated carotenoids have been tabulated.[165,166] Since hydrogenation limits the structural characterization of carotenoids, analysis by HPLC should prove to be increasingly valuable.

Acyclic Isoprenoid Hydrocarbons

Acyclic isoprenoid hydrocarbons are another important class of naturally occurring compounds which are believed to be derived from acyclic terpenes. Their widespread occurrence in geological materials such as ancient sediments and petroleum make them important as biological markers. The analysis of these acyclic isoprenoids can reveal much information about the diagenesis and maturation of sedimentary organic matter. Like the terpenes, the acyclic isoprenoids can have head-to-head, head-to-tail, and tail-to-tail linkages. Isoprenoids ranging from C_{10} to C_{40} have been identified in petroleum and in sediments of different geological periods. Albaiges[167] recently identified head-to-tail homologs up to C_{45}.

The C_{18}, C_{19}, and C_{20} isoprenoids are postulated to be derived from the phytol side chain of chlorophyll.[168,169] Being major constituents, the analysis of the C_{18} species, norpristane (2,6,10-trimethylpentadecane), the C_{19} species, pristane (2,6,10,14-tetramethylpentadecane), and the C_{20} species, phytane (2,6,10,14-tetramethylhexadecane) have received strong attention. Their analysis is further complicated by the presence of stereoisomers (Fig. 12). Using the stationary phases, Apiezon L, $C_{87}H_{176}$, or Carbowax 20M, pristane and phytane are eluted before the n-alkanes C_{17} and C_{18}, respectively, whereas using the nonpolar silicone phase OV-101, pristane, and phytane elute after these n-alkanes.[170,171] The ratios of n-C_{17}/pristane and n-C_{18}/phytane have been used as indices of degradation of oil

---

[165] R. F. Taylor and M. Ikawa, this series, Vol. 67, p. 233.
[166] R. F. Taylor and B. H. Davies, *J. Chromatogr.* **103**, 327 (1975).
[167] J. Albaiges in J. R. Maxwell, and A. G. Douglas, eds., "Advances in Organic Geochemistry," p. 19. Pergamon, Oxford, 1979.
[168] J. R. Maxwell, C. T. Pillinger, and G. Eglinton, *Q. Rev., Chem. Soc.* **4**, 571 (1971).
[169] W. G. Meinschein, E. S. Barghoorn, and J. W. Schopf, *Science* **145**, 262 (1964).
[170] F. C. Trusell, *in* "Chromatography in Petroleum Analysis" (K. H. Altgelt and T. H. Gouw, eds.), Chromatogr. Sci. Ser., Vol. 11, p. 91. Dekker, New York, 1979.
[171] H. Borwitzky and G. Schomburg, *J. Chromatogr.* **240**, 307 (1982).

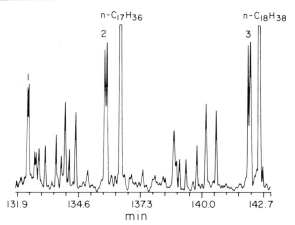

FIG. 12. Abridged chromatogram of diastereomers of norpristane (1), pristane (2), and phytane (3) in a petroleum cut. Column: 140 m × 0.27 mm i.d. coated with $C_{87}H_{176}$. In the overall run the column temperature was programmed from 100 to 230° at 0.8°/min. The relevant portion illustrated represents the temperature range 132–143°. The average linear velocity of hydrogen carrier was 35 cm/sec at 230°. From H. Borwitzky and G. Schomburg, J. Chromatogr. **240**, 307 (1982).

spills during the weathering process.[172,173] Since the ratios obtained are dependent on the resolution of the isoprenoid peaks,[174] it is important to use high-resolution GC systems. While the isoprenoid hydrocarbons are much more stable than the terpenes, their analysis is complicated by the fact that they occur in very complex mixtures.

In order to obtain a better quantitation of the isoprenoids, it may be desirable to use preseparation steps. The aromatic fraction of the sample may be separated on silica gel leaving the saturated fraction to be analyzed.[175] The n-alkanes can be removed by adduction with urea to concentrate the sample further.[176] Stationary phases such as squalane,[175] polyphenyl ether,[176] and Apiezon L[170,177] are commonly used. The limitations of polyphenyl ether and squalane as stationary phases have been discussed earlier in the chapter.

The use of capillary columns has permitted the resolution of the dia-

---

[172] M. Ehrhardt and M. Blumer, *Environ. Pollut.* **3**, 179 (1972).
[173] M. Blumer, M. Erhardt, and J. H. Jones, *Deep-Sea Res.* **20**, 239 (1973).
[174] O. C. Zafiriou, *Anal. Chem.* **45**, 952 (1973).
[175] K. E. H. Gohring, P. A. Schenck, and E. D. Engelhardt, *Nature (London)* **215**, 503 (1967).
[176] J. M. Gilbert, I. M. R. DeAndrade Bruning, D. W. Nooner, and J. Oro, *Chem. Geol.* **15**, 209 (1975).
[177] W. D. MacLeod, Jr., *J. Gas Chromatogr.* **6**, 591 (1968).

stereomers of various acyclic isoprenoids. The separation of acyclic isoprenoid alcohol isomers as their (+)-*trans*-chrysanthemates has been studied by Rowland and Maxwell.[178] Maxwell and co-workers used capillaries with polar stationary phases, BDS (butanediol succinate polyester), DEGS (diethylene glycol succinate), and DEGS-PEGS (polyethylene glycol succinate) (3:1) to achieve the partial separations of the diastereomers of pristane[179–182] and phytane.[183] Recently, Borwitzky and Schomburg[171] investigated the effect of elution temperature and stationary phase polarity on the resolution of the diastereomers of the C_{18}–C_{20} isoprenoids (norpristane, pristane and phytane). They found that lower temperatures produced better selectivities, and that Carbowax 20M more easily resolved the diastereomers than did the nonpolar stationary phases.

Isoprenoids ranging from C_{15} to C_{30} have been identified as the major constituents in the neutral lipids of various methanogenic and thermoacidophilic bacteria.[184,185] The minor constituents included the isoprenoids ranging from C_{14} to C_{20}. Due to the similarity between the isoprenoid content of bacteria studied and the occurrence of these compounds in sediment and petroleum, it was suggested that many isoprenoids isolated from geological samples may have been synthesized by bacteria.[185] The complex isomeric mixtures of isoprenoids in geological samples have necessitated the use of GC-MS for the identification of these compounds.[185,186]

---

[178] S. J. Rowland and J. R. Maxwell, *J. Chromatogr. Sci.* **21**, 298 (1983).
[179] J. R. Maxwell, R. E. Cox, R. G. Ackman, and S. N. Hooper, *Adv. Org. Geochem., Proc. Int. Meet., 5th., 1971* p. 277 (1972).
[180] R. E. Cox, J. R. Maxwell, R. G. Ackman, and S. N. Hooper, *Can. J. Biochem.* **50**, 1238 (1972).
[181] R. L. Patience, S. J. Rowland, and J. R. Maxwell, *Geochim. Cosmochim. Acta* **42**, 1871 (1978).
[182] A. S. Mackenzie, R. L. Patience, and J. R. Maxwell, *Geochim. Cosmochim. Acta* **44**, 1709 (1980).
[183] R. L. Patience, D. A. Yon, G. Ryback, and J. R. Maxwell, *Adv. Org. Geochem., Proc. Int. Meet. Org. Geochem., 9th, 1979* p. 287 (1980).
[184] T. G. Tornabene, T. A. Langworthy, G. Holzer, and J. Oró, *J. Mol. Evol.* **13**, 73 (1979).
[185] G. Holzer, J. Oró, and T. G. Tornabene, *J. Chromatogr.* **186**, 795 (1979).
[186] J. Albaiges, J. Borbon, and M. Gassiot, *J. Chromatogr.* **204**, 491 (1981).

[7] High-Performance Liquid Chromatography of Carotenoids

By MANFRED RUDDAT and OSCAR H. WILL III

While carotenoids were among the first substances separated by chromatography, they are apparently among the last ones to be separated by HPLC. Isolation of carotenoids usually involves processing the extract in at least two chromatographic steps, adsorption chromatography on an open column of alumina[1] or magnesium oxide/Hyflo-Super-Cel[2] followed by TLC[3,4] or in older methods by paper chromatography.[5] As relatively large volumes of solvents must be evaporated in these procedures, the repeated exposure of the inherently unstable carotenoids to conditions favoring partial isomerization and/or oxidative changes is unavoidable.

TLC separates carotenoids easily and rapidly into esthetically pleasing, highly visible, colored, sharp bands; recovery, however, is rarely quantitative and often impossible as the compounds begin to bleach before the support layer is scraped off the plate for qualitative and quantitative analysis. In HPLC these difficulties are avoided because the two chromatographic steps are combined; the saponified extract can be applied directly to the HPLC column. Carotenoids are eluted from the HPLC column as pure compounds in relatively small volumes available for further analysis. Spectrophotometric scanning for qualitative analysis can also be achieved on line with a stop-flow system,[6] and the carotenoids can also be quantitated with a calibrated recording integrator.

Gas chromatography (GC) rivals HPLC in resolution, reproducibility, and quantitation, but carotenoids require perhydrogenation before use in GC. Inherent thermal instability of the long carbon chain of carotenoids makes GC impractical.[7] As HPLC methods for carotenoids are now available, it is the analytical method of choice for separation, quantitation, and structural characterization of the naturally occurring and synthetic carotenoids as well as the most important metabolic product of β-carotene, vitamin A.

[1] S. C. Kushwaha, M. Kates, and J. W. Porter, *Can. J. Biochem.* **53**, 816 (1976).
[2] K. A. Buckle and F. M. Rahman, *J. Chromatogr.* **171**, 385 (1979).
[3] A. Hager and T. Meyer-Bertenrath, *Planta* **69**, 198 (1966).
[4] G. Britton and T. W. Goodwin, this series, Vol. 18, Part C, p. 654.
[5] Z. Sesták, *Photosynthetica* **14**, 239 (1980).
[6] M. Zakaria, K. Simpson, P. R. Brown, and A. Krstulovic, *J. Chromatogr.* **176**, 109 (1979).
[7] R. F. Taylor and M. Ikawa, this series, Vol. 67, p. 233.

The majority of the carotenoids are tetraterpenoids synthesized from eight isoprene pyrophosphate units, which are joined head-to-tail except for the head-to-head condensation at the center of the molecule. Hydrocarbon carotenoids are called carotenes, and oxygenated carotenoids, xanthophylls. Structures of most of the carotenoids discussed in this chapter are presented in Fig. 1. The commonly accepted trivial carotenoid name will be used throughout, accompanied at first mention by the semisystematic name recommended by the IUPAC–IUB Commission on Biochemical Nomenclature.

This chapter describes HPLC of carotenes, xanthophylls, vitamin A, butylated hydroxytoluene, a frequently used antioxidant, and ergosterol, a major component of fungal carotenoid extracts.

HPLC Equipment

An assortment of commercially available high-performance liquid chromatographs are suitable for carotenoid HPLC. We used an instrument assembled from components and generously placed at our disposal by Dr. F. J. Kézdy, Department of Biochemistry, University of Chicago. Assembling a chromatograph from components has the advantage of selecting parts over which one would have no control in a commercial instrument.

Two ISCO high pressure pumps (Instrument Specialty Corp., Lincoln, NE) were controlled by an ISCO DialaGrad programmer, model 384, for the production of gradients. Operating pressure was measured with a Viatran pressure transducer, model 108 D721 (Vinton Corp., Buffalo, NY), connected to an ISCO pressure monitor, model 1590. Samples were injected through a Rheodyne 7010 injector with a 48 μl loop (Rheodyne, Inc., Berkeley, CA). The injector was equipped with a magnetic switch to activate the Hewlett-Packard recorder/integrator.

Carotenoids were chromatographed on the following column packings: Ultrasphere ODS 5 μm, 25 cm × 4.6 mm i.d. (Altex Scientific Inc., Berkeley, CA); IBM Instruments Octadecyl, 250 mm × 4.5 mm i.d., 5 μm (IBM Instruments, Inc., Danbury, CT); Ultrasphere Octyl 5 μm, 25 cm × 4.6 mm i.d. (Altex Scientific); and Ultrasphere Si 5 μm, 25 cm × 4.6 mm (Altex Scientific). All columns were stainless steel and connected to an Altex prefilter.

The column eluent was monitored with a Perkin-Elmer (Norwalk, CT) LC-55 spectrophotometer (variable wavlength from 190 to 750 nm). A

FIG. 1. Structure of carotenoids. **I**, Phytoene; **II**, phytofluene; **III**, ζ-carotene; **IV**, neurosporene; **V**, lycopene; **VI**, β-zeacarotene; **VII**, γ-carotene; **VIII**, β-carotene; **IX**, α-carotene; **X**, zeaxanthin; **XI**, lutein.

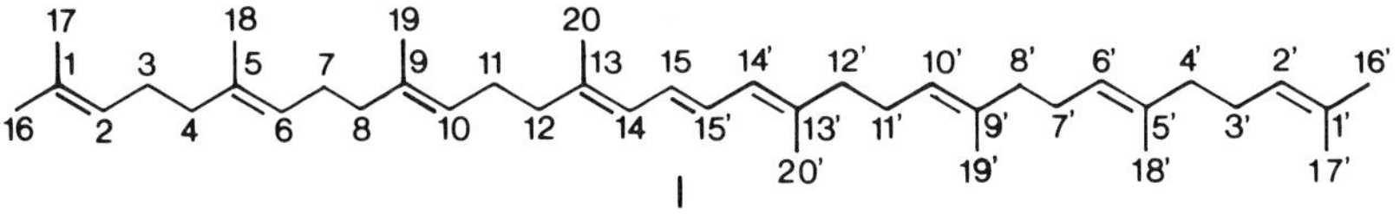

I

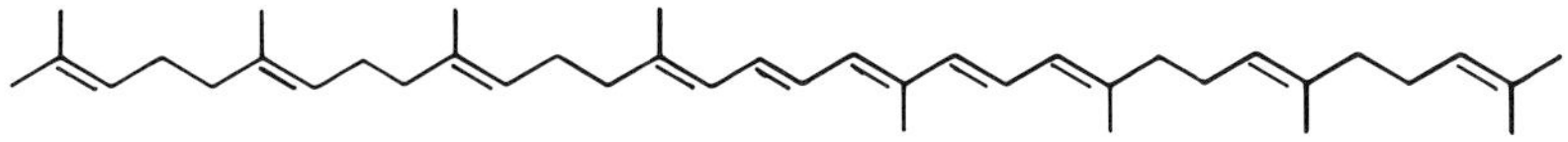

II

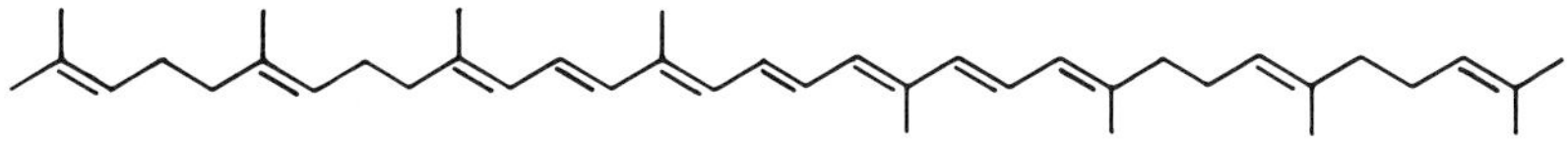

III

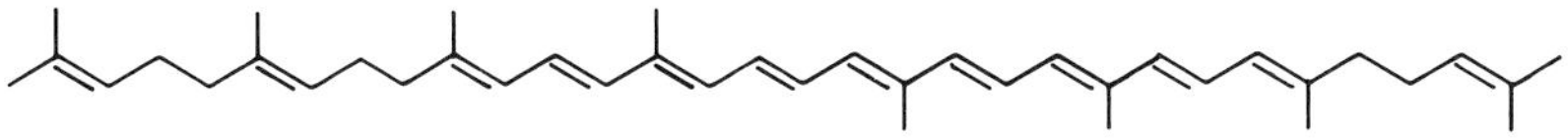

IV

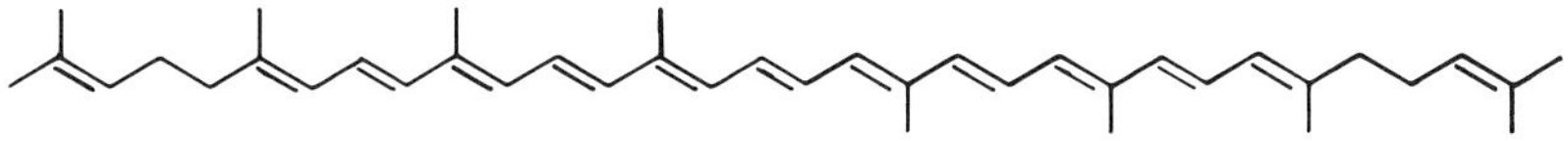

V

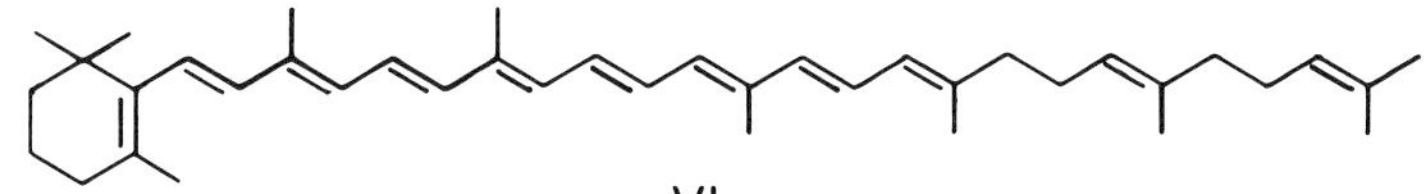

VI

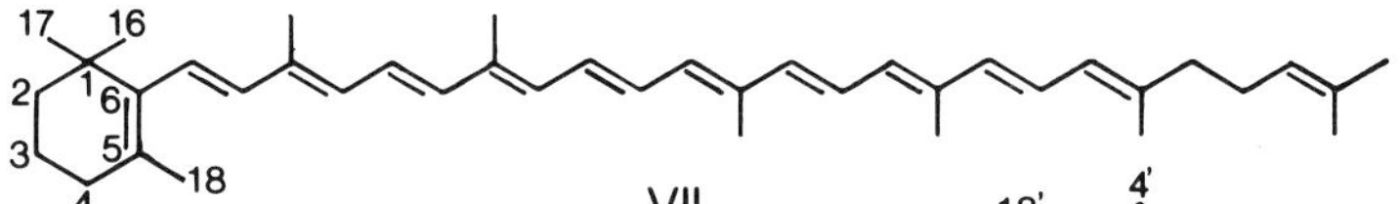

VII

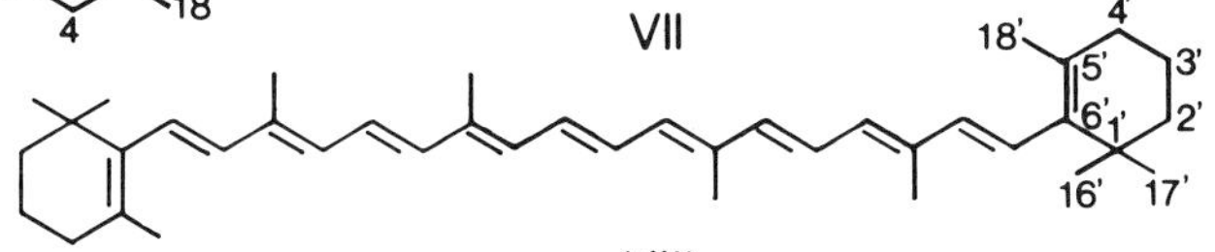

VIII

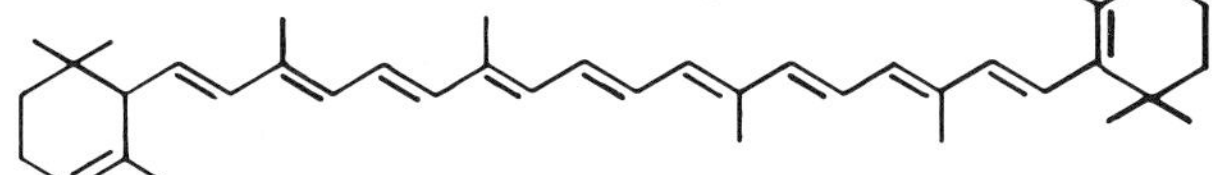

IX

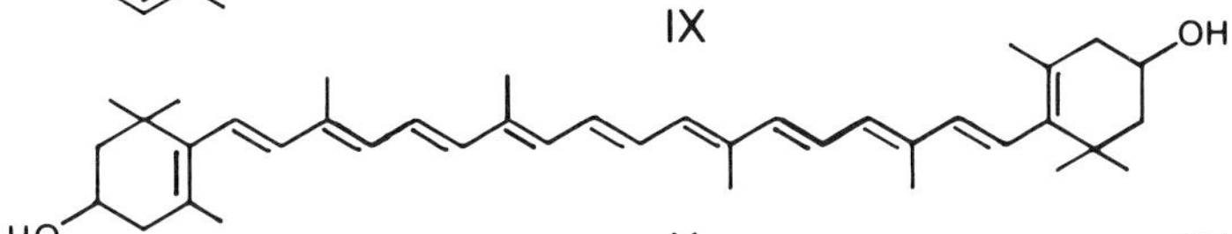

X

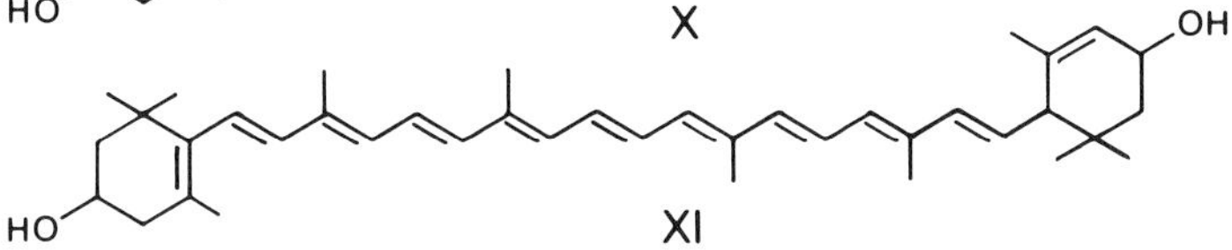

XI

high-pressure microcuvette was used in the spectrophotometer. The eluent of colored carotenoids was monitored at 425, 450 or 475 nm; that of colorless carotenoids, at 283, 330 nm; and of retinoids, at 315 or 340 nm. Elution profiles were recorded, the peak areas integrated, and the retention times printed with a Hewlett-Packard HP3390A reporting integrator (Avondale, PA).

A needle-valve stream-splitter upstream from the check valve in the waste line close to the spectrophotometer-cuvette allowed collection of desired fractions. Samples were collected from the ascending and descending slope of an absorption peak to assess the purity of the carotenoid under the absorption peak. Electronic absorption spectra for identification and quantitative determination of carotenoids[8] were recorded with an Aminco DW spectrophotometer (Silver Spring, MD) or a Perkin-Elmer Lambda 5 spectrophotometer.

Acetonitrile, 2-propanol, 1-propanol, methanol, hexane, isooctane, acetone, and tetrahydrofuran were HPLC grade from Burdick and Jackson (Muskegon, MI) and ethanol and dimethyl sulfoxide were from Fisher Scientific (Fair Lawn, NJ). Solvents were filtered through a Pyrex glass frit and degassed.

Carotenoids

Samples of β-carotene (β,β-carotene), β-zeacarotene (7′,8′-dihydro-β,ψ-carotene), and zeaxanthin [(3R,3′R)-β,β-carotene-3,3′-diol] were gifts from Hoffmann-LaRoche (Basel, Switzerland). Lutein [(3R,3′R, 6′R)-β,ε-carotene-3,3′-diol] and zeaxanthin [(3R,3′R)-β,β-carotene-3,3′-diol] were gifts of Dr. G. Marinetti, the University of Rochester. Lycopene (ψ,ψ-carotene), α-carotene [(6′R)-β,ε-carotene], β-carotene, retinol, and retinyl acetate were purchased from Sigma Chemical Company (St. Louis, MO).

Carotenoids were extracted from shake culture-grown sporidia of 28 mutant strains of *Ustilago violacea*[9] and from ripe tomatoes obtained at a local market. Carotenoid extraction followed established isolation procedures.[10–12] Standard precautions were taken in handling carotenoids to protect them from exposure to light, oxygen, acid and heat.

[8] S. Liaaen-Jensen and A. Jensen, this series, Vol. 23, p. 586.
[9] O. H. Will, III, M. Ruddat, E. D. Garber, and F. J. Kézdy, *Curr. Microbiol.* **10**, 57 (1984).
[10] G. Britton, this volume [5].
[11] B. H. Davies, *in* "Chemistry and Biochemistry of Plant Pigments" (T. W. Goodwin, ed.), Vol. 2, p. 38. Academic Press, New York, 1976.
[12] M. Ruddat and E. D. Garber, *in* "Secondary Metabolism and Differentiation in Fungi" (J. W. Bennett and A. Ciegler, eds.), p. 95. Dekker, New York, 1983.

Operating Conditions

Carotenoids were always injected dissolved in the solvent system used for elution; in gradient elution, the initial phase of the solvent gradient was used. The injection volume was 15 to 45 μl.

The solvent system, 2-propanol–acetonitrile–water (30:63:7 v/v/v), used in most of our investigations of carotenoids with ODS columns, was selected after exploring several other solvents in various combinations and ratios, including 2-propanol in combination with methanol, ethanol, or dimethyl sulfoxide (DMSO), respectively; methanol or dichloromethane in combination with acetonitrile; DMSO in combination with methanol, ethanol, acetonitrile, or tetrahydrofuran, respectively; and finally tetrahydrofuran/ethanol. The more polar solvent usually contained up to 15% water. While some of these solvents gave good separations, 2-propanol–acetonitrile–water was generally superior.

Columns were flushed with solvent between runs until the pressure returned to the initial level and the recorder-trace was stable. Reproducibility of separation and retention times depends on this equilibration phase.

The column was surrounded by a water-jacket to control column temperature. While the retention times were lowered at 27, 37, or 50° and increased at 15°, separation of carotenoids was not appreciably improved; consequently, chromatography was at 21°.

The lower limit of carotenoid detection depends on detector sensitivity, noise level, and column efficiency. In our system, 2–10 ng of carotenes was routinely detected. Peak shape remained the same over a large range of concentrations. High reproducibility of retention times and peak areas was obtained with a variation coefficient of less than 5% for retention times. With different columns, even from the same manufacturer, differences in retention times may occur.

Separation of Carotenoids by Reverse-Phase Chromatography

In reverse-phase chromatography, polar compounds elute earlier than nonpolar ones. Xanthophylls elute, therefore, before carotenes. Lycopene as the most polar carotene will elute as the first carotene, but after the xanthophylls, and phytoene as the last carotene.

Carotene Extracts from Ustilago violacea. Excellent separation of carotenes from saponified extracts of sporidia of *U. violacea* was obtained with ODS columns from Altex or IBM, which were eluted with a shallow concave gradient from 30:70 to 55:45 of 2-propanol and acetonitrile/water (9:1). Lycopene, γ-carotene (β,ψ-carotene), β-carotene, and

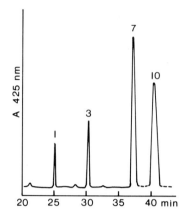

FIG. 2. Chromatogram of an extract of the orange strain of *U. violacea* obtained with an Ultrasphere ODS 5 μm (Altex) column. Eluent: 2-propanol and acetonitrile/water (9:1) in a gradient from 30:70 to 55:45. Flow rate 40 ml/hr; detection at 425 nm, phytoene at 283 nm. 1, Lycopene; 3, γ-carotene; 7, β-carotene; 10, phytoene.

phytoene (7,8,11,12,7′,8′,11′,12′-octahydro-ψ,ψ-carotene) were separated from an extract of the orange strain (2.D 37291 S) of *U. violacea* (Fig. 2). The eluent was monitored at 425 nm.

The area under the peaks in the recorder graph does not correspond to the amount of carotene, because the maximal absorption of the separated

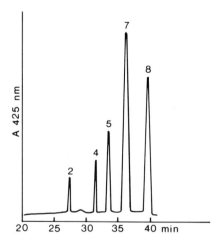

FIG. 3. Chromatogram of an extract of the yellow strain of *U. violacea* obtained with an Ultrasphere ODS 5 μm (Altex) column. Condition as in Fig. 2. 2, Neurosporene; 4, ζ-carotene; 5, β-zeacarotene; 7, β-carotene; 8, *cis*-β-zeacarotene.

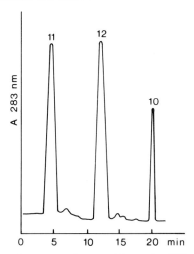

FIG. 4. Chromatogram of an extract from a white strain of *U. violacea* obtained with an Ultrasphere ODS 5 μm (Altex) column. Eluent: 2-propanol and acetonitrile/water (9:1), gradient from 50:50 to 65:35. Flow rate 40 ml/hr, detection at 283 nm. 10, Phytoene; 11, butylated hydroxytoluene; 12, ergosterol.

carotenes does not coincide with the fixed wavelength at which they were detected and the absorption coefficients of individual carotenes differ. Calibration of the recorder/integrator with carotene standards, by plotting the ratios of the peak areas of extracted carotenes against their standards, provides quantitation.

In the yellow strain (1.C2) of *U. violacea* five carotenes were separated and identified: neurosporene (7,8-dihydro-ψ,ψ-carotene), ζ-carotene (7,8,7'8'-tetrahydro-ψ,ψ-carotene), β-zeacarotene, β-carotene and *cis*-β-zeacarotene (Fig. 3). The eluents were monitored at 425 and 283 nm. The effluent from the separation of extracts from a white strain (2.C419) monitored at 283 nm contained phytoene, ergosterol, and butylated hydroxytoluene (Fig. 4). Butylated hydroxytoluene was added to all extracts as an antioxidant and served also as convenient external standard to monitor changes between runs. Even though ergosterol was precipitated by low temperature, its concentration remained relatively high in the extract.

In the 2-propanol–acetonitrile–water gradient, the carotenes eluted at 48 to 50% 2-propanol; the gradient, however, was routinely started at 30% 2-propanol to remove polar components from the extract and, more importantly, to check the extract for xanthophylls. In chromatographing extracts for phytoene, the gradient was started at 50% 2-propanol and gradually increased to 65% 2-propanol, thus saving time by eluting phytoene earlier (Figs. 2 and 4).

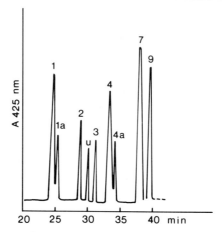

FIG. 5. Chromatogram of a tomato extract (carotene fraction) obtained with an IBM ODS 5 μm column. Conditions as in Fig. 2. Detection at 425 nm, phytofluene at 330 nm. 1, Lycopene; 1a, cis-lycopene; 2, neurosporene; 3, γ-carotene; 4, ζ-carotene; 4a, cis-ζ-carotene; 7, β-carotene; 9, phytofluene; u, unidentified.

Carotene Extract of Tomatoes. The carotene fraction of a tomato extract was chromatographed on an IBM ODS-column with 2-propanol–acetonitrile–water as the mobile phase (Fig. 5). Lycopene, neurosporene, γ-carotene, ζ-carotene, β-carotene, and phytofluene (7,8,11,12,7',8'-hexahydro-ψ,ψ-carotene), as well as *cis*-lycopene and *cis*-

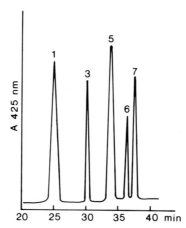

FIG. 6. Separation of carotene standards on an Ultrasphere ODS 5 μm (Altex) column. Conditions as in Fig. 2. 6, α-Carotene.

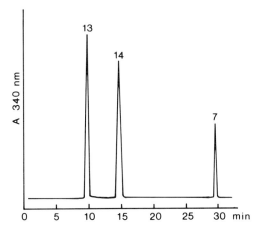

FIG. 7. Separation of standards of retinol, retinyl acetate and β-carotene on an Ultrasphere ODS 5 μm (Altex) column. Eluent: isocratic 5:95 of 2-propanol and acetonitrile–water (9:1), followed by a gradient reaching final ratio of 60:40. Flow rate 40 ml/hr, detection at 370 nm. 13, Retinol; 14, retinyl acetate; 7, β-carotene.

ζ-carotene, were well resolved. Phytofluene was monitored at 330 nm; all other carotenes at 425 nm.

Carotene Standards and Retinoids. Commercially available carotene standards were separated on an Altex ODS-column (Fig. 6). The retention times of carotenes from extracts and those of the standards were identical. The efficiency of the ODS column is indicated by the separation of α- and β-carotene.

The importance of β-carotene as the immediate precursor for vitamin A is well appreciated. HPLC of retinol and β-carotene on reverse-phase column was therefore included. Retinol, retinyl acetate, and β-carotene, all standards, were separated on an Altex ODS-column by eluting first isocratically with 2-propanol–acetonitrile–water 5:85.5:9.5, followed by a gradient in which 2-propanol was increased to 60% to elute β-carotene (Fig. 7). HPLC of retinol and its metabolites and synthetic derivatives has been thoroughly investigated.[13–15]

Xanthophylls. Standards of lutein and zeaxanthin were separated isocratically on an ODS-column from Altex (Fig. 8). The mobile phase was 2-propanol–acetonitrile–water 5:85.5:9.5.

Isocractic separation is advantageous in that the reequilibration time is eliminated. This can be, however, a disadvantage with samples of un-

[13] A. M. McCormick, J. L. Napoli, and H. F. DeLuca, this series, Vol. 67, p. 220.
[14] C. R. Broich, L. E. Gerber, and J. W. Erdman, Jr., *Lipids* **18**, 253 (1983).
[15] A. L. Ross, *Anal. Biochem.* **115**, 324 (1981).

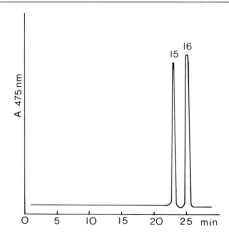

FIG. 8. Separation of lutein and zeaxanthin with an Ultrasphere ODS 5 μm (Altex) column. Eluent: 2-propanol–acetonitrile–water 4:85.5:9.5, isocratic. Flow rate 40 ml/hr, detection at 475 nm. 15, Lutein; 16, zeaxanthin.

known composition, where it may become difficult to know if the last compound has been eluted.

C_8 *columns.* Octylsilane columns have lower hydrophobicity than ODS-columns and therefore give decreased retention times, thereby saving time in the separation. However, for the separation of carotenes, we found that the hydrophobicity of a C_8 Altex column was insufficient. Similar observations were made by Zakaria et al.[6] with a LiChroscorp RP-8 (Brownlee, Santa Clara, CA) stationary phase. Octyl-columns are, however, well suited for xanthophyll chromatography. Neoxanthin [(3S,5R,6R,3'S,5'R,6'S)-5',6-epoxy-6,7-didehydro-5,6,5',6'-tetrahydro-β,β-carotene-3,5,3'-triol], violaxanthin [(3S,5R,6S,3'S,5'R,6'S)-5,6,5'6'-diepoxy-5,6,5',6'-tetrahydro-β,β-carotene-3,3'-diol], lutein 5,6-epoxide (5,6-epoxy-5,6-dihydro-β,ε-carotene-3,3'-diol), antheraxanthin (5,6-epoxy-5,6-didehydro-β,β-carotene-3,3'-diol), and lutein were well resolved on a LiChrosorp RP-8 column; and the retention times were about 50% less than that of an ODS-column, while the resolution was nearly the same.[16]

CN-column. Cyanopropylsilane bonded to microparticulate silica (Spherisorb S 5-CN) was successfully used as stationary phase to separate isomers of synthetic rhodoxanthin (4',5'-didehydro-4,5'-retro-β,β-carotene).[17] With hexane-isopropyl acetate-acetone (76:17:7), 6,6'-*cis*-rhodoxanthin was separated from 6-*cis*- and all-*trans*-rhodoxanthin with retention times of 9, 12 and 16 min, respectively.

[16] T. Braumann and L. H. Grimme, *Biochim. Biophys. Acta* **637**, 8 (1981).
[17] G. Englert and M. Vecchi, *J. Chromatogr.* **235**, 197 (1982).

STATIONARY AND MOBILE PHASES FOR CAROTENOID HPLC

Stationary phase	Mobile phase	Carotenoid source	Reference
Partisil-PXS-10/250DS-2 (Whatman)	8.5% chloroform–acetonitrile	Tomato carotenes	Zakaria et al.[a]
μBondapak C$_{18}$ (Waters)	8% chloroform–acetonitrile		Zakaria et al.[a]
Nucleosil 10 C$_{18}$ (Macherey-Nagel)	Methanol–H$_2$O (3:1), tetrahydrofuran–acetonitrile (1:1)	Daffodil carotene, polyprenoid alcohols	Beyer et al.[b]
Sil 60-RP 18 (Riedel-de Haen)	Methanol–acetonitrile (25:75) H$_2$O	Algal and spinach xanthophylls and carotenes	Braumann and Grimme[c]
Partisil C$_{18}$ (Whatman)	Methanol–acetone–H$_2$O (67.5:25:7.5)	Chloroplast carotenoids	Wellburn et al.[d]
Waters C$_{18}$ 10 μm	100% acetonitrile	Carotenes, suspension culture cells	Fosket et al.[e]
Sulpecosil LC-18 (Sulpelco)	Methanol–acetonitrile–chloroform (47:47:6)	Serum carotenes, retinoids	Broich et al.[f]
Ultrasphere-ODS 5 μm (Beckman)	88% acetonitrile–tetrahydrofuran (3:1) and 12% methanol–1% ammonium acetate (3:2)	Plasma carotenes	Peng and Beaudry[g]
Hypersil-ODS 5 μm (Shandon)	Methanol–H$_2$O (8:2), methanol–acetone (8:2)	Algal carotenoids and chlorophylls	Mantoura and Llewellyn[h]
Spherisorb S 5-CN 5 μm	n-hexane–isopropyl acetate–acetone (76:17:7)	Rhodoxanthin isomers	Englert and Vecchi[i]
Alox-T 5 μm alumina (Merck)	Dichloromethane–hexane (35:65)	Algal carotenes	Ben-Amotz et al.[j]

[a] M. Zakaria, K. Simpson, P. R. Brown, and A. Krstulovic, *J. Chromatogr.* **176**, 109 (1979).
[b] P. Beyer, K. Kreuz, and H. Kleining, *Planta* **150**, 435 (1980).
[c] T. Braumann and L. H. Grimme, *Biochim. Biophys. Acta* **637**, 8 (1981).
[d] A. R. Wellburn, D. C. Robinson, and F. A. M. Wellburn, *Planta* **154**, 259 (1982).
[e] D. E. Fosket, D. N. Radin, and M. Guiltinan, in "Biosynthesis and Function of Plant Lipids" (W. W. Thomson, J. B. Mudd, and M. Gibbs, eds.), Monogr., p. 195. Am. Soc. Plant Physiol., Rockville, Maryland, 1983.
[f] C. R. Broich, L. E. Gerber, and J. W. Erdman, Jr., *Lipids* **18**, 253 (1983).
[g] Y.-M. Peng and J. Beaudry, *J. Chromatogr.* **273**, 410 (1983).
[h] R. F. C. Mantoura and C. A. Llewellyn, *Anal. Chim. Acta* **151**, 297 (1983).
[i] G. Englert and M. Vecchi, *J. Chromatogr.* **235**, 297 (1982).
[j] A. Ben-Amotz, A. Katz, and M. Avron, *J. Phycol.* **18**, 529 (1982).

Separation of Carotenoids by Normal-Phase Chromatography

HPLC of carotenes on an adsorption column (Ultrasphere Si) with hexane or with the less polar isooctane as mobile phase did not satisfactorily separate carotenes. The least polar carotenes eluted with the solvent front, and the elution times of the more polar carotenes were insufficient. Similar results with silica adsorption columns have been reported by Fiksdahl et al.[18]

Excellent separation of cis-ζ-carotene, β-carotene, β-zeacarotene, trans-ζ-carotene, γ-carotene, and neurosporene extracted from the W_3Bu mutant of *Euglena gracilis* var. *bacillaris,* however, was obtained with an alumina column, Spherisorb A5Y.[19] The mobile phase consisted of a gradient of hexane–methyl *t*-butyl ether–acetonitrile from 97:3:0.08 to 60:40:0.2 and finally to 20:80:2.

In contrast to carotenes, xanthophylls appear to separate well on silica adsorption columns.[18]

Comments and Conclusions

Reverse-phase HPLC is the method of choice for carotenoid separation. It is rapid and selective with high reproducibility and recovery. On-line spectrophotometric scanning capability affords direct, convenient identification and assessment of the purity of the separated compounds. The hydrophobic interactions among nonpolar stationary phase, solute, and mobile phase in reverse-phase chromatography are much weaker than the ionic or polar forces in straight-phase chromatograhy, thus reducing loss of carotenoids or artifact formation during reverse-phase HPLC.

The assortment of commercially available reverse-phase columns is bewilderingly large. Because the degree of bonding nonpolar alkylsilanes to silanol groups varies with manufacturers, differences in resolution may be expected. Careful evaluation of the column is, therefore, required. Stationary phases and mobile phases commonly used in carotenoid HPLC are summarized in the table.

Finally detection of different carotenoids with similar retention times can often be attained, provided that their absorption at different wavelengths does not interfere, as in phytofluene and β-carotene. While the pure compounds are not isolated, qualitative and quantitative analysis is still achieved.

---

[18] A. Fiksdahl, J. T. Mortensen, and S. Liaaen-Jensen, *J. Chromatogr.* **157,** 111 (1978).
[19] F. X. Cunningham, Jr. and J. A. Schiff, *Plant Physiol.* **72,** Suppl., 155 (1983); personal communication.

[8] Isolation and Assay of Dolichol and Dolichyl Phosphate

By W. LEE ADAIR and R. KENNEDY KELLER

The isolation and characterization of long-chain prenols[1] and their monophosphate derivatives are complicated by several factors. First, they are present in small quantities in most tissues examined (5–100 µg/g wet weight). As a result, precise analysis requires the inclusion of a radioactive tracer to monitor purification and correct for losses. Second, the prenols occur naturally in a variety of derivatized states, including acyl and phosphomonoesters of the free alcohols as well as the many glycosylated forms involved in heteroglycan biosynthesis. Since simultaneous analysis of all these derivatives would be a complex task, it is convenient to convert them to a few easily assayable species, namely the free alcohol and the monophosphate ester. For animal tissues this is generally carried out by direct saponification,[2] which completely dissolves the tissue and converts all the dolichol derivatives to dolichol and dolichyl phosphate. With plant tissues (which often contain large amounts of cellulose), it is preferable to extract the prenols with organic solvents prior to saponification. Once extracted, the purification and assay of the prenols and prenyl phosphates have proved a challenge for chromatographers and chemists alike.

The chemistry of the long chain prenols is typical for that of allylic and saturated primary alcohols with regards to the hydroxyl end of the molecule. That is, the prenols may be reacted with acyl anhydrides and chlorides to form esters.[3] In addition, they can undergo oxidation to form prenals with a variety of oxidants, such as chromium trioxide-pyridine,[4] pyridinium chlorochromate,[5] and manganese dioxide.[6] It should be noted that treatment with manganese dioxide forms the basis of a convenient method to distinguish between polyprenol and dolichol, since only the former compound is susceptible to oxidation. An alternative procedure is

---

[1] Nomenclature: in this chapter we have used the term "polyprenol" to mean a fully unsaturated polyprenyl alcohol. The number of isoprene units in the molecule is indicated following a dash. Thus, nonadecaprenol would be called polyprenol-19. Derivatives of polyprenol are indicated, where appropriate, with the exception of the 2,3-dihydro derivative which is called "dolichol." "Prenol" is used to refer to *both* dolichol and polyprenol.

[2] J. Burgos, F. W. Hemming, J. F. Pennock, and R. A. Morton, *Biochem. J.* **88,** 470 (1963).

[3] R. K. Keller and W. L. Adair, *Biochim. Biophys. Acta* **489,** 330 (1977).

[4] R. W. Keenan and M. Kruszek, *Anal. Biochem.* **69,** 504 (1975).

[5] W. L. Adair and S. Robertson, *Biochem. J.* **189,** 441 (1980).

[6] O. Samuel, Z. Hachimi, and R. Azerad, *Biochimie* **56,** 1279 (1974).

to separate the two types of prenols by high-pressure liquid chromatography on silica,[7] a technique which is most effective when a single isoprenolog of each prenol is chromatographed. Should relatively large quantities of a given type of prenol be available, spectroscopic analysis (IR, NMR) can be successfully applied.[2,8,9]

With regards to the chemistry of the monophosphate esters of polyprenol and dolichol, both are stable to treatment with strong alkali, allowing the use of saponification conditions during purification without serious losses. In contrast, mild acid degrades polyprenyl phosphate while dolichyl phosphate is stable to treatment even with strong acid.[10]

The glycosylated derivatives of dolichyl and polyprenyl phosphate are labile to alkali and acid.[11] Both treatments degrade all these compounds to the monophosphate form, with the important exception of dolichyl phosphomannose, which undergoes an elimination reaction to form mannose 2-phosphate and dolichol[12] in the presence of base.

Methods

Thin-Layer Chromatography. Chromatography is performed on plastic-backed plates of silica gel 60 (0.25 mm thickness) from EM Labs. Tanks are equilibrated at least 1 hr before use. Detection of lipids is achieved by placing the plates in a TLC tank to which iodine crystals have been added. Alternatively, plates can be sprayed with anisaldehyde reagent and heated.

High-Pressure Liquid Chromatography. An instrument capable of maintaining back pressures of 5000 psi is required. An ultraviolet monitor with variable wavelength detection is preferred; however, a fixed wavelength monitor at 214 nm is adequate. The monitor is connected to a strip-chart recorder through an on-line integrator for quantitative analysis. We use a Laboratory Data Control Constametric IIG instrument with a Spectromonitor II variable wavelength detector, but any comparable instrument will suffice. For analysis of products from metabolic labeling studies, an on-line radioactivity monitor (e.g., Flow-One by Radiomatic Instruments) is convenient.

Most commercial columns are adequate for the separations reported below; however we have found the Brownlee analytical cartridge columns to be especially useful. We use 5-μm, 22-cm columns to which are affixed

---

[7] R. K. Keller, G. D. Rottler, and W. L. Adair, *J. Chromatogr.* **236**, 230 (1982).

[8] J. Feeney and F. W. Hemming, *Anal. Biochem.* **20**, 1 (1967).

[9] W. C. Breckenridge, L. S. Wolfe, and N. M. K. Ng Ying Kin, *J. Neurochem.* **21**, 1311 (1973).

[10] J. F. Wedgwood, C. D. Warren, and J. L. Strominger, *J. Biol. Chem.* **249**, 6316 (1974).

[11] C. D. Warren and R. W. Jeanloz, this series, Vol. 50 [8].

[12] C. D. Warren, I. Y. Liu, A. Herscovics, and R. Jeanloz, *J. Biol. Chem.* **250**, 8069 (1975).

3-cm guard cartridges in a single assembly. With this sytem, any significant loss of resolution or increase in back pressure can be easily remedied by insertion of a new guard cartridge.

Solvents. All solvents are HPLC grade. "Reagent alcohol" is a specially prepared solvent for HPLC use from Fisher Scientific. The composition is ethanol/methanol/isopropanol (90/5/5). All solvent mixtures are passed through a filter-degasser (Lazar Research Labs) prior to use and stored in 4-liter bottles.

Evaporations. Solvent removal is achieved using a 12-place evaporator (N-Evap, Organomation Associates) with the water bath set at 40° and argon or nitrogen as the flushing gas.

Preparation of Labeled Compounds

Preparation of [^{3}H]Polyprenol and [^{3}H]Dolichol

Principle. Dolichol or polyprenol is oxidized to the corresponding aldehyde by appropriate reagents. Reduction of the purified aldehyde product with NaB^{3}H$_4$ yields the desired labeled compounds.

Procedures. The procedures described here are those reported by Adair et al.[13] and Adair and Robertson.[5] In the oxidation of polyprenol, care must be taken to avoid cis–trans isomerization of the α-isoprene residue of the aldehyde product. Studies have shown that this side reaction is more prominent when using the chromium trioxide reagent and that it is enhanced by irradiation with UV light.[13] (Indeed, advantage can be taken of this side reaction as a convenient method to prepare the α-*trans*-polyprenol from the naturally occurring cis compound.) A second problem is the occurrence of an appreciable amount of 1,4-addition when sodium borotritide is used,[14] thereby generating tritiated dolichol as a side product. Although the use of lithium aluminum hydride avoids 1,4-addition,[14] high specific activity LiAl^{3}H$_4$ is not commercially available and the lability of this reagent makes it unattractive for routine laboratory use.

[1-^{3}H]Polyprenol. Polyprenol (2 mg) is dissolved in 100 μl hexane to which 10 mg MnO$_2$ is added. After stirring 60 min at 4° in a sealed, amber Reacti-vial (Pierce), the mixture is diluted with 2 ml hexane and centrifuged to remove the oxidizing reagent. Thin-layer chromatography of an aliquot of the supernatant on fluorescent dye-impregnated plates using chloroform as a solvent yields a single UV-positive spot (the allylic aldehyde) at $R_f = 0.8$. Spraying with anisaldehyde reagent shows traces of unreacted polyprenol in addition to the major product polyprenal. If any

---

[13] W. L. Adair, N. Cafmeyer, and R. K. Keller, *J. Biol. Chem.* **259**, 4441 (1984).
[14] M. R. Johnson and B. Rickborn, *J. Org. Chem.* **35**, 1041 (1970).

cis–trans isomerization occurs during workup the α-*trans*-polyprenal will appear as an additional UV-positive band migrating slightly slower than the α-cis compound. Purification is effected by applying the mixture to a 1 × 3-cm column of alumina (Brockman grade 3, equilibrated with hexane) and eluting the polyprenal with hexane/ether (9/1). Fractions of 1 ml each are collected and aliquots monitored by thin-layer chromatography. After evaporation of the eluting solvent, the aldehyde is dissolved in 0.25 ml 1,2-dichloroethane and treated with 100 μl NaB^3H_4 (100 mCi, 2–4 Ci/mmol) in alkaline ethanol (100 μl 1 N NaOH per 10 ml ethanol) for 90 min at room temperature. The reaction is stopped with 10 μl concentrated acetic acid and applied to a 1 × 3-cm column of alumina as before, eluting the [1-^{3}H]polyprenol with hexane/ether (4/1). Thin-layer chromatography in chloroform yields a single band with anisaldehyde spray and radiochromatogram scanning (R_f = 0.4), comigrating with authentic polyprenol (available from Sigma Chemical Co.). High-pressure liquid chromatography of such preparations, according to the method of Keller *et al.*,[7] shows small amounts (generally less than 10%) of [^{3}H]dolichol.

[1-^{3}H]Dolichol. Dolichol (2 mg) is dissolved in 80 μl dry dichloromethane and stirred with 1 mg pyridinium chlorochromate. After 60 min thin-layer chromatography (chloroform) indicates nearly quantitative conversion to a faster moving band (R_f = 0.8). Should the oxidation be incomplete, another addition of pyridium chlorochromate is made and the reaction stirred a further 30 min. The sample is diluted with 300 μl dichloromethane and applied to a 1 × 5-cm column of Florisil equilibrated and eluted with dichloromethane. Fractions of about 1 ml are monitored by thin-layer chromatography; those containing the aldehyde product are pooled and evaporated to a small volume. Reduction with NaB^3H_4 and final purification of [1-^{3}H]dolichol follows the protocol outlined above for the preparation of [1-^{3}H]polyprenol.

Following purification, the radiolabeled samples are evaporated to dryness, dissolved in benzene, and transferred to 2-ml glass storage vials (Pierce). The vials are flushed with argon, capped with Teflon-lined septa, and stored in the refrigerator. The purity should be checked periodically by reversed phase HPLC. Over time, some degradation to polar derivatives has been observed to occur. Repurification can be achieved using C_{18} Sep-Pak columns (Waters Associates). The sample is applied in methanol/ethanol (75/25), rinsed with methanol/ethanol (50/50), and eluted with methanol/ethanol (25/75).

Preparation of Dolichyl [^{32}P]Phosphate and Polyprenyl [^{32}P]Phosphate

Principle. The ^{32}P-labeled phosphomonoesters of dolichol and polyprenol can be prepared chemically from the free alcohol and inorganic

[^{32}P]phosphate by a modification of the procedure of Cramer et al.,[15] in which trichloroacetonitrile serves as the condensing agent.

Procedure. In this procedure the prenol is reacted with a slight stoichiometric excess of inorganic [^{32}P]phosphate in the presence of trichloroacetonitrile. This method has the advantage of producing only the monophosphate derivative and thus avoids the chromatographic purification protocol required in the procedure of Kandutsch et al.[16] Dolichyl [^{32}P]phosphate can also be prepared chemically using [^{32}P]POCl$_3$ by the method of Keenan.[17] However, this procedure, while giving excellent yields, is limited by the relatively low specific activity of ^{32}P product that has been attained (0.04 Ci/mol). In addition, since POCl$_3$ cannot successfully phosphorylate allylic alcohols (W. L. Adair, unpublished data), the POCl$_3$ method is not useful for preparing ^{32}P-labeled polyprenyl phosphate.

Dolichyl [^{32}P]Phosphate and Polyprenyl [^{32}P]Phosphate. Polyprenol or dolichol (1.0 mg, 0.77 μmol) is dissolved in 50 μl ethylene dichloride containing trichloroacetonitrile (2 μmol). Bis(triethylammonium)[^{32}P]-phosphate (1.0 μmole, 1.0 mCi)[18] in 50 μl acetonitrile is then added in a single portion. The reaction is then heated in a sealed vial at 50° for 1 hr. After cooling to room temperature, the reaction mixture is diluted with 2 ml chloroform, 1 ml methanol, and 0.75 ml water. Following centrifugation, the lower phase is applied to a 1 × 2-cm column of DEAE-cellulose (acetate form) equilibrated with chloroform/methanol (2/1). After rinsing with 10 ml chloroform/methanol (2/1), the product is eluted with 10 ml 0.3 M ammonium acetate in chloroform/methanol (2/1). One-quarter volume of water is added and the resulting lower phase, containing the labeled product, is collected after centrifugation. Thin-layer chromatography (chloroform/methanol/water, 65/35/4) of an aliquot of the washed eluate shows a single radioactive peak (R_f = 0.4) comigrating with standard dolichyl phosphate. The average yield is about 3–4%, based on radioactivity. The final product is dissolved in chloroform/methanol (2/1), aliquoted into glass vials and stored under argon at −70°.

Preparation of ^{3}H-Labeled Dolichyl Phosphate

Principle. In this method [^{3}H]dolichol is reacted with POCl$_3$ in the presence of triethylamine and the adduct hydrolyzed to produce the desired monophosphate product.

---

[15] F. Cramer, W. Rittersdorf, and W. Bohm, *Chem. Ber.* **654**, 180 (1961).
[16] A. A. Kandutsch, H. Paulus, E. Levine, and K. Bloch, *J. Biol. Chem.* **239**, 2504 (1964).
[17] R. W. Keenan, R. A. Martinez, and R. F. Williams, *J. Biol. Chem.* **257**, 14817 (1982).
[18] One micromole H$_3$PO$_4$ and 1 mCi H$_3$^{32}PO$_4$ (carrier free, HCl free) are taken to dryness in a 0.3-ml Reacti-vial. After an addition of 2 μmol triethylamine in 50 μl acetonitrile, the vial is capped and then vortexed well.

Procedure. Although the procedure outlined above for the preparation of dolichyl [^{32}P]phosphate can be employed for the preparation of [1-^{3}H]dolichyl phosphate, a higher yield can be attained using a modification of the procedure of Danilov and Chojnacki[19] which is described below. One drawback of this method is that it cannot be used to prepare tritiated polyprenyl phosphates: they must be prepared by other means, such as the procedure outlined above.

[1-^{3}H]Dolichyl Phosphate. [1-^{3}H]Dolichol (10^6–10^9 dpm) is dissolved in 25 μl benzene and added to a solution containing 25 μl POCl$_3$, 5 μl triethylamine, and 25 μl benzene. After incubating 1 hr at room temperature, the reaction is stopped by the addition of 0.5 ml of a solution of acetone/water/triethylamine (88/10/2). The mixture is allowed to stand overnight (to complete the hydrolysis of the chlorophosphate derivative) and then evaporated to a small volume. To the residue is added 2 ml chloroform, 1 ml methanol, and 0.75 ml water. The lower phase obtained after centrifugation is washed twice with 50% aqueous methanol containing 1 M H$_3$PO$_4$ and taken to dryness. The phosphoric acid is included in the washes to facilitate extraction of the excess triethylamine, which if not removed will interfere with subsequent chromatographic steps. The washed product is dissolved in 3 ml chloroform/methanol (2/1) and purified by ion exchange chromatography as described above for the ^{32}P-labeled derivative. Overall yield is about 75%.

Isolation and Assays of Dolichols and Polyprenols

The original procedure outlined by Burgos *et al.*[2] is used for the initial saponification and extraction of prenols. Various column techniques can then be employed for further purification prior to HPLC. These include chromatography on Florisil,[20] Lipidex,[21] alumina,[2] or gel filtration using Fractogel 6000[3] (a polyvinyl acetate resin originally sold by Merck, but subsequently discontinued), and Sephadex LH-20.[4] Both adsorption and reversed-phase high-pressure liquid chromatography have been used successfully, the former having the advantage of yielding a single peak, the latter having the advantage of displaying the distribution of isoprenologs.

The procedure which follows (Scheme 1) is used routinely in our laboratory for the isolation and quantitation of dolichol and dolichyl phosphate from rat liver, but can be adapted to any tissue.

[19] L. L. Danilov and T. Chojnacki, *FEBS Lett.* **131**, 310 (1981).
[20] K. K. Carroll, A. Vilim, and M. C. Woods, *Lipids* **8**, 246 (1973).
[21] T. Mankowski, T. Jankowski, T. Chojnacki, and P. Franke, *Biochemistry* **15**, 2125 (1976).

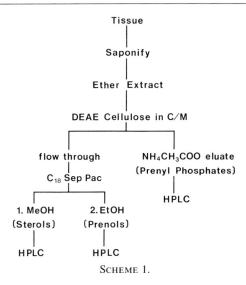

SCHEME 1.

Isolation of Dolichol from Animal Tissues

Procedure. To 1 g tissue (wet weight) in a 18 × 150-mm screw capped test tube is added 1 ml 60% KOH, 2 ml 0.25% pyrogallic acid in methanol, and 10^5 cpm of [^{3}H]dolichol. The solution immediately turns dark brown in color. The tube is sealed with a Teflon-lined plastic cap and is placed in a boiling water bath for 1 hr, during which time complete dissolution of the tissue occurs. After incubation, the sample is cooled to room temperature and extracted three times with 3 ml diethyl ether.[22] The extracts (total volume about 9 ml) are pooled in a 40-ml glass conical centrifuge tube with a Teflon-lined cap (Scientific Products C3985-40; Teflon lined cap is T1358-4) and washed once with an equal volume of 5% acetic acid. After centrifugation for 2 min at 1000 rpm, the upper phase is removed and taken to dryness. The residue, which sometimes contains small quantities (<100 μl) of water is dissolved in 2 ml chloroform/methanol (2/1) and applied to a DEAE-cellulose column (2 ml packed resin in a 10-ml disposable polypropylene syringe, fitted with a precut Whatman GF/C filter to retain the resin). The column allows all dolichol, cholesterol, and other neutral lipids to pass through while retaining dolichyl phosphate.[23] The

---

[22] Petroleum ether has also been used to extract saponification reactions. It has the advantage of not producing peroxides upon storage. Also, there may be fewer polar contaminants extracted into the upper phase.

[23] This step can be omitted if quantitation of dolichol only is desired, since it is highly unlikely that any of the compounds which are retained by the DEAE-cellulose column copurify with dolichol.

collected drop-through is treated with 0.25 volumes 0.88% KCl, vortexed, and centrifuged. The lower phase is evaporated to dryness and dissolved in 2 ml methanol. The sample is then applied to a C_{18} Sep-Pak column (Waters) equilibrated in methanol.[24] The bulk of the sterols pass through the column unretarded while dolichol is retained.[25] After washing with 10 ml methanol, dolichol is eluted in good yield with 10 ml reagent alcohol. The alcohol fraction is then taken to dryness in preparation of the sample for HPLC analysis.

The chief limitation of Sep-Pak columns is that sample size is generally restricted to less than 10 g wet weight of starting material. In those instances where large quantities of tissue are being processed for preparation of milligram quantities of prenols, purification can be achieved using large columns of alumina,[2] Florisil,[20] or Lipidex 5000,[21] the latter resin having the advantage of resolving isoprenolog species.

Isolation of Polyprenol from Plant Tissue

Procedure. Although direct saponification of plant tissue followed by filtration and solvent extraction has been used by many laboratories, we have found it more convenient to extract the tissue prior to saponification. Thus, 2.5 g of chopped plant leaves is suspended in 15 ml acetone and pulverized with a Polytron homogenizer in a well-ventilated hood. The suspension is filtered and the residue and filtrate are retained. The residue is suspended in 20 ml petroleum ether (bp 30–50°) and rehomogenized and filtered as before. Following a second petroleum ether extraction, the organic extracts are pooled and an appropriate radiolabeled internal standard (e.g., [^{3}H]polyprenol-19) is added. The sample is then treated with 50 ml water. After vigorous shaking, the upper phase is collected and dried over anhydrous Na_2SO_4. The sample is then flash-evaporated and saponified by treatment with 10 ml of methanol/isopropanol/water (2/1/1), 8.0 g KOH, and 0.5 g pyrogallic acid in a boiling water bath. We routinely carry out the saponifications in a 40-ml screw-capped conical centrifuge tube. After 2 hr, the reaction is cooled to room temperature, diluted with 3 ml water, and extracted three times with 15 ml petroleum ether. The organic layers are pooled, backwashed with 50 ml water, and dried over Na_2SO_4. After flash-evaporation, the residue is taken up in 1 ml hexane and chromatographed on 1 g of alumina (Brockman grade 3), packed in a

---

[24] Many investigators apply the total nonsaponifiable fraction directly to HPLC. However, we have observed that analysis of samples in this manner generally results in higher levels calculated for dolichol, indicating that contaminants comigrating with dolichol must be present.

[25] T. K. Wong and W. J. Lennarz, *J. Biol. Chem.* **257,** 6619 (1982).

Pasteur pipet, and equilibrated with hexane. After application of the sample, the column is rinsed with hexane and then sequentially eluted with 5 and 8 ml portions of hexane/diethyl ether (4/1) and (2/1), respectively. Fractions of about 2 ml are collected and aliquots taken for analysis by scintillation counting or thin-layer chromatography (chloroform). Those fractions containing the bulk of the polyprenol are pooled, flash-evaporated, and dissolved in the appropriate solvent for analysis by HPLC. Prior purification using the C_{18} Sep-Pak method described above for dolichol is to be avoided in this case, since many plants contain prenols with less than 15 isoprene units. These are not effectively retained by the Sep-Pak column, even when applied in various mixtures of methanol and water.

Assays of Total Dolichol and Polyprenol

Assay by High-Pressure Liquid Chromatography. HPLC of dolichol and polyprenol, as mentioned above, can be carried out on silica or octadecyl-silica, depending on the information desired. For those instances in which direct quantitation as a single peak is preferred, we use silica HPLC with 0.3% reagent alcohol in hexane as a solvent. The dried sample is dissolved in 0.25 ml of the HPLC solvent, and 0.1 ml is drawn up into a 0.1-ml syringe. Any turbidity in the sample to be injected can usually be cleared by heating 30 sec in a 50° water bath. Fifty microliters is taken for liquid scintillation counting and the remaining 50 μl is injected into the HPLC. The elution position of dolichol and polyprenol is dependent on their chain length.[7] For polyprenol-19 and dolichol-19, k' values of approximately 7 and 8, respectively, are obtained at a flow rate of 1 ml/min (Fig. 1a). Back pressures are on the order of 300 psi. Detection of prenols is by on-line UV monitoring at 210 nm. The lower limit of detection is about 10 ng using a high quality detector. By injecting fixed volumes (e.g., 50 μl using a 100 μl sample loop) of different dilutions of the standardized prenol solution (see below), a standard curve of peak height (or area) vs micrograms prenol is prepared. The level of prenol in the original sample is then calculated from the following equation:

$$\frac{\mu\text{g prenol}}{\text{g tissue}} = \frac{^3\text{H cpm prenol added/g}}{^3\text{H cpm injected}} \times \frac{\text{peak ht. of sample}}{\text{peak ht.}/\mu\text{g prenol}}$$

If the isoprene distribution of the prenol sample is desired, reversed-phase chromatography is carried out. The system is analogous to that described above for silica, except that 5 μm C_{18} cartridges are used. The solvent system is isopropanol/methanol (1/1). At a flow rate of 1.0 ml/min,

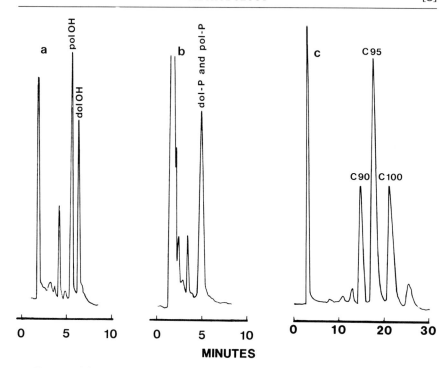

FIG. 1. High-pressure liquid chromatography of long-chain prenols and prenyl phosphates. Adsorption chromatography on silica of (a) polyprenol-19 and dolichol-19 and (b) polyprenyl-19 phosphate and dolichyl-19 phosphate. (c) Reversed-phase chromatography of dolichyl phosphate from pig liver. Values above peaks refer to number of carbon residues in individual isoprenologs. Flow rate on all runs is 1.0 ml/min. Ultraviolet monitoring is at 210 nm. Solvent systems are described in text.

the isoprenologs elute within 20 min with baseline resolution, as shown in Fig. 1c. Column back pressures are on the order of 2000–3000 psi. The above solvent system has proven convenient for separating isoprenologs containing 17–25 isoprene units. For shorter length species better resolution is obtained by decreasing the ratio of isopropanol to methanol. The isoprenolog peaks will appear as doublets if the sample (e.g., many plant tissues) contains both dolichols and polyprenols since these compounds are only partially resolved by reversed-phase HPLC.[26,27] Confirmation of both types of prenols in the sample can be achieved by subjecting one of the isoprenolog doublets to adsorption HPLC, which yields baseline reso-

[26] R. K. Keller, E. Jehle, and W. L. Adair, *J. Biol. Chem.* **257,** 8985 (1982).
[27] K. Ravi, J. W. Rip, and K. K. Carroll, *Lipids* **19,** 401 (1984).

lution of the two species, or by oxidation with MnO_2, which selectively oxidizes only the α-unsaturated polyprenol.

The purification procedure described above is generally satisfactory for the isolation of prenols in pure form from most tissues. If HPLC shows contaminants interfering with prenol analysis, then further purification is necessary. Samples are chromatographed on 1.5 × 30-cm columns of Sephadex LH-20 equilibrated in reagent alcohol. Prenols elute in approximately 25 ml and should be of sufficient purity for HPLC analysis.

The analytical HPLC columns described above can handle samples containing up to 1 mg prenol. HPLC analysis of up to 5 mg prenols can be achieved using 10-cm-diameter preparative columns in either the adsorption (silica) or reversed-phase (C_{18}) mode employing the solvents described above.

Assay by Derivatization. The development of HPLC analysis has superceded the use of derivatization reactions for the analysis of dolichol and polyprenol. However, derivatization is still the method of choice for the standardization of prenol solutions to be used for HPLC analysis. The procedure which follows is from the original acetylation reaction described by Keller and Adair[3] and is based on the formation of a stable derivative using a reagent with known specific activity. [^{3}H]Prenol is added in tracer quantities to determine the amount of starting material converted to product.

To 0.1-ml Reacti-vials (Pierce) are added various estimated amounts of prenol (1–10 μg) and 300,000 dpm [^{3}H]prenol. The samples are taken to dryness and 20 μl [^{14}C]acetic anhydride (10 μCi, 1 μmol) is added. The vials are capped with Teflon-lined septa and heated at 70° in a hood overnight. Slight loss of volume will be observed. After cooling, the samples are opened, evaporated to dryness in the hood, and dissolved in 10–20 μl chloroform/methanol (2/1). Thin-layer chromatography is then carried out in hexane/ether (8/2). Alongside the sample are spotted standards (5 μg each) of prenol and prenyl acetate (available from Sigma Chemical). After development, radioscanning shows three peaks: (1) at the origin, corresponding to excess unreacted ^{14}C material, (2) at $R_f = 0.2$, corresponding to any unreacted [^{3}H]prenol, and (3) at $R_f = 0.7$, corresponding to the double-labeled prenyl acetate. The latter two peaks should align with the corresponding standards. The prenyl acetate region is either subjected to sample oxidation or scraped from the plate into a 13 × 100-mm disposable test tube and eluted with 5 ml diethyl ether. If the sample is eluted, the eluant is then transferred to a scintillation vial, dried, and counted for ^{3}H and ^{14}C in a detergent-containing scintillation fluid (e.g.,

ACS, Amersham). The formula below is used to calculate the amount of prenol in the original sample.

$$\text{nanomoles prenol} = \frac{[^3H]\text{prenol added (dpm)}}{^3H \text{ dpm in derivative}} \times \frac{^{14}C \text{ dpm in derivative}}{^{14}C \text{ acetic anhydride sp. act. (dpm/nmol)}}$$

The above calculation assumes that the mass of [^{3}H]prenol is negligible. We routinely use specific activities in the range of 1–10 Ci/mmol, so that 300,000 dpm corresponds to about 14–140 pmol. Whether this is negligible in the above calculation depends on the amount of prenol derivatized. Correction for the mass of [^{3}H]prenol is easily achieved by carrying out a derivatization reaction to which no unlabeled prenol is added. It should be noted that the absolute specific activity of the [^{3}H]prenol can also be calculated from this reaction.

Isolation and Analysis of Free Unesterified Prenol and Prenyl Fatty Acid Esters

Procedure. Any of the common procedures using organic solvents can be employed for the initial extraction. These include the methods of Folch et al.,[28] Radin,[29] and Bligh and Dyer.[30] To the organic extracts are added about 10^5 dpm each of [^{3}H]prenol and [^{3}H]prenyl palmitate (prepared from [^{3}H]prenol and palmitoyl chloride[3]). Following any washing steps the sample can be chromatographed on Florisil for the separation of free and esterified prenol.[20] The free prenol fraction is saponified, purified, and assayed as described above. The ester fraction has been analyzed directly by reversed phase HPLC but cleaner preparations can be obtained if the sample is subjected to the saponification and purification procedure described above. Thus with the latter procedure, the prenyl esters will be analyzed as free prenol.

Isolation and Assay of Dolichyl Phosphate

Although the great majority of studies indicate that dolichyl phosphate is the prenyl phosphate involved in glycosyl transfer in eucaryotes, the possibility remains that the α-unsaturated polyprenyl phosphates may serve this function in some species, particularly plants and lower organisms. Accordingly, and because of the possibility that polyprenyl phosphate is a metabolic precursor for dolichyl phosphate,[31] it is desirable to

[28] J. Folch, M. Lees, and G. H. Sloane-Stanley, *J. Biol. Chem.* **226**, 497 (1957).
[29] N. S. Radin, this series, Vol. 72 [1].
[30] E. G. Bligh and W. J. Dyer, *Can. J. Biochem. Physiol.* **37**, 911 (1959).
[31] C. Vigo and W. L. Adair, *Biosci. Rep.* **2**, 835 (1982).

have available procedures which allow isolation and assay of both of these long chain prenyl phosphates. Solvent systems have not yet been developed which allow resolution of polyprenyl phosphate and dolichyl phosphate by HPLC. It is therefore necessary to carry out indirect determinations. For example, advantage can be taken of the lability of polyprenyl phosphate to acid: analysis (e.g., by HPLC) of the acid-untreated and treated samples yields values for polyprenyl phosphate by difference. Alternatively, the prenyl phosphate fraction may be dephosphorylated using phosphatases.[32,33] The resulting product may then be subjected to silica HPLC, which resolves polyprenol from dolichol (see above). The disadvantage of this method is that it assumes that the phosphatases work equally well on dolichyl and polyprenyl phosphates.

In theory, it should be possible to analyze for total dolichyl phosphate by carrying out extractions with chloroform/methanol (2/1) and chloroform/methanol/water (10/10/3) since all known dolichol-containing compounds should be soluble in one or both of these solvents. These extracts, when pooled and subjected to treatment with acid and base, should contain the total tissue dolichyl phosphate as free dolichyl phosphate. However, experiments in our laboratory have clearly demonstrated that the levels of dolichyl phosphate obtained from this procedure are significantly less than that obtained by direct saponification of the tissue. Studies have shown that the final insoluble residue after 10/10/3 extract contains a substantial amount of dolichyl phosphate (2–3 μg/g in rat liver). Although the basis for this finding is not yet apparent, it is clear that in order to carry out analysis of total dolichyl phosphate (minus dolichyl phosphomannose which is broken down to free dolichol[12]), tissue saponification and extraction are preferred to direct organic solvent extraction with subsequent acid–base treatments.

Isolation of Total Dolichyl Phosphate

Procedure. The procedure described is for the analysis of dolichyl phosphate; if the sample is suspected of containing polyprenyl phosphate, it must be subjected to acid hydrolysis or enzymatic dephosphorylation (see above).

The tissue is treated as described above for dolichol (see Scheme 1). ^{32}P- or ^{3}H-labeled dolichyl phosphate is added to the saponification mixture to monitor purification and correct for yield. Following ether extraction and washing, the sample is dried, dissolved in chloroform/methanol (2/1), and applied to DEAE-cellulose as described above. The column is

---

[32] D. D. Carson and W. J. Lennarz, *J. Biol. Chem.* **256**, 4679 (1981).
[33] H. Fujii, T. Koyama, and K. Ogura, *Biochim. Biophys. Acta* **712**, 716 (1982).

washed with 10 ml each of chloroform/methanol (2/1), chloroform/acetic acid (3/1), chloroform/methanol (2/1), and chloroform/methanol (2/1) containing 0.1 M ammonium acetate. The last fraction, containing the labeled dolichyl phosphate, is treated with 0.25 volumes water, vortexed, and centrifuged. The lower phase is underpipetted into a fresh tube and taken to dryness for subsequent derivatization or HPLC analysis (see below).

Assays of Dolichyl and Polyprenyl Phosphate

Assay by High-Pressure Liquid Chromatography. HPLC of dolichyl phosphate, like dolichol, can be carried out in the adsorption or reversed phase modes. For adsorption chromatography on silica, we use the same column and conditions as described above for dolichol except that the solvent employed is hexane/isopropanol/1.4 M H_3PO_4 or H_2SO_4 (965/35/0.5). A typical chromatographic run is shown in Fig. 1b. Because of its highly apolar nature, dolichyl phosphate elutes early from this system (6 min) compared to other phospholipids.[34] Quantitation is achieved by preparing a curve generated from injection of standardized solutions (see below). If it is desired to use the HPLC-purified dolichyl phosphate in further studies (e.g., derivatization) in which the phosphoric acid would interfere, the sample can be treated with 1.5 volumes of isopropanol/water (890/585). The resulting upper phase will contain all the dolichyl phosphate while the phosphoric acid quantitatively extracts into the lower phase.

For the reversed-phase HPLC of dolichyl phosphate, the system developed by Chaudhary *et al.*[35] is employed. For preparation of solvent, 1 g crystalline H_3PO_4 (Fluka) is dissolved in 1 liter of isopropanol/methanol (1/1). The column and conditions are the same as for reversed-phase analysis of dolichol. As with dolichol, reversed-phase chromatography results in baseline separation of the isoprenolog species (Fig. 1c).

Assay by Derivatization. This procedure is based on that described by Keller *et al.*[36] and involves the formation of a stable derivative of dolichyl phosphate with [^{14}C]phenylchloroformate.

$$\text{Dol}-\text{O}-\overset{\overset{\text{O}}{\|}}{\underset{\underset{\text{O}^-}{|}}{\text{P}}}-\text{O}^- + \text{Cl}-\overset{\overset{\text{O}}{\|}}{\text{C}}-\text{O}-\text{C}_6\text{H}_5 \longrightarrow \text{Dol}-\text{O}-\overset{\overset{\text{O}}{\|}}{\underset{\underset{\text{O}^-}{|}}{\text{P}}}-\text{O}-\overset{\overset{\text{O}}{\|}}{\text{C}}-\text{O}-\text{C}_6\text{H}_5$$

---

[34] R. K. Keller, D. Armstrong, F. C. Crum, and N. Koppang, *J. Neurochem.* **42,** 1040 (1984).

[35] N. Chaudhary, D. J. Freeman, J. W. Rip, and K. K. Carroll, *Lipids* **17,** 558 (1982).

[36] R. K. Keller, J. W. Tamkun, and W. L. Adair, *Biochemistry* **20,** 5831 (1981).

(The preparation of the radiolabeled derivatizing reagent, a simple one-step procedure, is described in the original publication.) The purified dolichyl phosphate fraction (usually from DEAE-cellulose chromatography) is taken to dryness in a polypropylene centrifuge tube. To the residue is added 0.1 ml chloroform, 0.01 ml 7 mM triethylamine in chloroform, and 0.01 ml (10 nmol, about 500,000 dpm) [^{14}C]phenyl chloroformate in hexane. The sample is taken to dryness at room temperature and dissolved in a small volume (10 μl) of chloroform/methanol (2/1) and spotted on a silica thin-layer plate. The choice of developing solvent is based on the ambient humidity: with low relative humidity we use chloroform/methanol (7/1) while with high relative humidity acetone is employed. In a separate lane is spotted a standard, prepared from a reaction using unlabeled materials. Radiochromatogram scanning of the sample lane after development reveals a peak at the origin (unreacted dolichyl phosphate), a peak at $R_f = 0.5$ (the derivative), and two faster moving peaks near the solvent front which are derived from unreacted [^{14}C]phenyl chloroformate. The derivative peak can either be subjected to sample oxidation[36] or scraped and eluted with 1.0 ml hexane/isopropanol/water (2/6/1.5). From the ^{3}H dpm (derived from the [^{3}H]dolichyl phosphate added to the original sample) and the ^{14}C dpm, the nanomoles of dolichyl phosphate in the sample is determined:

$$(\text{nmol Dol-P/g}) = \frac{^3\text{H dpm added/g}}{^3\text{H dpm in derivative}} \times \frac{^{14}\text{C dpm in derivative}}{[^{14}\text{C]phenyl chloroformate sp. act. (dpm/nmol)}}$$

If it is deemed necessary to substract the mass of the [^{3}H]dolichyl phosphate tracer added to the original sample, the procedure is analogous to that described above for dolichol derivatization.

The [^{14}C]phenyl chloroformate assay is rapid and allows more samples to be processed per unit time than HPLC analysis, since many derivatives can be applied to TLC plates in a single development. The sensitivity is comparable, since the specific activity of commercially available [^{14}C]phenol (the starting material for the [^{14}C]phenyl chloroformate synthesis) is on the order of 22,000–150,000 dpm/nmol.

One possible disadvantage of derivatization is that the partially purified sample may contain contaminants which interfere with the phenyl chloroformate reaction. Such interference would indicate HPLC as the method of choice for quantitation of these samples.

[9] Invertebrate Carotenoproteins

By P. F. ZAGALSKY

Noncovalent association of carotenoids with protein is common throughout the invertebrates, providing the multitudinous colors (blue, green, purple, red etc.) of tissues which may include, in addition to the external surface of the animal, the blood, eggs, ovary, hypodermis, and stomach wall. The nature, distribution, and properties of these pigments have been the subject of several reviews.[1-6] In brief, two types of complex may be distinguished: those (Type A) in which carotenoid is associated stoichiometrically with a simple protein or glycoprotein and those (Type B), usually less stable, in which carotenoid is associated with a lipo(glyco)protein. The latter are commonly found in the eggs, ovaries, and blood, while the former are more usual at the external surface, within the carapace of Crustacea and skin of Echinodermata, for example.

The more frequently occurring, but far from exclusive, carotenoid found in the combinations is astaxanthin (3,3'-dihydroxy-β,β-carotene-4,4'-dione). The two types of complex are typified by the extensively studied lobster pigments, crustacyanin,[7-20] the blue astaxanthin protein of

---

[1] D. F. Cheesman, W. L. Lee, and P. F. Zagalsky, *Biol. Rev. Cambridge Philos. Soc.* **42**, 131 (1967).
[2] H. Thommen, in "Carotenoids" (O. Isler, ed.), Chapter 8. Birkhaueser, Basel, 1971.
[3] P. F. Zagalsky, *Pure Appl. Chem.* **47**, 103 (1976).
[4] W. L. Lee, ed., "Carotenoproteins in Animal Coloration." Dowden, Hutchinson & Ross, Inc., Stroudsburg, Pennsylvania, 1977.
[5] G. Britton, G. M. Armitt, S. Y. M. Lau, A. K. Patel, and C. C. Shone, in "Carotenoid Chemistry and Biochemistry" (G. Britton and T. W. Goodwin, eds.), p. 237. Pergamon, Oxford, 1982.
[6] P. F. Zagalsky, *Oceanis* **9**, 73 (1983).
[7] G. Wald, N. Nathanson, W. P. Jencks, and E. Tarr, *Biol. Bull. (Woods Hole, Mass.)* **95**, 249 (1948).
[8] P. F. Zagalsky and D. F. Cheesman, *Biochem. J.* **89**, 21p (1963).
[9] W. P. Jencks and B. Buten, *Arch. Biochem. Biophys.* **107**, 511 (1964).
[10] D. F. Cheesman, P. F. Zagalsky, and H. J. Ceccaldi, *Proc. R. Soc. London, Ser. B* **164**, 130 (1966).
[11] R. Kuhn and H. Kühn, *Angew. Chem.* **78**, 979 (1966).
[12] R. Kuhn and H. Kühn, *Eur. J. Biochem.* **2**, 349 (1967).
*[13] M. Buchwald and W. P. Jencks, *Biochemistry* **7**, 844 (1968).
[14] R. Quarmby, D. A. Norden, P. F. Zagalsky, H. J. Ceccaldi, and R. Daumas, *Comp. Biochem. Physiol. B* **56B**, 55 (1977).
[15] D. B. Gammack, J. H. Raper, P. F. Zagalsky, and R. Quarmby, *Comp. Biochem. Physiol. B* **40B**, 295 (1971).

the carapace (Type A), and ovoverdin,[16,20–28] the green astaxanthin lipoglycoprotein (lipovitellin) of the ovary and eggs (Type B). While earlier studies have concentrated mainly on the carotenoproteins of Arthropoda, more recently detailed investigations have been made of those of Coelenterata,[17,28,29] including Hydrocorallina[30–32] and Echinodermata.[5,33,34]

The red astaxanthin glycoprotein of the freshwater South American snail, *Pomacea canaliculata*, has been analyzed thoroughly; the reader is referred to the paper of Cheesman[35] for practical information concerning this not easily accessible pigment.

Lobster Carapace Carotenoproteins

A number of distinct carotenoproteins occur in the calcified layer of lobster carapace and may be extracted following decalcification. The predominant carotenoprotein is the blue α-crustacyanin, λ_{max} 632 nm.[9] This pigment dissociates reversibly at low ionic strength to form α'-crustacyanin, λ_{max} 595 nm.[10] The latter changes on standing into another form,

---

[16] V. R. Salares, N. M. Young, H. J. Bernstein, and P. R. Carey, *Biochim. Biophys. Acta* **576**, 176 (1979).
[17] P. F. Zagalsky, *Comp. Biochem. Physiol. B* **71B**, 235 (1982).
[18] P. F. Zagalsky and R. Jones, *Comp. Biochem. Physiol. B* **71B**, 237 (1982).
[19] P. F. Zagalsky, *Comp. Biochem. Physiol. B* **73B**, 997 (1982).
[20] B. Renstrøm, H. Ronneberg, G. Borch, and S. Liaaen-Jensen, *Comp. Biochem. Physiol. B* **71B**, 249 (1982).
[21] K. G. Stern and K. Salomon, *Science* **86**, 310 (1937).
[22] K. G. Stern and K. Salomon, *J. Biol. Chem.* **122**, 461 (1938).
[23] R. Kuhn and N. A. Sörensen, *Ber. Dtsch. Chem. Ges.* **71**, 1879 (1938).
[24] R. Kuhn and N. A. Sörensen, *Z. Angew. Chem.* **51**, 465 (1938).
[25] T. W. Goodwin and S. Srisukh, *Biochem. J.* **45**, 268 (1949).
[26] H. J. Ceccaldi, D. F. Cheesman, and P. F. Zagalsky, *C. R. Seances Soc. Biol. Ses Fil.* **160**, 587 (1966).
[27] R. A. Wallace, S. L. Walker, and P. V. Hauschka, *Biochemistry* **6, 1582 (1967).
[28] P. F. Zagalsky, D. F. Cheesman, and H. J. Ceccaldi, *Comp. Biochem. Physiol.* **22**, 851 (1967).
[29] P. F. Zagalsky and P. J. Herring, *Philos. Trans. R. Soc. London, Ser. B* **299**, 289 (1977).
[30] H. Rønneberg, G. Borch, D. L. Fox, and S. Liaaen-Jensen, *Comp. Biochem. Physiol. B* **62B**, 309 (1979).
[31] H. Rønneberg, D. L. Fox, and S. Liaaen-Jensen, *Comp. Biochem. Physiol. B* **64B**, 407 (1979).
[32] H. Berger, H. Rønneberg, G. Borch, and S. Liaaen-Jensen, *Comp. Biochem. Physiol. B* **71B**, 253 (1982).
[33] A. Elgsaeter, J. D. Tauber, and S. Liaaen-Jensen, *Biochim. Biophys. Acta* **530**, 402 (1978).
[34] C. C. Shone, G. Britton, and T. W. Goodwin, *Comp. Biochem. Physiol. B* **62B**, 507 (1979).
[35] D. F. Cheesman, *Proc. R. Soc. London, Ser. B* **149**, 571 (1958).

β-crustacyanin, of similar size but with λ_{max} at 585–590 nm.[9–11] β-Crustacyanin, representing about 20% of the carapace carotenoprotein, may be resolved into a number of spectrally similar components.[14] Removal of the carotenoid prosthetic groups from either α- or β-crustacyanin results in reversible dissociation into apoprotein subunits of molecular weight ~20,000, about half the size of β-crustacyanin.[9–15]

The apoprotein of *Homarus gammarus* crustacyanin consists of five electrophoretically distinct components which may be classified in two groups, based on size, amino acid composition, and peptide mapping.[14,19] β-Crustacyanin is composed of associations of pairs of the apoproteins, one from each group, with each apoprotein binding a single astaxanthin.[11–13,14]

α-Crustacyanin is an octamer of β-crustacyanin-sized units[11–13,15] and is apparent in electron micrographs as a double tetrameric structure.[18] The specificity of the carotenoid–protein interaction has been investigated in several laboratories by reconstitution of pigments from apoprotein and synthetic carotenoids.[3,5,20,36] The mode of binding of astaxanthin has been studied by resonance Raman spectroscopy.[16]

About 10–20% of the carapace carotenoprotein is present in another form, γ-crustacyanin, λ_{max} 625 nm, different from α-crustacyanin in molecular shape and charge but identical in molecular weight and amino acid composition.[13]

In addition to the crustacyanins, a yellow, unstable pigment, λ_{max} 409 nm,[9] is located on the outer surface of the carapace.[37,38] The optical properties of the pigment, molecular weight ~90,000, result from exciton coupling between the 16–20 astaxanthin prosthetic groups stacked in a card-pack manner,[13,38,39] with twisting[40] or slipping[41] between carotenoids in the pile. The pigment is polydisperse on electrophoresis and has two groups of subunits, one of which is common to α-crustacyanin.[13,42]

Properties

The crustacyanins, and other Type A carotenoproteins, are relatively stable complexes over the pH range 5–8.5 and are crystallizable.[10,17] Dilute solutions of α- and β-crustacyanin (0.1 mg/ml) lose about 3% of absorbance at their respective λ_{max} on exposure to bright sunlight for a

[36] W. L. Lee and P. F. Zagalsky, *Biochem. J.* **101**, 9c (1966).
[37] M. L. Mackenthun, R. D. Tom, and T. A. Moore, *Nature (London)* **278**, 861 (1979).
[38] V. R. Salares, N. M. Young, H. J. Bernstein, and P. R. Carey, *Biochemistry* **16**, 4751 (1977).
[39] M. Buchwald and W. P. Jencks, *Biochemistry* **7**, 834 (1968).
[40] K. E. Van Holde, J. Brahms, and A. M. Michelson, *J. Mol. Biol.* **12**, 726 (1965).
[41] T. Takemura, P. K. Das, G. Hug, and R. S. Becker, *J. Am. Chem. Soc.* **100**, 2631 (1978).
[42] P. F. Zagalsky, *Comp. Biochem. Physiol. B* **71B**, 243 (1982).

day; in contrast, some preparations of ovoverdin under the same conditions become almost colorless. Dissociation of α-crustacyanin to the β-form takes place in solutions of low ionic strength (<0.075 M NaCl), and also in solutions containing 10% dioxane or 3 M urea.[9–11,43] Formation of β-crustacyanin occurs spontaneously, at a slow rate, during storage of the α-form in buffer or under ammonium sulfate and in electrophoresis using low ionic strength or discontinuous buffer systems. At high concentrations of denaturing agents of the "urea–guanidinium" class, crustacyanin tends to give a yellow denaturation product, but to give a red pigment with the "hydrophobic" class of denaturants.[9] During desalting on Sephadex G-25, α-crustacyanin initially precipitates then dissociates to the purple, α'-form. Freeze-drying is best carried out from a volatile buffer (3 mM NH$_4$Ac).[10] The reddish, hygroscopic powder redissolves to give the original pigment, usually with some β-crustacyanin, but is denatured on prolonged exposure to light, air, and, particularly, acid vapors.

In the author's laboratory, the crustacyanins have been stored, in the cold and dark, under 60% ammonium sulfate saturation for 2 or more years and in solution in 1 M KCl–20 mM potassium phosphate buffer, pH 7.0, or other suitable buffer,[9] for many months, with little alteration. The carapace pigments may be preserved *in situ* by freezing the whole animal for short periods but cleaned carapace reddens on prolonged storage in the deep-freeze.

The characteristics of the absorption spectra of the crustacyanins and yellow pigment are given in the table. The values reported for $A^{1\%}_{\lambda_{max}}$ of α- and β-crustacyanin in the visible region lie between 54 and 65.[11,13]

Studies on the lobster carotenoproteins reported from the United States and Canada invariably refer to *Homarus americanus* and those of European workers to the North Sea lobster, *Homarus gammarus*. It has generally been assumed that the carotenoproteins of the two species are identical. While their properties are closely similar, we have recently found some important differences (Zagalsky and Tidmarsh, in press). The pigments of *H. americanus* from Canadian waters bind more firmly to DEAE-cellulose and are eluted at higher ionic strengths in stepwise elution. There are differences also in the electrophoretic mobility of the crustacyanins and in the apocrustacyanin subunit compositions of the separate species.

Extraction

Decalcification of the finely ground carapace is necessary for extraction of the carotenoproteins. Agents that have been used for this purpose

---

[43] L. Wahlgren-Brännström and M. Baltscheffsky, *Acta Chem. Scand., Ser. B* **33**, 613 (1979).

SPECTRAL PROPERTIES OF LOBSTER CARAPACE CAROTENOPROTEINS AND ASTAXANTHIN[a]

Pigment	λ_{max} (nm)	$\nu^{1/2}$[b] (cm^{-1})	ε_{max}[c] ($\times 10^5$)	f[d]	Other max (nm)	$A_{vis/UV}$
α-Crustacyanin	632	4200	1.25	2.60	370, 320, 278	3.3
β-Crustacyanin	587	4300	1.16	2.54	360, 315,[e]278	3.3
γ-Crustacyanin	625	4200	1.19	2.52	366, 318, 278	3.3
Yellow	409	3200	1.00	1.91	300,[f] 255[f]	5.7
Astaxanthin						
In pyridine	492	4200	1.12	2.35		
In hexane	472	4200	1.24	2.60		

[a] Table taken from M. Buchwald and W. P. Jencks, *Biochemistry* **7**, 844 (1968). Reprinted with permission from *Biochemistry*. Copyright 1968 American Chemical Society.
[b] Half-band width.
[c] Molar extinction coefficient, based on astaxanthin content.
[d] Oscillator strength.
[e] Shoulder.
[f] Broad peak.

include 0.6 M ammonium sulfate,[44] 18% hexametaphosphate, pH 5.5,[11,12] and 0.5 M EDTA, pH 7.5.[9,13] Excess decalcifying agent is used and the pH is not allowed to rise above 8.5. The shell pieces may be satisfactorily powdered in the wet state with an agate mortar of a rotary type,[44] or in a rotary ball mill,[14] without denaturation of the pigments, providing care is taken to avoid heating. An initial grinding of the shell in an electric coffee grinder assists efficient powdering.

It is advisable to carry out the preparation and purification of the carotenoproteins in the cold and under weak illumination.

Procedure. Lobsters are killed by decapitation or by freezing at $-15°$. The carapace is removed and thoroughly freed from the underlying hypodermis by scrubbing under a cold tap. The uncalcified layer is removed by breaking the carapace and tearing free the adhering layer. The carapace is left overnight in a cold room spread on newspaper to dry. The blue parts of the shell are broken into small pieces and ground in short (10 sec) bursts in a Moulinex coffee grinder. This process is performed in a cold room, transferring the shell to a sieve (30 meshes to the inch) after grinding and allowing the grinder to cool between bursts; excessive heating must be avoided. It is advisable to wear a protective mask over the mouth

[44] H. J. Ceccaldi and B. H. Allemand, *Recl. Trav. Stn. Mar. Endoume* **35**, 3 (1964).

and nose to avoid inhalation of powder. The sieved shell is immediately transferred to 4 liters of 0.3 M boric acid brought to pH 6.8 with solid Tris, and ground for 16 hr at 4° in a rotary ball mill (5 liter capacity).[14] The suspension is filtered on a Büchner funnel through a layer of Celite Hyflo-Super-Cel (Koch Light), washed consecutively with borate-Tris and water, and the dry cake stirred overnight in 5 liters of 10% EDTA, pH 7.5 (25 g shell/liter). The mixture is refiltered through a layer of the filter aid and the shell residue reextracted for 48 hr with 5 liters of 10% EDTA, pH 7.5.

The pigments are precipitated from the blue filtrates, brought to pH 7.5 with 2 M HCl, at 50% ammonium sulfate saturation, redissolved in 250–500 ml of 0.2 M potassium phosphate buffer, pH 7.0, and fractionally precipitated between 30 and 50% ammonium sulfate saturation.

The shell residue may be reground with borate–Tris and reextracted with EDTA to yield some additional pigment.

Finally, the shell residue is mixed with an equal amount of filter aid, packed into a glass column, and washed (15 ml/hr) with cold 1 M KCNS. Residual pigment is eluted from the column as a purple band. The most concentrated pigment fractions are diluted with 10 volumes of water concentrated in an ultrafiltration cell (YM-10 membrane, Amicon) to a small volume and dialyzed against 50 mM potassium phosphate buffer, pH 7.0 overnight. Alternatively, the shell residue is stirred with 1 M KCNS (or 6 M urea) for 30 min and the pigments are precipitated at 50% ammonium sulfate saturation from the filtrate, after 10-fold dilution with water, and dialyzed as above. Additional α-, β-, and γ-crustacyanin can be recovered in this manner.[10,45] Separation of α- and γ-crustacyanin by adsorption onto a short column of DEAE-cellulose and elution with 0.25 M phosphate buffer, pH 7.0, is advisable before further purification.

Separation of Pigments

Resolution of the crustacyanins and yellow protein is achieved by DEAE-cellulose chromatography using linear salt gradients. The following procedure of Buchwald and Jencks[13] is suitable for 2 g protein; smaller columns and gradient volumes are used for lesser amounts of material.[38,45] This method applies for the pigments of *H. americanus*. Other separation methods described in this chapter have been developed for the proteins of *H. gammarus* and may require slight modification for the resolution of the proteins of *H. americanus*.

DEAE-Cellulose Chromatography: Gradient Elution. The pigments under ammonium sulfate are collected by centrifugation and dissolved in

[45] R. M. Quarmby, Ph.D. Thesis, London University (1971).

50 mM potassium phosphate buffer, pH 7.0. After thorough dialysis against several changes of the same buffer, the protein is applied to a column (4.5 × 38 cm) of DEAE-cellulose, previously washed with 0.5 M and 50 mM phosphate buffer, pH 7.0. β-Crustacyanin is not absorbed and is removed by washing the column with 50 mM buffer until the effluent is only faintly purple; it is stored under 60% ammonium sulfate saturation. The adsorbed pigments are then eluted with a linear gradient from 0 to 1.0 M KCl in 50 mM potassium phosphate buffer, pH 7.0, of 5 liters total volume. Following elution of the crustacyanins and the commencement of elution of yellow protein, the gradient is discontinued and the latter eluted with 1 M KCl in the 50 mM buffer.

The blue pigment fraction (α- and γ-crustacyanin) is precipitated at 50% saturation with ammonium sulfate, collected by centrifugation, dissolved in the 50 mM phosphate buffer, and dialyzed against the same. This fraction is rechromatographed on DEAE-cellulose in the same manner as described above, except that the gradient is 0 to 0.5 M KCl in the 50 mM phosphate buffer. The pigments are eluted in the order β-, γ-, and α-crustacyanin. The α- and γ-fractions are precipitated at 50% ammonium sulfate saturation and purified individually by stepwise DEAE-cellulose chromatography.

Stepwise Elution. The α-crustacyanin fraction in 50 mM potassium phosphate buffer, pH 7.0, is applied to a column of DEAE-cellulose (3.2 × 40 cm) equilibrated with the same buffer and washed free of β- and γ-crustacyanin with 50 mM and 0.15 M phosphate buffer, pH 7.0, respectively. α-Crustacyanin is then eluted with 0.25 M potassium phosphate buffer, pH 7.0. The central fractions with the highest ratio of visible to ultraviolet absorption ($A_{632/280}$) are pooled and stored as a precipitate under 50% ammonium sulfate.

The γ-crustacyanin fraction is purified by the same procedure and is eluted with the 0.15 M buffer.

After prolonged storage, α- and γ-crustacyanin are repurified to remove any β-crustacyanin formed by dissociation, either by repetition of the stepwise DEAE-cellulose chromatography or by gel filtration on Sephadex G-75, or Ultrogel AcA-44 (LKB), equilibrated with 0.1 M phosphate buffer, pH 7.0.[14]

Yellow Protein. The purification of the unstable yellow pigment is difficult to reproduce and gives low yields.[13]

Following dialysis against 0.1 M potassium phosphate buffer, pH 7.0, the pigment is rechromatographed on a DEAE-cellulose column (3 × 30 cm), with a linear gradient from 0.1 M phosphate buffer to 0.2 M phosphate buffer containing 1 M KCl, pH 7.0, and a total volume of 0.6 liter.

The pigment with $A_{409/280}$ of ~5–6 is stored as a precipitate under 60% ammonium sulfate.

Preparation of the α-Crustacyanin by Stepwise DEAE-Cellulose Chromatography

The following method[10] may be employed to obtain the α-crustacyanin of *H. gammarus* in reasonable purity from the extracted pigments following ammonium sulphate fractionation (see Extraction Procedure).

Procedure. The pigments in 0.15 M phosphate buffer, pH 6.9, are passed through a short column (2 × 3.5 cm) of DEAE-cellulose (DE-32; Whatman) to remove strongly adsorbing proteins (including yellow protein). The eluant is then passed through a column (3.5 × 30 cm, suitable for 0.5 g protein), so that the α-crustacyanin saturates 50–70% of the column. After washing with two bed volumes of the 0.15 M buffer to remove unadsorbed β- and γ-crustacyanin, the top few centimeters of DEAE-cellulose, blue-green in color, are removed by stirring and suction. The α-crustacyanin is eluted with 0.25 M phosphate buffer, pH 7.0, and fractions of $A_{632/280}$ greater than 3.0 are pooled and stored as a precipitate under 50% ammonium sulfate.

Remaining crustacyanin and yellow protein are recovered by leaving the exchanger overnight in 1 M KCl–50 mM phosphate buffer, pH 7.0.

Yield. About 40 mg α-crustacyanin is obtained per lobster,[12] about 1 mg/g powdered carapace.[10]

Separation of α- and γ-Crustacyanin on a Porath Column

The crustacyanins can conveniently be separated by electrophoresis on a Porath column (LKB, Sweden) using cellulose as supporting medium and 0.1 M Tris–4 mM EDTA–15 mM boric acid, pH 8.0, as electrophoretic buffer.[14]

Procedure. Yellow protein and impurities are removed by passage of the extracted pigments in 0.25 M phosphate buffer, pH 7.0, through a short column of DEAE-cellulose (DE-32).

Samples containing up to 150 mg protein in electrophoretic buffer in a volume not exceeding 15 ml are applied and run at 40–50 mA for 36–48 hr, the cooling water being kept at 4°. The column is eluted with electrophoretic buffer at a flow rate of 7 ml/hr. The order of elution of the crustacyanins (α-, β-, γ-) ensures complete separation of the α- and γ-pigments. Fractions of α- or γ-crustacyanin of $A_{vis/UV}$ greater than 3.0 are pooled, concentrated by ultrafiltration (YM-10 membrane), and any contaminating β-crustacyanin removed by gel filtration on Sephadex G-75 or ion exchange, as above.

Apoprotein Preparation

The apoproteins of crustacyanin and other Type A carotenoproteins have been prepared by (1) repeated precipitation with acetone,[35] (2) acetone treatment followed by ether extraction of the carotenoid,[20,31] (3) treatment with dimethylformamide (DMF) followed by separation of carotenoid and apoprotein on a column of Sephadex LH-20 (Pharmacia),[46] and (4) treatment of the pigment with 6 M urea and separation of the apoproteins on DEAE-cellulose.[14] The acetone precipitation method usually requires the presence of salts (KCNS or, less harmful,[47] LiCl) soluble at high acetone concentration to prevent loss of protein in the acetone[10,29]; complete removal of carotenoid is not attained and some aggregation of the apoprotein is unavoidable. The DMF and acetone–ether procedures are effective in giving a practically colorless apoprotein. The latter method is rapidly performed and leads to considerably less denaturation (see Fig. 2) than acetone precipitation. The DMF treatment is limited to small scale preparations of apoprotein, as required for reconstitution studies.

Acetone–Ether Method (adapted from Renstrøm et al.[20]). Crustacyanin (4–50 mg, spectrophotometrically determined amount) in 2 ml 50 mM phosphate buffer, pH 7.0, is mixed with 3 ml acetone (an equal volume is employed for some carotenoproteins[20]) in a stoppered tube. Diethyl ether (10 ml) is added and the mixture inverted several times during a period of 5–10 min. The ether layer, containing the extracted carotenoid, is removed by decantation or suction. Repetition of the ether extraction a further three times leaves a faintly purple aqueous phase. A colorless apoprotein preparation, for reconstitution studies, may be obtained by repeated addition of 2 ml acetone to the aqueous phase, made to 2 ml with water, and further (3–4) ether extractions. All the above procedures are carried out at 0°.

Residual acetone is removed by dialysis of the apoprotein (8 mm Cuprophan tubing, Medicell International), filtered through glass fiber paper, for about 3 hr against 50 mM phosphate buffer, pH 7.0, by rotary evaporation in the cold,[20] or under a gentle stream of nitrogen.

The apoprotein prepared by the above method, or by precipitation with acetone, may be separated from aggregated material and remaining α- or β-crustacyanin by gel filtration (see below); passage through a short

---

[46] G. Britton, G. M. Armitt, S. Y. M. Lau, A. K. Patel, and C. C. Shone, personal communication (1983).
[47] D. R. Robinson and W. P. Jencks, *J. Am. Chem. Soc.* **87**, 2470 (1965).

column of DEAE-cellulose equilibrated with 50 mM phosphate buffer, pH 7.0, removes α- but not β-crustacyanin.

DMF Procedure of Britton et al.[46] Carotenoprotein (50–100 μl) (2–20 mg/ml) in 50 mM sodium phosphate buffer, pH 7.5, is treated at 0° with 3 volumes of DMF. The apoprotein is separated from carotenoid on a Sephadex LH-20 column (3 × 0.75 cm) equilibrated with DMF:50 mM phosphate buffer, pH 7.5 (75/25, v/v). The apoprotein is eluted in the void volume, determined with Blue Dextran, in about 0.5 ml, and may be stored at −20° for reconstitution studies.

Reconstitution

α-Crustacyanin may be reconstituted from freshly prepared apoprotein and astaxanthin.[5,10,20] The ability to give the α-form is gradually lost on storage of the apoprotein in solution in a refrigerator or freezer or as a precipitate under 80% ammonium sulfate. Aged preparations and freeze-dried apoprotein give β-crustacyanin and a product of apoprotein size with λ_{max} 565 nm. A similar loss of reconstitution ability has been noted for other carotenoproteins[5] and rhodopsin.[48] Combination of astaxanthin with mixtures of the individual apoprotein units, separated by gel filtration or ion-exchange, gives, in the author's hands, only β-crustacyanin and apoprotein, λ_{max} 565 nm.

The specificity of carotenoid attachment has been studied for a number of carotenoproteins.[3,5,20,29,30,32,36] The following methods, giving good reconstitution of α-crustacyanin from freshly prepared apoprotein and astaxanthin, are carried out under dim light. The choice of method depends on the nature of both the carotenoid and apoprotein under investigation.

It is essential to confirm the identity of carotenoids in reformed pigments; release with dichloromethane is advised.[49]

Acetone Method. The procedure of Cheesman,[34] utilizing the solubility of apoprotein and carotenoid in acetone–water mixtures, has been applied with modifications.[10,29,30,32,36] The carotenoid in acetone is added, at 0°, to the apoprotein (2 mg/ml) in 50 mM phosphate buffer, pH 7.0, until a homogeneous phase is obtained (at ∼50% acetone for astaxanthin). The amount of carotenoid is at least 25% in excess of that derived from the carotenoprotein in the preparation of the apoprotein. Acetone is removed on a rotary evaporator under reduced pressure in the cold and the solution dialyzed against the 50 mM phosphate buffer.[30,32] Alternatively, the mix-

---

[48] E. W. Abrahamson and S. E. Ostroy, *Prog. Biophys. Microbiol.* **17**, 181 (1967).

[49] F. G. Pilkiewicz, M. J. Pettei, A. P. Yudd, and K. Nakanishi, *Exp. Eye Res.* **24**, 421 (1977).

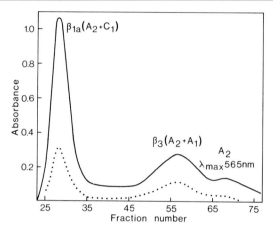

FIG. 1. Separation of pigments reconstituted from apocrustacyanin subunits and astaxanthin on DEAE-cellulose (DE-32). Column (1.0 × 12 cm) equilibrated at 5° with 5 mM phosphate buffer, pH 6.8. Apocrustacyanin subunit A_2 (~5 mg), A_1 (~1 mg), and C_1 (~3 mg), separated by DEAE-cellulose chromatography, recombined with astaxanthin (cellulose method). Reconstituted pigment applied in 5 mM phosphate buffer, pH 6.8. Elution buffer: linear concentration of gradient phosphate buffer, pH 6.8 (500 ml 5 mM–500 ml 0.1 M). Flow rate: 24 ml/hr. Fraction volume: 5 ml. Absorbance at 585 nm (———); absorbance at 280 nm (-----).

ture is diluted with 50 mM phosphate buffer containing 10 mM MgSO$_4$ to 10–15% acetone concentration and dialyzed against 0.1 M phosphate buffer–10 mM MgSO$_4$, pH 7.0, initially in the cold to remove acetone, then at room temperature. Free carotenoid is removed by centrifugation,[32] or by passage through a column of Celite [or CaCO$_3$–Celite, 1/1 (w/w)] equilibrated with 0.1 M phosphate buffer, pH 7.0.[36,45] The reconstituted pigments are separated by gel filtration on Sephadex G-75, ion-exchange (Fig. 1) or electrophoresis[3,36]; β-crustacyanin and pigments of apocrustacyanin size are not adsorbed onto DEAE-cellulose from 50 mM phosphate buffer, pH 7.0.

Detergent Method.[36] Carotenoids are dissolved in acetone, ether, dichloromethane, or other suitable solvent, taken to dryness in a round bottomed flask on a rotary evaporator, and dissolved in 50 mM phosphate buffer, pH 7.0, containing 5% Triton X-100. The carotenoid is added to the apoprotein (2 mg/ml) in 50 mM phosphate buffer, pH 7.0, to give a detergent concentration not exceeding 0.5% and left, or dialyzed against the phosphate buffer, in the dark at room temperature overnight. The amount of carotenoid added is twice that derived from the carotenoprotein in the preparation of the apoprotein. Reconstitution of apocrustacyanin with astaxanthin is practically instantaneous.

The solution is passed through a column (5 × 0.8 cm) of DEAE-cellulose, which is washed with 50 mM phosphate buffer, pH 7.0. Pigments of the α- or γ-type are adsorbed and eluted with 0.15 and 0.25 M phosphate buffer, pH 7.0, respectively. Excess carotenoid in detergent and β- and apo-type pigments are not adsorbed. The latter pigments may be separated by dialysis of the effluent against 1 mM phosphate buffer, pH 7.0, application to a column of DEAE-cellulose, and elution with a suitable concentration gradient of phosphate buffer (as in Fig. 1); detergent and excess carotenoid are unadsorbed.

DMF Method of Britton et al.[46] The carotenoid (100 μg) is dissolved in dichloromethane and adsorbed onto silicic acid (100 mesh) in a short column (3 × 0.5 cm) equilibrated with redistilled light petroleum. The solvent is displaced by nitrogen and the dried, bound carotenoid vortexed with 1 ml of DMF/50 mM sodium phosphate buffer–10 mM MgSO$_4$, pH 7.5 (75/25, v/v). The resulting solution is hand-pumped through a sintered glass funnel (porosity 3). The emerging carotenoid is added to the apoprotein at 0°, dissolved in the same DMF mixture, in approximately 1.5/1 M proportions. The solution is dialyzed in the cold, first against 50 mM phosphate–10 mM MgSO$_4$, then against 50 mM phosphate buffer, pH 7.0. Reconstituted pigments are separated on DEAE-cellulose, as above.

Cellulose Procedure.[29,45] Powdered cellulose is washed thoroughly with solutions in the following order: acetone, water, 0.25 M phosphate buffer, pH 7.0, 2% EDTA, pH 7.5, water, acetone. The cellulose is air-dried and solutions of the carotenoid (0.1–0.3 mg/ml) is acetone, dichloromethane, or other suitable solvent (2.5 ml) are stirred with the cellulose to make a slurry in a flask. The solvent is evaporated under a stream of nitrogen and the cellulose washed with ice-cold deaerated water. A solution (8 ml) of the apoprotein (~1 mg/ml) in 50 mM phosphate buffer, pH 7.0, is stirred for 24 hr at room temperature with the carotenoid-adsorbed cellulose. The slurry is filtered and reconstituted pigments isolated, as described above.

Celite, used as described for retinol-binding protein,[50] is more suitable than cellulose for hydrophobic carotenoid derivatives.

Isolation of Apoprotein Subunits

Gel Filtration. Apocrustacyanin (2–5 ml; 5–10 mg/ml) is separated on columns (52 × 2.5 cm) of Sephadex G-75 (superfine) (Pharmacia) equilibrated either with 50 mM phosphate buffer, pH 6.8, for subsequent ion-exchange chromatography, or with 25 mM imidazole–HCl, pH 7.4, for isolation of the subunits by chromatofocusing. The apoprotein is eluted in

---

[50] De W. S. Goodman and A. Raz, *J. Lipid Res.* **13**, 338 (1972).

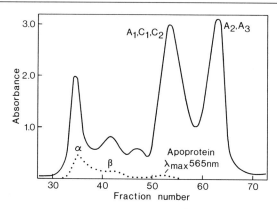

FIG. 2. Separation of apocrustacyanin in Sephadex G-75 (superfine). Column (52 × 2.5 cm) equilibrated at 5° with 25 mM imidazole–HCl, pH 7.4. Sample: 3 ml apocrustacyanin (~15 mg/ml) in equilibration buffer, prepared by the acetone–ether procedure. Flow rate: 6 ml/hr. Fraction volume: 3 ml. Absorbance at 280 nm (———); absorbance at 585 nm (----).

two fractions containing, in order of elution, apoproteins C_1, C_2, and A_1 (Type I subunits) and A_2 and A_3 (Type II subunits), respectively[14] (Fig. 2). Complete resolution of the two types of subunit may be achieved by concentrating each fraction to 2–5 ml by ultrafiltration (YM5 membrane; Amicon) and repetition of the gel filtration.

Ion-Exchange Chromatography. The two apoprotein subunit fractions are pooled, concentrated to 5–10 ml in an ultrafiltration cell (YM5 membrane), dialyzed against water, and ~20 mg applied to a column (1.8 × 40 cm) of DEAE-cellulose (DE-32; Whatman). The subunits are eluted in the order A_1, A_2, then A_3 with two consecutive linear concentration gradients of phosphate buffer, pH 6.8 (500 ml 5 mM–500 ml 45 mM for A_1 and A_2; 250 ml 35 mM–250 ml 0.25 M for A_3). Subunits C_1 and C_2 are not adsorbed and are subsequently separated, C_1 ahead of C_2, on a column (1.7 × 27 cm) of CM-cellulose (CM-52; Whatman) using a linear concentration gradient from 0 to 70 mM phosphate buffer, pH 6.8, in a total volume of 1.0 liter.

The apoproteins may similarly be resolved following dissociation of α-crustacyanin with 6 M urea, with the inclusion of 6 M urea in the gradients.[14]

A more rapid, and as satisfactory, resolution of the apoproteins is made by using the technique of chromatofocusing below.

Chromatofocusing. Buffers and exchanger are thoroughly degassed and the procedures recommended in the Pharmacia booklet "Chromatofocusing" followed.

The pooled apoprotein fractions from above (5 ml; 10–20 mg/ml) in 25 mM imidazole–HCl, pH 7.4, are applied to a column (1.0 × 27 cm) of Polybuffer Exchanger PBE94, equilibrated with the imidazole buffer. Prior to the addition of the sample, 5 ml Polybuffer 74, diluted 1/7 (v/v) and adjusted to pH 4.5 with 5 M HCl is passed through the column, followed by 1 ml of the imidazole buffer. The sample is washed into the column with 2 ml imidazole buffer and the subunits eluted with diluted Polybuffer 74, pH 4.5, in the order A_1, A_2, and A_3.

Subunits C_1 and C_2, not adsorbed on the exchanger, are concentrated to about 4 ml by ultrafiltration (YM5 membrane) and dialyzed against 25 mM ethanolamine–HCl, pH 9.4, in Cuprophan tubing (8 mm) for 6 hr. The sample is applied to a column (1.0 × 27 cm) of exchanger PBE94 equilibrated with 25 mM ethanolamine–HCl, pH 9.4. The subunits are eluted in the order C_2 followed by C_1 with Polybuffer 96 diluted 1/9 (v/v), pH 8.3.

The separations, carried out in a cold room, are shown in Fig. 3. Suitable fractions are pooled and the subunits precipitated at 80% ammonium sulfate saturation. The proteins are washed on the centrifuge with 80% ammonium sulfate to remove polybuffer, dialyzed against distilled water, and freeze-dried.

Separation of β-Crustacyanins

The β-crustacyanins are separated by DEAE-cellulose chromatography with linear concentration gradients of phosphate buffer pH 6.8, and elute in the order β_{1b}, β_{1a}, β_{2b}, β_{2a}, β_3, and β_4.[14]

The apoprotein compositions of the β-forms, determined by electrophoresis in starch or acrylamide gels containing 6 M urea (see later, Fig. 5), are β_{1a} ($A_2 + C_1$), β_{1b} ($A_2 + C_2$), β_{2a} ($A_3 + C_1$), β_3 ($A_2 + A_1$), and β_4 ($A_3 + A_1$).[14] Components β_{1a+b}, β_{2a}, and β_4 predominate and are formed in the approximate relative proportions 5/2/1 on dialysis of α-crustacyanin against water.[14]

Procedure. The β-crustacyanin fraction (see Separation of Pigments) is concentrated by ultrafiltration to 2–5 ml and applied to a column (2.5 × 52 cm) of Sephadex G-75 (superfine) equilibrated with 50 mM phosphate buffer, pH 6.8. Fractions with $A_{585/280}$ greater than 2.5 are concentrated and dialyzed against water. The β-crustacyanins (~35 mg) are partially separated on a column of DEAE-cellulose (1.8 × 12 cm) with a linear gradient (from 0 to 0.1 M) phosphate buffer, pH 6.8, in a total volume of 1 liter. The crustacyanins elute in the order β_{1a+b}, β_{2a+b}, β_3, and β_4. Fractions β_{1a+b} and β_{2a+b} are rechromatographed on DEAE-cellulose columns (1.8 × 12 cm) with shallower phosphate buffer gradients (500 ml 5 mM–

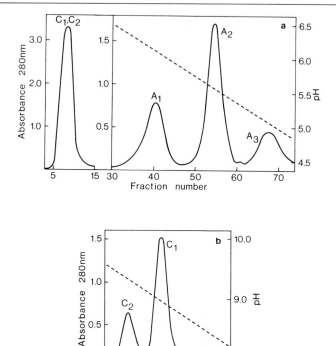

FIG. 3. Chromatofocusing of apocrustacyanin on exchanger PBE 94. (a) Column (1.0 × 27 cm) equilibrated at 5° with 25 mM imidazole–HCl, pH 7.4. Sample: 5 ml apocrustacyanin (~7 mg/ml) in equilibration buffer prepared by the acetone–ether procedure and separated on Sephadex G-75 (superfine) (Fig. 2). Elution buffer: Polybuffer 74 diluted 1/7 (v/v) and adjusted to pH 4.5 with 5 M HCl. Flow rate: 9 ml/hr. Fraction volume: 3 ml. (b) Column equilibrated at 5° with 25 mM ethanolamine–HCl, pH 9.4. Sample: 3 ml apocrustacyanin subunits C_1 and C_2 [eluting in void volume, (a) above; ~5 mg/ml in equilibration buffer]. Elution buffer: Polybuffer 96 diluted 1/9 (v/v), pH unadjusted (pH 8.3). Flow rate: 9 ml/hr. Fraction volume: 3 ml.

500 ml 25 mM and 500 ml 9 mM–500 ml 30 mM, respectively) to resolve the major components β_{1a} and β_{2a}.

β-Crustacyanins formed by dissociation of α-crustacyanin with 3 M urea are fractionated as above with the inclusion of 3 M urea in the gradients.[14]

Resolution of the β-crustacyanins is also accomplished satisfactorily by chromatofocusing on Exchanger PBE94 (Fig. 4).

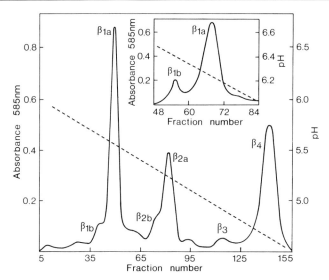

FIG. 4. Chromatofocusing of β-crustacyanin on exchanger PBE 94. Column (1.0 × 27 cm) equilibrated at 5° with 25 mM histidine–HCl, pH 6.2. Sample: 5 ml β-crustacyanin mixture (~3 mg/ml) in equilibration buffer. Elution: 25 ml of equilibration buffer followed by Polybuffer 74 diluted 1/7 (v/v) and adjusted to pH 4.5 with 5 M HCl. Flow rate: 9 ml/hr. Fraction volume: 3 ml. Inset: separation of $β_{1a}$- and $β_{1b}$-crustacyanin. Column equilibrated with 25 mM imidazole–HAc, pH 7.4. Sample in equilibration buffer. Elution: Polybuffer 96 diluted 1/12 (v/v) and adjusted to pH 6 with glacial acetic acid.

Minimum Molecular Weight

The amount of protein associated with a single carotenoid prosthetic group is estimated by determination of the protein and carotenoid contents of the purified carotenoproteins. The protein content may be determined in a number of ways and the choice of method may be dictated by availability of material; the methods are listed in order of reliability.

Protein Content

Direct Weighing. Freeze-dried protein is either directly weighed[35,51] or the protein content of a solution of the pigment is determined from the 280 nm absorption and $A_{280}^{1\%}$ value.[12] The latter is determined as follows: a volume of a salt-free solution of the pigment of known 280 nm absorption is freeze-dried and the protein dried at 105° to constant weight. For water-

---

[51] P. F. Zagalsky and P. J. Herring, *Comp. Biochem. Physiol. B* **41B**, 397 (1972).

insoluble pigments the absorption at 280 nm in a salt solution is measured prior to dialysis of a known volume against water and freeze-drying.

From Nitrogen Content.[13,51] A sample of the carotenoprotein in water or buffer (not Tris) is totally digested with concentrated sulfuric acid and the nitrogen content determined by the ninhydrin method of Jacobs.[52] The protein content is derived by measurement of the nitrogen content of the freeze-dried protein; for simple proteins this value may be calculated from the amino acid composition[13] or a value of 15.7%, measured for crustacyanin,[10,12,13] may be taken.

By Protein Determination. In order of sensitivity, the protein content may be determined by the modified Folin–Lowry method of Peterson,[53] biuret protein assay,[46] or, less reliably,[54] microtannin turbidity measurement.[46,55] Proteins of low aromatic acid content are likely to give anomalous color development in the Folin–Lowry assay.[54] A reasonable estimate (25,000) for the minimum molecular weight of α-crustacyanin has been obtained by the author, with crystalline serum albumin as standard, utilizing the modified Folin–Lowry assay; the carotenoid does not interfere with the assay. The carbohydrate content of glycoproteins must, in addition, be established by suitable means[33–35,56,57] for the minimum molecular weight to be derived.

By Measurement of Peptide Bond Absorption. The absorption at 215 nm of an optically clear solution of the carotenoprotein, obtained by centrifugation at high speed, is compared with that of a similarly treated standard solution of bovine serum albumin.[51] The value obtained (30,000) for the minimum molecular weight of α-crustacyanin by the author using this method exceeds the lowest reported value[12] by about 10,000. This may, in part, result from a carotenoid contribution to the absorption in the ultraviolet.[58]

Carotenoid Content. The carotenoid is liberated from the carotenoprotein with organic solvent (acetone, ethanol, or dichloromethane) and transferred to a solvent in which the molar extinction coefficient at absorption maximum is known[12,16,30,34,35]; astaxanthin has a molar absorbance coefficient of 1.12×10^5 in pyridine at the absorption maximum, 492 nm.[39]

[52] S. Jacobs, *Nature (London)* **183**, 346 (1959).
[53] G. L. Peterson, *Anal. Biochem.* **83**, 346 (1977).
[54] F. Layne, this series, Vol. 3, p. 447.
[55] S. Mejbaum-Fatzenellenbogen and W. J. Dobryszycka, *Clin. Chim. Acta* **4**, 515 (1959).
[56] P. F. Zagalsky, D. F. Cheesman, and H. J. Ceccaldi, *Comp. Biochem. Physiol.* **22**, 851 (1967).
[57] D. de Chaffoy de Courcelles and M. Kondo, *J. Biol. Chem.* **255**, 6727 (1980).
[58] G. Englert, F. Kienzle, and K. Noack, *Helv. Chim. Acta* **60**, 1209 (1977).

The method of Cheesman[35] avoids carotenoid loss in transference: a sample (5–50 μl) of the carotenoprotein in water is added to 1 ml pyridine in a spectrophotometric cell, mixed, and immediately read at 492 nm against pyridine to which the same volume of water has been added.[13]

The absorbance coefficient of astaxanthin in 95% pyridine is unaltered from that in the pure solvent.[39] Carotenoproteins insoluble in distilled water, but soluble in 0.1 M Tris–HCl or 50 mM phosphate buffers, are dissociated by addition of an equal volume of pyridine and the molar extinction coefficient at the absorption maximum taken as 1.12×10^5. Dissociation of α-crustacyanin in 0.1 M Tris–HCl, pH 7.5, with an equal volume of either pyridine or 5% sodium dodecyl sulfate (SDS) leads to practically identical absorbancies at the absorption maxima, 495 and 490 nm, respectively. There is a decrease in the absorbance at the maxima of about 11% compared to that of the carotenoprotein at 632 nm, as for dissociation of the pigment in 95% pyridine.[13]

There is a less rapid loss of color in the SDS solution than in pyridine and this method of dissociation is preferred for astaxanthin proteins. Some other carotenoids, however, aggregate in this detergent.[59,60]

Polyacrylamide Gel Electrophoresis (PAGE)

PAGE in the presence and absence of 0.1% SDS or 6 M urea, in rods or slabs, is carried out routinely[61,62] in the discontinuous buffer system of Laemmli.[63] The methods of preparation of gels and electrophoresis are as described previously in this series,[64] but with the following stock solutions.[65]

Stock Solutions

Solution A. 1.5 M Tris–HCl, pH 8.8
Solution B. 0.5 M Tris–HCl, pH 6.8
Solution C. Stock acrylamide solutions for separating gels:
 (1) 30 g acrylamide and 0.9 g bis(N,N'-methylenebis-acrylamide) in 100 ml of water
 (2) 30 g acrylamide and 2 g bis in 100 ml of water
Solution D. Stock acrylamide solution for stacking gels 30 g acrylamide and 0.8 g bis in 100 ml of water

---

[59] S. Takagi, K. Takeda, K. Kameyama, and T. Takagi, *Agric. Biol. Chem.* **46**, 2035 (1982).
[60] S. Takagi and K. Takeda, *Agric. Biol. Chem.* **46**, 2217 (1982).
[61] L. Shuster, this series, Vol. 22, p. 412.
[62] O. Gabriel, this series, Vol. 22, p. 565.
[63] U. K. Laemmli, *Nature (London)* **227**, 680 (1970).
[64] B. Dewald, J. T. Dulaney, and O. Touster, this series, Vol. 32, p. 82.
[65] U. K. Laemmli and M. Favre, *J. Mol. Biol.* **80**, 575 (1973).

Solution E. 4 mg riboflavin in 100 ml of water
Solution F. 50 g sucrose in 100 ml of water
Solution G. 90 mg sodium persulfate in 10 ml of water, prepared freshly
Solution H. Running buffer. 30 g Tris and 144 g glycine in 5 l of water. For SDS-PAGE, 5 g SDS is included in the buffer
Solution I. 10 g SDS in 100 ml of water

Separating Gel Solution. Thirty milliliters of solution contains 7.5 ml of solution A, the desired volume of solution C, 7.5 µl TEMED (N,N,N',N'-tetramethylethylenediamine), and 1 ml of solution G. Of solution I 0.3 ml is included for SDS-containing gels and 10.8 g urea (Analar) for 6 M urea gels. Solutions for acrylamide gradients are made similarly but with 0.5 ml of solution G and 3.75 µl of TEMED.

The gels are allowed to age at room temperature overnight.

Spacer Gel Solution. Thirty milliliters of solution is made with 7.5 ml of solution B, 3 ml (rods) or 4 ml (slabs) of solution D, 3.5 ml solution E, 6 ml of solution F, and 30 µl TEMED. Of solution I 0.3 ml is included for SDS-containing gels. Solution F is replaced by 10.8 g urea for 6 M urea-containing gels.

The spacer gel is polymerized by placing under a fluorescent light for 30 min.

Sample Preparation. Pigments and apoprotein are in 0.125 M Tris–HCl, pH 6.8–5% sucrose–0.001% bromophenol blue (BPB).

Samples for urea gels are in 6 M urea–1 mM EDTA–0.125 M Tris–HCl, pH 6.8–0.001% BPB (solution J) and are heated at 37° for 5 min, or 100° for 1 min,[29] to dissociate the pigments into subunits.

For electrophoresis in SDS the proteins and molecular weight markers are in 2% SDS–0.125 M Tris–HCl, pH 6.8–5% (w/v) 2-mercaptoethanol–5% sucrose–0.001% BPB (solution K) and are heated at 100° for 5 min.

Carotenoproteins resolved in PAGE are excised from gels and incubated at 37° with (1) 20 ml of solution J for 1 hr, or (2) 20 ml of solution K for 100 min. The pieces of gel are placed on the spacer gel of appropriate SDS or 6 M urea–acrylamide gels (rods or slabs) and covered with incubation buffer for electrophoresis.

Comparison of the sizes of apoproteins or of their peptide maps (see later) can be performed on a single sample of carotenoprotein excised from an acrylamide gel. The gel piece is incubated with solution J (as above) and the apoproteins are separated in a 6 M urea-containing rod gel. The rod is sliced longitudinally and incubated with solution K (as above) for size comparison, or with solution L (below) at 37° for 100 min for peptide mapping. For electrophoresis in the perpendicular direction, the slice is placed in close contact with the spacer gel of the appropriate SDS slab and covered with solution K or, for peptide mapping, with a layer of solution L containing 0.8% agarose (kept molten in a closed vessel at 56°).[66]

Gel pieces are stored overnight at 5° wrapped in cling film and resting on a cork in a beaker over a layer of water.

Separation of Pigments. Resolution of α- and γ-crustacyanins is achieved in 5% acrylamide ($T = 5.2$ and $C = 2.9$ in the notation of Hjértén[67]) made with stock solution $C_{(1)}$.

The β-crustacyanins and apoproteins are separated in 7.5% acrylamide gels [$T = 7.7$ and $C = 2.9$; stock solution $C_{(1)}$] (Fig. 5a).

Carotenoproteins of larger size[29] are separated in gels containing 0.5% agarose–3% acrylamide.[68]

Following dissociation of the pigments in 6 M urea, apoprotein components are examined in 5% acrylamide gels [$T = 5.2$, $C = 2.9$; stock solution $C_{(1)}$] containing 6 M urea (Fig. 5b and c). Alternatively, the pigments are in 0.125 M Tris–HCl, pH 6.8–5% sucrose–0.001% BPB and electrophoresis is interrupted for 30 min for dissociation of the proteins when the BPB front reaches the separating gel.

Separation of apocrustacyanin subunits, Type I, and Type II, is made in 12.5% acrylamide gels [$T = 12.9$ and $C = 2.9$; stock solution $C_{(1)}$] containing 0.1% SDS, or, with increased resolution, in 15% acrylamide 0.1% SDS gels [$T = 16.0$ and $C = 6.3$; stock solution $C_{(2)}$].

Starch gels, made with or without 6 M urea, have also been utilized for separation of the crustacyanins and apoproteins.[10,14]

Molecular Size of Apoproteins. Apparent molecular sizes[69] are determined in SDS-slabs made with a linear concentration gradient of acrylamide.[70] Molecular weight marker 12,300–78,000 (BDH) (cytochrome c, myoglobin, chymotrypsinogen A, ovalbumin, serum albumin, and ovotransferrin) is used as standard.

The gradient is formed from solutions (70 ml of each for 4 plates, 13.8 × 8.2 × 0.25 cm) containing 10% acrylamide [$T = 10.7$ and $C = 6.3$; stock solution $C_{(2)}$] and 15% acrylamide [$T = 16.0$ and $C = 6.3$; stock solution $C_{(2)}$], respectively, using Pharmacia peristaltic pump P-3 and gel slab casting apparatus GSC-8, with an overlay of 20% ethanol–0.1% SDS, or isobutanol–water (l/l, v/v). The separating and spacer gels are 12.0 and 1.8 cm, respectively, in length. A gradient produced from 12.5% acrylamide

---

[66] C. Bordier and A. Crettol-Järvinen, *J. Biol. Chem.* **254**, 2565 (1979).
[67] S. Hjértén, *Arch. Biochem. Biophys., Suppl.* **1**, 147 (1962).
[68] A. C. Peacock and C. W. Dingman, *Biochemistry* **7**, 668 (1968).
[69] D. M. Neville Jr. and H. Glossmann, this series, Vol. 32, p. 92.
[70] P. Lambin, *Anal. Biochem.* **85**, 114 (1978).

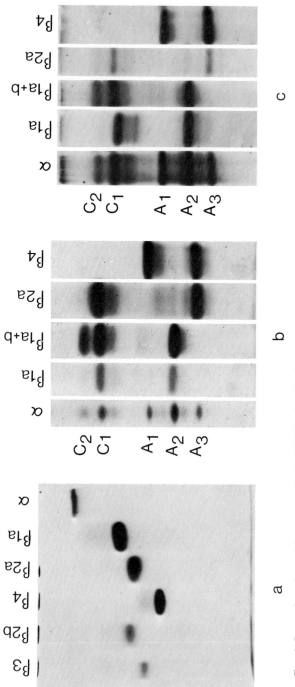

FIG. 5. Separation of α- and β-crustacyanins in PAGE and their subunit compositions in 6 M urea gels. (a) Separation of α- and β-crustacyanins in 7.5% acrylamide gel; (b) 5% acrylamide gel containing 6 M urea. Samples: α- and β-crustacyanins excised from 5 and 7.5% acrylamide gels, respectively, and incubated with solution J (see Sample Preparation); (c) 5% acrylamide gel containing 6 M urea. Samples: α- and β-crustacyanin in 0.125 M Tris–HCl, pH 6.8–5% sucrose–0.001% BPB. Electrophoresis interrupted for 30 min to allow dissociation into subunits (see Separation of Pigments).

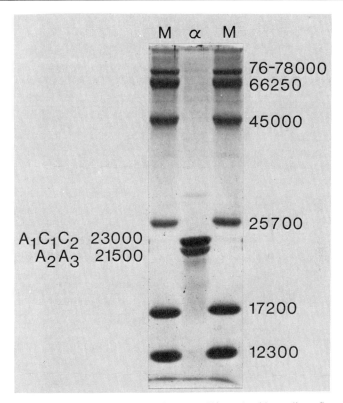

FIG. 6. SDS–PAGE of apocrustacyanin in a 10–15% acrylamide gradient. Samples: α,α-crustacyanin excised from a 5% acrylamide gel and incubated with solution K (see Sample Preparation); M, BDH molecular weight marker 12,300–78,000.

$[T = 12.9$ and $C = 2.9$; solution $C_{(1)}]$ and 20% acrylamide $[T = 20.6$ and $C = 2.9$; solution $C_{(1)}]$ gives better resolution of the molecular weight markers but poorer separation of the apoproteins.

Estimates of molecular size are obtained from plots of percentage mobility relative to chymotrypsinogen A (R_m[71]) against $\log_{MW}$ for the set of marker proteins (12.5–20% acrylamide gradient gels) or for the four smaller marker proteins (10–15% acrylamide gradient gels) (Fig. 6). The apparent sizes of Type I and Type II apocrustacyanins, 23,000 and 21,500, respectively, are in reasonable agreement with estimates (22,000 and 19,000, respectively) derived at a single acrylamide concentration using a continuous phosphate buffer system.[14]

Staining. Gels are stained in 0.1% Coomassie Brilliant Blue R-250 in

[71] W. L. Zahler, this series, Vol. 32, p. 70.

50% methanol, 10% acetic acid for 1–2 hr and destained in a solution of 5% methanol and 10% acetic acid.[72]

Comparison of Apoproteins by Peptide Mapping. The method of Cleveland et al.,[72] as modified by Bordier and Crettol-Järvinen,[66] for peptide mapping following limited proteolysis in 0.1% SDS is followed.

Procedure. Plates (13.8 × 8.2 × 0.25 cm) are made with both the spacer gel (length 4 cm; $T = 3.1$ and $C = 2.9$; stock solution D) and separating gel [length 8 cm; $T = 16.0$ and $C = 6.3$; stock solution $C_{(2)}$] containing 0.1% SDS and 1 mM EDTA.

Apoproteins are separated in 5% acrylamide–6 M urea gels and longitudinal slices incubated with 0.1% SDS–1 mM EDTA–0.125 M Tris–HCl, pH 6.8–0.001% BPF (solution L) and fixed within a layer of agarose above the spacer gel (see Sample Preparation).

Enzyme (50 μg) (*Staphylococcus* V_8 protease, chymotrypsin, activated papain[73]) in 0.5 ml solution L containing 5% sucrose is placed onto the agarose layer. Subtilisin is used with EDTA omitted from all solutions.

The costacking of protease and apoproteins is performed at 50 V and 20°. To extend the period of proteolysis, electrophoresis is interrupted for up to 3 hr when the BPB front reaches the separating gel; electrophoresis is continued at 70 V overnight (Fig. 7).

Staining. Gels are stained overnight in 0.05% Coomassie Brilliant Blue R-250 in 25% isopropanol, 10% acetic acid, and destained in 10% methanol, 7% acetic acid.

Comparison of Peptide Maps. A quantitative estimate of the similarity between peptide maps of pairs of the apoproteins[19] is made by deriving an index of similarity.[74] An index of dissimilarity is first calculated. This is defined as the total number of dissimilar bands (η) divided by the total number of bands for both proteins (N) expressed as a percentage [(η/N) × 100]. Subtraction of the index of dissimilarity from 100 gives the similarity index [$100 - (\eta/N) \times 100$]. The relatedness of the proteins is best expressed graphically in the form of a dendrogram.[74]

Crystallization

α-Crustacyanin (and other carotenoproteins[29,75]) may be crystallized from ammonium sulphate, in the form of needles,[10] or, more reprodu-

---

[72] D. W. Cleveland, S. G. Fischer, M. W. Kirschner, and U. K. Laemmli, *J. Biol. Chem.* **252**, 1102 (1977).
[73] R. Arnon, this series, Vol. 19, p. 226.
[74] R. Lawson, D. H. Davies, J. Casey, and S. A. Mohammad, *Lab. Pract.* **29**, 1065 (1980).
[75] P. F. Zagalsky, H. J. Ceccaldi, and R. Daumas, *Comp. Biochem. Physiol.* **34**, 579 (1970).

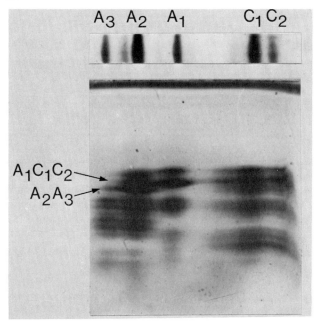

FIG. 7. Peptide mapping of apoprotein subunits of α-crustacyanin by PAGE following limited proteolysis in 0.1% SDS with subtilisin. Horizontal gel: 5% acrylamide gel containing 6 M urea. Sample: α-crustacyanin excised from a 5% acrylamide gel and incubated with solution J (see Sample Preparation). Vertical gel: 15% acrylamide gel containing 0.1% SDS. Sample: horizontal gel sliced longitudinally, incubated with solution L (see Sample Preparation) and overlayed with 50 μg subtilisin. PAGE interrupted for 3 hr to prolong digestion. Taken from P. F. Zagalsky, *Comp. Biochem. Physiol. B* **73B**, 997 (1982).

cibly, as hexagonal plates,[29] from aqueous polyethylene glycol (PEG) by the method of McPherson.[76]

Procedure. α-Crustacyanin (50 μl; 1–2 mg/ml) in 40 mM Tris–HCl, pH 6.9–4% PEG 6000 (BDH Ltd.), is placed in conical plastic tube endings (4 × 4 mm) held in a cork support. This is floated in an enclosure (10 × 10 × 5 cm) on a reservoir (20 ml) of 20% PEG 6000 and kept in the dark at room temperature. The whole procedure is carried out under sterile conditions. Crystals are formed within 1–2 weeks.

General Extraction Procedures for Carotenoproteins, Type A

Calcified tissues are ground with 0.3 M borate–Tris, pH 6.8, as for the extraction of crustacyanin. The powdered shell is collected by centrifuga-

---

[76] A. McPherson, Jr., *J. Biol. Chem.* **251**, 6300 (1976).

tion and suspended overnight in excess of decalcifying agent (0.5 M EDTA, pH 7.5 or 18% hexametaphosphate, pH 5.5). If the carotenoprotein is insoluble in the decalcifying solution,[75] the shell residue is filtered and suspended in 50 mM phosphate buffer, pH 7.0. It has been found that, in some cases,[75] only distilled water is effective in solubilizing the carotenoprotein from the decalcified shell.

Drastic procedures, such as extraction with a chaotropic agent[77] (1 M KCNS) or 6 M urea followed by dilution and dialysis,[10] may be employed as a last resort, but may not result in the isolation of native pigment.

Some carotenoproteins from noncalcified tissues are readily solubilized with phosphate or other buffer solutions, 10% NaCl, or with water alone[29,78]; others require repeated freezing and thawing in buffer for extraction.[51] The maintenance of a reasonably high ionic strength during extraction and purification is necessary for those proteins dissociating at low ionic strength[75]; a pH of solutions between 5 and 8.5 is advisable to preserve the native carotenoid–protein linkage.[1] Carotenoproteins with specific anion-binding sites, such as those of *Velella*,[29] require a saturating concentration of anion throughout procedures; alteration in quaternary structure or adsorption onto Sephadex and Agarose columns may otherwise result. The *Velella* pigments also show a tendency to aggregate in solution by disulfide bond formation, but redissolve if 20 mM mercaptoethanol is present. The inclusion of mercaptoethanol (or dithiothreitol) in extraction buffers should be tried for inextractable pigments.

Carotenoproteins of Echinodermata

The carotenoproteins of the dorsal surface of starfish are more difficult to free from associating contaminants than those of Crustacea and exhibit unusual properties of solubility and stability that distinguishes them from the latter pigments. Asteriarubin,[33] the blue carotenoprotein of *Asterias rubens*, has been extensively characterized.[5,33,34] Purple and blue pigments have been isolated from *Marthasterias glacialis* and studied in some detail.[5]

The water-soluble asteriarubin, λ_{max} 570 nm, is glycoprotein (0.5–1% carbohydrate) essentially devoid of tyrosine and tryptophan and, thereby, of ultraviolet absorption at 280 nm.[33] A mixture of carotenoids, 7,8,7',8'-tetrahydroastaxanthin, 7,8-dihydroastaxanthin, and astaxanthin, is bound to the latter protein with the acetylenic derivatives greatly predominating[33,34]; the carotenoproteins of *M. glacialis* have, likewise, a mixture

---

[77] W. H. Sawyer and J. Puckbridge, *J. Biol. Chem.* **248**, 8429 (1973).
[78] H. J. Ceccaldi and P. F. Zagalsky, *Comp. Biochem. Physiol.* **21**, 435 (1967).

of carotenoids, mainly astaxanthin (predominating) and adonirubin (3-hydroxy-β,β-carotene-4,4'-dione).[5] Asteriarubin has a single carotenoid binding site and dissociates into four identical subunits of molecular weight 11,000 in SDS–PAGE, but remains intact on removal of the carotenoid with acetone.[34] The specificity of the carotenoid–protein association has been investigated using synthetic carotenoid derivatives and part of the amino acid sequence has been determined.[5]

Properties.[5,46] Asteriarubin changes irreversibly in color to red in low ionic strength buffer (20 mM phosphate, pH 7.5) over a period of 30 days at 5°; this alteration is complete in 60 hr in the presence of 17 mM NaCl. The pigment is completely stable, however, in 20 mM phosphate buffer–47 mM NaCl between pH 5 and 8 and remains unbleached after a week in full sunlight. The carotenoprotein is insoluble in 67 mM NaCl–20 mM phosphate buffer, pH 7.5, and may be stored as a precipitate in this solution for several months at 5°. The pigment may be freeze-dried without apparent alteration, although differences in the ability to reconstitute result. Solutions in buffer are not stable to freezing.

The purple and blue carotenoproteins of *M. glacialis* change in color, irreversibly, within an hour at room temperature in 20 mM phosphate buffer, pH 7.5, but at 0° and between pH 4 and 8.5 are stable for several days (blue pigment) or months (purple pigment). In contrast to the purple carotenoprotein, the blue pigment is only soluble at low ionic strength and at a pH greater than 7.5, and is insoluble in 50 mM NaCl–20 mM phosphate buffer, pH 7.5. These solubility properties are employed to effect in the purification of the blue carotenoprotein.[79] The proteins are best preserved at 5° in 50 mM NaCl–20 mM phosphate buffer, pH 7.5, in which they are stable for several months. Both proteins are unstable to freezing and cannot be freeze-dried without alteration.

Purification of Asteriarubin.[46] Dorsal skins from 50 starfish (~540 g) are dissected and washed in cold water. They are homogenized, intermittently over a period of 30 min to avoid heating, with 300 ml ice-cold 0.1 M sodium phosphate buffer–10 mM EDTA, pH 7.5. The homogenate is centrifuged at 300 g for 30 min. The pellet is rehomogenized, recentrifuged, and the combined supernatants are adjusted to pH 5 to precipitate the carotenoprotein. It is important to omit this step if the pigment is to be used for reconstitution studies. The red–blue precipitate is dissolved in water and diluted to give a conductivity measurement below 3 mS on a conductivity meter. The carotenoprotein is adsorbed onto a wide bed of DEAE-cellulose (11 × 9 cm, in a sintered glass funnel, porosity 1), previously equilibrated with 20 mM phosphate buffer, pH 7.5. The DEAE-

---

[79] G. M. Armitt, Ph.D. Thesis, Liverpool (1981).

cellulose is then washed with 1 liter of this buffer to remove orange lipid material. The blue carotenoprotein is scraped from the top of the bed, transferred into another sintered glass funnel and washed with a further liter of 20 mM phosphate buffer. The greater part of the carotenoprotein is then eluted in two fractions with 20 mM phosphate buffer, pH 7.5, containing 50 mM and 0.1 M NaCl, respectively. These are pooled and the carotenoprotein is precipitated at 43% ammonium sulfate saturation. The precipitate, collected by centrifugation, is redissolved in 50 ml of 20 mM phosphate buffer, pH 7.5 and reprecipitated at 43% ammonium sulfate saturation.

Gel filtration on a column (1.4 × 90 cm) of Sephadex G-100, or Ultrogen AcA-54, equilibrated with 50 mM NaCl–20 mM phosphate buffer, pH 7.5, further ammonium sulfate precipitation at 43% saturation, and repeated gel filtration and DEAE-cellulose chromatography are necessary to obtain the carotenoprotein in a state homogeneous in PAGE.

A yield of about 36 mg purified carotenoprotein, with $A_{570/280}$ ratio of 7–7.5 is obtained.

Carotenoproteins of Hydrocorallina

Water-soluble carotenoproteins have been extracted from the colored (red, pink, yellow, purple red) aragonite skeletons of corals of the order Hydrocorallina.[30–32] Only that of *Allopora californica,* alloporin,[30] has been purified and studied in detail.[30,32] The pigment, λ_{max} 545 nm, binds a single astaxanthin, has an $A_{545/280}$ ratio of 4.1, and an $A_{545}^{1\%}$ value of 20.4. The protein has four subunits of molecular weight 17,000 and the specificity of combination between carotenoid and protein has been studied. *Stylaster* and *Distichopora* species have carotenoproteins binding both astaxanthin and zeaxanthin; the absorption spectra may be bathochromically or hypsochromically shifted compared to astaxanthin in acetone.[31]

It is noteworthy that the carotenoproteins are stable to 1.25 M NaOH at 70° in the finely ground coral.[31]

General Extraction Procedure of Rønnenberg et al.[31]

The corals are ground and washed repeatedly and sequentially with acetone, aqueous NaOH (1.25 M, 70°), boiling ethanol, and distilled water to remove nonskeletal material. The cleaned coral is treated with Na$_2$ EDTA · 2H$_2$O (3.7 g/g coral) in distilled water (250 ml) overnight. Solid particles are removed by filtration and the aqueous extract is concentrated by ultrafiltration (Diaflo membrane PM30). The aqueous concentrate is repeatedly diluted with distilled water and reconcentrated. Alloporin is purified by gel filtration on Sephadex G-100 equilibrated with 0.1 M triethanolamine, pH 7.0.

Carotenoproteins, Type B

Carotenoid–Lipovitellin Complexes of Decapods

A large part of the carotenoid present in decapod eggs and developing ovaries is associated with the main storage high density lipoglycoprotein (crustacean lipovitellin[27]), providing distinctive and vivid colorations (yellow, purple, green, blue). The primary carotenoid is usually astaxanthin which is often, particularly for blue and green complexes, selectively bound. In some species, however, all the carotenoids present in the tissues are found with the lipovitellin and in others a carotenoid other than astaxanthin may predominate.[1,3] The complexes are not necessarily of a stoichiometric nature and the carotenoid content may be determined by the amount available at the time of synthesis and deposition of the lipovitellin.[27]

Few extensive chemical or structural studies on the proteins have been reported. The lipovitellins, including ovoverdin of *Homarus* sp., of six decapod species have similar size ($\sim 3.5 \times 10^5$) and lipid content (30%) and are devoid of protein-bound phosphorus; $A_{280}^{0.1\%}$ values lie between 0.76 and 1.39.[27] Estimates of the number of astaxanthin molecules bound in ovoverdin range from 1 to $2^{16,21-24,26}$; the subunits of the pigment are of large size (105,000, 95,000, and 78,000) in SDS–PAGE.[16] The carbohydrate contents of ovoverdin and two other decapod lipovitellins are low (2–4%).[28]

Properties

The purple, blue, and green complexes especially are noted for the extreme lability of the absorption spectrum to the action of light, heat above 30°, low ionic strength, metal ions, surface denaturation (shaking), storage, and, for some,[27,81] pH values above 7.[1,3] Alterations in solubility behavior and precipitability with ammonium sulfate occur on storage of the extracted or unextracted lipovitellins and aggregation of the proteins may take place during normal manipulative procedure.[27] The lipovitellins cannot be freeze-dried without denaturation nor preserved by freezing. Treatment with acetone or alcohol leads to irreversible cleavage of the carotenoid–lipovitellin; the rupture of the linkage by heating is, however, usually reversible at temperatures not exceeding 60° provided salt is present.[1,16,22]

The unstable nature of the complex and hyperchromic alteration in spectrum on denaturation,[23,27] as well as possible nonstoichiometry, may partly explain the differences recorded in the spectral characteristics of ovoverdin. The positions of the two absorption bands at ~460 and 640 nm have variously been reported to lie between 460–476 and 637–675 nm, respectively, and the values of $A_{460/640}$ and $A_{278/460}$ given between 1.5 and 3.0 and 1.5 and 5.2, respectively.[16,20,22–27] Some of the discrepancies may also be the result of the use of separate species of lobster (*H. americanus* and *H. gammarus*) or of the eggs or ovaries being at different stages of development. The position of the long wavelength band is temperature dependent and shifts bathochromically, progressively, and reversibly, as the temperature is lowered from 25 to 5° with the appearance of fine structure at the lower temperature.[80] The pigment, less stable in the purified state,[22] fades considerably on dialysis against water in the dark and fine structure of the 640 nm band, evident in some preparations,[26] is lost. The broad 640 nm band is the more sensitive to heat, light, and dialysis; it may be selectively eliminated by shaking with ether, producing a new band at 480 nm.[16] Resonance Raman has been usefully employed to show that the two absorption bands of ovoverdin arise from distinct carotenoid-binding sites.[16]

Ovoverdin isolated from *H. gammarus* shows differences in absorption spectrum and subunit composition from those reported for *H. americanus*.[80]

Extraction and Purification

The instability of the pigments renders it difficult to make preparations in a reproducible manner and to carry out prolonged experiments on the same material. The use of fresh material, minimal exposure to light and air, maintenance of pH between 5 and 7, and freedom of solutions from metal ions is favored to minimize alteration during purification.[27] Methods involving ion-exchange are not, in general, recommended and have been found unsatisfactory for the separation of some lipovitellins.[20,26] Mild methods, such as density gradient centrifugation[81] and gel filtration,[27] are preferred.

A modification[26] of the original methods of Kuhn and Sörensen[23,24] for the rapid isolation of ovoverdin using calcium phosphate gel as adsorbant in the place of alumina and, in addition to ammonium sulfate fractiona-

---

[80] P. F. Zagalsky, *Comp. Biochem. Physiol. B*, in press.
[81] W. Fyffe and J. D. O'Connor, *Comp. Biochem. Physiol. B* **47B**, 851 (1974).

tion, removal of impurities with DEAE-cellulose and CM-cellulose, results in an electrophoretically homogeneous product. The purified ovoverdin, however, has altered solubility characteristics; although soluble in water, it precipitates at low ionic strength (1.25×10^{-3} M phosphate buffer, pH 7) unlike the native pigment,[27] but reminiscent of amphibian lipovitellin.[82]

A general and reproducible purification method for decapod lipovitellins has been described by Wallace et al.,[27] and is given below; ovoverdin isolated from *H. americanus* according to this procedure has reported absorption maxima at 460–465 nm and 637–645 nm[16,27] and an $A_{280}^{1\%}$ value of 0.85.[27]

Procedure. Eggs or mature ovaries are washed thoroughly with water and dispersed with a tight-fitting Teflon homogenizer in 3 volumes of 0.5 M NaCl–5 mM EDTA, pH 5.0. The mixture is centrifuged to remove debris and fatty material and the supernatant filtered through a Whatman No. 1 filter. Two volumes of saturated ammonium sulfate are added, slowly with stirring, to the filtrate. The precipitate, collected by centrifugation, is stable at 0–5° for about a month when overlaid with fresh 67% saturated ammonium sulfate.

Further purification of the lipovitellin is achieved by fractional precipitation between 45 and 65% ammonium sulfate saturation. The precipitate is solubilized by the addition of several volumes of water and the solution dialyzed against at least 20 volumes of 45% saturated ammonium sulfate. Any insoluble material is removed by centrifugation and four-sevenths volume of saturated ammonium sulfate slowly added to give a final concentration of 65% saturation. The precipitate is collected by centrifugation and dissolved in and thoroughly dialyzed against 3–4 changes of 0.5 M NaCl–5 mM EDTA, pH 5.0. Any insoluble material appearing after dialysis is removed by centrifugation. The yield of purified lipovitellin is of the order of 3–5 g/100 g ovary or eggs.

Gel filtration on columns (2.5 × 90 cm) of 6–8% Agarose may be used as a final purification step for small samples of protein (5–25 mg in 1–2 ml) with 0.5 M NaCl–5 mM EDTA, pH 5.0, as equilibration buffer.[16,27]

Storage

The purified pigments may be stored for short periods under nitrogen at 0–5° as precipitates under 65% saturated ammonium sulfate.[22–24,27] The precipitates fade with time with a concomitant increase in material insoluble at low ionic strength.[27] Ovaries may be stored in the deep freeze covered with 0.5 M NaCl– 5 mM EDTA, pH 5.0[16]; prolonged storage (for

[82] R. A. Wallace, *Biochim. Biophys. Acta* **74,** 505 (1963).

more than a month), however, is undesirable and leads to "aged" material salting out at a lower ammonium sulfate concentration.[27]

Carotenoprotein–Lipovitellin Complexes of Anostracans

In several species of anostracans canthaxanthin (β,β-carotene-4,4'-dione) is found associated with the storage lipoglycoprotein (lipovitellin) of the eggs.[3,83–85] The complexes are packaged in membrane-enclosed yolk platelets, together with certain enzymes, nucleic acids, and diguanosine nucleotides.[83] Analysis of the platelets of the brine shrimp, *Artemia salina* (L.), has been made following isolation by sucrose gradient centrifugation and by a nonaqueous procedure[83]; the platelets of the freshwater anostracan, *Branchipus stagnalis* (L.), have also been analyzed.[84] The mode of binding of canthaxanthin in both pigments has been investigated by resonance Raman spectroscopy.[86,87]

Properties

The absorption spectra of the orange-yellow canthaxanthin–lipovitellins of *Artemia*[85] and *Branchinecta*[88] have absorption maxima above and below that of the free carotenoid with a main absorption maximum in the 360 nm region and additional maxima in the 460 and 600 nm regions; the blue complex of *Branchipus* rapidly turns red at room temperature in the absence of chloride ions.[84] Polar organic solvents such as acetone or ethanol liberate the carotenoid from the complexes.

Detailed analysis of the composition and structure of the lipovitellin of *Artemia* has been reported.[57,85] The protein is a tetramer of two subunits of molecular weight 190,000 and 68,000 in SDS–PAGE. The subunits are present in equimolar amount and have been compared by peptide mapping.[57] The lipovitellin of *Branchipus* is similar to that of *Artemia* in composition and structure and both complexes bind 3–4 canthaxanthin molecules.[57,84] An alkaline protease (nonserine type) is associated with the *Artemia* lipovitellin at the gastrula stage and causes degradation of the larger subunit during embryonic development.[85]

[83] A. H. Warner, J. G. Puodziukas, and F. J. Finamore, *Exp. Cell Res.* **70**, 365 (1972).
[84] P. F. Zagalsky and B. M. Gilchrist, *Comp. Biochem. Physiol. B* **55B**, 195 (1976).
[85] D. de Chaffoy, J. Heip, L. Moens, and M. Kondo, *in* "The Brine Shrimp *Artemia*" (G. Persoone, P. Sorgeloos, O. Roels, and E. Jaspers, eds.), Vol. 2, p. 379. Universa Press, Wetheren, Belgium, 1980.
[86] P. F. Zagalsky, B. M. Gilchrist, R. J. H. Clark, and D. P. Fairclough, *Comp. Biochem. Physiol. B* **74B**, 647 (1983).
[87] P. F. Zagalsky, B. M. Gilchrist, R. J. H. Clark, and D. P. Fairclough, *Comp. Biochem. Physiol. B* **75B**, 163 (1983).
[88] B. M. Gilchrist and P. F. Zagalsky, *Comp. Biochem. Physiol. B* **76B**, 885 (1983).

The absorption spectra of anostracan carotenoid–lipovitellin complexes are, in general, more stable than those of the decapod Crustacea. The pigments may be stored under 65% ammonium sulfate in the cold for several months without alteration.

Isolation of Artemia Canthaxanthin–Lipovitellin (De Chaffoy et al.)[57,85]

All operations are carried out at 0–4° in quartz doubly distilled water (QBW) and under dim light.

The encysted gastrulae are hydrated in QBW at room temperature for 1 hr or at 0° overnight. The hydrated embryos are homogenized in QBW at 0° with a Virtis homogenizer at 55,000 rpm for 3 min. The homogenate is filtered through a glass wool column and centrifuged at 10,200 g for 2–4 min at 4°. The pellet is resuspended in QBW and recentrifuged as above. This operation is repeated until a clear supernatant is obtained. The lipovitellin is solubilized at 0° by gentle stirring in 50 mM Tris–HCl, pH 9, containing 1 M NaCl for 1–2 hr and the insoluble material removed by centrifugation. The supernatant is filtered through Whatman No. 1 filter paper and then chromatographed on Sepharose 6B using solubilization buffer.

Further purification of the lipovitellin is achieved by chromatography on DEAE-cellulose. The sample is diluted with 50 mM Tris–HCl, pH 9, to 0.02 M NaCl, applied to the column, and eluted with a gradient of NaCl (0.02 to 0.12 M) in the same buffer. The lipovitellin is eluted at 0.08 to 0.09 M NaCl after a small peak at 0.03 to 0.04 M NaCl.

The lipovitellins of the freshwater anostracans, *Branchipus*[80] and *Branchinecta*,[88] may be extracted and purified by gel filtration as above. In these cases the eggs are briefly homogenized in a Potter homogenizer. The DEAE-cellulose step is omitted for the *Branchipus* pigment; the protein adsorbs irreversibly to both DEAE-cellulose and Con A-Sepharose.[84]

Acknowledgments

The author thanks Dr. G. Britton and colleagues, Biochemistry Department, Liverpool University, and Dr. B. M. Gilchrist, Zoology Department, Bedford College, for helpful discussion and criticism of the manuscript.

[10] Separation of Mevalonate Phosphates and Isopentenyl Pyrophosphate by Thin-Layer Chromatography and of Short-Chain Prenyl Phosphates by Ion-Pair Chromatography on a High-Performance Liquid Chromatography Column

By PETER BEYER, KLAUS KREUZ, and H. KLEINIG

Phosphorylated intermediates in the pathway leading from mevalonate to geranylgeranyl pyrophosphate are rather difficult to separate because of their divergent polar and lipophilic properties, the very common simultaneous occurrence of mono- and pyrophosphates, and their tendency to adsorb unspecifically.

The paper chromatographical methods already described for the separation of isopentenyl pyrophosphate from its precursors achieve only partial resolution of the compounds.[1] We describe here a two-dimensional thin-layer system which is able to separate clearly mevalonate, mevalonate phosphate, mevalonate pyrophosphate, and isopentenyl pyrophosphate.

For the separation of short-chain prenyl phosphates HPLC methods are now applicable due to the introduction of ion-pairing. This offers the great advantages of reversed-phase methods (partition chromatography), and may be an alternative tool to the thin-layer systems on silica gel already described,[2,3] where overlapping of mono- and pyrophosphates and/or chain length homologs is sometimes observed. A method is presented here which allows the separation of geranyl, farnesyl, and geranylgeranyl mono- and pyrophosphates very rapidly.

Thin-Layer Chromatographical Separation of Mevalonate Phosphates and Isopentenyl Pyrophosphate

The *in vitro* enzymatic assays using radioactively labeled mevalonate as a biosynthetic precursor are extracted with chloroform/methanol according to conventional methods, e.g., that described by Bligh and Dyer[4] in order to remove lipidic material. The aqueous phase is briefly evaporated for removal of the remaining methanol and subsequently lyophi-

---

[1] T. T. Tchen, this series, Vol. 5, p. 489.
[2] I. Takahashi, K. Ogura, and S. Seto, *J. Biol. Chem.* **255**, 4539 (1980).
[3] S. S. Sofer and H. C. Rilling, *J. Lipid Res.* **10**, 183 (1969).
[4] E. S. Bligh and W. J. Dyer, *Can. J. Biochem. Physiol.* **37**, 911 (1959).

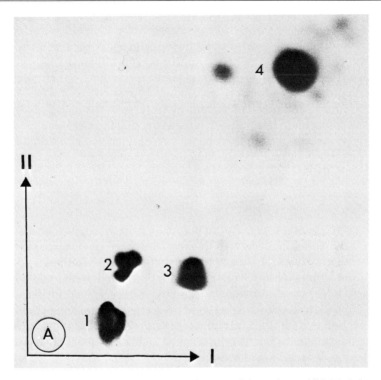

FIG. 1. Two-dimensional thin-layer separation on cellulose plates of ^{14}C-labeled mevalonate pyrophosphate (1), mevalonate phosphate (2), isopentenyl pyrophosphate (3), and mevalonate (4). I, II, first, second dimension; A, application point.

lized. The sample is dissolved in water and applied to a precoated cellulose thin-layer plate. The first dimension is developed with 1-propanol/NH$_3$ (25% aqueous solution)/H$_2$O (6:3:1) and the second dimension with 2-butanol/acetic acid/H$_2$O (40:10:13). A separation of the compounds is shown in Fig. 1. They can be detected by use of a radio scanner or by autoradiography and quantified by scintillation counting. Possibly occurring geranylphosphates and higher homologs in the water phase after extraction do not interfere with the separation of mevalonate phosphate, mevalonate pyrophosphate, and isopentenyl pyrophosphate, but may overlap with free mevalonate. For quantitative estimation of the latter in this case, the conversion of mevalonate into the lactone form and its separation according to Young and Berger[5] is recommended.

Our method does not allow the separation of isopentenyl pyrophos-

---

[5] N. L. Young and B. Berger, this series, Vol. 71, p. 498.

phate from dimethylallyl pyrophosphate. For this purpose the compounds may be eluted from the thin-layer plate with water and converted to free alcohols by treatment with alkaline phosphatase and separated by radio gas chromatography [e.g. 3% poly(A), 103 on 100/120 mesh Gas Chrom Q in a glass column 6 ft, 1/8 in., column temperature 70°, isothermal operation, helium flow 30 ml/min].

Significant decomposition of acid hydrolysis of the allylic pyrophosphates may occur in the second dimension of chromatography. Thus, recovery of these compounds would not be optimal and consequently the ion-pairing chromatography procedure described below is preferable for these compounds.

Ion-Pair Chromatography of Prenyl Phosphates

The principle of ion-pair chromatography is to transfer molecules with charged groups, in our case with ionized phosphate groups, into undissociated, externally neutral ion complexes by addition of a complex-forming agent to the mobile phase, in our case a tetrabutylammonium salt. Thus, it is possible to separate compounds exhibiting highly acidic or basic groups on the HPLC reversed-phase column as their complexes appear much less polar and are retained on the phase without exhibiting the dissociation equilibrium between the protonated and deprotonated form. This would lead to large tailing of peaks which otherwise could only be avoided by applying extreme pH values.

The Ion-Pair Reagent

The reagent (stock solution) contains tetrabutylammonium sulfate, 6.8 g, K_2HPO_4, 5.5 g, adjusted to pH 8.0 with KOH, in a total volume of 100 ml H_2O. The reagent is filtered through Millipore HA filters (0.45 μm). The separation of the prenyl phosphates is performed in a gradient system. The polar solvent contains 10 ml of the stock solution and 490 ml distilled water, pH 8.0. The solvent is filtered over Millipore HA filters (0.45 μm). For the nonpolar solvent 10 ml of the stock solution is added to 490 ml methanol. Precipitating phosphate is removed by centrifugation and subsequent filtration on Millipore FH filters (0.5 μm). This solvent is further diluted 1:4 with methanol. The polar solvent contains 4 mM, the nonpolar solvent 1 mM of the tetrabutylammonium reagent.

Formation of the Standards

Generally, prenyl phosphates to be analyzed from *in vitro* incubations are radioactively labeled. Therefore, and for convenient detection using a

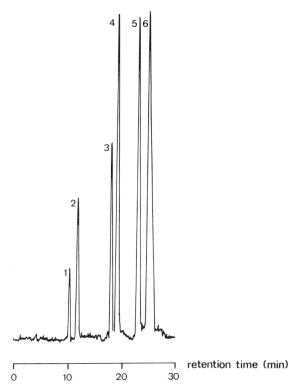

FIG. 2. Separation of synthetically ^{32}P-labeled prenyl mono- and pyrophosphates using a C_{18} reversed-phase column with the ion-pair gradient. 1, Geranyl pyrophosphate; 2, geranyl monophosphate; 3, farnesyl pyrophosphate; 4, farnesyl monophosphate; 5, geranylgeranyl pyrophosphate; 6, geranylgeranyl monophosphate.

radio column chromatography monitor (see below) ^{32}P-labeled standards are prepared from geraniol, farnesol, and geranylgeraniol according to published methods.[6,7]

The Separation System

For separation a HPLC system (Waters) equipped with a C_{18} reversed-phase column (μBondapak C_{18}, Waters) and a radio column chromatography monitor with a glass scintillator cell is used. The column is developed using a 15-min linear gradient ranging from 15% nonpolar solvent/85%

---

[6] G. Popják, R. H. Cornforth, J. W. Cornforth, R. Ryhage, and D. S. Goodman, *J. Biol. Chem.* **237**, 56 (1962).
[7] P. V. Bhat, L. M. de Luca, and M. L. Wind, *Anal. Biochem.* **102**, 243 (1980).

polar solvent to 70% nonpolar solvent/30% polar solvent at a flow rate of 1.3 ml/min. A typical separation of geranyl, farnesyl, and geranylgeranyl mono- and pyrophosphates is shown in Fig. 2. Under these conditions the decrease in polarity of the eluent is paralleled by a decrease in the concentration of ion-pair reagent, which allows the resolution of the mono- and pyrophosphates of the higher homologs. Ionic contaminants in higher concentration may interfere with the separation. After use, the HPLC unit should be rinsed thoroughly with distilled water.

Acknowledgment

This work was supported by Deutsche Forschungsgemeinschaft.

[11] Reversed-Phase High-Performance Liquid Chromatography of C_5 to C_{20} Isoprenoid Benzoates and Naphthoates

By TSAN-HSI YANG, T. MARK ZABRISKIE, and C. DALE POULTER

The major building reaction in the isoprenoid biosynthetic pathway is the sequential 1'-4 condensation of isopentenyl pyrophosphate with an allylic pyrophosphate to generate the five-carbon homolog of the allylic substrate.[1] Compounds with up to four isoprenoid units are very common, and metabolites with chain lengths clustering around 10 and 20 units are also widespread. Direct analysis of isoprenoids by liquid chromatography is limited by the lack of an intense UV chromaphore in many compounds. In addition, small samples of lower molecular weight isoprenoids are difficult to handle because of their high volatility. Since many isoprenoids occur naturally as alcohols or as acetate, phosphate, or carboxylate derivatives that can easily be converted to alcohols, aromatic esters of isoprenoid alcohols are excellent candidates for analysis. This chapter describes the high pressure liquid chromatographic properties for benzoate and naphthoate esters of linear five- to twenty-carbon isoprenoid alcohols on C_{18} reversed-phase columns.

[1] C. D. Poulter and H. C. Rilling, *Acc. Chem. Res.* **11**, 307 (1978).

Experimental Procedures

Materials

Dimethylallyl alcohol, isopentenol, 3-methyl-1-butanol, linalool, geraniol, benzoyl chloride, 2-naphthoic acid, 4-N,N-dimethylaminopyridine, and dicyclohexylcarbodiimide were obtained from Aldrich Chemical Co. E,E-Farnesol and nerol were provided by Givaudan Corp. Citronellol was obtained from Matheson, Coleman, and Bell. Fluorinated alcohols were available from a previous study in this laboratory.[2] Geranylgeraniol was provided by Professor R. M. Coates.

Benzoyl chloride, pyridine, and all solvents were freshly distilled onto 5-Å molecular sieves before use. N,N-Dimethylaminopyridine was recrystallized from chloroform. All other materials were used without further purification.

Synthesis of Isoprenoid Esters

Benzoates. Isoprenoid benzoates were prepared by treatment of the alcohols with a 10- to 50-fold excess of benzoyl chloride in dry diethyl ether at room temperature. Pyridine was used to neutralize the hydrochloric acid liberated during esterification. This procedure cleanly esterifies the alcohols in high yield without rearrangement.

In a typical esterification, 30 nmol of alcohol was dissolved in a solution of 150 μl of anhydrous diethyl ether and 50 μl of pyridine in a dry 5-ml test tube. Benzoyl chloride (300 nmol) was added, the tube was sealed with a plastic plug, and the resulting mixture was allowed to stir for 12 hr at room temperature. Diethyl ether (1 ml) was added, and the reaction mixture was washed in succession with 0.5 ml portions of 1 N hydrochloric acid (twice), water, and saturated sodium bicarbonate. The phases were agitated vigorously on a Vortex mixer, and the heavier aqueous layer was removed with a Pasteur pipet. The organic layer was dried with magnesium sulfate and then passed through a Pasteur pipet containing activity II aluminia. Solvent was removed with a gentle stream of dry nitrogen, and the residue was dissolved in acetonitrile for analysis. Conversions of alcohols to benzoates were typically greater than 85%. Large excesses of benzoyl chloride were used to overcome adventious moisture in the solvent or sample.

Naphthoates. Isoprene naphthoates were prepared by the method of Steglich[3] for the esterification of carboxylic acids. This procedure utilizes

---

[2] C. D. Poulter, P. L. Wiggins, and T. L. Plummer, *J. Org. Chem.* **46**, 1532 (1981).
[3] B. Neises and W. Steglich, *Angew. Chem., Int. Ed. Engl.* **7**, 522 (1978).

the acid directly and can be used to prepare benzoates as well. It is preferable when the samples to be esterified are fairly clean and does not require a separate step to activate the acid moiety before esterification.

In a typical esterification of 0.1 mmol of alcohol, 0.2 mmol of 2-naphthoic acid, and 1 ml of dry methylene chloride were placed in a 5-ml round bottomed flask equipped with a magnetic stirrer. The opening was sealed with a rubber septum, and the reaction was run in an atmosphere of dry nitrogen to exclude moisture. The alcohol was added by syringe, and the syringe was used to add two 0.2 ml portions of methylene chloride. 4-N,N-Dimethylaminopyridine (3–10 μmol) was added, and the flask was placed in an ice-water bath before 2 mmol of dicyclohexylcarbodiimide was added. The flask was kept in the ice-water bath for 10 min and then allowed to warm to room temperature. The progress of the reactions was followed by thin-layer chromatography. The R_f values of the esters were approximately 0.6 for elution with 4:1 hexane/ethyl acetate. The reactions were normally complete within 3–5 hr.

When the alcohol was consumed, the contents of the flask were passed through a plug of glasswool in a Pasteur pipet into a test tube. The plug was rinsed with two "plug volumes" of methylene chloride. The organic layer was washed in succession with 1 ml portions of 0.5 N hydrochloric acid, water, and saturated sodium bicarbonate using a Vortex mixer. The organic layer was dried over magnesium sulfate and passes through a plug of glass wool as described before. Solvent is removed at reduced pressure. The residue is extracted twice with 1 ml portions of hexane. The washings are combined, and solvent is removed at reduced pressure. The esters are purified by flash chromatography[4] using 95:5 hexane/ethyl acetate. Yields typically range from 50 to 78%.

Chromatography

A Waters high-pressure liquid chromatographic system consisting of a U6K injector, Model 6000A and M45 pumps, a RCM-100 radial compression module, a Model 441 detector, a Model 720 system controller, and a Model 730 data module were used. All separations were on Waters Radial PAK A columns (C_{18}, 10 μm, 8 mm × 10 cm). Doubly deionized water and acetonitrile (Merck, chromatographic grade) were passed through type HA (0.45 μm) and type FA (0.5 μm) filters (Millipore Corp.), respectively, and degassed under vacuum just prior to use. Samples were dissolved in acetonitrile, and 5–25 μl portions were injected.

---

[4] W. C. Still, M. Kahn, and A. Mitra, *J. Org. Chem.* **43**, 2923 (1978).

TABLE I
Capacity Factors (k') for Isoprenoid Benzoates on a Radial Pak C_{18} (10 μm) Reversed-Phase Column[a]

Benzoate	Volume fraction acetonitrile (ϕ)							
	0.3	0.4	0.5	0.6	0.7	0.8	0.9	1.0
C_5								
Dimethylallyl (**1**)	117	34.6	12.7	5.53	2.79	1.45	0.69	0.39
Isopentenyl (**2**)	102	28.6	10.8	4.78	2.74	1.21	0.61	0.31
3-Methyl-1-butyl (**3**)	229	58.7	21.1	8.48	3.87	1.95	0.96	0.52
C_{10}								
Geranyl (**4**)			76.3	25.5	10.5	4.56 $(1.67)^b$ $(5.20)^c$	1.77	0.83
Neryl (**5**)			58.7	19.7	8.32	3.56	1.74	0.76
Linalyl (**6**)			41.5	14.2	6.26	2.65	1.34	0.63
Citronellyl (**7**)			102	31.3	12.0	5.13	2.31	1.02
3,7-Dimethyl-1-octyl (**8**)			209	67.3	20.8	8.48	3.66	1.37
Z-3-Fluoromethyl-7-methyl-2,6-octadien-1-yl (**9**)					10.2	4.32	1.76	0.97
Z-3-Difluoromethyl-7-2,6-octadien-1-yl (**10**)					14.2	6.02	2.10	1.05
Z-3-Trifluoromethyl-7-2,6-octadien-1-yl (**11**)					17.1	7.15	2.84	1.36
C_{15}								
Farnesyl (**12**)				110	30.2	14.0	4.52	1.68
C_{20}								
Geranylgeranyl (**13**)					140.8	43.6	11.4	3.31

[a] Flow rate, 5 ml/min; temperature, 21°.
[b] Measured on the same column after 13 months.
[c] Measured on a new column.

Correlations of Capacity Factors

Capacity factors (k') for isoprenoid derivatives were determined from Eq. (1), where t_r and t_0 are the respective retention times of the aromatic esters and methanol, a t_0 marker.[5] Methanol was detected at 205 nm, and its retention time did not vary over the range of solvent compositions used in this study. Capacity factors for benzoates are listed in Table I, and those for naphthoates, in Table II.

[5] P. Jandera and J. Churacek, *Adv. Chromatogr.* **19**, 125 (1981).

TABLE II
CAPACITY FACTORS (k') FOR ISOPRENOID NAPHTHOATES ON A RADIAL PAK C_{18} (10 μm) REVERSED-PHASE COLUMN[a]

Naphthoate	Volume fraction acetonitrile (ϕ)				
	0.5^b	0.6^b	0.7^b	0.8^c	0.9^d
C_5					
Dimethylallyl (**14**)	15.7	6.19	2.77	1.25	0.53
Isopentenyl (**15**)	14.3	5.74	2.59	1.17	0.49
C_{10}					
Geranyl (**16**)	74.1	22.5	8.35 $(7.52)^e$	3.11 $(2.91)^e$	1.19
Neryl (**17**)	69.0	21.5	8.06 $(7.20)^e$	2.99	1.16
Citronellyl (**18**)	98.9	29.4	10.9	3.85	1.53
3,7-Dimethyl-1-octyl (**19**)	169.7	47.8	17.0	5.81	2.22
C_{15}					
Farnesyl (**20**)		77.9	24.8	7.65 $(6.91)^e$	2.52

[a] Temperatures, 21°.
[b] Flow rate, 7 ml/min.
[c] Flow rate, 4 ml/min.
[d] Flow rate, 2 ml/min.
[e] Measurements taken on the same column after 33 days.

$$k' = (t_r - t_0)/t_0 \tag{1}$$

The values listed in Tables I and II were reproducible to within ±2% for up to a week when the column was in continuous use. Over longer periods, however, the values slowly decreased, with changes as large as 10% for geranyl and neryl naphthoate after 33 days (see Table II). More substantial decreases in k' were noted as columns continued to age. Capacity factors for the benzoates listed in Table I were measured shortly after a column was first used while those for the naphthoates were measured on the same column 13 months later. A comparison of values in Tables I and II indicates that the capacity factors for the naphthoate esters are only slightly larger than those for benzoates in spite of a difference of four carbon atoms in the ester moieties. Reinjection of geranyl benzoate gave a k' in 80% acetonitrile which was only 36% of the former value. Upon installation of a new C_{18} column, geranyl benzoate gave a capacity factor of 5.20, close to the initial value of 4.56. Although capacity factors decreased substantially as the column aged, resolution degraded only slightly during the same period, and the elution "patterns" for the esters did not change. These results do, however, emphasize the need for frequent calibration of the chromatograms using standard samples. Measure-

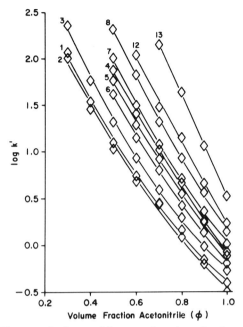

FIG. 1. A plot of log capacity factors (k') versus the volume fraction of acetonitrile (ϕ) for isoprenoid benzoates **1-8** and **12-13**. Lines through the experimental points were computer drawn using Eq. (3) and the parameters listed in Table III.

ments for the benzoates (Table I) were taken in a 1-week period, and those samples run early in the protocol were reevaluated at the end to insure that the values were reproducible. A similar procedure was followed for the naphthoates.

Composition of Solvent

Plots of log k' versus the volume fraction of acetonitrile ϕ (Fig. 1) show noticeable curvature for the less highly retained benzoate homologs which were examined over a wide range of solvent compositions. The degree of curvature is substantially less for the more highly retained benzoates. Similar, although less pronounced, behavior was observed for the naphthoates as shown in Fig. 2. Although many plots of log k' versus solvent composition in reversed-phase liquid chromatography published to date are linear, several groups have reported systems that show distinct curvature.[6] In these instances good fits of log k' versus solvent composi-

[6] B. L. Karger, J. R. Gant, A. Hartkope, and P. H. Weiner, *J. Chromatogr.* **128**, 65 (1982).

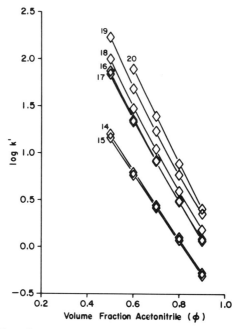

FIG. 2. A plot of log of capacity factors (k') versus the volume fraction of acetonitrile (ϕ) for isoprenoid naphthoates **14–20**. Lines through the experimental points were computer drawn using Eq. (3) and the parameters listed in Table III.

tion were obtained empirically with a quadratic function of the type shown in Eq. (2).

$$\log k' = A\phi^2 + B\phi + C \tag{2}$$

Using a nonlinear least-squares analysis, the data listed in Tables I and II also gave excellent fits to Eq. (3) for the values for A, B, and C listed in Table III. Wells and Clark[7] recently reported a similar study of a series of benzamides in acetonitrile/water and methanol/water over a wide range of solvent compositions. As ϕ approached zero (pure water), they found that curvature increased substantially, and their data fit neither a linear nor a quadratic model. One should, therefore, exercise caution in extrapolation of our data to values of ϕ outside of the experimentally determined range even when care has been taken to minimize changes in k' with increasing age of the columns.

[7] M. J. M. Wells and C. R. Clark, *J. Chromatogr.* **235**, 31 (1982).

TABLE III
PARAMETERS FOR ISOPRENOID BENZOATES AND
NAPHTHOATES FROM NONLINEAR LEAST-SQUARES
FIT TO EQ. (3)

Compound	A	B	C
1	1.95	−5.98	3.65
2	1.71	−5.68	3.50
3	2.30	−6.67	4.11
4	1.12	−5.57	4.38
5	1.17	−5.46	4.18
6	1.68	−6.10	4.24
7	1.51	−6.20	4.70
8	1.25	−6.19	5.08
9	1.58	−5.82	3.93
10	1.39	−5.93	4.22
11	1.40	−5.80	4.20
12	1.31	−6.54	5.48
13	0.83	−6.89	6.59
14	0.50	−4.35	3.24
15	0.43	−4.24	3.17
16	1.59	−6.67	4.79
17	1.63	−6.70	4.78
18	1.60	−6.76	4.97
19	1.52	−6.81	5.24
20	1.55	−7.30	5.72

Number of Isoprene Units

Plots of log k' versus the number of isoprene units (n) for the homologous series of linear allylic terpenoid benzoates (n = 1–4) exhibited excellent linearity in accord with Martin's rule[6] (Eq. (3), see Fig. 3). The slope (m), determined by a least squares analysis, is a measurement of the

sensitivity of capacity factors to isoprene number. The values for m decrease from 0.65 in 0.6 volume fraction (ϕ) of acetonitrile in water to 0.30 in pure acetonitrile. Similar behavior was found for the related isoprenoid naphthoates as shown in Fig. 4. For the naphthoates, m decreased from 0.55 to 0.34 as ϕ increased from 0.6 to 0.9. It was possible to evaluate capacity factors for solvents where $0.3 \leq \phi \leq 1.0$ with benzoate derivatives where n = 1. However, the working range shrank rapidly as

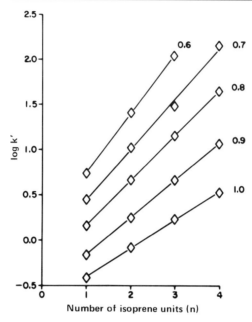

FIG. 3. A plot of log k' versus the number of isoprene units (n) in benzoate esters at different volume fractions (ϕ) of acetonitrile in water.

isoprene residues were added to the chain, and measurements for geranylgeranyl benzoate were confined to $0.7 \leq \phi \leq 1.0$.

$$\log k' = mn + b \tag{3}$$

Topology and Substitutents

Capacity factors for isomeric esters and esters with similar molecular weights varied substantially. Three general trends were noted for the compounds listed in Tables I and II. Capacity factors of compounds with the same general shape (skeletal and stereochemistry of double bonds where appropriate) increased substantially as unsaturation was removed. This trend is illustrated by comparisons of values shown in Fig. 1 for 3-methyl-1-butyl benzoate (**3**) with the unsaturated C_5 esters **1** and **2** or for 3,7-dimethyl-1-octyl benzoate (**8**) with C_{10} esters **4** and **7**. Similar behavior was found for C_{10} naphthoates as **16**, **18**, and **19** shown in Fig. 2. For both C_{10} benzoates and naphthoates, k' increased as each double bond was removed from the C_{10} chain. Within a family of isomers, increased branching or a double bond geometry that produced a more compact structure (i.e., cis versus trans) resulted in a decrease in k'. This trend is

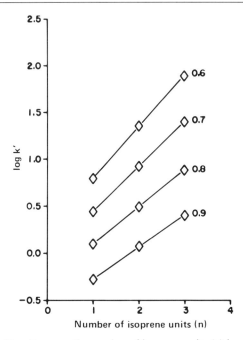

FIG. 4. A plot of log k' versus the number of isoprene units (n) in naphthoate esters at different volume fractions (ϕ) of acetonitrile in water.

noticeable in the series geranyl (**4**), neryl (**5**), and linalyl (**6**) benzoates. Similar relationships for the degree of unsaturation and the stereochemistry of double bonds are seen for the capacity factors of fatty acids.[8] Branching in isomeric compounds is also reported to result in major changes in k' by reducing the hydrophobic effect of individual carbons in a molecule.[8]

Introduction of fluorine into the methyl group at C-3 of geraniol produced little change in k' despite a substantial increase in molecular weight (see Fig. 5). Successive replacement of the second and third hydrogens in the C-3 methyl by fluorine resulted in a small, but regular, increase in k'.

General Considerations

Analysis and separation of terpene alcohols are difficult because many of the compounds of interest lack an intense UV chromaphore and many are volatile. These problems can be circumvented by conversion of the alcohols to their corresponding benzoate or naphthoate esters. These

[8] N. Tanaka and E. R. Thornton, *J. Am. Chem. Soc.* **99**, 7300 (1977).

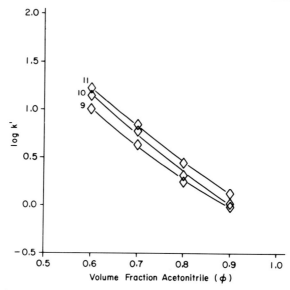

FIG. 5. A plot of log of capacity factors (k') versus the volume fraction of acetonitrile (ϕ) for isoprenoid benzoates **9–11**. Lines through the experimental points were computer drawn using Eq. (3) and the parameters listed in Table III.

derivatives have low volatilities, and small amounts of samples can be handled without significant losses. The molar extinction coefficients (ε) of the aromatic ester chromaphores provide excellent sensitivity for UV detection. We were able to detect the benzoate esters, $\varepsilon(214 \text{ nm})$ 7500 liter mol^{-1} cm^{-1}, at 0.1 μM levels when interference from background contaminates was low. A considerable improvement was found for naphthoate esters. Both the shift to longer wavelengths and the higher extinction coefficients, $\varepsilon(236 \text{ nm})$ 65,200 liter mol^{-1} cm^{-1}, permit detection of 50–100 nM concentrations of the derivatives. Radioactive compounds can be detected at much lower concentrations by collecting fractions and determining the radioactivity by liquid scintillation spectrometry.

Incremental increases in molecular weight per additional isoprene unit are sufficiently large so that the capacity factors for C_5, C_{10}, C_{15}, and C_{20} isoprenoids, with one exception, do not overlap. There is, however, considerable variation within a family as illustrated by the C_{10} series of benzoates. At one extreme, capacity factors for linalyl benzoate (**6**) are very close to those found for 3-methyl-1-butyl benzoate (**3**), the most highly retained member of the C_5 family. At the other extreme, values of k' for

3,7-dimethyl-1-octyl benzoate (8) are approximately 70% of those for farnesyl benzoate (12).

When analyzing mixtures containing C_5, C_{10}, C_{15}, and C_{20} isoprenoids, we found excellent resolution could be obtained by gradient elution chromatography. Because of the parallel behavior noted in Figs. 1, 2, and 5, once the order of elution is established for one set of conditions, the ratio of acetonitrile/water can be altered without disturbing the relative order in which the aromatic esters eluted. However, as mentioned previously, capacity factors do change substantially as the C_{18} columns aged and frequent calibrations with standard samples are essential.

Summary

Benzoate and naphthoate esters of C_5–C_{20} isoprenoid alcohols are excellent derivatives for reversed-phase liquid chromatography. Nonlinear responses of log k' to changes in the composition of the mobile phase are observed on Radial PAK C_{18} columns. In general, the capacity factors (k') increased with increases in carbon content and decreased with increases in the number of double bonds, branching, and Z double bond content. The large extinction coefficients of the aromatic ester moieties especially the 2-naphthoates, provide high sensitivity for UV detection.

Acknowledgments

We wish to thank the National Institutes of Health, Grants GM 25521 and GM 21328, for support of this research.

Section II

Sterol Metabolism

[12] Lecithin–Cholesterol Acyltransferase and Cholesterol Transport

By CHRISTOPHER J. FIELDING

Lecithin–cholesterol acyltransferase (LCAT; EC 2.3.1.43) catalyzes the transesterification of lecithin and cholesterol in plasma:

Lecithin + cholesterol → lysolecithin + cholesteryl ester

The reaction of the highly purified enzyme with synthetic substrates shows a broad reactivity with acyl acceptors including water,[1] lysolecithin itself,[2] long chain alcohols,[3] and thiols such as thiocholesterol.[4] Acyl donors include lecithin itself and demethylated lecithin analogs.[5] The specificity of LCAT with purified lipids probably represents in considerable measure a property of the physical structure of the substrate lipid surface rather than a specificity of the enzyme active site, since dipalmitoyllecithin, not an effective substrate as a pure lipid, is fully reactive when dispersed within a native plasma lipoprotein.[3] The reaction of purified LCAT is highly or totally dependent upon the presence of a lipoprotein apoprotein (apo A-I),[6] which is a component of a native substrates of LCAT.

After several reports of partial purifications of human LCAT from plasma (the only substantial source of the activity), the first isolation of LCAT was described in 1976 by Albers and colleagues.[7] Since that time a number of other procedures have been reported.[1,8–14] The chemical and

---

[1] L. Aron, S. Jones, and C. J. Fielding, *J. Biol. Chem.* **253,** 7220 (1978).
[2] P. V. Subbaiah, J. J. Albers, C. H. Chen, and J. D. Bagdade, *J. Biol. Chem.* **255,** 9275 (1980).
[3] K. Kitabatake, U. Piran, Y. Kamio, Y. Doi, and T. Nishida, *Biochim. Biophys. Acta* **573,** 145 (1979).
[4] L. Aron and C. J. Fielding, unpublished results.
[5] C. J. Fielding, *Scand J. Clin. Lab. Invest.* **33,** Suppl. 137, 15 (1974).
[6] C. J. Fielding, V. G. Shore, and P. E. Fielding, *Biochem. Biophys. Res. Commun.* **46,** 1493 (1972).
[7] J. J. Albers, V. G. Cabana, and Y. D. B. Stahl, *Biochemistry* **15,** 1084 (1976).
[8] J. J. Albers, J. Lin, and G. P. Roberts, *Artery* **5,** 61 (1979).
[9] J. Chung, D. A. Abano, G. M. Fless, and A. M. Scanu, *J. Biol. Chem.* **254,** 7456 (1979).
[10] G. Suzue, C. Vezina, and Y. L. Marcel, *Can. J. Biochem.* **58,** 539 (1980).
[11] K. S. Chong, L. Davidson, R. G. Huttash, and A. G. Lacko, *Arch. Biochem. Biophys.* **211,** 119 (1981).
[12] C. H. Chen and J. J. Albers, *Biochem. Med.* **25,** 215 (1981).
[13] C. E. Matz and A. Jonas, *J. Biol. Chem.* **257,** 4541 (1982).
[14] V. Mahadevan and L. A. Soloff, *Biochim. Biophys. Acta* **752,** 89 (1983).

physical properties of the purified proteins that have been reported in these studies are very similar, as have been the general principles involved in the purification itself. Major purification is usually achieved taking advantage of the low affinity of LCAT for hydroxylapatite.[1,3,7–9,11–15] Additional steps in common use involve the use of dextran sulfate[3] or ultracentrifugation[1,7–11,13] to remove contaminating plasma lipoproteins; hydrophobic chromatography on phenyl-Sepharose,[12] dodecylamine agarose,[11] Affigel-Blue agarose,[9,13] or HDL-agarose[7] (covalent complex of high-density lipoprotein and agarose) to remove bulk plasma protein including albumin, and DEAE-ion exchange chromatography[3,9–10,13,14] or anti-apolipoprotein D affinity chromatography[1,2] to separate LCAT from cholesteryl ester transfer protein. The level of LCAT protein in human plasma determined by specific immunoassay is about 6 μg/ml.[16]

Interest in the activity of LCAT has recently come into increased prominence because of the role of this enzyme *in vivo* in maintaining the balance of cholesterol mass between cell membranes and their surrounding fluid medium. LCAT forms part of a reaction chain[17] by means of which cholesterol of cellular origin is removed into the medium, esterified by LCAT, transferred in ester form to other lipoproteins by the activity of cholesteryl ester transfer proteins, and removed for irreversible degradation in the liver by various receptor-dependent and other pathways. Such centripetal transport is balanced by peripheral synthesis. At least 50% of the cholesteryl esters synthesized by LCAT may be derived from cell membrane cholesterol under conditions approximate to those of the vascular bed.[18] A chapter[19] in an earlier volume described in detail a purification method for LCAT from human plasma and a synthetic assay of lecithin–cholesterol liposomes activated for LCAT activity into apolipoprotein A-I. Very similar assays have been reported in connection with the purification procedures described above.[1,3,7–15] The present account will describe the newer analytical methods used for the study of the LCAT reaction in its biological context.

Cholesterol Mass Assay

Principle. Plasma or assay medium that contain LCAT and free and esterified cholesterol are incubated at 37° for 30–60 min. The increase in

[15] C. J. Fielding and P. E. Fielding, *FEBS Lett.* **15,** 355 (1971).
[16] J. J. Albers, J. L. Adolphson, and C. H. Chen, *J. Clin. Invest.* **67,** 141 (1981).
[17] C. J. Fielding and P. E. Fielding, *Med. Clin. North Am.* **66,** 363 (1982).
[18] P. E. Fielding, P. M. Davison, M. A. Karasek, and C. J. Fielding, *J. Cell Biol.* **94,** 350 (1982).
[19] Y. Doi and T. Nishida, this series, Vol. 71, p. 753.

medium cholesteryl ester, or decrease in medium free cholesterol, is measured enzymatically, using cholesterol esterase and cholesterol oxidase.[20]

Reagents

Surfal A nonionic detergent,[21] 10% v/v solution in isopropanol, reagent grade

0.1 M phosphate buffer (pH 7.8) containing 4 mM ethylenediaminetetraacetic acid

4-Hydroxy-3-methoxyphenylacetic acid, 4 mg/ml in distilled water. Horseradish peroxidase (150–250 units/mg), 3 mg/ml in distilled water

Cholesterol oxidase [from *Nocardia*, suspension in 1 M $(NH_4)_2SO_4$] (20 units/mg protein) diluted 10-fold with 0.1 M phosphate–EDTA (pH 7.8)

Cholesterol esterase [50 units/mg suspension in 3.2 M $(NH_4)_2SO_4$] (Boehringer), diluted 5-fold with phosphate buffer-EDTA

Assay. Freshly drawn blood containing 0.05 vol of 0.2 M sodium citrate (pH 7.0) is immediately centrifuged (2000 g, 0–2°, 30 min) to obtain plasma,[22] which is diluted with 0.15 M NaCl (5- to 20-fold) to a free cholesterol concentration of 20–80 μg/ml. For the assay of purified LCAT activity in synthetic assay media, a similar medium concentration of free cholesterol is used. The plasma or assay medium is brought to 1 mM with disodium–EDTA (pH 7.4), and 5–8 portions of 0.1 ml are taken by a micropipet dispensing device[23] into 0.9 ml of 0.15 M NaCl. This solution is then mixed with 1.0 ml of methanol (reagent grade). The remaining diluted plasma or assay medium is incubated at 37°, and samples (0.1 ml) are taken at intervals into 0.15 M NaCl and methanol, as initially. The samples in aqueous methanol are stored in 15-ml glass tubes[24] with Teflon-lined caps at −20°. One milliliter of chloroform (reagent grade) is added to the samples at the time of assay, and the tubes are vigorously mixed for 2 × 1 min intervals. The lower phase is then separated by centrifugation (2000 g, 10 min) and 0.2 ml samples are transferred by dispensing micropipette into 10 × 75-mm disposable glass culture tubes. The solvent is then evaporated just to dryness on a sand-bath under a stream of water-pumped N_2.

[20] J. G. Heider and R. L. Boyett, *J. Lipid Res.* **19**, 514 (1978).
[21] P-L Biochemicals, Milwaukee, Wisconsin.
[22] To maintain linear esterification rates in plasma, the blood should be drawn and cooled as expeditiously as possible and the assay carried out within 1–2 hr.
[23] Disposable glass micropipets should have an accuracy of 0.25% the stated volume; a suitable microdispenser is available from Drummond Scientific Co., Broomhall, Pennsylvania.
[24] The glass tubes have been cleaned of any organic contaminants by heating through the self-cleaning cycle of a domestic oven.

Fifty microliters of detergent solution in isopropanol is added, and the evaporated lipid is fully dissolved by mixing. Of the phosphate–EDTA buffer 1.0 ml is then added, the solution remixed, and 150 μl of 4-hydroxy-3-methoxyphenylacetic acid solution and 5 μl of peroxidase solution are then added. After remixing, the solution is then incubated for 10 min at 37°. Cholesterol and cholesteryl ester standards containing 1–10 μg of cholesterol are also carried through the procedure. An initial reading is then made by fluorimeter (excitation 32.5 nm, emission 41.5 nm). Fifteen microliters of the diluted cholesterol oxidase solution is then added, and the tubes are incubated for 30 min at 37°. A second reading in then made with the same fluorimeter settings. Finally diluted cholesterol esterase is added, and the tubes are incubated for a further 30 min at 37°. A final fluorimeter reading is then made. The difference between the first and second fluorimeter readings represents the loss of medium free cholesterol mediated by LCAT activity during the period of incubation; this decrease is inhibited (>97%) by incorporation of 1.5 mM dithiobis(2-nitrobenzoic acid) (DTNB) into the assay medium. Under the conditions described, 1 μg of cholesterol gives an optical density of about 1.0, and reproducibility should be < ±1.0% about the mean. Normal human plasma contains LCAT activity at a level of 20–40 μg of cholesterol esterified/ml/hr.[25]

A confirmation of the rate of LCAT activity is obtained from the difference between the second and third fluorimeter readings, which represents the increment of esterified cholesterol during the assay under the same conditions. In the absence of other reactions (e.g., the presence of cell membranes supplying or utilizing medium cholesterol) the rates of medium LCAT activity are equivalent.[26]

The Lipoprotein Origin of Cholesterol Esterified by LCAT

Principle. In the course of incubation of LCAT and plasma lipoproteins, nonsubstrate lipoproteins (low- and very low-density lipoproteins)

---

[25] P. E. Fielding, C. J. Fielding, R. J. Havel, J. P. Kane, and P. Tun, *J. Clin. Invest.* **71,** 449 (1983). Comparable values have been reported with a cholesterol mass assay in which automated injection gas–liquid chromatography was used to quantitate initial and final medium free cholesterol concentration [Y. L. Marcel and C. Vezina, *Biochim. Biophys. Acta* **306,** 497 (1973)]. Comparative study of the GLC and enzymatic methods in this laboratory shows similar reproducibility. However, the convenience of the enzymatic method for the considerable number of assays involved in the determination of LCAT activities is obviously much greater.

[26] It has been reported that cholesteryl arachidonate is not readily hydrolyzed by cholesterol esterase; however, an esterase preparation now commercially available (Boehringer) appears to be fully active with such esters.

are precipitated by magnesium chloride–dextran sulfate, and the loss of free cholesterol from the supernatant solution is compared with the total free cholesterol consumed by the LCAT reaction.

Reagents

Magnesium chloride, analytical reagent, 2 M solution in distilled water

Dextran sulfate, sodium salt, from dextran with M_w 5 × 10^5, 20 mg/ml

Reagents for free cholesterol assay, as above

Assay. Fresh plasma is diluted 2- to 5-fold with saline and brought to 1 mM disodium EDTA (pH 7.4). Initial samples of diluted plasma are taken for determination of total plasma free cholesterol as described above. An additional 2 ml sample is collected into ice water, and mixed with one-tenth volume of MgCl$_2$ solution. One-tenth volume of dextran sulfate solution is then added. The mixture is placed in ice water for 30 min, then centrifuged (2000 g, 30 min. 0–2°) to precipitate low- and very low-density lipoproteins.[27]

Samples of supernatant (5–8 0.1 ml) are taken into 0.15 M NaCl and 1.0 ml of methanol is added. The remaining diluted plasma is incubated at 37° for 30–60 min, and samples are again taken for total and MgCl$_2$–dextran sulfate soluble cholesterol.[28]

The cholesterol content of these samples is determined fluorimetrically, as described above. The proportion of total free cholesterol esterified that is supplied by high-density lipoprotein is calculated as decrease in MgCl$_2$–dextran sulfate soluble free cholesterol/decrease in total plasma free cholesterol, after correction is made for the dilution factors involved in the addition of the precipitating agents. The free cholesterol supplied by low- and very low-density lipoproteins, precipitated by MgCl$_2$–dextran sulfate, represents the difference between total free cholesterol esterified, and the free cholesterol loss from the soluble fraction. In normal plasma the major part of free cholesterol for LCAT activity (>80%) is supplied by VLDL and LDL[28] (Fig. 1).

Utilization of Cell Membrane Cholesterol for the LCAT Reaction

Principle. LCAT utilizes cell membrane cholesterol for esterification in the presence of a cholesterol carrier containing apolipoprotein A-I. In

---

[27] In the case of assay media containing high concentrations of triglyceride-rich lipoproteins, higher speeds of centrifugation are required to remove the flocculant precipitate. These can be conveniently obtained in an ultracentrifuge with 2-ml microadapters, such as those fitting the Beckman 40.3 rotor, and a speed of 20,000 rpm, 20 min at 0–2°.

[28] C. J. Fielding and P. E. Fielding, *J. Biol. Chem.* **256**, 2102 (1981).

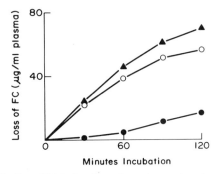

FIG. 1. The contribution of high-density vs low- + very low-density lipoproteins on the free cholesterol utilized for the LCAT reaction. Triangles, total loss of unesterified cholesterol from plasma; open circles, loss of free cholesterol from low- + very low-density lipoproteins; closed circles, loss of unesterified cholesterol from high-density lipoproteins. The result indicates clearly the relative inactivity of bulk high density lipoprotein as a source of substrate for the LCAT reaction. From ref. 28.

the presence of both cell membranes and plasma lipoproteins, the proportion of cholesterol for esterification derived from each source is determined by the difference in cholesterol decrement in an assay medium containing LCAT in the presence and absence of cell membranes.

Assay. Normal skin fibroblasts are grown to near confluence in 6-cm plastic culture dishes in 10% v/v fetal calf serum. The dishes contain 8–10 μg cell cholesterol/dish under these conditions. Freshly drawn blood is collected into one-twentieth volume of 0.2 M sodium citrate solution in ice water, and plasma is obtained by centrifugation (2000 g, 30 min, 0–2°). Fibrin formation is prevented by the removal of plasma fibrinogen on a column of immobilized antibody to human fibrinogen covalently coupled to agarose. The product (whose protein concentration is determined relative to that of fibrinogen-free native plasma) is then diluted to a concentration of 1.25% v/v relative to native plasma with phosphate-buffered saline containing Ca^{2+} (PBS).[29] The dishes of cells are washed four times with PBS-albumin (4 mg/ml) and four times with PBS, and 3 ml of diluted plasma is added to each of five dishes containing cells and five empty dishes rinsed in the same way. Three milliliters of diluted plasma is added to each dish, a 1 ml sample taken into a 15 ml conical glass tube with Teflon-lined screw cap, kept on ice, and the remaining 2 ml of medium in the dishes is incubated (60 min, 37°) in a tissue culture incubator. At the end of the incubation period a second 1 ml sample of medium is with-

---

[29] Phosphate-buffered saline containing $CaCl_2$, 0.1 g/liter; KCl, 0.2 g/liter; KH_2PO_4, 0.2 g/liter; $MgCl_2 \cdot 6H_2O$, 0.6 g/liter; NaCl, 8.0 g/liter; Na_2HPO_4, $7H_2$, 2.16 g/liter.

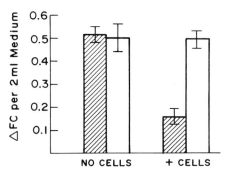

FIG. 2. Determination of cholesterol net transport between cultured normal fibroblasts and plasma. The loss of medium free cholesterol during incubation at 37° (hatched bars) is determined in empty culture dishes (no cells) and in the presence of fibroblasts (+ cells). The gain in medium esterified cholesterol under the same conditions (open bars) is also determined. The presence or absence of cell membrane cholesterol does not affect the rate of LCAT activity, as shown by the equivalent increase in ester cholesterol; however, less of the initial medium cholesterol is consumed by the reaction because of the sparing effect of cholesterol derived from the cell membranes. The difference in free cholesterol consumption in the presence and absence of the cells therefore represents the contribution of cell membrane cholesterol to the LCAT reaction (from ref. 30). (Detailed validation of the cholesterol balance technique applied to cultured cells is contained in ref. 25.)

drawn from each dish of cells and empty dish into another 15-ml conical glass tube. One milliliter of methanol (reagent grade) is added to each, and the stoppered tubes are stored at $-20°$ before determination of medium free cholesterol content. The mass of medium free and esterified cholesterol is determined enzymatically by fluorimeter, as described above for the LCAT mass assay.

Since the rate of LCAT activity is the same in the presence or absence of cells,[30] the contribution of cell membrane cholesterol to total substrate utilized by LCAT represents the difference in the decrement of free cholesterol decrease in the medium during incubation in the presence and absence of the cells.

Cholesterol net transport (cells → medium) per dish = [(initial FC of medium from empty dishes − final FC of medium from empty dishes) − (initial FC of medium from dishes containing cells − final FC of dishes containing cells)] × 2, since each incubated dish contains 2 ml of diluted plasma medium. Net transport for normal plasma under these conditions is about 0.4 μg FC transported (cells → medium) per hr, for plasma at a concentration of 1.25% v/v. Since native plasma contains about 400 μg/ml FC, approximately equal masses of cell membrane and plasma lipopro-

[30] C. J. Fielding and P. E. Fielding, *Proc. Natl. Acad. Sci. U.S.A.* **78**, 3911 (1981).

tein cholesterol are present in the assay under these conditions (Fig. 2). Since LCAT activity is not reduced by dilution under these conditions, net transport per ml plasma is approximately (0.4)/(0.0125 × 2) or 16 µg FC/ml native plasma/hr; total LCAT is 20–40 µg FC esterified/ml native plasma/hr under these conditions. This result indicates that at equivalent concentrations of FC, cell membranes are highly efficient at competing with plasma lipoproteins in donating FC for esterification by LCAT. The proportions of cell membrane and lipoprotein cholesterol selected for this assay are similar to those in small blood vessels.[18] Similar values for cell-to-plasma cholesterol transport have been reported for different cell species in culture; vascular endothelial cells appear to be most efficient in donating cholesterol to the LCAT reaction. The transport of cholesterol from cells to plasma is reduced, or even reversed, in plasma derived from humans with several pathological conditions, while LCAT activity is normal or increased.[25]

Acknowledgments

The research in this chapter was supported by grants from the National Institutes of Health (Arteriosclerosis SCOR HL 14237 and HL 23738).

[13] Assay of Cholesteryl Ester Transfer Activity and Purification of a Cholesteryl Ester Transfer Protein

By YASUSHI OGAWA and CHRISTOPHER J. FIELDING

Pioneering studies by Nichols and co-workers beginning in 1964[1,2] clearly established that during the incubation of plasma there took place a reciprocal exchange of cholesteryl ester in high-density lipoproteins (HDL) and triglyceride from very low-density lipoproteins (VLDL) and that such exchange was independent of the lecithin : cholesterol acyltransferase (LCAT) reaction. Other studies subsequently showed that only those lipoproteins containing apolipoprotein A-I, the cofactor of the LCAT reaction, were significant substrates for LCAT.[3] Nevertheless the major part of cholesteryl esters derived from the LCAT reaction were found in VLDL and its metabolic product, low-density lipoprotein

[1] C. S. Rehnborg and A. V. Nichols, *Biochim. Biophys. Acta* **84**, 596 (1964).
[2] A. V. Nichols and L. Smith, *J. Lipid Res.* **6**, 206, (1965).
[3] C. J. Fielding and P. E. Fielding, *FEBS Lett.* **15**, 355 (1971).

(LDL).[4] These findings, taken together, clearly implied the presence of a factor in plasma catalyzing the transfer of cholesteryl esters from the site of esterification by LCAT to the major cholesterol ester acceptors, VLDL and LDL. The first significant purification of such a transfer activity was achieved by Zilversmit and colleagues.[5] Subsequent studies in several other laboratories have taken advantage of the affinity of this transfer protein for concanavalin A-agarose[6,7] and its stability to isoelectric and chromatofocusing.[8] This chapter describes first, assay procedures for this activity, in whole plasma or in purified form; and second, conditions for the purification from plasma of a complex of LCAT, transfer activity, and apolipoprotein D (a minor protein of HDL); of a complex containing only transfer activity and apolipoprotein D but not LCAT; and of transfer activity containing neither LCAT nor apolipoprotein D.

Assay of Cholesteryl Ester Transfer Activity from Human Plasma

Three types of assay have been developed, representing different principles of transfer. In the first, the movement of radioactive cholesteryl ester between Sepharose-linked and soluble populations of synthetic liposomes is determined.[9] In the second, movement of labeled cholesteryl ester between labeled and unlabeled native lipoproteins is assayed, and the products separated after incubation either on the basis of their flotation properties[6] or of their differential affinity for heparin-Sepharose.[10] Finally, transfer can be assayed in terms of chemical mass transport between HDL and VLDL + LDL, with precipitation by Mg^{2+} and dextran sulfate to distinguish between donor and recipient lipoprotein fractions.[11,12]

Cholesteryl Ester Transfer between Synthetic Liposomes

Principle. Liposomes of phospholipid containing 5% w/w phosphatidylethanolamine and 2.5% w/w [³H]cholesteryl oleate are complexed co-

---

[4] J. A. Glomset, K. R. Norum, and W. King, *J. Clin. Invest.* **49**, 1827 (1970).
[5] D. B. Zilversmit, L. B. Hughes, and J. Balmer, *Biochim. Biophys. Acta* **409**, 393 (1975).
[6] T. Chajek and C. J. Fielding, *Proc. Natl. Acad. Sci. U.S.A.* **75**, 3445 (1978).
[7] J. Ihm, J. L. Ellsworth, B. Chataing, and J. A. K. Harmony, *J. Biol. Chem.* **257**, 4818 (1982).
[8] R. E. Morton and D. B. Zilversmit, *J. Lipid Res.* **23**, 1058 (1982).
[9] T. Chajek, L. Aron, and C. J. Fielding, *Biochemistry* **19**, 3673 (1980).
[10] J. L. Ellsworth, L. McVittie, and R. L. Jackson, *J. Lipid Res.* **23**, 653 (1982).
[11] P. E. Fielding, C. J. Fielding, R. J. Havel, J. P. Kane, and P. Tun, *J. Clin. Invest.* **71**, 449 (1983).
[12] C. J. Fielding, G. M. Reaven, and P. E. Fielding, *Proc. Natl. Acad. Sci. U.S.A.* **79**, 6365 (1982).

valently to CNBr-activated Sepharose. The complex is incubated with soluble liposomes containing phospholipids without cholesteryl ester, in the presence of cholesteryl ester transfer activity. The rate of activity is determined from the radioactivity in the soluble liposomes from which the labeled liposome-Sepharose complex has been removed by low speed centrifugation.

Reagents

Egg lecithin containing 5% w/w dioleyl phosphatidylethanolamine dispersed in distilled water (0.8 mg phospholipid/ml)

Egg lecithin containing 5% w/w dioleyl phosphatidylethanolamine and 2.5% w/w [^{3}H]cholesteryl oleate ($\sim 10^5$ dpm/μg)[13]

Plasma or purified transfer protein (see following section)

Assay. Approximately 4 ml of liposomes, dispersed by hand-shaking, is converted to single-walled vesicles in a French pressure cell[14] for three cycles of 30 sec each at a pressure of 20,000 lb/in.^{2} at room temperature. These particles form a single population of particles with a mean diameter of about 200 Å. To complex the vesicles containing [^{3}H]cholesteryl ester to agarose, 3.2 mg of phospholipid in aqueous dispersion is mixed with 0.5 g dry weight equivalent of Sepharose-CNBr which has been washed with 1 mM HCl then suspended in 0.1 M NaHCO$_3$. The mixture is then mixed by rotation or gentle shaking for 24 hr in the cold, filtered on sintered glass, then resuspended in 0.1 M ethanolamine (pH 7.4) to block any reactive groups unsubstituted with phospholipid amino groups. The Sepharose-phospholipid vesicle complex is then washed three times with distilled water on sintered glass, and resuspended in distilled water. Recovery is 40–50% on the basis of added lipid phosphorous and cholesteryl ester radioactivity. The incubation medium is made up of 50–100 μg/ml of soluble liposome phospholipid, the same mass of phospholipid complexed to Sepharose, and also contains 50 mg of recrystallized human albumin in a volume of 4 ml (pH 7.4) with 10 mM Tris–HCl buffer (pH 7.4). The reaction is initiated by the addition of about one unit (1 μg cholesteryl ester transferred/hr) of purified transfer activity or an equivalent of crude material. An initial sample (2 ml) is taken, and centrifuged at 1000 g (10 min). Multiple samples of supernatant solution (0.2 ml) are taken for analysis of radioactivity. The remaining assay mixture is then incubated for up to 30 min at 37°. At the end of incubation the labeled Sepharose–

---

[13] Prepared from oleyl chloride and 1,2-^{3}H-cholesterol by the method of D. Deykin and D. S. Goodman, *J. Biol. Chem.* **237**, 3649 (1962).

[14] R. L. Hamilton, J. Goerke, L. S. S. Guo, M. C. Williams, and R. J. Havel, *J. Lipid Res.* **21**, 981 (1980).

phospholipid complex is again separated by centrifugation (1000 g, 10 min) and supernatant radioactivity is determined. The rate of transfer in the absence of added transfer protein is <1% of the catalyzed rate.

Cholesteryl Ester Transfer between Isolated Plasma Lipoproteins

Principle. The unidirectional flux of cholesteryl ester between HDL and LDL is assayed by incubating HDL labeled biologically with [^{3}H]cholesteryl ester with unlabeled LDL in the presence of cholesteryl ester transfer protein. The lipoproteins are separated after incubation by ultracentrifugation, and the radioactivity in the floating LDL fraction is determined.

Reagents

[1,2-^{3}H]Cholesterol (>50 mCi/mmol)
Dithiobis(2-nitrobenzoic acid) (DTNB) (14 mM in 0.1 M sodium phosphate buffer, pH 7.4)
10% w/v recrystallized human serum albumin in 0.15 M NaCl, brought to pH 7.4 with 0.01 N NaOH
1.5 M Tris–HCl buffer (pH 7.4)

Assay. Preparation of ^{3}H-labeled HDL. Plasma is prepared from freshly drawn blood in 0.01 M sodium citrate by centrifugation (2000 g, 30 min). Repurified radiolabeled [^{3}H]cholesterol (0.5 mCi) in benzene solution (this volume [12]) is dried under nitrogen, dissolved in 0.5 ml of absolute ethanol, and added through a 26-gauge hypodermic needle from a glass syringe under the surface of 50 ml of stirred plasma. One-tenth volume of DTNB solution is added to inhibit LCAT activity[15] and the mixture is then incubated for 60 min at 37°. Under these conditions the unesterified cholesterol moiety of the plasma lipoproteins becomes labeled to a high specific activity ($\sim 10^4$ dpm/μg). The inhibition of LCAT is then reversed by addition of 2-mercaptoethanol to a final concentration of 20 mM, and the plasma is incubated for a further 60 min at 37°. ^{3}H-labeled cholesteryl esters are generated by LCAT and distributed to the major plasma lipoprotein fractions by cholesteryl ester transfer activity. HDL is recovered from the plasma by differential preparative ultracentrifugation.[16] The plasma is brought to a density of 1.08 g/ml with solid KBr and centrifuged at 40,000 rpm for 24 hr at 4° in the 40.3 rotor of a Beckman preparative ultracentrifuge. The infranatant solution is recovered, brought to a density of 1.21 g/ml with KBr, and centrifugation carried out

[15] K. T. Stokke and K. R. Norum, *Scand. J. Clin. Lab. Invest.* **27**, 21 (1971).
[16] R. J. Havel, H. A. Eder, and J. H. Bragdon, *J. Clin. Invest.* **34**, 1345 (1955).

for 48 hr under the same conditions. The supernatant HDL is diluted and recentrifuged at d 1.08 g/ml, then finally recovered at d 1.21 g/ml. To remove endogenous cholesteryl ester transfer activity, the HDL is dialyzed against 0.15 M NaCl to remove KBr, brought to 1 mM MgCl$_2$, 1 mM MnCl$_2$, 1 mM CaCl$_2$ by addition of concentrated solutions, and passed through a column (0.7 × 3 cm) of concanavalin A–Sepharose covalent complex, equilibrated in the same solution. Transfer activity is retained on the column[6] and the nonadsorbed ^{3}H-labeled HDL used in the assay described below.

Preparation of Unlabeled LDL. Unlabeled plasma containing 0.05% w/v EDTA is adjusted to a solution density of 1.019 g/ml with solid KBr then centrifuged at 40,000 rpm for 24 hr at 4° in the 40.3 rotor of a Beckman preparative ultracentrifuge. The bottom fraction is adjusted to 1.063 g/ml with KBr and centrifuged under the same conditions. The separation is repeated between the two same density limits.

The cholesteryl ester content of the ^{3}H-labeled HDL and unlabeled LDL is determined enzymatically.[17]

Assay. The assay of transfer activity is carried out in the 2-ml cellulose tubes used with polyethylene adaptors fitting the Beckman 40.3 ultracentrifuge rotor (5/16″ × 1-15/16″). Each assay contains 50 µg of ^{3}H-labeled HDL esterified cholesterol, and 300 µg of unlabeled LDL esterified cholesterol, together with 50 µl of 14 mM DTNB, 30 µl of 1.5 M Tris–HCl buffer (pH 7.4), 50 µl of 10% serum albumin, and 0.15 M NaCl in a final volume of 300 µl. The reaction is initiated by the addition of up to 200 µl of sample or buffer blank. The tubes are incubated at 37° for up to 3 hr, and the transfer reaction stopped by rapid cooling in ice water. The density of the reaction mixture is brought to 1.063 g/ml by addition of 1.5 ml of KBr solution (density 1.082 g/ml). The tubes are capped and centrifuged at 40,000 rpm for 24 hr at 4°. After centrifugation, the tube is sliced to recover the upper quarter of volume, and the LDL lipid is extracted with equal volumes of methanol and chloroform after mixing. Of the chloroform phase 250 µl is fractionated by thin-layer chromatography on silicic acid layers on glass plates developed with hexane : ethyl ether : acetic acid 83 : 16 : 1 v/v. The region containing cholesteryl ester (R_f 0.9) is recovered and its radioactivity is determined.

Assay of Cholesteryl Ester Mass Transport between HDL and VLDL + LDL in Plasma—Direct Method

Principle. The decrease in HDL cholesteryl ester mass during incubation is measured under conditions where the *de novo* synthesis of cholesteryl esters by LCAT is inhibited.

---

[17] J. G. Heider and R. L. Boyett, *J. Lipid Res.* **19**, 514 (1978).

Reagents

Dithiobis(2-nitrobenzoic acid) (DTNB), 14 mM in 0.1 M phosphate buffer, pH 7.4
0.3 M sodium EDTA, pH 7.4
2 M MgCl$_2$
Dextran sulfate (M_{av} 5 × 10^5), 20 mg/ml in distilled water

Assay. Blood is drawn into sodium citrate solution and plasma obtained by centrifugation as described above. The fresh plasma is used immediately in the incubation described below. (*Note.* This assay determines transport down a chemical potential gradient between HDL and VLDL + LDL; the activity will not be observed in stored or preincubated plasma.)

Fresh ice-cold plasma is mixed with three volumes of cold NaCl and brought to 10 mM Tris–HCl and 1 mM sodium EDTA by addition of the concentrated salt solutions. One-tenth volume of DTNB solution is added. A 2-ml portion of the diluted plasma–DTNB is then mixed with 0.05 ml of MgCl$_2$ and 0.05 ml of dextran sulfate solution. The remainder of the plasma–DTNB is transferred to a 37° water bath and incubated for 30–60 min. A second 1-ml portion is taken, cooled in ice, and mixed with MgCl$_2$ and dextran sulfate solutions as initially. The initial and final samples are each placed 30 min in ice, then centrifuged for 30 min at 0–2° and 2000 rpm to sediment the precipitated VLDL and LDL.[18] About 1.5 ml of the clear supernatant is transferred to a disposable 10 × 75-mm glass tube and pentuplicate 200 μl portions transferred to 0.9 ml 0.15 M NaCl in glass conical tubes. One milliliter of absolute methanol is added, and the tubes are capped with Teflon-lined screw caps.[19] The samples are stored at −20° before assay. The cholesteryl ester content of the supernatant (HDL) fraction of the precipitated plasma is determined enzymatically, as described in detail in the preceding chapter [12].

The decrease in HDL cholesteryl ester observed during incubation at 37° represents the rate of transfer of cholesteryl ester from HDL to the precipitated VLDL and LDL. In normal plasma, this rate is about 75% of the total cholesteryl ester synthesis rate catalyzed by the LCAT reaction, or 15–20 μg ester cholesterol transferred per ml plasma per hr. The same principle has been used to determine the net transport of cholesteryl ester from isolated HDL to LDL, with purified cholesteryl ester transfer activities.[20]

[18] Complete precipitation of VLDL + LDL from hypertriglyceridemic plasma requires higher centrifugal forces. A satisfactory separation can be obtained by centrifugation at 20,000 rpm for 30 min in the 40.3 rotor of a Beckman preparative ultracentrifuge, with 2-ml adaptors.
[19] The tubes have been cleaned of organic contaminants by heating through the self-cleaning cycle of a domestic oven.
[20] L. de Parscau and P. E. Fielding, unpublished experiments.

Assay of Cholesteryl Ester Mass Transfer—Indirect Method

Principle. The change in HDL cholesteryl ester during incubation at 37° is measured under conditions where LCAT activity continues to synthesize cholesteryl esters at a linear rate. The rate of transfer of cholesteryl esters to VLDL and LDL is given by the difference between the total increase in plasma cholesteryl esters and the increase in HDL cholesteryl esters, measured after the precipitation of VLDL and LDL with $MgCl_2$ and dextran and sulfate.

Assay. Plasma obtained as described in the preceding assay from citrated fresh blood is diluted 3-fold with cold 0.15 M NaCl, then brought to 10 mM Tris–HCl, 1 mM sodium EDTA (pH 7.4) by addition of the concentrated salts. An initial 2-ml portion of diluted buffered plasma is taken. Pentuplicate 0.2 ml samples are taken for the determination of initial cholesteryl ester content, as described above. A second 2 ml sample is mixed with 0.1 ml of $MgCl_2$ solution and 0.1 ml of dextran sulfate solution, and placed on ice for 30 min. The clear supernatant, obtained after centrifugation of the flocculent precipitate of VLDL and LDL at 2000 rpm for 30 min[18] is transferred and sampled (5 × 200 μl) into screw-capped glass tubes as described above for the direct method. The remainder of the diluted plasma is incubated for 60 min at 37°, cooled in ice water, then sampled directly for determination of total and HDL cholesteryl ester, exactly as for the initial samples. The determination of free and esterified cholesterol by the enzymatic method is as described in this volume [12].

The total synthesis rate of plasma cholesteryl ester (which is equivalent to the rate of decrease of plasma free cholesterol) is determined from the measurements with unfractionated plasma. The rate of accumulation of cholesteryl ester in HDL (measured as the insoluble cholesteryl ester after precipitation with $MgCl_2$ and dextran sulfate) is subtracted to determine the rate of transfer to VLDL and LDL. Appropriate correction is made for the dilution by $MgCl_2$ and dextran sulfate solutions.

Purification of Cholesteryl Ester Transfer Activity from Human Plasma

Reagents

74% w/v $MgCl_2$ solution in distilled water
36% w/v sodium phosphotungstate solution in distilled water
Phenyl-Sepharose (Pharmacia)
Calcium phosphate gel in 0.001 M phosphate buffer (pH 7.4)
Concanavalin A–Sepharose covalent complex (Pharmacia) crosslinked with glutaraldehyde[21]

---

[21] K. P. Campbell and D. H. MacLennan, *J. Biol. Chem.* **256,** 4626 (1981).

Gentamycin sulfate
Histidine–HCl buffer (25 mM) (either pH 5.5 or 5.8)
PBE 94 chromatofocusing resin and polybuffer 74 (Pharmacia)

Method. Transfer activity is isolated from human plasma freshly collected into 0.05 volumes of 0.26 M sodium citrate solution (pH 7.0) and cooled immediately in ice-water. When assayed with [^{3}H]cholesteryl ester-labeled HDL as described below, normocholesterolemic human plasma has a transfer activity of 10–30 μg ester cholesterol transferred (to LDL) per ml plasma per hr.

Step 1: Precipitation of Plasma Lipoproteins.[22] Blood plasma (100 ml) is obtained from blood by centrifugation (2000 g, 30 min) at 2–4°. Concentrated MgCl$_2$ solution (0.05 volumes) is added dropwise with stirring and then 0.05 volumes of sodium phosphotungstate solution. The flask containing the plasma is kept on ice for 20 min, then the flocculant precipitate (which contains VLDL, LDL, and the major part of HDL) is removed by centrifugation (20,000 rpm, 20 min) at 4° in the Type 30 rotor of a Beckman preparative ultracentrifugation. Subsequent steps are carried out at cold-room temperature (4°).

Step 2: Phenyl-Sepharose Chromatography. A column of phenyl sepharose (2.5 × 13 cm) is equilibrated with 3 M NaCl solution containing 0.01% w/v sodium azide, gentamycin sulfate (50 μg/ml) and 0.05% w/v sodium ethylenediaminetetraacetic acid (EDTA) (pH 7.4). The supernatant from step 1 is brought to 3 M with solid NaCl, and azide, EDTA, and gentamycin to the same concentrations as those in the column buffer (it has been found that the complete binding of LCAT and transfer activities for copurification requires loading at high salt concentrations). The supernatant is then loaded on the column, and the column is washed with 0.15 M NaCl containing azide, EDTA, and gentamycin as previously, until the protein concentration of the eluate decreases to <0.03 OD$_{280}$ units (this requires approximately 1200 ml of saline solution). LCAT and transfer activities are then eluted with distilled water, passed through the column at a flow rate of 1.5 ml/min. LCAT activity is purified about 100-fold, and cholesteryl ester transfer activity about 600-fold, relative to the original plasma (see the table).[23] A plasma inhibitor of transfer activity has been identified[24] whose removal may account for the part of the increase in transfer specific activity at this step. The peak of eluted proteins is pooled and may be stored in this form at 2–4° for at least 1 week without detectable loss of either activity.

[22] M. Bartholome, D. Niedmann, H. Wieland, and D. Seidel, *Biochim. Biophys. Acta* **664,** 327 (1981).
[23] O. H. Lowry, N. J. Rosebrough, A. L. Farr, and R. J. Randall, *J. Biol. Chem.* **193,** 265 (1951).
[24] R. E. Morton and D. B. Zilversmit, *J. Biol. Chem.* **256,** 11992 (1981).

COPURIFICATION OF LCAT AND CHOLESTERYL ESTER TRANSFER ACTIVITIES
FROM HUMAN PLASMA

Fraction	Specific activity[a]		Purification (fold)		Yield (%)	
	LCAT	CE transfer	LCAT	CE transfer	LCAT	CE transfer
Plasma	0.44	0.07	1.0	1.0	100	100
Phenyl-Sepharose	40.00	43.4	93	620	49	313
Hydroxylapatite	2960	1309	6727	18700	32	85
Concanavalin A	3216	1398	7310	19971	10	29

[a] Microgram cholesterol esterified or transferred per milligram protein[23] per hour. LCAT was assayed with synthetic egg lecithin–cholesterol liposomes activated with apolipoprotein A-1.[16] Cholesteryl ester transfer was assayed as the unidirectional flux of ^{3}H-labeled cholesteryl ester in HDL into unlabeled LDL.[6]

Step 3: Calcium Phosphate Gel Chromatography. The pool from step 2 is brought to 12 mM sodium phosphate solution[25] (pH 7.4) by addition of concentrated buffer, and to 0.01% w/v sodium azide. The protein solution is then added to calcium phosphate gel suspended in the same buffer in the ratio 1 ml packed gel per 5 OD_{280} units, kept on ice for 20 min, and then centrifuged (1500 g, 4° for 5 min). The protein concentration of the supernatant decreases to about 0.05 OD_{280} units (if a higher protein concentration remains, the calcium phosphate step is repeated under the same conditions). The following step removes a small further proportion of impurities, but is not required for the preparation of highly active material.

Step 4: Concanavalin A-Sepharose Chromatography. The gel supernatant is concentrated by ultrafiltration (Amicon YM-10 membrane). Remaining phosphate buffer is then removed by 2–3 cycles of dilution and reconcentration from distilled water containing 0.01% w/v azide (pH 7.0) (this is required to prevent precipitation by phosphate ions of the divalent cations equilibrated with concanavalin A in the following chromatographic step). The protein sample is then brought to 1 mM Tris–HCl (pH 7.4), 1 mM MgCl$_2$, 1 mM MnCl$_2$, 1 mM CaCl$_2$, and then applied to a column of concanavalin A–Sepharose, which had been cross-linked with glutaraldehyde as described by Campbell and MacLennan,[21] to prevent leaching of concanavalin A. The column (0.7 × 13 cm), is preequilibrated in the same buffer of Tris–HCl and divalent cations. The protein solution is added and the column is washed with Tris–divalent cation solution. The bound LCAT and transfer activity are eluted with 0.2 M methyl manno-

[25] Individual batches of this reagent have been found to differ slightly in their affinity for LCAT and transfer protein; the appropriate phosphate concentration for an individual lot can be determined using a gradient of phosphate (1–20 mM).

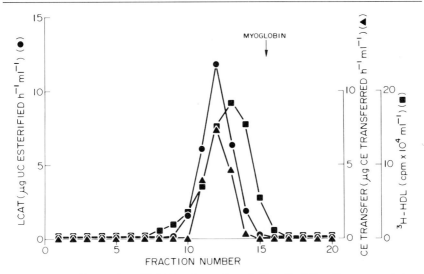

FIG. 1. Cosedimentation of LCAT and transfer activities. Centrifugation was carried out for 31 hr in a Spinco SW 41 rotor in a linear gradient of 5–25% w/v sucrose (6 ml). Either the concanavalin A eluate containing LCAT and transfer activity (see the table, step 4), total plasma HDL (1.08–1.21 g/ml), or recrystallized muscle myoglobin was applied in a 200 μl total volume. Fractions were collected from the bottom of the gradient. Circles, LCAT activity; triangles, transfer activity (assayed as the transfer from ^{3}H-labeled HDL to LDL); squares, HDL.

side containing 1 mM Tris–HCl (pH 7.4) and 0.01% sodium azide, at a flow rate of 0.3–0.4 ml/min. LCAT and transfer activity are coeluted at final fold-purification from plasma of 7000–8000 and 20,000 respectively, in yields of approximately 10% and 30% (see the table). The apolipoprotein D content of the purified fraction is 32 ± 10% by specific radial immunoassay.[26] During sedimentation velocity ultracentrifugation (Fig. 1) LCAT, cholesteryl ester transfer activity, and apolipoprotein D cosediment at a rate slightly faster than that of bulk plasma HDL. The activities can also be copurified (without further increase in specific activity) by pH gradient chromatography on hydroxylapatite (Fig. 2).

Separation of LCAT and Cholesteryl Ester Transfer Activities. LCAT has been isolated by several procedures from human plasma (see this volume [12]). It can also be isolated from the associated LCAT and transfer activities of the preceding step, by chromatofocusing column chromatography. The eluate from concanavalin A–Sepharose (step 4) is pooled and applied to a column (0.7 × 13 cm) of chromatofocusing resin equilibrated with 25 mM histidine buffer (pH 5.8). The column is washed with 40 ml of the same buffer, then eluted with Polybuffer 74 (pH 3.5) diluted

[26] P. E. Fielding and C. J. Fielding, *Proc. Natl. Acad. Sci. U.S.A.* **77**, 3327 (1980).

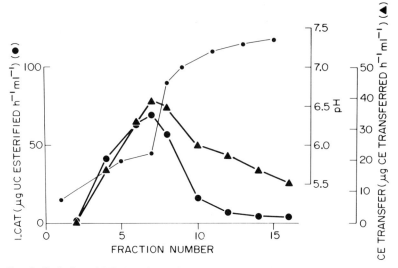

FIG. 2. Coelution of LCAT and transfer activity from a pH gradient on hydroxylapatite. The mixed LCAT-transfer activity was applied in 50 mM sodium acetate buffer (pH 4.5), 90 mM NaCl, to a column (1 × 15 cm) containing two parts hydroxylapatite and one part crystalline microcellulose in 15 mM sodium phosphate buffer, pH 7.4. Circles, LCAT activity; triangles, transfer activity, assayed as described in Fig. 1; small circles, pH of individual fractions.

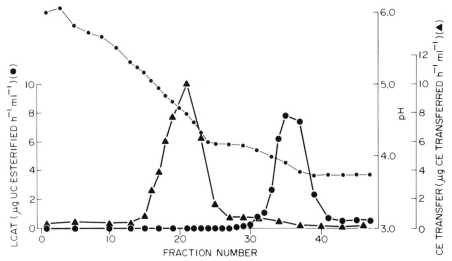

FIG. 3. Chromatofocusing of the complex of LCAT and transfer activities from concanavalin A–Sepharose chromatography. Fractionation was carried out on a column (0.7 × 13 cm) of Polybuffer exchanger 94 (Pharmacia) equilibrated in 25 mM histidine–HCl buffer (pH 5.8). Enzyme solution (10.5 ml) (step 4, see the table) was applied and the column was washed with 40 ml of histidine buffer. Polybuffer 74 diluted 8-fold with distilled water (final pH 3.5) (100 ml), was then passed through the column. Circles, LCAT activity; triangles, cholesteryl ester transfer activity, assayed as in Fig. 1; small circles, pH of individual fractions.

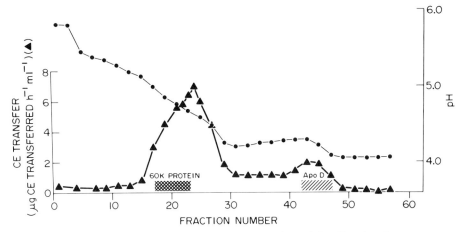

FIG. 4. Chromatofocusing of cholesteryl ester transfer activity. Fractionation was carried out as in Fig. 3 except that the resin was equilibrated in 25 mM histidine–HCl (pH 5.5) and eluted with Polybuffer 74 diluted 10-fold with distilled water (final pH 4.0). Triangles, cholesteryl ester transfer activity (assayed with ^{3}H-labeled HDL and unlabeled LDL); small circles, pH of individual fractions. The major proteins present in the individual fractions were identified by silver staining of polyacrylamide slab gels developed in Tris–glycine buffer (pH 8.3) in the presence of 0.1% w/v sodium dodecyl sulfate.[27] Samples were reduced before development with 60 mM 2-mercaptoethanol.

10-fold with distilled water, at a flow rate of 0.3 ml/min. As shown in Fig. 3, LCAT and transfer activities are completely separated and elute at pH 3.9 and 4.6, respectively. Apolipoprotein D is associated with both peaks of activity. Further purification of the transfer activity is achieved by chromatography with a shallow pH gradient. The fraction that contains transfer activity in distilled water containing 0.01% sodium azide is applied to the chromatofocusing column equilibrated in 25 mM histidine–HCl buffer (pH 5.5) and 10-fold diluted Polybuffer (pH 4.0) is used as eluant. As shown in Fig. 4, cholesteryl ester transfer activity is eluted in two peaks. In the first peak, transfer activity partially overlaps with a protein of molecular weight of ~60,000, determined by subsequent SDS slab gel chromatography[27]; a minor protein component with a molecular weight of ~70,000 can also be seen by sensitive silver strain.[28] The second peak contains as the major detectable protein, a species of molecular weight ~33,000, comigrating with authentic apolipoprotein D; it contains no detectable molecular weight 60,000 protein.

Acknowledgments

The research in this chapter was supported by grants from the National Institutes of Health (Arteriosclerosis SCOR HL 14237 and HL 23738).

[27] U. K. Laemmli, *Nature (London)* **227**, 680 (1970).
[28] B. R. Oakley, D. R. Kirsch, and N. R. Morris, *Anal. Biochem.* **105**, 361 (1980).

[14] Cholesterol Acyltransferase

By JEFFREY T. BILLHEIMER

$$\text{Fatty Acyl-CoA + Cholesterol} \longrightarrow \text{Cholesterol Ester + CoA}$$

Acyl-CoA : cholesterol acyltransferase (ACAT; EC 2.3-1.26) catalyzes the intracellular esterification of cholesterol in mammals, and a similar enzyme is present in yeast[1] and *Heliothis zea*.[2] A particulate enzyme, located primarily in the endoplasmic reticulum, ACAT, through sterol esterification, maintains the content of cellular free cholesterol within limits needed for proper membrane function. Two recent reviews deal with ACAT and its role in cholesterol metabolism.[3,4]

Assay Method

Principle

ACAT activity present in microsomes is commonly measured by a radiochemical assay in which [^{14}C]oleoyl-CoA is converted to [^{14}C]cholesterol ester. After the reaction has been terminated, radiolabeled cholesterol ester is separated from substrates and other labeled products by thin-layer chromatography and counted in a scintillation counter.

ACAT activity is sensitive to compounds such as detergents that disrupt the microsomal membrane. This complicates the assay since oleoyl-CoA itself is a strong detergent which can inhibit enzymatic activity. In addition the endogenous sterol present in the microsomes is not adequate for optimal ACAT activity and an exogenous source of cholesterol must be supplied.[5] Again, the use of organic solvents or strong detergents for the suspension of cholesterol must be avoided because of their affect on the enzyme.[6] A systematic study of ACAT activity in rat liver has led to the development of conditions which optimize cholesterol ester formation.[5]

---

[1] S. Taketani, T. Nishino, and H. Katsuki, *Biochim. Biophys. Acta* **575**, 148 (1979).
[2] J. T. Billheimer, D. M. Tavani, and K. S. Ritter, *Comp. Biochem. Physiol. B* **76B**, 127 (1983).
[3] A. A. Spector, S. N. Mathur, and T. L. Kaduce, *Prog. Lipid Res.* **18**, 31 (1979).
[4] T. Y. Chang and G. M. Doolittle, *in* "Enzymes" (P. Boyer, ed.), Vol. 16, p. 523. Academic Press, New York, 1983.
[5] J. T. Billheimer, D. M. Tavani, and W. R. Nes, *Anal. Biochem.* **111**, 331–335 (1981).
[6] S. K. Erickson, M. A. Shrewsbury, C. Brooks, and D. J. Meyer, *J. Lipid Res.* **21**, 930 (1980).

Reagents

Assay buffer: 0.1 M K-phosphate (pH 7.4) containing 1 mM glutathione
Fatty-acid free bovine serum albumin: 20 mg/ml in assay buffer
Cholesterol: recrystallized from acetone, 1 mg/ml in cyclohexane
[^{14}C]Cholesterol: 1000 dpm/μl in cyclohexane, 50 Ci/mol
Triton WR-1339: 10% solution in acetone
[^{14}C]Oleoyl-CoA: 5000 dpm/nmol, 1 mM in 0.01 M K-phosphate, pH 6.0
Cholesteryl oleate: 1 mg/ml in cyclohexane
Chloroform : methanol: 2 : 1, v/v

Substrate Solutions

Oleoyl-CoA is hydroscopic, therefore the concentration of solutions should be determined using its molar extinction coefficient at 256 nm (15,600 liters/mol). Oleoyl-CoA is labile and should be stored at $-80°$ under nitrogen. A new lot of labeled oleoyl-CoA should be checked in an enzyme assay against previous stock. Low ACAT activity may be due to oxidation or decomposition products of the substrate. Purity of labeled oleoyl-CoA can be determined by reversed-phase liquid chromatography according to the procedure of Baker and Schooley.[7]

Cholesterol or other sterols are suspended in assay buffer with the aid of Triton WR-1339. For ten assays, cholesterol (200 μg) is placed in a conical test tube and the cyclohexane evaporated. Triton WR-1339, 600 μg, is added and the sample warmed to ensure that the sterol is completely dissolved. Assay buffer (at 60°), 1 ml, is added, the solution vortexed, and the acetone removed by evaporation under nitrogen. The resulting solution is transparent; turbidity indicates incomplete dispersion of the sterol and will lead to lower ACAT activity.

Procedure

Microsomes are prepared from rat liver by differential centrifugation and washed with assay buffer prior to use.[5] The standard assay medium contains 100 μg of microsomal protein, 20 μl of the [^{14}C]oleoyl-CoA solution (100 μM, 5000 dpm/nmol), 50 μl of fatty acid-free bovine serum albumin (1 mg), 100 μl of cholesterol solution (52 nmol cholesterol in 0.3% Triton WR-1339), and assay buffer to a final volume of 200 μl. The medium excluding oleoyl-CoA is preincubated for 30 min at 37°, and the reaction is initiated by the addition of labeled oleoyl-CoA. After 10–15

---

[7] F. C. Baker and D. A. Schooley, this series, Vol. 72 [5].

min, the reaction is stopped by the addition of 4 ml of chloroform/methanol (2/1, v/v). Colesteryl oleate, 10 µg, and [^{14}C]cholesterol (10,000 dpm) are added as a carrier and internal standard, respectively. After allowing 30 min for lipid extraction, 0.8 ml of 0.88% KCl is added, the chloroform layer is removed, and the solvent evaporated under nitrogen. The residue is resuspended in 80 µl of chloroform, and the cholesterol esters are isolated by thin-layer chromatography on Gelman ITLC-SA polysilicic acid gel-impregnated glass fiber sheets using a petroleum ether, diethyl ether, acetic acid (85 : 15 : 0.5) solvent system. This system separates cholesteryl oleate (R_f 0.90) from other radioactive products, triglycerides (R_f 0.74), oleic acid (R_f 0.55), and cholesterol (R_f 0.28). The spots corresponding to cholesterol and cholesterol esters are visualized by spraying the sheet with a solution of 10% phosphomolybdic acid in 90% ethanol and heating for 2 min at 100°; the spots are cut out and placed in scintillation vials for counting. (*Note:* If [^{3}H]cholesteryl oleate is used as an internal standard, the spots are visualized with iodine vapor since phosphomolybdic acid interferes with dual-label counting.) An assay with boiled microsomes is used as a control. The specific activity of ACAT is expressed as nanomoles of cholesteryl oleate formed per minute per milligram of protein.

Comments

In addition to liver,[5] this assay has been utilized to quantitate ACAT activity in rat ovary,[8] corn earworm (*Heliothis zea*),[2] and yeast.[9]

Serum albumin is required in the assay medium to prevent the high concentration of oleoyl-CoA (100 µM) from inhibiting ACAT presumably by disrupting the microsomal membrane. Lichtenstein and Brecher[10] have shown that albumin binds oleoyl-CoA and thus acts as a reservoir, and albumin addition limits the concentration of free oleoyl-CoA during the assay. The molar ratio of albumin to oleoyl-CoA is critical with a 1 : 1 ratio being optimal. In the absence of albumin, the initial concentration of oleoyl-CoA is limited to about 15 µM and ACAT activity is reduced by 60%. This is probably due to the presence of a second enzyme in the microsomes, acyl-CoA hydrolase which hydrolyzes oleoyl-CoA at a rate an order of magnitude greater than esterification.[11] The free fatty acids thus generated are also inhibitory in the absence of albumin.[12]

[8] D. M. Tavani, T. Tanaka, J. F. Strauss, III, and J. T. Billheimer, *Endocrinology* **111,** 794 (1982).
[9] J. T. Billheimer, J. Adler, and S. Dorsey, *Fed. Proc. Fed. Am. Soc. Exp. Biol.* **43,** 1902 (1984).
[10] A. H. Lichtenstein and P. Brecher, *J. Biol. Chem.* **255,** 9098 (1980).
[11] R. K. Berge, *Biochim. Biophys. Acta* **574,** 321 (1979).
[12] D. S. Goodman, D. Deykin, and T. Shiratori, *J. Biol. Chem.* **230,** 1325 (1964).

Early investigations revealed that ACAT activity in liver increased as the microsomal content of free cholesterol increased, suggesting that membrane-bound cholesterol was limiting.[13] The addition of exogenous cholesterol in Triton WR-1339 increased ACAT activity by 3.5-fold (0.72 versus 0.21 mmol min^{-1} mg^{-1}) and this increase was shown to be due to the esterification of exogenous cholesterol.[5] A similar increase in ovarian ACAT activity is also observed upon addition of cholesterol in Triton WR-1339, and the activities are consistent with observed rates of cholesteryl ester synthesis *in vivo*.[8] Maximal cholesteryl oleate formation is obtained upon addition of 20 μg of cholesterol per assay tube utilizing a 30-min preincubation period.[5,8] Triton WR-1339 has also been used to deplete the pyridine nucleotide concentration in microsomes without changing their morphology or affecting the oxidative demethylation of sterols.[14] Most commonly employed nonionic detergents (e.g., Triton X-100) that have a hydrophile–lipophile balance value between 12 and 15, disrupt the membrane, and inhibit ACAT.[6,15] Neither are organic solvents such as acetone useful in the suspension of cholesterol since sterol precipitation occurs upon addition of the organic solution to the aqueous reaction mixture and the solvent itself may inhibit ACAT. Cholesterol-rich liposomes[16] and a cholesterol serum albumin emulsion[17] have also been utilized successfully as a source of exogenous cholesterol.

Labeled fatty acids can also be utilized to follow sterol esterification in the presence of ATP and coenzyme A by taking advantage of microsomal acyl-CoA synthetase.[12] Similarly, labeled cholesterol has also been used, however, the data are usually not quantitative because either the dilution by endogenous cholesterol or the extent of equilibration is not known.

Solubilization of ACAT

ACAT has been solubilized using deoxycholic acid and partially purified from pig liver[18] and has been solubilized from both rat liver[16] and Erhlich ascites cells[19] using Triton X-100. In all cases, the detergent had to be removed and ACAT reconstituted into an artificial membrane for optimal activity to be observed. Although the method of reconstitution varied extensively among the three groups,[16,18,19] each method demonstrated that ACAT is very sensitive to its microenvironment. ACAT ac-

[13] S. Hashimoto and S. Dayton, *Biochim. Biophys. Acta* **573**, 354 (1979).
[14] W. L. Miller, M. E. Kallafer, J. L. Gaylor, and C. V. Delwicke, *Biochemistry* **6**, 2673 (1967).
[15] R. W. Egan, *J. Biol. Chem.* **254**, 4442 (1976).
[16] K. Suckling, G. S. Boyd, and C. G. Smellie, *Biochim. Biophys. Acta* **740**, 154 (1982).
[17] F. J. Field and S. N. Mathur, *J. Lipid Res.* **24**, 409 (1983).
[18] G. M. Doolittle and T. Chang, *Biochemistry* **21**, 674 (1982).
[19] S. N. Mathur and A. A. Spector, *J. Lipid Res.* **23**, 692 (1982).

tivity in artificial liposomes is affected by the polar head group of the phospholipid, by the fatty acyl chains of the phospholipid, and by the phospholipid protein ratio. Also, cholesterol, in addition to being a substrate, can affect ACAT activity by changing the fluidity of the membrane.

Properties of ACAT

The microsomal enzyme from liver,[5] yeast,[9] and *Heliothis zea*[2] can be stored frozen ($-80°$) from one to several weeks without losing activity while the activity in microsomes from rat ovary[8] appears to be labile to freezing. ACAT from rat liver shows a broad pH optimum around neutrality and enzymic assays are generally run between pH 7.1 and 7.5.[12]

Substrate Specificity

In liver, Goodman *et al.* found oleoyl-CoA to be the best substrate followed by palmitoyl-, stearoyl-, and linoleoyl-CoA.[12] A part of this selectivity may be due to the presence of acyl-CoA hydrolase in the microsomal preparations. For this enzyme, the preferred substrate is palmityl-CoA which is hydrolyzed about eight times faster than oleoyl-CoA.[11] The apparent K_m for oleoyl-CoA is 55 μM in the assay outlined above, however, use of kinetic parameters such as K_m is of limited value due to the physical properties of oleoyl-CoA in solution and the addition of serum albumin to the assay.

The suspension of cholesterol in Triton WR-1339 promotes the use of exogenous cholesterol for esterification and has allowed investigations into the sterol specificity of ACAT.[20] In rat liver, ACAT shows a high degree of specificity toward cholesterol, its normal substrate. A 3β-hy-

droxyl group is required for esterification to occur. ACAT is especially sensitive to changes in the side chain of the sterol. Alkylation at C-24, found in phytosterols, reduces esterification by at least 80%. Increasing or decreasing the length of the side chain from that found in cholesterol also reduces esterification. Within the sterol nucleus, the presence of a gemdimethyl group at C-4 (e.g., lanosterol) reduces esterification to below de-

[20] D. M. Tavani, W. R. Nes, and J. T. Billheimer, *J. Lipid Res.* **23**, 774 (1982).

tectable limits. The presence of unsaturation in the β-ring nor the positioning of the double bond at C-5, when present (as in cholesterol) is not critical for esterification by ACAT; both cholestanol, the saturated analog of cholesterol, and lathosterol, the Δ^7-isomer of cholesterol, show appreciable esterification. Field and Mathur,[17] have shown that sitosterol is only esterified one-sixtieth as well as cholesterol by intestinal ACAT.

Ergosterol, the predominant sterol in yeast, contains a 24-β methyl group. In yeast microsomes, the rate of acyl-CoA-dependent esterification of ergosterol is similar to that of cholesterol. However, phytosterols which contain 24-α substituents are esterified only one-fifth as well as cholesterol.[9]

Inhibitors

As mentioned above, detergents and organic solvents which perturb the membrane structure inhibit ACAT activity. Detergents which have been shown to inhibit ACAT at concentrations above their critical micelle concentration are cholate, deoxycholate, Triton X-100, Tween 20, Tween 80, and Nonidet P-40.[6] Cholesterol complexing agents such as digitonin and filipin also inhibit ACAT activity.[6] Ethanol, acetone, and butanol all inhibit ACAT activity about 40% at a final concentration of 1% (v/v).[6]

The thio-blocking agent, 5,5'-dithiobis(2-nitrobenzoic acid) (0.15 mM) inhibited intestinal ACAT by 50%.[21]

A number of other compounds have been shown to inhibit ACAT activity *in vitro* in various tissues; prostaglandins of PGE and PGF families, skin, 50% inhibition about 5×10^{-7} M[22]; local anesthestics such as lidocaine, liver, 50% inhibition 1.5 mM,[23] the tranquilizer, chlorpromazine, aorta, 50% inhibition at 0.1 mM,[24] the hypocholesterolemic agent, clofibrate, aorta, 50% inhibition 5 mM.[25]

A number of naturally occurring and synthetic sterols and steroidal hormones have been tested for their effect on mammalian ACAT activity and the results are summarized in the table.[20,26–29] Naturally occurring sterols, such as sitosterol, which themselves are not esterified to any extent can competitively inhibit the esterification of exogenously supplied

[21] K. R. Norum, P. Helgerud, and A. C. Lilljequist, *Scand. J. Gastroenterol.* **16**, 401 (1981).
[22] V. A. Ziboh and M. A. Dreize, *Biochem. J.* **152**, 281 (1975).
[23] F. P. Bell and E. V. Hubert, *Biochim. Biophys. Acta* **619**, 302 (1980).
[24] F. P. Bell, *Exp. Mol. Pathol.* **38**, 336 (1981).
[25] P. I. Brecker and A. V. Chobanian, *Circ. Res.* **35**, 692 (1974).
[26] A. H. Lichtenstein and P. Brecker, *Biochim. Biophys. Acta* **751**, 340 (1983).
[27] E. R. Simpson and M. F. Burkhart, *Arch. Biochem. Biophys.* **200**, 79 (1980).
[28] J. L. Goldstein, J. R. Faust, J. H. Dygos, R. J. Chorvat, and M. S. Brown, *Proc. Natl. Acad. Sci. U.S.A.* **75**, 1877 (1978).
[29] J. T. Billheimer, *J. Lipid Res.* **25**, 412 (1984).

INHIBITION OF ACAT ACTIVITY BY VARIOUS STEROLS AND STEROIDS

Inhibit esterification of exogeneous cholesterol	Inhibit esterification of endogenous cholesterol	No inhibition
Sitosterol[a]	Z-17(20)-Dehydrocholesterol[b]	3-Epicholesterol[a]
Stigmosterol[a]	20-Methylcholesterol[b]	7-Ketocholesterol[c]
Ergosterol[a]	Progesterone[a,d,e]	Estriol[e]
Lanosterol[a]	7-Keto-20-oxacholesterol[c]	Corticosterone[e]
Cycloartenol[a]	Pregn-5-en-3β-ol[a]	Cortisol[e]
	Androsterone[d]	17α-Hydroxyprogesterone[e]
	Dihydrotestosterone[d]	
	Testosterone[e]	
	Pregnenolone[e]	
	5α-Dihydroprogesterone[e]	
	Deoxycorticosterone[e]	

[a] D. M. Tavani, W. R. Nes, and J. T. Billheimer, *J. Lipid Res.* **23,** 774 (1982).
[b] J. T. Billheimer, *J. Lipid Res.* **25,** 412 (1984).
[c] J. L. Goldstein, J. R. Faust, J. H. Dygos, R. J. Chorvat, and M. S. Brown, *Proc. Natl. Acad. Sci. U.S.A.* **75,** 1877 (1978).
[d] A. H. Lichtenstein and P. Brecher, *Biochim. Biophys. Acta* **751,** 340 (1983).
[e] E. R. Simpson and M. F. Burkhart, *Arch. Biochem. Biophys.* **200,** 79 (1980).

cholesterol but have little effect on the esterification of endogenous cholesterol.[20] Those compounds which inhibit the esterification of endogenous cholesterol have one or two similarities in structure; either a shortened side chain of three carbons or less or a change in the configuration at C-20 from that found in cholesterol. These compounds all inhibit ACAT activity by 50% at concentrations of 100 μM or less. Their mode of inhibition is not known. In the case of progesterone the inhibition is reversible; progesterone does not bind to the microsomes and no inhibition is observed in isolated microsomes.[26] Each of the four steroids (e.g., estriol) which either shows no, or only slight, inhibition at a concentration of 100 μM has an extra oxygen in its structure which may reduce its inhibitory activity.

Polyoxyethylated cholesterol, a water-soluble compound, inhibits ACAT activity in fibroblasts by 50% at about 25 μM.[30]

Recently a series of compounds have been synthesized by Sandoz, Inc. which inhibit ACAT activity. One of these, N-(1-oxo-9-octadecenyl)-DL-tryptophan(z) ethyl ester (compound 57-118), has shown to be a competitive inhibitor of intestinal ACAT with respect to acyl-CoA.[31] How-

[30] C. H. Fung and A. K. Khachadurian, *J. Biol. Chem.* **255,** 676 (1980).
[31] J. G. Heider, C. E. Pickens, and L. A. Kelly, *J. Lipid Res.* **24,** 1127 (1983).

ever, these inhibitors apparently do not affect triglyceride, phospholipid, or retinol ester synthesis from acyl-CoA.[31a]

Activation

Suckling *et al.*[32] have shown that ACAT activity is increased by an ATP-dependent process which is consistent with protein phosphorylation. Whether ACAT or another membrane component is phosphorylated is not known.

Upon the addition of exogenous cholesterol in the form of liposomes, a preincubation period is required for the uptake of cholesterol by microsomes before esterification can take place.[16] Nonspecific lipid transfer protein (sterol-carrier protein$_2$) increases the rate of this transfer process.[33]

[31a] A. C. Ross, C. J. Go, J. G. Heider, and G. H. Rothblat, *J. Biol. Chem.* **259**, 815 (1984).
[32] K. E. Suckling, E. F. Strange, and J. M. Pietschy, *FEBS Lett.* **151**, 111 (1983).
[33] J. M. Trzaskos and J. L. Gaylor, *Biochim. Biophys. Acta* **751**, 52 (1983).

[15] Sterol Carrier Protein

By MARY E. DEMPSEY

The protein originally called squalene and sterol carrier protein (SCP) was discovered during attempts to purify rat liver membrane-bound enzymes catalyzing late steps in cholesterol biosynthesis, i.e., from squalene to cholesterol.[1,2] SCP is now known to be a highly abundant, ubiquitous protein having multifunctional roles in the regulation of lipid metabolism and transport.[3,4] One of these roles is its activation of liver membrane-bound enzymes catalyzing cholesterol synthesis from lanosterol.[5] The spectrophotometric assay for SCP activity described in this chapter is based on the latter functional property, i.e., activation of membrane-bound Δ^7-sterol Δ^5-dehydrogenase.[3] Rat liver SCP also undergoes a dramatic diurnal change in amount, varying from 1 to 7 mg/g liver and

[1] M. C. Ritter and M. E. Dempsey, *J. Biol. Chem.* **252**, 5381 (1971).
[2] M. E. Dempsey, this series, Vol. 15, p. 505.
[3] M. E. Dempsey, K. E. McCoy, H. N. Baker, A. Dimitriadou-Vafiadou, T. Lorsbach, and J. B. Howard, *J. Biol. Chem.* **254**, 1867 (1981).
[4] M. E. Dempsey, *Curr. Top. Cell. Regul.* **24**, 63 (1984).
[5] M-K. H. Song and M. E. Dempsey, *Arch. Biochem. Biophys.* **211**, 523 (1981).

peaking during the dark period of a 12-hr dark, 12-hr light cycle.[6] The large scale, two-step purification scheme for SCP outlined here was developed by taking advantage of the unusual abundance of SCP in the cytosolic fraction during the dark period of the diurnal cycle.[3,6]

Assay Method

Principle. The conversion of Δ^7-cholestenol (5α-cholest-7-en-3β-ol) to $\Delta^{5,7}$-cholestadienol (cholest-5,7-dien-3β-ol) by membrane-bound Δ^7-sterol Δ^5-dehydrogenase requires SCP, oxygen, and NAD as cofactors. In this assay the drug, AY-9944, is used to block conversion of $\Delta^{5,7}$-cholestadienol to cholesterol, resulting in accumulation of the dienol.[7] The dienol has a distinctive ultraviolet light absorption pattern with a major peak at 281.5 mm.[8] The assay is performed under conditions where dienol formation is directly proportional to SCP concentration.

Reagents

SCP, homogeneous protein (150 μg/ml) or liver cystolic fraction (15 mg/ml) in 0.1 M potassium phosphate buffer, pH 7.4

Δ^7-Cholestenol (102 mg) (Steraloids, Inc.) and AY-9944 (0.3 mg) (Ayerst Laboratories Inc.) stock solution in 600 ml propylene glycol

Δ^7-Sterol Δ^5-dehydrogenase (1.5–1.8 mg/ml) in phosphate buffer, pH 7.4; the microsomal fraction (Fraction C) prepared from homogenates of female rat livers and washed as described previously[2]

NAD, 10 mM in potassium phosphate buffer, pH 7.4

Ethanol, 95% (v/v)

KOH, pellets

Cyclohexane, spectral grade

Procedure. SCP assays are performed in 1.5 × 10-cm tubes coated with polyethylene glycol.[9] The stock solution of Δ^7-cholestenol/AY-9944 (0.2 ml; final concentration in assay 58 μM and 120 mM, respectively) is added to 1.1 ml of 0.1 M potassium phosphate buffer, pH 7.4, containing 1 mM NAD and SCP (10–150 μg homogeneous protein or 0.1–15 mg of the cytosolic fraction). An identical reaction mixture, but without SCP, is

[6] D. M. McGuire, C. D. Olson, H. C. Towle, and M. E. Dempsey, *J. Biol. Chem.* **259,** 5368 (1984).

[7] D. Dvornik, M. Kraml, J. Dubuc, M. Givner, and R. Guadry, *J. Am. Chem. Soc.* **85,** 3309 (1963).

[8] M. E. Dempsey, J. D. Seaton, G. J. Schroepfer, Jr., and R. W. Trockman, *J. Biol. Chem.* **239,** 1381 (1964).

[9] K. J. Kramer, P. E. Dunn, R. C. Peterson, H. L. Seballos, L. L. Sanburg, and J. H. Law, *J. Biol. Chem.* **251,** 4979 (1976).

TABLE I
LARGE-SCALE PURIFICATION OF RAT LIVER SCP

Fraction	Volume (ml)	Total protein (mg)	Protein yield (%)	Total activity (units[a])	Yield of activity (%)	Specific activity (units/mg)	Purification factor (fold)
105,000 g Supernatant	100	1,960	100.0	20,972	100	10.7	1.0
Sephadex G-75, combined fractions (107–117, Fig. 1)	150	133	8.5	13,778	65	103.6	9.6
DEAE-Cellulose, combined fractions (50–62, Fig. 2)[b]	80	117	7.5	13,829	65	118.2	11.0

[a] A unit of SCP activity is the increase over the microsomal enzyme fraction without SCP of 1 nmol $\Delta^{5,7}$-cholestadienol synthesized in 45 min.

[b] Fractions 15–40, Fig. 2 (i.e., DEAE-peak I) are devoid of SCP functional activity after a second passage through the DEAE-cellulose column to remove traces of SCP.

also prepared. Each mixture is incubated at 37° for 15 min in a covered Dubnoff shaker. The microsomal enzyme fraction (0.2 ml) is then added and the incubation continued for 45 min. The reaction is stopped by adding 1.5 ml ethanol and two KOH pellets. Sterols are extracted into 3 ml of cyclohexane by vigorous mixing for 1 min. The organic layer is removed and scanned from 250 to 310 nm in a recording spectrophotometer. Each extract is read against a cyclohexane blank prepared from an identical reaction mixture to which ethanol–KOH is added prior to addition of the microsomal enzyme fraction. The amount of $\Delta^{5,7}$-cholestadienol synthesized is determined from the difference in absorbance between 321 and 281.5 nm and $\varepsilon = 11.8\ \text{m}M^{-1}\ \text{cm}^{-1}$ for $\Delta^{5,7}$-cholestadienol in cyclohexane at 281.5 nm.[8] The level of Δ^7-sterol Δ^5-dehydrogenase activity exhibited by the microsomal enzyme fraction without SCP is less than 1 nmol/45 min and cannot be detected in many preparations.[10] A unit of SCP activity is defined as the increase over the microsomal enzyme fraction without SCP of 1 nmol $\Delta^{5,7}$-cholestadienol synthesized in 45 min. The specific activity is expressed as units per milligram SCP (Table I).

The assay is linear up to 60 min and proportional to homogeneous SCP up to 10 units. Duplicate analyses vary by less than ±2%. The assay is a modification of the one described previously[2] and gives identical results to the more time-consuming isotopic derivative assays.[2] Lanosterol may also be used in place of Δ^7-cholestenol in the assay.[5] Interferences by extractable ultraviolet light absorbing substances in crude samples are

[10] M. C. Ritter and M. E. Dempsey, *Proc. Natl. Acad. Sci. U.S.A.* **70**, 265 (1973).

avoided by the preparation of a cyclohexane extract of an inactivated reaction mixture which is used as a blank during the scan of an extract of an identical active mixture.

Purification of Squalene and Sterol Carrier Protein (SCP)

The following procedure for purification of SCP is highly reproducible and yields 100–125 mg homogeneous SCP.[3,11] The procedure also purifies to homogeneity another fatty acid binding protein (DEAE-peak I), which is inactive in cholesterol synthesis.[3] All operations are performed at 4°. Buffer pH values are also measured at 4°. Protein is located in chromatographic fractions by measuring absorbance at 280 nm. Protein concentrations of the cytosolic fraction are determined by the method of Lowry et al.[12] using bovine serum albumin as standard. Solutions containing homogeneous SCP are quantitated using an ε (1% at 280 nm) of 3.4.[3] An ε (1% at 280 nm) of 8.6 is used to quantitate homogeneous DEAE-peak I.[3] Polyacrylamide gel electrophoresis in SDS [10% (w/v) acrylamide and 0.1% (w/v) SDS at pH 7.2] is performed according to Weber and Osborn[13] in the presence of 8 M urea.[3] Polyacrylamide gel electrophoresis in 7.5% (w/v) acrylamide gels at pH 4.3 is carried out as described by Shuster[14] in the presence of 8 M urea.[3]

Preparation of Liver Supernatant Fraction. The liver supernatant (cytosolic) fraction is prepared in potassium phosphate buffer, pH 7.4, from female rats (150–200 g), Sprague–Dawley strain, subjected to a 12-hr dark, 12-hr light cycle for 2 weeks prior to sacrifice.[2,6] They are allowed free access to Purina rat chow and water and are killed at 6 hr into the dark period.[2,6] All glassware is coated with polyethylene glycol.[9]

Separation of Liver Cytosolic Proteins. The supernatant from the 105,000 g centrifugation (100 ml; 1.8 to 2.0 g of protein) is applied to a column of Sephadex G-75 (5.5 × 100 cm) equilibrated with 30 mM Tris–HCl buffer pH 9.0. Column fractions are assayed for protein and SCP activity (Fig. 1). Fractions in the low-molecular-weight region of the column containing SCP activity (e.g., fractions 107–117, Fig. 1) are examined by polyacrylamide gel electrophoresis in SDS and 8 M urea[3,12] and at pH 4.3 in 8 M urea.[3,14–16] Fractions containing only two low-molecular-

---

[11] T. Ishibashi and K. Bloch, *J. Biol. Chem.* **256**, 12962 (1981).
[12] O. H. Lowry, N. J. Rosebrough, A. L. Farr, and R. J. Randall, *J. Biol. Chem.* **193**, 265 (1951).
[13] K. Weber and M. Osborn, *J. Biol. Chem.* **244**, 4406 (1969).
[14] L. Shuster, this series, Vol. 22, p. 412.
[15] If column fractions are monitored by polyacrylamide gel electrophoresis at pH 9.0, only SCP (pI 7.0) is observed; DEAE-peak I (pI 9.0) does not enter the gel, giving a false

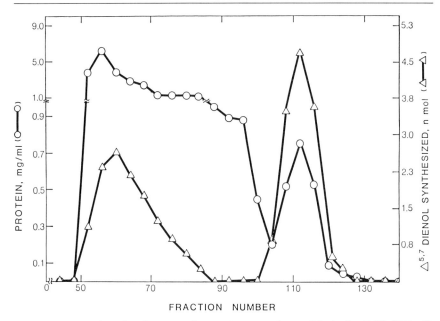

FIG. 1. Separation of rat liver cystolic proteins on a column of Sephadex G-75. SCP with the highest specific activity is present in the low-molecular-weight region of the column; it is associated with only one other protein. SCP present in the high-molecular-weight region is associated with lipid and several other proteins. (From Dempsey et al.[3] by permission of the *Journal of Biological Chemistry*.)

weight proteins (e.g., fractions 107–117, Fig. 1) are combined (140–150 ml, 125–135 mg protein) and applied to a DEAE-cellulose column (2.7 × 10 cm) equilibrated with 30 mM Tris–HCl, pH 9.0 buffer. The flow rate is 30 ml/hr and 7–8 ml fractions are collected. The column is washed with 30 mM Tris–HCl, pH 9.0, buffer until the first protein peak (a mixture of DEAE-peak I and a low level of SCP) is completely eluted (e.g., fraction 43, Fig. 2). The buffer is then changed to 30 mM Tris–HCl, pH 9.0, buffer containing 60 mM NaCl to elute the SCP peak (Fig. 2). The fractions containing DEAE-peak I can be rechromatographed on the DEAE-cellulose column to remove traces of SCP. Homogeneity of SCP and DEAE-

---

picture of the homogeneity of SCP. Although SCP and DEAE-peak I are nearly the same molecular weight (14,000), they are separated by SDS gel electrophoresis where SCP exhibits an M_r of 16,000, probably due to conformational changes and bound fatty acids.

[16] SCP functional activity occurs in both the high- and low-molecular-weight regions of the Sephadex G-75 column; however, SCP with the highest specific activity is present in the low-molecular-weight region and is accompanied by only one other protein (DEAE-peak I).

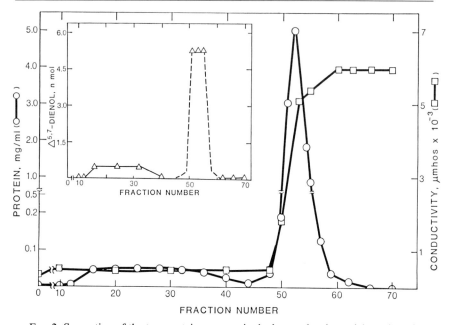

FIG. 2. Separation of the two proteins present in the low-molecular-weight region of the Sephadex G-75 column (Fig. 1) by chromatography on a column of DEAE-cellulose. SCP activity associated with the less abundant protein (DEAE-peak I; fractions 15–40) is due to contamination with SCP, which can be removed by rechromatography. Homogeneous DEAE-peak I is devoid of SCP functional activity. (From Dempsey et al.[3] by permission of the *Journal of Biological Chemistry*.)

peak I is verified by polyacrylamide gel electrophoresis at pH 4.3 in 8 M urea and analytical isoelectrical focusing.[3,14–16] Both proteins are stored dilute (>400 µg/ml) in 30 mM Tris–HCl buffer, pH 9.0 and in polyethylene glycol coated tubes at −70°.

Properties

Stability. Lyophilization or repeated freezing and thawing of SCP causes irreversible loss of SCP activity, as well as irreversible aggregation and precipitation of the protein from solution. Dilute solutions of SCP are stable for 6 months when frozen in liquid nitrogen and stored at −70°.

Structure and Physical Constants. Both homogeneous SCP and DEAE-peak I contain approximately 1 mol each of tightly or covalently bound long-chain fatty acids.[3] Thus, *both* SCP and DEAE-peak I are fatty acid binding proteins and are *both* measured by assay techniques that rely solely on fatty acid binding.[17] Neither SCP nor DEAE-peak I has associated carbohydrate, phosphate, amino sugar, sterol, or other lipid moieties.

The amino acid sequence of rat liver SCP has been determined by chemical and cloned cDNA analysis.[18,19] Rat liver SCP is a single polypeptide chain of 127 amino acids (M_r 14,184; pI 7.0) (Table II). The amino-terminal methionine is acetylated. Liver SCP is not made in a pro-form, i.e., it does not contain a cleavable signal peptide sequence or an internal signal sequence equivalent.[6,18]

Secondary structure predictions indicate that liver SCP has a moderate to high helical content (40% α-helix).[18] Other studies showed that the single free sulfhydryl group is partially buried in the native molecule and not required for functional activity with membrane-bound enzymes catalyzing cholesterol synthesis.[3] The absence of tryptophan residues causes an unusual and diagnostic ultraviolet absorption spectrum of SCP with an absorption maximum at 278 nm and substantial phenylalanine fine structure.[3] Rat liver SCP mRNA is abundant in liver and intestine.[6,8] In fact, it is the most abundant mRNA species in small intestinal mucosa.[18] Because of this unusual abundance, Gordon et al.[18] were readily able to determine the sequence of liver SCP mRNA isolated from intestine. With regard to the structure of other SCP molecules, the amino acid composition of homogeneous human liver SCP is strikingly similar to that of rat liver SCP (Table II). Furthermore, the amino acid composition of yeast SCP is also similar to the liver molecules, with the exception of a higher abundance of glycine and proline in yeast SCP. Related work in the author's laboratory suggests that the amino acid composition and structure of SCP molecules is remarkably conserved throughout nature.

The amino acid composition of the fatty acid binding protein (DEAE-peak I) (M_r 14,000; pI 9.0) that accompanies liver SCP during initial purification steps is distinct from that of SCP but similar to that of the nonspecific lipid transfer protein isolated by several groups,[20] retinol binding protein,[21] SCP_2,[22] and intestinal fatty acid binding protein.[23]

[17] R. K. Ockner, J. A. Manning, and J. P. Kane, *J. Biol. Chem.* **257**, 7873 (1982).
[18] J. I. Gordon, D. H. Alpers, R. F. Ockner, and A. W. Strauss, *J. Biol. Chem.* **258**, 3356 (1983).
[19] K. Takahashi, S. Odani, and T. Ono, *FEBS Lett.* **140**, 63 (1982).
[20] B. J. H. M. Poorthuis, J. J. C. Glatz, R. Akeroyd, and K. W. A. Wirtz, *Biochim. Biophys. Acta* **665**, 256 (1981).
[21] D. E. Ong and F. Cytil, *J. Biol. Chem.* **253**, 828 (1978).
[22] B. J. Noland, R. E. Arebalo, E. Hansbury, and T. J. Scallen, *J. Biol. Chem.* **255**, 4282 (1980).
[23] D. H. Alpers, A. W. Strauss, R. K. Ockner, and J. I. Gordon, *Proc. Natl. Acad. Sci. U.S.A.* **81**, 313 (1984).

TABLE II
AMINO ACID COMPOSITION OF SCP MOLECULES

Amino acid	Rat liver[a]	Human liver[b]	Yeast[b]
Ala	2	4	13
Arg	2	2	5
Asp, Asn	11	12	20
CM-Cys	1	1	1
Glu, Gln	17	22	15
Gly	12	12	20
His	2	2	4
Ile	9	10	8
Lys	17	16	14
Met	6	3	2
Phe	6	8	10
Pro	2	2	11
Ser	6	7	12
Thr	12	12	11
Trp	0	0	0
Tyr	3	2	5
Val	12	10	16
Total residues	127	133	177

[a] From cloned cDNA sequence.[18]
[b] By amino acid analysis.[3,4]

Occurrence, Abundance, and Subcellular Distribution. The data in Table III show the striking abundance of SCP in a variety of rat tissues and serum. SCP is especially abundant in tissues active in lipid metabolism, e.g., liver, adrenal, intestine, ovary, and cardiac muscle. Also there is a higher level (30%) of SCP in female rat tissues versus male tissues. These data (Table III) also document the unusual diurnal variation in amount exhibited by these tissues. The occurrence of SCP in serum is unexpected, since SCP is not made in a pro-form and lacks carbohydrate moieties, common characteristics of exported proteins.

SCP is also present in a variety of mammalian tissues, in addition to those of the rat, e.g., human liver, serum, brain, kidney, leukocytes, and tumors, bovine liver and adrenal, and mouse liver. It appears to be ubiquitous, occurring in lower forms of life, e.g., protozoa, yeast, fungi, and bacteria. SCP was also detected in a variety of cultured cell systems, e.g., human fibroblasts, rat hepatocytes, mouse macrophages, Chinese hamster ovary fibroblasts, 3T3 adipocytes, mouse L1 cells, and many human and mouse malignant cell lines.

TABLE III
AMOUNT OF SCP IN RAT TISSUES DURING THE
DIURNAL CYCLE

Tissue (soluble fraction)	SCP, amount (% of total protein)	
	Dark[a]	Light[b]
Liver	14.6	1.4
Adrenal	14.6	1.6
Intestine	14.2	1.2
Ovary	14.0	1.1
Cardiac muscle	13.0	0.6
Kidney	9.8	0.6
Lung	6.8	0.5
Brain	2.4	1.0
Spleen	1.8	0.6
Testes	1.5	0.6
Skeletal muscle	1.2	0.8
Adipose	1.0	0.5
Serum	0.4	0.1

[a] Six hours into dark period.
[b] Zero hour of light period.

With regard to subcellular distribution, SCP is distributed between the cytosolic (60%) and mitochondrial (30–35%) fractions of liver and adrenal cells.[24] The remainder of SCP is associated with the nuclear fraction. Little or no SCP is associated with the microsomal fraction. In cardiac muscle there is an even distribution of SCP between the cytosol and mitochondria. Nearly all mitochondrial SCP in liver and adrenal is present in the inner mitochondrial membrane.[24] The role of mitochondrial SCP is not completely understood. In the adrenal it appears involved in movement and conversion of cholesterol to steroid hormones; in caridac muscle it may be involved in transport of fatty acids for energy production.

Functions. Regarding known functions of SCP—first, SCP, as mentioned previously, is an absolute requirement for activation of liver membrane-bound enzymes catalyzing cholesterol synthesis from lanosterol.[5] In addition to SCP, the cofactor requirements for synthesis of cholesterol from lanosterol are NAD, NADPH, and oxygen.[5] Another liver protein, supernatant protein factor (SPF), was shown by Caras and Bloch[25] to

[24] O. M. Conneely, D. R. Headon, C. D. Olson, F. Ungar, and M. E. Dempsey, *Proc. Natl. Acad. Sci. U.S.A.* **81**, 2970 (1984).
[25] I. W. Caras and K. Bloch, *J. Biol. Chem.* **254**, 11816 (1979).

function with the membrane-bound enzyme system catalyzing the two-step conversion of squalene, a 30-carbon atom, unsaturated hydrocarbon, to lanosterol, a 30-carbon atom sterol. Thus, only two protein activators are required for synthesis of cholesterol from squalene, SPF, and SCP. Two groups[22,26] reported that another protein, called SCP_2 (DEAE-peak I), functions in sterol biosynthesis; however, the presence of SCP in their preparations is likely and has not been excluded. SCP also functions in the metabolism of cholesterol to bile acids[27] and in adrenal steroidogenesis.[24]

The second known function of SCP is that it has a high affinity for long-chain fatty acids and is considered to be an intracellular fatty acid transport protein, like albumin, the plasma carrier for fatty acids.[3,18] SCP also specifically activates membrane-bound enzymes catalyzing long-chain fatty acid metabolism, e.g., fatty acyl-CoA ligase and acyl-CoA: cholesterol acyltransferase (ACAT).[28,29] Several workers showed marked correlations between the level of SCP and rates of fatty acid uptake and utilization for triglyceride and phospholipid synthesis during a variety of physiological manipulations.[30,31] These and other authors also studied anionic drug, dye, heme, and bilirubin binding and intracellular transport by SCP.[32,33]

The third known property of SCP, no doubt related to its functional role in regulation of lipid metabolism and transport, is that SCP undergoes a rapid turnover *in vivo*. Its half-life was estimated to be less than an hour.[6] As shown in Table III, SCP undergoes a remarkable diurnal change in amount, peaking during the dark period.[6] The diurnal variation in SCP amount also occurs in a variety of rat tissues actively catalyzing lipid metabolism and in serum (Table III). Related studies indicated that although the total amount of, for example, liver SCP varies dramatically, the cellular distribution remains about the same. The large changes in SCP amount in liver and many other tissues (Table III) during the diurnal cycle are reflections of changes in the relative synthetic rate of SCP and the degradative rate is probably constant.[6] Related findings indicate food

[26] J. M. Trzaskos and J. L. Gaylor, *Biochim. Biophys. Acta* **751**, 52 (1983).
[27] G. Grabowski, K. E. McCoy, G. C. Williams, M. E. Dempsey, and R. F. Hanson, *Biochim. Biophys. Acta* **441**, 380 (1976).
[28] H. A. Daum and M. E. Dempsey, *Fed. Proc., Fed. Am. Soc. Exp. Biol.* **38**, 311 (1979).
[29] H. A. Daum and M. E. Dempsey, *Fed. Proc., Fed. Am. Soc. Exp. Biol.* **39**, 1643 (1980).
[30] S. Mishkin, L. Stein, G. Fleischner, Z. Gatmaitan, R. Gluck, and I. M. Arias, *Am. J. Physiol.* **228**, 1634 (1975).
[31] R. K. Ockner, D. A. Burnett, N. Lysenko, and J. A. Manning, *J. Clin. Invest.* **64**, 172 (1979).
[32] B. Ketter, E. Tipping, J. F. Hackney, and D. Beale, *Biochem. J.* **155**, 511 (1976).
[33] D. Trulzsch and I. M. Arias, *Arch. Biochem. Biophys.* **209**, 433 (1981).

consumption and related hormonal events, rather than photoperiod, are primarily responsible for the diurnal pattern in amount and synthesis of SCP.[34] When the functional activity of liver SCP mRNA was measured during the dark–light cycle, no significant changes were found in SCP mRNA levels. Since SCP mRNA is available for translation at all times, the diurnal variation in liver SCP synthesis must involve mechanisms regulating the efficiency of translation of SCP mRNA.[6] These mechanisms are currently under investigation.

Acknowledgments

The author is grateful to Carol Olson for her excellent technical assistance. These studies were supported by the American Cancer Society (BC-378) and the National Heart, Lung, and Blood Institute (HL-28176).

[34] P. A. Stewart, C. D. Olson, and M. E. Dempsey, *Fed. Proc., Fed. Am. Soc. Exp. Biol.* **43**, 1719 (1984).

[16] Epoxide Hydrolases in the Catabolism of Sterols and Isoprenoids

By BRUCE D. HAMMOCK, DAVID E. MOODY, and ALEX SEVANIAN

Background Information

Epoxide hydrolases are enzymes which add water to three-membered cyclic ethers known as epoxides, oxiranes, or oxides of olefinic or aromatic compounds. The products of these reactions are diols (aliphatic systems) or dihydrodiols (aromatic systems). These epoxide ether hydrolases (EC 3.3.2.3) were formerly known as epoxide hydrases or hydratases (EC 4.2.1.63). The hydration leads to a product much more polar than the precursor epoxide. Epoxide hydration, in most cases, represents a detoxification or inactivation reaction, but there are cases where the reaction leads to a more toxic material or is involved in the biosynthesis of a biologically active compound. Numerous reviews are available on these enzymes.[1–4]

[1] F. Oesch, *Xenobiotica* **3**, 305 (1972).
[2] A. Y. H. Lu and G. T. Miwa, *Annu. Rev. Pharmacol. Toxicol.* **20**, 513 (1980).
[3] B. D. Hammock, S. S. Gill, S. M. Mumby, and K. Ota, *in* "Molecular Basis of Environ-

There is evidence for at least four forms of epoxide hydrolase in vertebrate systems. One of these forms hydrates leukotriene A_4 and will not be further discussed. There is evidence that steroids and isoprenoids are at least incidental if not physiologically significant substrates for the remaining three forms of epoxide hydrolase. Most of the research in this area has been focused on the microsomal form which is responsible for the hydration of arene oxides. It is clear that this form can hydrate some steroids,[5] but since its assay and purification have been reported in this series, it will not be discussed.[6,7] Some of the assay procedures discussed here can be modified for the analysis of the microsomal enzyme. The solubilized form of the microsomal epoxide hydrolase which occurs in preneoplastic and neoplastic lesions in the liver can be assayed by modifications of the techniques used for the analysis of the membrane bound form of the microsomal epoxide hydrolase. It is important that this "preneoplastic antigen" not be confused with the cytosolic epoxide hydrolase discussed below.[8,9]

Methods for the analysis of two other forms will be presented. The predominantly cytosolic form hydrates a variety of epoxides on aliphatic systems with the epoxides of fatty acids being very rapidly hydrolyzed.[10,11] Although epoxides of isoprenoid compounds such as the epoxides of squalene or lanosterol are more slowly metabolized, these compounds have very low K_m values indicating that the enzyme may have a physiological role in the metabolism of such compounds either produced as by-products of steroid biosynthesis or as dietary components.[3] A partition method using the juvenile hormones for the analysis of the cytosolic epoxide hydrolase from vertebrates is first presented. Subsequently simpler methods are presented based on model substrates.

Evidence is good that there is a microsomal epoxide hydrolase which specifically hydrolyzes the α and β epoxides of cholesterol, but is different from the microsomal epoxide hydrolase which metabolizes arene ox-

---

mental Toxicity" (R. S. Bhatnager, ed.), p. 229. Ann Arbor Sci. Publ., Ann Arbor, Michigan, 1980.

[4] F. P. Guengerich, *Rev. Biochem. Toxicol.* **4**, 5 (1982).

[5] U. Bindel, A. Sparrow, H. Schmassmann, M. Golan, P. Bentley, and F. Oesch, *Eur. J. Biochem.* **97**, 275 (1979).

[6] A. Y. H. Lu and W. Levin, this series, Vol. 52, p. 193.

[7] T. M. Guenthner, P. Bentley, and F. Oesch, this series, Vol. 77, p. 344.

[8] W. Levin, A. Y. H. Lu, P. E. Thomas, D. Ryan, D. E. Kizer, and M. J. Griffin, *Proc. Natl. Acad. Sci. U.S.A.* **75**, 3240 (1978).

[9] M. J. Griffin and K. I. Noda, *Cancer Res.* **40**, 2768 (1980).

[10] S. S. Gill and B. D. Hammock, *Biochem. Biophys. Res. Commun.* **89**, 965 (1979).

[11] N. Chacos, J. Capdevila, J. R. Falck, S. Manna, C. Martin-Wixtrom, S. S. Gill, B. D. Hammock, and R. W. Estabrook, *Arch. Biochem. Biophys.* **223**, 639 (1983).

ides and substrates such as 16,17-epoxyandrosten-3-one.[5,12,13] Thus, a method for the analysis of this enzyme using cholesterol epoxide is presented.

Epoxide hydration in insects has received much less attention and several reviews cover the available information.[14,15] However, it appears in the few species so far studied that there are several hydrolases predominantly associated with the microsomal fraction which hydrate a number of substrates including cyclodiene insecticides. It is interesting that the most active epoxide hydrolase so far purified has come from an insect.[16] The assays presented here and elsewhere for vertebrate systems certainly need to be applied to insect systems. However, in relation to this volume the activity of primary importance is the one which hydrates the terpenoid insect juvenile hormones as originally reported by Slade and Zibitt.[17] In this regard an assay for epoxide hydrolase activity using the natural juvenile hormones of insects is presented.

Assay with Juvenile Hormone

General Considerations

The partition assay for epoxide hydrolase activity described here is more tedious than some of the partition methods described below. However, use of these substrates led to the discovery of the cytosolic epoxide hydrolase in vertebrates, and they represent commercially available model compounds for the hydration of terpenoid epoxides. Entomologists are fortunate in having these compounds available for studying epoxide hydrolase activity with intrinsec substrates. For such studies, the only viable alternatives to this method are a modification of the TLC methods discussed in this volume [32] and with regard to cholesterol epoxide below, or HPLC or GLC methods which are quite laborious.

As discussed in this volume [32] it is important to ensure that epoxide hydration is the only pathway involved in the system being investigated. In the vertebrates studied, most of the carboxyesterase activity on the

---

[12] T. Watabe, M. Kanai, M. Isobe, and N. Ozawa, *J. Biol. Chem.* **256**, 2900 (1981).
[13] W. Levin, D. P. Michaud, P. E. Thomas, and D. M. Jerina, *Arch. Biochem. Biophys.* **220**, 485 (1983).
[14] B. D. Hammock and G. B. Quistad, *Prog. Pestic. Biochem.* **1**, 1 (1983).
[15] B. D. Hammock, *in* "Comprehensive Insect Physiology, Biochemistry and Pharmacology" (G. A. Kerkut and L. I. Gilbert, eds.), Chapter 12. Pergamon, Oxford, 1984 (in press).
[16] C. Mullin and C. F. Wilkinson, *Insect Biochem.* **10**, 681 (1980).
[17] M. Slade and C. H. Zibitt, *in* "Insect Juvenile Hormones: Chemistry and Action" (J. J. Menn and M. Beroza, eds.), p. 155. Academic Press, New York, 1972.

juvenile hormones is in the microsomes while in the Lepidoptera studied it is in the cytosol. The epoxide hydrolase activity is in the opposite fraction in each case. If esterase activity is a problem, O,O-diisopropyl phosphorofluoridate (DFP), or O,O-diethyl-O-p-nitrophenyl phosphate (paraoxon) are suggested as inhibitors. For insects O-ethyl-S-phenylphosphoramidothioate (EPPAT) and 3-octylthio-1,1,1-trifluoro-2-propanone (OTFP) should be added to the list.[18,19] Since glutathione reacts very poorly with trisubstituted, aliphatic epoxides, formation of the conjugates of the juvenile hormones is usually not a problem.

Enzyme Preparation

In vertebrate tissue the enzyme referred to as cytosolic epoxide hydrolase is primarily located in the cytosol fraction, but similar enzyme activity has been described in particulate fractions.[20–22] Therefore, initial studies should consider the ratio of soluble to particulate activity. For common rodents the cytosolic fraction can be prepared by standard protocols. Liver is homogenized in sodium phosphate buffer pH 7.4 with an ionic strength of 0.2 M, and the cytosolic fraction is prepared from the postmitochondrial supernatant. While the ionic strength and pH of the buffer are important for reproducible assays, we have found other buffers, such as 0.25 M sucrose, 10 mM Tris–HCl (pH 7.4), appropriate for tissue preparation. Recently, the purification of cytosolic epoxide hydrolase from rabbit,[23] human,[24] and mouse[25] liver has been described. These references should be consulted for preparation of purified enzyme. Although this discussion applies to murine epoxide hydrolase, it represents a good starting point for optimizing assay conditions in insects.

For studies of epoxide hydration in tissues of lepidopteran (*Trichoplusia ni*) tissues were homogenized in sodium phosphate buffer at pH 7.4 with an ionic strength of 0.2 M with 0.01% phenylthiourea to inhibit tyrosinases. Activity was either examined in a 20,000 g supernatant with esterase inhibitors or in a postmitochondrial pellet. Assay conditions should be optimized for the particular study planned.

[18] T. C. Sparks and B. D. Hammock, *Pestic. Biochem. Physiol.* **14**, 290 (1980).
[19] T. C. Sparks, B. D. Hammock, and L. M. Riddiford, *Insect Biochem.* **13**, 529 (1983).
[20] S. S. Gill and B. D. Hammock, *Biochem. Pharmacol.* **30**, 2111 (1981).
[21] T. M. Guenthner, U. Vogel-Bindel, and F. Oesch, *Arch. Toxicol., Suppl.* **5**, 365 (1982).
[22] F. Waechter, P. Bentley, F. Bieri, W. Staubli, A. Vozkl, and H. D. Fahimi, *FEBS Lett.* **158**, 225 (1983).
[23] F. Waechter, M. Merdes, F. Bieri, W. Staubli, and P. Bentley, *Eur. J. Biochem.* **125**, 457 (1982).
[24] P. Wang, J. Meijer, and F. P. Geungerich, *Biochemistry* **21**, 5769 (1982).
[25] S. S. Gill, *Biochem. Biophys. Res. Commun.* **112**, 763 (1983).

Principle

The enzyme is assayed by the conversion of the epoxides of juvenile hormone I, II, or III to their corresponding diols. The method is described for JH I, but is equally applicable to any of the commercially available compounds. Following the enzyme reaction, the majority of the JH is extracted with a hydrocarbon solvent (isooctane or dodecane is used to reduce volatility) while the majority of the product diol remains in the aqueous phase. The proportion of the diol in the aqueous phase is increased by incorporating methanol.[26] General principles of partition assays are described in more detail elsewhere.[27,28]

Reagents

The juvenile hormones such as JH I [methyl-(2E,6E,10-cis)-10,11-epoxy-3,11-dimethyl-7-ethyl-2,6-tridecadienoate] are available as racemic mixtures with a tritium in the 10 position from New England Nuclear while the unlabeled materials are available from many suppliers including Calbiochem and Sigma. The labeled and unlabeled materials should be mixed to give the desired substrate concentration (see this volume [32]). For routine work a concentration of 1×10^{-4} M in absolute ethanol containing approximately 10,000–20,000 cpm per microliter is useful. For developing a standard curve JH diol can be prepared chemically[17] or enzymatically as discussed below. The only variation would be to extract the aqueous methanol several times with pentane to remove all traces of JH. Varying ratios of JH and JH diol can be prepared and subjected to the partition assay for the development of a standard curve. In addition one needs methanol (ethylene or propylene glycols can be substituted), isooctane, or other hydrocarbon solvent, silicon or polyethylene glycol coated tubes, and a scintillation solution capable of handling aqueous solvents.

Assay Procedure

Place 100 μl of enzyme solution in a round bottom 6 × 50 or 10 × 75-mm glass tube treated with a 1% solution of polyethylene glycol (PEG 20,000) to reduce substrate binding to the glass. Equilibrate the tube to 30 or 37.5° (insect or vertebrate) and add substrate to initiate the reaction. The substrate is added with a 50-μl Hamilton syringe in a repeating dispenser designed to give 1 μl yielding a final substrate concentration of 1×10^{-6} M (this is just below the critical micellar concentration for JH I). The

---

[26] B. D. Hammock and T. C. Sparks, *Anal. Biochem.* **82**, 573 (1977).
[27] M. T. Bush, this series, Vol. 77, p. 353.
[28] B. D. Hammock and R. N. Wixtrom, *Adv. Biochem. Pharmacol.* (in press).

reaction is terminated (usually at 0 to 30 min) by the addition of 200 µl of methanol. Isooctane (250 µl) is then added with a repeating dispenser and the mixture vortexed vigorously and then centrifuged. Under these conditions one can expect over 76% of the diol to be in the aqueous phase with less than 5% contamination by the epoxide. Initially one can sample 100 µl of both phases and determine radioactivity by liquid scintillation counting, but for routine work only the aqueous phase needs to be sampled.

Most systems give only a small, consistent level of quenching so that cpm can be used in calculations. One should run the control where methanol, but no isooctane is added to provide the maximum number of counts, determine machine background, and do a zero time experiment to yield the assay background. Spontaneous hydration of these substrates is very small. The percentage of diol formed can be determined from a standard curve by regression or by simultaneous equations.

Assay with *trans*- and *cis*-Stilbene Oxide

General Considerations

A specific and rapid partition assay for cytosolic epoxide hydrolase recently has been devised which uses *trans*-stilbene oxide (TSO) as substrate.[29,30] The geometric isomer, *cis*-stilbene oxide (CSO), can be used to assay the common microsomal epoxide hydrolase. This assay will also be considered at this time. The use of these geometrical isomers to compare the cytosolic and microsomal epoxide hydrolase is attractive since they have similar physical properties. The stilbene oxides can conjugate with glutathione. Tissues which contain significant amounts of glutathione such as rat liver cytosol, should be dialyzed overnight before use. Otherwise, the principles for preparation of the cytosolic fraction are the same as discussed previously. Microsomes can be prepared by conventional methods. The principles of the partition assay are similar as those discussed for juvenile hormone. The differentiation of the cytosolic and microsomal activities depends upon substrate specificity, pH, and subcellular fractions used. Cytosolic epoxide hydrolase is therefore defined as the conversion of TSO to its diol at pH 7.4 using cytosolic fractions. Microsomal epoxide hydrolase is defined as the hydrolysis of CSO to its diol at pH 9.0 using microsomal fractions.

[29] S. S. Gill, K. Ota, and B. D. Hammock, *Anal. Biochem.* **131,** 273 (1983).
[30] D. E. Moody, D. N. Loury, and B. D. Hammock, submitted.

Reagents

The preparation of TSO and CSO and their tritiated counterparts has been previously described in detail and will not be repeated here.[29] The substrates should be dissolved in ethanol at a concentration of 5 mM with enough radiolabeled substrate to give 10,000–20,000 cpm per microliter. The tissue to be assayed should be suspended in 76 mM sodium phosphate (pH 7.4) for the cytosolic epoxide hydrolase or in 100 mM Tris–HCl (pH 9.0) for the microsomal epoxide hydrolase. Dodecane or isooctane (with or without previous addition of up to 25% methanol) is used for the extraction, and scintillation solution appropriate for aqueous solutions is needed.

Assay Procedure for Cytosolic Epoxide Hydrolase

Place 100 μl of enzyme preparation in a round bottomed 10 × 75 or 13 × 100-mm glass tube and preincubate for 1 min at 37°. Initiate the reaction by adding 1 μl of [^{3}H]TSO at 5 mM using a 50-μl Hamilton syringe with a repeating dispenser. The reaction is stopped by adding 200 μl of dodecane or isooctane followed by vigorous vortexing. Centrifuge the tubes briefly to ensure phase separation and then withdraw a 50-μl aliquot of the aqueous phase for counting. Blanks are run simultaneously by adding buffer instead of enzyme. Standards are prepared by adding 1 μl of the radiolabeled substrate directly to the scintillation cocktail. Activity is calculated using the assumption that 92% of the diol is left in the aqueous phase. This value may be increased by previous addition of up to 25% methanol in the extracting solvent.

Assay Procedure for Microsomal Epoxide Hydrolase

The procedure for the microsomal epoxide hydrolase is essentially the same as that described above, so only the differences will be pointed out. A similar volume of enzyme is used, but it must be diluted in 100 mM Tris–HCl (pH 9.0). After equilibration of the enzyme 1 μl of [^{3}H]CSO at 5 mM is added to start the reaction. All the following procedures are identical to those described above.

Assay with Cholesterol Epoxide

General Considerations

It is now known that a distinct enzyme(s) exists in the microsomes of vertebrates which specifically hydrates Δ^5-steroid epoxides. This enzyme has been shown to be distinct from the more commonly studied microsomal epoxide hydrolase by studies on its substrate specificity, pH optima, inhibition, induction, and immunochemical properties.[12,13,31,32] While studies comparing this enzyme with cytosolic epoxide hydrolase are few, little or no activity for hydrolysis of the Δ^5-steroids in the cytosol has been found suggesting that these two enzymes are also distinct.[32] Assays for cholesterol epoxide hydrolase have been described using isotope dilution as detected by GC-MS,[33,34] HPLC, or GC analysis of column separated extracts,[35] capillary GC,[34] and TLC separation of radiolabeled substrate and product.[13,32] The GC-MS assay is quite precise and sensitive and can be useful for assays when the recovery of activity is low or when radiolabeled substrates are not available. The most convenient and still relatively sensitive assay is the TLC method, and such an assay will be discussed below.

Enzyme Preparation

Cholesterol epoxide hydrolase is commonly assayed using microsomal fractions. Sodium cholate detergent extracts have also been found appropriate, and while not studied, postmitochondrial supernates, tissue homogenates, and cell suspensions may also be appropriate. Microsomes can be prepared by any conventional means, but before assay they should be suspended in 100 mM potassium phosphate (pH 7.4), as both the buffer used and its pH have been shown to effect activity.[13,32] At this time a method for purifying the cholesterol epoxide hydrolase has not been described.

Principle

The enzyme is assayed by the conversion of cholesterol epoxide (other 5,6-epoxysteroids may also be used) to the corresponding cholesterol triol (5,6-glycol) using radiolabeled substrate. The products formed then are separated from the substrate by TLC, both recovered by scraping, and the radiolactivity quantitated. This method was first described by Jerina *et al.*[36] for the assay of a number of epoxides, and subsequently has been modified specifically for the cholesterol epoxide hydrolase.

[31] L. Aringer and P. Eneroth, *J. Lipid Res.* **15,** 389 (1974).
[32] A. Sevanian, R. A. Stein, and J. F. Mead, *Biochim. Biophys. Acta* **614,** 489 (1980).
[33] N. L. Petrakis, L. D. Gruenke, and J. C. Craig, *Cancer Res.* **41,** 2563 (1981).
[34] J. Gumulka, J. S. Pyrek, and L. L. Smith, *Lipids* **17,** 197 (1982).
[35] L.-S. Tsai and C. A. Hudson, *J. Am. Oil Chem. Soc.* **58,** 931 (1981).
[36] D. M. Jerina, P. M. Dansette, A. Y. H. Lu, and W. Levin, *Mol. Pharmacol.* **13,** 342 (1977).

Reagents

The synthesis of unlabeled or radiolabeled cholesterol epoxide and cholesterol triol from cholesterol can be achieved by a few different methods which will not be described further.[12,13,31,32] It is important that the relative amounts of the α- and β-epoxide are known for their rates of spontaneous and enzymatic hydrolysis are different. Therefore, pure isomers are the preferred substrates. The delivery solvent for the cholesterol epoxide is also critical and maximal activity can be achieved by using acetonitrile leading to a final concentration of 6.25%. Siliconized test tubes are recommended to prevent adherance of the substrate. TLC separation has most commonly been carried out using Whatman LK5DF silica gel plates.

Assay Procedure

Add 150 μl of microsomes (200–500 μg/ml) in 100 mM potassium phosphate (pH 7.4). Preincubate microsomes for 1 min at 37° and then add cholesterol epoxide (final concentration 20–100 μM) dissolved in enough acetonitrile to give a final concentration of 6.25%. After 2–20 min incubation at 37° the reaction is terminated by addition of 50 μl of tetrahydrofuran containing cold carrier substrate and product. Use of unlabeled standards is particularly important if low amounts of product are expected. Aliquots (up to 70 μl) are then quickly transferred to the preabsorbant zone of the silica gel plate (Whatman LK5DF). The plate is then developed in benzene–ethyl acetate (3:2). Spots can then be visualized with iodine vapor and are then scraped for quantitation by scintillation counting.

[17] Biosynthesis and Interconversion of Sterols in Plants and Marine Invertebrates

By L. J. Goad

The broad outline of the sterol biosynthetic pathway leading to plant sterols is now understood[1] although some facets remain to be clarified. However, many different sterols, varying in the structure of the side chain, the position of double bonds and the presence of methyl groups at

---

[1] L. J. Goad, *in* "Lipids and Lipid Polymers in Higher Plants" (M. Tevini and H. K. Lichtenthaler, eds.), p. 46. Springer-Verlag, Berlin and New York, 1977.

C-4 and C-14, have been characterized from higher plant and algal sources.[2] Therefore it is apparent that the phytosterol biosynthetic pathways operating in different plant families may vary significantly in some details. For example, it is now recognized that several mechanisms are operative in different families of higher plants and algae for the introduction of the alkyl group at C-24 which is characteristic of most phytosterols.[3] Thus it can be anticipated that the identification of the sterols and the elucidation of the sterol biosynthetic pathway operating in various species of higher plants and algae may prove to be of phylogenetic and perhaps chemotaxonomic value.

A rich array of sterols has been found in the marine environment. Some of the sterols have modified ring systems while many others possess exotic side chain structures exhibiting additional alkyl group substituents at carbons other than C-24 and which are not encountered in terrestrial species.[4-6] Many of the sterols, although found in invertebrates, are probably derived via the food chain from an algal producer. These "unusual" marine sterols present intriguing challenges regarding both the organisms of their origin and the biosynthetic routes which lead to their production.

This chapter describes some of the methodology available for studies on plant and marine invertebrate sterol biosynthesis and metabolism. The many research groups with interests in sterol biochemistry have developed a wide range of excellent analytical techniques to suit their own particular needs. It is not possible to describe all these techniques in this chapter and only some of them are cited where appropriate. The methods described here are largely the ones which we have employed in our work but they are not claimed to have any particular merits when compared to the alternative techniques available.

Choice and Preparation of Labeled Precursors

General Comments. The choice of a suitable labeled precursor will be of course determined mainly by the particular problem under investigation. So far radioactively labeled precursors have been utilized for most studies on plant and invertebrate sterol biosynthesis. However, for some investigations on plant sterol biosynthesis, particularly on mechanistic

---

[2] W. R. Nes and M. C. McKean, "Biochemistry of Steroids and other Isoprenoids". University Park Press, Baltimore, Maryland, 1977.

[3] L. J. Goad, F. F. Knapp, J. R. Lenton, and T. W. Goodwin, *Lipids* **9**, 582 (1974).

[4] F. J. Schmitz, in "Marine Natural Products" (P. J. Scheuer, ed.), Vol. 1, p. 241. Academic Press, New York, 1978.

[5] C. Djerassi, *Pure Appl. Chem.* **53**, 873 (1981).

[6] L. J. Goad, *Pure Appl. Chem.* **53**, 837 (1981).

problems, stable isotopes are superior,[7,8] and they are now being increasingly used as suitable labeled compounds and sensitive analytical instrumentation (GLC-MS, NMR) become more widely available.

In studies designed to determine the ability of an organism to synthesize sterols *de novo* or to obtain labeled intermediates for isolation and identification the obvious precursors are ^{14}C-labeled acetic acid or mevalonic acid. These compounds can be purchased commercially and as they are water soluble they can be easily administered to both plants and marine invertebrates. However, the percentage incorporation of [^{14}C]acetic acid into sterol is often considerably lower than that observed with [2-^{14}C]mevalonic acid. This may be due to any of several reasons which include (1) the acetate must be converted into acetyl-CoA by a thiokinase which may be a rate limiting step; (2) excessive dilution of the [^{14}C]acetic acid with an endogenous pool of acetate (or acetyl-CoA); (3) utilization of the acetyl-CoA by competing metabolic pathways; (4) the step catalyzed by 3-hydroxy-3-methyglutaryl-CoA (HMG-CoA) reductase could be rate limiting in the organism under study. However, while the use of [2-^{14}C]mevalonic acid can be advantageous in order to obtain higher incorporations of radioactivity into compounds and so facilitate their characterization, this may give a false impression that sterol synthesis is proceeding at a considerably faster rate than is in fact the true physiological situation. Also, the introduction into an organism of a significant amount of labeled exogenous precursor, such as mevalonic acid, which enters the biosynthetic pathway after a primary regulatory point (e.g., HMG-CoA reductase) could lead to a distortion of the flux of precursors through the pathway. This may lead to an abnormal accumulation of an intermediate at a secondary control point or even to a diversion of material resulting in increased production of by-products on side pathways. While this phenomenon may be used to advantage for the identification of some metabolites it again does not necessarily give a true picture of the normal physiological process.

Much attention has been paid to the problem of accurately measuring the true absolute rates of sterol synthesis in mammalian tissues and [^{14}C]octanoate and particularly tritiated water have been advocated as the best labeled substrates for such measurements.[9,10] However, methods for measuring absolute rates of sterol synthesis in plant tissues or invertebrate animals have not been reported to date.

[7] S. Seo, Y. Tomita, and K. Tori, *J. Chem. Soc., Chem. Commun.* p. 329 (1978).
[8] S. Seo, A. Uomori, Y. Yoshimura, and K. Takeda, *J. Am. Chem. Soc.* **105**, 6343 (1983).
[9] J. M. Dietschy and J. D. McGarry, *J. Biol. Chem.* **249**, 52 (1974).
[10] J. M. Anderson and J. M. Dietschy, *J Lipid Res.* **20**, 740 (1979).

Most of the labeled sterol precursors which are required to investigate sterol interconversions are not available commercially and consequently they must be synthesized by the user. In general ^{14}C-labeled compounds are preferable to ^{3}H-labeled compounds. The former can be detected more readily by autoradiography and thin-layer chromatography radioscanning at the low levels of radioactivity often incorporated into metabolites produced in incubations with plant or invertebrate preparations. With ^{3}H-labeled compounds there is also the problem that, depending upon the labeling position and the biotransformations to which the compound is subjected, the ^{3}H may become labile in some metabolites thus possibly resulting in partial or even total loss of ^{3}H and a consequent underestimation of incorporation into products. However, tritium-labeled sterols have been very extensively used because the preparation of a ^{3}H-labeled sterol is usually more straightforward, often permits the preparation of compounds of higher specific activity, requires smaller amounts of starting compounds for the synthesis, and it is often less costly than the preparation of the corresponding ^{14}C-labeled compound. Wherever possible the chemical synthesis of a radiolabeled sterol is to be preferred rather than preparation by isolation from plant or animal tissues incubated with a precursor such as [^{14}C]acetate or [2-^{14}C]mevalonate. Radiochemical purity is more likely to be assured by the former method.

When precursors labeled with the stable isotopes ^{2}H or ^{13}C are used the products are analyzed by either mass spectrometry or NMR spectroscopy. Despite the high sensitivity of the mass spectrometer some problems of detection can arise if large amounts of the endogenous product sterol are present in the organism under study. This may cause excessive dilution of the small quantity of a deuterium-labeled product yielded by a low incorporation of precursor and it can be a particular problem with investigations using higher plant tissues.

Outlines of some of the methods available for the preparation of labeled sterols and other precursors are given below. The methods given and the references cited may refer to the labeling of particular compounds. However, appropriate modifications of the described procedures should permit the preparation of other labeled sterols to suit specific needs. The safe handling of the high levels of radioactivity required for these procedures demands adequate radiochemical laboratory facilities and precautions to prevent personal and laboratory contamination.

^{3}H-Labeled Species of Mevalonic Acid. Mevalonic acid (MVA) labeled stereospecifically with ^{2}H or ^{3}H at C-2, C-4, or C-5 has been used to investigate various steps in the biosynthesis of sterols and other terpenoids which involve introduction or isomerization of double bonds or

hydride migrations.[11,12] When using the ^{3}H-labeled forms of MVA the procedure requires admixture of the appropriate [^{3}H]MVA with [2-^{14}C]MVA to give a ^{3}H : ^{14}C ratio usually in the range 5 : 1 to 10 : 1. Following incubation of this dual labeled MVA with plant or animal preparations the products are isolated and their ^{3}H : ^{14}C ratios determined and converted to ^{3}H : ^{14}C atomic ratios based upon the number of labeled carbon atoms known to be incorporated into the compound in question. This ratio then reveals if retention or loss of tritium atoms has occurred during biosynthesis and by suitable degradative procedures the retained tritium atoms can be located. When investigating sterol synthesis it is usually advantageous to isolate labeled squalene as a reference compound since this will have a ^{3}H : ^{14}C atomic ratio of 6 : 6. Caution is sometimes required in the interpretation of results obtained with (2R)-[2-^{3}H$_1$] and (2S)-[2-^{3}H$_1$]MVA since some loss and the scrambling of tritium label in the C-2 position can occur due to the reversible action of the isopentenyl pyrophosphate-dimethylallyl pyrophosphate isomerase.[12]

Methods are available for the synthesis of (2R)- and (2S)-[2-^{3}H$_1$]MVA, (4R)- and (4S)-[4-^{3}H$_1$]MVA,[13] and (5R)-[5-^{3}H$_1$]MVA.[14] Methods have also been described for the synthesis of [3,4-^{13}C$_2$]MVA[13,16] and [2,3-^{13}C$_2$]MVA.[17]

Squalene. Methods have been published for the chemical synthesis of [^{14}C]squalene and [^{3}H]squalene.[18] It is more conveniently prepared biosynthetically by incubating either [2-^{14}C]MVA or [^{3}H]MVA with a rat liver homogenate under anaerobic conditions.[19] The labeled squalene is then isolated from the nonsaponifiable lipid by column chromatography on alumina (Brockman grade III) and by preparative TLC on silica gel.

2,3-Oxidosqualene. This can be prepared[20] from labeled squalene by treatment with N-bromosuccinimide in aqueous glyme to provide the 2,3-bromohydrin which is purified by preparative TLC (silica gel, ethyl ace-

[11] G. Popják and J. W. Cornforth, *Biochem. J.* **101**, 553 (1966).
[12] L. J. Goad, in "Naturally Occuring Compounds Formed Biologically from Mevalonic Acid" (T. W. Goodwin, ed.), p. 45. Academic Press, London, 1970.
[13] R. H. Cornforth and G. Popják, this series, Vol. 15, p. 359.
[14] R. J. H. Williams, G. Britton, J. M. Charlton, and T. W. Goodwin, *Biochem. J.* **104**, 767 (1967).
[15] G. Popják, this series, Vol. 15, p. 393.
[16] J. A. Lawson, W. T. Colwell, J. I. DeGraw, R. H. Peters, R. L. Dehn, and M. Tanabe, *Synthesis*, p. 279 (1975).
[17] J. I. DeGraw and I. Uemura, *J. Labelled Compd. Radiopharm.* **16**, 547 (1979).
[18] R. G. Nadeau and R. P. Hanzlik, this series, Vol. 15, p. 346.
[19] T. T. Tchen, this series, Vol. 6, p. 505.
[20] E. E. van Tamelen and T. J. Curphey, *Tetrahedron Lett.* p. 121 (1962).

tate–benzene, 5:95). The bromohydrin is then converted into labeled (3R,S)-2,3-oxidosqualene by treatment with ethanolic base and purified by preparative TLC (silica gel, ethyl acetate–hexane, 5:95). An alternative method has been published for the production of 2,3-[4-^{3}H]oxidosqualene.[18]

4,4-Dimethylsterols. Methods are available for the preparation of ^{14}C-labeled 4,4-dimethylsterols which require partial degradation of the side chain and its subsequent reconstruction by the Wittig reaction using appropriate ^{14}C-labeled reagents. Thus, [26,27-^{14}C$_2$]cycloartenol[21] and [28-^{14}C]24-methylenecycloartanol[22] have been prepared for biosynthetic studies.

A simple procedure can be used to introduce tritium into the C-2 position of 4,4-dimethylsterols which requires only a small amount of starting compound and that is of general applicability to many sterols irrespective of ring or side chain structure.[22,23] The 3β-hydroxy group of the sterol is first oxidized with chromic acid to yield the 3-oxo compound which is purified by preparative TLC (silica gel, chloroform–ethanol, 98:2). Tritium is then introduced at C-2 by basic keto–enol exchange in the presence of tritiated water. Two methods can be used for performing this exchange. In the method developed by Klein and Knight[24] the 3-oxo compound is dissolved in benzene–hexane (1:1) and applied to a small column of basic alumina previously deactivated by addition of tritiated water of high specific activity. The 3-oxo compound is left absorbed on the column for a few hours to permit exchange and then eluted and reduced with lithium aluminum hydride to restore the 3β-hydroxy group. Finally the 4,4-[2-^{3}H$_2$]dimethylsterol is purified by preparative TLC to remove any 3α-hydroxy compound produced in the reduction step.

Alternatively the introduction of tritium at C-2 of the 3-oxo compound can be achieved in an enclosed system[23] using a tritiated water–dioxane–potassium hydroxide mixture contained in a closed breakseal tube. At the end of the exchange time the tritiated water–dioxane mixture is transferred via a vacuum system using standard procedures[25] to a second breakseal tube cooled to $-70°$ which can then be sealed and thus permitting recovery of the tritiated water and its reuse for further labeling reactions. The ^{3}H-labeled 3-oxo compound is extracted with petroleum ether and washed with a large volume of water to remove traces of alkali and so

[21] L. Cattel, G. Balliano, F. Viola, and O. Caputo, *Planta Med.* **38**, 112 (1980).
[22] R. Heintz and P. Benveniste, *J. Biol. Chem.* **249**, 4267 (1974).
[23] J. R. Lenton, H. Hall, A. R. H. Smith, E. L. Ghisalberti, H. H. Rees, L. J. Goad, and T. W. Goodwin, *Arch. Biochem. Biophys.* **143**, 664 (1971).
[24] P. D. Klein and J. C. Knight, *J. Am. Chem. Soc.* **87**, 2657 (1965).
[25] "Guide for Users of Labelled Compounds." Radiochemical Centre, Amersham, England, 1972.

prevent exchange of the tritium label. The product is then reduced with lithium aluminum hydride or sodium borohydride to yield the 3β-hydroxy compound which is purified by preparative TLC as outlined above.

When using 4,4-[2-^{3}H$_2$]dimethylsterols it should be noted that the tritium is potentially labile since the biological C-4 demethylation reactions require the production of a 3-oxo intermediate. This exposes the tritium atoms at C-2 to possible keto–enol exchange although this did not appear to be a significant problem in our particular plant studies[23] using these 2-^{3}H$_2$-labeled sterols.

For studying reactions such as the C-24 alkylation step or the opening of the 9β,19-cyclopropane ring in plant sterol biosynthesis which do not involve an oxidation of the 3β-hydroxy group a 4,4-[3α-^{3}H$_1$]dimethyl (or 4α-methyl) sterol can be utilized.[22] The sterol is first oxidized with Jones' reagent to give the 3-oxo compound. The 3β-hydroxy group is then restored by reduction with sodium [^{3}H]borohydride to introduce the 3α-^{3}H followed by purification of the 3β-hydroxy sterol by preparative TLC.

For some investigations sterols labeled at C-24 with tritium have proved useful and methods are described in the literature for the synthesis of [24-^{3}H]cycloartenol[21] and [24-^{3}H]lanosterol.[26]

4α-Methylsterols. Compounds such as 24-methylenelophenol, 24-ethylidinelophenol, and cycloeucalenol have been labeled[23] with tritium by the exchange technique with the 3-oxo derivative as described above for the 4,4-dimethylsterols. However, in these cases, an additional tritium atom is introduced at C-4 to give a 4α-[2,4-^{3}H$_3$]methylsterol. 4α-[3α-^{3}H$_1$]Methylsterols can also be prepared as indicated above.

A method has been developed for the introduction of tritium into the side chain of some sterols with a 24-methylene group such as cycloeucalenol.[23] Treatment of the cycloeucalenol with osmium tetroxide in pyridine followed by addition of aqueous sodium metabisulfite produces the 24,28-dihydroxy derivative which can be purified by preparative TLC (silica gel, CHCl$_3$–MeOH, 100:7). The diol is then cleaved by excess sodium periodate in aqueous dioxane to produce the 24-oxosterol. This compound can be labeled with tritium at C-23 and C-25 by basic exchange in the presence of tritiated water as described above. The cycloeucalenol is then reconstituted by a Wittig reaction using triphenylphosphine methylene bromide and purified by preparative TLC.

4-Demethylsterols. 4-Demethylsterols can be labeled with ^{14}C at positions C-4[27] or C-3[28] and although [4-^{14}C]cholesterol is produced commer-

---

[26] K. H. Raab, N. J. De Souza, and W. R. Nes, *Biochem. Biophys. Acta* **152**, 742 (1968).
[27] T. F. Gallagher and B. Belleau, *J. Am. Chem. Soc.* **73**, 4458 (1951).
[28] P.Ph.L. Ott, A. C. Besemer, and W. H. J. M. Wientjens, *J. Labelled Compd.* **6**, 111 (1970).

cially it is generally not a particularly convenient method for the preparation of labeled phytosterols. 4-Demethylsterols with a double bond in the side chain can be synthesized from appropriate precursors using a ^{14}C-labeled Wittig reagent.[29] For example [28-^{14}C]ergosta-5,7,22,24(28)-tetra-en-3β-ol, [28-^{14}C]ergosta-8,24(28)-dien-3β-ol, and [28-^{14}C]ergosta-7,24-(28)-dien-3β-ol can be synthesized from the corresponding 24-oxosterol precursor.[30] The Grignard reaction can also be employed to introduce labeling into the sterol side chain. For example, this method has been used[31] for the synthesis of (24S)-24-[26-^{14}C]ethylcholesta-5,25-dien-3β-ol and 24-[26-^{14}C]ethylcholesta-5,24-dien-3β-ol. 3β-Tetrahydropyranosyl-(24S)-24-ethyl-27-nor-cholest-5-en-25-one, obtained by degradation of clerosterol [(24S)-24-ethylcholesta-5,25-dien-3β-ol] is reacted with a Grignard reagent generated from [^{14}C]methyl iodide to yield 3β-tetrahydropyranosyl-(24S)-24-[26-^{14}C]ethylcholest-5-en-25-ol which is then converted into the 3β,25-diol by treatment with HCl in tetrahydrofuran. Acetylation of the labeled diol gives 3β-acetoxy-(24S)-24-[26-^{14}C]ethylcholest-5-en-25-ol which is dehydrated by POCl$_3$ to produce a mixture of (24S)-24-[26-^{14}C]ethylcholesta-5,25-dien-3β-yl acetate (~70%) and 24-[26-^{14}C]ethylcholesta-5,24-dien-3β-yl acetate (~30%) which are readily separated by preparative TLC on AgNO$_3$-silica gel developed with pure chloroform. The 3β-hydroxysterols are then obtained by saponification of the acetates.

4-Demethylsterols with the nuclear double bond in the Δ^7- or Δ^8-position can be easily labeled with tritium. Base catalyzed exchange with tritiated water using the 3-oxo derivative of the sterol obtained by Jones oxidation followed by sodium borohydride reduction to restore the 3β-hydroxy group yields the 2,4-^{3}H$_4$-labeled sterol.[30,32] Alternatively the 3α-^{3}H-labeled sterol may be obtained from the 3-oxosteroid by reduction with tritium-labeled sodium borohydride.[22,33]

With Δ^5-sterols base catalyzed exchange of the 3-oxo derivative results in isomerization of the double bond into the conjugated Δ^4-position. The Δ^5-bond can be restored by refluxing the labeled 3-oxo-Δ^4-steroid with isopropenyl acetate and H$_2$SO$_4$ followed by addition of dry sodium acetate and removal of the solvent under vacuum.[34] This produces the Δ^{3,5}-enol acetate which can be reduced with sodium borohydride in aqueous ethanol followed by reflux with ethanol–HCl to destroy any 3β-

---

[29] M. Fryberg, A. C. Oehlschlager, and A. M. Unrau, *Tetrahedron* **27**, 1261 (1971).
[30] M. Fryberg, A. C. Oehlschlager, and A. M. Unrau, *J. Am. Chem. Soc.* **95**, 5747 (1973).
[31] C. Largeau, L. J. Goad, and T. W. Goodwin, *Phytochemistry* **16**, 1925 (1977).
[32] F. F. Knapp, L. J. Goad, and T. W. Goodwin, *Phytochemistry* **16**, 1683 (1977).
[33] M. Devys, A. Alcaide, and M. Barbier, *Phytochemistry* **8**, 1441 (1969).
[34] M. J. Thompson, O. W. Berngrubber, and P. D. Klein, *Lipids* **6**, 233 (1971).

hydroxy-Δ^4-sterol and so yield a mixture of 3β- and 3α-hydroxy-Δ^5-sterol. The 3β-hydroxy-Δ^5-sterol is purified from this mixture by column chromatography or preparative TLC.

Δ^5-Sterols can also be labeled with tritium at the C-7 or C-6 positions by a simple procedure which can be performed with milligram quantities of sterol.[32,35] The method can be exemplified by reference to cholest-5-en-3β-ol (cholesterol) but it is equally applicable to other Δ^5-sterols of varying side chain structure. The cholest-5-en-3β-ol is first converted via the 3β-tosylate into the *i*-sterol (3α,5α-cyclocholestan-6β-ol). This compound is then oxidized by Jones' reagent to yield 3α,5α-cyclocholestan-6-one. Base catalyzed exchange of this 6-oxo steroid in the presence of tritiated water introduces ^{3}H at the C-7 position to yield 3α,5α-[7-^{3}H$_2$]cyclocholestan-6-one which is reduced with sodium borohydride to produce 3α,5α-[7-^{3}H$_2$]cyclocholestan-6β-ol.[32] Rearrangement of the latter compound by reflux with zinc acetate in glacial acetic acid gives [7-^{3}H$_2$]cholest-5-en-3β-yl acetate or alternatively treatment of the labeled *i*-sterol with dioxane containing H$_2$SO$_4$ gives [7-^{3}H$_2$]cholest-5-en-3β-ol.[36] To avoid handling the large amounts of radioactivity required when tritiated water is used for the C-7 exchange reaction an alternative reduction procedure may be used to introduce tritium at C-6 of the sterol.[35] Reduction of the 3α,5α-cyclocholestan-6-one with sodium [^{3}H]borohydride produces 3α,5α-[6-^{3}H]cyclocholestan-6β-ol which can then be rearranged to give [6-^{3}H]cholest-5-en-3β-ol.

Deuterium-Labeled Precursors. The most used deuterium-labeled precursor for plant sterol biosynthetic studies has been [C^{2}H$_3$]methionine. Use of this substrate has permitted the elucidation of the C-24 alkylation mechanism operating in various algae, fungi, and higher plants. A method for the synthesis of [C^{2}H$_3$]methionine has been described by Dolphin and Endo.[37] S-Adenosyl[C^{2}H$_3$]methionine can be prepared from this deuterium-labeled methionine by incubation with yeast (*Saccharomyces cerevisiae*).[38]

Any of the methods described above for the introduction of tritium into sterol precursors can be simply modified to allow the synthesis of deuterium-labeled compounds. A combination of the two procedures described above using *i*-sterol intermediates for the introduction of labeled hydrogen at C-6 and C-7 has been described[36] for the synthesis of Δ^5-6,7,7-^{2}H$_3$-labeled sterols.

[35] M. A. Palmer, L. J. Goad, and T. W. Goodwin, *Nouv. J. Chim.* **2**, 401 (1978).
[36] L. J. Goad, M. A. Breen, N. B. Rendell, M. E. Rose, J. N. Duncan, and A. P. Wade, *Lipids* **17**, 982 (1982).
[37] D. Dolphin and K. Endo, *Anal. Biochem.* **36**, 338 (1970).
[38] N. L. A. Misso and L. J. Goad, *Phytochemistry* **22**, 2473 (1983).

Incubation of Labeled Precursors with Plant and Marine Invertebrate Organisms and Tissues

Algae. The uptake of particular labeled precursors (both water soluble and lipophilic) from the growth medium by some species of algae can be very poor. Water-soluble precursors may be simply added to the culture medium employing aseptic conditions either at the time of initial inoculation with cells or at any time required during the culture period. Labeled sterols can be added as a solution in a small volume of ethanol (typically 0.1–0.2 ml of ethanol solution added to 100–200 ml of growth medium).[32]

Vascular Plants. Water-soluble precursors may be administered by one of several methods depending upon the nature of the plant material and the particular study in hand. With some seeds addition of the precursor together with the water used to moisten the seeds at initiation of germination has allowed uptake of the labeled compound.[39] Excised shoots can be placed with the cut ends in an aqueous solution of the precursor and uptake encouraged by locating the shoots in a gentle air stream to facilitate transpiration; additional water is added to the shoots as needed during long incubations. To permit use of whole plants the precursor can be administered by wick feeding. A cotton wick is introduced into a small hole cut in the stem of the plant and one end placed in a solution of the precursor thus allowing its continuous uptake.[40]

Incubation with small amounts of leaves or other plant tissue may be conveniently performed with chopped preparations.[41] The plant material (1–10 g) is cut into strips about 1–2 mm wide with a razor blade, and the tissue then moistened with a small volume of water or appropriate buffer containing the radioactive precursor. The use of plant tissue cultures has proved invaluable in many areas of plant biochemistry[42] and they will take up many water soluble compounds. Cell suspension cultures of a variety of plant species have been used very effectively for studies on plant sterol biosynthesis and metabolism by Benveniste and his colleagues.[43]

The effective administration of water-insoluble radioactively labeled sterols to higher plant tissues in order to examine their conversion into the major 4-demethyl sterols has proved to be rather difficult. A low incorporation of any substrate into product may of course be the result, in part at least, of a slow biosynthetic rate. However, poor uptake and transport of

---

[39] D. J. Baisted, E. Capstick, and W. R. Nes, *Biochemistry* **1**, 537 (1962).
[40] R. D. Bennett, E. Heftmann, W. H. Preston, and J. R. Hann, *Arch. Biochem. Biophys.* **103**, 74 (1963).
[41] L. J. Goad and T. W. Goodwin, *Biochem. J.* **99**, 735 (1966).
[42] J. J. MacCarthy and P. K. Stumpf, this series, Vol. 72, p. 754.
[43] P. Fonteneau, M. A. Hartmann, and P. Benveniste, *Plant Sci. Lett.* **10**, 147 (1977).

water-insoluble labeled sterol precursors to the site of sterol biosynthesis can undoubtedly be a factor with plant tissues. Various methods have been devised in an attempt to overcome this practical problem.

The radioactive compound can be applied to a leaf surface as a solution in acetone.[44] The leaves are then carefully sprayed with a petroleum ether solution of a silicone oil; this is done slowly and carefully to permit evaporation of the petroleum ether and to ensure deposition of the oil as a thin film on the leaf surface. The absorption of the oil into the leaf tissue then facilitates the uptake of the radioactive sterol precursor.

Some plants may absorb labeled sterols if the cut ends of the shoots are immersed in an aqueous solution containing the precursor as an emulsion.[45] The emulsion is prepared by adding 4 ml of a solution of the labeled compound in 0.5% Tween 20 in acetone to 8 ml of 0.1% Tween 20 in water. The acetone is then evaporated under a stream of nitrogen and an emulsion obtained by vigorous agitation or sonication of the mixture.

The use of relatively rapidly growing plant material, such as developing seedlings or plant tissue cultures, offers an opportunity to maximize uptake and conversion of a precursor into the major sterols. Labeled sterols have been added to the growth medium of plant tissue cultures and shown to be absorbed and metabolized.[46] Similarly, germinating seedlings have been found to utilize labeled sterols applied to the surface of the seeds prior to germination.[47] Isolated barley embryos cultured on a suitable medium have also been used to demonstrate some steps in sterol biosynthesis.[31,48] Barley *(Hordeum vulgare)* seeds are dehusked and sterilized by careful treatment with 50% H_2SO_4 for 1.5 hr. The seeds are thoroughly washed with 10 × 500 ml volumes of sterile water and the embryos removed from the seeds under aseptic conditions. (Note, although barley is sufficiently robust for this H_2SO_4 treatment, other Gramineae seeds cannot be sterilized in this way, dilute sodium hypochlorite solution can be used as an alternative for surface sterilization.) The barley embryos are cultured at 23° in covered dishes (9.0 cm diameter × 5.0 cm deep) or conical flasks containing 4.0 ml of Rappaports mineral medium supplemented with 5.0 mM sodium L-glutamate, 5.0 mM L-asparagine, 20 mM sucrose, and 10 μM gibberellic acid. Labeled sterol, added in 0.1 ml

---

[44] R. D. Bennett and E. Heftmann, *Phytochemistry* **4**, 475 (1965).
[45] G. F. Gibbons, L. J. Goad, T. W. Goodwin, and W. R. Nes, *J. Biol. Chem.* **246**, 3967 (1971).
[46] M. J. E. Hewlins, J. D. Ehrhardt, L. Hirth, and G. Ourisson, *Eur. J. Biochem.* **8**, 184 (1969).
[47] F. Nicotra, F. Ronchetti, G. Russo, G. Lugaro, and M. Casellato, *J. Chem. Soc., Chem. Commun.* p. 889 (1977).
[48] J. R. Lenton, L. J. Goad, and T. W. Goodwin, *Phytochemistry* **14**, 1523 (1975).

solvent, is incorporated into this medium by sonication in the presence of 0.05% Triton X-100. Considerable coleoptile development occurs during a 6- to 8-day culture period.

Marine Invertebrates. Water-soluble substrates can be injected as an aqueous solution into the body cavity or mass of tissue of most marine invertebrates.[49] With some animals the injected material may pass from the animal into the surrounding sea water during incubation and be lost if the aquarium is maintained with circulating filtered sea water. To minimize this problem it may be advisable to maintain the animal in the smallest volume of aerated, noncirculating, sea water which is consistent with keeping the animal in a healthy condition during the incubation period. Isolated tissues of the echinoderm, *Asterias rubens,* will incorporate radioactivity into squalene and sterols when incubated at 20° with [2-^{14}C]mevalonic acid dissolved in 0.1 M sodium phosphate buffer pH 7.8.[49]

Radioactive sterols can be administered to many marine invertebrates by injection into the body cavity as an emulsion in 0.2 ml of 5% (w/v) aqueous Tween 80[50] or dissolved in 0.05 ml propylene glycol–ethanol (1 : 1).[51] For an investigation of sterol metabolism by a sponge an ethanolic solution (2 ml) of [4-^{14}C]cholesterol was added to the aquarium sea water (50 liters) and it was adequately absorbed and metabolized by the animal.[52]

An effective way of administering radioactive sterols to *Asterias rubens* (Asteroidea, Echinodermata) is to introduce the labeled compound dissolved in diethyl ether into one-half of a small gelatin capsule. After evaporation of the solvent the capsule is capped and embedded in a slit made in the adductor muscle of a suitable "opened" mollusc such as *Chlamys opercularis.* When this preparation is placed in the aquarium an *A. rubens* can usually be induced to feed upon the mollusc and in so doing it efficiently digests the gelatin capsule and absorbs the labeled sterol together with the other digestive products from the mollusc flesh.

Use of Algal and Plant Cell-Free Preparations for Studies on the Later Stages of Sterol Biosynthesis

Presqualene Pyrophosphate to Squalene. The preparation of a cell free homogenate from *Pisum sativum* seeds catalyzing this interconver-

[49] A. G. Smith and L. J. Goad, *Biochem. J.* **146**, 25 (1975).
[50] A. G. Smith and L. J. Goad, *Biochem. J.* **146**, 35 (1975).
[51] S-I. Teshima, A. Kanazawa, and H. Miyawaki, *Comp. Biochem. Physiol. B* **63B**, 323 (1979).
[52] M. De Rosa, L. Minale, and G. Sodano, *Experientia* **31**, 408 (1975).

sion has been described.[53] Seeds (50) germinated for 1–5 days are ground (4°) with 0.1 mM potassium phosphate buffer, pH 7.4 (50 ml) containing 0.4 M sucrose, 5 mM glutathionine, and 5 mM MgCl$_2$. After filtration through cheesecloth and rehomogenization (Thomas homogenizer) the homogenate is centrifuged for 10 min at 40,000 g to give the supernatant used for incubations. Incubations contain (final volume 11 ml) 40,000 g supernatant, [^{3}H]presqualene pyrophosphate (1.6 × 10^5 cpm) emulsified with 0.1% Tween 80 in 0.1 M phosphate buffer pH 7.4, 0.25 mM NADPH, 0.25 mM NADP, 2 mM glucose 6-phosphate, and 1.5 units glucose-6-phosphate dehydrogenase. The incubations are conducted anaerobically at 25° for 1.5 hr are stopped by addition of 10% (w/v) KOH in methanol–water (85 : 15) and 50 μg of carrier squalene is added. This is followed by saponification and isolation of the labeled squalene by preparative TLC on silica gel developed with petroleum ether.

2,3-Oxidosqualene-Cycloartenol Cyclase. Active cell-free preparations catalyzing this reaction have been obtained from the chrysophyte alga *Ochromonas malhamensis*,[54,55] leaves of bean *(Phaseolus vulgaris)*,[56] and a tissue culture of *Nicotinia tobacum*.[57]

The *O. malhamensis* preparation is obtained by harvesting cells from a 6- to 7-day culture (1 liter), washing with cold 0.1 M potassium phosphate buffer pH 7.4, resuspending in the same buffer (20 ml), and passing the paste through a French pressure cell. The resulting homogenate can be centrifuged at 12,000 g for 20 min to give a supernatant suitable for incubations. Alternatively it can be subjected to differential centrifugation to obtain an active 105,000 g "microsomal" preparation and further purified by ammonium sulfate precipitation and Sephadex gel filtration. The incubation mixture reported[55] for maximal cyclase activity consists of (in a total volume of 1 ml): enzyme preparation (1 mg of protein), potassium phosphate buffer, pH 7.4 (10 μmol), KCl (350 μmol), sodium deoxycholate (1 mg), dithiothreitol (5 μmol), 2,3-[^{3}H] oxidosqualene (7.5 × 10^4 dpm, the amount will depend upon specific activity) emulsified in 0.1 ml of 0.3% Tween 80 in 0.01 M potassium phosphate buffer, pH 7.4. The mixture is incubated for 30 min at 37° and the reaction stopped by addition of 10% (w/v) KOH in aqueous methanol and carrier 2,3-oxidosqualene (1 mg) and cycloartenol (1 mg) added. After saponification the 2,3-oxidosqualene and cycloartenol are obtained by preparative TLC on silica gel.

Cycloeucalenol—Obtusifoliol Isomerase. This is the microsomal en-

[53] G. H. Beastall, H. H. Rees, and T. W. Goodwin, *FEBS Lett.* **28**, 243 (1972).
[54] H. H. Rees, L. J. Goad, and T. W. Goodwin, *Biochim. Biophys. Acta* **176**, 892 (1969).
[55] G. H. Beastall, H. H. Rees, and T. W. Goodwin, *FEBS Lett.* **18**, 175 (1971).
[56] H. H. Rees, L. J. Goad, and T. W. Goodwin, *Tetrahedron Lett.* p. 723 (1968).
[57] R. Heintz and P. Benveniste, *Phytochemistry* **9**, 1499 (1970).

zyme responsible for opening the 9β,10-cyclopropane ring of cycloeucalenol to yield the Δ⁸-compound obtusifoliol. Methods for obtaining an enzymatically active cell-free preparation and assay conditions have been described using a tissue culture of bramble *(Rubus fruticosus)*[22] and coleoptiles of maize *(Zea mays)*[58] as the plant sources.

S-Adenosylmethionine-Cycloartenol Methyltransferase. Cell-free preparations demonstrating this enzyme activity have been obtained from algae[59] and higher plant tissues.[38,43,60]

Algal cells *(Scenedesmus obliquus* or *Trebouxia* sp. 213/3) are harvested by centrifugation and washed with cold 0.1 M sodium phosphate buffer, pH 7.6. The cells are resuspended in this buffer (5 g wet cell per 7 ml buffer), 35 ml portions of the suspension are mixed with glass beads (0.25–0.30 mm diameter, 10 g) and shaken at 0–2° for 5 min in a Braun Cell Homogenizer (4000 oscillations per minute). Centrifugation at 4000 g for 20 min gives a supernatant which is enzymatically active. The enzyme is associated with the cell supernatant and microsomal membrane fractions. A crude microsomal material can be obtained if required by centrifugation at 20,000 g for 20 min followed by recentrifugation of the supernatant at 105,000 g for 1 hr and the pellet suspended in an appropriate volume of 0.1 M sodium phosphate buffer, pH 7.6. The enzyme incubation mixture contains homogenate (5.0 ml), $MgCl_2$ (40 μmol, glutathione (20 μmol), cycloartenol (0.25 mg emulsified by sonication in 0.5 ml of 1% Tween 80 in water), and S-adenosyl-L-[$^{14}CH_3$]methionine (0.5μCi) in a final volume of 6.3 ml. After incubation at 33° the reaction is terminated by addition of 10% (w/v) KOH in aqueous 75% ethanol, the mixture refluxed, and the nonsaponifiable lipid extracted. The labeled product sterol is obtained by preparative TLC on silica gel developed with chloroform. The sterol products obtained with *S. obliquus* and *Trebouxia* sp. homogenates are a mixture of 24-methylenecycloartanol (75%) and cyclolaudenol (25%). Methods for identifying these sterols are given later.

An active microsomal preparation exhibiting S-adenosylmethionine-cycloartenol methyltransferase activity has been obtained from maize coleoptiles[43] and shoots.[38] Maize shoots harvested 4–6 days after germination of seeds are ground (1 g shoots per 1 ml medium) at 5° in a mortar and pestle with 0.1 M Tris–HCl buffer pH 8.0, 0.5 M mannitol, 1 mM EDTA, 10 mM mercaptoethanol, 0.5% bovine serum albumin, and solid polyvinylpyrrolidone (1 mg/ml of medium). After filtration through gauze the filtrate is centrifuged at 1000 g for 5 min to remove debris and recentrifuged at 10,000 g for 10 min. Centrifugation at 105,000 g for 60 min gives a pellet which is resuspended in 0.1 M Tris–HCl buffer pH 7.4, 4 mM

---

[58] A. Rahier, L. Cattel, and P. Benveniste, *Phytochemistry* **16,** 1187 (1977).
[59] Z. A. Wojciechowski, L. J. Goad, and T. W. Goodwin, *Biochem. J.* **136,** 405 (1973).
[60] M. A. Hartmann and P. Benveniste, *Phytochemistry* **17,** 1037 (1978).

MgCl$_2$, and 2 mM mercaptoethanol to provide the material for incubations. The incubation mixture contains in a final volume of 20 ml the microsomal preparation (18 ml), cycloartenol (5 μmol emulsified in 1.5 ml of 1% aqueous Tween 80) and S-adenosyl[^{14}CH$_3$]methionine (5 μmol, 2.5 μCi). The mixture is held at 30° for up to 6 hr and the reaction terminated by saponification (10% KOH in aqueous ethanol). After addition of carrier cycloartenol, cyclolaudenol, and 24-methylenecycloartanol the labeled products can be isolated by preparative TLC.

Extraction, Isolation, and Characterization of Radioactive Sterol Metabolites

Lipid Extraction. If it is desired to determine the incorporation of radioactivity from administered labeled compounds into both the steryl esters and the free sterols then the total lipid must be extracted. With plant material total lipid can be extracted by maceration of fresh tissue with acetone. Filtration and reextraction of the plant debris two or three more times with acetone is usually sufficient to remove the bulk of the lipid and the effectiveness of the extraction can be judged roughly by the removal of pigment (chlorophyll and carotenoids). The pooled acetone extracts are reduced substantially in volume by rotary evaporation, water added, and the lipid extracted in petroleum ether or petroleum ether–diethyl ether (1:1). The combined solvent extract is washed with several volumes of water, dried over anhydrous sodium sulfate, filtered, and the solvent removed by rotary evaporation to leave the total lipid.

The total lipid can be extracted from marine invertebrate tissues by homogenization with chloroform–methanol (2:1) following the Bligh and Dyer[61] procedure. Sterol conjugates such as steryl glycosides and steryl sulfates, as well as steryl esters and free sterols can be obtained by chloroform–methanol (2:1) extraction of lyophilized tissue.

If steryl esters are not specifically required it is often more convenient to obtain the nonsaponifiable lipid for subsequent chromatography. The total lipid extract can be saponified by reflux for 2–3 hr with 10% potassium hydroxide in 85% aqueous ethanol followed by dilution and extraction several times with petroleum ether or petroleum ether–diethyl ether (1:1). The combined solvent extracts are washed several times with water, dried over anhydrous sodium sulfate, filtered, and evaporated to dryness. Alternatively, the nonsaponifiable lipid can be obtained by direct saponification of chopped plant or marine invertebrate tissue by the above procedure, filtering through glass wool, prior to solvent extraction.

Addition of Carrier Sterols. It is often advantageous to add a few milligrams of appropriate unlabeled carrier sterols to a lipid extract to aid

---
[61] E. G. Bligh and W. J. Dyer, *Can. J. Biochem. Physiol.* **37,** 911 (1959).

the location and identification of labeled metabolites during chromatographic procedures. However, if carrier sterols are to be used some consideration needs to be given to the amount and the point of addition in the isolation procedure. Excessive amounts of added unlabeled carrier to a metabolite produced in low radioactive yield may reduce the specific activity to a point which makes some techniques impracticable. For example resolution on preparative GLC could be adversely effected due to column overloading.

Chromatography. A preliminary rough separation of the lipid into fractions enriched with different sterol types can be achieved on columns of silicic acid[62] or alumina. For convenience we have favored alumina deactivated to Brockmann grade III.[41] The lipid is applied to the column in a petroleum ether solution and the column eluted stepwise with petroleum ether and petroleum ether–diethyl ether mixtures. For optimum resolution the column loading should be up to 10–20 mg lipid/g of alumina,[63] the glass column diameter is chosen such that the alumina height : diameter ratio is approximately 5 : 1 and the fraction volumes are 100 ml/10 g of alumina. Fractions are eluted with petroleum ether (hydrocarbons, including squalene), petroleum ether–diethyl ether (98 : 2) (steryl esters, 2,3-oxidosqualene), petroleum ether–diethyl ether (94 : 6) (4,4-dimethylsterols), petroleum ether–diethyl ether (91 : 9) (4α-methylsterols), petroleum ether–diethyl ether (85 : 15 or 80 : 20) (4-demethylsterols), diethyl ether (more polar materials). Although particular compounds are concentrated in the fractions indicated some overlap of compounds into the preceding and following fractions occurs, and further purification of the required materials can be achieved by TLC or by HPLC.

For preparative TLC on silica gel the sterols can be applied in cyclohexane and the plates developed with chloroform–ethanol (98 : 2). Bands of material are located by light spraying with 0.01% rhodamine 6G in acetone or 0.005% berberine in ethanol and viewing under a UV lamp. R_f values may vary somewhat but they are in the order squalene 0.90, 4,4-dimethylsterols 0.4, 4α-methylsterols 0.30, 4-demethylsterols 0.20. After location bands are scraped and compounds eluted with dry diethyl ether by filtering under reduced pressure through a small sintered glass funnel.

The above procedures will separate sterols only according to the degree of methyl group substitution at C-4. Further separation into individual sterols can be achieved by use of argentation chromatography, gas liquid chromatography, or HPLC. Sterols differing by the degree of unsat-

---

[62] E. Hansbury and T. J. Scallen, *J. Lipid Res.* **21**, 921 (1980).
[63] When large amounts of lipid are being chromatographed these proportions are not convenient and higher loadings of sample onto the column can be tolerated to yield fractions enriched in particular compounds for subsequent more careful chromatography.

uration and position of double bonds in either the ring system or the side chain can be separated after acetylation (pyridine-acetic anhydride) by preparative TLC on 10% $AgNO_3$-silica gel plates. A number of solvent systems have been recommended but we find that the best resolution is achieved with rigorously purified chloroform prepared freshly just before use. A small amount of chloroform is washed exhaustively with water to remove ethanol (added as a stabilizer to prevent phosgene formation), dried over anhydrous calcium chloride, and fractionally distilled. (Note that this purified chloroform is now subject to oxidation to produce toxic phosgene and adequate precautions such as good ventilation should be taken when using this solvent.) Steryl acetates are applied in cyclohexane solution to the $AgNO_3$-silica gel plates. The separations are rather sensitive to overloading and best separations are achieved if no more than 10 mg of steryl acetate is applied to a 0.5-mm-thick 20 × 20-cm plate. Bands are located by light spraying with rhodamine 6 G or berberine solutions (see above) and viewing under UV light. Bands are scraped and eluted with dry diethyl ether. If desired, to prevent any possible contamination of the steryl acetate with silver nitrate, a small amount of alumina can be placed in the small sintered glass funnel used for the elution step. Although pure chloroform is satisfactory for the separation of many steryl acetate mixtures the addition of up to 3% diethyl ether gives significantly better resolution of the more unsaturated or polar steryl acetates.[64] The R_f values of steryl acetates on this system can be somewhat variable but Table I lists typical R_f values recorded for a selection of 4-demethylsteryl acetates[64] while Table II presents the R_f values of some 4,4-dimethylsteryl acetates[65] to give an indication of the separations which can be achieved. It is apparent from Table I that preparative $AgNO_3$-silica gel TLC of a labeled steryl acetate mixture cannot be relied upon to provide a radiochemically pure individual steryl acetate. The acetates of many potential sterol metabolites have rather similar R_f values and must be purified further by HPLC or preparative GLC prior to crystallization to constant specific radioactivity with appropriate carrier or alternatively they must be subjected to derivative formation for characterization. A rigorous chromatographic procedure for the isolation of individual sterols from the complex mixtures which are found in marine invertebrates has been described by Djerassi and colleagues.[66]

High-Performance Liquid Chromatography. HPLC is becoming increasingly important as a means of isolating and purifying sterols from

---

[64] I. Rubinstein, Ph.D. Thesis, University of Liverpool, 1973.
[65] N. L. A. Misso, Ph.D. Thesis, University of Liverpool, 1982.
[66] S. Popov, R. M. K. Carlson, A. Wegmann, and C. Djerassi, *Steroids* **28**, 699 (1976).

TABLE I
Thin-Layer Chromatographic Migration of Some Steryl Acetates on 10% AgNO$_3$-Silica Gel Developed with Chloroform–Diethyl Ether (97:3)[64]

Compound	R_f
(24R)-24-Methylcholesta-5,7,22E-trien-3β-yl acetate	0.06
(24R)-24-Ethylcholesta-5,7,22E-trien-3β-yl acetate	0.06
Cholesta-5,7-dien-3β-yl acetate	0.08
(24R)-24-Ethylcholesta-5,22E,25-trien-3β-yl acetate	0.11
24-Methylcholesta-5,24(28)-dien-3β-yl acetate	0.12
24-Methyl-5α-cholesta-8,14,22E-trien-3β-yl acetate	0.12
24-Methylcholesta-5,8,22E-trien-3β-yl acetate	0.13
(24S)-24-Methylcholesta-5,25-dien-3β-yl acetate	0.20
(24R)-24-Methylcholesta-5,7,9,(11),22E-tetraen-3β-yl acetate	0.20
(24R)-24-Ethyl-5α-cholesta-7,22E,25-trien-3β-yl acetate	0.21
24-Norcholesta-5,22Z-dien-3β-yl acetate	0.21
24-Methyl-5α-cholestan-7,24(28)-dien-3β-yl acetate	0.22
Cholesta-5,24-dien-3β-yl acetate	0.24
Cholesta-5,22Z-dien-3β-yl acetate	0.29
24-Norcholesta-5,22E-dien-3β-yl acetate	0.30
Cholesta-8,14-dien-3β-yl acetate	0.30
24-Ethylcholsta-5,24(28)Z-dien-3β-yl acetate	0.30
5α-Cholesta-7,24-dien-3β-yl acetate	0.31
(24S)-24-Ethylcholesta-5,25-dien-3β-yl acetate	0.32
Cholesta-5,22E-dien-3β-yl acetate	0.33
24-Methylcholesta-5,23E-dien-3β-yl acetate	0.35
(24R)-24-Methylcholesta-5,22E-dien-3β-yl acetate	0.36
(24S)-24-Methylcholesta-5,22E-dien-3β-yl acetate	0.36
5α-Cholest-9(11)-en-3β-yl acetate	0.37
Cholesta-5,9(11)-dien-3β-yl acetate	0.37
24-Ethylcholesta-5,24(28)E-dien-3β-yl acetate	0.37
24-Ethylcholesta-5,23E-dien-3β-yl acetate	0.39
24-Nor-5α-cholesta-7,22E-dien-3β-yl acetate	0.40
24-Ethyl-5α-cholesta-7,24(28)Z-dien-3β-yl acetate	0.40
24-Ethyl-5α-cholesta-7,24(28)E-dien-3β-yl acetate	0.44
24-Methylcholesta-5,24-dien-3β-yl acetate	0.44
5α-Cholesta-7,22E-dien-3β-yl acetate	0.45
24-Ethylcholesta-5,24-dien-3β-yl acetate	0.47
(24R)-24-Methyl-5α-cholesta-7,22E-dien-3β-yl acetate	0.50
(24S)-24-Methyl-5α-cholesta-7,22E-dien-3β-yl acetate	0.50
(24S)-24-Ethylcholesta-5,22E-dien-3β-yl acetate	0.51
(24R)-24-Ethylcholesta-5,22E-dien-3β-yl acetate	0.51
(24R)-24-Ethyl-5α-cholesta-7,22E-dien-3β-yl acetate	0.58
(24S)-24-Ethyl-5α-cholesta-7,22E-dien-3β-yl acetate	0.58
Cholest-5-en-3β-yl acetate	0.59
(24R)-24-Methylcholest-5-en-3β-yl acetate	0.59

TABLE I (continued)

Compound	R_f
(24R)-24-Ethylcholest-5-en-3β-yl acetate	0.59
(24S)-24-Ethylcholest-5-en-3β-yl acetate	0.59
Cholest-4-en-3β-yl acetate	0.61
5α-Cholest-7-en-3β-yl acetate	0.61
(24R)-24-Methyl-5α-cholest-7-en-3β-yl acetate	0.61
(24R)-24-Ethyl-5α-cholest-7-en-3β-yl acetate	0.61
5α-Cholestan-3β-yl acetate	0.65

plant[67] and marine sources.[66] The excellent papers by Hansbury and Scallen[62] and by DiBussolo and Nes[68] should be consulted for the details of the use of silicic acid and reversed-phase HPLC columns for the resolution of a wide range of 4,4-dimethyl-, 4α-methyl-, and 4-demethylsterols. With the three step procedure described by Hansbury and Scallen[62] for the separation of mammalian sterol mixtures a preliminary chromatography on a short column of silicic acid is employed to provide the 4,4-methyl and 4α-methylsterols as a combined fraction and the 4-demethylsterols as a second fraction. Each of these fractions is then subjected to reversed-phase HPLC. Elution is with acetonitrile–water (88:12) and gives some individual sterols and mixtures of others. The latter mixtures are then acetylated and separated into individual compounds by silicic-acid HPLC with i-octane–cyclohexane–toluene (5:3:2) as solvent. This paper also describes the separation of squalene and 2,3-oxidosqualene by reversed-phase HPLC; elution is with acetonitrile. The system described by DiBussolo and Nes[68] for the separation of sterols employs C_{18} or C_8 reversed-phase columns eluted with acetonitrile or acetonitrile–methanol (95:5), respectively. If steryl acetates have been isolated by $AgNO_3$-silica gel TLC they require saponification (reflux with 8% potassium hydroxide in 85% aqueous ethanol) before reversed-phase HPLC separation by the above systems. However, some of the critical mixtures of steryl acetates which cochromatograph on $AgNO_3$ silica gel TLC (Table I) can be resolved by reversed-phase HPLC without prior saponification to the free sterol. Thus 24-methylcholest-5-en-3β-yl acetate and 24-ethylcholest-5-en-3β-yl acetate are cleanly resolved on an Ultrasphere 5ODS column eluted with methanol–water (95:5).[69] This system is also effective for the

[67] H. H. Rees, P. L. Donnahey, and T. W. Goodwin, J. Chromatogr. 116, 281 (1976).
[68] J. M. DiBussolo and W. R. Nes, J. Chromatogr. Sci. 20, 193 (1982).
[69] N. L. A. Misso and L. J. Goad, Phytochemistry 23, 73 (1983).

TABLE II
SEPARATION OF SOME
9β,19-CYCLOPROPANE-4,4-DIMETHYLSTERYL ACETATES ON 10%
AgNO$_3$-SILICA GEL DEVELOPED TWICE WITH
CHLOROFORM–DIETHYL ETHER (98 : 2)[65]

Compound	R_f
24-Methyl-9β,19-cyclolanost-24(28)-en-3β-yl acetate (24-methylenecycloartanyl acetate)	0.47
(24S)-24-Methyl-9β,19-cyclolanost-25-en-3β-yl acetate (cyclolaudenyl acetate)	0.53
9β,19-Cyclolanost-24-en-3β-yl acetate (cycloartenyl acetate)	0.69
Lanosta-8,24-dien-3β-yl acetate (lanosteryl acetate)	0.69
24-Methyl-9β,19-cyclolanost-23-en-3β-yl acetate (cyclosadyl acetate)	0.72
24-Methyl-9β,19-cyclolanost-24-en-3β-yl acetate (24-methylcycloartenyl acetate)	0.78

separation of other mixtures such as 24-methylcholesta-5,23E-dien-3β-yl acetate, 24-ethylcholesta-5,25-dien-3β-yl acetate, and 24-ethylcholesta-5,24(28)Z-dien-3β-yl acetate.[69] The cis–trans isomeric sterols 24-methylcholesta-5,23E-dien-3β-yl acetate and 24-methylcholesta-5,23Z-dien-3β-yl acetate can be separated by reversed-phase chromatography on Lichrosorb RP18 eluted with methanol.[69]

Radioactivity detectors are now commercially available for monitoring HPLC effluents for radioactive compounds. However, with the low radioactivities often associated with the sterol metabolites recovered from plant and invertebrate incubations an alternative to the use of such instruments is the collection and radioassay of fractions from the HPLC column. After passage through the HPLC detector, the column eluant is collected at 30-sec or 1-min intervals and each radioassayed by liquid scintillation counting. A histogram plot then permits the comparison of radioactive peaks with the individual sterols as revealed by the HPLC detector record.

Gas–Liquid Chromatography. GLC is a well established and valuable method for sterol identification and quantitation. For biosynthetic studies it is a valuable aid for verifying the identities of compounds purified by TLC and HPLC methods. Preparative GLC of sterols has also proved a useful technique for the identification of radioactive sterol metabolites.[32] Conventional packed columns are employed and a splitter is fitted at the effluent end of the column. It is usually most convenient to arrange for about 1/10 to 1/25 of the effluent to pass to the GLC detector. The remainder of the column effluent passes to an exchangeable glass melting point capillary tube held at ambient temperature. The tube is replaced at 1-min intervals and eluting sterols or steryl acetates condense in the tube. They can then be transferred to vials for radioassay by liquid scintillation counting. A histogram plot of radioactivity is constructed on the FID detector record to identify labeled compounds.[32] This technique is not suitable for sterol samples with a low specific radioactivity. Column overloading and loss of resolution of peaks may result from the relatively large amounts of sterol which must be injected to permit trapping of adequate amounts of radioactivity for reliable assay.

Derivative Formation. Some combinations of sterols are difficult to separate by the chromatographic procedures outlined above and unambiguous identification of the labeled component(s) of such a mixture may require formation of suitable, separable, derivatives.

Cycloartenol and lanosterol, which are inseparable by TLC on silica gel or $AgNO_3$-silica gel, have been separated by formation of the mono- and diepoxide, respectively, for biosynthetic studies on plant sterols.[70] Similarly, epoxide formation has been used to give separable derivatives of radioactively labeled 5α-cholestan-3β-ol, 5α-cholest-7-en-3β-ol, and cholest-5-en-3β-ol.[50] The sterol mixture is dissolved in chloroform (1–2 ml) and m-chloroperbenzoic acid (1.1 M equivalent) added. The mixture is stirred for 3 hr, diethyl ether (25 ml) added, and the solution washed with dilute solutions of sodium nitrite and sodium chloride, followed by water. After drying over anhydrous sodium sulfate the solvent is removed to leave the sterol epoxides. The mixture is separated by preparative TLC on silica gel developed with chloroform–ethyl acetate (65:35) to yield unchanged 5α-cholesta-3β-ol (R_f 0.65), $7\alpha,8\alpha$-epoxy-5α-cholestan-3β-ol (R_f 0.52), and $5\alpha,6\alpha$-epoxycholestan-3β-ol (R_f 0.45).

Some Δ^7 and Δ^5-steryl acetates which are inseparable by $AgNO_3$-silica gel TLC (Table I) have an additional double bond in the side chain and epoxidation may then prove to be an unsuitable method of separation. In such cases chromic acid oxidation of the 3β-hydroxysterol (obtained by saponification of the steryl acetate) may be used to advantage. A Δ^7-sterol will yield a 3-oxo-Δ^7 derivative while a Δ^5-sterol will produce a 3,6-dioxo-Δ^4 compound.[69,71] The sterol mixture is dissolved in diethyl ether (10 ml) and 0.2 ml chromic acid solution (100 mg of potassium dichromate in 0.6 ml of water and 0.1 ml of concentrated H_2SO_4) added. The reaction mixture is stirred at room temperature for 2 hr excess water added, and the products extracted into diethyl ether. The solvent extract is washed

---

[70] P. Benveniste, L. Hirth, and G. Ourisson, *Phytochemistry*, **5**, 45 (1966).
[71] L. J. Goad and T. W. Goodwin, *Eur. J. Biochem.* **7**, 502 (1969).

with saturated sodium chloride solution, sodium bicarbonate solution, and finally water before drying over anhydrous sodium sulfate. The products are separated by TLC on silica gel developed with chloroform–ethanol (98 : 2) to yield the 3,6-dioxo-Δ^4-steroid (R_f 0.4; UV λ_{max}^{EtOH} 253 nm) and the 3-oxo-Δ^7-steroid (R_f 0.6, UV λ_{max}^{EtOH} 221 nm).

Cyclolaudenol (Δ^{25}) and 24-methylenecycloartanol [$\Delta^{24(28)}$] have been reported as the products the S-adenosylmethionine-cycloartenol methyltransferase catalyzed reaction in two algae[59] and a higher plant.[38] These two compounds are rather difficult to separate by chromatography although very careful TLC of the acetates on $AgNO_3$-silica gel (Table II) does achieve some separation. To identify these compounds a procedure has been employed which involves osmium tetroxide oxidation of the terminal methylene group of the two steryl acetates.[59] Osmium tetroxide (75 mg) in pyridine (3.5 ml) is added to the steryl acetates (100 mg) and the mixture left at room temperature for 18 hr. Water (2 ml) and sodium metabisulfite (310 mg) are added and the mixture stirred for 2 hr. After addition of water (50 ml) the reaction mixture is extracted with diethyl ether (3 × 50 ml). This produces a mixture of isomeric 25,26-diols and 24,28-diols from cyclolaudenyl acetate and 24-methylenecycloartanyl-acetate, respectively. These diols are separated by TLC on silica gel (chloroform-methanol, 92 : 8).[59] Treatment of the respective diols (50 mg) in dioxane (5 ml) with potassium metaperiodate (50 mg) in water (2.5 ml) with stirring for 18 hr yields 24-methyl-25-oxo-26-norcycloartanyl acetate (derived from cyclolaudenol) and 24-oxocycloartanyl acetate (derived from 24-methylenecycloartanol) which can be purified by TLC on silica gel (chloroform–methanol, 97 : 3).

Acknowledgments

The author thanks the Science and Engineering Research Council for financial support and his many colleagues for their valuable contributions to the work conducted in Liverpool.

[18] Yeast Sterols: Yeast Mutants as Tools for the Study of Sterol Metabolism

By LEO W. PARKS, CYNTHIA D. K. BOTTEMA, RUSSELL J. RODRIGUEZ, and THOMAS A. LEWIS

Introduction

Yeast Mutants

The yeast sterol mutants have been valuable for obtaining unusual yeast sterols, for studying the mechanism and regulation of ergosterol biosynthesis, and for studying the effects of altered sterol structures on cell physiology and physical properties of cell membranes. It is the purpose of this contribution to discuss sterol mutants with regards to their selection, maintenance, and utility. A comprehensive review of yeast sterols has been published, and may be consulted for sterol structures, biosynthetic pathways, and nomenclature.[1]

In general two types of yeast mutants have been used in studies on sterol metabolism: sterol auxotrophs and sterol mutants. Sterol auxotrophs must be supplied an exogenous source of sterol for growth. Sterol auxotrophy can arise by three different mechanisms: (1) enzymatic defects early in the ergosterol biosynthetic pathway, (2) enzymatic defects resulting in heme deficiency, and (3) growth of cells under anaerobic conditions.

Sterol mutants result from enzymatic defects late in ergosterol biosynthesis. Although unable to synthesize ergosterol, these organisms accumulate sterols which can satisfy the minimal structural features for sterol function, and thus do not require supplementation with exogenous sterol for aerobic growth. Sterol mutants have been obtained for almost every major biosynthetic step after cyclization of squalene by virtue of resistance to various antifungal compounds.

Sterol mutants may also be mimicked biochemically. Induced sterol defects result from the use of metabolic inhibitors of ergosterol biosynthesis, such as azasterol or imidazole antifungals, in wild-type cells. However, results from experiments with biochemically induced sterol defects must be interpreted cautiously as these inhibitors may affect other cellular phenomena.

[1] L. W. Parks, *CRC Crit. Rev. Microbiol.* **6**, 301 (1978).

Natural Auxotrophy

A natural auxotropy for ergosterol exists in wild-type cells when they are grown under strictly anaerobic conditions. Morpurgo et al.[2] showed that sterol deprivation under anaerobic conditions resulted in cell death instead of the usual growth stasis that is seen in yeast during starvation for many auxotrophic requirements. Inoculating an aerobically grown culture into anaerobic media, in the absence of sterols, stops growth after 5 to 7 generations, and the organisms begin to die 24 hr thereafter. Cells produced during that growth interval lack significant internal cellular structures. Extreme care must be taken to exclude molecular oxygen during these experiments since even very small amounts elicit sterol synthesis. Using the anaerobic technique, the structural features of sterols that are essential for satisfying the needs of the anaerobic cell have been defined. Proudlock et al.[3] reported that the molecule required was a planar sterol (α-configuration at C-5) with a long alkyl side chain and a C-3-OH group. More recent data[4] indicate that the C-3-OH group must be of β-configuration and a methyl group at C-24 is essential.

While the anaerobic procedure has resulted in interesting data on the structural features of the sterol, it imposes a very difficult constraint for physiological experimentation. To avoid this, sterol auxotrophs have been isolated. Sterol auxotrophs allow manipulation of ergosterol levels to demonstrate quantitative relationships of sterols to cellular function under aerobic growth conditions.

Methods

Sterol Mutant Isolation

Mutagenesis of yeast can be done by several methods including UV irradiation, nitrous acid, or ethylmethane sulfonate (EMS). Treatment with EMS has been described previously by Fink[5] and is one of the most commonly used techniques because this potent mutagen is convenient to use and gives high mutant yields. By varying the time of EMS treatment and/or concentration of EMS, it is possible to manipulate the conditions for the desired results. Single mutations are easily obtained by treating with EMS for 1 to 2 hr at concentrations of 1–3%. Enrichment for double

---

[2] G. Morpurgo, G. Serlupi-Crescenzi, G. Tecce, F. Valente, and D. Venetacei, *Nature (London)* **201,** 897 (1964).
[3] J. W. Proudlock, L. W. Wheeldon, D. J. Jollow, and A. W. Linnane, *Biochim. Biophys. Acta* **152,** 434 (1968).
[4] W. R. Nes, B. C. Sekula, W. D. Nes, and J. H. Adler, *J. Biol. Chem.* **253,** 6218 (1978).
[5] G. R. Fink, this series, Vol. 17, p. 59.

mutations can be achieved by treating overnight with 0.5% EMS or using higher concentrations (5%).

Although EMS treatment is usually done in liquid, we have successfully mutagenized on YTD agar plates (0.5% yeast extract, 1.0% tryptone, 2.0% dextrose, 1.5% Bacto agar) with wells containing decreasing concentrations of EMS for mutagenesis.[6] Mutagenesis plates are prepared by adding 50 μl of 5, 3, 1, and 0.5% (v/v) EMS to four wells 5 mm in diameter placed 90° apart and 2 cm from the edge of 10-cm plates. Plates are equilibrated for 16 hr to allow the mutagen to diffuse prior to inoculation. The plates are inoculated with wild-type haploid cells grown to early stationary phase in liquid YTD. Cells are diluted to 10^6 per ml, and 0.5 ml of this dilution is spread uniformly on the EMS plates. Twenty four hours after inoculation, the EMS plates are replicated onto selective media. After 3 days of incubation at 28°, an average of 5 resistant colonies per plate is found.

For the selection of sterol mutants, antimycotics are usually employed. By selecting for resistance to drugs which inhibit ergosterol biosynthesis, it is possible to obtain mutants with defects in the pathway. Such antimycotics, however, usually give poor results. Selection with polyene antibiotics has been more successful. For *Saccharomyces,* nystatin is the preferred polyene antibiotic because it very effectively binds ergosterol, destroying cellular membranes.[7] Nystatin does not bind other sterols as effectively. Thus, mutants in the sterol biosynthetic pathway which accumulate sterols other than ergosterol are more resistant to nystatin binding. By replicating the EMS plates to nystatin-containing plates, it is possible to obtain mutants with defects in the enzymatic steps of sterol biosynthesis.[6] Nystatin plates are prepared by adding nystatin in dimethylformamide or ethanol to Wickerham's complete synthetic media[8] with 2% dextrose just prior to pouring the plates (final concentration nystatin is 15 units/ml). Nystatin plates with YTD can also be used, but the concentration of nystatin should be increased to 30 units/ml. Independent mutants are streaked on YTD plates for single clones and, after overnight growth, are replicated to 2 nystatin plates. Duplicate plates are necessary to avoid false readings. At 24 and 48 hr, the nystatin plates are compared, and those clones with good growth on both plates are transferred for sterol analysis. It is good practice, however, to rescreen clones for nystatin resistance prior to analysis since nystatin is fairly unstable, being sensitive to both heat and light. Analyzing the sterol composition of

[6] L. W. Parks, C. McLean-Bowen, C. K. Bottema, F. R. Taylor, R. Gonzales, B. W. Jensen, and J. L. Ramp, *Lipids* **17,** 187 (1982).

[7] A. W. Norman, A. M. Spielvogel, and R. G. Wong, *Adv. Lipids Res.* **14,** 127 (1976).

[8] L. J. Wickerman, *J. Bacteriol.* **52,** 292 (1946).

the mutants involves extraction of the sterols, separation from other lipid classes, and identification.

Characterization of Sterol Mutants

Extraction of Sterols. Sterols may be extracted by several methods including (1) alkaline saponification, (2) chloroform–methanol extraction, or (3) dimethyl sulfoxide (DMSO)–hexane extraction.

1. Alkaline saponification: Cells are acid labilized by resuspending in 20 volumes 0.1 N HCl, and steaming for 20 min. After steaming, the cells are washed in distilled water. (Acid labilization is not essential, but does greatly increase the final sterol yield.[9]) Up to 2 g wet weight cells are resuspended in 2 ml 60% KOH in water, 3 ml methanol, 2 ml 0.5% pyrogallol in methanol with boiling chips, and refluxed for 1–2 hr at 60–80°. After cooling, the saponification mix is extracted by the addition of 10 ml hexane. The phases may be separated by centrifugation at 500 g for 1 min. The top hexane layer is removed to a clean tube and the cells extracted 2 more times with hexane. Any emulsions occurring may be dispersed by the addition of 0.5 ml methanol.

2. Chloroform–methanol extraction: Cells are resuspended in distilled water (1 ml/g wet weight cells) and broken with glass beads in a cell homogenizer with CO_2 cooling. Four milliliters methanol is added per 1.6 ml cell suspension and and vortexed. Two milliliters chloroform is added to the supernatant and vortexed, followed by the addition of 2 ml saline (0.9% Nacl) with vortexing. The phases are separated by centrifugation at 500 g for 5 min. The top layer is removed and the chloroform layer washed 2 times by adding 2 ml 2 M KCl and vortexing. Again the layers are separated by centrifugation and the top layer removed. In tubes containing emulsions, the suspension may require incubation at 37° for 1–3 hr. The addition of 0.5 ml methanol may also be necessary.

3. DMSO–hexane extraction: Up to 2 g wet weight cells are pelleted and frozen. The cell pellet is lyophilized and minced to a powder. DMSO is added to just saturate the pellet and the tubes steamed for 1 hr. After cooling, 10 ml hexane is added and the solution vortexed. Ten milliliters of methanol–water (1:4 v/v) is added and vortexed. The suspension is centrifuged at 500 g for 5 min. The hexane layer is removed to a clean tube and the pellet reextracted 2 more times with hexane.

While all of these methods are efficient, alkaline saponification is usually the method of choice since it is the most rapid for screening large numbers of mutants. It can only be used, however, for identification of

[9] R. A. Gonzales and L. W. Parks, *Biochim. Biophys. Acta* **489**, 507 (1977).

total sterol. In those cases where the composition of both the free sterol and the steryl ester fractions is being analyzed, it is necessary to use one of the other methods. For quantitating sterols with respect to other lipid classes (e.g., phospholipids), the chloroform–methanol extraction is used.

Purification of Sterols. Sterols are purified from the lipid extract by either thin-layer chromatography (TLC) or high-pressure liquid chromatography (HPLC). For HPLC the lipid extract is evaporated under nitrogen with gentle heating or rotoevaporated to dryness, resuspended in isopropanol, and run on a HPLC column as described elsewhere in this volume.[10] For TLC the lipid extract is evaporated, resuspended in either diethyl ether or chloroform–methanol (4 : 1 v/v), and streaked on silica gel plates (EM Laboratories Silica Gel 60 F-254, 20 cm length). Plates of various widths may be used depending on the amount of lipid present in the extract. Wide plates may also be sectioned for running several samples by using a sharp needle or razor blade to make lanes in the silica gel. Overloading of the plates should be avoided since it hinders separation. Plates are chromatographed in isopropyl ether : glacial acetic acid (90 : 4 v/v) up to 12 cm from the baseline, dried, and then chromatographed (to the top of the plate) in hexane : diethyl ether : glacial acetic acid (90 : 10 : 1 v/v/v) as described by Skipski and Barclay.[11] The sterol band may be visualized by iodine. Short wavelength UV light may also be used for visualization if any conjugated double bonds are present. Free sterols are removed from the plate by scraping the sterol-containing silica gel from the plate, powdering the gel, and placing the silica gel in test tubes for extraction. The silica gel is extracted (3–5 times) by vortexing with 3–10 ml of diethyl ether or chloroform–methanol (4 : 1 v/v). The solvent is best removed between extractions if the silica gel is pelleted by centrifugation at 500 g for 5 min. The solvent supernatant is run over Na_2SO_4 columns to remove any water or silica gel in the sample, and then evaporated under nitrogen with gentle heating. Silica gel-containing steryl esters is scraped into test tubes for direct saponification. The conditions for steryl ester saponification are as described above with the exception that 6% KOH in methanol is used. After the esters are hydrolyzed, they are processed as free sterols.

Individual sterols may be further purified when necessary by HPLC[10] or by additional TLC. Derivatization of sterols is often useful, particularly acetylation. Sterol acetates are more stable and allow separation on silver nitrate-impregnated plates. Sterols are acetylated by resuspending in 0.5 ml pyridine and 1.0 ml acetic anhydride and incubating 1–2 hr at 25°.

[10] R. J. Rodriquez and L. W. Parks, this volume [2].
[11] V. P. Skipski and M. Barclay, this series, Vol. 14, p. 530.

Silver nitrate plates are made by placing the silica gel plates in a 10% solution of silver nitrate in methanol:water (3:1 v/v) for 5 sec and drying in a 100° oven for 0.5–1.0 hr. If the plates are not used immediately, they must be stored, desiccated, in the dark. Sterol acetates may be separated by using the solvent system of cyclohexane:benzene (65:35 v/v).

Sterol Analysis. Sterols are identified by a variety of techniques. UV-spectroscopy scans from 310 to 210 nm of sterol dissolved in redistilled hexane give characteristic peaks for sterols with conjugated double bonds. Retention indices for HPLC[10] and gas liquid chromatography (GLC) also provide means of identification. Sterols dissolved in diethyl ether or chloroform can be separated on a 3% SE-30 stationary phase column at 235°. The column is 6 ft × 4 mm (internal diameter) and is run with a nitrogen flow rate of 35 ml/min. By using a Varian Model 2740 chromatograph equipped with a CDS 111 recording integrator and a flame ionization detector, we can determine the relative carbon numbers for the unknown sterol from the following formula: $C_x = C_a + [\log t_x - \log t_a]/(\log t_b - \log t_a)] (C_b - C_a)$, where C_a, C_b, C_x are the relative carbon numbers of the standards a and b and the unknown x, and t_a, t_b, t_x are the respective retention times. The standards are long chain alkanes of 26, 28, 32, 34, 36 carbons with relative carbon numbers of 2600, 2800, 3200, 3400, 3600. The choice of standards should be such that $t_a < t_x < t_b$. By determining the relative carbon numbers of sterol standards, unknown sterols may be identified.

Gas chromatography coupled to mass spectroscopy (GC-MS) is generally necessary for sterol identification, especially when standards are not available. For GC-MS sterols may be separated on a 7% OV-101-packed column at 70 eV. Using the mass ion peaks, it is possible to fingerprint the unknown sterol.[12] While these techniques are usually sufficient, it is occasionally necessary to use infrared spectroscopy (IR) or nuclear magnetic resonance (NMR).

We have quantitated sterols by several means. Using [³H]acetic anhydride, sterols may be acetylated, purified by TLC, and quantitated by radioactivity.[13] The major drawback to this method is variability in recovery. We have also used a modified Liebermann–Burchard assay,[14] but only in those cases where the individual sterols have been first purified and the extinction coefficients are known.[1] Quantitation by UV is only possible for those sterols with conjugated double bonds. The preferable method is to add a sterol standard of known concentration during sterol extraction for determining extraction efficiency, and quantitating the puri-

---

[12] R. B. Bailey and L. W. Parks, *J. Bacteriol.* **127**, 606 (1975).
[13] C. A. McLean-Bowen and L. W. Parks, *J. Bacteriol.* **143**, 1325 (1981).
[14] P. R. Starr and L. W. Parks, *J. Cell. Comp. Physiol.* **59**, 107 (1962).

fied yeast sterol fraction with respect to this sterol standard by GLC.[15] GLC quantitation is very accurate if an internal standard (e.g., cholestane) of known concentration is added to the sample just prior to injection and the peak areas of the internal standard, the sterol extraction standard, and the sterol sample are compared by integration.

Sterol Auxotroph Isolation

Two approaches have been taken in obtaining yeast sterol auxotrophs: (1) isolation by virtue of nystatin resistance using the methods described above, and (2) selection of strains deficient in heme biosynthesis.[16] For our studies, we have modified auxotrophic strains obtained from the above procedures for better growth characteristics. This was accomplished by first mating strain RW2 (Berkeley stock culture collection) with a wild-type strain, and selecting an auxotrophic segregant (FY1) for nonflocculant growth.[17] Standard procedures for yeast mating, sporulation of diploids, and dissection of asci are described by Sherman *et al.*[18] FY1 grows well on ergosterol but very poorly on cholesterol. This strain was used to select a strain (FY3) for adequate growth on cholesterol.[17] FY3 was further modified to derive an auxotroph (RD5) unable to desaturate sterols at the C-5(6) position.[19] RD5 can be obtained by two methods: (1) EMS mutagenesis of FY3 as described above, and (2) mating FY3 to a sterol mutant defective in the C-5(6) desaturase enzyme.[19]

Strains and Culture Conditions

The sterol mutants used in our studies (designated as *erg* mutations) have been described in detail.[1,6] Sterol auxotrophs FY1, FY3, and RD5 have also been characterized.[17,19] FY1 is a heme mutant and can synthesize sterol if provided δ-aminolevulinic acid (alv). FY3 has an additional mutation in 2,3-oxidosqualene cyclase and cannot synthesize sterol even in the presence of alv. FY3 has been used for experiments dealing with sterol uptake,[20] esterification of sterols,[21] and the mode of action of imidazole and azasterol antimycotics.[22] To better understand the bulk mem-

---

[15] C. K. Bottema and L. W. Parks, *Lipids* **15**, 987 (1980).
[16] J. M. Haslam and A. A. Astin, this series, Vol. 56, p. 558.
[17] F. R. Taylor and L. W. Parks, *Biochem. Biophys. Res. Commun.* **95**, 1437 (1980).
[18] F. Sherman, G. R. Fink, and J. B. Hicks, "Methods in Yeast Genetics." Cold Spring Harbor Press, Cold Spring Harbor, New York, 1981.
[19] R. J. Rodriguez and L. W. Parks, *Arch. Biochem. Biophys.* **225**, 861 (1983).
[20] L. F. Salerno and L. W. Parks, *Biochim. Biophys. Acta* **752**, 240 (1983).
[21] F. R. Taylor and L. W. Parks, *J. Biol. Chem.* **256**, 13048 (1981).
[22] F. R. Taylor, R. J. Rodriguez, and L. W. Parks, *Antimicrob. Agents Chemother.* **23**, 515 (1983).

brane and high specificity sparking requirements for sterols we have further modified FY3 to obtain RD5.[19] All strains used in mutant selection and experimentation are haploid. Sterol mutants are routinely cultured in rich medium (YTD) containing 0.5% yeast extract, 1.0% tryptone, and 2.0% dextrose. When experiments involving mitochondria are performed, the sterol mutants are grown in the same medium except dextrose is replaced by 95% ethanol (2.0%). Sterol auxotrophs are grown in defined media[19] and supplemented with unsaturated fatty acids (oleate : palmitoleate, 4 : 1 v/v) and sterol. Both lipid supplements are added from stock solutions in tyloxapol–95% ethanol (1 : 1 v/v) to final concentrations of 100 and 5 μg/ml, respectively. All cultures are grown at 27–30°.

Of particular concern in mutant isolation are the choices of the haploid parental yeast strains and the polyenes for the selective medium. We have observed considerable variation in the classes of mutants obtained in different wild-type strains that we have used.[6] For example, the C-24 methyltransferase mutants were the most abundant type found by Lynn Miller. In contrast, we have never seen a C-24 methyltransferase mutant type from our wild-type strain, 3701B. The most abundant type of mutant in this organism is that containing a defect in the Δ^5-desaturation of the B ring of the sterol nucleus. The basis of this difference remains obscure, but "preferred" mutant classes are seen in a variety of different parental strains. Similar concern should be addressed in the selection of the polyene being used. Different organisms seem to yield different preferred mutant types with the different polyenes being employed. In addition, the levels of sensitivity to the polyenes within the different parentals is seen.

While this section is concerned with sterol mutants in *Saccharomyces*, it is important to note that a variety of eucaryotic organisms with altered sterol metabolism have been obtained by selecting for resistance to various fungicides. *Penicillium*,[23] *Fusarium*,[24] *Ustilago*,[25] *Neurospora*,[26–28] *Aspergillus*,[29] *Arthroderma*,[30] and *Candida*[31,32] have yielded sterol mutants using procedures of the type we describe here.

[23] M. A. DeWaard, H. Groeneweg, and J. G. M. Van Nisterlrooy, *Neth. J. Plant Pathol.* **88**, 99 (1982).
[24] G. Defago and H. Kern, *Physiol. Plant Pathol.* **22**, 29 (1983).
[25] P. Leroux and M. Gredt, *C. R. Hebd. Seances Acad. Sci., Ser. III* **296**, 191 (1983).
[26] M. Grindle and R. Farrow, *Mol. Gen. Genet.* **165**, 305 (1978).
[27] A. M. Johnston, I. R. Aarouson, and C. E. Martin, *Biochim. Biophys. Acta* **713**, 512 (1982).
[28] M. Grindle, *Mol. Gen. Genet.* **130**, 81 (1974).
[29] M. R. Siegel and Z. Solel, *Pestic. Biochem. Physiol.* **15**, 222 (1981).
[30] D. H. Howard and N. Dabrowa, *Sabouraudia* **17**, 35 (1979).
[31] J. M. T. Hamilton-Miller, *J. Gen. Microbiol.* **73**, 201 (1972).
[32] A. M. Pierce, H. D. Pierce, A. M. Unrau, and A. C. Oehlschlager, *Can. J. Biochem.* **56**, 135 (1978).

Maintenance of Mutants

Maintenance of the sterol mutants is often difficult since resistance to polyene antibiotics can be relatively unstable in yeast. Stocks are maintained on solid media containing 1.5% Bacto agar (described above) at 4° and frequently transferred (1–2 months). Storing on filter paper or silica gel[5] is recommended for long periods of time. Constructing diploids of mutants with wild-types can be very useful since the diploids are very stable and may be sporulated to regain the original mutant.

Utilization of Mutants

Sterol Auxotrophs

The first mutants isolated which had an aerobic requirement for sterols were later shown to possess lesions in porphyrin biosynthesis.[33] These mutants showed pleiotrophic requirements for sterol, unsaturated fatty acid, and methionine[34–36] due to the inability to assemble hemoprotein biosynthetic enzymes. Auxotrophic requirements of these mutants can be satisfied by heme supplements such as hemin, protoporphyrin IX, or δ-aminolevulinic acid (alv). Studies of these mutants have shown that they accumulate lanosterol,[17] they can use cholesterol to replace ergosterol as the sterol supplement,[35] and they are more resistant to nystatin than the wild-type strains.[35] The effect of increasing concentrations of alv to restore increasing levels of ergosterol in one heme mutant, ole-3 (*hem1*; alv synthetase), was exploited to observe the quantitative effect of ergosterol on mitochondrial membranes.[37]

Heme mutations impose restraints on studying sterols in yeast similar to those of anaerobiosis; when an absolute requirement for sterol exists, much of the cell's oxygen-dependent metabolism cannot occur. Mutants used to study the physiological roles of sterol under aerobic conditions should ideally be defective only in sterol synthesis. Tight, nonconditional mutations in sterol biosynthesis (prior to lanosterol synthesis) have been isolated only in a *hem* background.[17,38] The inability to segregate viable *erg* mutants from these strains has shown that the *hem* mutation is re-

---

[33] R. A. Woods, H. K. Sanders, M. Briquet, F. Foury, B. Drysdale, and J. R. Mattoon, *J. Biol. Chem.* **250**, 9090 (1975).
[34] M. A. Resnik and R. K. Mortimer, *J. Bacteriol.* **92**, 597 (1966).
[35] F. Karst and F. Lacroute, *Physiol. Veg.* **11**, 563 (1973).
[36] E. G. Gollub, P. Trocha, P. K. Liu, and D. B. Sprinson, *Biochem. Biophys. Res. Commun.* **56**, 471 (1974).
[37] A. M. Astin and J. M. Haslam, *Biochem. J.* **166**, 275 (1977).
[38] E. G. Gollub, K.-P. Liu, J. Dayan, M. Adlersberg, and D. B. Sprinson, *J. Biol. Chem.* **252**, 2846 (1977).

quired for viability of tight *erg* mutations.[38] By observing the effect of alv on one *hem*1 *erg*7 mutant, FY3, it appears that the uptake of sterol is inhibited by the Hem$^+$ condition.[39]

The isolation of one *erg* mutation outside a *hem* background was reported in which one leaky squalene epoxidase mutant was found after analysis of 2000 strains.[40] Conditional mutants, found to be temperature sensitive in their sterol requirement, have since been isolated.[41] These mutants were also observed to have a complicated phenotype in that sporulation of diploids produced from crosses with wild-type cells gave poor spore viability, and in some cases, an additional auxotrophy was required for viability of *erg*-containing spores. One of these auxotrophies (*aux*$_{32}$) can be replaced by a *hem*1 mutation.[39]

Using sterol auxotrophs, it is possible to study the specificity of sterol uptake under aerobic conditions. A preference has been demonstrated for molecules with $\Delta^{5,7,22}$ unsaturations as well as the C-24 β-methyl group.[21] We have shown that cells preferentially utilize ergosterol over cholesterol, and in the presence of both sterols, ergosterol diminishes both uptake and esterification of cholesterol. Sterol uptake is dependent on the metabolic state of the cell, and occurs only during cellular growth.[20]

In yeast, sterols are found in free form and esterified to long chain fatty acids. Although free sterols are available for membrane synthesis, esterified sterols are not incorporated into cell membranes and are stored in lipid droplets in the cytosol. Different specificities for esterification may provide the cell a mechanism for controlling the types of sterols in membranes. By making different sterol combinations available to sterol auxotrophs we have observed preferential deposition of ergosterol-like sterols into the free sterol fraction and preferential esterification of nonergosterol-like sterols.[21]

With the availability of sterol auxotrophs, it is also possible to study the specificity of sterols for satisfying the growth requirements of yeast. Using highly purified substrates, we have found evidence for at least two major classes of sterol functions.[19,42] We have designated these as bulk membrane and high-specificity (sparking) functions. While a variety of sterols and stanols can satisfy the bulk requirement, only those having a C-5(6) unsaturation or those capable of being desaturated at C-5 fulfill the high-specificity requirement. Analysis of sterol mutants reported to lack

---

[39] T. A. Lewis, F. R. Taylor, and L. W. Parks, in preparation.
[40] F. Karst and F. Lacroute, *Biochem. Biophys. Res. Commun.* **59**, 370 (1974).
[41] F. Karst and F. Lacroute, *Mol. Gen. Genet.* **154**, 269 (1977).
[42] R. J. Rodriguez, F. R. Taylor, and L. W. Parks, *Biochem. Biophys. Res. Commun.* **106**, 435 (1982).

C-5(6)-desaturase activity provides support for the sparking hypothesis. All of the presumed sterol mutants defective in this enzyme were found by HPLC analysis to contain small amounts of ergosterol indicating some C-5(6)-desaturate activity.[19] The amount of ergosterol in these mutants is approximately equal to that necessary to satisfy the sparking requirement in the auxotrophs.

A comparison of the sterol requirements of auxotrophy for sterols in aerobic mutants and in wild-type cells grown anaerobically reveals ambiguity with regard to the structural features of the sterol molecule. The data of Nes et al.[4] are clear in that the methylated side chain is essential for growth of yeast under anaerobic conditions. Our data show that under aerobic conditions the methyl group at C-24 of ergosterol is not required for growth, but does allow for greater ease of desaturation at the C-5(6).[19] How sterol physiology differs under these conditions remains an intriguing question.

The ability to feed sterol auxotrophs a large variety of sterols and stanols provides a unique opportunity to examine the effects of changing sterol structure on cellular physiology. By physical studies (e.g., fluorescence polarization, differential scanning calorimetry, nuclear magnetic resonance) and enzymatic analyses (e.g., Arrhenius kinetics of membrane proteins) of isolated cellular fractions from sterol auxotrophs, it is also possible to ascertain how inidividual sterol structural features affect membrane properties.

Sterol Mutants

Sterol mutants may be utilized to obtain unusual sterols. This is of particular value for assaying of ergosterol metabolic enzymes. These mutants are resistant to polyenes and lack at least one of the enzymes following the formation of lanosterol.[43] This relationship between sterol metabolism and polyene resistance was first observed by Lynn Miller (personal communication). In every case, altered sterols are produced, usually with a mixture of sterol types being found. A variety of mutants have now been characterized,[44,45] and cross-resistance to other polyenes observed.[46]

Because these organisms accumulate ergosterol intermediates, the mutants have been used extensively in studying the biosynthetic pathway

[43] S. A. Henry, in "The Molecular Biology of the Yeast *Saccharomyces*. Metabolism and Gene Expression" (J. N. Strathern, E. W. Jones, and J. R. Broach, eds.), p. 101. Cold Spring Harbor Press, Cold Spring Harbor, New York, 1982.
[44] R. A. Woods, *J. Bacteriol.* **108,** 69 (1974).
[45] M. Bard, *J. Bacteriol.* **111,** 649 (1972).
[46] S. W. Molzahn and R. A. Woods, *J. Gen. Microbiol.* **72,** 339 (1972).

of ergosterol.[47–49] However, mutations in the latter stages of ergosterol synthesis do not prevent modifications which occur subsequent to the effected step. For example, in the mutants lacking the methyltransferase function, zymosterol (the substrate for the methyltransferase) is the predominant sterol, but minor sterols, $\Delta^{5,7,24}$ and $\Delta^{5,7,22,24}$-cholesterol derivatives, are also present. Predictions of the nature of the enzymatic defects have been made based on the predominant intermediates accumulated, but have been verified by enzymatic analyses in only a very few instances.[50,51]

Sterol mutants have also been used to elucidate sterol structural features preferable to the cell. Sterol mutants obtained give an indication of the advantages of the various modifications preceding ergosterol formation. Although some sterol mutants have growth characteristics indistinguishable from wild-type, certain mutants have very different growth properties, are less readily isolated, and are less stable. For example, it is difficult to maintain a double mutant defective in C-14 demethylation and C-5(6) desaturation.[52] Such double mutants tend to revert to C-14 demethylase competence, retaining the defect in C-5(6) desaturation. It appears that although the major sterol of the double mutant [14α-methylergosta-8,24(28)-dienol] is acceptable, demethylation at C-14 is preferred. We have also observed that mutants which accumulate lanosterol require a more satisfactory sterol for growth and that there is a selective advantage in these mutants for an additional defect in lanosterol formation.[17] All of these observations suggest that some sterol structures are preferred physiologically and give a selective advantage to the organism.

Because unusual sterols are produced, the sterol mutants can also be utilized for a variety of studies aimed at deciphering the effect of sterol alterations on various cellular functions. The effects of the altered sterol composition have been studied by analysis of membrane fluidity,[13,53] stability,[54,55] permeability,[56–58] resistance to lysis,[59] and cellular growth.[60,61]

---

[47] D. H. R. Barton, A. A. L. Gunatilaka, T. R. Jarman, D. A. Widdowson, M. Bard, and R. A. Woods, *J. Chem. Soc., Perkin Trans. 1* p. 88 (1975).

[48] T. R. Jarman, A. A. L. Gunatilaka, and D. A. Widdowson, *Bioorg. Chem.* **4,** 202 (1975).

[49] D. H. R. Barton, J. E. T. Corrie, D. A. Widdowson, M. Bard, and R. A. Woods, *J. Chem. Soc., Perkin Trans. 1* p. 1326 (1974).

[50] R. B. Bailey, L. Miller, and L. W. Parks, *J. Bacteriol.* **126,** 1012 (1976).

[51] W. D. Neal and L. W. Parks, *J. Bacteriol.* **139,** 1375 (1977).

[52] F. R. Taylor, R. J. Rodriguez, and L. W. Parks, *J. Bacteriol.* **155,** 64 (1983).

[53] C. D. K. Bottema, C. A. McLean-Bowen, and L. W. Parks, *Biochim. Biophys. Acta* **743,** 235 (1983).

[54] C. A. McLean-Bowen, and L. W. Parks, *Lipids* **17,** 662 (1982).

We have observed by fluorescence polarization, for instance, that the membranes of sterol mutants undergo a phase transition which does not occur in the ergosterol-containing membranes of wild-type strains.[13,53] Using sterols and phospholipids from the sterol mutants, we have made reconstituted model liposomes and demonstrated that these phase transitions are a result of the unusual sterol composition of the mutants.[62] Lees et al.[55] showed that the membrane order is higher in membranes of sterol mutants compared to wild-type membranes. As a result of altering the sterol composition, the kinetics of membrane enzymes in the sterol mutants are also different from the kinetics of wild-type membrane enzymes.[13,53,63] Sterol mutants show increased sensitivity to ethanol and detergents.[59] The results from these studies generate another important library of information for use in studying the physiological functions of yeast sterols.

Summary

Yeast mutants defective in ergosterol synthesis are valuable tools for investigating sterol metabolism. Both sterol mutants and sterol auxotrophs have been utilized in determining what sterol structural features are required for yeast cell viability. Both types of mutants can also be studied to ascertain how changes in sterol structure affect membrane properties.

Other aspects of sterol metabolism, such as the specificity of sterol esterification, have been elucidated by the sterol auxotrophs. In broader applications, interrelationships between sterol metabolism and other cellular functions (e.g., heme metabolism) may also be examined with these mutants. By analyzing the lipid composition of the sterol mutants, on the other hand, much of the ergosterol biosynthetic pathway has been delineated. The unusual sterols of the mutants can also be obtained to develop

---

[55] N. D. Lees, M. Bard, M. D. Kemple, R. A. Haak, and F. W. Kleinhans, *Biochim. Biophys. Acta* **553**, 469 (1979).
[56] F. W. Kleinhans, N. D. Lees, M. Bard, R. A. Haak, and R. A. Woods, *Chem. Phys. Lipids* **23**, 143 (1979).
[57] C. A. McLean-Bowen, and L. W. Parks, *Chem. Phys. Lipids* **29**, 137 (1981).
[58] M. Bard, N. D. Lees, L. S. Burrows, and F. W. Kleinhans, *J. Bacteriol.* **135**, 1146 (1978).
[59] N. D. Lees, S. L. Lofton, R. A. Woods, and M. Bard, *J. Gen. Microbiol.* **118**, 209 (1980).
[60] L. W. Parks, E. D. Thompson, and R. B. Bailey, *J. Am. Oil. Chem. Soc.* **51**, 522A (1974).
[61] E. D. Thompson and L. W. Parks, *J. Bacteriol.* **120**, 779 (1974).
[62] C. D. K. Bottema, R. J. Rodriguez, and L. W. Parks, in preparation.
[63] E. D. Thompson and L. W. Parks, *Biochem. Biophys. Res. Commun.* **57**, 1207 (1974).

assays for the enzymes involved in ergosterol synthesis. Thus, by utilizing mutants, the simple eukaryotic system of yeast may be extended to explore the entire field of sterol metabolism and its relationship to cellular physiology.

Acknowledgments

This work was supported in part by grants from the National Science Foundation (PCM-8306625) and the U. S. Public Health Service (AM-05190).

[19] Enzymatic Dealkylation of Phytosterols in Insects

By YOSHINORI FUJIMOTO, MASUO MORISAKI, and NOBUO IKEKAWA

Since the first rigorous demonstration of the dealkylation of ergosterol into Δ^{22}-cholesterol in the German cockroach *Blattella germanica*,[1] several reports have appeared on the conversion of phytosterols into cholesterol in a variety of insects. In general, phytophagous insects utilize phytosterols, e.g., sitosterol (**1**), campesterol, (**2**) and stigmasterol, through the metabolic conversion thereof into cholesterol (**3**). However, carnivorous insects require cholesterol as a nutritional sterol. The dealkylation reaction is one of essential metabolism in insects which have no capacity of *de novo* sterol biosynthesis.[2]

For the deethylation of sitosterol (**1**), Svoboda *et al.* identified fucosterol (**7**) and desmosterol (**5**) as intermediates around 1970.[3] In 1972, we proposed fucosterol 24,28-epoxide (**4**) as the missing link between **7** and **5**.[4] Successive studies, including the proof of hydrogen migration from C-25 to C-24 during the dealkylation, have established the pathway as shown in Scheme 1 (R = CH$_3$).[5,6] Then we focused our attention on determination of which isomer of the 24,28-epoxide (see Fig. 1) is the

[1] A. J. Clark and K. Bloch, *J. Biol. Chem.* **234**, 2583 (1959).
[2] M. Morisaki, Y. Fujimoto, A. Takasu, Y. Isaka, and N. Ikekawa, in "Metabolic Aspects of Lipid Nutrition in Insects" (T. E. Mittler and R. H. Dadd, eds.), p. 17. Westview Press, Boulder, Colorado, 1983; J. A. Svoboda and M. J. Thompson, *ibid.* p. 1; J. A. Svoboda, J. N. Kaplanis, W. E. Robbins, and M. J. Thompson, *Annu. Rev. Entomol.* **20**, 205 (1975).
[3] J. A. Svoboda, M. J. Thompson, and W. E. Robbins, *Nature (London), New Biol.* **230**, 57 (1971); J. A. Svoboda and W. E. Robbins, *Experientia* **24**, 1131 (1968).
[4] M. Morisaki, H. Ohtaka, M. Okubayashi, N. Ikekawa, Y. Horie, and S. Nakasone, *J. Chem. Soc., Chem. Commun.* p. 1275 (1972).
[5] Y. Fujimoto, N. Awata, M. Morisaki, and N. Ikekawa, *Tetrahedron Lett.* p. 4335 (1974).
[6] Y. Fujimoto, M. Morisaki, and N. Ikekawa, *Biochemistry* **19**, 1065 (1980).

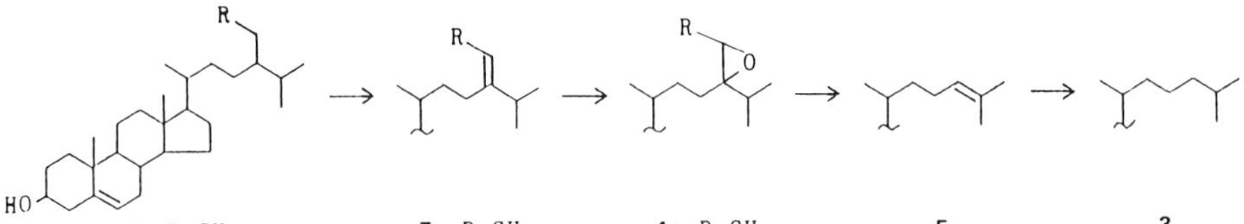

SCHEME 1. Mechanism of phytosterol dealkylation in insects.

intermediate in this dealkylation. Results of *in vivo* experiments using the silkworm *Bombyx mori* are summarized as follows. (1) The larvae are able to utilize (24R,28R)-epoxide (**4a**) and (24S,28S)-epoxide (**4b**), but not the other two isomers (**4cd**) to meet their nutritional requirement for sterol. (2) Epoxides **4a** and **4b** are converted to cholesterol at similar rates. (3) Both **4a** and **4b** are metabolically produced from labeled fucosterol. (4) Both **4a** and **4b**, but not **4cd** are isolated and identified from fifth instar larvae. These and *in vitro* (see Fig. 3) results might lead to the conclusion that both epoxides **4a** and **4b** are intermediates of the dealkylation pathway in the silkworm.[6,7] Somewhat different selectivity of the 24,28-epoxide were reported for *Tenebrio molitor*.[8]

For the demethylation of campesterol (**2**), the intermediacy of 24-methylenecholesterol (**8**) and desmosterol (**5**) has been reported by Svoboda *et al.*[8] However, 24-methylenecholesterol epoxide (**6**) was shown to be an inadequate nutrient for the silkworm growth. In order to obtain a further insight into this problem, we have synthesized deuterated compounds, [24-^{2}H]-, [25-^{2}H]-, [23,23,25-^{2}H$_3$]campesterols and 24-[23,23,25-^{2}H$_3$]methylenecholesterol and examined their metabolism in the silkworm larvae. Gas chromatography–mass spectrometry analysis of metabolically produced cholesterols showed that the cholesterol molecules that came from the four deuterated compounds mentioned above contain zero, one, three, and three deuterium atom in that order. In addition, the desmosterol molecules from the latter two deuterium precursors were shown to have three deuterium atoms. These results indicated that the C-24 hydrogen is lost, while the C-23 and C-25 hydrogens are retained during campesterol demethylation: the C-25 hydrogen must migrate, probably to C-24.[9] Similar results were recently obtained with *Tenebrio molitor*.[10]

[7] N. Ikekawa, Y. Fujimoto, A. Takasu, and M. Morisaki, *J. Chem. Soc., Chem. Commun.* p. 709 (1980).

[8] J. A. Svoboda, M. J. Thompson, and W. E. Robbins, *Lipids* **7**, 156 (1972).

[9] S. Maruyama, Y. Fujimoto, M. Morisaki, and N. Ikekawa, *Tetrahedron Lett.* **23**, 1701 (1982).

[10] F. Nicotra, F. Ronchetti, G. Russo, and L. Toma, *J. Chem. Soc., Perkin Trans. 1* p. 787 (1983).

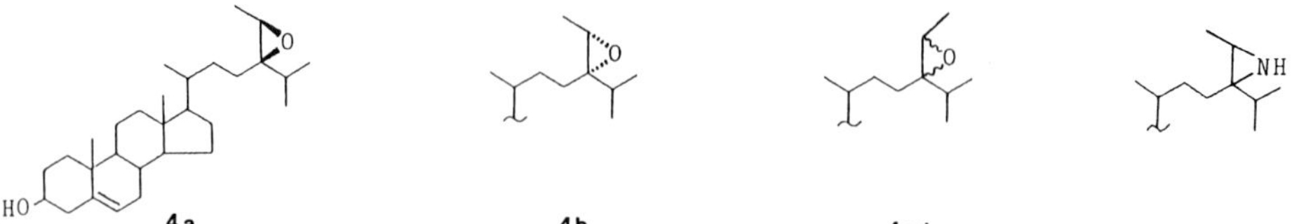

FIG. 1. Structure of 24,28-epoxide stereoisomers (**4a–d**) and imine (**9**).

Thus, campesterol (**2**) would be dealkylated through 24-methylenecholesterol 24,28-epoxide (**6**).

For the deethylation of stigmasterol, similar migration of hydrogen from C-25 to C-24[11] and the intermediacy of cholesta-5,22,24-triene-3β-ol and desmosterol[12] have been observed so far.

Among the four steps of enzymatic reaction (see Scheme 1), active cell free preparation has been prepared for the steps of the epoxide (**4**) → desmosterol (**5**) (prepared from *Bombyx mori*,[6,13] *Schistocerca gregaria*,[14] and *Spodoptera littoralis*[14]) and desmosterol (**5**) → cholesterol (**3**) (prepared from *Manduca sexta*[15] and *Bombyx mori*[6]).

Synthesis of Substrate

A 1:1 mixture of (24R,28R)- and (24S,28S)fucosterol 24,28-epoxide (**4ab**) was prepared as follows. A solution of fucosterol acetate (9.1 g) and *m*-chloroperbenzoic acid (5.2 g) in chloroform (300 ml) was stirred for 5 min at 0°. The mixture was washed with 1 N NaOH and then with water and dried over Na_2SO_4. After removal of the solvent, the product was chromatographed on silica gel. The fraction eluted with hexane–benzene (1:3) afforded the epoxide acetate (6.9 g). The acetate (370 mg) was treated with 5% KOH–methanol (10 ml) and tetrahydrofuran (10 ml) at room temperature for 1 hr. Extraction with ether, washing with brine, drying over Na_2SO_4, and removal of solvent gave crude epoxide, which was chromatographed on silica gel. Elution with benzene afforded crystalline **4ab** (280 mg).

[11] Y. Fujimoto, M. Kimura, A. Takasu, F. A. M. Khalifa, M. Morisaki, and N. Ikekawa, *Tetrahedron Lett.* **25**, 1501 (1984).
[12] J. A. Svoboda, R. F. N. Hutchins, M. J. Thompson, and W. E. Robbins, *Steroids* **14**, 469 (1969); J. A. Svoboda and W. E. Robbins, *Experientia* **24**, 2461 (1968).
[13] N. Awata, M. Morisaki, and N. Ikekawa, *Biochem. Biophys. Res. Commun.* **64**, 157 (1975).
[14] H. H. Rees, T. G. Davies, L. N. Dinan, W. J. S. Lockley, and T. W. Goodwin, in "Progress in Ecdysone Research" (J. A. Hoffmann, ed.), p. 125. Elsevier/North-Holland Biomedical Press, Amsterdam, 1980.
[15] J. A. Svoboda, M. Womack, M. J. Thompson, and W. E. Robbins, *Comp. Biochem. Physiol.* **30**, 541 (1969).

The corresponding [3α-³H]epoxide was obtained as follows. Fucosterol 24,28-epoxide (**4ab**) (3 mg) was oxidized with chromic oxide (5 mg)–pyridine (8 μl) complex in dichloromethane (0.3 ml) at room temperature for 15 min. The crude products obtained after usual work up using ether as an extracting solvent, were dissolved in isopropanol (50 μl) and added to the solution of sodium [³H]borohydride (~8 mCi) (New England Nuclear, 7 Ci/mmol) in isopropanol (30 μl). After standing at room temperature for 1 hr, unlabeled sodium borohydride (1 mg) was added to complete the reaction. The product was chromatographed on silica gel affording 3α-³H-labeled **4ab** (1.1 mg, 101 Ci/mol) by elution with benzene.[13]

Prepration of Cell-Free Extracts

Fifty guts of the fifth instar larvae of silkworms were removed in a chilled Bucher's medium[16] at 5°. The guts were ground with sea sand in a motar containing 10 ml of the same medium. The homogenate was transferred to a test tube by aid of Bucher's medium (final volume was adjusted to 50 ml) and centrifuged at 1500 g for 10 min at 4°. The resulting supernatant was distributed into 10 test tubes (each 5 ml, protein concentration 5.0 mg/ml was estimated by Lowry–Folin method) and used for incubation.

The cell-free extracts can be stored frozen for at least 1 week without significant loss of activity.

Although a detail of subcellular distribution of the enzyme activity must await further investigation, it was preliminarily observed that the 20,000 g supernatant performed the conversion as effectively as the 1500 g supernatant did. It should also be noted that the insect gut does not seem to be the obligatory site of dealkylation.[13]

Incubation and Analysis of Incubation Products

Two assay methods, a tracer assay and a gas–liquid chromatographic (GLC) assay, have been developed. The GLC method is more convenient when more than 50 μg of a substrate epoxide is incubated.

Tracer Assay. Five milliliters of the extracts was mixed with dimethylformamide (50 μl) solution of [3α-³H]epoxide (**4ab**) (10 μg, 2.24 μCi) and the mixture was incubated in air at 30° for 2 hr. The enzyme reaction was terminated by the addition of methanol (5 ml), lipids were extracted with ether (3 × 20 ml), and then saponified with 5% KOH–methanol (10 ml) under reflux for 1 hr. To the nonsaponifiable fraction (recovery of the radioactivity is approximately 90%) a few mg of cholesterol was added as

[16] N. L. R. Bucher and K. McGarrahan, *J. Biol. Chem.* **222**, 1 (1956).

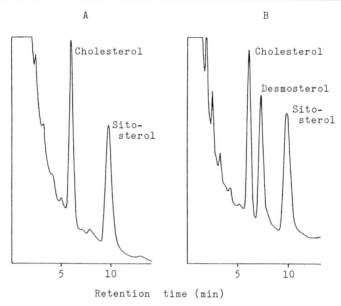

FIG. 2. GLC analysis of desmosterol formed by incubation with cell-free extracts. (A) Background sterols in cell-free extracts; (B) sterols after incubation of fucosterol 24,28-epoxide (**4ab**) (200 μg).

a carrier. This mixture was treated with benzoyl chloride (50 μl) in pyridine (1 ml) at room temperature overnight. By the usual work-up a mixture of benzoates was obtained, which was diluted with a carrier (a few milligrams of **4ab**-benzoate). Column chromatography on silica gel and elution with hexane–benzene (2:1) afforded a "sterol fraction" which contained cholesterol benzoate and desmosterol benzoate. Radioactivity of an aliquot of the fraction was determined with a Packard 3320 liquid scintillation counter using a solution of toluene containing 0.4% DPO and 0.03% dimethyl-POPOP. The conversion yield (23% in the typical run) into the sterol fraction in which radioactivity resided in desmosterol benzoate exclusively was determined after correction for the recovery in the extraction stage.[13]

Gas–Liquid Chromatographic Assay. To the extracts (5 ml) was added acetone (100 μl) solution of the epoxide (**4ab**) (200 μg) and the mixture was incubated at 30°. After being shaken for 2 hr, the mixture was extracted with 10 ml of ethyl acetate twice using a Vortex mixer (sterol recovery is approximately 90%). The combined organic phase was evaporated to dryness under reduced pressure. The residue was dissolved in a small amount of ethyl acetate and an aliquot was analyzed by GLC (1%

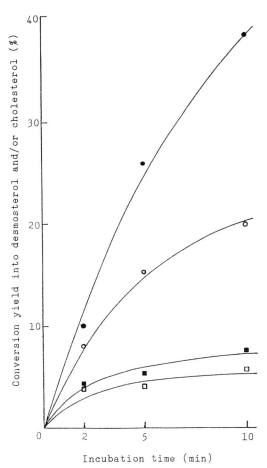

FIG. 3. Time course of formation of desmosterol and/or cholesterol by incubation of fucosterol 24,28-epoxide stereoisomer (**4a** or **4b**) with cell-free extracts (20,000 g sup). (●) (24R,28R)-Epoxide (**4a**) plus NADPH; (○) (24R,28R)-epoxide; (■) (24S,28S)-epoxide (**4b**) plus NADPH; (□) (24S,28S)-epoxide.

SUBSTRATE SPECIFICITY OF EPOXIDE ANALOGS

Substrate epoxide	(%) Conversion yield into desmosterol
Fucosterol 24,28-epoxide (**4ab**)	32
Fucosterol 24,28-epoxide benzoate (**4ab**-benzoate)	8
24-Methylenecholesterol 24,28-epoxide (**6**)	16

OV-17 on Shimalite W (AW-DMCS), 1.5 m × 4 mm i.d., oven temperature 270°). The amount of desmosterol observed was calculated by using a calibration curve. A typical GLC trace is illustrated in Fig. 2.[17]

Substrate Specificity and Comments

Although the substrate specificity of this enzyme has not been systematically explored, the following data have been obtained so far. The cell free extracts may catalyze the first two steps, i.e., conversion of sitosterol (**1**) to fucosterol (**7**)[17] and fucosterol (**7**) to fucosterol 24,28-epoxide (**4**)[6] under a modified incubation condition in a low yield.[6] However, Δ^{24}-reductase activity [conversion of desmosterol (**5**) to cholesterol (**3**)] was easily observed after addition of NADPH.[6]

The substrate specificity of the enzyme was examined for (24R,28R)- and (24S,28S)-epoxides (**4a** and **4b**) with or without NADPH using a 20,000 g supernatant. The results are illustrated in Fig. 3, which indicates that **4a** is a superior substrate to **4b**.[6] Conversion yield into desmosterol of the other epoxides was compared using GLC assay method (see the table).[17]

When the imine (**9**),[18] a structural analog of fucosterol 24,28-epoxide (**4**), was added to the cell-free extracts at the same level with the epoxide, the conversion of the epoxide (**4ab**) to desmosterol (**5**) was completely blocked.[13]

[17] Y. Fujimoto and N. Ikekawa, unpublished work.
[18] Y. Fujimoto, M. Morisaki, and N. Ikekawa, *Steroids* **24**, 367 (1974).

[20] Side-Chain Cleavage of Cholesterol by Gas Chromatography–Mass Spectrometry in a Selected Ion Monitoring Mode

By Masuo Morisaki, Mikio Shikita, and Nobuo Ikekawa

The first step of steroid hormone production is the carbon–carbon bond cleavage between C-20 and C-22 of cholesterol. The reaction is catalyzed by the cytochrome *P*-450 specific for side-chain cleavage (SCC) of cholesterol in the presence of adrenodoxin, adrenodoxin reductase, and a NADPH-generating system. Three moles each of molecular oxygen and NADPH is required for the conversion of 1 mol of cholesterol to pregnen-

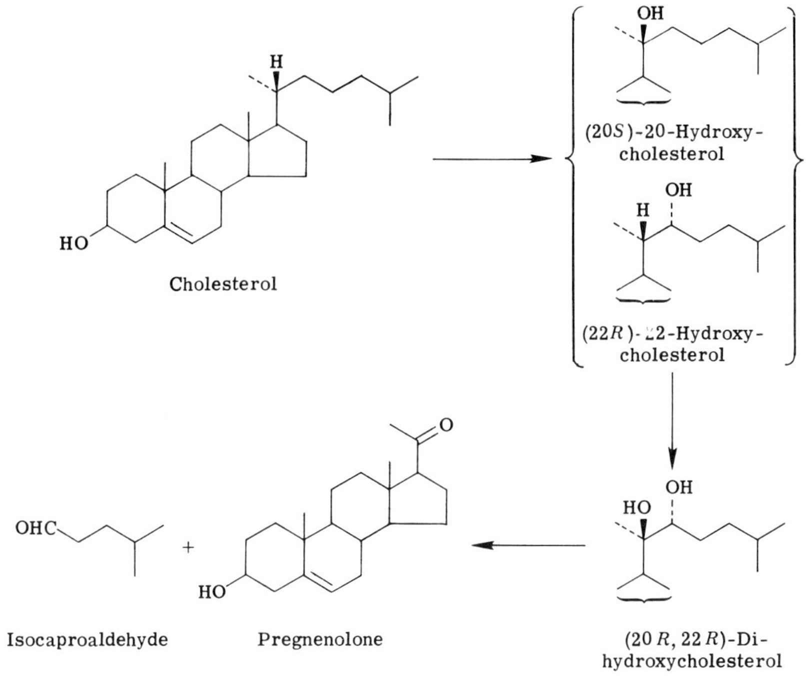

FIG. 1. Pathway of side-chain cleavage of cholesterol.

olone and isocaproaldehyde. As the intermediates of this conversion, (20R,22R)-20,22-dihydroxycholesterol, (22R)-22-hydroxycholesterol, and (20S)-20-hydroxycholesterol have been proposed (Fig. 1). The mechanism of SCC of cholesterol has been studied mainly by the use of radioactive substrates (^{3}H or ^{14}C),[1] but their preparation and a precise analysis of the reaction products require rather complex procedures. Analysis by gas chromatography–mass spectrometry (GC-MS) of incubation products from nonlabeled substrates or the substrates labeled with a stable isotope (^{2}H, ^{13}C, or ^{18}O) may be an attractive alternative for the method to investigate SCC of cholesterol. The usefulness of the GC-MS method is exemplified by the work of Burstein et al.[2] By GC-MS of (20R,22R)-20,22-dihydroxycholesterol formed from cholesterol under $^{18}O_2$ and $^{16}O_2$, they demonstrated that the oxygen atoms of the 20- and 22-hydroxyl groups were derived from distinct oxygen molecules. The result indicates sequential hydroxylation of cholesterol.

[1] This series, Vol. 15 [27].
[2] S. Burstein, B. S. Middleditch, and M. Gut, J. Biol. Chem. **250**, 9028 (1975).

The drawback of the GC-MS method, its low sensitivity of detection as compared to radioactive assay, can be overcome by increasing the size of the incubation volume as well as by utilizing the selected ion monitoring (SIM) method. In this technique, a mass spectrometer is used as a detector of the gas chromatograph. One or several ion species that are characteristic of the compound under analysis are monitored. If one m/z value is going to be monitored (single ion monitoring), the magnetic field is adjusted to focus that m/z value using a standard sample. If several ions are sequentially monitored in rapid succession (multiple ion monitoring), an m/z value of the GC column background is set as a reference and the preselected ions are alternately focused through an accelerating voltage alternator. In this way, detection sensitivity increases approximately by a thousand times compared to GC with a flame ionization detector.

Incubation

Cytochrome P-450$_{scc}$,[3] adrenodoxin,[4] and adrenodoxin reductase (EC 1.18.1.2, ferredoxin-NADP$^+$ reductase)[4] are prepared as described. Cholesterol or its analogs (1–10 μg dissolved in 10 μl of dimethylformamide) is placed in 0.37 ml of 50 mM potassium phosphate buffer pH 7.2, containing 0.1 nmol of cytochrome P-450$_{scc}$, 2 nmol of adrenodoxin, five dinitrophenol indophenol units of adrenodoxin reductase, 5 mM cysteine, and 1 mM EDTA. After 20 min of preincubation, 0.13 ml of NADPH-generating system (1.15 μmol of glucose 6-phosphate, 85 nmol of NADP, and 0.5 ng of glucose-6-phosphate dehydrogenase) is added and the mixture is incubated at 37° for 5–10 min. The enzymatic reaction is terminated by addition of 5 ml of dichloromethane followed by vigorous agitation with a Vortex mixer for 10 sec.

Determination of Pregnenolone[5]

Internal standard [17,21,21,21-^{2}H$_4$]pregnenolone is prepared as follows. To a mixture of anhydrous dioxane (10 ml) and deuterium oxide (Merck, 10 ml), is added slowly sodium metal (250 mg) and then pregnenolone (Steraloids, 100 mg). After refluxing for 30 min, the mixture is extracted with diethyl ether (50 ml), washed with water, dried over magnesium sulfate, and evaporated to dryness. Recrystallization from ethyl acetate gives the deuterated pregnenolone (75 mg), mp 187–190°. The mass spectrum shows m/z 320 (M$^+$), 302 (M-H$_2$O), and 287 (M-H$_2$O-Me).

[3] This series, Vol. 37 [25].
[4] This series, Vol. 52 [13].
[5] M. Morisaki, C. Duque, N. Ikekawa, and M. Shikita, *J. Steroid Biochem.* **13**, 545 (1980).

For determination of pregnenolone, [^{2}H$_4$]pregnenolone (200 ng) is added to the dichloromethane used for extraction of the incubation product (vide supra). The extract is concentrated by solvent evaporation and transferred into a small test tube (4 cm × 0.4 mm i.d.). The solvent is evaporated, leaving the residue at the bottom of test tube. To the residue is added trimethylsilylimidazole (20–30 μl) to convert pregnenolone to the TMS ether. A 2 μl-portion of this solution is injected into a Shimadzu-LKB 9000S gas chromatograph–mass spectrometer equipped with a multiple ion detector. The column (1 m × 3 mm i.d.) is packed with 1.5% OV-17 on Shimalite W (80–100 mesh) and operated at 250° with helium carrier gas at a flow rate of 30 ml/min. The temperature of the injection port and separator is set at 280 and 290°, respectively. The ions of m/z 298 and m/z 302 (corresponding to molecular ion minus trimethylsilanol) are used to monitor pregnenolone and [^{2}H$_4$]pregnenolone, respectively. The ratio between heights of the peaks m/z 298 and m/z 302 of mixtures of different amounts of standard pregnenolone together with a fixed amount of standard [^{2}H$_4$]pregnenolone is linearly related with the amount of pregnenolone. Using this calibration curve, pregnenolone in incubation product is determined (Fig. 2). The minimum amount of pregnenolone to be determined by this method is 0.1 ng with an error of 3%.

Analysis of Hydroxylated Cholesterol Derivatives[6]

The intermediates of cholesterol side-chain cleavage are accumulated in the incubation medium only in minute amounts, because they are instantly transformed into pregnenolone. Therefore, a highly sensitive method such as SIM is needed for their identification. In the following way, all the three hydroxylated choelsterols, (20S)-20-*hydroxycholesterol*, (22R)-22-hydroxycholesterol, and (20R,22R)-20,22-dihydroxycholesterol, can be identified from the incubation product of 10 μg of cholesterol (Fig. 3). The crude extract of the incubation mixture is applied to a silica gel plate (3 × 7 cm) and developed with *n*-hexane/ethyl acetate (2 : 1). The hydroxycholesterols are located at the zone of R_f 0.2–0.5 and are extracted from the silica gel by dichloromethane. The residue obtained by solvent evaporation is heated with trimethylsilyl imidazole (20 μl) in a sealed tube at 80° for 1 hr. A 2 μl-portion of the resulting solution is analyzed in the same manner as described for pregnenolone except for a column temperature of 268° and the monitoring ions: m/z 173 for (22R)-22-hydroxycholesterol bis-TMS ether, m/z 201 for (20S)-20-hydroxycholesterol bis-TMS ether, and m/z 461 for (20R,22R)-20,22-dihydroxycholesterol tris-TMS ether.

[6] M. Morisaki, S. Sato, N. Ikekawa, and M. Shikita, *FEBS Lett.* **72**, 337 (1976).

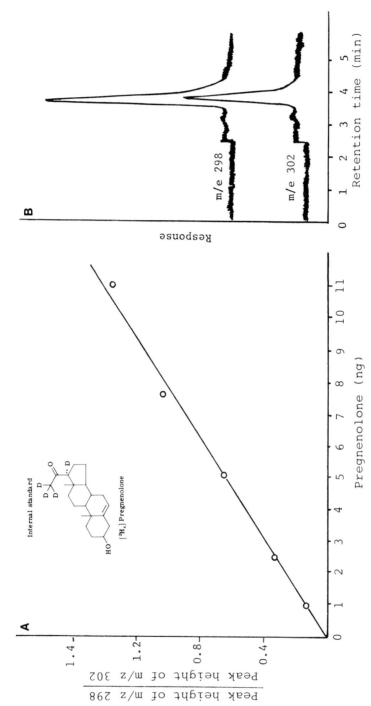

FIG. 2. Determination of pregnenolone by GC-MS on SIM mode. (A) The ratio between peak heights at *m/z* 298 and *m/z* 302 in SIM of the TMS ethers of different standard mixtures of pregnenolone with a fixed amount (10 ng) of [$^2H_4$]pregnenolone. (B) SIM of the extract of incubation mixture of (20*R*)-20-isopentylpregn-5-en-3β-ol (compound 10 in Fig. 7).

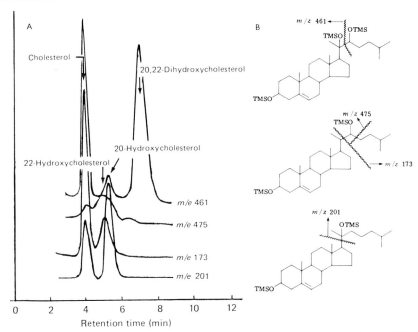

FIG. 3. Detection of hydroxylated cholesterols on incubation of cholesterol. (A) SIM of 20- and 22-hydroxy- and 20,22-dihydroxycholesterol (in the form of the TMS ethers) which were found by incubation (2.5 min) of cholesterol (10 μg) with P-450$_{scc}$. (B) Mass spectra fragmentation pattern of the three hydoxylated cholesterol TMS ethers.

TABLE I
INCUBATIONS OF CHOLESTEROL ANALOGS WITH CYTOCHROME P-450$_{scc}$

Substrates	Incubation products Analytical forms (monitored fragment ion)	Yield (%)
Cholest-5-en-3β-ol	Pregn-5-en-20-on-3β-yl TMS ether (m/e 298)	32
Cholest-5-en-3α-ol	Pregn-5-en-20-3β-yl TMS ether (m/e 298)	12
Cholest-5-en-3β-yl palmitate	Pregn-5-en-20-3β-yl TMS ether (m/e 298)	<0.1
Cholest-5-en-3β-yl sulfate	Pregn-5-en-20-on-3β-yl TMS ether (m/e 298)	<0.1
Cholest-5-en-3β-yl acetate	Pregn-5-en-20-on-3β-yl acetate (m/e 298)	<0.1
4,4-Dimethylcholest-5-en-3β-ol	4,4-Dimethylpregn-5-en-20-on-3β-yl TMS ether (m/e 287)	<0.1
5α-Cholestan-3β-ol	5α-Pregn-20-on-3β-yl TMS ether (m/e 300)	<1
5β-Cholestan-3β-ol	5α-Pregn-20-on-3β-yl TMS ether (m/e 300)	<0.1
Cholest-4-en-3-one	Pregn-4-ene-3,20-dione (m/e 314)	3

Under these conditions, some of the stereoisomers of the hydroxylated cholesterol derivatives are separated from each other. Thus, the method described above also is useful for elucidation of some stereochemical problems. Retention times of the TMS ethers in minutes are (20R)-20-hydroxycholesterol, 14.6; (20S)-20-hydroxycholesterol, 15.2; (22S)-22-hydroxycholesterol, 13.6; (22R)-22-hydroxycholesterol, 14.4; (20S,22R)-20,22-dihydroxycholesterol, 18.6; (20S,22S)-20,22-dihydroxycholesterol, 18.6; (20R,22R)-20,22-dihydroxycholesterol, 19.6; (20R,22S)-20,22-dihydroxycholesterol, 20.4. A better separation can be obtained on an open tubular glass capillary column (SE-30, 30 m × 0.3 mm i.d.) at 290° (Fig. 4).

Analysis of Cleavage Products of Cholesterol Analogs Modified in A/B Ring[7]

Side-chain cleavage products from cholesterol analogs modified in an A/B-ring can be analyzed by SIM on the fragment ions indicted in Table I. Incubation and SIM analysis are carried out in the same manner as described for pregnenolone except progesterone is analyzed without derivatization. Some esters of cholesterol, e.g., sulfate, acetate, or palmitate, can also be examined by a prior acidic or basic hydrolysis of the incubation products followed by SIM analysis of the resulting pregnenolone.

Analysis of Possible Cleavage Product of Cholesterol Analogs Containing a Vicinal Diol Function in the Side Chain[8]

Although (20R,22R)-20,22-dihydroxycholesterol is a good substrate for cytochrome $P\text{-}450_{scc}$, its analogs, e.g., the 22,23-diol, 23,24-diol, 24,25-diol, and 25,26-diol are transformed into pregnenolone only in minute or negligibly small amounts.[5] It can be postulated that these compounds are cleaved not on the C-20,22 bond but the respective vincinal diol portions to yield the corresponding aldehyde or ketone. These carbonyl compounds, if produced, can be identified by SIM in the following manner. The incubation mixture is extracted with diethyl ether (3 ml). The ether extract is dried over magnesium sulfate and filtered. The filtrate is stirred with addition of lithium aluminum hydride (20 mg) at 20° for 5 min. Excess of reagent is destroyed by addition of moist ether (5 ml). The residue is removed by decantation and the ether layer is washed with 1 N hydrochloric acid (2 ml) and water (2 × 2 ml), dried over magnesium

---

[7] M. Morisaki, C. Duque, K. Takane, N. Ikekawa, and M. Shikita, *J. Steroid Biochem.* **16**, 101 (1982).

[8] A. Calzolari, M. Morisaki, and N. Ikekawa, unpublished.

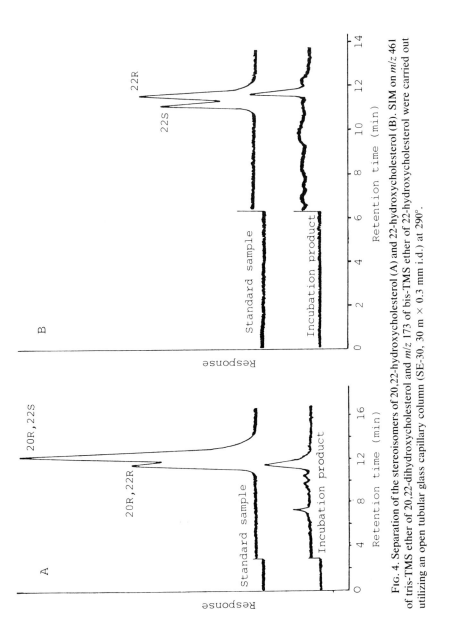

FIG. 4. Separation of the stereoisomers of 20,22-hydroxycholesterol (A) and 22-hydroxycholesterol (B). SIM on m/z 461 of tris-TMS ether of 20,22-dihydroxycholesterol and m/z 173 of bis-TMS ether of 22-hydroxycholesterol were carried out utilizing an open tubular glass capillary column (SE-30, 30 m × 0.3 mm i.d.) at 290°.

TABLE II
SIM ANALYSIS OF SCC PRODUCT FROM THE VICINAL DIOLS[a]

Substrate	Product	Analytical form	Selected ions
R^1–CH(OH)–CH(OH)–iPr	R^1–CHO	R^2–CH$_2$OTMS	m/z 386
R^1–CH(OH)–CH(OH)–CH(CH$_3$)–	R^1–CHO	R^2–CH$_2$OTMS	m/z 400
R^1–...–C(OH)(CH$_3$)–CH$_2$OH	R^1–...–CHO	R^2–...–CH$_2$OTMS	m/z 414, 375
R^1–...–C(OH)(CH$_3$)–CH$_2$CH$_2$OH	R^1–...–C(=O)–	R^2–...–OTMS	m/z 442, 403

[a] R^1 = cholesterol-3-ol moiety (3β-hydroxy-Δ5); R^2 = cholesterol-3-OTMS moiety (3β-OTMS-Δ5)

sulfate, and evaporated to dryness. The residue in a small test tube is added with trimethylsilylimidazole (20 μl) and a 2-μl portion of this solution is analyzed by SIM on the ions indicated in Table II.

Application

For understanding the mechanism of cholesterol SSC, identification of the intermediates is of crucial importance. For this purpose, the SIM method is extremely useful. For example, the recently proposed 20(22)-dehydrocholesterol and the corresponding 20,22-epoxide were excluded as intermediates of the cholesterol SCC, because pregnenolone was not

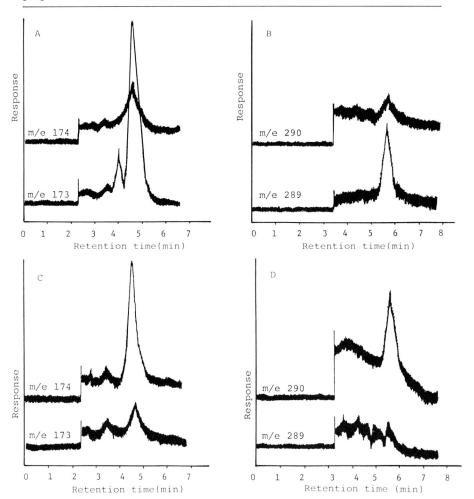

FIG. 5. SIM of 22-hydroxycholesterol (A,C) and 20,22-dihydroxycholesterol (B,D) as their TMS ethers, produced on incubation of (22R)-[22-^{2}H]cholesterol (A,B) and (22S)-[22-^{2}H]cholesterol (C,D) with adrenocortical mitochondria acetone powder.

detectable in any significant amount by SIM in the incubation of these compounds with the SCC enzyme system.[9]

The stereochemical course of C-22 hydroxylation of cholesterol, which is an obligatory step in cholesterol SCC, was clarified by incubation

[9] M. Morisaki, K. Bannai, N. Ikekawa, and M. Shikita, *Biochem. Biophys. Res. Commun.* **69**, 481 (1976).

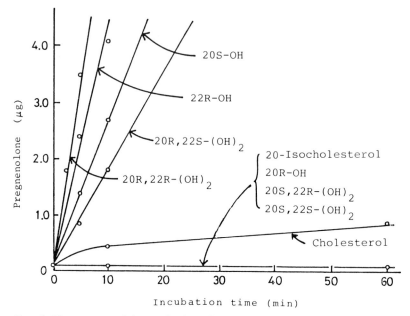

FIG. 6. Time course of the production of pregnenolone when various C-20 and C-22 stereoisomers of cholesterol derivatives (10 μg each) were incubated with $P\text{-}450_{scc}$.

of cholesterol and (20S)-20-hydroxycholesterol stereospecifically labeled with deuterium at 22-pro-R or 22-pro-S positions, and the subsequent SIM analysis of the incubation products. The SIM on m/z 173 and 174 for (22R)-22-hydroxycholesterol bis-TMS ether, and on m/z 289 and 290 for (20R,22R)-20,22-dihydroxycholesterol tris-TMS ether revealed that the 22-hydroxylation occurs by stereospecific displacement of the 22-pro-R hydrogen with retention of configuration (Fig. 5).[10]

Isoelectrofocusing of a highly purified cytochrome $P\text{-}450_{scc}$ showed a single peak of enzyme activity that was assayed by SIM quantitation of the pregnenolone produced by incubation of either cholesterol, (20S)-20-hydroxycholesterol, (22R)-22-hydroxycholesterol, or (20R,22R)-20,22-dihydroxycholesterol. This result is one piece of evidence that a single species of enzyme catalyzes all three steps of the reaction, i.e., 20-hydroxylation, 22-hydroxylation, and SCC between C-20 and C-22.[11]

Studies on the substrate specificity of cytochrome $P\text{-}450_{scc}$ are important not only for information on the cholesterol SCC mechanism but for

[10] C. Duque, M. Morisaki, N. Ikekawa, and M. Shikita, *Tetrahedron Lett.* p. 4479 (1979).
[11] C. Duque, M. Morisaki, N. Ikekawa, and M. Shikita, *Biochem. Biophys. Res. Commun.* **82**, 179 (1978).

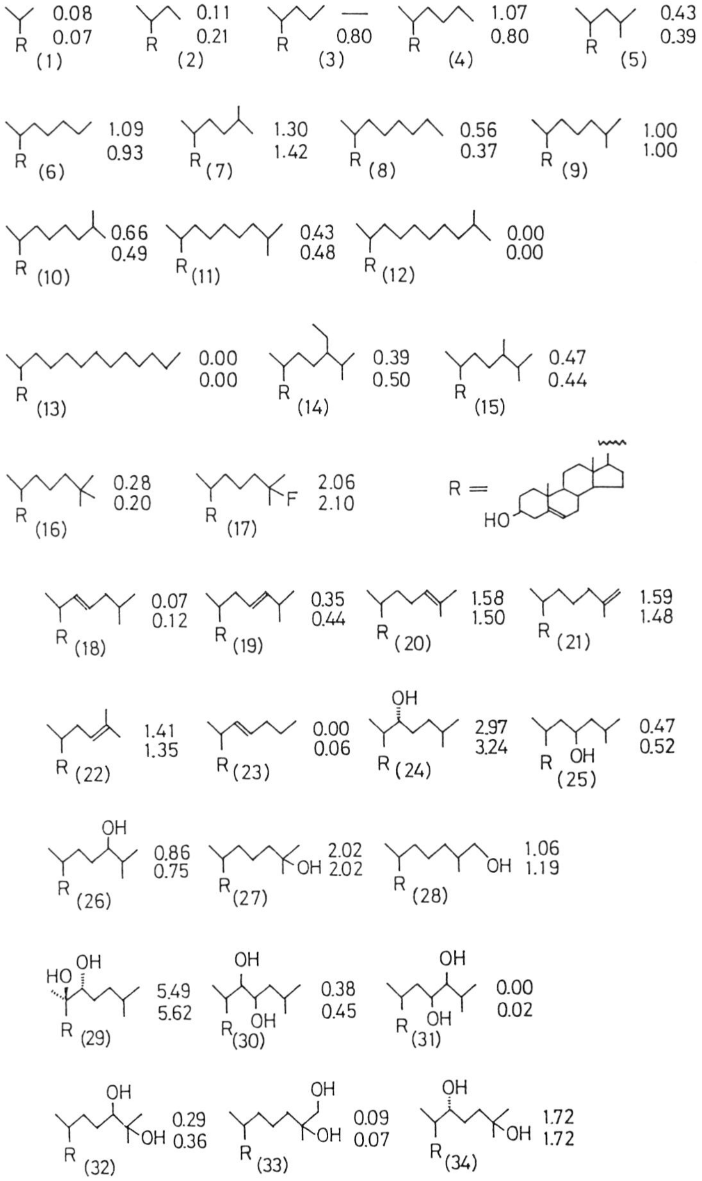

FIG. 7. Cleavage of C-20,22 bond of various cholesterol analogs with modified side chains by $P\text{-}450_{scc}$. The amounts of pregnenolone produced in 5 min (top) and 10 min (bottom) incubation of these compounds are presented on the right hand side of each compound in the yield relative to the amount of pregnenolone produced from cholesterol (0.147 nmol in 5 min and 0.210 nmol in 10 min).

basic considerations of topography of the active site of the enzyme. About 50 analogs of cholesterol have been chemically synthesized, and their efficiency as the substrate for cytochrome P-450$_{scc}$ was examined by SIM analysis of the cleavage products.[5-8,12] It was found that 20-isocholesterol, (20R)-20-hydroxycholesterol, (20S,22R)- and (20S,22S)-20,22-dihydroxycholesterol are not converted to pregnenolone at all, whereas (22R)- and (22S)-22-hydroxycholesterol and (20R,22R)- and (20R,22S)-20,22-dihydroxycholesterol are all good substrate for cytochrome P-450$_{scc}$ (Fig. 6). These results suggest that this enzyme requires strict stereospecificity around C-20 of the substrate, although it is generous about the configuration at C-22. Figure 7 indicates that cholesterol analogs either with 5 to 10 saturated carbon side chain, with a double bond at C-24 or C-25, or with hydroxyl group(s) at C-22, C-24, C-25, C-26, or C-22,25, are all effectively converted to pregnenolone. The C-20,22 bonds of 3-epicholesterol and cholest-4-en-3-one are also cleaved in a moderate efficiency (Table I). However, saturation of the 5,6-double bond, introduction of the 4,4-dimethyl group, or esterification of the 3-hydroxyl group completely prohibits the cleavage reaction. The side-chain analogs with a shorter (C_3 or C_4) or longer (C_{11} or C_{13}) carbon length also are not substrates for cytochrome P-450$_{scc}$ (Fig. 7).

[12] R. Goto, M. Morisaki, and N. Ikekawa, *Chem. Pharm. Bull.* **31**, 3528 (1983).

[21] Purification from Rabbit and Rat Liver of Cytochromes P-450 Involved in Bile Acid Biosynthesis

By Stefan Andersson, Hans Boström, Henry Danielsson, and Kjell Wikvall

The conversion of cholesterol into the primary bile acids cholic acid and chenodeoxycholic acid in the liver involves several hydroxylations of the steroid nucleus as well as of the side chain.[1] The first and rate-limiting step is the 7α-hydroxylation of cholesterol. In the formation of cholic acid a 12α-hydroxylation occurs at an early stage. The degradation of the C_{27}-steroid side chain starts with a 26-hydroxylation. The side chain may also be hydroxylated in the C-25 position in an alternative pathway. The 7α-, 12α-, and 25-hydroxylations are catalyzed by microsomal monooxygenase systems involving cytochromes P-450 and NADPH-cytochrome

[1] H. Danielsson and J. Sjövall, *Annu. Rev. Biochem.* **44**, 233 (1975).

P-450 reductase.[2,3] The 26-hydroxylation is catalyzed by a mitochondrial monooxygenase system involving cytochrome P-450, ferredoxin, and ferredoxin reductase.[4,5] The microsomal cytochromes P-450 participating in bile acid biosynthesis show a high degree of specificity and at least the 7α-hydroxylation of cholesterol is under metabolic control. Also the 12α-hydroxylation has been ascribed a regulatory role in determining the ratio between cholic acid and chenodeoxycholic acid.[1]

This chapter describes methods for purification of cytochrome P-450 fractions active in hydroxylations in bile acid biosynthesis from rat and rabbit liver.

Assay Methods

Cytochrome P-450 Concentration. The concentration of cytochrome P-450 in purified preparations as well as in microsomes is determined from the CO difference spectrum between 450 and 490 nm of the reduced sample using the extinction coefficient of 91 mM^{-1} cm^{-1}.[6]

Protein Concentration. Protein concentrations are determined as described by Lowry *et al.*[7] using crystalline bovine serum albumin as standard.

Gel Electrophoresis. Polyacrylamide gel electrophoresis is performed in the presence of sodium dodecyl sulfate using Pharmacia gradient slab gels, PAA 4/30. The electrophoresis buffer is at pH 7.4 and contains 40 mM Tris–Cl, 20 mM sodium acetate, 2 mM EDTA, and 0.2% (w/v) sodium dodecyl sulfate. Electrophoresis is carried out in an electrophoresis apparatus GE-4 II (Pharmacia) as follows. After preelectrophoresis 70 V for 1 hr without samples, the protein samples (1–5 μg), pretreated as described by Laemmli,[8] are applied and electrophoresis is carried out first at 200 V for 10 min and then at 150 V for 3–6 hr. Staining and destaining of the gels are performed as described by Haugen *et al.*[9]

Incubation Conditions and Analyses of Incubation Mixtures. The catalytic activities of the purified cytochromes P-450 toward various C_{27}-steroids are analyzed in reconstituted systems. The microsomal cytochromes P-450 are reconstituted with NADPH–cytochrome P-450

[2] H. Boström and K. Wikvall, *J. Biol. Chem.* **257**, 11755 (1982).
[3] R. Hansson and K. Wikvall, *Eur. J. Biochem.* **125**, 423 (1982).
[4] R. Sato, Y. Atsuta, Y. Imai, S. Taniguchi, and K. Okuda, *Proc. Natl. Acad. Sci. U.S.A.* **74**, 5477 (1977).
[5] J. I. Pedersen, I. Björkhem, and J. Gustafsson, *J. Biol. Chem.* **254**, 6464 (1979).
[6] T. Omura and R. Sato, *J. Biol. Chem.* **239**, 2379 (1964).
[7] O. H. Lowry, N. J. Rosebrough, A. L. Farr, and R. J. Randall, *J. Biol. Chem.* **193**, 265 (1951).
[8] U. K. Laemmli, *Nature (London)* **227**, 680 (1970).
[9] D. A. Haugen, T. A. van der Hoeven, and M. J. Coon, *J. Biol. Chem.* **250**, 3567 (1975).

reductase and dilauroylglycero-3-phosphorylcholine. The mitochondrial cytochrome P-450 is reconstituted with ferredoxin and ferrodoxin reductase from bovine adrenal mitochondria. Sources and synthesis of the various labeled C_{27}-steroids have been described elsewhere.[10,11]

7α-Hydroxylation of cholesterol catalyzed by a microsomal cytochrome P-450 is assayed as follows. Cytochrome P-450, 0.05–0.5 nmol, is preincubated at 37° for 2 min with 1–3 units of NADPH–cytochrome P-450 reductase, 25 µg of dilauroylglycero-3-phosphorylcholine, and 1–5 µmol of dithiothreitol in a volume of 0.85 ml of 50 mM Tris-acetate buffer, pH 7.4, containing 20% glycerol and 0.1 mM EDTA. [4-^{14}C]Cholesterol (61 Ci/mol), 25 nmol in 25 µl of acetone, and 0.75 mg of Triton X-100 are then added. The reaction is started by addition of 1 µmol of NADPH. The mixture (final volume, 1 ml) is incubated at 37° for 5–20 min. The reaction is terminated by addition of 3 ml of ethanol, 96% (v/v), and the incubation mixture is extracted twice with hexane/ethyl acetate (1/1, v/v). The organic phase is evaporated and the residue subjected to thin-layer chromatography. The developing solvent is toluene/ethyl acetate (2/3, v/v).

12α-Hydroxylation of 5β-cholestane-$3\alpha,7\alpha$-diol and 25-hydroxylation of 5β-cholestane-$3\alpha,7\alpha,12\alpha$-triol catalyzed by microsomal cytochromes P-450 are assayed as follows. Cytochrome P-450, 0.1–0.5 nmol, is preincubated at 37° for 2 min with 2–4 units of NADPH–cytochrome P-450 reductase and 25 µg of dilauroylglycero-3-phosphorylcholine before addition of 5β-[7β-^{3}H]cholestane-$3\alpha,7\alpha$-diol (500 Ci/mol) or 5β-[7β-^{3}H]cholestane-$3\alpha,7\alpha,12\alpha$-triol (500 Ci/mol), 125–250 nmol in 25 µl of acetone, and 1–2 µmol of NADPH in a total volume of 1 ml of 0.15 M phosphate buffer, pH 7.4, containing 20% glycerol, 0.1 mM EDTA. In incubations with 5β-cholestane-$3\alpha,7\alpha,12\alpha$-triol and rat liver microsomal cytochrome P-450 the phospholipid is omitted since it has an inhibitory effect on the 25-hydroxylase activity.[12] The mixture is incubated at 37° from 10 to 20 min. The reaction is terminated by addition of 3 ml of ethanol, 96% (v/v). The incubation mixture is acidified and extracted twice with ether. The organic phase is washed with water until neutral, the solvent is evaporated, and the residue subjected to thin-layer chromatography. The developing solvent is ethyl acetate/trimethylpentane (4/1, v/v) for 5β-cholestane-$3\alpha,7\alpha$-diol and ethyl acetate/trimethylpentane/acetic acid (50/50/15, v/v/v) for 5β-cholestane-$3\alpha,7\alpha,12\alpha$-triol. The chromatoplates are analyzed by radioactive scanning. The products are then analyzed by radio-gas chromatography.[11,13]

[10] R. Hansson and K. Wikvall, *Eur. J. Biochem.* **93**, 419 (1979).
[11] I. Björkhem and J. Gustafsson, *Eur. J. Biochem.* **36**, 201 (1973).
[12] S. Andersson, I. Holmberg, and K. Wikvall, *J. Biol. Chem.* **258**, 6777 (1983).
[13] I. Björkhem, H. Danielsson, and K. Wikvall, *J. Biol. Chem.* **251**, 3495 (1976).

26-Hydroxylation of 5β-cholestane-3α,7α,12α-triol catalyzed by mitochondrial cytochrome P-450 is assayed as follows. Cytochrome P-450, 0.025–0.25 nmol, is incubated at 37° for 20 min together with 2 nmol of ferredoxin, 0.2 nmol of ferredoxin reductase, 5β-[7β-^{3}H]cholestane-3α,7α,12α-triol, 250 nmol in 25 μl of acetone, and 1 μmol of NADPH in a total volume of 1 ml of 50 mM Tris-acetate buffer, pH 7.4. The reaction is terminated by addition of 3 ml of ethanol, 96% (v/v). The incubation mixture is acidified and extracted twice with ether. The organic phase is washed with water until neutral, the solvent is evaporated and the residue subjected to thin-layer chromatography. The developing solvent is ethyl acetate/trimethylpentane/acetic acid (50/50/15, v/v/v). The chromatoplates are analyzed by scanning for radioactivity. The product is identified by thin-layer chromatography. The identity of the product can be confirmed by radio-gas chromatography.[11,13]

Preparation of Microsomes and Mitochondria from Rat and Rabbit Liver

Male rats of the Sprague–Dawley strain, weighing 250–350 g, and male New Zealand rabbits, weighing 2–3 kg, are used. The animals are untreated or treated with cholestyramine, 3% (w/w), in the diet for 1 week.[2]

Microsomes. The microsomal fraction is prepared from a 20–33% (w/v) liver homogenate in 0.25 M sucrose containing 1 mM EDTA and 10 mM Tris-acetate buffer, pH 7.4. The homogenate is centrifuged at 20,000 g for 20 min. The 20,000 g precipitate is resuspended in half the volume of the sucrose buffer, homogenized, and recentrifuged at 20,000 g for 20 min. The 20,000 g precipitate is used for preparation of the mitochondrial fraction as described below. The combined 20,000 g supernatant fluids are centrifuged at 100,000 g for 70 min. The microsomal pellet is suspended in half the original volume of 0.1 M potassium pyrophosphate buffer, pH 7.4, containing 1 mM EDTA, homogenized, and recentrifuged at 100,000 g for 70 min. The resulting microsomal pellet is suspended in the appropriate buffer used for solubilization of the various microsomal cytochrome P-450 preparations.

Mitochondria. The mitochondrial fraction is prepared from the 20,000 g precipitate obtained in the preparation of rabbit liver microsomes. The precipitate is suspended in a volume of 0.25 M sucrose, containing 10 mM Tris-acetate buffer, pH 7.4, and 1 mM EDTA, corresponding to a 10% (w/v) liver homogenate. The suspension is homogenized and centrifuged at 800 g for 10 min. The 800 g supernatant fluid is centrifuged at 6500 g for 20 min. The mitochondrial pellet obtained is resuspended and recentrifuged at 6500 g for 20 min three times. If not processed immediately the mitochondria are suspended in the sucrose buffer and stored under nitrogen at −20°.

Purification Procedure

The procedures are carried out at 4° and all buffers contain 20% (v/v) glycerol unless otherwise stated. Phosphate buffer is used as the potassium salt and Tris buffer as the acetate salt.

Octylamine-Sepharose 4B is prepared from 350 ml of cyanogen bromide-activated Sepharose 4B (from Pharmacia), previously washed with 1 mM HCl, and 350 ml of a 2 M octamethylenediamine solution adjusted to pH 10 by concentrated HCl. The mixture is shaken for 16 hr, transferred to a Büchner funnel, and washed extensively first with water and then with 0.1 M phosphate buffer, pH 7.25 (without glycerol) before equilibration.

Hydroxylapatite (BioGel HTP from Bio-Rad) is mixed with an equal volume of Whatman CF-1 cellulose powder before equilibration.

NADPH–cytochrome P-450 reductase is purified to electrophoretic homogeneity from liver microsomes of phenobarbital-treated rabbits and rats as described by Yasukochi and Masters.[14] The specific activity of the preparations is 50–55 units/mg of protein. The reductase activity is assayed according to Masters et al.[15] at 30° using 0.3 M phosphate buffer, pH 7.4. One unit corresponds to reduction of 1 μmol of cytochrome c/min.

Ferredoxin and ferredoxin reductase are purified to electrophoretic homogeneity from bovine adrenal mitochondria as described elsewhere.[16] Ferredoxin and ferredoxin reductase concentrations are determined as described by Huang and Kimura[17] and by Chu and Kimura.[18]

Purification of a Cytochrome P-450 from Rat Liver Microsomes Catalyzing 7α-Hydroxylation of Cholesterol

Step 1. Polyethylene Glycol Fractionation of Cholate-Solubilized Preparation. Liver microsomes from 50–100 cholestyramine-treated rats (500–1000 g of liver) are suspended in approximately 100 ml of 50 mM Tris buffer, pH 7.4, containing 1 mM EDTA. The protein concentration is

---

[14] Y. Yasukochi and B. S. S. Masters, *J. Biol. Chem.* **251**, 5337 (1976).
[15] B. S. S. Masters, C. H. Williams, Jr., and H. Kamin, this series, Vol. 10, p. 565.
[16] K. Wikvall, *J. Biol. Chem.* **259**, 3800 (1984).
[17] J. J. Huang and T. Kimura, *Biochemistry* **12**, 406 (1973).
[18] J.-W. Chu and T. Kimura, *J. Biol. Chem.* **248**, 2089 (1973).

adjusted to 15 mg/ml with the same buffer. The suspension is mixed with 0.2 M Tris buffer, pH 7.4, containing 1 mM EDTA to a final protein concentration of 6 mg/ml. A 10% (w/v) sodium cholate solution is added dropwise with stirring to give a final concentration of 1.8% (w/v). The mixture is stirred for 30 min. A 50% (w/v) solution of polyethylene glycol 6000 is added dropwise to a concentration of 6% and stirring is continued for 10 min. The mixture is centrifuged at 20,000 g for 15 min. Polyethylene glycol is added to the supernatant fluid to a final concentration of 15% and the mixture is centrifuged at 20,000 g for 15 min. The resulting precipitate is suspended in approximately 500 ml of 0.1 M phosphate buffer, pH 7.25, containing 1 mM EDTA and 0.7% (w/v) cholate.

Step 2. Octylamine-Sepharose Column Chromatography. The red solution is applied to an octylamine-Sepharose column (4 × 45 cm) equilibrated with 0.1 M phosphate buffer, pH 7.25, containing 1 mM EDTA and 0.7% cholate. The column is washed with the equilibrating buffer containing 0.46% cholate until the eluate is colorless. Cytochrome *P*-450 is eluted with the equilibrating buffer containing 0.37% cholate and 0.06% Emulgen 913. Fractions of 20 ml are collected. Cytochrome *P*-450 is eluted as one peak with this buffer. The fractions on the downward slope of the peak with an absorbance of 416 nm of 0.4 to 0.05 are pooled and used for further purification.

Step 3. Hydroxylapatite Column Chromatography. The pooled fractions from the octylamine-Sepharose column are diluted with 4 volumes of 20% glycerol and applied to a hydroxylapatite column (3 × 20 cm) equilibrated with 10 mM phosphate buffer, pH 7.25, containing 0.1 mM EDTA. The column is first washed with 1 liter of 40 mM phosphate buffer, pH 7.25, containing 0.1 mM EDTA and 0.2% Emulgen 913 and then with the same buffer containing 80 mM phosphate until no absorbance at 416 nm is detectable. Cytochrome *P*-450 active in cholesterol 7α-hydroxylation is then eluted with 0.2 M phosphate buffer, pH 7.25, containing 0.1 mM EDTA and 0.2% Emulgen 913.

Step 4. DEAE-Sepharose Column Chromatography. The 0.2 M phosphate eluate from the hydroxylapatite chromatography is concentrated by ultrafiltration (using an Amicon Diaflo PM 30 membrane) to approximately 20 ml and then diluted with 1 volume of 20% glycerol. The sample is dialyzed for 36 hr against 2 liters of 5 mM phosphate buffer, pH 7.6, containing 0.1 mM EDTA, 0.1% Emulgen 913, 0.2% cholate, and 0.5 mM dithiothreitol with two changes of the dialysis buffer. The dialyzate is applied to a DEAE-Sepharose column (1.5 × 10 cm) equilibrated with the same buffer. After washing the column with the equilibrating buffer until no absorbance at 416 nm is detectable, cytochrome *P*-450 active in choles-

TABLE I
Purification of a Cytochrome P-450 from Liver Microsomes of Cholestyramine-Treated Rats Active in 7α-Hydroxylation of Cholesterol

	Protein (mg)	Cytochrome P-450 (specific content) (nmol/mg protein)	Cytochrome P-450 (yield) (%)
Microsomes	12,000	0.75	100
Octylamine-Sepharose and hydroxylapatite	20	5.3	1.2
DEAE-Sepharose and hydroxylapatite	6	3.0	0.2

terol 7α-hydroxylation is eluted with 50 mM phosphate buffer, pH 7.6, containing 0.1 mM EDTA, 0.2% Emulgen 913, 0.2% cholate, and 0.5 mM dithiothreitol at a flow rate of 0.3 ml/min. The peak fractions are pooled.

Step 5. Hydroxylapatite Column Chromatography to Remove Detergent (Emulgen 913). The 50 mM phosphate eluate from DEAE-Sepharose chromatography is diluted with 4 volumes of 20% glycerol and applied to a small hydroxylapatite column (1 × 4 cm) equilibrated with 10 mM phosphate buffer, pH 7.25, containing 0.1 mM EDTA. The column is washed with the equilibrating buffer until no absorption at 276 nm (due to Emulgen 913) is detectable. Cytochrome P-450 is then eluted with 0.5 M phosphate buffer, pH 7.4, containing 0.1 mM EDTA and 0.1% cholate. The resulting cytochrome P-450 fraction is dialyzed overnight against 50 mM phosphate buffer, pH 7.4, containing 0.1 mM EDTA and stored under nitrogen at $-20°$.

Table I summarizes the purification of the cytochrome P-450 active in cholesterol 7α-hydroxylation. The final cytochrome P-450 preparation shows one protein band with a $M_r = 51,000$ when subjected to gel electrophoresis in the presence of sodium dodecyl sulfate.

The catalytic properties are summarized in Table V.

Purification of a Cytochrome P-450 from Rat Liver Microsomes Catalyzing 25-Hydroxylation of C_{27}-Steroids

Step 1. Solubilization of Microsomes with Sodium Cholate. Liver microsomes from 25 untreated rats (250–350 g of liver) are suspended in 0.1 M phosphate buffer, pH 7.25, containing 1 mM EDTA to a final protein concentration of 4 mg/ml. A 10% sodium cholate solution is added dropwise with stirring to give a final concentration of 0.7% (w/v). The mixture is stirred for 30 min and then centrifuged at 100,000 g for 70 min. The supernatant fraction, containing cytochrome P-450, is used for further purification.

Step 2. Octylamine-Sepharose Column Chromatography. The red supernatant fluid is applied to an octylamine-Sepharose column (3 × 35 cm) equilibrated with 0.1 M phosphate buffer, pH 7.25, containing 1 mM EDTA and 0.7% cholate. The column is washed with the equilibrating buffer containing 0.46% cholate until the eluate is colorless. Cytochrome P-450 is then eluted as one peak with the equilibrating buffer containing 0.37% cholate and 0.06% Emulgen 913. Fractions of 20 ml are collected. The peak fractions with an absorbance at 416 nm higher than 0.4 are pooled.

Step 3. Hydroxylapatite Column Chromatography. The cytochrome P-450 fraction eluted from the octylamine-Sepharose column is diluted with 4 volumes of 20% glycerol and applied to a hydroxylapatite column (3 × 20 cm) equilibrated with 10 mM phosphate buffer, pH 7.25, containing 0.1 mM EDTA. The column is washed with 40 mM phosphate buffer, pH 7.25, containing 0.1 mM EDTA and 0.2% Emulgen 913 until no absorbance at 416 nm is detectable. Cytochrome P-450 active in C_{27}-steroid 25-hydroxylation is then eluted with 80 mM phosphate in the buffer. The fractions showing absorbance at 416 nm higher than 0.25 are pooled and concentrated to approximately 20 ml by ultrafiltration using an Amicon Diaflo PM 30 membrane. The concentrated sample is dialyzed for 36 hr against 2 liters of 5 mM phosphate buffer, pH 7.8, containing 0.1 mM EDTA and 0.2% Emulgen with two changes of the dialysis buffer.

Step 4. DEAE-Sepharose Column Chromatography. The dialysate is diluted with an equal volume of 20% glycerol containing 0.2% Emulgen 913 and applied to a DEAE-Sepharose column (2.2 × 10 cm) equilibrated with the dialysis buffer. The cytochrome P-450 fractions active in 25-hydroxylation of C_{27}-steroids are eluted with the equilibrating buffer. The fractions showing one major protein band (M_r = 51,000) upon gel electrophoresis are pooled and further purified.

Step 5. Octyl-Sepharose Column Chromatography. The phosphate concentration of the cytochrome P-450 from the DEAE-Sepharose chromatography is adjusted to 0.1 M with 1 M phosphate, pH 7.25. The fraction is treated with Amberlite XAD-2 (16 g wet weight/100 ml of enzyme solution). The mixture is stirred for 20 min and filtered through glass wool. The filtrate is diluted with 3 volumes of 0.1 M phosphate buffer, pH 7.25, containing 0.1 mM EDTA. Sodium cholate is added to a final concentration of 0.7% and the sample is applied to an octyl-Sepharose column (1.2 × 8 cm) equilibrated with 0.1 M phosphate buffer, pH 7.25, containing 0.1 mM EDTA and 0.7% cholate. The column is washed with

TABLE II
PURIFICATION OF CYTOCHROME P-450 FROM LIVER MICROSOMES OF
UNTREATED RATS ACTIVE IN 25-HYDROXYLATION OF C_{27}-STEROIDS

	Protein (mg)	Cytochrome P-450 (specific content) (nmol/mg protein)	Cytochrome P-450 (yield) (%)
Microsomes	3200	0.8	100
Octylamine-Sepharose	208	3.8	31
Hydroxylapatite	32	6.7	8.4
DEAE-Sepharose	10.5	11.0	4.5
Octyl-Sepharose and hydroxylapatite	3.4	15.6	2.1

the equilibrating buffer containing 0.46% cholate until no absorbance at 416 nm is detectable. Cytochrome P-450 active in 25-hydroxylation of C_{27}-steroids is eluted with 0.37% cholate and 0.2% Emulgen 913 in the buffer at a flow rate of 0.3 ml/min. The peak fractions are pooled.

Step 6. Hydroxylapatite Column Chromatography to Remove Detergent (Emulgen 913). The cytochrome P-450 fraction is diluted with 20% glycerol and chromatographed on a small hydroxylapatite column according to the same procedure as described for the cytochrome P-450 active in cholesterol 7α-hydroxylation. The resulting cytochrome P-450 fraction is dialyzed overnight against 50 mM phosphate buffer, pH 7.4, containing 0.1 mM EDTA and stored under nitrogen at −20°.

Table II summarizes the purification of the cytochrome P-450 active in 25-hydroxylation of C_{27}-steroids. The final cytochrome P-450 preparation shows one protein band with $M_r = 51,000$ when subjected to gel electrophoresis in the presence of sodium dodecyl sulfate.[12]

The catalytic properties are summarized in Table V.

Purification of Cytochromes P-450 (P-450 LM_4I and P-450 LM_4II) from Rabbit Liver Microsomes with Different Substrate Specificity toward C_{27}-Steroids

Liver microsomes from 5 cholestyramine-treated rabbits (corresponding to 400–500 g of liver) are used for preparation of electrophoretically homogeneous cytochrome P-450 LM_4. The procedure used initially is the one described previously in this series by Coon *et al.*[19] for purification of cytochrome P-450 LM_4 from β-naphthoflavone-treated rabbits.

[19] M. J. Coon, T. A. van der Hoeven, S. B. Dahl, and D. A. Haugen, this series, Vol. 52C, p. 109.

TABLE III
PURIFICATION OF TWO CYTOCHROMES P-450 (P-450 LM$_4$I AND P-450 LM$_4$II) FROM LIVER MICROSOMES OF CHOLESTYRAMINE-TREATED RABBITS WITH DIFFERENT SUBSTRATE SPECIFICITY TOWARD C_{27}-STEROIDS

	Protein (mg)	Cytochrome P-450 (specific content) (nmol/mg protein)	Cytochrome P-450 (yield) (%)
Microsomes	3000	1.3	100
Cytochrome P-450 LM$_4$	13	15.2	5
Cytochrome P-450 LM$_4$ I	1.5	18.8	0.7
Cytochrome P-450 LM$_4$ II	2.8	10.8	0.8

Isolation of Cytochromes P-450 LM$_4$ I and P-450 LM$_4$ II. Two hundred nanomoles of cytochrome P-450 LM$_4$ is dialyzed overnight against 20 volumes of 0.1 M phosphate buffer, pH 7.25, containing 1 mM EDTA and 0.7% (w/v) sodium cholate. The dialyzate is applied to an octylamine-Sepharose column (1 × 15 cm) equilibrated with the dialysis buffer. Elution of cytochrome P-450 from the column is performed in three steps by changing the detergent composition of the equilibrating buffer. The buffer contains 0.45% (w/v) cholate (step 1), 0.37% (w/v) cholate and 0.06% (w/v) Emulgen 913 (step 2), and 0.2% (w/v) Emulgen 913 (step 3). A major peak of cytochrome P-450 is eluted with the buffer containing 0.37% cholate and 0.06% Emulgen. Additional cytochrome P-450 is eluted in a minor peak with the buffer containing 0.2% Emulgen. The early fractions of the major peak are pooled to give cytochrome P-450 LM$_4$ I and the fractions in the minor peak eluted with 0.2% Emulgen are pooled to give cytochrome P-450 LM$_4$ II. Depending on the binding capacity of the octylamine-Sepharose, cytochrome P-450 LM$_4$ I is also eluted in some experiments with the buffer containing 0.45% cholate (step 1).

The cytochrome P-450 LM$_4$ I and LM$_4$ II fractions are diluted with 2 volumes of 20% glycerol and separately applied to small hydroxylapatite columns (1 × 5 cm) equilibrated with 33 mM phosphate buffer, pH 7.4, containing 0.1 mM EDTA. The columns are washed with the equilibrating buffer until no absorbance at 276 nm is detectable. Cytochrome P-450 is then eluted with 0.5 M phosphate in the buffer. The cytochrome P-450 LM$_4$ I and P-450 LM$_4$ II preparations are dialyzed against 20 volumes of 50 mM phosphate buffer, pH 7.4, containing 0.1 mM EDTA and stored under nitrogen at $-20°$.

Table III summarizes the purification of cytochromes P-450 LM$_4$ I and P-450 LM$_4$ II. Cytochromes P-450 LM$_4$ I and P-450 LM$_4$ II show the same apparent molecular weight (M_r = 53,000) upon gel electrophoresis in the presence of sodium dodecyl sulfate. The fractions differ in amino acid composition with respect to content of arginine, lysine, proline, and glutamine plus glutamic acid.[2]

Table V summarizes the catalytic properties of the two cytochrome P-450 fractions from rabbit liver.

Purification of Cytochrome P-450 from Rabbit Liver Mitochondria Active in 26-Hydroxylation of C_{27}-Steroids

Liver mitochondria from 10–15 untreated rabbits (800–1500 g of liver) are suspended in 0.25 M sucrose containing 10 mM Tris buffer, pH 7.4, and 1 mM EDTA and centrifuged at 100,000 g for 1 hr. It is possible to use mitochondria that have been frozen (cf. above under Preparation of Microsomes and Mitochondria from Rat and Rabbit Liver).

Step 1. Treatment of Mitochondria with Hypotonic Phosphate Buffer and Sonication. The mitochondria collected after centrifugation are suspended in 10 mM phosphate buffer, pH 7.4 (without glycerol), to a protein concentration of 25 mg/ml, homogenized, and allowed to stand at 4° overnight. The mitochondrial suspension is sonicated for a total period of 3 min at intervals of 15 sec using a Branson sonifier (cell disrupter B 15) at setting 8 of the output control. The sonicate is homogenized and centrifuged at 100,000 g for 1 hr.

Step 2. Solubilization of Mitochondrial Membrane with Sodium Cholate. The mitochondrial membrane pellet is suspended in 0.1 M phosphate buffer, pH 7.25, containing 1 mM EDTA, to a protein concentration of 4 mg/ml. A 10% sodium cholate solution is added with stirring to give a final cholate concentration of 0.8% (w/v) and stirring is continued for 1 hr. The mixture is centrifuged at 100,000 g for 1 hr.

Step 3. Octylamine-Sepharose Column Chromatography. The 100,000 g supernatant fluid, containing cytochrome P-450, is applied to an octylamine-Sepharose column (3 × 40 cm) equilibrated with 0.1 M phosphate buffer, pH 7.25, containing 1 mM EDTA and 0.7% cholate. The column is washed with the equilibrating buffer containing 0.5% cholate until the eluate is colorless. Cytochrome P-450 is eluted with 0.4% cholate and 0.08% Emulgen 913 in the buffer. Fractions of 10 ml are collected. The fractions with absorbance at 416 nm higher than 0.15 are pooled.

Step 4. Hydroxylapatite Column Chromatography. The cytochrome P-450 fraction obtained from octylamine–Sepharose chromatography is diluted with 3 volumes of 20% glycerol and applied to a hydroxylapatite column (3 × 20 cm) equilibrated with 25 mM phosphate buffer, pH 7.25, containing 0.1 mM EDTA. The column is washed with 35 mM phosphate buffer, pH 7.25, containing 0.1 mM EDTA and 0.2% Emulgen 913 until the eluate is colorless. Cytochrome P-450 is eluted with 0.15 M phosphate in the buffer. The fractions with absorbance at 416 nm higher than 0.2 are pooled and dialyzed overnight against 5 liters of 5 mM phosphate buffer, pH 7.4, with one change of the dialysis buffer.

Step 5. DEAE-Cellulose Column Chromatography. Sodium cholate is added to the dialyzate to a final concentration of 0.1% (w/v). The sample is diluted with one volume of 10 mM phosphate buffer, pH 7.4, containing 0.05% (w/v) Emulgen 913 and 0.1% (w/v) cholate, stirred for 10 min, and applied to a Whatman DE-52 cellulose column (1.5 × 30 cm) equilibrated with the same buffer. Cytochrome P-450 is first eluted with the equilibrating buffer and then with 400 ml of a linear KCl-gradient (0–0.3 M) in the equilibrating buffer. Fractions of 10 ml are collected. The chromatography results in three peaks of cytochrome P-450 as measured by absorbance at 416 nm. The first peak is eluted with the equilibrating buffer, the second and the major peak in the beginning and the third peak in the middle of the KCl gradient. The fractions with absorbance at 416 nm of 0.09 or higher in the second, major peak are pooled and dialyzed overnight against 5 liters of 10 mM phosphate buffer, pH 7.4, containing 0.1 mM EDTA.

Step 6. Hydroxylapatite Column Chromatography to Remove Detergent (Emulgen 913). The dialysate is applied to a small hydroxylapatite column (1 × 5 cm) equilibrated with the dialysis buffer. The column is washed with the dialysis buffer and cytochrome P-450 is eluted as described for the cytochrome P-450 active in cholesterol 7α-hydroxylation. The cytochrome P-450 is dialyzed overnight against 50 volumes of 50 mM phosphate buffer, pH 7.4, containing 0.1 mM EDTA.

Table IV summarizes the purification of cytochrome P-450 from rabbit liver mitochondria active in 26-hydroxlation of C_{27}-steroids. The final cytochrome P-450 preparation shows one protein band (M_r = 53,000) upon gel electrophoresis in the presence of sodium dodecyl sulfate.[16]

The catalytic properties are summarized in Table V.

Properties

Table V summarizes the catalytic properties as well as the apparent minimum molecular weights and heme contents of the purified cytochromes P-450.

The cytochrome P-450 preparations show a high degree of specificity with respect to the hydroxylations in bile acid biosynthesis. The presence

TABLE IV
PURIFICATION OF A MITOCHONDRIAL CYTOCHROME P-450 FROM LIVER
MITOCHONDRIA OF UNTREATED RABBITS ACTIVE IN 26-HYDROXYLATION OF
C_{27}-STEROIDS

	Protein (mg)	Cytochrome P-450 (specific content) (nmol/mg protein)	Cytochrome P-450 (yield) (%)
Solubilized cytochrome P-450 (cholate extract)	850	0.18	100
Octylamine-Sepharose and hydroxylapatite	8.5	4.9	25
DE-52 cellulose (Pool 2)	0.6	10.2	4.0

of more than one C_{27}-steroid hydroxylase activity in some preparations might indicate that these preparations are not homogeneous but contaminated to some extent with each other.

Although the preparations are apparently homogeneous upon gel electrophoresis several of them show a heme content which is less than maximal. This is probably due to heme losses in the purification procedure.[19] The phenomenon is especially obvious with the cholesterol 7α-hydroxylating and with the mitochondrial species of cytochrome P-450. It should

TABLE V
CATALYTIC PROPERTIES OF CYTOCHROMES P-450 PURIFIED FROM RAT AND RABBIT LIVER

	M_r	Cytochrome P-450 (nmol/mg protein)	Hydroxylation			
			7α	12α	25	26
			(pmol/nmol cytochrome P-450 × min)			
Cytochrome P-450$_{7\alpha}$ (rat liver microsomes)	51,000	3.0	7,500	50	≤1	≤1
Cytochrome P-450$_{25}$ (rat liver microsomes)	51,000	15.6	≤1	≤1	2,600	≤1
Cytochrome P-450 LM$_4$ I (rabbit liver microsomes)	53,000	18.8	≤1	160	120	≤1
Cytochrome P-450 LM$_4$ II (rabbit liver microsomes)	53,000	10.8	80	50	50	≤1
Cytochrome P-450$_{26}$ (rabbit liver mitochondria)	53,000	10.2	≤1	≤1	≤1	18,500

be pointed out in this connection that the preparations with less than maximal content are not less catalytically active than the preparation with maximal heme content. In fact, the preparation from microsomes with the lowest heme content, cytochrome $P\text{-}450_{7\alpha}$ from rat liver, is the one which shows the highest increase in hydroxylase activity—about 200-fold. Also, the 26-hydroxylase activity of the mitochondrial cytochrome $P\text{-}450$ is up to 1000 times higher than that of mitochondria.

[22] Biosynthesis and Metabolism of Ecdysteroids and Methods of Isolation and Identification of the Free and Conjugated Compounds

By H. H. REES and R. E. ISAAC

Overall Pathways of Biosynthesis and Metabolism

Insect molting hormones (collectively referred to as ecdysteroids) occur not only in the tissues of immature insects but also in ovaries of adult females and in embryos. Although significant amounts of ecdysteroids apparently do not generally occur in adult male insects, low levels of such steroids have been reported in males of a few species. Recent reviews are available on the ecdysteroid endocrine system in immature[1,2] and adult/embryonic[2-4] stages of insect development.

According to the classical scheme of insect postembryonic development, ecdysteroids in immature stages (larvae and pupae) should be secreted by the paired prothoracic glands. That larval prothoracic glands do secrete ecdysone, but not 20-hydroxyecdysone, has been demonstrated for several insect species (for review, see ref. 5). Following secretion into

[1] G. Richards, *Biol. Rev. Cambridge Philos. Soc.* **56**, 501 (1981).
[2] H. H. Rees, *in* "Biosynthesis of Isoprenoid Compounds" (J. W. Porter and S. L. Spurgeon, eds.), Vol. 2, p. 463. Wiley, New York, 1983.
[3] J. A. Hoffmann, M. Lagueux, C. Hetru, M. Charlet, and F. Goltzené, *in* "Progress in Ecdysone Research" (J. A. Hoffmann, ed.), p. 431. Elsevier/North Holland Biomedical Press, Amsterdam, 1980.
[4] H. H. Rees and R. E. Isaac, *in* "Biosynthesis, Metabolism and Mode of Action of Invertebrate Hormones" (J. A. Hoffmann and M. Porchet, eds.), p. 181. Springer-Verlag, Berlin and New York, 1984.
[5] H. H. Rees, *in* "Comprehensive Insect Physiology, Biochemistry and Pharmacology" (G. A. Kerkut and L. I. Gilbert, eds.) Vol. 7, p. 249. Pergamon, Oxford, 1984 (in press).

the hemolymph, the ecdysone, which is primarily regarded as a prohormone, undergoes hydroxylation at C-20 in several, but not all, peripheral tissues, yielding the generally more active hormone, 20-hydroxyecdysone. The source of the appreciable ecdysteroid production in abdomens of *Tenebrio* pupae[6] is unclear; very low levels of ecdysteroid production have also been reported in abdomens of certain larval species.[5]

Molting hormone titers, nowadays generally determined by radioimmunoassay, exhibit distinct peaks at specific stages during development in larvae and pupae. These fluctuating titers are apparently controlled by several factors including changes in the rates of biosynthesis, inactivation (further metabolism), and excretion of the hormones, in addition to the extent of their sequestration (storage).

In adult females of most insect species investigated, ecdysteroids are largely confined to the ovaries, where they are synthesized.[3] At least in the migratory locust, *Locusta migratoria,* this synthesis occurs in the follicle cells.[7] The amounts of ecdysteroids occurring in ovaries are generally far higher than those in larvae.[3,4] It has been shown in many, but not all, species investigated that the ovarian ecdysteroids occur largely as polar conjugates, hydrolyzable with crude hydrolytic enzymes (often described as an "arylsulfatase preparation") from the snail, *Helix pomatia.* In most species investigated the ecdysteroid conjugates are largely passed into the oocytes and occur in the newly laid eggs. In the desert locust, *Schistocerca gregaria,* the conjugates in newly laid eggs have been identified as the 22-phosphate derivatives.[8,9] Hydrolysis of these phosphates by the crude *Helix* arylsulfatase preparation is accounted for by the presence of appreciable phosphatase activity. In many species of insects, distinct peaks in the titer of immunoreactive free ecdysteroids during embryogenesis have been detected, there being an apparent correlation between the timing of the peaks and cycles of embryonic cuticulogenesis/membrane formation (see ref. 4). At least in *S. gregaria,* it has been demonstrated that the ecdysteroid 22-phosphates may serve as inactive storage forms of hormone, being hydrolyzed to release free ecdysteroids by a phosphatase enzyme preparation from developing embryos.[4,10] However, contribution to the increasing hormone titers during embryogenesis from *de novo* synthesis or metabolism of precursors also present in the eggs is not pre-

---

[6] J. P. Delbecque, J. Delachambre, M. Hirn, and M. De Reggi, *Gen. Comp. Endocrinol.* **35**, 436 (1978).

[7] F. Goltzené, M. Lagueux, M. Charlet, and J. A. Hoffmann, *Hoppe-Seyler's Z. Physiol. Chem.* **359**, 1427 (1978).

[8] R. E. Isaac, M. E. Rose, H. H. Rees, and T. W. Goodwin, *J. Chem. Soc., Chem. Commun.* p. 249 (1982).

[9] R. E. Isaac, M. E. Rose, H. H. Rees, and T. W. Goodwin, *Biochem. J.* **213**, 533 (1983).

[10] R. E. Isaac, F. P. Sweeney, and H. H. Rees, *Biochem. Soc. Trans.* **11**, 379 (1983).

cluded. Decrease in the immunoreactive ecdysteroid titer in the closed system of the eggs during embryogenesis occurs by further metabolism (hormone inactivation).

Peaks of immunoreactive, free ecdysteroids have also been detected in adult females of ticks, with formation of nonpolar esters of the hormones decreasing the free hormone titer.[11,12] In the cattle tick, *Boophilus microplus,* these partly characterized long chain acyl esters of ecdysone are transferred to the newly laid eggs, where they greatly predominate over free hormones. Also in this species, it has been shown that the esters may be hydrolyzed by a homogenate of developing eggs, releasing free hormone.[11] Thus, it is possible that the esters may represent inactive storage forms of ecdysone, releasing active, free hormone by enzymatic hydrolysis during embryogenesis. In these ticks, the ecdysteroid acyl esters may function in an analogous manner to the phosphate esters in the locust, *S. gregaria.*

Whereas in immature stages of insects, the major ecdysteroid is generally 20-hydroxyecdysone accompanied by smaller amounts of ecdysone, in ovaries/eggs the converse usually holds; in the latter cases appreciable amounts of 2-deoxyecdysone also generally occur. However, in certain Hemiptera and Heteroptera, the C_{28}-steroid, makisterone A, is the major ecdysteroid (for references, see ref. 13).

The occurrence of ecdysteroids is not confined to the Arthropod phylum (Insecta, Crustacea, Myriapoda, and Arachnida), but they also occur in mollusks, annelids, nematodes, and platyhelminths (trematodes and cestodes) in some cases in particularly low concentrations.[2,14-17]

Biosynthesis

The biosynthesis of ecdysteroids has been investigated mostly in insects and has been reviewed recently.[2,5,18] Possible diverging pathways of

[11] K. P. Wigglesworth, D. Lewis, and H. H. Rees, *Arch. Insect Biochem. Physiol.* (1984), in press; presented at VIth European Ecdysone Workshop, Szeged, Hungary (1983).
[12] J-L. Connat, P. A. Diehl, and M. Morici, presented at VIth European Ecdysone Workshop, Szeged, Hungary (1983).
[13] J. M. Gibson, M. S. I. Majumder, A. H. W. Mendis, and H. H. Rees, *Arch. Insect Biochem. Physiol.* **1,** 105 (1983).
[14] A. H. W. Mendis, M. E. Rose, H. H. Rees, and T. W. Goodwin, *Mol. Biochem. Parasitol.* **9,** 209 (1983).
[15] A. H. W. Mendis, H. H. Rees, and T. W. Goodwin, *Mol. Biochem. Parasitol.* **10,** 123 (1984).
[16] P. Nirde, G. Torpier, M. L. De Reggi, and A. Capron, *FEBS Lett.* **151,** 223 (1983).
[17] H. H. Rees and A. H. W. Mendis, in "Biosynthesis, Metabolism and Mode of Action of Invertebrate Hormones" (J. A. Hoffmann and M. Porchet, eds.), p. 338. Springer-Verlag, Berlin and New York, 1984.
[18] L. I. Gilbert, W. Goodman, and W. E. Bollenbacher, *Int. Rev. Biochem.* **14,** 1 (1977).

ecdysteroid biosynthesis are summarized in Fig. 1, which is a composite of results obtained from studies utilizing different species and experimental systems.

The distal precursor of C_{27} ecdysteroids is cholesterol (1), which undergoes transformations in the nucleus before side-chain hydroxylations. 7-Dehydrocholesterol (2) is a possible, but not established, early intermediate in the pathway, which may proceed via a 3-oxo-Δ^4-steroid (3). There is evidence supporting the involvement of $3\beta,14\alpha$-dihydroxy-5β-cholest-7-en-6-one (6), although the exact sequence of subsequent hydroxylations is uncertain. There may be more than one sequence operating even in a single species at diffferent stages of development or in different species. In immature stages of some, but not all, insect species, the steroid (7) possessing a complete ecdysone nucleus may be an intermediate (6 → 7 → 8 → 9 → 10). However, the occurrence, interconversion, and metabolism of 2-deoxyecdysteroids in ovaries/eggs suggest that a pathway involving hydroxylation at C-2 as a final step may operate in those cases (6 → 11 → 12 → 9 → 10). In either case, the apparent sequence of side-chain hydroxylations is C-25 followed by C-22. However, the isolation of steroids 13 and 14 together with 12, 9, and 10 from ovaries/eggs of certain species suggests that branched pathways may operate in the terminal steps of 20-hydroxyecdysone (10) formation. In ovarian synthesis, the exact stage of conjugation (phosphorylation) is uncertain, although it must be a late process, since 22-hydroxylation itself is a late step.

Inactivation

From the limited information available at present, it appears that the metabolic routes (inactivation) of ecdysteroids in larvae and pupae are very similar to those operating in developing eggs/embryos. Comprehen-

---

FIG. 1. Possible diverging pathways of ecdysteroid biosynthesis from cholesterol. The scheme is based on the combined results obtained utilizing different species of insects. (1) Cholesterol, (2) 7-dehydrocholesterol, (3) 3-oxo- Δ^4-sterol, (4) 3β-hydroxy-5β-sterol, (5) 3β-hydroxy-5β-cholest-7-en-6-one, (6) $3\beta,14$-dihydroxy-5β-cholest-7-en-6-one, (7) $2\beta,3\beta,14$-trihydroxy-5β-cholest-7-en-6-one (22,25-dideoxyecdysone), (8) $2\beta,3\beta,14,25$-tetrahydroxy-5β-cholest-7-en-6-one (22-deoxyecdysone), (9) $(22R)$-$2\beta,3\beta,14,22,25$-pentahydroxy-5β-cholest-7-en-6-one (ecdysone), (10) $(22R)$-$2\beta,3\beta,14,20,22,25$-hexahydroxy-$5\beta$-cholest-7-en-6-one (20-hydroxyecdysone), (11) $3\beta,14,25$-trihydroxy-5β-cholest-7-en-6-one (2,22-dideoxyecdysone), (12) $(22R)$-$3\beta,14,22,25$-tetrahydroxy-5β-cholest-7-en-6-one (2-deoxyecdysone), (13) $3\beta,14,20,25$-tetrahydroxy-5β-cholest-7-en-6-one (2,22-dideoxy-20-hydroxyecdysone), (14) $(22R)$-$3\beta,14,20,22,25$-pentahydroxy-5β-cholest-7-en-6-one (2-deoxy-20-hydroxyecdysone).

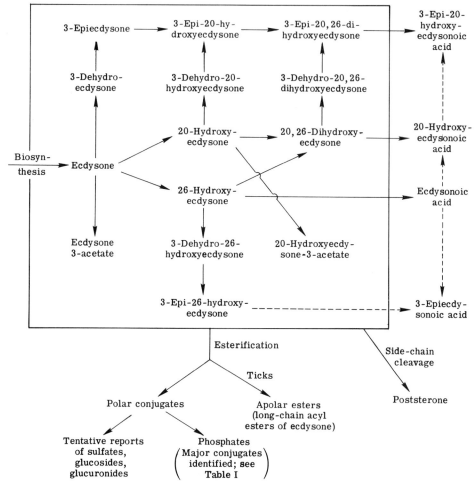

FIG. 2. Summary of composite possible metabolic routes of ecdysteroids occurring in various insect species (modified and extended from ref. 19). The dashed arrows represent likely reactions. The ecdysteroid acetates have been detected mainly, but not entirely, as phosphate conjugates.

sive reviews of ecdysteroid metabolism are available.[19,20] A composite scheme of possible metabolic routes of ecdysteroids occurring in various insect species is shown in Fig. 2. All the metabolites of ecdysone and 20-

[19] J. Koolman, *Insect Biochem.* **12**, 225 (1982).
[20] R. Lafont and J. Koolman, in "Biosynthesis, Metabolism and Mode of Action of Invertebrate Hormones" (J. A. Hoffmann and M. Porchet, eds.), p. 196. Springer-Verlag, Berlin and New York, 1984.

hydroxyecdysone have much reduced biological (molting hormone) activity compared to the parent hormones[21,22] and, thus, represent inactivation products.

It appears that ecdysteroid 26-oic acids formed via 26-hydroxy compounds are prominent ecdysteroid inactivation products in many species of insects, and may, indeed, be universal.[22,23] Reports of the formation of 3-epiecdysteroids have been prominent in Lepidoptera, but are not confined by any means to that order. There is evidence that 3-epiecdysteroid formation from 3β-hydroxy compounds occurs via the corresponding 3-dehydro compounds.[24] There is little doubt that several early reports of the formation of 3-dehydroecdysteroids from [^{3}H]ecdysone on the basis of silica thin-layer chromatographic mobility warrant reinvestigation in view of the fact that the corresponding 3-acetate derivatives cochromatograph in this system.[25]

A prominent metabolic route of free ecdysteroids in most insect species at various stages of development is the formation of polar conjugates. The only conjugates to be characterized, as yet, by physical techniques are various phosphate esters (see Table I), which have been isolated mainly from locust eggs. Reports of the 22-phosphate esters of ecdysteroids not bearing an acetate group appear to be largely limited to inactivation of ovarian ecdysteroids so far. It appears that ecdysteroid acetates occur primarily as conjugates (phosphates) and have been detected in larvae[26,27] as well as in developing eggs. The sequence of acetylation and phosphorylation in the formation of the acetylecdysteroid phosphates indicated in Table I has not been demonstrated experimentally. Reports of the occurrence of sulfate, glucoside, and glucuronide ecdysteroid conjugates, with the exception of a few of those for sulfates, are based upon hydrolysis with enzymes, which were probably impure.

In adult females of the cattle tick, *B. microplus*, formation of polar ecdysteroid conjugates is apparently a very minor pathway, the major metabolic (inactivation) products being apolar long-chain acyl esters of

[21] R. Bergamasco and D. H. S. Horn, in "Progress in Ecdysone Research" (J. A. Hoffmann, ed.), p. 299. Elsevier/North Holland Biomedical Press, Amsterdam, 1980.

[22] R. Lafont, C. Blais, P. Beydon, J.-F. Modde, U. Enderle, and J. Koolman, *Arch. Insect Biochem. Physiol.* **1**, 41 (1983).

[23] R. E. Isaac, N. P. Milner, and H. H. Rees, *Biochem. J.* **213**, 261 (1983).

[24] N. P. Milner and H. H. Rees, presented at VIth European Ecdysone Workshop, Szeged, Hungary (1983).

[25] R. E. Isaac, H. H. Rees, and T. W. Goodwin, *J. Chem. Soc., Chem. Commun.* p. 594 (1981).

[26] J. M. Gibson, R. E. Isaac, L. N. Dinan, and H. H. Rees, *Arch. Insect Biochem. Physiol.* **1**, 385 (1984).

[27] G. Tsoupras, B. Luu, C. Hetru, J.-F. Muller, and J. Hoffmann, *C.R. Hebd. Seances Acad. Sci., Ser. III* **296**, 77 (1983).

ecdysone.[11] Similar apolar derivatives of 20-hydroxyecdysone as well as of ecdysone have been detected in another tick species.[12]

It appears that side-chain cleavage of ecdysteroids at the C-20–C-22 bond with formation of poststerone is a minor pathway in insects. However, such a metabolite may have escaped detection in studies utilizing [23,24-^{3}H$_2$]ecdysone, since the label would be located in the cleaved side-chain fragment.

Experimental Approaches

When considering pathways of biosynthesis and metabolism of compounds, the unequivocal establishment of the intermediacy of any postulated compound necessitates the fulfilment of a number of criteria, which were suggested originally by Frantz and Schroepfer for intermediates in cholesterol biosynthesis.[28] The compound should be (1) isolated from tissues and its structure established, (2) formed biologically from the distal precursor, (3) converted biologically into the final product, (4) formed biologically from the postulated immediate precursor, (5) converted biologically into the next postulated compound in the scheme, and (6) in addition, each individual step in the pathway should be demonstrated directly. Furthermore, establishment of the quantitative importance of a proposed pathway necessitates much additional information.

During the isolation of steroids, particularly radioactively labeled ones, from tissues, it is mandatory to ascertain their purity, for example, by HPLC on more than one type of column, and where possible by derivatization (see subsequent sections).

Biosynthesis

Cholesterol as Precursor. For certain types of study, radioactive cholesterol has been incorporated *in vivo* into ecdysteroids (primarily 20-hydroxyecdysone and ecdysone) at times of increasing ecdysteroid titers. In larval systems the percentage incorporations (usually within the range 0.015–0.12%) are low,[2,5] primarily because ecdysteroids occur only in small amounts, and also owing to the presence of large cholesterol pools in insects. However, we have taken advantage of the comparatively large synthesis of ecdysteroids (phosphates) in ovaries of *S. gregaria* to effect a much greater incorporation (0.5–1.0% of the injected substrate) of [^{14}C]- or [^{3}H]cholesterol into ecdysteroids (mainly ecdysone and 2-deoxyecdysone.)[29] For this. radioactive cholesterol is injected into maturing adult female *S. gregaria* before the increase in ecdysteroid phosphate titer and

[28] I. D. Frantz and G. J. Schroepfer, *Annu. Rev. Biochem.* **36**, 691 (1967).
[29] L. N. Dinan and H. H. Rees, *Insect Biochem.* **11**, 255 (1981).

the incorporation into ecdysteroids examined either in dissected ovaries, or in newly laid eggs, which contain the bulk of the ovarian ecdysteroid conjugates. In general, during incorporation of radioactive cholesterol into ecdysteroids, labeled intermediates do not accumulate, except certain 2-deoxyecdysteroids in ovarian systems.[3,5] In fact, several 2-deoxyecdysteroids which are presumably biosynthetic intermediates have been isolated from ovaries/eggs in sufficient quantities for characterization by physical techniques.

Although studies on the incorporation of [^{3}H]- or [^{14}C]cholesterol into ecdysteroids have provided only limited information on biosynthetic intermediates, much valuable data have been obtained utilizing various stereospecifically ^{3}H-labeled cholesterol species in admixture with [4-^{14}C]cholesterol. In this way, the stereospecificities of Δ^7 bond formation, and of the C-2 and C-22 hydroxylation reactions have been established.[2,5] Similarly, utilization of [3α-^{3}H$_1$]-, [4α-^{3}H$_1$]-, and [4β-^{3}H$_1$]cholesterol species, each in admixture with [4-^{14}C]cholesterol, as ecdysteroid precursors demonstrated the specific removal of the 3α- and 4β-hydrogens of cholesterol with retention of the 4β-hydrogen.[30] This was interpreted in terms of the possible intermediacy of a 3-oxo-Δ^4-steroid intermediate (see Fig. 1).

By culturing *in vitro* prothoracic glands (ring glands in the case of Cyclorrhapha) or ovaries from several species in appropriate media, increase in ecdysteroid immunoreactivity has been observed, indicating the production of free ecdysteroids from endogenous precursors in these organs (see ref. 5). Apparent stimulation of ecdysteroid production by lipoprotein-containing hemolymph fractions has been reported in some cases. However, such *in vitro* systems have only met with limited success for incorporation of exogenous radioactive cholesterol.[5]

Putative Intermediates. Much information on the later stages of ecdysone biosynthesis, particularly the sequence of the side-chain hydroxylations, has been obtained by chemical synthesis of radioactive putative intermediates and estimating their efficacy as hormone precursors either in *in vivo* or *in vitro* systems.[2,5,18,31] In this way, the percentage incorporations of various compounds can be compared. The *in vitro* systems which have been primarily used in this approach are *Manduca sexta* prothoracic glands[18] and *Locusta migratoria* ovaries,[32] whereas larvae and pupae of

---

[30] H. H. Rees, T. G. Davies, L. N. Dinan, W. J. S. Lockley, and T. W. Goodwin, in "Progress in Ecdysone Research" (J. A. Hoffmann, ed.), p. 125. Elsevier/North-Holland Biomedical Press, Amsterdam, 1980.

[31] J. A. Svoboda, M. J. Thompson, W. E. Robbins, and J. N. Kaplanis, *Lipids* **13**, 742 (1978).

[32] C. Hetru, C. Kappler, J. A. Hoffmann, R. Nearn, L. Bang, and D. H. S. Horn, *Mol. Cell. Endocrinol.* **26**, 51 (1982).

various species at a stage of increasing ecdysteroids titer have been used for the *in vivo* work.

In this approach it is distinctly desirable to compare the percentage incorporation of each putative intermediate to that of cholesterol in the *same in vivo* or *in vitro* incubation, for example, by using the ^{3}H-labeled putative intermediate in admixture with [^{14}C]cholesterol. This overcomes variability between experiments, especially differences in the exact stages of development of the insects.

This approach suffers especially from possible differences between compounds in their permeability to the site of ecdysone synthesis, and furthermore, some compounds that are not true intermediates, may be metabolized by enzymes operating at low specificity. It is of paramount importance to take into account that it is the *combination* of the rates of reactions in a pathway that determine the overall rate of transformation of cholesterol into ecdysone. It is not sufficient to demonstrate that a potential intermediate is metabolized adequately to the final product, but it is mandatory to show that such a compound is formed from cholesterol at a significant rate (see ref. 5).

Metabolism

The rates of molting hormone inactivation (further metabolism) and excretion vary during insect development, both being apparently low during high ecdysteroid titer, whereas the reverse holds for periods of low hormone levels.[2,19] The metabolism of ecdysteroids has been investigated mainly by examination of the fate of ^{3}H-labeled ecdysteroids (usually synthetic [^{3}H]ecdysone) either *in vivo* or in isolated tissues and cell-free preparations *in vitro*. Besides [^{3}H]ecdysone, other labeled ecdysteroids, which have been isolated and purified from insects as metabolic products, have been utilized as substrates. In such cases, the purity and unequivocal identification of the substrate are important. The *in vivo* approach has certain limitations, including the following. The exogenous substrate may not penetrate to all normal sites of metabolism of endogenous hormone. For example, a negligible amount of [^{3}H]ecdysone injected into adult desert locusts, *S. gregaria*, is taken up by the ovaries and passed into the eggs. Furthermore, ecdysteroid-binding macromolecules have been demonstrated in hemolymph of several insect species.[33] Consequently, bound ecdysteroids may have a longer half-life than the free hormones, and the fate of exogenous free ecdysteroid may well differ from that produced endogenously and bound to macromolecules. Greater compartmentaliza-

---

[33] R. Feyereisen, *in* "Progress in Ecdysone Research" (J. A. Hoffmann, ed.), p. 325. Elsevier/North-Holland Biomedical Press, Amsterdam, 1980.

tion of endogenous hormone than exogenous material may also be envisaged.

The foregoing disadvantages of examination of the fate of exogenous [^{3}H]ecdysone *in vivo* may be largely circumvented by investigating the fate of ecdysteroids formed biosynthetically *in situ* from radioactive cholesterol. Such an approach, in principle, may be used following synthesis of ecdysteroids in larvae, pupae, or in adults. However, its success depends on efficient incorporation of the cholesterol into the ecdysteroid in the first place. The technique has been particularly valuable in following the fate of ovarian ecdysteroid phosphates in eggs of *S. gregaria* during embryogenesis,[4,29] especially since comparatively high incorporation of cholesterol into ecdysteroids is observed in adult females. This approach should reflect quantitative metabolic routes of endogenous ecdysteroids much more accurately than the administration of exogenous ^{3}H-labeled ecdysteroids. In fact, marked differences in the metabolism of molting hormones have been observed in *Pieris brassicae* pupae following injection of [^{3}H]ecdysone or 20-[^{3}H]hydroxyecdysone and after long-term [^{3}H]cholesterol incorporation.[34]

Isolation, Fractionation, and Characterization of Ecdysteroids

Investigation of ecdysteroids in invertebrates frequently requires analytical procedures for separation, identification, and quantification of closely related ecdysteroids, their precursors, as well as their inactivation products. For the absolute identification of an isolated steroid, it is necessary to characterize the compound by physicochemical techniques. However, it is often not possible to isolate sufficient quantities of material for such analyses. Ecdysteroids which are radioactively labeled or are present in such low amounts that RIA is required for their detection can be identified with some degree of confidence if cochromatography with authentic compounds is demonstrated in more than one chromatographic system. This approach can be coupled with appropriate chemical derivatization (e.g., acetate or acetonide formation) in the case of radioactive compounds.

In the following sections emphasis will be placed on methods used in our laboratory, although extensive reference to alternative procedures will also be made. Typical ecdysteroid isolation and characterization procedures utilized by the authors are summarized in Fig. 3. Aspects of the recent application of mass spectrometry and nuclear magnetic resonance (NMR) spectroscopy to ecdysteroid metabolism will be considered separately in the two subsequent sections.

[34] P. Beydon, J. Claret, P. Porcheron, and R. Lafont, *Steroids* **38**, 633 (1981).

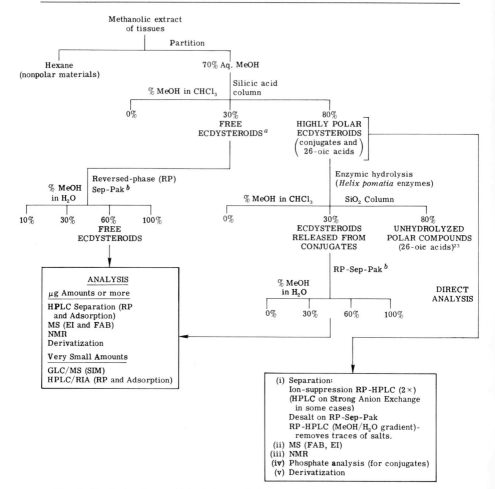

FIG. 3. Typical ecdysteroid isolation and characterization procedures used in the authors' laboratory. [a] Also contains any "long-chain acyl esters of ecdysteroids"[11] when present. [b] Generally omitted for large samples. Since it is important to avoid overloading the Sep-Pak cartridge, this step should be validated with each type of extract.

Extraction of Tissues

A typical procedure for extraction of free and highly polar ecdysteroids from tissues is described. The sample is macerated and extracted three times with methanol–water (7:3; 10 ml/g of material) and the residue extracted a further three times with methanol.[35] To remove apolar

---

[35] Other extraction procedures include homogenization in warm (60°) ethanol–H_2O (6:4), methanol, or acetonitrile[36–38] Extraction into water–chloroform (1:1) has also been employed, but it is not suitable for compounds less polar than ecdysone.[39]

material, extracts are partitioned between countersaturated hexane and methanol–water (7:3), and each phase backextracted with the respective counter phase. The aqueous methanol phase containing free and highly polar ecdysteroids is then evaporated to dryness under reduced pressure.

Group Separation

Ecdysteroids are conveniently divided into two groups, free ecdysteroids and highly polar ecdysteroids, on the basis of their polarity. Highly polar ecdysteroids are comprised of ecdysteroid conjugates, hydrolyzable with a crude mixture of hydrolase enzymes (often referred to as an "arylsulfatase preparation") from *H. pomatia*, and ecdysteroid 26-oic acids (Table I).[40–44]

Silicic Acid Column Chromatography.[45] The extract dissolved in methanol is adsorbed onto Celite (diatomaceous earth; Sigma Chemical Co., London) by evaporation of the solvent either under N_2 or reduced pressure. The coated Celite is then applied as a dry powder or as a suspension in chloroform to a silicic acid column (Kieselgel 60, 70–230 mesh; E. Merck A-G, Darmstadt, FRG) prepared in chloroform. Alternatively, small amounts of free ecdysteroids may be conveniently loaded on to the column by first dissolving in the minimum volume of methanol followed by addition of chloroform to yield a methanol/chloroform ratio of 1/19.[22] The lower solubility of polar ecdysteroid conjugates in this solvent may preclude application of the latter method to such compounds.

The column is developed sequentially with chloroform, 30% methanol in chloroform (which elutes the free ecdysteroid fraction), and either 80% methanol in chloroform, or methanol (for elution of the highly polar ecdysteroid fraction). The volume of each eluting solvent is 10 ml/g of adsorbent. The tentatively identified long-chain acyl esters of ecdysone

---

[36] E. D. Morgan and C. F. Poole, *Adv. Insect Physiol.* **12,** 17 (1976).
[37] D. H. S. Horn, *in* "Naturally Occurring Insecticides" (M. Jacobson and N. G. Crosby, eds.), p. 333. Dekker, New York, 1971.
[38] R. Lafont, G. Somme-Martin, B. Mauchamp, B. F. Maume, and J. P. Delbecque, *in* "Progress in Ecdysone Research" (J. A. Hoffmann, ed.), p. 45. Elsevier/North-Holland Biomedical Press, Amsterdam, 1980.
[39] R. Lafont, J.-L. Pennetier, M. Andrianjafintrimo, J. Claret, J.-F. Modde, and C. Blais, *J. Chromatogr.* **236,** 137 (1982).
[40] G. Tsoupras, C. Hetru, B. Luu, E. Constantin, M. Lagueux, and J. Hoffmann, *Tetrahedron* **39,** 1789 (1983).
[41] G. Tsoupras, B. Luu, and J. A. Hoffmann, *Steroids* **40,** 551 (1982).
[42] R. E. Isaac and H. H. Rees, *Biochem. J.* **221,** 459 (1984).
[43] R. E. Isaac, H. Desmond, and H. H. Rees, *Biochem. J.* **217,** 239 (1984).
[44] G. Tsoupras, C. Hetru, B. Luu, M. Lagueux, E. Constantin, and J. A. Hoffmann, *Tetrahedron Lett.* **23,** 2045 (1982).
[45] L. N. Dinan and H. H. Rees, *J. Insect Physiol.* **27,** 51 (1981).

TABLE I
ECDYSTEROID PHOSPHATES AND ECDYSTEROID
ACIDS ISOLATED FROM INSECT TISSUE AND
CHARACTERIZED BY PHYSICOCHEMICAL TECHNIQUES

Compound
Ecdysone 22-phosphate
2-Deoxyecdysone 22-phosphate
20-Hydroxyecdysone 22-phosphate
2-Deoxy-20-hydroxyecdysone 22-phosphate
3-(or 2)-Acetyl-20-hydroxyecdysone 22-phosphate[a]
3-(and 2)-Acetyldysone 22-phosphate
3-Acetylecdysone 2-phosphate
3-Acetyl-20-hydroxyecdysone 2-phosphate
20-Hydroxyecdysone 2-(or 3)-phosphate[a]
Edysone 2-phosphate[b]
3-Epi-2-deoxyecdysone 3-phosphate
Ecdysonoic acid
20-Hydroxyecdysonoic acid

[a] Isolated as metabolites of 20-hydroxyecdysone from *in vitro* tissue incubations.
[b] Isolated as a nonenzymatic degradation product of 3-acetylecdysone 2-phosphate during storage, but may also be present *in situ*.

detected as metabolites in adult female ticks also elute during this chromatographic procedure in the 30% methanol in chloroform fraction together with free ecdysteroids.[11]

Prepacked disposable silica cartridges (Sep-Pak silica cartridges; Waters Associates) are convenient for the work-up of multiple small samples and can be regarded as 1 g silica columns for calculating elution volumes.[13]

Small-Scale Reversed-Phase Column Chromatography. Free ecdysteroids may also be separated from highly polar ecdysteroids in small samples utilizing disposable cartridges packed with silica bonded to C_{18} hydrocarbon[13,39] (Sep-Pak C_{18} cartridges, Waters Associates).[46] However, in samples containing comparatively large amounts of polar ecdysteroid conjugates, some of these compounds may elute early in the free ecdysteroid fraction (see later). This method is also valuable for further clean-up of free ecdysteroids eluted from silicic acid columns prior to analysis by HPLC, RIA, HPLC-RIA, or GC-MS(SIM).

Extracts dissolved in methanol (100 μl) and adjusted to 10% methanol in water are loaded onto the activated cartridge using a further 1 ml of

[46] Bond Elut C_{18} (Analytichem International) and "Baker" C_{18} extraction columns (J. T. Baker Chemicals, B. V. Deventer, Holland) are comparable disposable columns.

10% methanol in water to ensure transfer of all the sample. The cartridge is then developed sequentially with 10% methanol in water (2 ml), 30% methanol in water (4 ml; highly polar ecdysteroid fraction), 60% methanol in water (6 ml; contains ecdysteroids of similar polarity to ecdysone and 20-hydroxyecdysone), and finally methanol (4 ml) to ensure the elution of any relatively nonpolar ecdysteroids.

Other similar procedures for the isolation of ecdysteroids on Sep-Pak C_{18} cartridges (Water Associates) have been described.[39,47]

Butanol–Water Partition. This is sometimes included in procedures for isolating free ecdysteroids, which partition into the butanol phase.[36,37] Initial countersaturation of solvents is essential, together with backextraction of each phase with the other solvent; sometimes a centrifugation step to break emulsions is necessary. Highly polar ecdysteroids can be recovered from the water phase,[47] but this is not always quantitative.[38,45]

Further Purification and Analysis

Free Ecdysteroid Fraction

Silica gel thin-layer chromatography. The growing awareness of the complexity of ecdysteroid mixtures in extracts of insect tissue has resulted in a general decline in the use of this low resolution method for ecdysteroid analysis. Silica gel TLC is, however, useful for the preliminary analysis of extracts to indicate the range of ecdysteroids present. This is particularly true for the analysis of labeled metabolites which can be detected readily by scanning for radioactivity. Chloroform–methanol (or chloroform–ethanol) (4 : 1) is a frequently used solvent system which separates ecdysone, 20-hydroxyecdysone, and the corresponding 26-hydroxy and 2-deoxy compounds (e.g., R_f values: ecdysone, 0.27; 20-hydroxyecdysone, 0.21; 26-hydroxyecdysone, 0.15; 20,26-dihydroxyecdysone, 0.11).[48] Highly polar ecdysteroids remain at the origin with this solvent system. Other solvent systems may be required for more specialized applications,[36,37] including the separation of 3-epiecdysteroids from the corresponding 3β-hydroxy compounds by continuous elution.[49]

Reversed-phase thin-layer chromatography. Thin-layer plates (silica bonded to C_{18} hydrocarbon or paraffin coated silica) developed in methanol–water (1 : 1) give good separation of 2-deoxyecdysone, ecdysone, and 20-hydroxyecdysone.[50,51] Resolution is inadequate for the separation of

---

[47] S. Scalia and E. D. Morgan, *J. Insect Physiol.* **28,** 647 (1982).
[48] J. Koolman, L. Reum, and P. Karlson, *Hoppe-Seyler's Z. Physiol. Chem.* **360,** 1351 (1979).
[49] L. N. Dinan and H. H. Rees, *Steroids* **32,** 629 (1978).
[50] I. D. Wilson, S. Scalia, and E. D. Morgan, *J. Chromatogr.* **212,** 211 (1981).
[51] I. D. Wilson, C. R. Bielby, and E. D. Morgan, *J. Chromatogr.* **242,** 202 (1982).

TABLE II
RETENTION VOLUMES OF ECDYSTEROIDS ON A
REVERSED-PHASE COLUMN[52] [a]

Ecdysteroid	Retention volume (ml)
Ecdysone	13.1
20-Hydroxyecdysone	9.4
3-Epiecdysone	13.6
3-Epi-20-hydroxyecdysone	9.9
3-Dehydroecdysone	13.5
3-Dehydro-20-hydroxyecdysone	9.9
26-Hydroxyecdysone	9.7
20,26-Dihydroxyecdysone	4.9
Ecdysone 3-acetate	16.0
Ecdysone 2-acetate	18.3
2-Deoxyecdysone	18.6
2-Deoxy-20-hydroxyecdysone	14.8
3-Epi-2-deoxyecdysone	20.0
3-Epi-2-deoxy-20-hydroxyecdysone	16.6
Ecdysone 22-phosphate	1.3
2-Deoxyecdysone 22-phosphate	2.0
Poststerone	9.6
Inokosterone	9.9
Ponasterone A	17.2

[a] Column: 15 cm × 4.6 mm i.d. Mobile phase: linear gradient of 40% methanol in water increasing to 80% methanol in water over 20 min; flow rate, 1 ml/min.

complex mixtures of ecdysteroid metabolites but such a system can be useful in certain circumstances to complement chromatography on silica gel TLC.

High-performance liquid chromatography (HPLC). The excellent resolution of HPLC allows the separation of complex mixtures of ecdysteroids. However, some compounds have similar or identical retention times (see Table II for examples) and, therefore, it is important that isolated ecdysteroids are shown to be homogenous or cochromatograph with authentic material on at least two systems which differ in their chromatographic principles (e.g., reversed-phase and adsorption).

1. *Reversed-phase HPLC.* Free ecdysteroids of a wide spectrum of polarity can be resolved on reversed-phase (C_{18}, C_8, or phenyl) columns eluted with a gradient solvent system (Table II). Typical solvent systems are linear gradient of 40 to 80% methanol in water[52]; linear gradient of 8 to

[52] R. E. Isaac, N. P. Milner, and H. H. Rees, *J. Chromatogr.* **246,** 317 (1982).

40% acetonitrile in 20 mM Tris–HClO$_4$, pH 7.5 or 8.5.[22,38,53] Superior resolution is obtained with 5-μm-diameter column packings [e.g. Ultrasphere ODS (Beckman) or Resolve C$_{18}$ (Waters Associates)] than with larger packing materials. For elution from reversed-phase columns of nonpolar long-chain acyl esters of ecdysone in free ecdysteroid-containing fractions from ticks, a linear gradient of 40% methanol in water to pure methanol has been used.[11]

A HPLC system incorporating a sensitive UV monitor set at 254 nm and isocratic elution can detect ecdysteroids in the low nanogram range. In practice, the estimation of ecdysteroids by HPLC in conjunction with UV detection is restricted to relatively clean samples because of the possible interference from other UV-absorbing compounds.

2. *Adsorption HPLC columns*. Silica columns separate a number of ecdysteroids,[38,53] but are deactivated with time by polar solvents such as methanol or water, which can result in inconsistent retention times and long equilibration periods on changing solvents. These problems are largely circumvented by using aminopropyl packing (e.g., APS-Hypersil, Shandon Southern Products) as an adsorption column.[54] This is often used in conjunction with reversed-phase HPLC for separation and identification of free ecdysteroids.[13,25,45,55] The aminopropyl column gives clear separation of 3-epi-, 3-dehydro-, and 3β-hydroxy-ecdysteroids, which are poorly resolved on reversed-phase columns.[54] Solvent systems (generally dichloroethane : 2-propanol : methanol mixtures) for the chromatography of numerous ecdysteroids on APS-Hypersil are described by Dinan *et al.*[54] This system has been used effectively for investigation of ecdysteroid titers and of ecdysone 3-epimerase activity in *Spodoptera littoralis*.[24]

High-performance liquid chromatography–radioimmunoassay (HPLC—RIA). A technique that is particularly useful in the characterization of very small amounts of ecdysteroids is the monitoring by RIA of effluent (e.g., in fractions collected every 1 min) from HPLC fractionation of samples.[14] In this method it is convenient to collect the effluent in small tubes suitable for carrying out the RIA directly and to remove the solvent under vacuum in a spinning rotor (e.g., using a Speedvac concentrator; Savant). It is possible to use both reversed-phase and adsorption (e.g.,

---

[53] R. Lafont, P. Beydon, B. Mauchamp, G. Somme-Martin, M. Andrianjafintrimo, and P. Krien, *in* "Regulation of Insect Development and Behaviour" (F. Sehnal, A. Zabza, J. J. Menn, and B. Cymborowski, eds.), p. 125. Wroclaw Technical University Press, Poland, 1981.

[54] L. N. Dinan, P. L. Donnahey, H. H. Rees, and T. W. Goodwin, *J. Chromatogr.* **205**, 139 (1981).

[55] R. E. Isaac, H. H. Rees, and T. W. Goodwin, *J. Chem. Soc., Chem. Commun.* p. 418 (1981).

APS-Hypersil) columns and to choose antisera exhibiting specificities for different parts of the ecdysteroid structure.[56,57] The chromatographic positions of peaks of immunoreactivity are characterized by comparison with those of authentic compounds (monitored by U.V. absorbance at 254 nm), chromatographed under the same conditions prior to the samples. It is mandatory to thoroughly "flush" the HPLC system with solvent and to carry out a blank chromatographic run with collection of fractions for RIA immediately preceding the chromatography of the biological samples to check for possible "memory effects" (ecdysteroid contamination) in the HPLC system.

Gas–liquid chromatography (GLC). Ecdysteroids are amenable to analysis by GLC with flame ionization and electron capture detection, once they have been converted into volatile trimethylsilyl derivatives.[36] Electron capture detection is extremely sensitive (picogram level) for the detection of ecdysteroid trimethylsilyl ethers. Derivatization of all the tertiary hydroxy groups in ecdysteroids necessitates fairly harsh conditions.[36] Ecdysteroids are usually reacted with silylating reagent[36,38,58,59] [N-trimethylsilylimidazole (TMSI) or N,O-bis(trimethylsilyl)trifluoroacetamide)] for long periods at elevated temperatures. A derivatization procedure which is used in the authors' laboratory is described in the following section. Both electron capture and flame ionization detection lack specificity and, therefore, the identification of ecdysteroids depends upon the prior removal of as many impurities as possible by careful clean-up procedures of biological extracts. For electron-capture detection, chlorinated solvents should be avoided in the work-up procedure.[60] GLC analysis of ecdysteroids is now generally not favored, except where it is coupled to mass spectrometry, owing to the need for derivatization and the requirement for extensively purified samples.

Gas–liquid chromatography–mass spectrometry (GLC-MS). Greater specificity and considerable structural information during analysis of silylated ecdysteroids is obtained by interfacing the gas–liquid chromatograph to a mass spectrometer. The mass spectrum obtained for each eluted component together with the retention time on the GLC column provide information to distinguish ecdysteroids[14,36,38] and their biosynthetic intermediates[58] from nonecdysteroid impurities.

[56] L. Reum and J. Koolman, *Insect Biochem.* **9**, 135 (1979).
[57] M. A. De Laage, M. H. Hirn, and M. L. De Reggi, this series, Vol. 84, p. 350.
[58] C. Hetru, M. Lagueux, B. Luu, and J. A. Hoffmann, *Life Sci.* **22**, 2141 (1978).
[59] E. D. Morgan and C. F. Poole, *J. Chromatogr.* **116**, 333 (1976).
[60] C. R. Bielby, A. R. Gande, E. D. Morgan, and I. D. Wilson, *J. Chromatogr.* **194**, 43 (1980).

1. *Preparation of fully silylated derivatives.* The method described[14] is based on previous work.[59] Ecdysteroid samples, after thorough desiccation in vacuum are heated under dry nitrogen with N-trimethylsilylimidazole (60 μl) in "Reacti vials" (Pierce) at 140° for 17hr. The reaction products are then diluted with 0.8 ml of highly purified[61] hexane–ethyl acetate (7:3), applied to a 2 g silicic acid column (E. Merck, A-G Darmstadt, G.F.R., Kieselgel 70–230 mesh) prepared in hexane–ethyl acetate (7:3) and eluted with 20 ml of the same solvent. The eluate is evaporated to dryness under reduced pressure and the derivatives dissolved in purified hexane[61] for GLC-MS analysis. Silylated glassware is used throughout this procedure.

Gas–liquid chromatography–mass spectrometry (selected ion monitoring) [GLC-MS (SIM)] or mass fragmentography. By monitoring a limited number of selected ions, which are highly diagnostic of ecdysteroid trimethylsilyl ethers [GLC-MS (SIM)] greatly increased sensitivity can be achieved compared to GLC-MS, where complete mass spectra are obtained. GLC-MS (SIM) can be used to analyze and quantify ecdysteroids down to 0.1–1 ng depending on the ecdysteroid nucleus (see later), provided that a suitable internal standard is employed.[38] Suitable ecdysteroid nuclear fragment ions for monitoring fully silylated derivatives of compounds containing the ecdysone- or 20-hydroxyecdysone-type nucleus are m/z 567 and 561, respectively.[14,38] The occurrence of mass fragmentogram peaks corresponding to such ions also indicates the nature of the steroid nucleus in unidentified ecdysteroids. Makisterone A or ponasterone A can be used as internal standards (m/z 561) provided that these compounds are not already present in the biological sample; both makisterone A and ponasterone A have been detected in plants and crustacea,[2,37] whereas makisterone A has been reported additionally in certain insect species (for references, see ref. 13). Alternatively, a known amount of authentic ecdysteroid silylated with deuterated TMSI may be used as internal standard, since the nuclear fragment monitored in this case occurs at 27 mass units higher (9 extra mass units per TMS group present) than when nondeuterated TMSI is employed.[38]

The retention times of various fully silylated authentic ecdysteroids on a 1% OV-1 column, relative to that of silylated makisterone A together with the ions monitored are given in Table III. Since 3-epiecdysteroids are not separated from the corresponding 3β-hydroxy compounds on this OV-1 column, it is convenient to complement the GLC-MS (SIM) analy-

---

[61] Hexane and ethyl acetate are prepared from HPLC grade solvents by elution through a neutral alumina column followed by redistillation.

TABLE III
RELATIVE RETENTION TIMES OF THE FULLY SILYLATED TRIMETHYLSILYL ETHERS OF SOME ECDYSTEROIDS ON A 1% OV-1 COLUMN[a,b]

Silylated ecdysteroid	Retention time relative to silylated Makisterone A	Ions monitored (m/z)
Ponasterone A	0.54	561
Ecdysone	0.63	567
3-Epiecdysone	0.63	567
20-Hydroxyecdysone	0.79	561
3-Epi-20-hydroxyecdysone	0.79	561
Inokosterone	0.85	561
Makisterone A	1.00	561
26-Hydroxyecdysone	0.95[c]	567
20,26-Dihydroxyecdysone	1.24	561
Podecdysone A	1.17	561

[a] From ref. 14.
[b] GC-MS (SIM) of silylated ecdysteroids was carried out on a Pye-Unicam 204 gas–liquid chromatograph coupled via a single-stage all glass jet separator to a VG Micromass 7070F mass spectrometer linked to a Finnigan Incos data system. The conditions were 1.5 m × 2 mm glass column packed with 1% (w/w) OV-1 silicone phase on Gas Chrom Q (100–120 mesh); injector temperature, 300°; column, 280°; jet separator, 280°; inlet pressure of helium carrier gas, 2 kg cm^{-2}.
[c] M. Cleator and H. H. Rees (unpublished results).

sis of small amounts of compounds by their HPLC-RIA fractionation on an adsorption system, such as APS-Hypersil, which clearly resolves such pairs of ecdysteroids.[54]

1. *Quantification of ecdysone and 20-hydroxyecdysone by GLC-MS (SIM)*.[14] A standard curve is obtained by derivatizing mixtures of equimolar amounts of ecdysone and 20-hydroxyecdysone containing 10 pg to 100 ng of each compound and 5 ng of makisterone A (internal standard). Peak areas corresponding to fragment ions m/z 567 for ecdysone and m/z 561 for 20-hydroxyecdysone are expressed relative to that of makisterone A (m/z 561) and plotted against the original mass of ecdysteroid silylated. By adding internal standard (5 ng of makisterone A) to the biological samples (usually containing 5–10 ng of ecdysteroid) and obtaining peak areas for m/z 567 and 561 the amount of ecdysone and 20-hydroxyecdysone present in the sample can be calculated from the standard curve. The ion at m/z 561 in the spectrum of hexa-TMS-20-hydroxyecdysone is more

abundant (~10-fold) than the ion m/z 567 for the spectrum of penta-TMS-ecydsone, resulting in a better sensitivity for 20-hydroxyecdysone.

Derivatization reactions. The formation of ecdysteroid derivatives can provide structural information and can indicate the degree of purity of a sample. In addition, derivatization can aid purification of ecdysteroids from a biological extract by converting the steroid into a less polar compound, which may then be separated from polar impurities by chromatography.

1. *Acetylation.* The degree of acetylation depends upon the number and accessibility of the hydroxy groups present in the steroid and the reaction time.[37] A mixture of ecdysteroid acetates is usually obtained, the more fully acetylated products requiring longer reaction times. For example, 20-hydroxyecdysone acetylated at 20° for 20hr gives mainly 2,3,22-triacetate with some 2,3,22,25-tetraacetate derivative.[62] The position of free, accessible hydroxy groups may be indicated by comparing the acetylated products of an unknown sample with those of reference ecdysteroids.

1a. *Reaction procedure.* Samples are dissolved in the minimum amount of pyridine, to which is added acetic anhydride (final ratio of pyridine to acetic anhydride, 2:1).[62] The reaction is terminated by the addition of ice, and after standing for 30 min the solvent is evaporated under reduced pressure, adding ethanol as an azeotrope to remove pyridine. The resulting ecdysteroid acetates are readily soluble in solvents which are appreciably less polar than methanol, e.g., ethyl acetate, acetone, and dichloromethane.

Ecdysteroid mono-, di-, tri-, and tetraacetates are conveniently separated by silica TLC,[37] developing with chloroform–ethanol (19:1). Within these groups of compounds the 2- and 3-monoacetates as well as the 2,22- and 3,22-diacetates are not resolved in this system, but both of these pairs are readily separated by reversed-phase HPLC[63] (see Table II).

2. *Acetonide formation.* Ecdysteroids possessing vicinal *cis*-diol groups can form acetonides. Mono-, di-, and triacetonides are formed by ecdysone, 20-hydroxyecdysone, and 20,26-dihydroxyecdysone, respectively. Any free accessible hydroxy groups remaining after acetonide formation can be acetylated by the aforementioned procedure. Of course, the $2\beta,3\alpha$-hydroxy groups of 3-epiecdysteroids do not form acetonides.

2a. *Reaction procedure.*[62] Samples (up to 3 mg) are dissolved in anhydrous acetone (1 ml; dried over molecular sieve 4A) to which is added

---

[62] M. N. Galbraith and D. H. S. Horn, *Aust. J. Chem.* **22**, 1045 (1969).
[63] D. R. Greenwood, L. N. Dinan, and H. H. Rees, *Biochem. J.* **217**, 783 (1984).

phosphomolybdic acid (~200 μg) and the mixture stirred at ambient temperature for 20 to 30 min; it is important that fresh phosphomolybdic acid is used. The reaction is terminated by adding the mixture to 1-butanol (20 ml), which is then washed with saturated sodium bicarbonate (20 ml) followed by water (20 ml). The butanol layer, containing the acetonide derivative and any unreacted ecdysteroids, is then evaporated to dryness under reduced pressure. The reaction products are then analysed on silica TLC developed with chloroform–methanol (19:1). Acetonide derivatives are less polar than the parent ecdysteroid and, therefore, have greater mobility on this TLC system.

3. *Methylation of carboxylic acids.* Ecdysonoic acid and 20-hydroxyecdysonoic acids are methylated by reaction with diazomethane. The less polar methylated derivatives show greater retention than the parent free acids on reversed-phase HPLC. Using the same chromatographic conditions as in Fig. 4, the retention volumes for the methyl esters of 20-hydroxyecdysonoic acid and ecdysonoic acid are 41.5 and 49.5 ml, respectively.[23] Some free ecdysteroid-26-oic acids may also be detected on HPLC; this may indicate incomplete methylation, or more likely, some hydrolysis of the ester during the work-up. Alternatively, the methylated products may be separated by HPLC on a silica column, using dichloromethane–2–propanol–water (125:25:2) as eluting solvent.[22]

3a. *Reaction procedure.* Diazomethane (highly explosive) is prepared from *N*-methyl-*N*-nitroso-*p*-toluene sulfonamide ("Diazald," Aldrich). The preparation and storage of diazomethane are carried out with the utmost care as directed by the manufacturers. Diazomethane in diethyl ether is added in excess to a methanolic solution (0.3 ml) of the ecdysteroid-26-oic acids (<1 mg). After 10 min at 22°, the diazomethane together with solvent is evaporated under a stream of N_2.[23] The reaction products are dissolved in methanol for HPLC analysis.

Highly Polar Ecdysteroid Fraction. Analysis of the highly polar ecdysteroid fraction may be carried out either on the intact compounds or on the ecdysteroids released from hydrolysis of the conjugates with *Helix* hydrolases (see Fig. 3).

Analysis of enzymatically hydrolyzed conjugates. Samples are dissolved in 1 ml of 0.1 *M* MES [2-(*N*-morpholino)ethanesulfonic acid] buffer, pH 5.5, containing 0.5 mg[64] (12 units; one unit will hydrolyze 1 μmol of nitrocatechol sulfate in 1hr at pH 5.0 and 37°) of a so-called "arylsulfatase preparation" from *H. pomatia* (Sigma Chemical Co., Lon-

---

[64] The *H. pomatia* enzyme preparation contained less than 50 pg ecdysone equivalents per mg as determined by RIA.[14] Appropriate controls are required whenever RIA is used to measure ecdysteroids released by hydrolysis as the immunoreactivity of different batches of *H. pomatia* enzyme can vary.

don) for 17 hr at 37°. MES buffer can be replaced with sodium acetate or sodium citrate buffers, pH 5.5, but not phosphate buffer, which inhibits the hydrolyzing enzyme. Reactions are terminated by the addition of 3 ml of chilled ethanol and the resulting protein precipitate after centrifugation is extracted with 3 ml of methanol. The combined alcoholic extracts are evaporated to dryness, adsorbed onto Celite, and subjected to chromatography on a silicic acid column as described earlier. The free ecdysteroid fraction now containing the ecdysteroids released from conjugation is then analyzed as for the ecdysteroids which were present in the free form.

For the hydrolysis of ecdysteroid conjugates in relatively clean or small extracts, the incubation volume can be reduced to 100–500 μl. The reaction is stopped by the addition of an equal volume of chilled methanol and after centrifugation, aliquots of the supernatant are injected directly onto a reversed-phase HPLC system for the analysis of released ecdysteroids. This latter method is recommended for the analysis of ecdysteroid acetates liberated by the hydrolysis of the phosphate conjugate, since the faster work-up procedure and the omission of a silicic acid column step minimizes migration and elimination of the acetate group.[25]

Separation and analysis of the intact highly polar ecdysteroid fraction. 1. *Silica gel TLC.* Ecdysteroid-26-oic acids can be separated from polar ecdysteroid conjugates by TLC on silica gel developing with a polar solvent system, ethyl acetate–ethanol–H_2O (20:80:10).[22,65,66] The highly polar metabolites[67] of ecdysone that occur during embryogenesis of *S. gregaria,* namely ecdysonoic acid (R_f 0.42) and 3-acetylecdysone 2-phosphate (R_f 0.25), are resolved on this system. Since the recovery of the polar compounds, in particular conjugates, from the silica may be low, the usefulness of this technique is probably limited to analysis of radioactive compounds detected by scanning for radioactivity.

2. *High voltage paper electrophoresis.* Polar ecdysteroid conjugates (acidic) can be separated from ecdysteroid-26-oic acids by high voltage paper electrophoresis using 1 *M* acetic acid–pyridine buffer, pH 2.5.[22,65,66] At this pH the acidic ecdysteroid conjugates migrate toward the anode while ecdysonoic and 20-hydroxyecdysonoic acids remain at or near the origin.[22] Again, this separation technique is probably most useful in studies involving radioactive metabolites.

3. *HPLC of ecdysteroid phosphates and ecdysteroid-26-oic acids.* These compounds are amenable to HPLC on reversed-phase columns provided that the anionic charge is first neutralized by ion-suppression

---

[65] J. Koolman, J. A. Hoffmann, and P. Karlson, *Hoppe-Seyler's Z. Physiol. Chem.* **354,** 1043 (1973).
[66] A. Sannasi and P. Karlson, *Zool Jahrb., Abt. Allg. Zool. Physiol. Tiere* **78,** 378 (1974).
[67] R. E. Isaac and H. H. Rees, *Insect Biochem.* (1984), in press.

TABLE IV
RETENTION VOLUMES FOR SOME ECDYSTEROID PHOSPHATES
ON PARTISIL-SAX[a,b]

Ecdysteroid	Retention volume (ml)
Ecdysone 22-phosphate	16
Ecdysone 2-phosphate	31
2-Deoxyecdysone 22-phosphate	19
3-Acetylecdysone 2-phosphate	28
20-Hydroxyecdysone 22-phosphate	31

[a] Compounds were chromatographed on a Partisil-SAX column (Whatman; 25 cm × 4.6 mm i.d.) with a mobile phase of 0.1 M ammonium acetate at a flow rate of 2 ml/min.
[b] R. E. Issac and H. H. Rees (unpublished results).

with a suitable buffer (e.g., 20 mM Tris–HClO$_4$, pH 8.5[38]; 20 mM sodium acetate, pH 5.5[8]; 20 mM potassium phosphate, pH 5.7[38]; 20 mM sodium citrate, pH 6.5[9]) or by an ion-pairing agent (e.g., tetrabutylammonium phosphate[52,53,68]). These agents are added to the aqueous component of the mobile phase. The isolation of the ecdysteroid conjugate as a salt with a simple metal counter ion obviates subsequent complications in the interpretation of mass and NMR spectra. For this reason the use of Tris or tetrabutylammonium cation is preferably avoided when compounds are to be analyzed by these physical techniques.

The use of ion-suppression reversed-phase HPLC for the separation of highly polar ecdysteroids extracted from developing eggs of *S. gregaria* is illustrated in Fig. 4. Greater resolution of unresolved components (e.g., peaks 5 and 6 in Fig. 4) can be achieved either by rechromatography on the same column under appropriate isocratic elution conditions, or by using a longer (50 cm) reversed-phase column[42] [e.g. Partisil ODS-3, 50 cm × 9.4 mm i.d. (Magnum 9); Whatman]. The occurrence of nonenzymatic acetyl migration and deacetylation in acetylecdysteroid phosphates during their isolation makes the analysis more complex.[42]

The HPLC of ecdysteroid phosphates on a strong anion exchange column serves to complement reversed-phase chromatography in the analysis and identification of ecdysteroid phosphates. The phosphate conjugates are eluted isocratically with a suitable strength salt solution, e.g., 0.02 M potassium phosphate buffer, pH 5.0 or 0.1 M ammonium acetate[9] (Table IV).

[68] S. Scalia and E. D. Morgan, *J. Chromatogr.* **238**, 457 (1982).

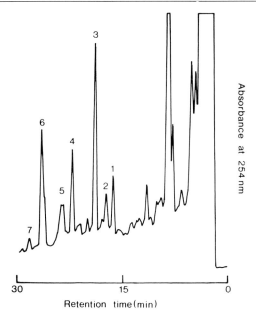

FIG. 4. Reversed-phase HPLC of highly polar ecdysteroids isolated from late-stage eggs of *S. gregaria*.[23,42,43] The UV absorbing peaks are 1, 20-hydroxyecdysone 22-phosphate; 2, 20-hydroxyecdysonoic acid; 3, ecdysone 22-phosphate; 4, ecdysonoic acid; 5, a mixture of 2-deoxy-20-hydroxyecdysone 22-phosphate, 3-acetyl-20-hydroxyecdysone 2-phosphate, and 3(or 2)-acetylecdysone 22-phosphate; 6, a mixture of 2-deoxyecdysone 22-phosphate and 3-acetylecdysone 2-phosphate; 7, 3-epiecdysone 3-phosphate. Chromatography was on a Partisil ODS-3 (25 cm × 4.6 mm i.d.; Whatman) eluted with a linear gradient of 10% methanol in 20 mM sodium citrate buffer, pH 6.5 to 70% methanol in the same buffer over a period of 30 min at a flow rate of 2 ml/min.

4. *Desalting of highly polar ecdysteroids.* Samples of highly polar ecdysteroids containing salts after collection from HPLC fractionation are desalted by loading onto reversed-phase Sep-Pak C_{18} cartridges[46] in 5 ml of water, salt being eluted with 5 ml of distilled water, and the ecdysteroids recovered by eluting with 5 ml of methanol.[9]

Mass Spectrometry

Electron Impact (EI)

Electron impact spectra of ecdysteroids contain many peaks which are of value for structural characterization.[37,69] EI mass spectrometry requires

[69] K. Nakanishi, *Pure Appl. Chem.* **25**, 167 (1971).

the sample to be ionized in the vapor phase and in the case of the polyhydroxylated ecdysteroids results in the sequential loss of water molecules with consequent very small or, more typically, lack of molecular ions.[37] Most ecdysteroids undergo characteristic side-chain cleavage either between C-20 and C-22 or between C-17 and C-20.

The EI spectrum of ecdysone includes distinctive peaks at m/z 348 [M − ($C_{22} \to C_{27}$)]$^+$, 300 [M − ($C_{20} \to C_{27}$)−H_2O]$^+$, and the side chain fragments m/z 99 [$C_6H_{11}O$]$^+$ and 81 [$C_6H_{11}O-H_2O$]$^+$. The spectrum of 20-hydroxyecdysone contains peaks at m/z 363 [M − ($C_{22} \to C_{27}$)]$^+$, 300 [M − ($C_{20} \to C_{27}$)−H_2O]$^+$, 99 [$C_6H_{11}O$]$^+$, and 81 ($C_6H_{11}O-H_2O$)$^+$. Characteristic ions are also observed in the EI spectra of 2-deoxy[70,71] and 26-hydroxy[72,73] derivatives of ecdysone and 20-hydroxyecdysone. Confirmation of the side-chain structure is obtained from the observation of prominent side-chain fragments,[37,69] e.g., the C_{28} ecdysteroid, makisterone A gives high abundance peaks at m/z 113 [$C_7H_{13}O$]$^+$ and 95 [$C_7H_{13}O-H_2O$]$^+$, 20,26-dihydroxyecdysone[72] gives peaks at m/z 133 [$C_6H_{13}HO_3$]$^+$ and 115 [$C_6H_{13}O_3-H_2O$)]$^+$, whereas 26-hydroxyecdysone[73] yields a side chain fragment at m/z 115 [$C_6H_{13}O_3-H_2O$]$^+$ (cf. foregoing side-chain fragments for ecdysone and 20-hydroxyecdysone). The EI spectra of ecdysteroid acetates include ions for the sequential loss of 60 mass units.[25,74]

EI mass spectra do not always indicate the molecular weight of an ecdysteroid because of the low abundance of the molecular ion. For a similar reason, EI mass spectrometry has been of little use in the identification of ecdysteroid phosphate conjugates. The polar phosphate group undergoes thermal elimination on heating the sample prior to ionization by a process which is probably analogous to the commonly observed loss of H_2O from the ecdysteroid molecule.[8,9]

Chemical Ionization (CI)

In chemical ionization (CI) mass spectrometry, ionization is achieved by the transfer of a proton from the reactant gas ions (ion plasma) to form quasi- (or pseudo)molecular ions [M + H]$^+$. These ions are generally

---

[70] M. N. Galbraith, D. H. S. Horn, E. J. Middleton, and R. J. Hackney, *J. Chem. Soc., Chem. Commun.* p. 83 (1968).

[71] Y. K. Chong, M. N. Galbraith, and D. H. S. Horn, *J. Chem. Soc., Chem. Commun.* p. 1217 (1970).

[72] M. J. Thompson, J. N. Kaplanis, W. E. Robbins, and R. T. Yamamoto, *J. Chem. Soc., Chem. Commun.* p. 650 (1967).

[73] J. N. Kaplanis, W. E. Robbins, M. J. Thompson, and S. R. Dutky, *Science* **180**, 307 (1973).

[74] R. E. Isaac, M. E. Rose, and H. H. Rees, unpublished results.

more prominent in the positive ion CI mass spectra[25,55,63] of ecdysteroids (and derivatives) than are molecular ions in the EI mass spectra of such compounds. Hence, of these two ionization techniques, CI frequently gives a better indication of the molecular weight of the compound. Common reactant gases in CI-mass spectrometry are NH_3, isobutane, and methane. The occurrence of relatively low abundance adduct ions (e.g., $[M + NH_4]^+$ in the case of ammonia, $[M + C_3H_7]^+$ and $[M + C_2H_5]^+$ in the case of isobutane) might cause confusion in the interpretation of CI mass spectra, especially the spectra of compounds (e.g., ecdysteroid acetates) which undergo elimination of H_2O and CH_3COOH. Another complication encountered in positive CI spectra of ecdysteroid acetates observed when NH_3 was employed as reactant gas is that the spectra of the monoacetate and diacetate anomalously resemble those expected of the diacetate and triacetate, respectively. Analogous intermolecular capture of an acetyl group has apparently been observed during mass spectrometry (EI) of acetylated monosaccharides.[75] However, such a problem is not observed during fast atom bombardment (FAB) mass spectrometry of ecdysteroid acetates (see following section). The paucity of fragment ions in CI spectra of ecdysteroids compared to those in the EI mode generally provides less structural information in the former case.

"In Beam" Ionization (Direct EI or Direct CI, Desorption EI or Desorption CI)[76]

In these techniques the molecules are vaporized directly into the ionizing beam (EI) or into reactant gas ions (CI) by rapid heating of the probe, which usually results in little thermal decomposition of the sample occurring before ionization. This method, used in the CI mode (direct CI or desorption CI) has given abundant quasimolecular ions, for example in the case of ponasterone A.[77] The methyl ester of 20-hydroxyecdysonoic acid submitted to a similar technique gave $[M + H]^+$ ions in low abundance.[22] Attempts to obtain molecular ions of ecdysteroid phosphates by direct CI failed due to the nonvolatile, thermally labile nature of these highly polar compounds.[74] However, molecular ions (low intensity) together with dehydrated ions have been observed in the case of certain

---

[75] H. Budzikiewicz, C. Djerassi, and D. H. Williams, "Structure Elucidation of Natural Products by Mass Spectrometry," Vol. II, p. 217. Holden-Day, San Francisco, California, 1964.

[76] M. E. Rose and R. A. W. Johnstone, "Mass Spectrometry for Chemists and Biochemists." Cambridge Univ. Press, London and New York, 1982.

[77] F. Lachaise, M. Goudeau, C. Hetru, C. Kappler, and J. A. Hoffmann, *Hoppe-Seyler's Z. Physiol. Chem.* **382**, 521 (1981).

ecdysteroid phosphates using the analogous technique, "direct EI" mass spectrometry with rapid heating of the sample probe.[27,41]

Fast Atom Bombardment (FAB).[76] In FAB mass spectrometry, ionization is achieved by the bombardment of the sample (either as a solid or in a nonvolatile solvent, e.g., glycerol) at ambient temperature with rapidly moving neutral atoms of gas (e.g., xenon). This "soft ionization" technique carried out at ambient temperature overcomes the difficulty in obtaining quasimolecular ions and, hence, the molecular weight of polar compounds, such as ecdysteroids and their derivatives including polar conjugates. Both positive and negative ion spectra may be recorded, but in our experience, the negative ion mass spectra are superior for the observation of quasi-molecular ions, $[M - H]^-$, of free ecdysteroids,[74] ecdysteroid acetates,[74] ecdysteroid phosphates,[8,9,42,43] and ecdysteroid acids.[23] In our laboratory, FAB mass spectra are recorded on a VG Micromass 70-70F spectrometer employing a primary atom beam of xenon with energy 8 keV for ionization. Samples, as solutions in methanol, are added to glycerol on the probe tip prior to FAB mass spectrometry.

In negative ion FAB mass spectra of ecdysteroids and their derivatives the only predominant ions present in the high mass region correspond to $[M - H]^-$ and a quasimolecular ion of the sodium salt. A low abundance peak at $[M - H - 16]^-$ is also generally seen (Fig. 5). The $[M - H]^-$ ion does not distinguish between an ecdysteroid monophosphate and an ecdysteroid monosulfate. However, prominent peaks at m/z 79 and 97 in the negative ion spectra of ecdysteroid phosphates correspond to $[PO_3]^-$ and $[H_2PO_4]^-$ ions, respectively[8,9] (Fig. 5). The sulfate ester of 20-hydroxyecdysone (synthesized by Dr. D. R. Greenwood, DSIR, N.Z.)[9] and sulfate esters of vertebrate steroids yield large peaks at m/z 80 and 97, corresponding to $[SO_3]^-$ and $[HSO_4]^-$ ions, respectively.[78] The stability of these oxyanions results in very large peaks in the negative ion mass spectra which are clearly diagnostic of the conjugating moiety.

Of the mass spectrometric techniques considered in this section, EI probably furnishes spectra yielding the most structural information on ecdysteroids (nucleus and side chain). However, for determination of molecular weights of unknown ecdysteroids, including polar phosphate conjugates. negative ion FAB furnishes unambiguous estimation of the molecular weights of all such compounds investigated so far. Of course, for determination of molecular formulas of compounds, accurate mass measurements by mass spectrometry are necessary.[76] As considered in this section, soft ionization techniques besides FAB can provide useful

---

[78] M. E. Rose, M. P. Veares, I. A. S. Lewis, and L. J. Goad, *Biochem. Soc. Trans.* **11,** 601 (1983).

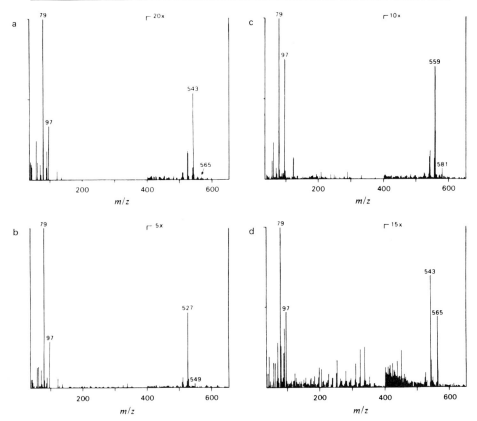

FIG. 5. Negative-ion FAB mass spectra of ecdysteroid conjugates from newly laid eggs of *Schistocerca gregaria* (from ref. 9). The negative ion FAB mass spectra of ecdysone 22-phosphate (a), 2-deoxyecdysone 22-phosphate (b), 20-hydroxyecdysone 22-phosphate (c), and 2-deoxy-20-hydroxyecdysone 22-phosphate (d) show distinct [M − H]⁻ ions at m/z 543, 527, 559, and 543, respectively.

spectra. Furthermore, another of these techniques, field desorption has furnished prominent quasimolecular ions for 3-dehydro-20-hydroxyecdysone.[79]

NMR Spectroscopy

For the recording of NMR spectra, free ecdysteroids are usually dissolved in [^{2}H$_5$]pyridine or [^{2}H$_4$]methanol, while the less polar ecdysteroid

[79] J. Koolman and K.-D. Spindler, *Hoppe-Seyler's Z. Physiol. Chem.* **358**, 1339 (1977).

TABLE V[a]
^{1}H CHEMICAL SHIFTS (ppm) OF SOME FREE ECDYSTEROIDS AND THEIR HIGHLY POLAR DERIVATIVES

Compounds	^{3}H signals					^{1}H signals[b]				Reference
	C-18	C-19	C-21	C-26/27	CH$_3$CO	C-2	C-3	C-22	C-7[c]	
Ecdysone	0.74 (s)	0.97 (s)	0.94, 0.97 (d, J 7 Hz)	1.20, 1.21 (2s)	—	3.85 (m, W$_{1/2}$ 21 Hz)	3.96 (m, W$_{1/2}$ 7 Hz)	3.61 (m, W$_{1/2}$ 16 Hz)	5.81 (d, J 2 Hz)	8,9
Ecdysone 22-phosphate	0.75 (s)	0.97 (s)	0.97, 1.00 (d, J 7 Hz)	1.18, 1.19 (2s)	—	3.86 (m, W$_{1/2}$ 21 Hz)	3.97 (m, W$_{1/2}$ 8 Hz)	4.20	5.81 (d, J 2 Hz)	8,9
3-Acetylecdysone 2-phosphate	0.73 (s)	1.00 (s)	0.93, 0.95 (d, J 6 Hz)	1.19, 1.20 (2s)	2.09 (s)	—	—	3.58 (m, W$_{1/2}$ 16 Hz)	5.82	43
Ecdysone 2-phosphate	0.72 (s)	0.97 (s)	0.93, 0.95 (d, J 6 Hz)	1.19, 1.20 (2s)	—	4.37 (m, W$_{1/2}$ 24 Hz)	4.18 (m, W$_{1/2}$ 8 Hz)	—	5.80	43
3-(or 2)-Acetylecdysone 22-phosphate	0.74 (s)	0.99 (s)	0.975, 0.99 (d, J 6 Hz)	1.17, 1.185 (2s)	2.06, 2.18 (s)[d]	—	—	—	5.82, 5.84[d]	42
Ecdysonoic acid	0.71 (s)	0.96 (s)	0.91 (d, J 10 Hz)	1.33 (s)	—	3.85 (m, W$_{1/2}$ 22 Hz)	3.95 (m, W$_{1/2}$ 9 Hz)	—	5.80	23
2-Deoxyecdysone	0.73 (s)	0.96 (s)	0.93, 0.95 (d, J 7 Hz)	1.19/1.20 (2s)	—	—	—	—	5.80 (d, J 2 Hz)	8,9
2-Deoxyecdysone 22-phosphate	0.74 (s)	0.96 (s)	0.96, 0.98 (d, J 8 Hz)	1.17, 1.19 (2s)	—	—	3.98 (m, W$_{1/2}$ 18 Hz)	4.20 (m, W$_{1/2}$ 22 Hz)	5.80 (d, J 2 Hz)	8,9
3-Epi-2-deoxyecdysone 3-phosphate	0.73 (s)	0.91 (s)	0.94, 0.96 (d, J 6 Hz)	1.19, 1.20 (2s)	—	—	4.08 (m, W$_{1/2}$ 20 Hz)	3.60 (m, W$_{1/2}$ 19 Hz)	5.80	42

Compound	Me-18	Me-19	Me-21	Me-26,27	OAc	H-2	H-3	H-22	H-7	Ref.
20-Hydroxyecdysone	0.89 (s)	0.96 (s)	1.19 (s)	1.19, 1.20 (2s)	—	3.84 (m, $W_{1/2}$ 21 Hz)	3.95 (m, $W_{1/2}$ 7 Hz)	3.33 (m, $W_{1/2}$ 14 Hz)	5.80 (d, J 2 Hz)	9
20-Hydroxyecdysone 22-phosphate	0.89 (s)	0.96 (s)	1.26 (s)	1.18, 1.19 (2s)	—	3.84 (m, $W_{1/2}$ 22 Hz)	3.95	4.07	5.80 (d, J 2 Hz)	9
3-Acetyl-20-hydroxyecdysone 2-phosphate	0.885 (s)	0.995 (s)	1.19 (s)	1.19, 1.20 (2s)	2.09 (s)	—	—	—	5.82	42
20-Hydroxyecdysone 2 (or 3)-phosphate[e]	0.91 (s)	1.15, 1.18[f]	1.22 (s)	1.22 (2s)	—	—	—	—	—	27
3-(or 2)-Acetyl-20 hydroxyecdysone 22-phosphate[e]	0.92 (s)	0.99 (s)	1.32 (s)	1.21, 1.22 (2s)	1.92	—	—	—	—	27, 41
2-Deoxy-20-hydroxyecdysone 22-phosphate	0.89 (s)	0.95 (s)	1.26 (s)	1.18, 1.20 (2s)	—	—	3.98 (m, $W_{1/2}$ 18 Hz)	4.07 (m, $W_{1/2}$ 22 Hz)	5.80 (d, J 2 Hz)	9
20-Hydroxyecdysonoic acid	0.86 (s)	0.95 (s)	1.16 (s)	1.32 (s)	—	3.85 (m, $W_{1/2}$ 22 Hz)	3.95 (m, $W_{1/2}$ 9 Hz)	—	5.80	23

[a] [$^2H_4$]Methanol was used as the solvent and TMS as a reference.
[b] Some proton signals were not observed because of low sample concentration.
[c] Only when this region of the spectrum is expanded are the doublets discernible.
[d] Acetyl migration resulted in a mixture of 2- and 3-acetyl derivatives which gives rise to two signals.
[e] These spectra were recorded under different conditions and, compared to the other spectra, all the chemical shifts appear to be displaced by about +0.03 ppm.
[f] This sample appears to be a mixture of two compounds.

acetates are more soluble in [^{2}H]chloroform. Suitable solvents for highly polar ecdysteroids (phosphates and 26-oic acids) are [^{2}H$_4$]methanol or D$_2$O. For comparison of spectra of free ecdysteroids and those of highly polar derivatives without observing differences due to solvent effects, [^{2}H$_4$]methanol has been used as a common solvent in a number of studies on the structural elucidation of ecdysteroid conjugates[8,9,27,41,43] and acids.[23]

The high sensitivity and spectral resolution obtained with high field Fourier transform NMR instruments (250–400 MHz for ^{1}H) allow spectra to be obtained with quantities as low as 50 μg for ^{1}H and 700 μg for ^{13}C nuclei, within acceptable accumulation times.

In the following sections, the ^{1}H and ^{13}C NMR resonances are given relative to tetramethylsilane at 0 ppm.

^{1}H NMR Spectroscopy

The CH_3 and in the case of some ecdysteroids (mainly acetate and acetonide derivatives) the CH signals have been assigned and are invaluable for structural elucidation. ^{1}H NMR spectroscopy of ecdysteroids has been reviewed by Horn[37] and Nakanishi[69] and many chemical shift values have been tabulated.[37] ^{1}H NMR data from the spectra of some recently identified ecdysteroids isolated from insects are given in Table V. The chemical shifts of the CH_3 resonances at C-18, C-21, C-26, and C-27 are, in the main, determined by the nature of the ecdysteroid side chain. Introduction of a hydroxy group at C-20 results in a large downfield shift of the C-18 and C-21 CH_3 signals relative to those of ecdysone. Hydroxylation of C-26 and further oxidation of this group to a carboxylic acid shifts the C-27 CH_3 downfield, while ecdysteroid 22-phosphates have the C-21 CH_3 signal occurring downfield from that of the nonconjugated steroid (see Table V). Differences in the chemical shift of the C-19 CH_3 due to changes in the A-ring structure are not always observed (e.g., absence of a 2-hydroxy group). However, in the spectra of 3-dehydro[49,79] and 3-epi-2-deoxy compounds,[55,62] the C-19 signal occurs at higher field (+0.05 to +0.07 ppm) relative to ecdysone or 20-hydroxyecdysone.

The CH resonances can be frequently discerned in high field Fourier transform spectra of even relatively small samples and can be important in pin-pointing esterified hydroxy groups. The broad CH OH (axial) resonance of the 2-H can be distinguished from the narrow CH OH (equatorial) signal of the 3-H; acetylation or phosphorylation of the hydroxy groups at these positions results in large downfield shifts of the signals (see Table V). Similarly, the CH OH signal of the 22-H of ecdysteroid 22-phosphates occurs downfield from the corresponding signal in the spec-

TABLE VI
13C CHEMICAL SHIFTS (δ) OF HYDROXY-SUBSTITUTED CARBON ATOMS OF FREE ECDYSTEROIDS AND ECDYSTEROID PHOSPHATES[a,b]

Compound	δ (ppm)[c]					
	C-2	C-3	C-14	C-20	C-22	C-25
Ecdysone	68.7	68.5	85.2	—	75.3	71.5
	(d)	(d)	(s)		(d)	(s)
Ecdysone 22-phosphate	68.7	68.6	85.4	—	79.3	71.8
	(d)	(d)	(s)		(d)	(s)
2-Deoxyecdysone 22-phosphate	—	65.6	85.6	—	79.1	71.8
		(d)	(s)		(d)	(s)
20-Hydroxyecdysone	68.7	68.6	85.3	77.9	78.4	71.3
	(d)	(d)	(s)	(s)	(d)	(s)
20-Hydroxyecdysone 22-phosphate	68.7	68.6	85.3	78.5	82.9	71.4
	(d)	(d)	(s)	(s)	(d)	(s)

[a] From. ref. 9.
[b] Spectra were recorded in methanol with TMS as a reference.
[c] The type of peak observed in off-resonance decoupled spectra is indicated below the chemical shift. —, a nonsubstituted carbon.

trum of the free ecdysteroid (+0.59 ppm for 20-deoxyecdysteroids and +0.74 ppm for 20-hydroxy compounds; see Table V). The axial C-3 proton of 3-epi-2-deoxyecdysone[55] and 3-epi-2-deoxyecdysone 3-phosphate (see Table V) is observed as a broad signal at δ 4.15 ppm (pyridine) and 4.08 ppm (methanol), respectively.

The quality of the spectra obtained with small quantities of sample is dependent upon the purity of the sample and the deuterated NMR solvent. In particular, ^{1}H signals due to nondeuterated methanol in [^{2}H]methanol solvent can mask the *H*COH signal of C-22. The use of only a trace of TMS as an internal reference is also important in obtaining a clean spectrum with small amounts of sample.

^{13}C NMR Spectroscopy

For proton noise decoupled spectra, relatively large quantities (~1 mg) of sample are required if long accumulation times are to be avoided. The ^{13}C nuclear magnetic resonance signals for all the carbon atoms in several free ecdysteroids have been assigned.[80] The signals (Table VI) in ^{13}C NMR spectra of ecdysteroid phosphate conjugates isolated from the

---

[80] J. Krepinsky, J. A. Findlay, B. Danieli, G. Palmisano, P. Beynon, and S. Murakami, *Org. Magn. Reson.* **10,** 255 (1977).

eggs of the locust, *S. gregaria,* have been assigned[8,9,23,42,43] by off-resonance decoupling and by comparison with the foregoing published data.[80] ^{13}C NMR spectra have also been used in the determination of the structures of polar ecdysteroid conjugates isolated from another locust species (see this volume [23]). The resonances of the hydroxy substituted carbons occur downfield (60–100 ppm), well clear of other carbon signals, and are easily assigned by off resonance decoupling. The introduction of a phosphate group shifts the signal of the esterfied carbon downfield (~ 4.0 ppm; see Table VI) and owing to ^{13}C–^{31}P coupling this signal is observed as a doublet or as a broad unresolved peak. Other changes are observed in the chemical shifts of carbons close to the phosphorylated carbon. For example, compared with ecdysone the corresponding 22-phosphate ester shows the following shifts: C-20 (−2.2 ppm), C-21 (+0.3 ppm), C-23 (−0.4 ppm), and C-24 (−1.24 ppm).[9]

^{31}P NMR Spectroscopy

The presence of phosphate in purified samples of ecdysteroid conjugates can be detected nondestructively by proton noise-decoupled ^{31}P NMR. The ^{31}P signal referenced to external H_3PO_4, of ecdysteroid 22-phosphates occurs at 1 to 3 ppm.[8,9,40,44] Quantification of the phosphate present in samples and, hence, the determination of the molar ratio of ecdysteroid (determined by UV absorbance at 242 nm; ε_{242} 12,400 in methanol) to phosphate can be carried out readily by the colorimetric assay of the inorganic phosphate released on acid hydrolysis (0.5 M HCl at 110°).[8,9]

Acknowledgments

Work from the authors' laboratory was supported by the Science and Engineering Research Council, Agricultural Research Council, the Wellcome Trust, and The Royal Society. We thank Dr. M. E. Rose and Mr. M. C. Prescott for mass spectra, and Drs. B. Mann (University of Sheffield), I. P. Jones (Bruker Spectrospin Ltd.), and I. H. Sadler (University of Edinburgh) for the NMR spectra.

[23] Ecdysone Conjugates: Isolation and Identification

By C. HETRU, B. LUU, and J. A. HOFFMANN

Following the isolation[1] and structural elucidation[2,3] of ecdysone, a series of investigations started on the biosynthetic and catabolic pathways of this steroid hormone. In 1970, Heinrich and Hoffmeister[4] proposed that in the blowfly *Calliphora erythrocephala*, ecdysone was inactivated as a glycoside conjugate (incorporation studies with [^{14}C]glucose). Several subsequent studies on a variety of insect species showed that enzymatic hydrolysis of highly polar ecdysteroid yielded free ecdysone and/or 20-hydroxyecdysone which led to the assumption that conjugation was a major catabolic pathway of ecdysteroids (review in Koolman[5]). On the basis of incorporation experiments of labeled molecules into ecdysteroid conjugates and of hydrolytic studies with defined enzymes, it was concluded that ecdysteroids were inactivated as sulfate, glucoside, glucuronide, or phosphate conjugates (references in review cited above). It is only very recently that an ecdysteroid conjugate was isolated from fecal material of *Locusta migratoria*[6] and *Schistocerca gregaria*[7] and identified by physical methods as a 2-phosphate 3-acetate of 20-hydroxyecdysone. By similar methods, Malphighian tubules–digestive tract complexes were shown in *L. migratoria* to convert 20-hydroxyecdysone to a 22-phosphate 2- (or) 3-acetate conjugate.[8]

Independently from the studies on inactivation of ecdysteroids during postembryonic development, Ohnishi and co-workers[9] had noted in 1977 that 2-deoxyecdysone was present in ovaries of *Bombyx mori* pupae as a polar compound presumed to be a conjugate. The observation that ova-

---

[1] A. Butenandt and P. Karlson, *Z. Naturforsch., B: Anorg. Chem., Org. Chem., Biochem., Biophys., Biol.* **9B**, 389 (1954).
[2] P. Karlson, H. Hoffmeister, H. Hummel, P. Hocks, and G. Spiteller, *Chem. Ber.* **98**, 2394 (1965).
[3] R. Huber and W. Hoppe, *Chem. Ber.* **98**, 2403 (1965).
[4] G. Heinrich and H. Hoffmeister, *Z. Naturforsch., B: Anorg. Chem., Org. Chem., Biochem., Biophys., Biol.* **25B**, 358 (1970).
[5] J. Koolman, *Insect Biochem.* **12**, 225 (1982).
[6] J. F. Modde, Thesis, Université Pierre et Marie Curie, Paris (1983) (submitted as as article for *J. Invertebr. Reprod. Dev.*)
[7] R. E. Isaac and H. H. Rees, *Commun. Int. CNRS Symp. Invertebr. Horm., 1983* (1983).
[8] G. Tsoupras, B. Luu, C. Hetru, J. F. Muller, and J. A. Hoffmann, *C.R. Hebd. Seances Acad. Sci., Ser. D* **296**, 77 (1983).
[9] E. Ohnishi, T. Mizuno, F. Chatani, N. Ikekawa, and S. Sakurai, *Science* **197**, 66 (1977).

ries contain large quantities of ecdysteroids predominently in conjugated form has since been confirmed in a variety of species which include *S. gregaria*,[10–13] *L. migratoria*,[14,15] and *Galleria mellonella*.[16] The observations that these conjugated ecdysteroids are synthesized inside the vitellogenic ovaries,[17,18] accumulate in the oocytes, and are obviously hydrolyzed during embryogenesis[11,14] have prompted studies toward the isolation and identification of these compounds, and several structures have recently been reported.[12,13,19–22] The present chapter will essentially deal with the methods of purification and structure determination of ecdysteroid conjugates.

Extraction and Purification

For the extraction of ecdysteroid conjugates the various authors mostly use polar solvents. Pure methanol,[22] various proportions of methanol–water mixtures,[13,21] or even pure water[6,23] are recommended in the literature. As a rule, the extraction by polar solvents is followed by partition destined to remove the majority of the lipophilic substances: either methanol–water/hexane partitions[13,21,24] or water/chloroform partitions[6,23] can be used.

The ecdysteroid conjugates recovered in the polar phase of these partitions can subsequently be submitted to liquid chromatography, either on a silicic acid column[6,13] or on a reverse-phase C_8 column.[21] It has been recommended[24] to partition the residues from the first partition step between water-saturated *n*-butanol and water: the free ecdysteroids are then

---

[10] I. D. Wilson and E. D. Morgan, *Insect Physiol.* **24**, 751 (1978).
[11] L. N. Dinan and H. H. Rees, *Insect. Physiol.* **27**, 51 (1981).
[12] R. E. Isaac, M. E. Rose, H. H. Rees, and T. W. Goodwin, *Chem. Commun.* p. 249 (1982).
[13] R. E. Isaac, M. E. Rose, H. H. Rees, and T. W. Goodwin, *Biochem. J.* **213**, 533 (1983).
[14] J. A. Hoffmann, M. Lagueux, C. Hetru, M. Charlet, and F. Goltzené, in "Progress in Ecdysone Research" (J. A. Hoffmann, ed.). Elsevier, Amsterdam, 1980.
[15] M. Lagueux, C. Sall, and J. A. Hoffmann, *Am. Zool.* **21**, 751 (1981).
[16] T. H. Hsiao and C. Hsiao, *Insect Physiol.* **25**, 45 (1979).
[17] M. Lagueux, M. Hirn, and J. A. Hoffmann, *Insect Physiol.* **23**, 109 (1977).
[18] F. Goltzené, M. Lagueux, M. Charlet, and J. A. Hoffmann, *Hoppe-Seyler's Z. Physiol. Chem.* **359**, 1427 (1978).
[19] R. E. Isaac, H. H. Rees, and T. W. Goodwin, *Chem. Commun.* p. 594 (1981).
[20] G. Tsoupras, C. Hetru, B. Luu, M. Lagueux, E. Constantin, and J. A. Hoffmann, *Tetrahedron Lett.* p. 2045 (1982).
[21] G. Tsoupras, C. Hetru, B. Luu, E. Constantin, M. Lagueux, and J. A. Hoffmann, *Tetrahedron* **39**, 1789 (1983).
[22] G. Tsoupras, B. Luu, and J. A. Hoffmann, *Science* **220**, 507 (1983).
[23] R. Lafont, J. L. Pennetier, M. Andrianjafintrimo, J. Claret, J. F. Modde, and C. Blais, *J. Chromatogr.* **236**, 137 (1982).
[24] S. Scalia and E. D. Morgan, *J. Chromatogr.* **238**, 457 (1982).

TABLE I
PURIFICATION OF ECDYSTEROID CONJUGATES BY HPLC: TECHNIQUES USED IN THE RECENT LITERATURE

Stationary phase	Mobile phase	Reference
ODS, 5μm Spherisorb (Clwyd, U.K.)	Linear gradient (12 min) from 25 to 50% methanol in 0.4 M ammonium acetate buffer (pH 7) in the continuous presence in the mobile phase of 0.003 M tetrabutylammonium hydroxide	24
ODS, Zorbax (Dupont) ODS, Ultrasphere (Beckman)	Linear gradient from 8 to 40% acetonitrile in 0.02 M Tris/perchloric acid buffer (pH 7.5) or in 0.05 M sodium citrate buffer (pH 6.1 or 3)	6
ODS, 5μm, Lichrosorb, (Merck)	Linear gradient from 12 to 44% acetonitrile in 0.2 M Tris–HCl buffer (pH 7.4)	25
ODS-3, Partisil (Whatman)	Isocratic elution (32 min) with 35% methanol in 0.2 M sodium acetate buffer (pH 5.5) Linear gradient (30 min) from 10 to 70% methanol in 0.02 M sodium citrate buffer (pH 6.5)	13
SAX, Partisil (Whatman)	Isocratic elution with 0.1 M ammonium acetate	13

recovered in the butanol phase, whereas the polar conjugates are present in the aqueous phase.

All investigations on ecdysteroid conjugates use HPLC for further purification, either on a preparative or on an analytical scale. Table I[25] gives the details on the stationary and mobile phases and the modes of elution used in recent studies.

We have summarized in Fig. 1 a complete procedure for extraction and purification which we recommend at present for the extraction of polar ecdysteroid conjugates.

Identification Procedures

The identification of ecdysteroid conjugates is usually by a combination of spectroscopic techniques: infrared and ultraviolet absorption, mass spectrometry, and ^{1}H, ^{13}C, ^{31}P nuclear magnetic resonance.

Infrared Spectroscopy

Infrared spectroscopy can be useful in the study of ecdysteroid conjugates. In the study of Tsoupras et al.[21] three bands of absorption are

[25] G. Tsoupras, Thesis, Université Louis Pasteur, Strasbourg (1982).

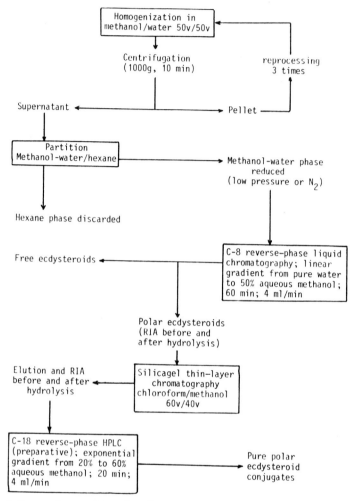

FIG. 1. Proposed extraction and purification procedures for polar ecdysteroid conjugates from insects or other invertebrate systems.

given for several ecdysteroid conjugates: two correspond to the ecdysteroid moiety [IR(KBr, cm^{-1}): 1640 for the C=O bond and 3400 for the O—H bonds]; a third band at 1050–1100 is indicative of the presence of a P—O—C bond.

Ultraviolet Spectroscopy

The presence of an α,β-ethylenic ketone in ecdysteroids explains the strong absorption at 250 nm in ecdysteroid conjugates.[13,21]

TABLE II
CHARACTERISTIC ^{1}H NMR DATA OF SOME RELEVANT ECDYSTEROIDS
AND ECDYSTEROID CONJUGATES

Compound	Chemical shifts of methyl groups					Solvent	Reference
	C-18	C-19	C-21	C-26	C-27		
Ecdysone	0.74	1.07	1.28	1.38	1.38	C_5D_5N	25
2-Deoxyecdysone	0.74	1.05	1.28	1.38	1.38	C_5D_5N	25
20-Hydroxyecdysone	1.19	1.05	1.54	1.38	1.38	C_5D_5N	25
22-Phosphate of ecdysone	0.70	0.96	0.92	1.20	1.20	D_2O	25
22-Phosphate of 20-hydroxyecdysone	0.92	0.98	1.32	1.21	1.21	D_2O	26
22-Phosphate of ecdysone	0.75	0.97	0.98	1.18	1.19	CD_3OD	12
22-Phosphate of 2-deoxyecdysone	0.73	0.96	0.97	1.18	1.18	CD_3OD	12
22-Adenosine monophosphate of ecdysone	0.72	0.96	0.92	1.20	1.20	D_2O	21

Nuclear Magnetic Resonance (NMR)

Nuclear magnetic resonance is a powerful tool for the identification of ecdysteroid conjugates. As a rule, ^{1}H NMR and ^{13}C NMR spectra are first taken of the intact pure ecdysteroid conjugate, after which the conjugate is subjected to enzymatic or chemical hydrolysis (see below, Table IV). The free ecdysteroid is then purified and analyzed by ^{1}H and ^{13}C NMR spectroscopy. The comparison between the spectra of the free identified ecdysteroid and the intact conjugated ecdysteroid usually allows inferences on the nature of the conjugating moiety.

The comparison of chemical shifts of certain signals in ^{1}H and ^{13}C NMR between the free and the conjugated ecdysteroid is useful in assigning the position at which the linkage between the conjugating moiety and the ecdysteroid has occurred. Table II[26] gives a certain number of characteristic chemical shifts in ^{1}H NMR of relevant ecdysteroids and ecdysteroid conjugates. In Table III the ^{13}C NMR signals of the 27 carbon atoms of several ecdysteroids are also presented and the multiplicity of the signals is indicated. These tables illustrate, for instance, that in the ^{1}H NMR spectra of the 22-phosphate of ecdysone, the signal of the C-21 methyl group is shifted downfield. In ^{13}C NMR, the signal of the carbon C-22 of 2-deoxyecdysone is shifted downfield, whereas the signals of C-20 and C-23 are upfield shifted. This sort of observations had led[13,25] to the conclusion that the conjugating moiety is linked to the ecdysteroids in the C-22 position in these compounds.

[26] G. Tsoupras, B. Luu, and J. A. Hoffmann, *Steroids* **40**, 551 (1982).

TABLE III
^{13}C NMR DATA OF ECDYSTEROIDS AND CONJUGATES[a]

Number of carbon atoms	A	B	C	D	E
1	37.90(t)	29.48(t)	29.50(t)	34.50(t)	34.00(t)
2	68.03(d)	29.07(t)	28.20(t)	32.10(t)	29.70(t)
3	68.03(d)	64.06(d)	66.04(d)	69.30(d)	74.60(d)[b]
4	31.80(t)	33.07(t)	32.80(t)	35.90(t)	33.60(t)
5	51.30(d)	51.61(d)	52.21(d)	57.40(d)	56.90(d)
6	203.20(s)	203.20(s)	210.00(s)	202.20(s)	207.80(s)
7	121.50(d)	121.30(d)	121.00(d)	121.50(d)	121.70(d)
8	165.60(s)	166.01(s)	170.00(s)	166.06(s)	169.60(s)
9	34.50(d)	34.40(d)	35.50(d)	34.20(d)	34.60(d)
10	38.60(s)	36.96(s)	37.60(s)	37.00(s)	37.20(s)
11	21.10(t)	21.00(t)	21.80(t)	21.00(t)	21.00(t)
12	31.40(t)	31.70(t)	32.10(t)	31.60(t)	31.70(t)
13	47.50(s)	48.00(s)	—	47.80(s)	48.00(s)
14	83.80(s)	84.01(s)	86.70(s)	83.90(s)	85.90(s)
15	32.40(t)	31.70(t)	31.70(t)	31.80(t)	32.20(t)
16	25.50(t)	25.50(t)	26.40(t)	35.80(t)	25.20(t)
17	48.30(d)	48.32(d)	48.76(d)	48.50(d)	48.50(d)
18	15.80(q)	15.80(q)	16.57(q)	16.00(q)	16.40(q)
19	24.40(q)	24.30(q)	24.40(q)	24.10(q)	23.80(q)
20	42.90(d)	42.99(d)	40.80(d)	43.20(d)	42.70(d)
21	13.60(q)	13.60(q)	13.60(q)	13.90(q)	13.50(q)
22	73.90(d)	73.90(d)	81.09(d)[b]	74.20(d)	75.50(d)
23	26.60(t)	26.70(t)	24.90(t)	26.80(t)	26.60(t)
24	42.40(t)	42.46(t)	41.24(t)	42.70(t)	41.50(t)
25	69.70(s)	69.60(s)	73.02(s)	69.80(s)	72.70(s)
26	30.01(q)	29.98(q)	29.00(q)	30.00(q)	29.10(q)
27	30.20(q)	30.20(q)	29.20(q)	30.20(q)	28.80(q)

[a] Ecdysone (A), 2-deoxyecdysone (B), 22-phosphate of 2-deoxyecdysone (C), 3-epi 2-deoxyecdysone (D), and 3-phosphate of 3-epi-2-deoxyecdysone (E). Spectra of A, B, and D were obtained in [^{2}H$_5$]pyridine, C and E in D$_2$O; δ in parts per million from trimethylsilane (internal standard); s, singlet; d, doublet; t, triplet, q, quadruplet. From Ref. 25.

[b] In proton-noise decoupled spectra, this signal appears as a doublet.

^{31}P NMR is mostly used to confirm the presence of a phosphate group in some conjugates.[13,21]

Mass Spectrometry

Most ecdysteroid conjugates are polar compounds and are not very volatile. In addition, they are probably unstable under various conditions. A variety of techniques of volatilization and ionization have been em-

TABLE IV
CONDITIONS OF ENZYMATIC HYDROLYSIS OF ECDYSTEROID CONJUGATES

Enzymes	Buffer systems	Conditions			Work up	References
		pH	Time	Temperature		
Arylsulfatase of *Helix pomatia* (Sigma)	0.1 M 4-morpholinoethanesulfonic acid	5.5	17 hr	37°	Addition of ethanol	13
Digestive juice of *Helix pomatia* (Koch-Light Laboratories Colnbrook, U.K.)	100 mM acetate	5.2	Overnight	37°	Injection into Sep-Pak C_{18} and elution with a methanol/water mixture	24
Digestive juice of *Helix pomatia* (Industrie Biologique Française, Clichy, France)	50 mM acetate	5.3	12 hr	37°	Addition of methanol	25
Digestive juice of *Helix pomatia* (Industrie Biologique Française, Clichy, France) and acid phosphatase from potatoes	50 mM sodium acetate	5.3	1 or 6 hr	30°	—	6

TABLE V
ECDYSTEROID CONJUGATES ISOLATED AND IDENTIFIED IN INSECTS

Compound	Reference
2-Acetate of ecdysone	6
2-Acetate of 20-hydroxyecdysone	6
3-Acetate of ecdysone	6,19
3-Acetate of 20-hydroxyecdysone	6
22-Adenosine monophosphate of 2-deoxyecdysone	20,21
22-Adenosine monophosphate of ecdysone	20,21
22-N^6-(Isopentenyl)adenosine monophosphate of ecdysone	22
2-Phosphate-3-acetate of ecdysone	6
2-Phosphate-3-acetate of 20-hydroxyecdysone	6
3-Phosphate of ecdysone	25
3-Phosphate of 3-epi-2-deoxyecdysone	20,21
3-Phosphate of 20-hydroxyecdysone	6
22-Phosphate 3- (or 2-)acetate of 20-hydroxyecdysone	26
22-Phosphate of 2-deoxyecdysone	12,13,21
22-Phosphate of 2-deoxy-20-hydroxyecdysone	13,26
22-Phosphate-2,3-diacetate of ecdysone	25
22-Phosphate of ecdysone	12,13,21
22-Phosphate of 20-hydroxyecdysone	13,26

ployed: chemical ionization,[12,21] laser ionization,[8] gold thread technique,[21] and fast heating technique.[25] Of special interest is the recently developed technique of fast atom bombardment, which allows the recording of the quasimolecular ion (e.g., for 22-phosphates of ecdysone, 20-hydroxyecdysone, 2-deoxyecdysone, and 2-deoxy-20-hydroxyecdysone, see Isaac et al.[13]).

Hydrolysis

Table IV lists several procedures which are commonly used for the enzymatic hydrolysis of ecdysteroid conjugates, as well as the operating conditions. *Helix pomatia* juice is the most widely used enzymatic mixture in this field; it should be emphasized that some *Helix* preparations contain ecdysteroids. Ecdysteroid conjugates can also be hydrolyzed chemically, e.g., by a 2 hr treatment with 0.1 N HCl.[25]

Concluding Remarks

Table V gives a list of ecdysteroid conjugates which have been isolated and subsequently identified by physicochemical methods. Most of these results have been obtained over the last 2 years and there can be no doubt that this list will rapidly increase.

With the exception of the acetate esters of ecdysteroids, the identified conjugates are all polar compounds (in fact, mainly phosphates). There is however increasing evidence for the existence of ecdysteroid conjugates of very low polarity: ticks for instance are capable of esterifying ecdysone with fatty acids.[27-29] For the extraction of such esters, ethanol can be recommended. Until more information becomes available, partitions as used in Fig. 1 for the purification of polar conjugates, should be employed with great care for the nonpolar conjugates. Esterases or K_2CO_3 treatment is preferred to the *Helix* digestive juice for hydrolysis of these products.

[27] P. Diehl, J. L. Connat, P. Vuilleme, M. Morici, and J. Bouvier, *Commun. Int. CNRS Symp. Invertebr. Horm., 1983* (1983).
[28] K. P. Wigglesworth and H. H. Rees, *Commun. Int. CNRS Symp. Invertebr. Horm., 1983* (1983).
[29] P. Diehl and R. Lafont, personal communication.

[24] Ecdysone Oxidase

By J. KOOLMAN

Ecdysone oxidase (EC 1.1.3.16) catalyzes the reaction shown in Fig. 1.[1] The enzyme is found in insects. It is probably part of the catabolic system that inactivates the molting hormones.[2] Products of the reaction of ecdysone oxidase are 3-dehydroecdysteroids. These metabolites do not accumulate *in vivo* but are converted further to 3α-ecdysteroids (3-epiecdysteroids) by action of a 3-dehydroecdysone reductase.[3,4] Recent experiments with the butterfly, *Pieris brassicae,* indicate that another reductase still exists which converts 3-dehydroecdysteroids into 3β forms.[5] 3α-Ecdysteroids have been isolated or detected as ecdysone metabolites in various stages of insects[6-9] and represent along with conjugates and ecdysonoic acids major catabolites of ecdysone.[10]

[1] J. Koolman and P. Karlson, *Eur. J. Biochem.* **89,** 453 (1978).
[2] J. Koolman, in "Actualités sur les Hormones d'Invertebres" (M. M. Durchon, ed.), p. 403. CNRS, Paris, 1976.
[3] P. Karlson and J. Koolman, *Insect Biochem.* **3,** 409 (1973).
[4] R. Lafont and J. Koolman, in "Biosynthesis, Metabolism and Mode of Action of Invertebrate Hormones" (J. A. Hoffman and M. Porchet, eds.), p. 196. Springer-Verlag, Berlin and New York, 1984.
[5] C. Blais and R. Lafont, *Hoppe-Seyler's Z. Physiol. Chem.* **365,** 809 (1984).
[6] J. N. Kaplanis, M. J. Thompson, S. R. Dutky, and W. E. Robbins, *Steroids* **34,** 333 (1979).
[7] J. N. Kaplanis, G. F. Weirich, J. A. Svoboda, M. J. Thompson, and W. E. Robbins, in

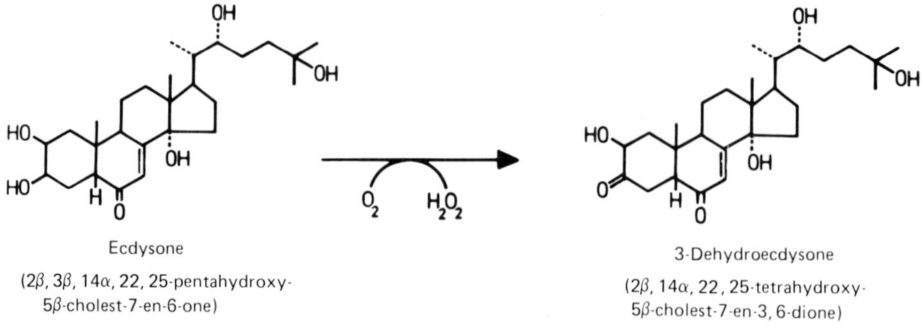

FIG. 1. Reaction catalyzed by ecdysone oxidase.

Ecdysone
(2β, 3β, 14α, 22, 25-pentahydroxy-5β-cholest-7-en-6-one)

3-Dehydroecdysone
(2β, 14α, 22, 25-tetrahydroxy-5β-cholest-7-en-3, 6-dione)

Though the reaction products of ecdysone oxidase have also been detected in a number of other species (*Locusta migratoria, Aeshna cyanea, Choristoneura fumiferana, Drosophila hydei, Tenebrio molitor, Gryllus bimaculatus, Drosophila melanogaster,* and *Pieris brassicae*),[10] biochemical studies on ecdysone oxidase are so far limited to the blowfly, *Calliphora vicina*.[1,11,12] In view of the large evolutionary variety among the arthropods, more work on other species is needed to establish the general occurrence and relevance of ecdysone oxidase. Of special interest is the question whether the activity of "ecdysone 3-epimerase"[13,14] (see also this volume [26]) results from the combination of the enzymes ecdysone oxidase and 3-dehydroecdysone reductase (3α-specific) as was discussed recently.[4,10]

Assay

The activity of ecdysone oxidase can be assayed by a radioassay or by an optical assay.

"Progress in Ecdysone Research" (J. A. Hoffmann, ed.), p. 163. Elsevier/North Holland Biomedical Press, Amsterdam, 1980.

[8] R. Lafont, G. Somme-Martin, B. Mauchamp, B. F. Maume, and J. P. Delbecque, in "Progress in Ecdysone Research" (J. A. Hoffmann, ed.), p. 45. Elsevier/North Holland Biomedical Press, Amsterdam, 1980.

[9] R. Lafont, J.-L. Pennetier, M. Andrianjafintrimo, J. Claret, J.-L. Modde, and C. Blais, *J. Chromatogr.* **236,** 137 (1982).

[10] J. Koolman, *Insect Biochem.* **12,** 225 (1982).

[11] J. Koolman and P. Karlson, *Hoppe-Seyler's Z. Physiol. Chem.* **356,** 1131 (1975).

[12] J. Koolman, *Hoppe-Seyler's Z. Physiol. Chem.* **359,** 1315 (1978).

[13] H. N. Nigg, J. A. Svoboda, M. J. Thompson, J. N. Kaplanis, J. N. Dutky, and W. E. Robbins, *Lipids* **9,** 971 (1974).

[14] R. T. Mayer, J. L. Durrant, G. M. Holman, G. F. Weirich, and J. A. Svoboda, *Steroids* **34,** 555 (1979).

The principle of the *radioassay* is the enzymatic conversion of radiolabeled ecdysone into 3-dehydroecdysone. The reaction mixture is then extracted, substrate and product are separated by TLC, and the proportion of remaining ecdysteroid is determined by radioscanning.[1,11] For the enzymatic reaction, closed 1.0 ml reaction vials containing 200 µl of reaction mixture are used. The mixture consists of 100 µl enzyme solution, 50 µl buffer (2.5 µmol sodium potassium phosphate, pH 7.0), and 50 µl substrate (0.5 nmol, 1 µCi tritiated ecdysone in water). The samples are incubated for 120 min at 25°. Then 600 µl methanol is added to stop the reaction. Precipitated protein is removed by centrifugation. The supernatant is concentrated by vacuum centrifugation at room temperature and finally transferred onto a thin-layer plate (silica gel type 60 F254, Merck Darmstadt). The width of each trace containing the sample is 3 cm. The plate is developed in chloroform/methanol (80:20, v/v) twice. Radioactivity on the plates is quantified by use of a TLC scanner (Berthold, Wildbad) adapted to a count integration unit.

In order to calculate the initial rate of enzyme activity (v) the formula

$$v = n_0/t \times \ln c_0/c$$

is used (n_0 = initial amount of ecdysone; t = period of incubation; c_0/c = quotient of the ecdysone concentrations at the beginning and the end of the reaction). A conversion of 1% ecdysone to 3-dehydroecdysone corresponds to an enzyme activity of 42×10^{-9} U. The sensitivity of the assay (lower limit: 2×10^{-7} enzyme units) can be increased by use of smaller amounts of substrate with concomitant increase of its specific radioactivity. Consecutive or side-reactions leading to other products than 3-dehydroecdysone are observed with unpurified homogenates only. A simple purification step of ecdysone oxidase by ammonium sulfate precipitation in most cases is sufficient for their suppression. The side products are usually more polar than ecdysone and thus are separated on TLC from less polar 3-dehydroecdysteroids. The rate of 3-dehydroecdysone formation can be used to calculate the enzyme activity in this case, too.

The *optical test* uses 2,6-dichloroindophenol as hydrogen acceptor.[1] The reaction is followed by recording the decrease of absorbance at 600 nm with a spectrophotometer. The temperature of the reaction in microcuvettes is 25°. The incubation mixture (total volume 1.20 ml) consists of 500 µl ecdysone (172.2 nmol), 100 µl 2,6-dichloroindophenol (60 nmol), 100 µl sodium, potassium phosphate (40 µmol, pH 6.5), and 500 µl ecdysone oxidase (at least 120 µU). On the basis of a molar absorption coefficient[15] of 2,6-dichloroindophenol at pH 7.0 of 19,100 M^{-1} 1 mU enzyme

---

[15] T. P. Singer and T. Cremona, this series, Vol. 9, p. 302.

corresponds to an absorbance change $A_{600} = 0.0159$ min^{-1} ($\lambda = 600$ nm, $d = 1.00$ cm). The optical test can be performed in the presence of oxygen because ecdysone oxidase nearly exclusively reacts with the dye.[11] The use of the assay is restricted to purified enzyme preparations because the sensitivity (limit: 10^{-4} enzyme units) is low compared to the radioassay (by a factor of about 500) and unpurified solutions often show a diaphorase activity which is not due to ecdysone oxidase. This can be controlled by an incubation with a substrate blank.

Preparation

Highest activities of ecdysone oxidase are detected in insect eggs and in pupae of holometabolous insects.[12] A suitable starting material for the isolation of ecdysone oxidase are young pupae of the blowfly, *Calliphora vicina* (1 to 3 days after pupariation). The pupae may be collected over a longer period and stored frozen at $-25°$.

The isolation of the enzyme is performed at $5°$. Pupae (300 g) are homogenized with an ultraturrax (Janke and Kunkel, Staufen, i.Br.) in 1 l Tris–HCl buffer (100 mM, pH 8.0). To inhibit phenol oxidase 2.4 g (13.6 mmol) ascorbic acid is added to the buffer. The homogenate is centrifuged at 70,000 g for 60 min. Ecdysone oxidase is found in the supernatant from which it is recovered by fractional precipitation with ammonium sulfate. The enzyme is found in the protein precipitate at 40–65% ammonium sulfate saturation (1.56–2.54 M). The protein is dissolved in buffer A (Tris–HCl 100 mM, pH 8.5) and desalted by gel chromatography on Sephadex G-25. For further purification DEAE-cellulose (DE 32, Whatman) is used. For a batch step 1 g ion exchanger per 50 mg protein is equilibrated with buffer A, mixed with the protein solution, allowed to stand for 30 min, then separated by centrifugation and washed twice with an equal volume of buffer A. Ecdysone oxidase remains in the supernatants which are combined. The enzyme preparation is equilibrated either by dialysis or gel chromatography with buffer B (10 mM Tris–HCl, pH 8.5) for the following anion exchange chromatography on a column filled with DE 32 (1 g/35 mg protein). The proteins are adsorbed to the ion exchanger, which subsequently is washed by 3 volumes of buffer B. With a gradient of sodium chloride (0–100 mM) in buffer B proteins are eluted from the column. Ecdysone oxidase is found in the fractions eluted with 5–35 mM sodium chloride. These fractions are pooled and concentrated by ultrafiltration. By gel chromatography on Sephadex G-25 the enzyme preparation is equilibrated with buffer C (sodium, potassium phosphate 10 mM, pH 6.5). The subsequent purification step uses a cation exchanger (CM 32, Whatman; 1 g/10 mg protein, equilibrated with buffer C). Ecdysone oxidase is adsorbed to the ion exchanger and eluted with a linear

TABLE I
PURIFICATION OF ECDYSONE OXIDASE FROM 300 g PUPAE OF THE BLUE BLOWFLY, *Calliphora vicina*[a]

Fraction	Protein (mg)	Specific enzyme activity[b] (μU/mg protein)	Purification	Recovery (%)
Ammonium sulfate precipitate (40–65% saturat.)	3,430	5.0	1	100
Supernatant of DE-32 batch adsorpt.	410	57.6	11.6	90
Fractions from chromatography on DE-32	23.8	744	149	75
Fraction from chromatography on CM-32	0.24	11,200	2,240	11

[a] J. Koolman and P. Karlson, *Hoppe-Seyler's Z. Physiol. Chem.* **356**, 1131 (1975).
[b] Determined by radioassay.

gradient of sodium chloride (0–500 mM) in buffer C. The enzyme is found in the fractions eluted with 50–110 mM sodium chloride. Table I gives a typical purification scheme of ecdysone oxidase.[11]

Although a purification factor of more than 2000 has been achieved the resulting enzyme preparation still showed several peptide bands in nondenaturing PAGE.[16] Thus a further purification of ecdysone oxidase is possible.

Characterization

Ecdysone oxidase is a relatively stable enzyme. It does not lose activity when repeatedly frozen and thawed. Under sterile conditions it retains its activity in aqueous solution at 0° for several days and frozen at −25° for more than 3 months. Precipitation with ammonium sulfate and dialysis do not decrease its activity. Lyophilization causes a slight (<10%) loss of activity.[1] The enzyme is stable in gels during nondenaturing PAGE and can be detected after electrophoresis by incubation of gel slices with radiolabeled ecdysone under conditions of the radioassay.

Ecdysone oxidase is not inhibited significantly by the following inhibitors of monophenol monooxygenase (phenol oxidase, tyrosinase, EC

[16] J. Koolman, unpublished (1978).

TABLE II
ECDYSONE OXIDASE IN HOMOGENATES OF TOTAL BLOWFLY LARVAE[a]

Fraction	Duration of centrifugation (min)	Enzyme activity[b]	
		(μU/g larvae)	(nU/mg protein)
Homogenate[c]	—	7.2	64
1,000 g supernatant	20	8.1	97
10,000 g supernatant	20	13.6	139
100,000 g supernatant	60	30.2	345
100,000 g pellet	60	3.2	129

[a] Few hours before pupariation. J. Koolman, *Hoppe-Seyler's Z. Physiol. Chem.* **359**, 1315 (1978).
[b] Determined by radioassay.
[c] The analysis of crude extracts may result in an underestimation of the enzyme reaction due to the presence of 3-dehydroecdysone reductases and small amounts of their cofactors. C. Blais and R. Lafont, personal communication (1983).

1.14.18.1): potassium cyanide, sodium azide, cysteine, reduced glutathione, 1-phenyl-2-thiourea, and ascorbic acid when tested in 1 mM concentration. Other enzyme inhibitors such as *p*-chloromercuribenzoate (tested in concentrations up to 1 mM), iodoacetamide (up to 1 mM), ethylenediaminetetraacetate (up to 100 μM), α-naphthoflavone (up to 1 mM), metyrapone (up to 100 μM), 3-aminotriazole (10 mM), and dimilin (1 mM) are also ineffective. Methanol inhibits ecdysone oxidase. Methanol (5 vol%) decreases the activity of the enzyme by 31%.[1]

The reaction catalyzed by ecdysone oxidase is irreversible.[1] Its pH optimum is 6.5 when tested with the optical assay in MOPS buffer [2-(*N*-morpholino)propanesulfonic acid/NaOH].[11] The K_m values of ecdysone and 20-hydroxyecdysone in the radioassay are 98 and 31 μM, respectively.[1] When tested with the optical assay the K_m for ecdysone is 41 μM and for 2,6-dichloroindophenol is 7 μM.[11] The differences in the K_m for the steroid substrate are due to use of different assay systems and may be explained by conformational influence of the artificial dye in the optical assay.[11]

By analogy to enzymes catalyzing similar reactions it may be assumed that the prosthetic group of ecdysone oxidase is a flavin. However no supportive evidence for this assumption is available.[1,11] The participation of iron ions is not so likely because the enzyme is active in the presence of chelators such as EDTA.

Ecdysone oxidase from blowfly larvae is a cytosolic enzyme. Most of its activity is present in the 100,000 g supernatant of homogenates (Table II). The enzyme has a molecular weight of 240,000 when analyzed by gel

TABLE III
ACTIVITY OF ECDYSONE OXIDASE WITH VARIOUS
STEROIDS AS SUBSTRATES[a]

Substrate[b]	Relative enzyme activity[c] (%)
Inokosterone	136
Makisterone	134
2-Deoxy-20-hydroxyecdysone	108
Ecdysone	100
20-Hydroxyecdysone	95
Cyasterone	90
2-Deoxyecdysone	85
5,20-Dihydroxyecdysone	45
22,25-Dideoxyecdysone	0
20-Hydroxycholesterol	0
7-Dehydrocholesterol	0
Cholesterol	0

[a] Koolman and Karlson, *Eur. J. Biochem.* **89,** 453 (1978).
[b] For the systematic names and formulas see this volume [22].
[c] Optical assay.

chromatography (Sephadex G-150).[11] The tissue distribution in blowfly larvae and the developmental profile of the enzyme activity have been reported. Major activities are found in the fat body and the digestive tract. The enzyme is found in high concentrations in insect eggs and pupae.[11]

Substrates

Ecdysone oxidase is relatively specific for its steroid substrate. Only ecdysteroids are oxidized (Table III). Structural requirements for a substrate are the α,β-unsaturated keto function at C-6 and several hydroxyl functions including the hydroxyls at C-3 and at C-22. On the other hand the enzyme is active with ecdysteroids lacking hydroxyls at C-2, C-20, and C-25 or having an additional hydroxyl at C-26 or a methyl at C-24.[1]

The alcoholic function of the steroid at C-3 is oxidized to a ketone by enzymatic transfer of two hydrogens to a second substrate (cosubstrate). The physiological electron acceptor most likely is molecular oxygen. There is no indication for other physiological electron acceptors. Certain dyes can serve as artificial electron acceptors *in vitro* (Table IV). They differ in the rate of stimulation of the reaction: ferricyanide < methylene blue < 2,6-dichloroindophenol < 5-methylphenazine methyl sulfate.[1]

TABLE IV
ACTIVITY OF ECDYSONE OXIDASE WITH VARIOUS HYDROGEN
ACCEPTORS AS COSUBSTRATES[a]

Cosubstrate	Concentration (mM)	Relative enzyme activity[b] (%)
None added	—	100
Nitrogen (exclusion of air)	—	16
Cytochrome c	0.25	57
FAD	0.50	91
NAD$^+$	0.50	96
NADP$^+$	0.50	107
FMN	0.50	110
Potassium ferricyanide	1.00	132
Methylene blue	1.00	142
2,6-Dichloroindophenol	1.00	178
5-Methylphenazine methyl sulfate	1.00	189

[a] Koolman and Karlson, *Hoppe-Seyler's Z. Physiol. Chem.* **356**, 1131 (1975).
[b] Radioassay.

Products

3-Dehydroecdysteroids are the products of the reaction of ecdysone oxidase with various ecdysteroids (Table III). These metabolites can be obtained either by enzymatic oxidation with ecdysone oxidase[17,18] or by chemical oxidation with platinum.[18–20] Their physicochemical properties (UV, IR, NMR, and MS spectra) have been reported.[17,18,20] The maximum of UV absorption of 3-dehydroecdysteroids in methanol is at 242–244 nm (ε of 3-dehydroecdysone = 10,360 M^{-1}).[17,18,20] In aqueous solution the maximum is shifted to longer wavelength (248–250 nm).

Separation and identification of 3-dehydroecdysteroids are possible by a combination of TLC, HPLC, color reactions, and derivative formation.

For TLC precoated silica gel plates are used impregnated with a dye fluorescent upon excitation at 254 nm. The plates are prerun in solvent and dried prior to use. They are developed in chloroform/methanol (80/20, v/v) twice. Because of the UV-absorbing 7-en-6-one structure

[17] P. Karlson, H. Bugany, H. Döpp, and G.-A. Hoyer, *Hoppe-Seyler's Z. Physiol. Chem.* **353**, 1610 (1972).
[18] J. Koolman and K.-D. Spindler, *Hoppe-Seyler's Z. Physiol. Chem.* **358**, 1339 (1977).
[19] K.-D. Spindler, J. Koolman, F. Mosora, and H. Emmerich, *J. Insect Physiol.* **23**, 441 (1977).
[20] L. N. Dinan and H. H. Rees, *Steroids* **32**, 629 (1978).

which by definition is common to all ecdysteroids, substrates as well as products of the enzymatic reaction can be localized visually on the plates with UV light at 254 nm.

Color reactions allow a more specific detection of ecdysteroids on TLC plates.

Reaction I: Spraying with vanillin-sulfuric acid reagent (1 g vanillin in 100 ml conc. sulfuric acid; after spraying, heating at 120° for 10 min)[17] gives red to turquoise colors depending on the structure of ecdysteroid.[1]

Reaction II: Spraying with ethanol–conc. sulfuric acid reagent (1:1, v/v) followed by 10 min at 100° gives a mostly blue fluorescence of ecdysteroids when the plate is irradiated with UV light at 366 nm.[21,22]

Reaction III: Upon spraying with tetrazolium reagent (4 g 2,3,5-triphenyltetrazolium chloride in 100 ml methanol mixed with 100 ml 1 N sodium hydroxide prior to use) and subsequent heating at 100° for 5–10 min, 3-dehydroecdysteroids become red while other ecdysteroids remain uncolored.[1,11]

Reaction IV: The plate is sprayed with a solution of 2,4-dinitrophenylhydrazine (100 mg in 100 ml ethanol to which 1 ml 36% hydrochloric acid is added). 3-Dehydroecdysteroids become yellow. Their parent steroids do not react.[1]

Reaction V: The plate is sprayed with sodium carbonate (20%, w/v in water) and dried. Subsequently Folin reagent (Merck, Darmstadt, diluted with 3 volumes water) is sprayed onto the plate. 3-Dehydroecdysteroids become blue; the parent ecdysteroids remain invisible.

While reagent I gives a positive reaction with many steroids and reaction II is specific for ecdysteroids in general, 3-dehydroecdysteroids are selectively detected by reactions III, IV, and V. Reagent III is sensitive to reducing substances such as the 2-ol-3-one structure of 3-dehydroecdysteroids. Reagent IV is able to detect reactive aldehydes and ketones (the 6-one function of ecdysteroids does not react because it is protected by the conjugated double bond) and reagent V is positive with phenol-like substances.

3-Dehydroecdysteroids are less polar than their parent ecdysteroids. The R_f value in TLC on silica plates is therefore increased. A list of R_f values is found in the literature.[1,23] The R_m function ($R_m = \log R_f^{-1} - 1$) changes by -0.4 when ecdysteroids are oxidized at C-3. Unfortunately, ecdysteroid acetates have a very similar R_f value to 3-dehydroecdysteroids and for this reason can be mistaken for 3-dehydroecdysteroids.[4]

[21] R. T. Mayer and J. A. Svoboda, *Steroids* **31**, 139 (1978).
[22] J. Koolman, *Insect Biochem.* **10**, 381 (1980).
[23] L. N. Dinan, P. L. Donnahey, H. H. Rees, and T. W. Goodwin, *J. Chromatogr.* **205**, 139 (1981).

TLC of ecdysteroids on reversed-phase silica gel plates has also been described.[24,25]

Separation and analysis of ecdysteroids can also be achieved successfully by HPLC on silica straight and reversed-phase columns as well as on silica gel derivatized with aminopropyl groups.[8,9,23,26–28] The chromatography of pure 3-dehydroecdysteroids gives reproducible results in most cases. The HPLC of 3-dehydroecdysteroids from biological extracts apparently is more difficult on the other hand. Often a serious loss of material or a considerable change in the eluting properties of 3-dehydroecdysteroids is encountered. This reflects a general instability of the compounds and may be especially due to chemical interaction of the reactive 2,3-ketol function of ring A with other components of the probe (such as amines,[29] oxidizing substances etc.). Chromatography of 3-dehydroecdysteroids at elevated temperature (50°) leads to improved separation in HPLC.[5] Kept in a refrigerator in methanolic solution 3-dehydroecdysteroids are stable for weeks only.

Derivative formation[30] can be used to identify 3-dehydroecdysteroids as acetates, acetonides, and other compounds. Upon reaction with acetone or 2,2-dimethoxypropane, parent ecdysteroids give 2,3-acetonides, whereas 3-dehydroecdysteroids cannot form these 2,3-acetonides. The number of possible acetates formed by reaction with aceticanhydride is diminished by one in 3-dehydroecdysteroids due to lack of the hydroxyl at C-3. In addition, the chromatographic properties of the acetates formed differ. Carbonyl reagents such as 2,4-dinitrophenylhydrazine react with 3-dehydroecdysteroids only. 3-Dehydroecdysone can also be separated and distinguished from ecdysone by ion exchange chromatography as a complex with boric acid or by affinity chromatography on boronic acid gel because only the vicinal 2,3-diol moiety of ecdysone can form cyclic diesters.[31]

The hormonal activity of 3-dehydroecdysone and 3-dehydro-20-hydroxyecdysone has been tested in several biological assays. The substances were found to be less (3/10–1/15) active than their parent compounds in the *Calliphora* bioassay[17,18] and two different *Drosophila* puff induction assays.[19,32]

[24] I. D. Wilson, C. R. Bielby, and E. D. Morgan, *J. Chromatogr.* **242**, 202 (1982).
[25] I. D. Wilson, S. Scalia, and E. D. Morgan, *J. Chromatogr.* **212**, 211 (1981).
[26] M. W. Gilgan, *J. Chromatogr.* **129**, 447 (1976).
[27] R. Lafont, G. Martin-Somme, and J.-C. Chambet, *J. Chromatogr.* **170**, 185 (1979).
[28] R. E. Isaac, N. P. Milner, and H. H. Rees, *J. Chromatogr.* **246**, 317 (1982).
[29] C.-G. Eriksson, L. Nordström, and P. Eneroth, *J. Steroid Biochem.* **19**, 1199 (1983).
[30] J. Koolman, L. Reum, and P. Karlson, *Hoppe-Seyler's Z. Physiol. Chem.* **360**, 1351 (1979).
[31] J. Koolman, unpublished (1980).
[32] G. Richards, *J. Insect Physiol.* **24**, 329 (1978).

The second product of the ecdysone oxidase reaction under physiological conditions is hydrogen peroxide which is difficult to detect due to the small amounts formed and due to contaminating catalase. The detection method that has proven to be most suitable uses the hydrogen peroxide formed by ecdysone oxidase to oxidize methanol to formaldehyde which in turn is quantified colorimetrically.[1]

[25] Ecdysteroid Carrier Proteins

By J. KOOLMAN

Blood proteins which are able to bind steroid hormones have been extensively studied in vertebrates[1,2] in which several serum proteins serve for this purpose. Transcortin is one example out of the group of carrier proteins with high affinity and low capacity for certain steroid hormones. Albumin is characteristic for the other type of carrier which nonspecifically binds various steroid hormones with low affinity and high capacity.[3]

Steroid-binding blood proteins have been found in invertebrates, also. Until now, the search for ecdysteroid-binding proteins in the phylum of arthropods has led to inconsistent results. No evidence for an ecdysteroid carrier was obtained in two crustacean species, *Pachygrapsus crassipes*[4] and *Orconectes limosus*[5] and three insect species, the silkmoths *Philosamia cynthia*,[6] *Antheraea polyphemus*,[7] and *Hyalophora cecropia*.[8] The possibility of a weak binding between hemolymph proteins and ecdysteroids, however, could not be excluded in any of these cases.[9]

From studies with other insect species a reversible binding of ecdysteroids to blood proteins was indicated. Preliminary experiments suggested binding in the bug *Pyrrhocoris apterus*,[10] the fruit fly *Drosophila hy-*

[1] U. Westphal, *Hoppe-Seyler's Z. Physiol. Chem.* **359**, 431 (1978).
[2] U. Westphal, *J. Steroid Biochem.* **19**, 1 (1983).
[3] U. Westphal, in "Steroid-Protein Interactions" (F. Gross, A. Labhart, T. Mann, L. T. Samuels, and J. Zander, eds.), p. 101. Springer-Verlag, Berlin and New York, 1971.
[4] E. S. Chang, B. A. Sage, and J. D. O'Connor, *Gen. Comp. Endocrinol.* **30**, 21 (1976).
[5] P. Kuppert, M. Büchler, and K.-D. Spindler, *Z. Naturforsch., C: Biosci.* **33C**, 437 (1978).
[6] H. Chino, L. I. Gilbert, J. B. Siddall, and W. Hafferl, *J. Insect Physiol.* **16**, 2033 (1970).
[7] L. Cherbas and P. Cherbas, *Biol. Bull.* (*Woods Hole, Mass.*) **138**, 115 (1970).
[8] T. A. Gorell, L. I. Gilbert, and J. Tash, *Insect Biochem.* **2**, 94 (1972).
[9] L. I. Gilbert and H. Chino, *J. Lipid Res.* **15**, 439 (1974).
[10] H. Emmerich, *J. Insect. Physiol.* **16**, 725 (1970).

dei,[11,12] and the blowfly *Calliphora vicina*.[13] The ecdysteroid-binding protein from hemolymph of the blowfly was recently identified as arylphorin.[14,15] Arylphorin (identical with calliphorin when isolated from *Calliphora*), the major storage protein of dipteran larvae, has a low affinity for ecdysteroids.

Binding of ecdysteroids to a hemolymph protein was also shown in the locust *Locusta migratoria*.[16,17] The protein has a high degree of specificity and a considerable affinity for 20-hydroxyecdysone (K_D 10^{-7} M). It is found in rather low concentrations in the circulatory system. In addition, indications for a second type of ecdysteroid-binding hemolymph protein were observed in locusts. This protein preferentially binds ecdysone.[18]

It might be expected that due to the huge variety among arthropods many different types of steroid-binding hemolymph proteins have evolved. The investigation of steroid-binding proteins in arthropods thus might turn out to be especially rewarding. In this chapter I will concentrate on measurement, isolation, and description of such carriers which are characterized by a low affinity for ecdysteroids. The chapter by Feyereisen[17] focuses on the binders of higher affinity.

Assay

Measurement of ecdysteroid carrier proteins is based on the affinity of the proteins for ecdysteroids. The measurement includes the separation of the small ecdysteroids from macromolecules and subsequently the detection and quantification of ecdysteroids in the macromolecular fraction.

Gel permeation chromatography, equilibrium dialysis, or electrophoresis can be used for the separation of free from bound ecdysteroids. A radioimmunoassay of unlabeled ecdysteroids or scintillation counting of radiolabeled ecdysteroids is then used to detect the steroids. For the identification of ecdysteroid binders, photoaffinity labeling is an alternative method. Measurement of the quenching of protein fluorescence

[11] H. Emmerich, *Z. Vergl. Physiol.* **68**, 385 (1970).
[12] F. M. Butterworth and H. D. Berendes, *J. Insect Physiol.* **20**, 2195 (1974).
[13] G. Thamer and P. Karlson, *Z. Naturforsch, B: Anorg. Chem., Org. Chem., Biochem., Biophys., Biol.* **27B**, 1191 (1972).
[14] L. Reum, G. Käuser, U. Enderle, and J. Koolman, *Z. Naturforsch., C: Biosci.* **37C**, 967 (1982).
[15] U. Enderle, G. Käuser, L. Reum, K. Scheller, and J. Koolman, in "The Larval Serum Proteins of Insects" (K. Scheller, ed.), p. 40. Thieme, Stuttgart, 1983.
[16] R. Feyereisen, M. Lagueux, and J. A. Hoffmann, *C. R. Hebd. Seances Acad. Sci.* **280**, 1709 (1975).
[17] R. Feyereisen, this volume [27].
[18] M. Charlet and J. A. Hoffmann, *Conf. Comp. Endocrinol., 11th, 1981*, Abstract, p. 86 (1981).

caused by steroids is another approach to study protein steroid interactions.

Gel permeation chromatography is an appropriate method to separate macromolecules from steroids.[19,20] Ecdysteroids, like other steroid hormones, have a relatively low molecular mass (ecdysone: 464) that allows the separation of free from protein-bound ecdysteroids by chromatography on suitable gels such as Sephadex G-25.[15] Following chromatography the ecdysteroids are detected by radioimmunoassay[21,22] or fluorometry.[23] The occurrence of ecdysteroids in fractions that contain macromolecules is an indication for the existence of binding proteins. However, as a nonequilibrium method the chromatography may not be suitable for weak steroid–protein complexes. These complexes may remain undetected although they might be of physiological significance.

Incubation of the hemolymph sample prior to gel chromatography (or prior to other separation methods) with radioactive ecdysteroids to label the carrier protein(s) has several limitations. (1) Only few ecdysteroids which potentially are bound to hemolymph carriers are commercially available in tritiated form. (2) The exchange rate between endogenous and added labeled ecdysteroid is unknown. (3) Extended *in vitro* incubations of radioactive ecdysteroid with hemolymph at elevated temperature can lead to the formation of artifactual metabolites. For instance, 3-dehydroecdysone is formed upon incubation of blowfly hemolymph with ecdysone due to the presence of traces of ecdysone oxidase.[24] This less polar ecdysone metabolite has a higher affinity than ecdysone for the carrier protein of blowflies and thus simulates the binding of the ecdysteroid applied. For this reason suitable control experiments which verify the identity of the tracer after binding to the protein (e.g., by HPLC) are mandatory.

Some of the limitations mentioned can be overcome. By HPLC of the steroid–carrier complex on suitable protein columns the speed of separation is increased which leads to a reduction of the dissociation of the complex.

The analysis of binding of radiolabeled ecdysteroid to low affinity binders is also possible by chromatography under steady-state conditions.[25] The latter method is a chromatography in the presence of a con-

[19] U. Westphal, this series, Vol. 15, p. 761.
[20] M. R. Sherman, this series Vol. 36, p. 211.
[21] L. Reum and J. Koolman, *Insect Biochem.* **2**, 135 (1979).
[22] M. A. Delaage, M. H. Hirn, and M. L. De Reggi, this series, Vol. 84, p. 350.
[23] J. Koolman, *Insect Biochem.* **10**, 381 (1980).
[24] J. Koolman, unpublished (1979).
[25] C. W. Burke, *Biochim. Biophys. Acta* **176**, 403 (1969).

stant concentration of ecdysteroid. While it leads to unequivocal results with low affinity carrier proteins the method is rather expensive because of the high amounts of labeled and unlabeled ecdysteroid needed.[24]

Equilibrium dialysis likewise can be used to detect binding of steroids by macromolecules. This method has been described extensively.[26] With low affinity binders (K_D of the protein–steroid complex above 10^{-4} M) the method, however, gives only poor results. This may explain the lack of evidence for ecdysteroid carriers in studies using this technique.

Photoaffinity labeling is a very powerful method to detect proteins which interact with steroids.[27] Enzymes as well as serum proteins and hormone receptors can be labeled.[28,29] The principle of the method is the conversion of the steroid into a radical by irradiation with light. The steroid radical then stabilizes by reaction with a molecule in its neighborhood. If the bound steroid is in direct contact to a protein, this protein is more likely to be covalently linked to the steroid than other unrelated proteins.

Photoaffinity labeling of ecdysteroid-binding proteins was introduced to link ecdysteroids covalently to their nuclear receptors.[30,31] The method is also useful for the detection of ecdysteroid-binding carrier proteins.[14,32] UV light at a wavelength around the minor absorption maximum of ecdysteroids at 315 nm converts the 7-en-6-one structure of the steroids into a radical which may combine with binding protein.

Method of Irradiation[14]

Ecdysteroids dissolved in phosphate-buffered saline (20 mM sodium phosphate, pH 7.3, 150 mM sodium chloride) are irradiated in the presence of proteins in a cylindrical cuvette or optical glass (type 165; Hellma, München) which has an optical length of 20 mm and a stirred volume of 3.0 ml. The high-power light source used (type LX 501 with a 450 W xenon bulb; Zeiss, Oberkochen) is part of a commercial cytophotometer and is also found as a component of Zeiss fluorometers. The light source produces a focused beam of light which passes through a 6-cm water filter to absorb the IR light portion. The UV light of short wavelength is ab-

[26] U. Westphal, in "Steroid-Protein Interactions" (F. Gross, A. Labhart, T. Mann, L. T. Samüels, and J. Zander, eds.), p. 43. Springer-Verlag, Berlin and New York, 1971.
[27] J. A. Katzenellenbogen, Biochem. Actions Horm. **4**, 1 (1978).
[28] F. Wold, this series, Vol. 46, p. 3.
[29] H. Bayley and J. R. Knowles, this series, Vol. 46, p. 69.
[30] H. Gronemeyer and O. Pongs, Proc. Natl. Acad. Sci. U.S.A. **77**, 2108 (1980).
[31] K. Schaltmann and O. Pongs, Hoppe-Seyler's Z. Physiol. Chem. **361**, 207 (1980).
[32] M. Cao, B. Duvic, D. Zachary, P. Harry, and J. A. Hoffmann, Insect Biochem. **13**, 567 (1983).

sorbed by the front part of the cuvette (special optical glass) and by an additional glass filter (object slide, type Resistance) put in front of the cuvette. Other filters may also be used.[14] The absorbance of these components (filter plus front of cuvette) is 0.5 at 316 nm, 1.0 at 307 nm, and 2.0 at 296 nm when measured against air. Irradiation of ecdysteroids in the presence of protein is normally performed for 30 min at 10° constant temperature (measured inside the cuvette). An efficient cooling system is required to compensate for the high energy input due to the light.

Optimal conditions were found to depend not only on the duration of irradiation but on several other variables such as the type of steroid irradiated, the optical density of the solution, and the optical system used (i.e., the intensity and the wavelength distribution of light and spectral characteristics of the filter). Since the steroid radical and its reaction products differ chemically from the parent molecule they can be separated, e.g., by TLC. The half-life of the ecdysteroid during irradiation is a useful parameter for the optimization of the method.[14]

Quantitative Analysis of Cross-Linkage[14]

The ratio of covalently linked ecdysteroid to total ecdysteroid is taken as yield of photoaffinity labeling. To determine this extent of cross-linkage the ecdysteroid used in the experiment has to be radiolabeled. The kinetics of photoaffinity labeling are analyzed by taking aliquots from the solution in the cuvette prior to and at various timed intervals during irradiation. After separation of proteins from free ecdysteroids by either of the methods described in the following paragraphs, TLC, precipitation with trichloroacetic acid, or charcoal treatment,[14] the percentage of protein-linked steroid is determined.

TLC is performed on silica gel plates (Type 60 F254; Merck, Darmstadt). The plates are developed twice with chloroform/methanol (80:20; v/v). Reference substances (unlabeled ecdysteroids) are cochromatographed to localize the ecdysteroids which remained unlinked with proteins during irradiation. The radioactivity on the plate is detected by radioscanning. Ecdysone has an R_f of 0.29 (R_f values of other ecdysteroids are found in the literature[33]), while protein-bound ecdysteroid is detected at the origin of the plate (R_f of less than 0.05).

Proteins are precipitated by addition of 3 vol of aqueous trichloroacetic acid (5%; w/v). Aliquots containing less than 1 mg protein are mixed with 50 μl albumin solution (20 mg/ml) prior to precipitation. After 2 hr standing at 5° the proteins are pelleted by centrifugation, washed with solvent twice, mixed with a suitable cocktail for liquid scintillation count-

---

[33] J. Koolman and P. Karlson, *Eur. J. Biochem.* **89**, 453 (1978).

ing of proteins, and counted. The resulting radioactivity is expressed as a percentage of the total radioactivity taken for the irradiation experiment.

Charcoal treatment of aliquots withdrawn from the irradiation experiment removes the free steroid portion by adsorption. An equal volume of dextran-coated charcoal suspension [5% charcoal, 0.5% dextran (T 500; Serva, Heidelberg) in water; w/v] is added, mixed by vortexing, and incubated for 30 min at 0° with agitation every 10 min. Radioactivity is determined in the supernatant after removal of the charcoal by centrifugation.

Nonquantitative methods have been described which are based on an irradiation of ecdysteroid–protein complexes present *in situ* or in gel slices. After irradiation the ecdysteroid portion is detected selectively by antibodies against ecdysteroids.[30,31,34]

Specificity of Photoaffinity Labeling in the Detection of Steroid-Binding Proteins. The general method to test whether the photoaffinity labeling selectively labels steroid-binding proteins is to increase the concentration of the steroid by about two orders of magnitude (which concomitantly leads to a decrease of specific radioactivity of the steroid). In this case steroid-binding proteins which have a limited capacity to bind steroids are not labeled significantly. Hence, covalent incorporation of radioactivity is reduced to the background level if the process of photoaffinity labeling is specific.

Unfortunately this test for specificity of the method does not work with steroid-binding proteins of high binding capacity. For this type of protein another control experiment may be used. Its principle is to demonstrate that upon irradiation the steroid preferentially reacts with the steroid-binding protein and not with other proteins.[14] The steroid is irradiated in the presence of either selected single proteins or in the presence of a mixture of several proteins. If photoaffinity labeling specifically labels the binding protein, this will show the highest specific radioactivity after cross-linking. Most conveniently, protein mixtures irradiated with radioactive steroid are separated by SDS–PAGE and analyzed for radioactivity by counting of gel slices.

Other control experiments include (1) the analysis of structural and functional integrity of the proteins under the conditions of irradiation (e.g., by electrophoresis or by an enzyme assay), and (2) the exclusion of a dark reaction of irradiated components by mixing the protein and steroid only after irradiation.

Fluorescence quenching can be used to demonstrate the interaction

---

[34] K. Schaltmann and O. Pongs, *Proc. Natl. Acad. Sci. U.S.A.* **79**, 6 (1982).

between proteins and small molecules.[35-37] Fortunately the arylphorins have a high content of aromatic amino acids and, for this reason, show considerable fluorescence (calliphorin: excitation at 280 nm, emission at 336 nm). Addition of ecdysteroids in aqueous or ethanolic solution quenches the fluorescence of the arylphorin of the blowfly due to radiationless deactivation of excited amino acids.[19] Although further studies are needed to evaluate the effects quantitatively, it is evident that the method allows determination of the dissociation constants of the ecdysteroid protein complexes. Its great advantage is that no radiolabeled steroids are needed.

Isolation

The methods used for the isolation of hemolymph proteins which are able to bind ecdysteroids with low affinity and high capacity reflect the relative abundance of the carriers. Arylphorin and similar proteins represent up to 80% of the hemolymph proteins depending on the stage of the insect. This allows their isolation in few steps. Conventional methods are used to purify arylphorin, including precipitation with salts, ultracentrifugation, gel chromatography, and ion exchange chromatography.[38] Quite by contrast a special immunological technique has been used to isolate selectively ecdysteroid-binding proteins from crude biological extracts. The application of this method for the isolation of an ecdysteroid carrier protein is described here.

Isolation of Ecdysteroid Binding Proteins by a Combination of Photoaffinity Labeling and Immunoadsorption.[14] Hemolymph is prepared at 0° by mass isolation from mature third-instar larvae of the blowfly. The activity of phenol oxidase is inhibited by addition of 100 μmol/ml phenylthiourea. Hemocytes are removed by centrifugation (10 min at 3000 g). In order to cross-link endogenous ecdysteroids by photoaffinity labeling with carrier protein, the hemolymph is diluted with 4 volumes phosphate-buffered saline and irradiated as described above. Subsequently ecdysteroids which did not react with protein are removed by exhaustive dialysis against saline.

For the isolation of ecdysteroid-linked protein(s) an immunoadsorbent is prepared from an ecdysteroid-specific antiserum.[39] The antiserum is

[35] N. A. Attallah and G. F. Lata, *Biochim. Biophys. Acta* **168**, 321 (1968).
[36] S. D. Stroupe, S. L. Cheng, and U. Westphal, *Arch. Biochem. Biophys.* **168**, 473 (1975).
[37] S. S. Lehrer and P. C. Leavis, this series, Vol. 49, p. 222.
[38] L. Levenbook and A. C. Bauer, *Insect Biochem.* **10**, 693 (1980).
[39] L. Reum, D. Haustein, and J. Koolman, *Z. Naturforsch., C: Biosci.* **36C**, 790 (1981).

obtained from rabbits by immunization. For preparation of the immunogen ecdysone is coupled at position C-6 in the steroid nucleus to bovine serum albumin.[21,39] From the resulting antiserum the γ-globulin fraction is isolated by 3-fold precipitation with 50% saturated (1.95 M) ammonium sulfate. The globulins are coupled to Sepharose 4B (Pharmacia, Freiburg; 9.4 mg protein per ml gel) by cyanogen bromide activation.[40] Remaining reactive groups are blocked by treatment with a surplus of ethanolamine. The adsorbent is stored at 4° in phosphate-buffered saline containing 0.1% (w/v) sodium azide.

A column (0.5 cm i.d. and 12 cm height) filled with 9.0 ml immunoadsorbent is used for adsorption of ecdysteroid-linked proteins. The immunoadsorbent should have the capacity to bind more than 5 nmol ecdysone. The protein solution which contains the irradiated hemolymph proteins (2.0 mg) is passed slowly through the column. The column is washed with phosphate-buffered saline until no protein (monitored by UV measurement at 280 nm) is detected in the eluate. The ecdysteroid-linked protein is eluted with 3.0 M sodium trichloroacetate and immediately dialyzed against saline. Of course the steroid binding assays do not function any more with the binding protein isolated by this method since the binding sites are at least partially blocked by the photochemical reaction and the protein is denatured to a large extent.

Properties

Among the proteins which have been identified as ecdysteroid carriers in invertebrates, arylphorin of blowfly larvae so far is the only protein studied in great detail.[41,42] Arylphorin is a fairly symmetrical, globular hexamer (M_r about 500,000) consisting of similar but not identical subunits. It dissociates reversibly into subunits at pH values above 6.5 while other factors also influence this process. Arylphorin has been isolated from blowfly larvae and purified to homogeneity.[38] Its amino acid composition is known as well as the carbohydrate and lipid content.[43,44] Moreover, the messenger RNAs have been isolated and translated *in vitro*.[45] The arylphorin genes of a number of insect species have been isolated,

[40] S. C. March, I. Parikh, and P. Cuatrecasas, *Anal. Biochem.* **60**, 149 (1974).
[41] L. Levenbook, *in* "The Larval Serum Proteins" (K. Scheller, ed.), p. 1. Thieme, Stuttgart, 1983.
[42] L. Levenbook, *in* "Comprehensive Insect Physiology, Biochemistry and Pharmacology" (G. A. Kerkut and L. I. Gilbert, eds.), Vol. 10, Chapter 7. Pergamon, Oxford (in press).
[43] E. A. Munn, A. Feinstein, and G. D. Greville, *Biochem. J.* **124**, 367 (1971).
[44] J. F. Kinnear and J. A. Thomson, *Insect Biochem.* **5**, 531 (975).
[45] C. E. Sekeris and K. Scheller, *Dev. Biol.* **59**, 12 (1977).

cloned, mapped, and are being sequenced presently.[46-48] Monographs and reviews on larval serum proteins covering various aspects of this group of proteins have appeared.[42,49,50]

Arylphorins are not the only group of hemolymph proteins able to bind ecdysteroids. Several other ecdysteroid carrier proteins were found in the hemolymph of various insect species as was discussed previously. However, few ecdysteroid carriers have been identified in arthopods in comparison with vertebrates. Ecdysteroid-binding proteins also have been found inside the cells. They differ from ecdysteroid receptors and belong to the group of vitellins.[51,52] Their precursors, the vitellogenins, are detected in the hemolymph. Whether these hemolymph proteins are also able to bind ecdysteroids awaits further elucidation.

[46] D. F. Smith, A. McClelland, B. N. White, C. F. Addison, and D. M. Glover, *Cell* **23**, 441 (1981).
[47] H. Schenkel, C. Myllek, M. König, P. Hausberg, and K. Scheller, *in* "The Larval Serum Proteins" (K. Scheller, ed.), p. 18. Thieme, Stuttgart, 1983.
[48] D. B. Roberts, *in* "The Larval Serum Proteins" (K. Scheller, ed.), p. 86. Thieme, Stuttgart, 1983.
[49] G. R. Wyatt and M. P. Pan, *Annu. Rev. Biochem.* **47**, 779 (1978).
[50] K. Scheller, ed. "The Larval Serum Proteins." Thieme, Stuttgart, 1983.
[51] M. Lagueux, P. Harry, and J. A. Hoffmann, *Mol Cell. Endocrinol.* **24**, 325 (1981).
[52] G. Käuser, J. Koolman, and P. Koch, *Verh. Dtsch. Zool. Ges.* p. 235 (1982).

[26] Ecdysone 3-Epimerase

By MALCOLM J. THOMPSON, GUNTER F. WEIRICH, and JAMES A. SVOBODA

Ecdysone 3-epimerase is an enzyme found in the cytosol of midgut cells of the tobacco hornworm, *Manduca sexta*. The enzyme converts insect molting hormones (3β-ecdysteroids) to their corresponding 3α-epimers in a reaction requiring oxygen and NADPH or NADH. These 3-epiecdysteroids are 1/10 to 1/15 as active in the house fly molting hormone assay as their corresponding 3β-epimers. When first discovered, ecdysone 3-epimerase was referred to as ecdysone dehydrogenase-isomerase.[1] In a study that extended the original work, the enzyme, because of

[1] H. N. Nigg, J. A. Svoboda, M. J. Thompson, J. N. Kaplanis, S. R. Dutky, and W. E. Robbins, *Lipids* **9**, 971 (1974).

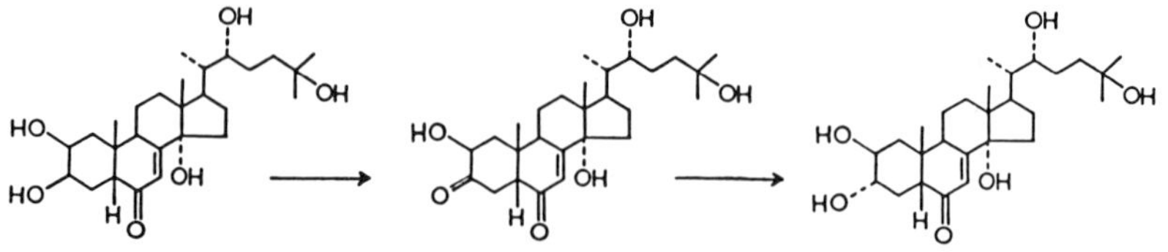

FIG. 1. Proposed scheme for the conversion of ecdysone to 3-epiecdysone.

its overall reaction, was called ecdysone 3-epimerase.[2] The two-step reaction (Fig. 1) proposed for the formation of the 3α-epimer[1] assumes the formation of a 3-dehydroecdysteroid as an intermediate. Such a scheme implies a dual-function enzyme or enzyme complex, with one component related to the ecdysone oxidase of flies[3] and the other to the 3α-hydroxysteroid oxidoreductase of mammals[4] and bacteria.[5]

In a recent study of sixth-instar larvae of the fall armyworm, *Spodoptera littoralis,* the incubation of undialyzed and dialyzed cytosolic fractions from the midgut with [^{3}H]ecdysone revealed not only conversion of ecdysone to 3-epiecdysone, but also evidence for the formation of 3-dehydroecdysone as an intermediate.[6]

This chapter describes procedures for the study of ecdysone 3-epimerase obtained from the 80,000 g supernatant of *M. sexta* midgut.

Enzyme Preparation

M. sexta larvae, reared to the late fifth-instar until they have stopped feeding, are immobilized by chilling. Then, their midguts are removed, placed in cold 10 mM Tris–HCl buffer (pH 7.4) that contains 0.3 M sucrose and 1.0 mM EDTA (buffer A), and cleansed of fat body, Malpighian tubules, tracheae, and food boluses. The cleaned midguts are blotted dry with tissue paper, weighed, minced with scissors, and then homogenized in a Ten Broeck homogenizer in four volumes of chilled buffer A. The pestle is operated first by hand until the tissue is broken up, and then for two strokes with a motor at approximately 650 rpm. The postmicrosomal supernatant is prepared by a sequence of centrifugations, described here by the average g forces and the time integrals of the

[2] R. T. Mayer, J. L. Durrant, G. M. Holman, G. F. Weirich, and J. A. Svoboda, *Steroids* **34,** 555 (1979).
[3] J. Koolman and P. Karlson, *Eur. J. Biochem.* **89,** 453 (1978).
[4] G. Verhoeven, W. Heyns, and P. DeMoor, *J. Steroid Biochem.* **8,** 731 (1977).
[5] B. A. Skalhegg, *Eur. J. Biochem.* **46,** 117 (1974).
[6] N. P. Milner and H. H. Rees, *Commun. Int. CNRS Symp. Invertebr. Horm., 1983,* Poster Abstract (1983).

squared angular velocities $[W = 981 \times 60(g \min)/r_{av}]$.[7] The times required for the initial centrifugations are reduced to a minimum by increasing the speed,[8] and some centrifugations consist only of acceleration and deceleration, with no plateau phase in between.

For the first centrifugation of the homogenate, the rotor is accelerated to 3000 g for a total of $W = 2.4 \times 10^7$ rad^2 sec^{-1}. The pellet is resuspended in two volumes of buffer A and recentrifuged as above. The combined supernatants are accelerated to 20,000 g for a total $W = 4.0 \times 10^8$ rad^2 sec^{-1}. The resulting supernatant is centrifuged at 28,000 g ($W = 9.0 \times 10^8$ rad^2 sec^{-1}) and then at 80,000 g ($W = 3.6 \times 10^{10}$ rad^2 sec^{-1}).

From the 80,000 g supernatant, approximately 90% of the enzyme is precipitated between 20 and 45% ammonium sulfate saturation. The protein precipitate is pelleted at 10,000 g for 15 min; the remaining supernatant has no measurable activity. The pellet is dissolved in 50 mM potassium phosphate buffer (pH 7.3) containing 0.1 mM EDTA (buffer B), at a ratio of 75 ml/liter of postmicrosomal supernatant. The solution is then dialyzed overnight against 100 volumes of buffer B (0–5°). The ecdysone 3-epimerase preparation thus obtained has 5 to 8 times the specific activity of the postmicrosomal supernatant. The protein concentration[9] is adjusted to 20–25 mg/ml, and the preparation is stored in 10-ml portions at $-20°$ until used.

Enzyme Assay

Each assay mixture contains enzyme preparation (3–5 mg protein), 20 μg (43 nmol) of [^{3}H]ecdysone (New England Nuclear, Boston, MA)[10] or 20-hydroxyecdysone in 5–10 μl methanol, 0.64 mM NADPH, and buffer B to a final volume of 1.0 ml. An NADPH-regenerating system is added (2.5 U of glucose-6-phosphate dehydrogenase, from bakers' yeast, Sigma, Type XV, and 6 mM glucose 6-phosphate). The mixtures are incubated in 20-ml glass scintillation vials mounted in a Dubnoff metabolic incubator (30°). The reactions are initiated by addition of ecdysteroid substrate and stopped by addition of 5.0 ml of methanol. Protein is sedimented by a 15-min centrifugation at 20,000 g, and the pellets are reextracted with another 5 ml of methanol. The methanol extracts of each assay are combined and evaporated under vacuum. The residues are dissolved in 1 ml of

---

[7] H. Beaufay and A. Amar-Costesec, *Methods Membr. Biol.* **6**, 1 (1976).

[8] L. L. Keeley, *Comp. Biochem. Physiol. B* **46**, 147 (1973).

[9] O. H. Lowry, N. J. Rosebrough, A. L. Farr, and R. J. Randall, *J. Biol. Chem.* **193**, 265 (1951).

[10] Mention of a company or proprietary product does not imply endorsement by the U.S. Department of Agriculture.

10% methanol in water, applied to a C_{18} SEP-PAK cartridge (Waters Associates, Milford, MA), and rinsed with an additional 1 ml of 10% methanol. Then, the cartridge is eluted with 5 ml each of 30, 60, and 100% methanol. The 60% methanol fraction contains the ecdysteroids and is analyzed by high-performance liquid chromatography (HPLC).

High-Performance Liquid Chromatography

Samples of the ecdysteroid fractions are analyzed by reversed-phase HPLC on a C_{18} μBondapak column (4.0 mm × 30 cm; Waters) or a C_{18} Radial Compression column (5.0 mm × 10 cm, 10 μm particle size; Waters) by isocratic elution with a methanol/water or an acetonitrile/water mixture. Typical ecdysone/3-epiecdysone separations are shown in Fig. 2.

Some apolar tritium-labeled impurities of the samples are retained by the columns at the solvent strength used for the separations. Because these compounds tend to bleed slowly from the columns, they can cause errors in subsequent analyses. A 5- to 10-min flush of the column with a more polar solvent (i.e., acetonitrile/water, 50/50, or methanol/water, 50/50) after each analysis is necessary to eliminate this problem.

The conversion rate of ecdysone or 20-hydroxyecdysone to its respective 3α-epimer is calculated from the peak heights or peak areas of the substrate and product on UV recordings (after calibraton of the HPLC system), or from radioactivity determinations of the appropriate effluent fractions. In our experience the isotopic method has been more reproducible for determining ecdysone/3-epiecdysone ratios than the UV-based method (involving either manual or electronic peak integration with the Waters Data Module 730).

Specificity

Ecdysone 3-epimerase converts ecdysone, 20-hydroxyecdysone, and 22-deoxyecdysone to their corresponding 3α-epimers, but does not convert 22,25-dideoxyecdysone.[1,2] These results suggest a requirement for an ecdysone nucleus and a 25-hydroxy group on the side chain of the substrate.

Properties

The apparent K_m and V_{max} for ecdysone are 17.0 ± 1.4 μM and 110.6 ± 14.6 pmol/min/mg protein, respectively; for 20-hydroxyecdysone, $K_m = 47.3 \pm 7.5$ μM, $V_{max} = 131.0 \pm 3.5$ pmol/min/mg protein.[2] The

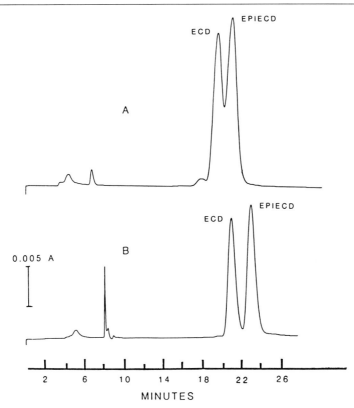

FIG. 2. Profiles of ecdysone (ECD) and 3-epiecdysone (EPIECD), 0.5 μg each, separated by reversed-phase HPLC on a C_{18} μBondapak column (4.0 mm × 30 cm) at 40°. Absorbance of the effluent at 254 nm was monitored with a Waters Model 441 absorbance detector and automatically recorded by a Shimadzu Model C-R1B recording integrator. (A) Solvent system, 36% methanol in water; flow rate, 0.48 ml/min. Ecdysone was eluted at 19.32 min; 3-epiecdysone, at 20.72 min. (B) Solvent system 20% acetonitrile in water; flow rate, 0.4 ml/min. Ecdysone was eluted at 20.91 min; 3-epiecdysone, at 22.88 min.

reaction is not reversible. Ecdysone inhibits the conversion of 20-hydroxyecdysone and vice versa, but in each case the inhibiting ecdysteroid is being epimerized.[2] 22,25-Dideoxyecdysone, which is not converted by the epimerase, does not inhibit the enzyme.

The pH optimum for the conversion of ecdysone to its 3-epimer is 7.3. Ca^{2+} and Mg^{2+} ions at 10 mM stimulate the reaction by 10% but inhibit the reaction at higher concentrations. K^+ (up to 100 mM) has little effect on the activity; Mn^{2+}, though without effect at 10 mM, inhibits the reaction at higher concentrations (100% inhibition at 50 mM).[2] The reaction is completely suppressed under anaerobic conditions.[2]

Comments

When the supernatant of *M. sexta* midgut is incubated with ecdysone, several reactions occur simultaneously. Ecdysone is converted to 3-epiecdysone, and both ecdysone and 3-epiecdysone are converted to conjugates, even without added cosubstrates or cofactors.[11] The cytosol contains enough cosubstrates and cofactors to sustain the reactions at substantial rates. This complex situation virtually precludes an accurate determination of the rate of the ecdysone/3-epiecdysone conversion.

To suppress the conjugate formation, the cosubstrate (ATP) and the cofactor (Mg^{2+}) of the reaction have to be removed from the crude supernatant by dialysis or gel filtration on Sephadex G-25.[11] These operations also remove endogenous supplies of NADPH and NADH and, consequently, allow a more precise quantification of the cofactor in the reaction mixture for the 3-epimerase.

[11] G. F. Weirich, M. J. Thompson, and J. A. Svoboda, unpublished observations (1983).

[27] 20-Hydroxyecdysone Binding Protein from Locust Hemolymph

By R. FEYEREISEN

The importance of vertebrate plasma proteins in the transport of low-molecular-weight ligands is well established. In addition to albumin binding of fatty acids, several specific binding proteins have been described which have an affinity for hormones (thyroxine, steroid hormones) and vitamins (A, B_{12}, D), etc. It is generally held that the free fraction of steroids in plasma is available to target tissues. Because binding of hormones to blood proteins may play a role in bioavailability and mode of action of hormones,[1] studies on hemolymph protein binding of the steroid molting hormone of arthropods were conducted as soon as radiolabeled ecdysone became available.[2-7] It was however soon found that 20-hy-

[1] U. Westphal, "Steroid-Protein Interactions." Springer-Verlag, Berlin and New York, 1971.
[2] H. Chino, L. I. Gilbert, J. B. Siddall, and W. Hafferl, *J. Insect Physiol.* **16**, 2033 (1970).
[3] H. Emmerich, *J. Insect Physiol.* **16**, 725 (1970).
[4] H. Emmerich, *Z. Vergl. Physiol.* **68**, 385 (1970).
[5] G. Thamer and P. Karlson, *Z. Naturforsch., B: Anorg. Chem., Org. Chem., Biochem., Biophys., Biol.* **27B**, 1191 (1972).
[6] T. A. Gorell, L. I. Gilbert, and J. Tash, *Insect Biochem.* **2**, 94 (1972).
[7] F. M. Butterworth and H. D. Berendes, *J. Insect Physiol.* **2**, 94 (1972).

droxyecdysone has a higher biological activity than ecdysone in several *in vitro* bioassays, and that 20-hydroxyecdysone is a major circulating form of the molting hormone is insect larvae. Renewed efforts to search for an ecdysteroid binding protein in insect hemolymph therefore used radiolabeled 20-hydroxyecdysone as ligand and this led to the discovery[8,9] of a specific binding protein in the hemolymph of the African migratory locust, *Locusta migratoria* L. (Orthoptera). This chapter will describe progress in the purification of the locust 20-hydroxyecdysone binding protein as well as some of its characteristics.

Biosynthesis of 20-[^{3}H]Hydroxyecdysone[8]

Because radiolabeled 20-hydroxyecdysone is not commercially available at the high specific activity necessary for binding studies, this radiolabeled ligand is biosynthesized from [23,24-^{3}H]ecdysone ($\approx$60 Ci/mmol) (available from New England Nuclear). Malpighian tubules are an important site of ecdysone conversion to 20-hydroxyecdysone in locusts *in vitro*[10] and can easily be dissected. This tissue is chosen in preference to fat body or midgut because nonlabeled (endogenous) ecdysone and 20-hydroxyecdysone present in Malpighian tubules can be more easily rinsed out prior to incubation, and this minimizes dilution of the radiolabeled ecdysteroids. Malpighian tubules of 50 late last larval stage locusts (*Locusta migratoria*) are rinsed for at least 2 hr at 4° in locust saline. Highest conversion rates of ecdysone to 20-hydroxyecdysone are observed when the hemolymph concentration of 20-hydroxyecdysone is highest i.e., during the last larval stage. The solvent containing 15 nmol of [23,24-^{3}H]ecdysone (900 μCi) is dried under a stream of nitrogen and the hormone is taken up by 5 ml of Landureau's medium (i.e., final ecdysone concentration $\approx$3 μM). Malpighian tubules are added to this medium and incubated at 37° for 6 hr with constant, gentle shaking. Ecdysteroids are extracted with methanol[11] and separated by thin-layer chromatography (2 developments in the solvent system chloroform : methanol, 4 : 1 v/v. Merck HF$_{254}$ precoated silica gel plates, 0.25 mm thickness). In our hands, 80% of the radioactivity recovered was 20-hydroxyecdysone.[8] Radioimmunoassay of small aliquots of [^{3}H]ecdysone and 20-[^{3}H]hydroxyecdysone using a ^{125}I-iodinated tracer[12] shows that there is no significant dilution of the biosyn-

[8] R. Feyereisen, *Experientia* **33**, 1111 (1977).
[9] R. Feyereisen, *in* "Progress in Ecdysone Research" (J. A. Hoffmann, ed.), p. 325. Elsevier/North-Holland Biomedical Press, Amsterdam, 1980.
[10] R. Feyereisen, M. Lagueux, and J. A. Hoffmann, *Gen. Comp. Endocrinol.* **29**, 319 (1976).
[11] J. A. Hoffmann, J. Koolman, P. Karlson, and P. Joly, *Gen. Comp. Endocrinol.* **22**, 90 (1974).
[12] M. L. DeReggi, M. H. Hirm, and M. A. Delaage, *Biochem. Biophys. Res. Commun.* **66**, 1307 (1975).

thesized 20-hydroxyecdysone by unlabeled hormone present in the tissue. Specific activity of the biosynthesized 20-[^{3}H]hydroxyecdysone can also be determined by HPLC. The procedure can be scaled down but the final ecdysone concentration must remain as high as possible.

Comments. In addition to its use in the search for ecdysteroid binding proteins in insect hemolymph, the biosynthesized 20-[^{3}H]hydroxyecdysone has been used in ecdysteroid metabolism studies,[13] and in research on cellular receptors for ecdysteroids in crustaceans[14,15] and insects.[16] Very useful techniques for the purification of ecdysteroids have been described, using Sep-Pak cartridges[17,18] and HPLC. Reversed-phase TLC plates have been recommended because they allow better recoveries of ecdysteroids when compared to normal phase silica TLC plates.[19] Rearing of *Locusta migratoria* is not allowed in the United States, but biosynthesis of 20-[^{3}H]hydroxyecdysone can be done using tissues of other insects, for instance *Manduca sexta*[20] Malpighian tubules or fat body (K. Dyer, personal communication).

Hemolymph Preparation

Hemolymph is collected after severance of a metathoracic leg and it is allowed to clot for at least 10 min. The supernatant obtained after centrifugation ("serum") does not darken if kept frozen ($-30°$) or if diluted with at least 1 vol 50 mM sodium phosphate buffer, pH 6.9. Clotting in *L. migratoria* is an interaction between factor(s) released by hemocytes and hemolymph proteins resulting in the precipitation of the abundant glycolipoprotein coagulogen[21] (shown to be identical to lipophorin[22]). A rapid centrifugation of blood cells from hemolymph collected from chilled insects prevents clotting and yields a stable, nondiluted "plasma."[21] According to Gellissen,[22] clotting can also be avoided by diluting hemolymph directly upon collection with an equal volume of ice cold 0.05 M Tris–HCl buffer, pH 8.3, containing 0.5 M NaCl, 2.5 mM reduced glutathione, and 2 mM EDTA (high salt concentration prevents the precipitation of lipophorin and vitellogenin).

[13] R. Lafont, P. Beydon, G. Somme-Martin, and C. Blais, *Steroids* **36**, 185 (1980).
[14] P. Kuppert, M. Buchler, and K. D. Spindler, *Z. Naturforsch., C: Biosci.* **33C**, 437 (1978).
[15] P. Kuppert, S. Wilhelm, and K. D. Spindler, *J. Comp. Physiol.* **128**, 95 (1978).
[16] K. Schaltmann and O. Pongs, *Proc. Natl. Acad. Sci. U.S.A.* **79**, 6 (1982).
[17] R. Lafont, J. L. Pennetier, M. Andrianjafintrimo, J. Claret, J. F. Modde, and C. Blais, *J. Chromatogr.* **236**, 137 (1982).
[18] R. D. Watson and E. Spaziani, *J. Liq. Chromatogr.* **5**, 535 (1982).
[19] I. D. Wilson, S. Scalia, and E. D. Morgan, *J. Chromatogr.* **212**, 211 (1981).
[20] D. S. King, *Gen. Comp. Endocrinol., Suppl.* **3**, 221 (1972).
[21] M. Brehelin, *Comp. Biochem. Physiol. B* **62B**, 329 (1979).
[22] G. Gellissen, *Naturwissenschaften* **70**, 45 (1983).

Assay of 20-Hydroxyecdysone Binding Protein

Binding of 20-[³H]hydroxyecdysone to *L. migratoria* hemolymph proteins was first shown by gel filtration on Sephadex G-100 or G-25. 20-[³H]Hydroxyecdysone was found associated with the macromolecular fractions after injection of [³H]ecdysone into larval and adult locusts,[23] and also after *in vitro* incubation of cell-free hemolymph with 20-[³H]hydroxyecdysone.[8] Although both approaches showed that labeled hormone competed with endogenous unlabeled hormone for a limited number of binding sites (because apparent binding percentage was inversely related to the endogenous hormone titer[8,23]), the technique of gel filtration is not suited for rapid screening of binding activity. Cao *et al.* recently utilized photoaffinity labeling to confirm the existence of a 20-hydroxyecdysone binding protein in locust hemolymph.[24] Two useful techniques for the assay of 20-hydroxyecdysone binding protein are described below.

Dextran-Coated Charcoal Adsorption.[9] Charcoal (Norit A, acid and base washed) is added to a 0.1% dextran (60–90.000) solution in 50 mM sodium phosphate buffer. The suspension (1% charcoal, 0.1% dextran) is adjusted to pH 6.9 and continuously mixed with a magnetic stirrer. Of this dextran-coated charcoal suspension 150 μl is mixed (5 sec, vortex) with an equal volume of buffer containing 1% bovine serum albumin and labeled 20-hydroxyecdysone (30 nM to 30 μM), then incubated for 5 min at 7° and centrifuged for 5 min. More than 99% of the labeled 20-hydroxyecdysone is adsorbed to the dextran-coated charcoal (pellet) provided that the suspension is freshly prepared (i.e., less than 48 hr under continuous stirring). Final concentrations of dextran and charcoal lower than 0.05 and 0.5%, respectively, or incubation times shorter than 5 min result in significant amounts of radioligand not being adsorbed under these conditions, and are thus not suitable for the detection of hemolymph protein binding. Cell-free hemolymph serum is diluted to 10 mg protein/ml with 50 mM sodium phosphate buffer, pH 6.9 and incubated for 60 min with 20-[³H]hydroxyecdysone (9 × 10⁴ cpm, 30 nM to 30 μM) in a total volume of 150 μl. Binding is measured on two 100-μl aliquots of the supernatant obtained after dextran-coated charcoal treatment as described above. Although this assay technique is rapid and easy to perform, it is a nonequilibrium technique which results in an underestimation of the actual binding of 20-[³H]hydroxyecdysone to hemolymph proteins. This can easily be shown by varying the incubation time in the presence of dextran-coated charcoal.[9] Rapid dissociation of the 20-[³H]hydroxyecdysone–

---

[23] R. Feyereisen, M. Lagueux, and J. A. Hoffmann, *C. R. Hebd. Seances Acad. Sci., Ser. D* **280,** 1709 (1975).

[24] M. Cao, B. Duvic, D. Zachary, P. Harry, and J. A. Hoffmann, *Insect Biochem.* **13,** 567 (1983).

protein complex also explains our failure to detect binding of 20-[³H]hydroxyecdysone after polyacrylamide gel electrophoresis or isoelectric focusing. Binding activity is however retained after polyacrylamide gel electrophoresis under nondenaturing conditions or after isoelectric focusing. The position of the binding protein on gels can thus be determined by performing binding assays on eluted proteins.[24]

Equilibrium Dialysis.[9] An ideal technique for binding studies is of course equilibrium dialysis because it does not have (by definition) the experimental constraints of nonequilibrium techniques such as charcoal adsorption or gel filtration. A number of dialysis instruments are available commercially (e.g., Dianorm, Bachofer). We have routinely used a Biologie Appliquée (Marseille, France) apparatus.[25] Sartorius SM 11533 or Schleicher and Schuell RC 53 membranes separate two cylindrical 200 μl chambers filled with 150 μl of the desired solutions. SGE syringes (Austin, Texas) fitted with repeating dispensers are used for filling and sampling the dialysis chambers. In routine assays, one chamber is filled with 20-[³H]hydroxyecdysone in 50 mM sodium phosphate buffer, pH 6.9, containing 1% bovine serum albumin, whereas the opposite chamber is filled with 150 μl cell-free hemolymph serum diluted to 10 mg protein/ml with buffer. Dialysis is allowed to proceed at 7° under constant agitation. The radioactivity of 100-μl aliquots from both chambers is assayed (binding of label to the dialysis membrane is negligible). Figure 1 shows that equilibrium is reached after about 2 hr, and therefore we have chosen a standard dialysis time of 3 hr (equilibrium is maintained for at least 12 hr).

Partial Purification of the Binding Protein

The 20-hydroxyecdysone binding protein from locust hemolymph has not yet been purified to homogeneity. Attempts to purify the protein have been described[9,24] and some salient features will be reported here. Collection of hemolymph can be done with the minimum of precautions because clotting does not reduce the apparent binding activity. Thus, locust lipophorin is not an ecdysteroid binding protein. Locust vitellogenin is likewise not an ecdysteroid binding protein, because vitellogenin is not present in larval hemolymph, and because removal of vitellogenin from adult female hemolymph leaves binding activity unchanged.

Sixty mililiters of hemolymph (2.28 g protein, from approximately 1000 mid-fifth larval stage locusts) yields about 50 ml of serum (1.85 g protein). Sepharose 4B gel filtration on a preparative scale (3.5 × 67-cm

[25] H. L. Cailla, G. S. Cros, E. J. P. Jolu, M. A. Delaage, and R. C. Depieds, *Anal. Biochem.* **56,** 383 (1973).

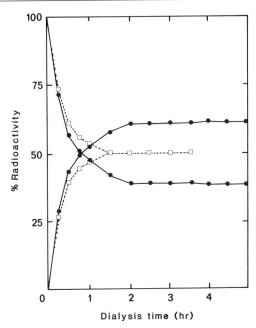

FIG. 1. Equilibrium dialysis of 20-[^{3}H]hydroxyecdysone (30 nM in a 1% BSA solution in 50 mM phosphate buffer pH 6.9) against *L. migratoria* partially purified binding protein (10 mg/ml) (●) or against a 1% BSA solution in the same buffer (□). Dialysis with constant agitation was at 7°. Radioactivity in both chambers was measured at various times of dialysis and expressed as a percentage of total radioactivity.

column loaded with 50 ml of serum and eluted with 50 mM phosphate buffer, pH 6.9; flow rate 25 ml/hr) removes the endogenous hormone and separates two peaks of binding activity. The first peak (proteins of ~280,000 MW) is quantitatively most important. Binding to proteins of the second, minor peak (proteins of less than 50,000 MW) is not saturable at 20-hydroxyecdysone concentrations up to 30 μM. Also, proteins of the second peak bind [^{3}H]ecdysone as well as 20-[^{3}H]hydroxyecdysone as opposed to proteins of the first peak (see below). The Sepharose 4B gel filtration results in an approximately 7-fold purification of the 20-hydroxyecdysone binding protein (first peak). Fractions of the first peak of binding activity (45 ml, 608 mg protein) are pooled and concentrated by ammonium sulfate precipitation (35–70% saturation). The protein pellet is taken up in 20 mM phosphate buffer, pH 7.4 and diafiltered in an Amicon cell. The concentrated binding protein (7.2 ml, 283 mg protein) is not bound by a DEAE-cellulose column eluted with 20 mM phosphate buffer, pH 7.4, but the eluting peak of binding activity (9.6 ml, 186 mg protein) is purified approximately 14-fold (binding activity is lost on CM-cellulose, even though all protein can be recovered). The partially purified 20-hydroxyecdysone binding protein is diluted and stored at $-30°$ in 50 mM phosphate buffer, pH 6.9. Whereas unpurified, undiluted cell-free hemolymph retains all its binding activity for over 12 months when stored at $-30°$, the partially purified binding protein is unstable, losing half of its binding activity after 44 days of storage. It would therefore seem advisable to protect partially purified binding protein with ligand (20-hydroxyecdysone) and to remove the ligand when necessary by prolonged charcoal treatment.

Cao et al. have described a purification scheme for the binding protein from adult female locusts.[24] They used adult locusts because endogenous ecdysteroid levels are low in adult female hemolymph. Hemolymph ecdysteroid concentrations equivalent to 0.5 μM ecdysone are measured by RIA in reproductively competent adult female L. migratoria.[26] In the first step of the Cao et al. procedure, 6 ml of serum is brought to pH 8.0 with 10 mM Tris and loaded on a DEAE-cellulose column (0.9 × 15 cm) equilibrated and eluted with a linear gradient of NaCl in the same buffer. The binding protein is eluted at approximately 80 mM NaCl and can be localized either by binding essay (dextran-coated charcoal adsorption) on eluted fractions, or by the behavior on this column of binding protein labeled with 20-[^{3}H]hydroxyecdysone after photoactivation (Fig. 2). The active (20-hydroxyecdysone binding) fractions from the anion-exchange column can be further purified by gel permeation on Ultrogel AcA 34 and by polyacrylamide gel electrophoresis.[24] Silver staining of the gels obtained in the last step reveal that the binding protein is nearly homogeneous, but no indications are given on purification factors or yields. The various approaches described above will certainly be useful in a future preparative scale purification of the binding protein.

20-Hydroxyecdysone Binding Protein Properties

The molecular weight of the binding protein is estimated at 280,000 in larvae[9] and 270,000 in adult females.[24] The protein may be a dimer of 133,000 subunits, and it has an isoelectric point of 5.6.[24] Similarities in specificity and molecular weight suggest that the larval and adult binding protein may be the same protein in L. migratoria.

The apparent association constant of the 20-hydroxyecdysone–protein complex was determined by equilibrium dialysis[9] at 7° and found to be

[26] M. Lagueux, M. Hirn, M. DeReggi, and J. A. Hoffman, C. R. Hebd. Seances Acad. Sci., Ser. D 282, 1187 (1976).

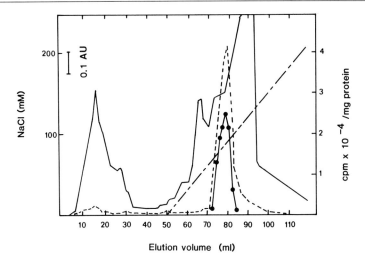

FIG. 2. DEAE-cellulose chromatography of *L. migratoria* serum proteins. Solid line: optical density measured at 280 nm (inset scale: 0.1 absorbance unit); dashed line: binding activity for 20-[^{3}H]hydroxyecdysone of the various eluted fractions; ●, radioactivity linked to proteins chromatographed after photoactivation of 20-[^{3}H]hydroxyecdysone in the presence of serum. Radioactivity expressed in cpm/mg of eluted protein. Operating conditions: 0.9 × 15-cm DEAE cellulose column, equilibrated with 10 mM Tris buffer (pH 8); 6 ml of serum in the same buffer was deposited and eluted in this buffer until fraction 50 after which elution occurred with a linear gradient from 0 to 200 mM NaCl (dashed straight line in the figure); flow rate: 0.2 ml/min; collection of 1 ml fractions. Adapted from M. Cao, B. Duvic, D. Zachary, P. Harry, and J. A. Hoffmann, *Insect Biochem.* **13**, 567 (1983).

$K_A = 1.03 \pm 0.05 \times 10^7 \, M^{-1}$, a value in good agreement with the apparent association constant estimated by nonequilibrium techniques. Sephadex G-25 gel filtration[8] of unpurified, cell-free hemolymph gave $K_A = 3.3 \times 10^6 \, M^{-1}$, and dextran-coated charcoal binding assays[24] with partially purified binding protein gave $K_A = 3.7 \times 10^6 \, M^{-1}$.

Specificity.[9] Specificity of the partially purified binding protein is best studied by equilibrium dialysis in a competition assay with unlabeled ecdysteroids. Stock solutions of ecdysteroids are prepared in methanol and a UV spectrum is recorded. Concentrations can then be calculated using the extinction coefficients listed in Table I. Methanol dilutions of stock ecdysteroid solutions are dried under a nitrogen stream and taken up in 500 μl of 50 mM sodium phosphate buffer, pH 6.9 containing 1% bovine serum albumin and a tracer amount (1.10^5 cpm) of 20-[^{3}H]hydroxyecdysone. Three 150-μl aliquots are dialyzed against 150 μl partially purified binding protein diluted to 10 mg/ml. Because determinations are performed at equilibrium, and because the Scatchard plot obtained with

TABLE I
Ultraviolet Absorption of Various Ecdysteroids in Methanol

Ecdysteroid	λ_{max}	ε^a
20-Hydroxyecdysone	242	12,400
2-Deoxy-20-hydroxyecdysone	243	12,100
Ponasterone A	244	12,400
Inokosterone	243	12,100
3-Dehydro-20-hydroxyecdysone	242	10,360[b]
Makisterone A	243	12,400
Calonysterone	292	7,850
Kaladasterone	300	10,800
Ecdysone	242	12,400
22-Isoecdysone	244	11,850
Muristerone A	232	8,900
2β,3β,14α-Trihydroxy-5β-cholest-7-en-6-one	243	12,200
2β,3β-Dihydroxy-5α-cholest-7-en-6-one	248	14,600

[a] Molar extinction coefficients are from D. H. S. Horn, in "Naturally Occurring Insecticides" (M. Jacobson and D. G. Crosby, eds.), p. 333. Dekker, New York, 1971; and from L. Canonica, B. Danieli, G. Ferrari, J. Krepinsky, and I. Weisz-Kincze, *Phytochemistry* **14**, 525 (1975).

[b] Not reported. This value for 3-dehydroecdysone may be an underestimate [P. Karlson, H. Bugany, H. Dopp, and G. A. Hoyer, *Hoppe-Seyler's Z. Physiol. Chem.* **353**, 1610 (1972)].

20-hydroxyecdysone binding data is linear ($n = 15$, $r = 0.9953$) indicating that only one type of equivalent binding sites is involved, it is legitimate to calculate the ratio of equilibrium constants of association (RAC) from competition data (displacement curves). If R is the ratio of free/bound tracer in the absence of competitor and RA is the ratio of the concentrations of 20-hydroxyecdysone/competitor required to decrease binding of the tracer by 50%, the RAC can be calculated according to the following formula[27] designed for equilibrium dialysis data:

$$RAC = \frac{K_A \text{ competitor}}{K_A \text{ 20-hydroxyecdysone}} = \frac{2(2R + 1)(RA)}{4R + 3 - RA}$$

Table II gives the RAC of 14 ecdysteroids of natural or synthetic origin. It is possible to compare pairs of ecdysteroids differing by structural features such as substituents or conformation and deduce the role of such

[27] M. A. Delaage, personal communication.

TABLE II
RATIO OF EQUILIBRIUM CONSTANTS OF ASSOCIATION
(RAC)

Ecdysteroid[a]	RAC[9]
20-Hydroxyecdysone	1.000
2-Deoxy-20-hydroxyecdysone	1.212
Ponasterone A	0.445
Inokosterone	0.425
3-Dehydro-20-hydroxyecdysone	0.387
Makisterone A	0.036
Calonysterone	0.033
Kaladasterone	0.015
Ecdysone	0.012
22-Isoecdysone	0.005
Muristerone A	0.002
$2\beta,3\beta,14\alpha$-Trihydroxy-5β-cholest-7-en-6-one	<0.001[b]
$2\beta,3\beta$-Dihydroxy-5α-cholest-7-en-6-one	<0.001[b]
Poststerone	<0.001[b]

[a] Structural formulas of ecdysteroids: C. Hetru and D. H. S. Horn, in "Progress in Ecdysone Research" (J. A. Hoffmann, ed.), p. 13. Elsevier/North-Holland Biomedical Press, Amsterdam, 1980.

[b] These compounds did not displace the tracer 20-hydroxyecdysone at the highest concentration tested (3.3×10^{-5} M).

features on binding. Such comparisons can be facilitated[28] by calculating the free energy of binding $\triangle G°$ of the ecdysteroid binding–protein complexes. The side-chain of 20-hydroxyecdysone contributes to more than 40% of the $\triangle G°$ and the C-20 hydroxyl group contributes alone to 30% of the $\triangle G°$ of binding. Thus, ecdysone is a poor ligand for the binding protein and this has been confirmed by other techniques as well.[8,24] The relative importance of the three side-chain hydroxyl groups is C-20 ≥ C-22 > C-25. Modifications of the side-chain, e.g., addition of a methyl group in C-24 (makisterone A) or the shift of a hydroxyl group from C-25 to C-26 or from C-22(R) to C-22(S), are all detrimental to binding. On the steroid nucleus, it appears that the C-2β hydroxyl group slightly decreases binding affinity, with the result that 2-deoxy-20-hydroxyecdysone (deoxycrustecdysone) is a better ligand than 20-hydroxyecdysone itself.

[28] M. E. Wolff, J. D. Baxter, P. A. Kollman, D. L. Lee, I. D. Kuntz, E. Bloom, D. T. Matulich, and J. Morris, *Biochemistry* **17**, 3201 (1978).

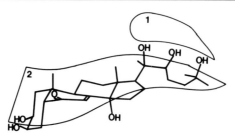

FIG. 3. 20-Hydroxyecdysone and possible features of the binding site of the hemolymph binding protein of L. migratoria. (1) Site(s) recognizing the C-20 hydroxyl group (and the side chain). (2) Postulated hydrophobic region.

The C-3β hydroxyl group is however preferred to a C-3 ketone which modifies the conformation of the A-ring. Hydroxylation of the C-ring (C-11α) under the plane of the steroid nucleus is very detrimental to binding. Interestingly, calonysterone is bound with an appreciable affinity, despite the considerable modifications of the ring structure. Indeed, the orientation of the C-7 ketone conjugated by C-5, C-8, and C-14 double bonds makes the ring structure almost planar as opposed to the bent orientation of the A-ring in all ecdysteroids with a cis-fused A/B-ring junction (5β-H). It thus appears that the presence of the C-20 hydroxyl group is quantitatively more important for binding activity than the conformation of the steroid ring structure. These interpretations are summarized in Fig. 3 which delineates the features which are important for binding of ecdysteroids to the hemolymph binding protein. In addition to a binding site (hydrogen bond acceptor?) recognizing the side chain and the C-20 hydroxyl group in particular, it is postulated that the hormone is presenting a long hydrophobic side which would fit in a similarly hydrophobic pocket of the binding site. This hypothesis would rationalize the effect of C-2β, C-11α, C-22(S), and C-26 hydroxyl groups, but needs to be confirmed when more ecdysteroids are tested on a homogeneously pure binding protein.

Concluding Remarks

Little is known about the biological function of the 20-hydroxyecdysone binding protein in locust hemolymph and more extensive work is clearly warranted. The specificity of this ecdysteroid–protein interaction is reminiscent of the specificity of ecdysteroid effects on biological systems in vitro[29] and the hemolymph binding protein may therefore serve as

[29] R. Bergamasco and D. H. S. Horn, in "Progress in Ecdysone Research" (J. A. Hoffmann, ed.), p. 299. Elsevier/North-Holland Biomedical Press, Amsterdam, 1980.

model for the study of ecdysteroid–protein interactions in cellular ecdysteroid receptors.

Binding of endogenous 20-hydroxyecdysone to hemolymph proteins occurs beyond any doubt in *Calliphora vicina*[30,31] and *L. migratoria* but variations of the ratio of free and bound hormone remain to be studied under various physiological conditions. Differences between the *C. vicina* and *L. migratoria* binding proteins are considerable: in the blowfly larva, 20-hydroxyecdysone is bound with little specificity to calliphorin, a major larval protein, whereas in locusts 20-hydroxyecdysone binds with a high degree of specificity to a less abundant protein. Such differences may be reflected in the respective physiological function(s) of these proteins.

Difficulties in research on ecdysteroid binding protein in insect hemolymph can be caused by the use of the wrong ligand (ecdysone has a very low affinity to the locust binding protein), by the high hemolymph concentrations of ecdysteroids which considerably dilute the tracer amounts of ligand, and by the very rapid dissociation of the ecdysteroid–protein complex which points to equilibrium techniques (equilibrium dialysis, microSephadex G-25 equilibrium assays[32] or steady state polyacrylamide gel electrophoresis[33]) as the methods of choice for future research. Photoaffinity labeling techniques are of course extremely useful for very limited experimental purposes[24,30,31] (and Koolman, this volume [25]) but because these techniques involve a covalent linkage between the protein and its ligand, they cannot be used to study the physiological role of the binding protein.

---

[30] L. Reum, G. Kauser, U. Enderle, and J. Koolman, *Z. Naturforsch., C: Biosci.* **37C,** 967 (1982).
[31] U. Enderle, G. Kauser, L. Reum, K. Scheller, and J. Koolman, in "The Larval Serum Proteins of Insects" (K. Scheller, ed.), p. 40. Thieme, Stuttgart, 1983.
[32] D. W. Payne and J. A. Katzenellenbogen, *Endocrinology* **105,** 743 (1979).
[33] V. Hansson, K. Purris, A. Attramadal, T. Varaas, and E. M. Ritzen, *J. Steroid Biochem.* **8,** 771 (1977).

[28] Ecdysone 20-Monooxygenase[1]

By GUNTER F. WEIRICH

Hydroxylation of ecdysone[2] to 20-hydroxyecdysone is generally considered to be the last step in the biosynthesis of the insect molting hormone. This conversion takes place in various tissues, e.g., fat body, Malpighian tubules, and midgut.[3] Depending on the species and tissue under study, ecdysone 20-monooxygenase can be found in mitochondria[4-6] and/or microsomes.[7,8] In both of these cellular components, the enzyme is a cytochrome P-450 monooxygenase. This chapter describes procedures for the study of ecdysone 20-monooxygenases in midgut mitochondria and microsomes of the tobacco hornworm, *Manduca sexta*.

[³H]Ecdysone Purification

[23,24-³H]Ecdysone (New England Nuclear),[9] to be used as substrate in the assay, is purified to ≥95% radiopurity by reversed-phase high-performance liquid chromatography (RP-HPLC) on a radial compression C_{18} column (Waters Associates; 8 × 100 mm, 10-μm particle size). [³H]Ecdysone is eluted isocratically with a mixture of methanol (Burdick & Jackson) and glass-distilled water (38/62) at 2 ml/min. After elution and collection of the [³H]ecdysone, the column is flushed with methanol/water (60/40) until effluent radioactivity is at background level. RP-HPLC

---

[1] EC 1.14.99.22.
[2] 2β,3β,14α,22R,25-Pentahydroxy-5β-cholest-7-en-6-one.
[3] M. J. Thompson, J. N. Kaplanis, G. F. Weirich, J. A. Svoboda, and W. E. Robbins, in "Regulation of Insect Development and Behaviour" (F. Sehnal, A. Zabza, J. J. Menn, and B. Cymborowski, eds.), p. 107. Wroclaw Technical University Press, Poland, 1981.
[4] P. Johnson and H. H. Ress, *Biochem. J.* **168**, 513 (1977).
[5] R. T. Mayer, J. A. Svoboda, and G. F. Weirich, *Hoppe-Seyler's Z. Physiol. Chem.* **359**, 1247 (1978).
[6] S. L. Smith, W. E. Bollenbacher, D. Y. Cooper, H. Schleyer, J. J. Wielgus, and L. I. Gilbert, *Mol. Cell. Endocrinol.* **15**, 111 (1979).
[7] R. Feyereisen and F. Durst, *Eur. J. Biochem.* **88**, 37 (1978).
[8] J. N. Kaplanis, G. F. Weirich, J. A. Svoboda, M. J. Thompson, and W. E. Robbins, in "Progress in Ecdysone Research" (J. A. Hoffmann, ed.), p. 163. Elsevier/North-Holland Biomedical Press, Amsterdam, 1980.
[9] Mention of a company or proprietary product does not imply endorsement by the U.S. Department of Agriculture.

in an acetonitrile/water system yielded unstable [^{3}H]ecdysone solutions and was, therefore, deemed unsuitable for purification.

Assay

Incubation

The assay mixture contains 0.64 mM NADP$^+$, 6.0 mM glucose 6-phosphate, 1.0 unit glucose-6-phosphate dehydrogenase (from bakers' yeast, Sigma, Type XV), and mitochondria (2–10 mg protein) or microsomes (2–6 mg protein) in 0.5 ml buffer solution (33 mM potassium phosphate, pH 7.8, 100 mM sucrose, 1 mM EDTA). To prevent losses of the mitochondrial enzyme activity, bovine serum albumin (Sigma, crystallized and lyophilized) is added to the assay mixture at 2.0 mg/ml. The reaction vessels are kept in a Dubnoff metabolic incubator at 30° and at 90–100 oscillations per min. After a 5-min preincubation, the reaction is started by the addition of 10 μg [^{3}H]ecdysone (4.6 Ci/mol) in 3–5 μl methanol. The reaction is stopped at an appropriate time by addition of 5.0 ml methanol. Protein is sedimented by a 15-min centrifugation at 20,000 g, and the pellets are reextracted with another 5 ml methanol.

Isolation and Determination of 20-[^{3}H]Hydroxyecdysone

The methanol extract from each incubation mixture is dried under vacuum, redissolved in 1.0 ml methanol/water (1/9), and applied to a C$_{18}$ SEP-PAK cartridge (Waters). The cartridge is rinsed in sequence with 1.0 ml of methanol/water (1/9) and 5 ml methanol/water (3/7). Ecdysone and 20-hydroxyecdysone are eluted with 6 ml methanol/water (6/4; >90% recovery), the eluate is dried under vacuum, the residue dissolved in a small volume of methanol, and analyzed by RP-HPLC. Ecdysone and 20-hydroxyecdysone are separated by isocratic elution with an acetonitrile (Burdick & Jackson)/water mixture from a μBondapak C$_{18}$ column (Waters Associates; acetonitrile/water, 20/80, 1 ml/min; retention times for ecdysone and 20-hydroxyecdysone, 19–22 and 8–10 min, respectively). Alternately, a Radial Compression C$_{18}$ column can be used (Waters Associates; 5 × 100 mm, 10-μm particle size; acetonitrile/water, 18/82, 1 ml/min; retention times for ecdysone and 20-hydroxyecdysone, 14–15 and 5–6 min, respectively). The absorbance of the effluent is monitored at 254 nm.

Some apolar tritium-labeled impurities of the samples are retained by the columns at the solvent strengths used for the separation. As these compounds tend to bleed slowly from the columns, they can cause errors in subsequent [^{3}H]ecdysone or 20-[^{3}H]hydroxyecdysone determinations.

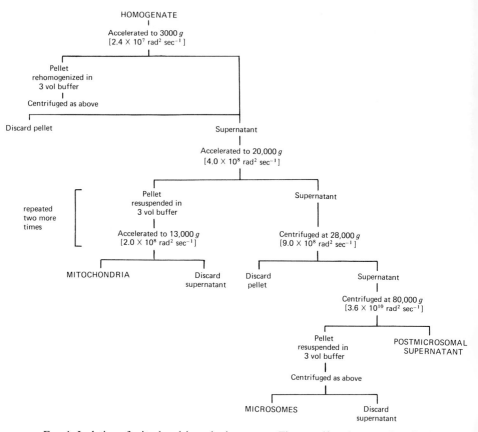

FIG. 1. Isolation of mitochondria and microsomes. The centrifugations are described by the average g forces and the time integrals of the squared angular velocities [$W = 981 \times 60\,(g\,\text{min})/r_{av}$], given in brackets [H. Beaufay and A. Amar-Costesec, *Meth. Membr. Biol.* **6**, 1 (1976)]. [Modified from G. F. Weirich and J. R. Adams, *Arch. Insect Biochem. Physiol.* **1**, 311 (1984).]

A 5–10 min flush of the column with acetonitrile/water (50/50) after each analysis is necessary to eliminate this problem. The conversion rate of ecdysone to 20-hydroxyecdysone is calculated from peak heights or peak areas of the two compounds on UV tracings (after calibration of the HPLC system), or from radioactivity determinations of the corresponding effluent fractions.

Comments

It is essential that assays of the mitochondrial enzyme system are performed in a hypotonic incubation mixture. In isotonic buffer solutions

(300 mM sucrose) mitochondria generally do not hydroxylate ecdysone, probably because intact mitochondrial membranes prevent its access to the enzyme system.[8] In assays of the mitochondrial ecdysone 20-monooxygenase, various hydrogen donors can be substituted for the NADPH-generating system (e.g., succinate, isocitrate, malate, NADH).[5] All of these substrates are linked to the generation of NADPH by mitochondrial pathways.[5] These pathways, however, do not yield predictable NADPH concentrations, and the substitute hydrogen donors are, therefore, generally not suitable for the determination of the monooxygenase activity. Mitochondria contain a pool of endogenous NADPH and/or other hydrogen donors that can sustain the ecdysone 20-hydroxylation without added cosubstrate(s).[8] The NADPH-generating system is added to optimize the NADPH supply in the assay mixtures. Microsomes, on the other hand, do not have permeability barriers for ecdysone. They have no endogenous pool of NADPH and cannot utilize other hydrogen donors.[10]

Isolation of Mitochondria and Microsomes

Fifth-instar larvae of *M. sexta* are collected on the first day of "wandering"[11,12] and cooled in ice. The midguts are dissected and freed from fat body, Malpighian tubules, trachea, and gut contents under ice-cooled 10 mM Tris–HCl buffer, pH 7.4, containing 300 mM sucrose and 1 mM EDTA (isolation buffer).

The cleaned and rinsed guts are then blotted dry on tissue, weighed, minced with scissors, and homogenized in 6 ml isolation buffer per gram tissue (6 "volumes") in a Ten Broeck homogenizer cooled in ice. The homogenizer is initially operated by hand until the tissue is broken up and suspended in the buffer. The homogenization is then completed by two strokes with the motor-driven pestle at approximately 650 rpm. The homogenate is fractionated as outlined in Fig. 1. In this scheme the duration of all centrifugations (except for the sedimentation of microsomes) is reduced to a minimum by increasing the speed,[13] and some centrifugations consist only of acceleration and deceleration, with no plateau phase in between. Mitochondrial and microsomal pellets are resuspended by gentle mixing with isolation buffer in glass-Teflon homogenizers.

Isolation of mitochondria and microsomes from insect tissues can also be accomplished by centrifugation in a sucrose density gradient.[7] Mito-

[10] G. F. Weirich, unpublished observation (1979).
[11] J. W. Truman and L. M. Riddiford, *J. Exp. Biol.* **60**, 371 (1974).
[12] H. F. Nijhout and C. M. Williams, *J. Exp. Biol.* **61**, 481 (1974).
[13] L. L. Keeley, *Comp. Biochem. Physiol. B* **46**, 147 (1973).

chondrial preparations contain a small proportion of microsomes even after repeated washings. To assess the extent of this contamination, NADPH–cytochrome c reductase activity has to be determined for both the isolated mitochondria and microsomes.[14]

Properties

For the mitochondrial enzyme system, the apparent K_m and V_{max} for ecdysone are 18.3 ± 6.8 μM and 46.6 ± 14.2 pmol/min/mg protein, respectively.[5] The microsomal system has not been characterized.

[14] G. F. Weirich and J. R. Adams, *Arch. Insect Biochem. Physiol.* **1**, 311 (1984).

[29] Measurement and Characterization of Ecdysteroid Receptors

By BECKY A. SAGE and JOHN D. O'CONNOR

Introduction

The modulation of cellular responses to steroid hormones by intracellular steroid receptor molecules has been the central component of the interpretive paradigm of steroid action since the early 1960s.[1–3] However, it was not until 1978 that the unequivocal demonstration of ecdysteroid receptors in cells and tissues of *Drosophila melanogaster* was forthcoming.[4,5] During the intervening years between the first demonstration of intracellular steroid receptors and the present time there have been four characteristics which have emerged as the sine qua non of receptors in general and a fifth characteristic particular to systems responding to steroid hormones. The first four characteristics of a receptor molecule are saturability, specificity, high affinity for the biological ligand, and the association of the biological response with the presence of the receptor. Paraphrasing this last point: in the absence of a high affinity, highly spe-

[1] E. V. Jensen and H. I. Jacobson, *Recent Prog. Horm. Res.* **18**, 307 (1962).
[2] W. D. Noteboom and J. Gorski, *Arch. Biochem. Biophys.* **111**, 559 (1965).
[3] R. J. B. King, J. Gordon, and D. R. Inman, *Endocrinology* **32**, 9 (1965).
[4] P. Maroy, R. Dennis, C. Beckers, B. A. Sage, and J. D. O'Connor, *Proc. Natl. Acad. Sci. U.S.A.* **75**, 6035 (1978).
[5] M. A. Yund, D. S. King, and J. W. Fristrom, *Proc. Natl. Acad. Sci. U.S.A.* **75**, 6039 (1978).

cific, saturable protein the hormone fails to elicit the typical response of "target" cells. A fifth characteristic which is particularly important for steroid receptors is the ability to bind to DNA. In particular, the current model of the mode of action of ecdysteroids formulated by Ashburner[6] and supported by data from numerous laboratories (for review, see Cherbas et al.[7]) places the primary action of the steroid receptor complex at the initiation of transcription of particular, tissue-specific genetic loci.

It is the purpose of this presentation to indicate the various biological sources in which ecdysteroid receptors have been demonstrated and to detail specifically manners in which nuclear and cytosolic receptors from the K_c cell line have been characterized.

Tissue Sources

To date there have been four tissue or cell types in which ecdysteroid receptors have been demonstrated and from which their partial kinetic and hydrodynamic characteristics have been obtained. A brief discussion of the data obtained from three of those sources is presented in this section so that some comparisons may be made to the more detailed presentation below of the isolation and characterization of ecdysteroid receptors of K_c cells.

Crustacean Hypodermis

Ecdysteroid receptors have been demonstrated in both the cytosol[8] and the nucleus[9] of the crayfish *Orconectes limosus*. The cytosol was obtained by homogenizing the integument in a glass Potter-Elvehjem homogenizer in ice cold Van Harreveld's saline[10] supplemented to a final concentration of 20% sucrose and 10 mM mercaptoethanol. Subsequently the homogenate was centrifuged twice (20 min at 20,000 g and 60 min at 120,000 g) and the resultant supernatant used as cytosol. An alternative buffer (5 mM HEPES, 120 mM NaCl, 5 mM mercaptoethanol, 10% sucrose pH 7.5) has subsequently been used for cytosol isolation without any discernible difference in kinetic constants.[9]

Nuclei were obtained by homogenizing diced integument from 2 or-

[6] M. Ashburner, C. Chihara, P. Meltzer, and G. Richards, *Cold Spring Harbor Symp. Quant. Biol.* **38**, 655 (1974).
[7] L. Cherbas, J. W. Fristrom, and J. D. O'Connor, in "Biosynthesis, Metabolism and Mode of Action of Invertebrate Hormones" (J. A. Hoffmann and M. Prochet, eds.). Springer-Verlag, Berlin and New York (in press).
[8] M. Londershausen and K.-D. Spindler, *Mol. Cell. Endocrinol.* **24**, 253 (1981).
[9] M. Londershausen, P. Kuppert, and K.-D. Spindler, *Hoppe-Seyler's Z. Physiol. Chem.* **363**, 797 (1982).
[10] A. Van Harreveld, *Proc. Soc. Exp. Biol. Med.* **34**, 428 (1936).

TABLE I
BINDING CONSTANTS OF CYTOPLASMIC AND
NUCLEAR PREPARATIONS FROM THE INTEGUMENT
OF *Orconectes limosus*[a]

Ecdysteroid	Cytoplasmic preparation (M)	Nuclear preparation (M)
Ecdysone	1.8×10^{-8}	6.8×10^{-9}
20-OH-ecdysone	2.7×10^{-9}	1.5×10^{-9}
Ponasterone A	6.0×10^{-11}	5.6×10^{-11}

[a] Data taken from ref. 9.

ganisms in 2 ml of 0.33 M sucrose, 4 mM CaCl$_2$ solution in a Dounce homogenizer (pestle A) and stroking 10 times.

The resulting homogenate was filtered through a 48-μm mesh nylon net and centrifuged 15 min at 1000 g at 4°. The pellet was resuspended in the original solution and layered on a 1.7 M column of sucrose containing 3 mM CaCl$_2$ and again centrifuged (1000 g, 15 min). The pellet was resuspended in the original solution and layered on 2.3 M sucrose, 3 mM CaCl$_2$ and centrifuged 2 times 45,000 g at 4°. The resultant pellet was washed twice in 0.33 M sucrose, 4 mM CaCl$_2$ and then suspended in a buffer containing 10 mM Tris, 25 mM NaCl, 25 mM KCl, 10 mM MgCl$_2$, 1 mM dithiothreitol, 0.33 M sucrose, at pH 7.4.

As an alternative method the very first nuclear pellet was resuspended in 0.33 M sucrose, 4 mM CaCl$_2$, 0.1% Triton X-100, and homogenized by 5 strokes of a Dounce homogenizer. The resultant homogenate was let stand for 15 min at 4° and then centrifuged at 1000 g for 15 min. The pellet was designated "Triton nuclei."

The binding characteristics of these preparations are illustrated in Table I. It should be pointed out that the circulating titer of 20-OH-ecdysone in the intermolt crustaceans in which it has been measured is appreciably higher than the K_d indicated in Table I.[11] This would suggest that the integumentary receptor is always "occupied" with steroid. It this is the case, the characteristic response of intermolt crustaceans to an injection of active ecdysteroids must be due to the mediation of a secondary set of empty receptors! Alternatively, the reported K_d values may be in error. Using a photoaffinity labeling system described below the crustacean ec-

[11] K.-D. Spindler, R. Keller, and J. D. O'Connor, *in* "Recent Progress in Ecdysone Research" (J. A. Hoffmann, ed.), p. 247. Elsevier/North-Holland Biomedical Press, Amsterdam, 1980.

dysteroid receptors have been reported to have a molecular weight of approximately 70,000 and an isoelectric point of 5.9.

Salivary Glands

Ecdysteroid binding moieties have been demonstrated in salivary glands of *Drosophila melanogaster* by Pongs and colleagues.[12,13] Glands are either hand dissected or mass isolated by the method of Cohen[14] and incubated for 30 min in a buffer, pH 7.2, containing 40 mM NaCl, 40 mM KCl, 2 mM KH$_2$PO$_4$/1 mM Na$_2$HPO$_4$, 1.2 mM MgCl$_2$, 1.2 mM Mg acetate, 1 mM CaCl$_2$, 80 mM sucrose, and 1 μM 20-OH-ecdysone. The glands were then washed in the buffer without 20-OH-ecdysone and irradiated in an air cooled cylindrical cuvette with an HBO 500 W/2 high-pressure mercury lamp [Schoffel, (TrappenKamp, Germany); Lamphouse: Oriel (Darmstadt, Germany)] at a distance of 5 cm through a 1-mm-thick WG320 long-pass filter for 5 min. Irradiation was performed at room temperature in a nitrogen atmosphere. After irradiation the glands were collected by short (1 min) centrifugation and disrupted by sonication for 5 min. The resulting homogenate is centrifuged[15] and the supernatant brought to 30% ammonium sulfate (w/v) and the precipitate is collected by centrifugation (20 min, 14,000 g).

Following unidimensional SDS–PAGE according to Laemmli[16] the protein was transferred to nitrocellulose filters (Western blot) and probed with an anti-ecdysteroid antiserum.

The results of this approach revealed a single band in the salivary gland preparation whose mobility suggested a molecular weight of 130,000. This was similar to, if not identical with, a similar band present in a population of K$_c$ cells prepared in similar fashion.

In contrast, when the above preparation from salivary glands is layered on a 5–20% sucrose gradient and centrifuged at 41,000 rpm for 18 hr at 4° in a Beckman SW41 rotor, there were 2 peaks of material which reacted with anti-ecdysteroid antiserum. One peak migrated at 6 S and the other at 9 S. (In this experiment each fraction from the gradient was absorbed on nitrocellulose filter paper and subsequently incubated with anti-ecdysone antiserum as indicated above.) To date there are no binding data available for salivary gland receptors.

[12] H. Gronemeyer and O. Pongs, *Proc. Natl. Acad. Sci. U.S.A.* **77**, 2108 (1980).
[13] K. Schaltmann and O. Pongs, *Proc. Natl. Acad. Sci. U.S.A.* **79**, 6 (1982).
[14] L. H. Cohen and B. V. Gotchel, *J. Biol. Chem.* **246**, 1841 (1971).
[15] The authors of Ref. 13 did not mention the g force used on the centrifuge model. The sonicated homogenate was centrifuged 20 min at 4600 RPM.
[16] U. K. Laemmli, *Nature (London)* **227**, 680 (1970).

Imaginal Disks

Early observations of Yund and Fristrom[17] suggested that imaginal disks possessed intracellular ecdysteroid receptors. However, the definitive demonstration was published using [^{3}H]ponasterone A as the ligand.[5] In this latter study 10,000–20,000 mass isolated imaginal disks[18] were either exposed to 0.2 nM concentrations of [^{3}H]ponasterone A in 1 ml Robbs medium[19] and then homogenized or homogenized first and the subcellular components incubated with 2–3 nM ponasterone A. In either case the disks were homogenized at a concentration of 75,000–100,000 disks/ml in a 10 mM Tris buffer (pH 7.4) containing 1.5 mM EDTA and 7 mM dithiothreitol. The cytoplasm fraction was the supernatant of centrifugation either at 100,000 g, 60 min or 17,000 g for 30 min. Apparently both supernatants yielded equivalent results. The description of the nuclear fraction was simply that it was prepared by the method of Williams and Gorski.[20] The separation of bound [^{3}H]ponasterone from free [^{3}H]ponasterone in the above preparations was accomplished by passing the material over a 1.3 × 35 cm Sephacryl S-200 column at a flow rate of 50 ml/hr at 4°. The elution buffer was similar to the homogenization buffer except that dithiothreitol was omitted and 100 mM KCl was included.

The kinetic data obtained revealed the K_d for both the nuclear and cytoplasmic preparations was $3-4 \times 10^{-9}$ M for ponasterone A. The concentration of unlabeled ecdysteroids necessary to displace 50% of [^{3}H]ponasterone when the receptor is labeled by incubation with 3 nM [^{3}H]ponasterone was 2 nM, 52 nM, and 72 μM for ponasterone A, 20-OH-ecdysone, and ecdysone, respectively. This is consistent with the concentration of these ecdysteroids necessary to produce 50% evagination in mass isolated imaginal disks. Unfortunately no hydrodynamic data or physical characterization of the receptor in imaginal disks is available.

Preparation of Crude Cytosolic and Nuclear Ecdysteroid Receptors from K_c Cells

Reagents and Materials

Buffers

Tris: 10 mM Tris, pH 7.2 at 22°
TM: 10 mM Tris, 5 mM MgCl$_2$, pH 6.9 at 22°
TMT: 10 mM Tris, 5 mM MgCl$_2$, 0.1% Triton X-100, pH 6.9 at 22°

---

[17] M. A. Yund and J. W. Fristrom, *Dev. Biol.* **43**, 287 (1975).
[18] J. W. Fristrom, *Results Probl. Cell Differ.* **5**, 109 (1972).
[19] J. A. Robb, *J. Cell Biol.* **41**, 876 (1969).
[20] D. Williams and J. Gorski, *Biochem. Biophys. Res. Commun.* **45**, 258 (1971).

TMK: TM buffer containing 0.8 M KCl

TE: 10 mM Tris, 1 mM Na$_2$EDTA, pH 7.3 at 22°

Reagents are stored at 4°, and all procedures described below are performed at 4° unless otherwise specified.

Cytosolic Receptors from Frozen Cells

Cells are harvested and resuspended in TM buffer. After a period of 10 min on ice, the cell suspension is centrifuged (2000 g, 10 min), and the cell pellet immediately frozen at −20 or −70°.

To obtain cytosol from frozen cells, the cells are thawed at 8° and then homogenized by 8 passes in a motor-driven Teflon/glass Potter-Elvehjem apparatus. The homogenate is centrifuged (Type 50 rotor, 100,000 g, 60 min), and the resulting supernatant is designated as a crude "high speed cytosol."

Nuclear Receptor Preparation

K$_c$ cells are harvested and resuspended in 10 vol of TMT buffer. After 15 min on ice, the cell suspension is homogenized by 20 passes in a Dounce homogenizer (A pestle). Crude nuclei are then collected by centrifugation (2000 g, 10 min). The nuclear pellet is washed once by resuspension in 4 vol of TM buffer. Following centrifugation (4000 g, 10 min), the nuclear pellet is extracted with an equal volume of TMK buffer by gentle mixing with a glass rod to resuspend the nuclei. The suspension is kept on ice for 10 min and then centrifuged (14,000 g, 10 min). The supernatant is desalted by centrifugation through Sephadex G-25 columns equilibrated with Tris buffer.[21] The G-25 columns have a maximum sample/bed volume ratio of 0.23. The desalted nuclear extract is centrifuged (25,000 g, 20 min), and the supernatant designated as a crude "nuclear receptor" preparation.

Whole Cell Receptor Preparations

Cytosol, nuclear, and whole cell receptor preparations can also be made concurrently from a singular source of cells which are either freshly harvested or stored at −20°.

Freshly harvested cells are resuspended in TM buffer and kept on ice for 10 min. The cell suspension is then centrifuged (2000 g, 10 min), and the cell pellet is homogenized by 8 passes in a motor-driven Teflon/glass Potter-Elvehjem apparatus followed by 20 passes in a Dounce homoge-

---

[21] M. W. Neal and J. R. Florini, *Anal. Biochem.* **55**, 328 (1973).

nizer (A pestle). A portion of the homogenate is immediately salt-extracted by the dropwise addition of 2.4 M KCl in TM buffer. The volume that is added should give a final salt concentration of 0.4 M KCl. After gentle stirring, the suspension is kept on ice for 10 min and then centrifuged (14,000 g, 10 min). The supernatant is then subjected to a desalting step as described above, and after a final centrifugation (25,000 g, 20 min) is designated as a "whole cell receptor" preparation from freshly harvested cells.

For cytosol and nuclear receptor preparations, the remainder of the whole cell homogenate is centrifuged (3000 g, 15 min), and the pellet and supernatant fractions are treated as follows. The pellet, consisting primarily of nuclei, is washed once with 4 vol of TM buffer and then extracted with an equal volume of TMK buffer. Subsequent desalting and centrifugation of this "nuclear receptor extract" is then performed as described above. The supernatant fraction of the whole cell homogenate is centrifuged a second time (Type 50 rotor, 100,000 g, 60 min), and the resulting supernatant designated as a "high speed cytosol" from freshly harvested cells.

Cytosol, nuclear, and whole cell receptor preparations can also be obtained from frozen cells using procedures similar to that described above for freshly harvested cells. One difference is that the cells are maintained at $-20°$ for 14–18 hr following the initial swelling/centrifugation step in TM buffer.

Ecdysteroid Receptor Binding Determinations

High speed cytosol, nuclear, and whole cell receptor preparations are radiolabeled with the ecdysteroid, [^{3}H]ponasterone A (sp. act. = 106 Ci/mmol). The preparation and purification of this radioligand have been previously described.[22] Receptor preparations are incubated with the radioligand (10^{-9} M) for 60–90 min at 22° or overnight at 4°. Bound ligand is routinely determined on 100 μl duplicate samples using the dextran-coated charcoal method of McGuire.[23] The accuracy of these determinations can be confirmed by an independent separation of the bound and free ligand using Sephadex G-25 gel filtration at 4°. The bound ligand determinations by both of these methods vary by less than 10%.

In some cases, the determination of ecdysteroid binding characteristics is facilitated by removal of the free ligand. This is accomplished using the centrifugation desalting method described above. This procedure

[22] B. A. Sage, M. A. Tanis, and J. D. O'Connor, *J. Biol. Chem.* **257**, 6373 (1982).
[23] W. McGuire, this series, Vol. 36, p. 248.

results in the removal of approximately 90% of the free ligand with only a minimal increase in sample volume.

Characteristics of Ecdysteroid Receptors in K_c Cells

A detailed report of the characterization of cytosol, nuclear, and whole cell ecdysteroid receptors in K_c cells has been previously presented.[22] However, a review of these characteristics is given below for purposes of comparison with data already presented.

The stability of cytosol and nuclear ecdysteroid receptors from K_c cells has been monitored at both 4 and 22°. In the continual presence of hormone both nuclear and cytosol preparations show a high degree of receptor binding stability at 4°. However, at 22°, the binding activity decreases markedly and after 20 hr more than half of the activity has been lost. Similar results are obtained if the receptor preparations are maintained at 4 and 22° without ecdysteroid present. At 4° the unloaded cytosol and nuclear ecdysteroid binding moieties are very stable, while at 22° a 50% loss of receptor binding activity occurs within 14 hr. This would seem to indicate that loss of binding activity at 22° is somewhat accelerated when receptor preparations are maintained without hormone.

Ecdysteroid receptor binding activity in both cytosol and nuclear preparations is reduced in the presence of 0.4 M KCl. This salt effect, however, can be reversed and the binding activity of the receptor preparation restored if the salt is removed from the cytosol or nuclear extracts using the centrifugation desalting method of Neal and Florini.[21]

The hydrodynamic properties of ecdysteroid receptors prepared from fresh and frozen K_c cell populations are given in Table II. From these data it is apparent that in fresh cell preparations both the cytosol and nuclear ecdysteroid receptors exhibit the same s value, namely 6.3 S. This stands in marked contrast, however, to the difference seen in the sedimentation properties of the cytosol and nuclear receptors prepared from frozen cells, where the s value for the cytosolic receptor has shifted from 6.3 to 4.2 S. The nature of this hydrodynamic change is still under investigation in our laboratory, but the data summarized in Table II are important in that they demonstrate the similarity of the hydrodynamic properties of the ecdysteroid receptor present in cytosol and nuclear preparations of freshly harvested K_c cells.

Table III summarizes the kinetic characteristics of the ecdysteroid binding moieties present in nuclear and cytosol preparations. The determinations demonstrate the saturability and competibility of the ecdysteroid receptors, as well as binding specificities that are consistent with

TABLE II
Sedimentation Properties of Ecdysteroid Receptors in Fresh and Frozen K_c Cell Populations[a]

Cell population	Receptor preparation	s value of ecdysteroid receptor activity (S)
Fresh	Whole cell	6.3
Fresh	Cytosol	6.3
Fresh	Nuclear	6.3
Frozen	Whole cell	6.3 and 4.2
Frozen	Cytosol	4.2
Frozen	Nuclear	6.3

[a] Velocity sedimentation analyses were performed on reorienting 10–40% sucrose gradients (TM buffer, Sorvall TV-865, 370,000 g, 2.5 hr, 4°). Molecular weight markers were [^{14}C]globin or protein markers prepared by reductive alkylation using [^{14}C]formaldehyde (G. E. Means, this series, Vol. 47, p. 469).

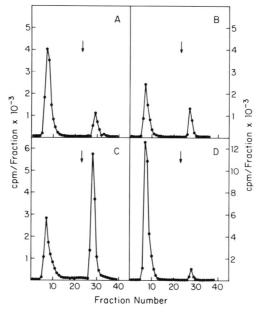

FIG. 1. Binding of ecdysteroid receptor preparations from K_c cells to DNA cellulose. Following incubation with [^{3}H]ponasterone A, samples were desalted to remove free ligand and layered onto 4.8 × 1-cm columns containing 1.0 g of DNA cellulose at 4°. The samples were allowed to percolate slowly through the column packing, and were initially washed with 3 bed volumes of TE buffer containing 50 mM KCl (fractions 1–22). The columns were then washed with 2 bed volumes of 0.5 M KCl in TE buffer. (Arrows indicate application of this 0.5 M KCl wash to the columns.) (A) Elution profile of 4.2 S cytoplasmic receptor preparation. (B) Elution profile of 6.3 S cytoplasmic receptor preparation. (C) Elution profile of 6.3 S nuclear receptor preparation. (D) Elution profile of a nuclear receptor preparation from a receptor-deficient variant K_c cell line.

TABLE III
KINETIC PROPERTIES OF CYTOSOL AND
NUCLEAR ECDYSTEROID RECEPTORS[a]

Binding parameter	K_c Cytosol	K_c nuclear extract
Saturation level [³H]ponasterone A	$1.2 \times 10^{-8}\ M$	$1.8 \times 10^{-8}\ M$
Binding sites/cell	1200	800
Equilibrium dissociation constants [K_d]		
Ponasterone A	$3.4 \times 10^{-9}\ M$	$4.2 \times 10^{-9}\ M$
20-OH-ecdysone	$2.4 \times 10^{-7}\ M$	$2.0 \times 10^{-7}\ M$
Ecdysone	$3.6 \times 10^{-5}\ M$	$2.7 \times 10^{-5}\ M$

[a] Cytosol receptor preparations were made from frozen cell populations. Scatchard [G. Scatchard, *Ann. N.Y. Acad. Sci.* **51,** 660 (1949)] analyses of the saturation kinetics of [³H]ponasterone A binding to ecdysteroid receptor preparations were used to derive the values for binding sites/cell and the K_ds of ponasterone A. The equilibrium dissociation constants for 20-OH-ecdysone and ecdysone were calculated from the ecdysteroid inhibition of [³H]ponasterone A receptor binding.

other biological responses.[24–26] Moreover, the high degree of similarity between the binding properties of the 4.2 S cytosolic receptor and the 6.3 S nuclear receptor is noteworthy, since it would seem to indicate a similarity in the hormone binding domains of these two forms of the receptor. These data, in conjunction with the similarity in hydrodynamic properties noted above, seem to provide strong evidence for a single molecular ecdysteroid binding moiety present in both cytosol and nuclear preparations of K_c cells, even though the cytosol form of the ecdysteroid receptor easily undergoes a size reduction or shape change if the cells are stored for any significant period of time at $-20°$.

Both nuclear and cytoplasmic receptor preparations have been examined for their ability to bind to DNA cellulose prepared by the method of Alberts and Herrick.[27] The results of these studies are illustrated in Fig. 1. Clearly the 6.3 S receptor from both nuclear and cytoplasmic preparations binds very effectively to DNA (Fig. 1B and C). In contrast, the 4.2 S

[24] A.-M. Courgeon, *Exp. Cell Res.* **74,** 327 (1972).
[25] J. W. Fristrom and M. A. Yund, in "Invertebrate Tissue Culture" (K. Maramorosch, ed.), p. 161. Academic Press, New York, 1976.
[26] L. Cherbas, C. D. Yonge, P. Cherbas, and C. M. Williams, *Wilhelm Roux's Arch. Dev. Biol.* **189,** 1 (1980).
[27] B. Alberts and G. Herrick, this series, Vol. 21, p. 198.

cytoplasmic receptor from frozen cells has a markedly lower affinity for the DNA cellulose (Fig. 1A). Thus, it would appear that either a change in shape or a reduction in size caused by freezing and characterized by the 6 S → 4 S transition masks or eliminates the receptor's DNA binding domain. The somewhat reduced binding efficiency of the 6 S cytosol receptor–hormone complex in comparison to the 6 S nuclear complex may be an artifact of the preparation.

Acknowledgment

The work was supported by a grant from the National Science Foundation.

Section III

Metabolism of Other Isoprenoids

[30] Dolichol Kinase, Phosphatase, and Esterase Activity in Calf Brain

By CARLOTA SUMBILLA and CHARLES J. WAECHTER

Dolichol Kinase

$$\text{Dolichol} + \text{CTP} \rightarrow \text{Dolichyl phosphate} + \text{CDP}$$

Particulate preparations from a variety of animal tissues have been shown to catalyze the CTP-dependent phosphorylation of dolichol.[1-4] Enzymatic studies with calf brain microsomes have demonstrated that the endogenous dolichyl [^{32}P]monophosphate, labeled by incubation with [γ-^{32}P]CTP, can be utilized for the biosynthesis of mannosylphosphoryldolichol,[2] glucosylphosphoryldolichol,[5] and N-acetylglucosaminylpyrophosphoryldolichol[2] (Fig. 1).

Assay Method

Principle. The properties of dolichol kinase activity in calf brain microsomes have been studied by following the enzymatic transfer of [^{32}P]phosphate from [γ-^{32}P]CTP to endogenous or exogenous dolichol.[2] Dolichol kinase activity has also been assayed in other systems by measuring the conversion of [^{3}H]dolichol to [^{3}H]dolichyl monophosphate chromatographically.[1,3,6,7]

Reagents

0.1 M Tris–HCl (pH 7.2)–0.25 M sucrose–1 mM EDTA–10 mM mercaptoethanol
0.1 M CaCl$_2$
500 mM UTP

---

[1] C. M. Allen, Jr., J. R. Kalin, J. Sack, and D. Verizzo, *Biochemistry* **17**, 5020 (1978).
[2] W. A. Burton, M. G. Scher, and C. J. Waechter, *J. Biol. Chem.* **254**, 7129 (1979).
[3] J. W. Rip and K. K. Carroll, *Can. J. Biochem.* **58**, 1051 (1980).
[4] W. A. Burton, J. J. Lucas, and C. J. Waechter, *J. Biol. Chem.* **256**, 632 (1981).
[5] M. G. Scher, W. A. Burton, and C. J. Waechter, *J. Neurochem.* **35**, 844 (1980).
[6] R. K. Keller, G. D. Rottler, N. Cafmeyer, and W. L. Adair, Jr., *Biochim. Biophys. Acta* **719**, 118 (1982).
[7] C. Gandhi and R. W. Keenan, *J. Biol. Chem.* **258**, 7639 (1983).

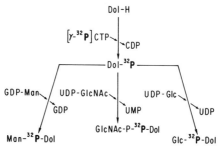

FIG. 1. Utilization of dolichyl phosphate formed by dolichol kinase for the synthesis of glycolipid intermediates.

400 μM [γ-^{32}P]CTP (400–600 cpm/pmol)
Dolichol (1 mg/ml) dispersed in 0.5% Nonidet P-40
Enzyme source

Preparation of Crude Microsomal Fraction from Calf Brain. Dolichol kinase, phosphatase, and esterase activity can be studied with a crude microsomal fraction from calf brain. In typical preparations two calf brains (approximately 650 g) are minced and suspended in an equal volume of 0.1 M Tris–HCl (pH 7.2)–0.25 M sucrose–1 mM EDTA containing 10 mM 2-mercaptoethanol (buffer A). All steps in the procedure are carried out at 0–4°. The tissue is homogenized by 13 strokes with pestle A (large clearance) in a 40 ml Dounce homogenizer (Kontes, Vineland, NJ). Two more volumes of buffer A are added and the homogenate is centrifuged at 600 g for 15 min. The supernate is saved and the pellet is resuspended in an equal volume of buffer A. The suspension is further homogenized by 6 strokes with pestle B in a Dounce homogenizer. Additional buffer is added and the homogenate is centrifuged again at 600 g for 15 min. The rehomogenization procedure is repeated and the low-speed supernates are pooled and centrifuged at 9000 g for 15 min. The pellets are discarded and the supernatant fluid is filtered through three layers of cheesecloth. The crude microsomal fraction is sedimented by centrifugation of the filtrate at 39,000 g for 15 min. The membranous pellet is resuspended in buffer A at a concentration of 20–40 mg protein/ml and used as enzyme. Contaminating myelin is removed by flotation on 0.1 M Tris–HCl (pH 7.2)–0.9 M sucrose–1 mM EDTA by centrifugation at 45,000 g for 1 hr in a Beckman SW 25.1 rotor.

Procedure. Typical reaction mixtures contain brain membranes (1 mg protein), 50 mM Tris–HCl (pH 7.2), 0.125 M sucrose, 0.5 mM EDTA, 20 mM UTP, 30 mM CaCl$_2$, 5 mM mercaptoethanol and 40 μM [γ-^{32}P]CTP (400 cpm/pmol) in a total volume of 0.1 ml. *In vitro* reactions are terminated by the addition of 20 vol of $CHCl_3$–CH_3OH (2 : 1). The membrane residue is sedimented by centrifugation and the lipid extract is saved. The membrane residue is washed twice with 1 ml of $CHCl_3$–CH_3OH (2 : 1) and the lipid extracts are pooled. Unreacted [γ-^{32}P]CTP and other ^{32}P-labeled water-soluble compounds are removed by washing the lipid extract with one-fifth volume of 0.9% NaCl–10 mM EDTA. The mixture is placed in ice for 5 min and centrifuged to produce two phases. The upper (aqueous) phase is discarded and the lower phase is washed three times with CH_3OH–0.9% NaCl (1 : 1) containing 5 mM EDTA. After removal of the upper phases the lower phase, containing the enzymatically synthesized dolichyl [^{32}P]monophosphate, is transferred to a scintillation vial. The organic solvent is evaporated by a stream of air, and the amount of radiolabeled dolichyl monophosphate formed is measured in a scintillation spectrometer after the addition of 10 ml of Liquiscint (National Diagnostics).

Properties of Membrane-Bound Dolichol Kinase

Inhibitors and Activators. The rapid hydrolysis of the radiolabeled nucleotide substrate presents a major technical difficulty when assaying dolichol kinase activity in crude microsomal preparations. This problem can be overcome by the addition of 20 mM UTP to the incubation mixture. Under these conditions [γ-^{32}P]CTP is effectively protected from enzymatic hydrolysis. The brain kinase activity is reduced by 1 mM CDP, presumably by end-product inhibition. CMP, ADP, GDP, or UDP have virtually no effect. The hen oviduct enzyme is inhibited similarly by CDP or dCDP.[4] The brain enzyme requires a divalent cation for maximal activity with Ca^{2+} (20–30 mM) providing a greater stimulation than Mn^{2+} or Mg^{2+}. Related to the Ca^{2+} requirement, activity can be restored to EGTA-washed microsomes from rat brain by the addition of calmodulin.[7] Calf brain dolichol kinase is also enhanced by the addition of exogenous dolichol presented as an aqueous dispersion in 0.07% Nonidet P-40. The latter observation indicates that the polyprenol substrate is present in subsaturating levels in these preparations.

pH Effect. Dolichol kinase activity is optimal at neutral pH.

Specificity. CTP was the most effective phosphoryl donor of three nucleotide substrates tested (Table I). An apparent K_m value = 6.5 μM has been calculated for CTP in the brain microsomal system. A similar K_m was found for dCTP. The observation that UTP is not inhibitory when present at a concentration in a 500-fold excess of [γ-^{32}P]CTP indicates that the 4-amino group of the pyrimidine moiety is important in the recognition of the phosphoryl donor. A CTP dependence has also been found in other

TABLE I
NUCLEOTIDE SPECIFICITY OF CALF BRAIN
DOLICHOL KINASE[a]

Labeled substrate (40 μM)	Dolichyl [^{32}P]phosphate formed (pmol/mg protein)
[γ-^{32}P]CTP	6.10
[γ-^{32}P]ATP	0.10
[γ-^{32}P]GTP	0.06

[a] W. A. Burton, M. G. Scher, and C. J. Waechter, *J. Biol. Chem.* **254**, 7129 (1979).

systems.[1,3,4] γ-Labeled CTP and dCTP both served as phosphoryl donors for the hen oviduct kinase.[4] The oviduct enzyme exhibited a slight preference for CTP, but both cytidine nucleotides might act as substrates *in vivo*. A comparison of various polyprenol substrates showed that the liver kinase preferred the 11–19 isoprenologs with saturated α-isoprene units.[6]

Solubilization. Active preparations of the calf brain kinase are not readily obtained by solubilization with Triton X-100 or sodium deoxycholate. However, a substantial fraction of the kinase activity can be solubilized by extracting salt-washed microsomes with the zwitterionic detergent, CHAPS (Fig. 2). Prior to treatment with detergent, brain microsomes are washed with 1 M NaCl to remove loosely bound extrinsic membrane proteins, and adsorbed nucleic acids and proteoglycans. All

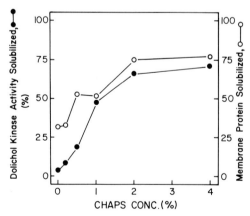

FIG. 2. Solubilization of dolichol kinase activity by extraction of salt-washed microsomes with CHAPS. CHAPS, 3-[(3-Cholamidopropyl)dimethylammonio]-1-propane sulfonate.

steps in the procedure are conducted at 0–4°. Microsomes (360 mg protein) are suspended in 40 mM Tris–HCl (pH 7.2), 0.1 M sucrose, 0.4 mM EDTA, 4 mM mercaptoethanol, and 1 M NaCl in a total volume of 140 ml. The membranes are suspended by 13 strokes of pestle A in a 40 ml Dounce homogenizer (Kontes) and placed in ice for 60 min. The washed membranes are sedimented by centrifugation at 100,000 g for 60 min. To solubilize dolichol kinase activity, the washed microsomes (4–5 mg protein/ml) are resuspended in 45 mM Tris–HCl (pH 7.2), 0.112 M sucrose, 0.45 mM EDTA, 4.5 mM mercaptoethanol, 1 M NaCl, and 10% glycerol containing 2% CHAPS by 16 strokes with pestle A in a 7 ml Dounce homogenizer. The detergent extract is placed on ice for 60 min. Approximately 65% of the dolichol kinase activity is recovered in the supernatant fluid after centrifugation at 100,000 g for 60 min. Dolichol kinase activity has also been solubilized by treating bovine liver microsomes with 0.5% sodium deoxycholate or 0.5% Triton X-100 containing 0.5 M NH$_4$Cl.[8]

Properties of CHAPS-Solubilized Dolichol Kinase. The properties of the soluble enzyme closely resemble those of particulate dolichol kinase. The soluble kinase utilizes either CTP or dCTP as the phosphoryl donor. The apparent K_m values for CTP and dCTP are 7.7 and 9.1 μM, respectively. CDP inhibits dolichol phosphorylation with an apparent K_i value of 121 μM. Kinetic analysis indicates that the inhibitory effect is competitive. Kinase activity is not affected by CMP, UMP, GMP, or AMP. Maximal activity is seen at neutral pH in the presence of 10 mM CaCl$_2$. The solubilized kinase phosphorylates a variety of exogenous polyprenols, including S-dolichol, R-dolichol, and undecaprenol (C$_{55}$). The brain enzyme exhibits a slight preference for long-chain (C$_{90-95}$) polyprenols with saturated α-isoprene units of the S-configuration.

Subcellular Distribution. Subellular fractionation of brain tissue indicates that the major site of dolichol kinase is the rough endoplasmic reticulum (Table II).[9] The specific activity in the heavy microsomal fraction was 15 times higher than in the light microsomal fraction. Dolichol kinase activity has also been found predominantly in microsomal fractions from liver.[3,6,10]

Possible Functions of Dolichol Kinase. The exact function(s) of dolichol kinase in animal tissues is still a subject for conjecture. The kinase may participate in a phosphorylation–dephosphorylation process that regulates the amount of dolichyl phosphate available for the synthesis of

---

[8] C. M. Allen, J. D. Muth, and N. Gildersleeve, *Biochim. Biophys. Acta* **712**, 33 (1982).

[9] M. G. Scher, G. H. DeVries, and C. J. Waechter, *Arch. Biochem. Biophys.* **231**, 293 (1984).

[10] T. Ekström, T. Chojnacki, and G. Dallner, *J. Lipid Res.* **23**, 972 (1982).

TABLE II
Distribution of Dolichol Kinase and Dolichyl Phosphate Phosphatase in Calf Brain Subcellular Fractions[a]

Subcellular fraction[b]	Dolichol kinase (pmol/mg/min)	Dolichyl phosphate phosphatase (nmol/mg/min)
Homogenate	0.08	1.0
Light microsomes	0.20(9)[c]	2.6(71)
Heavy microsomes	3.0(79)	1.0(16)
Mitochondria lysosomes	0.43(8)	0.3(3)
Synaptic plasma membrane	0.19(2)	1.8(9)
Myelin	0.15(2)	0.16(1)

[a] Data from M. G. Scher et al.[9]
[b] B. L. Roth, M. B. Laskowski, and C. J. Coscia, *J. Biol. Chem.* **256**, 10117 (1981).
[c] Data in parentheses represent percentages of total recovered activity.

glycolipid intermediates. Inductions of dolichol kinase activity have been found that correspond to increases in glycosylation activity in estrogen-treated oviducts[4] and developing sea urchin embryos.[11] If dolichyl pyrophosphate were sequentially dephosphorylated by the pyro- and monophosphatase after the transfer of the glucosylated oligosaccharide on the luminal side of the ER, the kinase could rephosphorylate the polyprenol after the hydroxyl terminus diffused back to the cytoplasmic surface. Studies with liver microsomes indicate that the active site of the kinase is asymmetrically located on the cytoplasmic side of the ER.[10,12] In addition to resphosphorylating various reserve pools of dolichol, the kinase could catalyze the terminal step in the *de novo* pathway for dolichyl phosphate biosynthesis if 2,3-didehydrodolichol were the substrate for the reductase acting on the α-isoprene unit.

Dolichyl Phosphate Phosphatase

$$\text{Dolichyl Monophosphate} + H_2O \rightarrow \text{Dolichol} + P_i$$

The enzymatic dephosphorylation of exogenous dolichyl monophosphate has been demonstrated in several eukaryotic systems.[13–18] Calf

[11] D. P. Rossignol, W. J. Lennarz, and C. J. Waechter, *J. Biol. Chem.* **256**, 10538 (1981).
[12] W. J. Adair, Jr. and N. Cafmeyer, *Biochim. Biophys. Acta* **751**, 21 (1983).
[13] G. S. Adrian and R. W. Keenan, *Biochim. Biophys. Acta* **575**, 431 (1979).
[14] J. F. Wedgwood and J. L. Strominger, *J. Biol. Chem.* **255**, 1120 (1980).

brain membranes have been shown to catalyze the dephosphorylation of endogenous and exogenous dolichyl phosphate.[16]

Assay Method

Principle. Dolichyl phosphate phosphatase activity in calf brain membranes can be conveniently assayed by monitoring the release of [^{32}P]phosphate from exogenous dolichyl [^{32}P]phosphate added as an aqueous dispersion in a variety of detergents. The polyisoprenyl phosphatase activity has also been assayed chromatographically by following the conversion of [^{3}H]dolichyl phosphate to [^{3}H]dolichol.[13,15,17]

Reagents

0.1 M Tris–HCl (pH 7)–0.25 M sucrose–1 mM EDTA–10 mM mercaptoethanol
0.4 mM dolichyl [^{32}P]monophosphate dispersed in 2% Triton X-100
Enzyme source

Preparation of Dolichyl [^{32}P]Phosphate. Dolichyl [^{32}P]phosphate is enzymatically labeled by incubating calf brain microsomes (0.65 mg protein), 50 mM Tris–HCl (pH 7.1), 0.125 M sucrose, 0.5 mM EDTA, 20 mM UTP, 5 mM mercaptoethanol, 30 mM CaCl$_2$, and 41 μM [γ-^{32}P]CTP (21.9 Ci/m mol) in a total volume of 0.05 ml at 37° for 40 min. UTP is included in the reaction mixture to prevent the rapid enzymatic hydrolysis of [γ-^{32}P]CTP. The reaction mixture is stopped by the addition of 20 vol of CHCl$_3$–CH$_3$OH (2:1). The mixtures are then sedimented by centrifugation and the lipid extract is saved. The membrane residue is washed again with 1 ml of CHCl$_3$–CH$_3$OH (2:1). The lipid extracts are pooled and washed with one fifth volume of 0.9% NaCl–10 mM EDTA to remove unreacted [γ-^{32}P]CTP and other water-soluble ^{32}P-labeled compounds. The lower phase is washed two more times with CHCl$_3$–CH$_3$OH–H$_2$O (3:48:47) containing 5 mM EDTA and the lower phase is dried under N$_2$. Contaminating glycerophosphatides and neutral glycerides are removed by mild alkaline methanolysis (4 ml of 0.1 N KOH in CH$_3$OH–toluene (3:1), 60 min, 0°). The methanolysate is neutralized by the addition of 0.4 ml of 1 N acetic acid and a two-phase system is produced by the addition of 5 ml of CHCl$_3$ and 1.4 ml of H$_2$O. Dolichyl [^{32}P]phosphate is recovered

---

[15] V. Idoyaga-Vargas, E. Belocopitow, A. Mentaberry, and H. Carminatti, *FEBS Lett.* **112**, 63 (1980).
[16] W. A. Burton, M. G. Scher, and C. J. Waechter, *Arch. Biochem. Biophys.* **208**, 409 (1981).
[17] J. W. Rip, A. Rupar, N. Chaudhary, and K. K. Carroll, *J. Biol. Chem.* **256**, 1929 (1981).
[18] D. P. Rossignol, M. Scher, C. J. Waechter, and W. J. Lennarz, *J. Biol. Chem.* **258**, 9122 (1983).

in the lower phase. The lower phase is dried and the ^{32}P-labeled lipid is redissolved in a small volume of $CHCl_3-CH_3OH-H_2O$ (10:10:3). The lipid is then applied to a DEAE-cellulose column (acetate form, 1.45 × 5.0 cm) equilibrated with $CHCl_3-CH_3OH-H_2O$ (10:10:3). Dolichyl [^{32}P]phosphate remains bound during elution of fatty acid methyl esters with the equilibration solvent and the same mixture containing 2 mM ammonium acetate. The ^{32}P-labeled phospholipid is then eluted with the equilibration solvent containing 20 mM ammonium acetate. Fractions containing dolichyl [^{32}P]phosphate are pooled and $CHCl_3$ and H_2O are added to achieve final proportions of $CHCl_3-CH_3OH-H_2O$ (2:1:0.6). The mixture is cooled and centrifuged to produce two phases. Dolichyl [^{32}P]phosphate is recovered in the lower phase. Dolichyl [^{32}P]phosphate synthesized by chemical procedures can also be used as substrate for phosphatase assays.[14,19]

Procedure. Standard reaction mixtures consist of brain membranes (1 mg protein), 50 mM Tris–HCl (pH 7.1), 0.125 M sucrose, 0.5 mM EDTA, and 0.2 mM dolichyl [^{32}P]phosphate (100 cpm/nmol) dispersed in Triton X-100 (final concentration = 1%) in a total volume of 0.1 ml. Enzymatic reactions at 37° are stopped by the addition of 24 vol of $CHCl_3-CH_3OH-0.9\%$ NaCl (2:1:0.6). The extraction mixture is placed in ice for 5 min and then centrifuged to produce two phases. The upper phase containing the enzymatically released [^{32}P]phosphate is transferred to a scintillation vial. The lower phase and the membrane residue at the interface are washed with approximately 1 ml of $CHCl_3-CH_3OH-0.9\%$ NaCl (3:48:47). The pooled upper phases are dried under a stream of air. The amount of [^{32}P]phosphate released from dolichyl [^{32}P]phosphate is measured in a scintillation spectrometer after the labeled product is redissolved in 1 ml of 1% SDS and 10 ml of Liquiscint (National Diagnostics).

Properties of Dolichyl Phosphate Phosphatase

Inhibitors and Activators. Dolichyl monophosphate phosphatase activity is inhibited by 10 mM Mn^{2+} (57%), F$^-$ (99%), or phosphate ions (75%). EDTA is slightly stimulatory with some preparations.

Specificity. While the precise specificity and number of the calf brain phosphatases have not been defined conclusively, the polyisoprenyl monophosphatase is considerably more sensitive to inhibition by fluoride and phosphate ions than phosphatidate phosphatase activity associated with the same membrane preparations.[16] The phosphatase hydrolyzing dolichyl monophosphate is also substantially more stable to heat inactivation at 60° than the general phosphatase activity hydrolyzing

---

[19] R. W. Keenan, R. A. Martinez, and R. F. Williams, *J. Biol. Chem.* **257**, 14817 (1982).

p-nitrophenyl phosphate. Glucose 6-phosphate, pyridoxal phosphate, NADP$^+$, AMP, methylene diphosphonate, and p-nitrophenyl phosphate had virtually no inhibitory effect on the dephosphorylation of dolichyl [^{32}P]monophosphate when added at equimolar concentrations. Consistent with this enzyme being a polyisoprenyl monophosphatase, it was inhibited 43% by retinyl monophosphate. The apparent K_m value for dolichyl monophosphate is 200 μM.

pH Effect. The calf brain phosphatase activity is optimally active at neutral pH (6.5–7.5).

Solubilization. After crude membrane preparations are initially washed with 1 M NaCl, approximately 30% of the dolichyl monophosphate activity can be solubilized by treatment with 1% Triton X-100.[16] Crude microsomes are first washed with 1 M NaCl as described for dolichol kinase. The salt-washed microsomes (70 mg protein) are suspended in 50 mM Tris–HCl (pH 7.1), 0.125 M sucrose, 0.5 mM EDTA, and 1% Triton X-100 in a total volume of 4 ml. The membrane suspension is further homogenized by 12 strokes with pestle B in a 7 ml Dounce homogenizer (Kontes) at 37°. The membrane suspension is then incubated at 37°. After 60 min the solubilized preparation is centrifuged at 100,000 g for 60 min at 4°. Approximately 30–40% of the dolichyl phosphate phosphatase activity is recovered in the supernatant fluid. The properties of the detergent-solubilized activity are essentially identical to the membrane-bound form of the enzyme. Dolichyl phosphate phosphatase activity has also been solubilized by treating rat liver microsomes with Triton X-100.[20]

Subcellular Distribution. The order of specific activities of the phosphatase in the various subcellular fractions is light microsomes > synaptic plasma membranes > heavy microsomes > mitochondria-lysosomes > myelin (Table II). It is currently not known if the dephosphorylation of dolichyl phosphate is catalyzed by a single enzyme in the various subcellular fractions. The highest specific activity is recovered in an axolemma-enriched fraction isolated by a separate procedure (Table III). The highest specific activity of the liver enzyme was also found to be associated with a plasma membrane fraction.[17]

Possible Functions of the Polyisoprenyl Phosphatase Activity. Operating in coordination with the CTP-dependent kinase, the phosphatase could play a regulatory role in controlling dolichyl phosphate levels. The polyisoprenyl phosphatase may also be involved in the *de novo* pathway for the biosynthesis of dolichyl monophosphate. If the free polyprenol were the preferred substrate for the enzyme catalyzing the reduction of the α-isoprenyl unit, the phosphatase could be required to dephosphorylate 2,3-didehydrodolichyl monophosphate.

---

[20] E. Belocopitow and D. Boscoboinik, *Eur. J. Biochem.* **125**, 167 (1982).

TABLE III
ENRICHMENT OF DOLICHYL PHOSPHATE PHOSPHATASE IN AXOLEMMA
PREPARATIONS FROM RAT BRAIN[a]

Enzyme activity	Axolemma-enriched fraction[b]	Mitochondria–lysosomes	Microsomes
Acetylcholinesterase (nmol/mg/min)	39.8	26.3	14.8
Cytochrome c oxidase (μmol/mg/hr)	0.42	23.2	2.9
Dolichol kinase activity (pmol/mg/min)	0.52	0.9	2.4
Dolichyl phosphate phosphatase activity (nmol/mg/10 min)	24.0	10.7	9.4

[a] Data from M. G. Scher et al.[9]
[b] G. H. DeVries and M. J. Lazdunski, *J. Biol. Chem.* **257,** 11684 (1982).

Dolichol Esterase

Dolichyl Oleate + H_2O → Dolichol + Oleic Acid

The presence of fatty acyl esters of dolichol in animal tissues is well documented.[21,22] If the esterified form of the polyprenol is a reserve pool that can be recruited for dolichyl phosphate synthesis, deacylation would occur prior to enzymatic phosphorylation by the CTP-dependent kinase. While the utilization of dolichyl esters for dolichyl phosphate biosynthesis has not been demonstrated, calf brain membranes have been shown to catalyze the enzymatic deacylation of dolichyl oleate.[23] A similar esterase activity recovered in cell-free extracts of intestinal mucosa is capable of hydrolyzing dolichyl palmitate.[24]

Assay Method

Principle. Dolichyl esterase activity has been assayed by incubating dolichyl [^{14}C]oleate with calf brain membrane preparations and following the release of [^{14}C]oleic acid. The appearance of [^{14}C]oleic acid can be

[21] P. H. W. Butterworth and F. W. Hemming, *Arch. Biochem. Biophys.* **128,** 503 (1968).
[22] C. A. Rupar and K. K. Carroll, *Lipids* **13,** 291 (1978).
[23] M. G. Scher and C. J. Waechter, *Biochem. Biophys. Res. Commun.* **99,** 675 (1981).
[24] R. W. Keenan, N. Rice, and G. S. Adrian, *Biochim. Biophys. Acta* **711,** 490 (1982).

measured by adapting the partition method of Khoo et al.[25] used for assay of triglyceride hydrolase activity.

Reagents

0.1 M Tris–HCl (pH 7.1)–0.25 M sucrose–1 mM EDTA
0.1 mM Dolichyl [^{14}C]oleate dispersed in 0.5% Triton X-100
Bovine serum albumin (10 mg/ml)
0.1 M MgCl$_2$
Enzyme source

Chemical Synthesis of Dolichyl [^{14}C]Oleate. The radiolabeled ester is synthesized by reacting [^{14}C]oleoyl chloride with dolichol. The [^{14}C]acyl chloride is formed by a reaction of [^{14}C]oleic acid (32.4 μmol, 4 μCi/μmol) with oxalyl chloride (97.2 μmol) in 0.2 ml of benzene for 2 hr at 23°. The reaction mixture is dried under N$_2$ and 97.2 μmol of oxalyl chloride dissolved in 0.2 ml of benzene is added. After an additional 2 hr at 23°, the mixture is dried under N$_2$ and dolichol (10.8 μmol), dissolved in 1.0 ml of anhydrous benzene, is added with pyridine (86.9 μmol). After 2 hr at room temperature the mixture is dried under N$_2$, and 2 ml of benzene and 1 ml of H$_2$O are added. The radiolabeled dolichyl ester is recovered in the organic phase.

Procedure. Typical reaction mixtures contain calf brain membranes (3–5 mg protein), 50 mM Tris–HCl (pH 7.1), 125 mM sucrose, 0.5 mM EDTA, bovine serum albumin (5 mg/ml), 20 mM MgCl$_2$, 0.03 mM dolichyl [^{14}C]oleate (3570 cpm/nmol) dispersed in Triton X-100 (final concentration = 0.04%) in a total volume of 0.2 ml. Following incubation at 37° for the desired period of time, the reaction is stopped by the addition of 2.5 ml of CHCl$_3$–CH$_3$OH–benzene (2:2.4:1) containing 0.1 mM oleic acid. The lipid extract is agitated with a Vortex mixer and the membrane residue is sedimented by centrifugation. The lipid extract is saved and the residue is washed again with 0.5 ml of the extraction solvent. The extracts are pooled and placed in ice for 5 min. After cooling, 0.3 ml of 0.33 N NaOH is added and two phases are formed by centrifugation. The upper phase, containing the enzymatically released [^{14}C]oleic acid, is removed and the lower phase is washed with 1.6 ml of synthetic upper phase. The pooled upper phases are transferred to scintillation vials, placed in a warm water bath, and dried by a stream of air. The amount of [^{14}C]oleic acid released from the ester in each incubation is measured by scintillation spectrometry after the addition of 10 ml of Liquiscint (National Diagnostics) and 0.5 ml of 1% SDS.

[25] J. C. Khoo, D. Steinberg, J. J. Huang, and P. R. Vagelos, *J. Biol. Chem.* **251**, 2282 (1976).

Properties of the Brain Esterase Activity. The calf brain esterase activity is optimally active at pH 7.5. This pH optimum distinguishes the dolichol esterase activity from cholesterol ester and triolein hydrolase activities associated with the same membrane preparations. The latter hydrolase activities are optimal at pH 5.0–5.5. Dolichol esterase activity is not affected by divalent cations or EDTA. The enzymatic hydrolysis of dolichyl [^{14}C]oleate is inhibited by *p*-hydroxymercuribenzoate or iodoacetamide, suggesting the presence of an essential sulfhydryl group(s). Esterase activity is also inhibited substantially when Triton X-100 or sodium taurocholate is added in excess of 0.06%. Apparent end-product inhibition is observed in the presence of 0.1 mM oleic acid (24%). A slight stimulation is produced by the addition of bovine serum albumin (5 mg/ ml), probably due to binding of fatty acid released by the enzymatic reaction.

[31] Purification and Properties of the Juvenile Hormone Carrier Protein from the Hemolymph of *Manduca sexta*

By RONALD C. PETERSON

Carrier proteins for juvenile hormone have been found in the hemolymph of a number of different insects from several orders. These carrier proteins are distinguished from lipoproteins and other lipid binding proteins by their high affinity and specificity for juvenile hormones. A major role of the carrier protein is the transport of juvenile hormone from the site of synthesis through the hemolymph to the target tissues. During this process the carrier protein protects the hormone from general esterases present in the hemolymph and other tissues. The juvenile hormone carrier protein from *Manduca sexta* (tobacco hornworm) was the first such protein described and is the best characterized member of this class of proteins.

Assay

The assay for juvenile hormone carrier protein from *M. sexta* is based on the observation that the protein binds to DEAE-ion exchange matrixes at low ionic strength.[1,2]

---

[1] K. J. Kramer, P. E. Dunn, R. C. Peterson, H. L. Seballos, L. L. Sanburg, and J. H. Law, *J. Biol. Chem.* **251,** 4979 (1976).

[2] K. J. Kramer, P. E. Dunn, R. C. Peterson, and J. H. Law, *in* "The Juvenile Hormones" (L. I. Gilbert, ed.), p. 327. Plenum, New York, 1976.

Materials

Methyl (2E,6E)-*cis*-10,11-epoxy-3,11-dimethyl-7-ethyl-2,6-tridecadienoate (JH I) (Calbiochem.) in hexane.[3]

[10-^{3}H]JH I, 13.5 Ci/mmol (New England Nuclear) in heptane.

DEAE-cellulose filter discs, DE-81 (Whatman). Discs (2.5 cm) are washed with 0.5 N HCl, distilled water, 0.5 M NaOH, distilled water, and extensively washed and equilibrated with 5 mM Tris–HCl, pH 8.3.

5 mM Tris–HCl, pH 8.3.

5 mM Tris–HCl, pH 8.3, with 1% Triton X-100.

All glassware coming into contact with the juvenile hormone or the carrier protein should be precoated with a 1% (w/v) solution of polyethylene glycol (PEG-20,000), rinsed extensively with deionized water, and dried at 110°.

Solutions of JH. Aqueous solutions of juvenile hormone of the desired specific activity (usually 1 Ci/mmol) are prepared by mixing labeled hormone with the appropriate amount of unlabeled hormone in a PEG-coated test tube, removal of the organic solvent by evaporation at 0° with a stream of N_2, and addition of an appropriate volume of 5 mM Tris–HCl, pH 8.3. The tube is mixed repeatedly on a Vortex mixer over a 1-hr period to aid solubilization and the final concentration is determined by measurement of radioactive JH in the solution.

Procedure

For assay of juvenile hormone binding activity, 10 μl of radioactive JH (10^{-6} M) is mixed with an equal volume of carrier protein solution or chromatography fraction in a PEG-coated glass microcentrifuge tube and the mixture is incubated for 30 min on ice. DEAE-cellulose filters equilibrated with 5 mM Tris–HCl, pH 8.3, are placed on a filter manifold and washed with 1 ml of ice-cold 1% Triton X-100 in 5 mM Tris–HCl, pH 8.3. Assay mixtures are transferred to the washed filter disc using PEG-coated glass capillary tubes. After a 1 min incubation on the filter, the filters are washed with 5 ml 1% Triton X-100 in 5 mM Tris–HCl, pH 8.3, using sufficient vacuum to allow a flow rate of 2.5 ml/min. Excess liquid is removed from the washed filters on the manifold with increased vacuum and the adsorbed radioactivity is measured using a liquid scintillation fluid prepared from 1 liter Triton X-100, 2 liters toluene, and 12 g Omnifluor (New England Nuclear).

---

[3] The purity of JH I should be monitored by gas–liquid chromatography and when necessary, further purified by high-performance liquid chromatography as described by W. Goodman, D. A. Schooley, and L. I. Gilbert, *Proc. Natl. Acad. Sci. U.S.A.* **75**, 185 (1978).

PURIFICATION OF JUVENILE HORMONE CARRIER PROTEIN FROM *Manduca sexta*

	Protein (mg)	Total binding (units)	Specific activity (units/mg)	Purification	Yield (%)
Reconstituted acetone powder[a]	3084	123	0.04		
Sephadex G-100	149	137[b]	0.92	21	100[b]
DEAE BioGel A	18.5	124	6.73	153	91
SP-Sephadex C-50	1.96	50.1	25.6	582	36

[a] Acetone powder of hemolymph from 144 larvae.
[b] Full recovery of binding activity after gel filtration was assumed. The assay of crude hemolymph or reconstituted acetone powder is complicated by the presence of lipoproteins and esterases that degrade the hormone.

One binding unit is defined as the amount of carrier protein that binds 1 nmol of JH when equilibrated with 1×10^{-6} JH I and assayed using the DEAE-filter method.

The DEAE-filter binding assay was designed to fit the properties of the carrier protein from *M. sexta*. If the carrier protein from another species is being investigated, it may be necessary to modify the assay conditions to ensure that the protein–hormone complex is quantitatively retained on the filter. Also, large variations in the pH and ionic strength of the buffer used in the assay mixture may affect the quantitative retention of the complex on the filter.

Purification

The carrier protein can be purified to homogeniety from the hemolymph of *M. sexta* larvae by three chromatographic steps.[4] Other procedures for purification of this protein have been described but these procedures result in a lower yield of protein.[1,5] Results of this purification scheme are shown in the table. All chromatography and ultrafiltration steps were done at 4°.

Animals. Eggs of *M. sexta* were generously provided by Dr. J. P. Reinecke, United States Department of Agriculture, Fargo, ND. Colonies are maintained as described by Kramer *et al.*[6]

[4] R. C. Peterson, P. E. Dunn, H. L. Seballos, B. K. Barbeau, P. S. Keim, C. T. Riley, R. L. Heinrikson, and J. H. Law, *Insect Biochem.* **12,** 643 (1982).
[5] W. Goodman, P. A. O'Hern, R. H. Zuagg, and L. I. Gilbert, *Mol. Cell Endocrinol.* **11,** 225 (1978).
[6] K. J. Kramer, L. L. Sanburg, F. J. Kézdy, and J. H. Law, *Proc. Natl. Acad. Sci U.S.A.* **71,** 493 (1974).

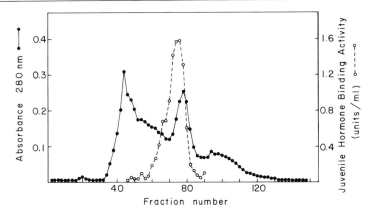

FIG. 1. Chromatography of juvenile hormone carrier protein on DEAE-Bio Gel A in 5 mM Tris–HCl, pH 8.0, 32 mM NaCl. Concentrated fractions from three Sephadex G-100 columns were applied to this column (1.5 × 96 cm, 169 ml). Fractions (4.0 ml) were analyzed for protein (A_{280}, ●) and JH binding activity ○. [Adapted from R. C. Peterson, P. E. Dunn, H. L. Seballos, B. K. Barbeau, P. S. Keim, C. T. Riley, R. L. Heinrikson, and J. H. Law, *Insect Biochem.* **12**, 643 (1982)].

Preparation of Hemolymph Acetone Powder. Larvae (fifth-instar larger than 6.5 g) are cooled in ice and hemolymph is obtained by cutting the first proleg with a fine scissors. The hemolymph (approximately 0.75 ml per larva) is collected into a 20-fold excess of cold (0°) acetone with constant stirring. The resulting suspension is recovered by gentle suction filtration and washed extensively with acetone. The resulting fine, light blue powder is stored desiccated at 4°.

Gel Filtration. Acetone powder from 40 to 50 larvae is reconstituted in 11 ml of cold 20 mM Tris–HCl, pH 8.3, 100 mM NaCl, 0.2 mM phenylthiourea, 0.02% NaN$_3$, and centrifuged at 3000 g for 10 min. The supernatant is applied to a column of Sephadex G-100 (Pharmacia) (2.0 × 150 cm) equilibrated with the same buffer and fractions (6.5 ml) are collected. The fractions containing JH carrier protein are pooled and frozen at $-20°$.

Ion Exchange Chromatography on DEAE-BioGel A (Bio-Rad Laboratories). The pooled fractions from several gel filtration columns (equivalent to 120–150 larvae) are concentrated and equilibrated with 5 mM Tris–HCl, pH 8.0, 32 mM NaCl by ultrafiltration using a UM-2 membrane (Amicon). The concentrate is applied to a column of DEAE-BioGel A (1.5 × 96 cm) equilibrated with 5 mM Tris–HCl, pH 8.0, 32 mM NaCl, and fractions (4 ml) of effluent are collected (Fig. 1).

Under these buffer and salt conditions the carrier protein is retarded on the DEAE-BioGel A column but does not bind tightly. This represents the best compromise between adequate recovery and resolution. The exact conditions for optimal resolution may vary with different lots of DEAE-BioGel A and it may be necessary to alter the salt concentration slightly for different lots of this material. We find it best to buy a large quantity of one lot and to clean and reuse it.

Ion Exchange Chromatography on SP—Sephadex (Pharmacia). The fractions from DEAE-BioGel A chromatography that contained juvenile hormone carrier protein are concentrated by ultrafiltration (as above) and equilibrated with 10 mM sodium acetate, 10 mM NaCl, pH 5.0. This concentrate is applied to a column of sulfopropyl-Sephadex C-50 (0.9 × 24 cm) equilibrated with 10 mM sodium acetate, pH 5.0, 10 mM NaCl, 5% (w/v) glycerol. The column is washed with three column volumes of buffer and a shallow linear gradient from 10 to 25 mM NaCl in 10 mM sodium acetate, pH 5.0, 5% glycerol is applied (total volume of gradient 50 ml). Following the gradient the column is washed with 25 mM NaCl, 10 mM sodium acetate, pH 5.0, 5% glycerol. The active fractions are pooled and the buffer exchanged with 5 mM Tris-HCl, pH 8.3 by ultrafiltration as above. This step is necessary because the protein loses activity on standing in acidic medium.

In order to achieve good yields from the SP-Sephadex, the column should be loaded to capacity. We judge this by an overload of about 10% of the activity eluting in the void volume. This material in the void volume can be added to a subsequent run to improve the overall yield. If the column is not loaded to capacity, much lower yields are obtained.

Comments. The hemolymph may also be collected directly into stirred buffer (20 mM Tris-HCl, 100 mM NaCl, 0.02% NaN$_3$, pH 8.3, saturated with phenylthiourea) using 0.4 ml buffer for each larva. The phenylthiourea is added to inhibit phenol oxidase activity in the hemolymph. The hemolymph is stored frozen at $-20°$. This preparation is thawed and centrifuged at 30,000 g for 20 min, and the supernatant applied to a Sephadex G-100 column.

Properties

Molecular Weight. The juvenile hormone carrier protein from the hemolymph of *M. sexta* has a molecular weight of 28,000 as determined by SDS gel electrophoresis and gel filtration chromatography.[1] The isoelectric point is 4.95.[1] The amino acid composition[1] and the sequence at the amino terminus[4] are known. The purified protein shows an absorbance of 1.02 mg^{-1} ml^{-1} at 278 nm and 5.01 mg^{-1} ml^{-1} at 230 nm.

Binding Site. The carrier protein binds one juvenile hormone molecule per protein monomer with a dissociation constant of 4×10^{-7} M. The binding is specific for juvenile hormone and requires both the 10,11 epoxide and the ester functions on the hormone for strong interaction.[2,7] The geometry of the double bonds and the stereochemistry of the epoxide function are also important determinants of specific binding.[7,8] The carrier protein–hormone complex is stable and can be subjected to gel filtration chromatography and gel electrophoresis under native conditions.[2,4,6]

Stability. The carrier protein is not stable under prolonged storage at −20 or 4° for periods longer than 1 month. The most satisfactory procedure is storage at 4° in 5 mM Tris–HCl, pH 8.3.

Acknowledgments

This work was done in the laboratory of Dr. John H. Law and was supported by grants from the National Science Foundation and the National Institute of General Medical Sciences.

[7] R. C. Peterson, M. R. Reich, P. E. Dunn, J. H. Law, and J. A. Katzenellenbogen, *Biochemistry* **16**, 2305 (1977).
[8] D. A. Schooley, B. J. Bergot, W. Goodman, and L. I. Gilbert, *Biochem. Biophys. Res. Commun.* **81**, 743 (1978).

[32] Analysis of Juvenile Hormone Esterase Activity

By BRUCE D. HAMMOCK and RICHARD M. ROE

Background

The terpenoid juvenile hormones (JHs) of insects control a variety of functions including metamorphosis and reproduction, and insects appear unique in having such terpenes as hormones.[1] The hormones identified to date are homologs and/or isomers of the 10,11-epoxide of methyl farnesoate. Thus, there are many possible sites of metabolic attack with hydration of the 10,11-epoxide and hydrolysis of the conjugated methyl ester representing hydrolytic pathways.[2] In several insect species it appears that degradation in addition to biosynthesis serve to regulate hormone titer.[3,4] For this and other reasons it is important to monitor metabolism of

[1] C. A. D. de Kort and N. A. Granger, *Annu. Rev. Entomol.* **26**, 1 (1981).
[2] M. Slade and C. H. Zibitt, in "Insect Juvenile Hormones: Chemistry and Action" (J. J. Menn and M. Beroza, eds.), p. 155. Academic Press, New York, 1972.
[3] B. D. Hammock and G. B. Quistad, *Prog. Pestic. Biochem.* **1**, 1 (1981).
[4] B. D. Hammock, in "Comprehensive Insect Physiology, Biochemistry and Pharmacology" (G. A. Kerkut and L. I. Gilbert, eds.), Vol. 7, Chapter 13. Pergamon, Oxford, 1984 (in press).

the JHs. This chapter will emphasize methods for monitoring the ester cleavage of the juvenile hormones.

Three of the known juvenile hormones are commercially available. Both the labeled and unlabeled materials have a high degree of isomeric purity *(E, E, cis)*, but are racemic at C-10 or C-10 and C-11. The radiolabeled materials have a tritium at C-10 introduced by reductive tritiation of a haloketone followed by separation of the erythro and threo compounds and base catalyzed cyclization to the epoxide. There are several published methods leading to radiolabeled hormones,[3-5] and of note are biosynthetic methods yielding optically active hormones. Assay methods involving determination of radioactive methanol produced from JH methyl esters have been discussed elsewhere.[3,4] A caution is that radioactive methanol is readily lost during the assay procedure, even from plastic scintilliation vials.

Metabolite Identification

In performing the esterase assays, it is important to confirm that ester hydrolysis is the only pathway of importance in the assay used. Confirmation of metabolite identity by spectral means is seldom warranted. Tentative identification can be provided by a chromatographic method (thin-layer or high-performance liquid chromatography, TLC or HPLC) with careful attention to recovery of radioactivity and precise cochromatography with authentic standards. TLC is adequate to separate the major metabolites of JH from most organisms (JH, JH acid, JH diol, JH diol acid, polar conjugates),[2,6] but there has been some debate regarding the relative positions of JH acid and diol on TLC and HPLC. Since the acid can exist in a protonated or anion form, the developing solvent, the type or lot of TLC plate used, and even the method of extraction used can cause JH acid to change its position relative to the diol. Since the epoxide of JH is very stable to acid, a trace of acetic acid added before extraction will improve the extraction efficiency of acidic metabolites and reduce trailing on TLC. There are a variety of procedures to lend support to the structural assignment of JH acid by simple chromatography. If the R_f values of the metabolite and standard are significantly reduced on a plate predeveloped with 5% triethylamine in ether, or on a plate developed in the usual solvent with a trace of ammonium hydroxide added, the metabo-

[5] R. C. Jennings and A. F. Hamnett, *in* "Juvenile Hormone Biochemistry" (G. E. Pratt and G. T. Brooks, eds.), p. 375. Elsevier/North-Holland Biomedical Press, Amsterdam, 1981.
[6] B. D. Hammock, J. Nowock, W. Goodman, V. Stamoudis, and L. I. Gilbert, *Mol. Cell. Endocrinol.* **3**, 167 (1975).

lite is likely to be an acid. On a 10×10-cm TLC plate one can develop the plate in the first dimension with a solvent such as hexane : ethyl acetate (2 : 1), then in the second dimension with the same solvent containing a few drops of ammonium hydroxide after exposing the plate to ammonia vapors. JH acid should show a tight spot of $R_f \sim 0.4$ in the first dimension but a low R_f and possible a trailing spot in the second dimension. JH acid can be reesterified with diazomethane or ethane,[7] and it should fail to react with n-butylboronic acid (a few drops of a 0.1% solution in ethyl acetate added just before spotting), while JH diol does react, and the resulting adduct has a high R_f.

Thus to evaluate the structure of a metabolite suspected to be JH acid, incubate the enzyme system (0.1–1 ml) with the substrate concentration to be used in the routine assay, using ~0.01–1 μCi of ^{3}H per incubation, terminate the reaction by adding sodium chloride to saturation and authentic standards, then extract three times with equal volumes of peroxide-free ethyl ether. Alternatively, ether : ethanol or ethyl acetate can be used to remove more polar materials and a drop of acetic acid can be added if the acid spot trails on TLC. Following drying over anhydrous sodium sulfate, the solvent is removed under a blanket of nitrogen and the residue spotted as a narrow horizontal band 2 cm wide, 1.5 cm from the bottom, and 1.5 cm from the left-hand side of a silica gel TLC (such as Brinkman Silica Gel 60, F_{254}, 250 μm). The plate is then developed in hexane : ethyl acetate (2 : 1). If a large amount of material is present, the solvent may be passed repeatedly through the origin and developed to a distance of ~5 cm, or a more polar solvent such as ethyl acetate or methanol may be used to develop the plate just past the origin. If the presence of lipids causes diffuse spots, the plate then may be developed 20 cm in toluene before using the hexane : ethyl acetate system. After developing in the first dimension the plate is rotated 90° and developed in the second dimension. A polar solvent such as toluene : propanol (4 : 1) is used to bring the original band into a tight spot by developing 5 cm. Then the plate is developed in toluene : propanol (10 : 1) for 20 cm.[6]

If oxidative processes are expected a nonpolar solvent such as hexane : ether (10 : 1) or petroleum ether : ether (5 : 1) can be used to separate JH from the corresponding diepoxide and triene. The radioactive materials can then be qualitatively evaluated by fluorography and the appropriate spots to be scrapped can be identified by placing the developed film under the TLC plate over a light box.[8] For subsequent assays, one-dimensional development normally can be used.

[7] H. M. Fales, T. M. Jaouni, and J. F. Babashak, *Anal. Chem.* **45**, 2302 (1973).
[8] K. Randerath, *Anal. Biochem.* **34**, 188 (1970).

Selection of Analytical Method

Once JH acid is established as the sole or major metabolite under the conditions to be routinely evaluated, many analytical methods can be employed. A convenient, continuous assay using model substrates offers many advantages, but the numerous controls which must be run to ensure that the same esterases which metabolize JH are metabolizing the model substrate dictate that the radioactive, natural JHs offer the most unambiguous methods. One can monitor metabolism by gas–liquid chromatography using either flame ionization detection or electron capture, by utilizing the limited ability of the conjugated ester to capture electrons. HPLC can be employed using the weak chromophore of JH ($\lambda_{max} \cong 217$ nm, $\varepsilon \cong 14770$)[9] or radiochemical detection.[3,4] By far the two most rapid techniques yet developed are the partition and TLC methods described below.

The TLC method offers the advantage of having a lower background, requiring less radioactivity, allowing one to monitor several metabolic reactions on a single plate and to scrape the plates for liquid scintillation counting (LSC) at leisure. With improvements in TLC scanners, this method could replace the partition method in speed and economy, but at the present time it is more laborious and expensive and the partition method is suggested.

The assays suggested were optimized for the esterase activities for *Manduca sexta* and *Trichoplusia ni*. The conditions of assay should be adjusted for the experiment in hand. Partition methods can also be adjusted to optimize conditions for higher speed, lower cost, lower background, etc. usually at the expense of another factor.

The Partition Assay Method

Principle. Esterase activity is assayed by the conversion of radiolabeled JH (^{3}H at C-10) to the JH acid and subsequent partitioning of the substrate and product by simple solvent extraction. Under the conditions of the assay >99% of the radiolabeled JH is extracted into a hydrocarbon phase while >99% of the radiolabeled JH acid remains in a basic aqueous–methanol phase.[10,11]

Enzyme Preparation. Insect whole hemolymph is collected and diluted for assay immediately (before clotting occurs) in 4° sodium phos-

---

[9] W. Goodman, D. A. Schooley, and L. I. Gilbert, *Proc. Natl. Acad. Sci. U.S.A.* **75**, 185 (1978).
[10] B. D. Hammock and T. C. Sparks, *Anal. Biochem.* **82**, 573 (1977).
[11] T. C. Sparks, B. D. Hammock, and L. M. Riddiford, *Insect Biochem.* **13**, 529 (1983).

phate buffer (pH 7.4, ionic strength 0.2 M with 0.01% phenylthiourea to inhibit tyrosinases). Insect plasma is obtained only in those specific cases where clotting is slow enough to allow collection and centrifugation at 1000 g for 5 min to selectively remove only insect hemocytes. The plasma (supernatant) is then immediately diluted in sodium phosphate buffer as above. Serum is obtained by allowing the hemolymph to thoroughly clot (the time varies with insect species and temperature) and then is centrifuged at 1000 g for 5 min. The serum (supernatant) is then immediately diluted for assay. Whole tissue or whole body preparations are obtained by homogenization in sodium phosphate buffer, centrifugation at 20,000 g for 15 min, and filtration of the 4° supernatant through glass wool previously washed with pentane and dried. The filtrate should contain over 90% of the activity found in the crude homogenate.

Substrate Preparation. Labeled and unlabeled JH homologs are available from a number of commercial sources. It is preferable to have a source that provides JH in solution in a sealed glass ampoule as opposed to neat JH in a rubber septum, sealed serum vial which is available from some sources. Unlabeled JH is available from Calbiochem-Behring in nanograde hexane and [10-^{3}H]JH from New England Nuclear at 10–20 Ci/mmol in toluene : hexane (4 : 1). JH III is used as substrate for routine monitoring of JH esterase activity because it is the least expensive and most soluble of the available JHs, but differences in the rates of metabolism of the homologs may occur. The unlabeled JH received is diluted with nanograde hexane to 2.5×10^{-2} M and stored as a stock solution at $-20°$. Labeled JH is also diluted with nanograde hexane to ~80,000 cpm/ μl and stored as a stock solution at $-20°$. A working substrate solution is formulated by transferring 20 μl of stock unlabeled and 80 μl of stock [^{3}H]JH to a 4 ml Teflon-sealed screw-capped vial. The toluene : hexane solvent is removed by gentle heat (~30 to 35°) while blanketing the vial with nitrogen. Extreme care should be taken to not use a strong flow of nitrogen. Once the solvent is evaporated just to dryness, 1 ml of absolute anhydrous ethanol is added to bring the JH to the final concentration of 5×10^{-4} M, with ~6400 cpm/μl. This substrate is stored at $-20°$, but is brought to room temperature before opening.

Assay Procedure

Step 1. Pipet 100 μl of phosphate buffer (blank, B), and 100 μl of each enzyme preparation (EP) into individual 10 × 75-mm disposable culture tubes in triplicate in an ice : water bath. *Caution:* Conical tubes are unacceptable since they reduce mixing during vortexing in step 5.

Step 2. One microliter of substrate solution is added to each tube with a Hamilton repeating dispenser equipped with a 50-μl syringe. The standard deviation of the amount of label added by this method is ±6%. It usually is unnecessary to clean the syringe needle between enzyme preparations. The final substrate concentration is 5×10^{-6} M.

Step 3. Once substrate is added to all tubes, they are immediately transferred to a 30° water bath, shaken vigorously for 30 sec, and incubated for various times (routinely 15 min). *Caution:* Vigorous shaking is essential for rapid temperature equilibration to 30°.

Step 4. After incubation at 30°, all tubes are transferred back to the ice:water bath, shaken vigorously for 30 sec, and then 50 μl of a methanol:water:concentrated ammonium hydroxide solution (10:9:1) is added. When this solution is added, all enzymatic JH esterase activity is irreversibly stopped. There obviously are many variations on this basic procedure. For instance one could initiate the reaction by addition of substrate or terminate the reaction by adding the basic methanol solution to the enzyme.

Step 5. Tubes can now be removed from the ice bath, 250 μl of isooctane or another hydrocarbon solvent added to each tube, and each tube vigorously mixed (Vortex) and centrifuged for 5 min at 1000 g. *Caution:* For an effective partition, it is necessary to vortex the tube several times until an emulsion forms.

Step 6. From each tube, remove 75 μl of the lower methanol:water phase with a Hamilton 100-μl syringe (a small air bubble should be extruded as the syringe passes through the organic layer) and count for 5 min in an aqueous counting solution. Also count by liquid scintillation in triplicate, aqueous counting solution alone (ACS_B) and aqueous counting solution plus 1 μl of working substrate (ACS_S). *Caution:* The assumption is made that there is no quenching due to the additon of 75 μl of the methanol:water phase from the enzyme preparations being assayed. If there is unequal quenching between samples and/or control vials, then controls (B, ACS_B, and ACS_S) must be redesigned to include the same quench factor or one must use dpm rather than cpm for the calculations. One can also determine maximum counts (ACS_S) by adding substrate to enzyme, adding basic methanol, and then counting a sample without prior extraction with isooctane.

Calculations. Percentage metabolism (%M) is

$$\%M = \left[\frac{(\bar{x}\text{cpm}_{EP} - \bar{x}\text{cpm}_B) \times 2}{\bar{x}\text{cpm}_{ACS_S} - \bar{x}\text{cpm}_{ACS_B}}\right] \times 100$$

In these equations $\bar{x}$ refers to the average counts per minute (cpm) indicated by the subscripts defined in the text. Enzyme dilutions and incubation times should be chosen so that data are accumulated in a region where %M is linearly dependent upon incubation time and protein concentration. When lower substrate concentrations are used, the linear relationship between %M and incubation time occurs over a shorter range. It is critical that this relationship be determined at all working substrate concentrations. If graphically the line does not pass through the origin one may correct for this "zero time" hydrolysis (d) by the following relationship.

Corrected percentage metabolism =

$$\left[\left\{\frac{(\bar{x}\text{cpm}_{EP} - (\bar{x}\text{cpm}_B) \times 2}{\bar{x}\text{cpm}_{ACS_S} - \bar{x}\text{cpm}_{ACS_B}}\right\} \times 100\right] - d$$

Activity per unit time per unit volume is

$n M$ JH metabolized/min/ml = (%M/100)(Dilution factor)(0.333)

The correction factor (0.333) is applicable for a final substrate concentration of $5 \times 10^{-6} M$ with a 15 min incubation time.

TLC Assay Method

Principle. Esterase activity is assayed by conversion of radiolabeled JH to the corresponding acid and separation of substrate and product following direct spotting of the reaction mixture on the cellulose prelayer of a TLC plate.

Assay Procedure

Step 4. Follow steps 1–3 as above. To terminate reaction remove 25 μl of the incubation mixture and mix with 25 μl of methanol (or tetrahydrofuran) containing unlabeled standards in a 6 × 50-mm glass tube. This procedure is important because simply spotting the material on the prelayer may not stop the reaction immediately. If spotting the reaction mixture directly stops the reaction, one may go to step 5 (below) using plates to which standards have already been applied. The juvenoid, 1-(4'-ethylphenoxy)-3,7-dimethyl-6,7-epoxy-*trans*-2-octene and its corresponding diol are routinely used in this laboratory as unlabeled standards. Alternatively one can use one of the commercial JH homologs and prepare JH acid by base hydrolysis or esterase action. One must be cautious since base hydrolysis may cause migration of the 2,3 double bond with loss of UV absorbance. A more reasonable approach is to use commerically available, UV dense standards such as benzophenone and diphenyl acetic acid. In this system JH I chromatographs slightly below benzophenone and slightly above phenol while JH I acid runs with diphenyl acetic acid and slightly above 2-chlorobenzoic acid. In one run one can determine the amount of standard to add to the methanol based on step 7 (10 mg/ml is a reasonable amount), and the relative positions of JH, JH acid, and the standards on a TLC plate using the TLC system peculiar to one's laboratory.

Step 5. In most cases it is possible to spot the aqueous methanol directly, but if streaking occurs one can agitate the tube and centrifuge out the protein before spotting. Spot the material in 25-μl aliquots on the cellulose prelayer of Whatman LK5DF silica gel plates (250 μm thickness). The aqueous material must neither contact the silica layer nor extend below the surface of the developing solvent when the TLC plate is placed in the tank.

Step 6. The TLC plate is air dried for at least 20 min and then developed using hexane : ethyl acetate (2 : 1) or toluene : propanol (20 : 1). If trailing occurs one can develop the plate through the cellulose layer using a more polar solvent, air dry, and then develop in the usual solvent.

Step 7. Remove the plate from the developing tank and air dry. Visualize the standard spots using a 254-nm lamp and mark them with a pencil. Spray the plate lightly with water to reduce radioactive dust, effects of static, and chemiluminescence and then scrape the appropriate regions corresponding to JH and JH acid into a scintillation vial. Counting solutions designed for aqueous samples effectively extract the moderately polar JH acid from the silica gel. Because of severe chemiluminescence problems with LK5DF plates, the vials should be held in the dark for several hours before counting. Silica gel exposed to UV light should also be counted to ensure absence of chemiluminescence.

Step 8. The amount of JH acid can be expressed as a percentage of the radioactivity recovered from the TLC plate and the enzyme activity calculated by slight modifications of the procedure discussed above.

[33] Cellular Juvenile Hormone Binding Proteins

By ERNEST S. CHANG

Juvenile hormones (JH) are unique sesquiterpene derivatives that regulate a number of developmental functions in insects.[1] Although the dramatic morphological effects of the hormone were documented many years

---

[1] N. A. Granger and W. E. Bollenbacher, in "Metamorphosis: A Problem in Developmental Biology" (L. I. Gilbert and E. Frieden, eds.), 2nd ed., p. 105. Plenum, New York, 1981.

ago,[2] it has only been relatively recently that preliminary information has been obtained concerning the molecular mechanism by which the hormone may exert its dramatic effects at the cellular level. Initial data indicate that JH may modulate differential gene expression in a manner similar to that proposed for steroid hormones; namely, that the hormone passively diffuses into cells and then is complexed with a cellular JH binding protein (CJHBP). Preliminary data also indicate that there are nuclear binding proteins for JH (see below). Whether or not the cytosolic proteins are able to translocate JH to the nucleus, or are simply artifacts of cytosol preparation (as suggested by some researchers) remains to be elucidated.

Data indicative of a cytosolic location of a CJHBP have been obtained from *Drosophila hydei* epidermis,[3] fat body and ovary of the cockroach, *Leucophaea maderae*,[4-8] and fat body of the locust, *Locusta migratoria*.[9] Our laboratory has concentrated on the K_c cell line of *Drosophila melanogaster*.[10-12] Established by Echalier and Ohanessian,[13] this cell line offers several advantages to the use of tissues obtained from whole insects. First, these cells can be grown in copious quantities in the absence of serum. Second, the cells are amenable to cloning. These features permit the isolation of billions of cells that have been grown in a defined medium without tedious dissections. In addition, there are no endogenous hormones in the culture medium. Thus any addition of radiolabeled ligand will not be subject to lowering of the specific activity and the cells would not show any preconditioned responses, such as down regulation of hor-

[2] V. B. Wigglesworth, *Q. J. Microsc. Sci.* **77**, 191 (1934).
[3] G. Klages, H. Emmerich, and M. G. Peter, *Nature (London)* **286**, 282 (1980).
[4] F. Engelmann, *in* "Insect Biology in the Future" (M. Locke and D. S. Smith, eds.), p. 311. Academic Press, New York, 1980.
[5] F. Engelmann, *in* "Juvenile Hormone Biochemistry" (G. E. Pratt and G. T. Brooks, eds.), p. 263. Elsevier/North-Holland Biomedical Press, Amsterdam, 1981.
[6] F. Engelmann, *Mol. Cell. Endocrinol.* **24**, 103 (1981).
[7] J. K. Koeppe, G. E. Kovalick, and M. C. Lapointe, *in* "Juvenile Hormone Biochemistry" (G. E. Pratt and G. T. Brooks, eds.), p. 215. Elsevier/North-Holland, Amsterdam, 1981.
[8] G. E. Kovalick and J. K. Koeppe, *Mol. Cell. Endocrinol.* **31**, 271 (1983).
[9] P. E. Roberts and G. R. Wyatt, *Mol. Cell. Endocrinol.* **31**, 53 (1983).
[10] E. S. Chang, T. A. Coudron, M. J. Bruce, B. A. Sage, J. D. O'Connor, and J. H. Law, *Proc. Natl. Acad. Sci. U.S.A.* **77**, 4657 (1980).
[11] J. D. O'Connor and E. S. Chang, *in* "Metamorphosis: A Problem in Developmental Biology" (L. I. Gilbert and E. Frieden, eds.), 2nd ed., p. 241. Plenum, New York, 1981.
[12] W. G. Goodman and E. S. Chang, *in* "Comparative Insect Physiology, Biochemistry and Pharmacology" (G. A. Kerkut and L. I. Gilbert, eds.), Vol. 7. Pergamon, Oxford, 1984 (in press).
[13] G. Echalier and A. Ohanessian, *In Vitro* **5**, 162 (1970).

mone receptors. Third, K_c cells respond to JH[14,15] (Figs. 1 and 2). Last, in the case of cell lines from *Drosophila,* the potential exists to use the extensive genetic information of this genus to help in the elucidation of JH action at the cellular level.

Culture Conditions and Cytosol Preparation

Drosophila melanogaster K_c cells may usually be obtained from the various research laboratories that are currently culturing them. They are not presently available from any commercial sources (such as the American Type Culture Collection). We maintain the cells in 1- and 3-liter spinner flasks (Bellco Glass) at a stirring speed of 60 rpm on a spinner flask stirring plate (Bellco Glass). Most general laboratory stirring plates heat the cultures and do not provide slow enough stirring rates. We have found that fastest growth occurs in the smaller flasks, perhaps due to a greater surface area to volume ratio for gas exchange. The flasks are kept in a refrigerated incubator at 25° in an air atmosphere.

The culture medium is D-20[13] except that the serum has been eliminated. The organic ingredients are from Sigma with the exception of Yeastolate (Difco). The inorganic ingredients are from Mallinckrodt. After final pH adjustment to 6.8, the medium (15 liters in a glass bottle) is kept in the cold overnight to permit settling of insoluble material. The following day, the medium is filtered by means of a 142-mm stainless-steel filter holder (Millipore). The medium is passed through two prefilters and a final 0.22 μm-filter (Millipore) by means of a peristaltic pump. One-liter bottles are filled while working in a laminar flow hood. The filtered medium is kept at room temperature for 1 day prior to refrigeration. Any contamination usually becomes noticeable at this time.

As an additional precaution, a small portion (approximately 100 ml) of medium is removed from the bulk medium prior to addition of the antibiotics. This portion is separately filtered prior to filtration of the bulk medium and then placed in an incubator. The absence of the antibiotics in this small portion will permit the rapid growth of any bacteria that may be penicillin or streptomycin sensitive. Medium is stored for several days prior to use so that any contamination will become apparent. Batches of medium are prepared as needed, but usually not less than once a month. Different batches are allowed to overlap, such that a new batch can be tested on a few cultures prior to expiration of the previous one. Lack of contamination and suitable cell growth are the parameters that are mea-

[14] E. S. Chang, A. I. Yudin, and W. H. Clark, Jr., *In Vitro* **18,** 297 (1982).
[15] L. Cherbas, personal communication.

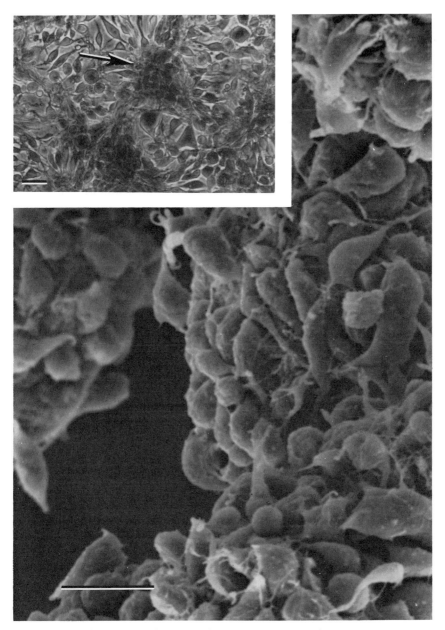

FIG. 1. Scanning electron micrograph of K_c cells exposed to 10^{-7} M 20-hydroxyecdysone without JH for 3 days. The cells have formed dense clusters and many have a fibroblastic shape with numerous filopodial extensions intertwined within the cell mass. Bar = 10 μm. Inset: Phase contrast micrograph of cells from the same culture. The dense areas represent clusters of cells (arrow). Bar = 10 μm. From unpublished work by E. S. Chang and A. I. Yudin.

FIG. 2. Scanning electron micrograph of K_c cells exposed to both 10^{-7} M 20-hydroxyecdysone and 10^{-6} M JH III for 3 days. The spindle-shaped cells (S) have developed long filopodial extensions, but have not formed the large cell aggregates seen in Fig. 1. Bar = 10 μm. Inset: Phase contrast micrograph of cells from the same culture. Bar = 10 μm. From unpublished work by E. S. Chang and A. I. Yudin.

sured for each new batch of medium. Other precautions that we take include duplication of our cultures in different incubators that are on separate electrical circuits. Clones are maintained in a staggered fashion, such that all of the cultures of a particular clone are never fed on the same day.

The clones currently used in our laboratory were started by dilution of wild type cells on irradiated feeder layers[16] and provided by Dr. J. D. O'Connor (University of California, Los Angeles).

Cells are harvested when they are at a density of $3-6 \times 10^6$ per ml, at which time they are in a logarithmic phase of growth. They are pelleted by centrifugation at 800 g for 3 min. They are then resuspended in TM buffer (10 mM Tris/5 mM MgCl$_2$, pH 6.9 at 4°). Cells are repelleted (800 g, 8 min) and then transferred as a slurry to a Dounce homogenizer (7 ml, Wheaton). Cells are homogenized with 20 strokes of the A pestle. The homogenate is then centrifuged at high speed (122,000 g, 90 min) to obtain the supernatant fraction designated as cell cytosol. If necessary, dilutions of the cytosol are made using TMK buffer (10 mM Tris/5 mM MgCl$_2$/150 mM KCl, pH 7.4 at 22°).

Assay of JH Binding Activity

Separation of the protein-bound radiolabeled ligand from the unbound hormone is necessary for all subsequent characterizations of the CJHBP. These methods include adsorption and pelleting of unbound hormone by dextran-coated charcoal,[3-6,10] precipitation and pelleting of the binding protein with polyethylene glycol (PEG),[7,8] and adsorption and pelleting of the binding protein with hydroxylapatite.[9] Although there are reports of anomalous results with dextran-coated charcoal (DCC),[9] and the concentration and timing of the assay are important,[6] we have found that this method provides the most consistent results for the *Drosophila* K_c system.

DCC is prepared according to the method of Kramer *et al.*[17] Activated charcoal (Sigma) is washed with 1 N HCl, followed by washes with water, 1% NaHCO$_3$ in water (w/v), and water until neutral. The washed charcoal is then dried and can be stored indefinitely. The charcoal is coated with dextran by dissolving 0.5 g of dextran (Sigma, MW = 80,700) in 100 ml of 0.01 M Tris/1.5 mM EDTA/3 mM NaN$_3$ (pH 7.3). One gram of charcoal is added to this solution and left at 4° overnight with gentle stirring. The

---

[16] C. Richard-Molard and A. Ohanessian, *Wilhelm Roux's Arch. Dev. Biol.* **181**, 135 (1977).
[17] K. J. Kramer, P. E. Dunn, R. L. Peterson, H. L. Seballos, L. L. Sanburg, and J. H. Law, *J. Biol. Chem.* **251**, 4979 (1976).

working solution of DCC is prepared by washing the above suspension 10 times in TMK buffer. The DCC is pelleted gently (200 g, 1 min) after each wash. The final working suspension is 1.25% DCC (w/v) in TMK.

Necessary precautions in working with JH are detailed by Prestwich et al.[18] These include the coating of all glassware with PEG, addition of JH in aqueous solutions, and the measurement of hormone stock solutions by spectrophotometry. Our standard JH binding assay consists of 100 μl of the protein and hormone(s) (radiolabeled with or without unlabeled JH) in TMK in 10 × 75-mm culture tubes. After 90 min, the tubes are placed on ice and 50 μl of the DCC solution is added. The mixture is vortexed and allowed to react for 15 min at 0°. When adding the DCC, the stock suspension is kept homogeneous by gently stirring it with a small stirring bar and a magnetic stirring plate. The solutions are then centrifuged (4800 g, 10 min) to pellet the charcoal. Seventy-five microliters is then removed for scintillation spectrometry.

Demonstration of a Binding Protein

For a CJHBP to be functioning as a receptor for the hormone, certain criteria must, at least, be met: (1) affinity for the hormone should be high, usually exceeding that of the serum carrier protein by an order of magnitude; (2) the binding sites should be saturable; (3) radiolabeled hormone should be removed by competition by excess unlabeled ligand; and (4) binding should be specific for the native hormone and the protein should not bind metabolites or inactive analogs. When these criteria have been addressed, then subsequent characterizations may be conducted on the binding protein.

Binding affinities are most commonly analyzed by the method of Scatchard.[19] These are usually constructed from the data obtained by measuring the amount of radiolabeled hormone ([^{3}H]JH I or III, available from New England Nuclear in racemic form) bound to a constant amount of cytosol or protein preparation (Fig. 3). Unbound hormone concentrations are essentially equal to the concentration of added hormone for these determinations. The data are then analyzed by plotting the amount of bound hormone divided by the amount of unbound (free) hormone versus the bound (Fig. 4). The Scatchard plot then enables an estimation of the binding activity. The negative inverse of the slope of the derived line (fit by least squares analysis) is the equilibrium dissociation constant

[18] G. D. Prestwich, J. K. Koeppe, G. E. Kovalick, J. J. Brown, E. S. Chang, and A. K. Singh, this volume [34].
[19] G. Scatchard, Ann. N. Y. Acad. Sci. **51**, 660 (1949).

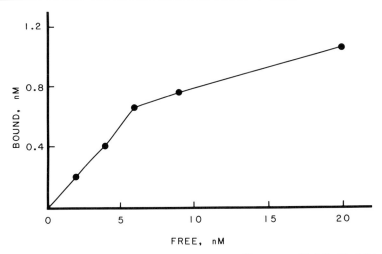

FIG. 3. Bound [³H]JH I as a function of the amount of hormone added (2–20 nM) to 12.5 µl of K_c cell cytosol in a total volume of 100 µl. Incubation was for 3 hr at 23°. Modified from Chang et al.[10]

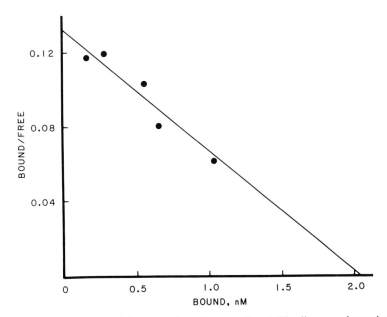

FIG. 4. Scatchard plot of the same data shown in Fig. 3. The line was drawn by the method of least squares and indicated a $K_d = 1.6 \pm 0.6 \times 10^{-8}$ M. Modified from Chang et al.[10]

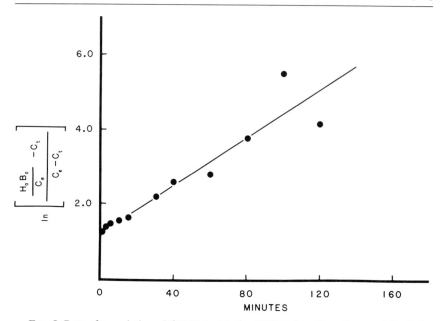

FIG. 5. Rate of association of [^{3}H]JH I with the CJHBP from K$_c$ cell cytosol. Radiolabeled hormone (10 nM) was incubated with a 50% solution of cytosol in TMK. Specific binding was determined by subtracting the amount of binding in a parallel series of tubes that were incubated with excess (10 μM) unlabeled JH I. The line was drawn by the method of least squares and indicated a $k_a = 1.3 \pm 0.1 \times 10^6\ M^{-1}\ \text{min}^{-1}$. H_0, the initial concentration of unbound hormone; B_0, the initial concentration of binding sites; C_e, the equilibrium concentration of the hormone-binding protein when saturated; C_t, the concentration of the hormone-binding protein complex at the designated time. Modified from Chang et al.[10]

(K_d). The x-intercept of the derived line is an estimate of the number of binding sites. Multicomponent curves are sometimes obtained, which may be indicative of several classes of binding sites (see below). Methods of resolving these multicomponent binding sites have been published.[20,21]

It is also useful to obtain kinetic data concerning the rate association (k_a) and rate dissociation (k_d) constants. These data can then be used as an alternative method to verify K_d (since $K_d = k_d/k_a$). They also provide information concerning the temporal stability of the hormone-binding protein complex. The k_a is determined by incubating cytosol with or without a 1000-fold excess of JH. The amount of binding is then measured at various timepoints. The difference in binding between the tubes incubated with or without excess JH is determined and is designated as the specifi-

[20] J. G. Nørby, P. Ottolenghi, and J. Jensen, *Anal. Biochem.* **102**, 318 (1980).
[21] D. Rodbard, P. J. Munson, and A. K. Thakur, *Cancer* **46**, 2907 (1980).

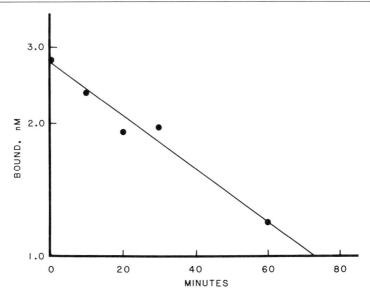

FIG. 6. Semi-log$_e$ plot of the rate of dissociation of [^{3}H]JH I from the CJHBP of K_c cytosol. Radiolabeled hormone (10 nM) was incubated with a 50% solution of cytosol in TMK buffer for 3 hr and then excess (10 μM) unlabeled JH I was added. At the designated times, aliquots were removed and assayed for bound hormone by means of the DCC assay. The values were corrected for nonspecific binding. A k_d value was determined (1.3 ± 0.2 × 10^{-2} min^{-1}) from the line drawn by the method of least squares. Modified from Chang et al.[10]

cally bound hormone. These data are then transformed using a plot of $\ln\{[(H_0B_0/C_e) - C_t]/(C_e - C_t)\}$ versus time, where H_0 is initial concentration of unbound hormone (added hormone); B_0 is initial concentration of binding sites; C_e is equilibrium concentration of the hormone-binding site complexes (the amount of bound hormone at equilibrium); and C_t is amount of hormone-binding site complexes at time t (Fig. 5). The line derived from regression analysis has a slope of $[(H_0B_0/C_e) - C_e]k_a$, from which k_a can be calculated.

The rate of dissociation of the hormone can also be determined by measuring the rate at which bound labeled hormone is displaced by excess unlabeled hormone. A plot of these data versus time (Fig. 6) results in a line with a slope equal to k_d. These data are obtained by first incubating cytosol with saturating amounts of radiolabeled hormone for an amount of time that results in maximal binding (determined from the rate association data). An excess amount of unlabeled hormone is then added. At various times afterward, aliquots of the incubation mixture are removed and assayed for binding. The data are corrected for nonspecific binding. The K_d

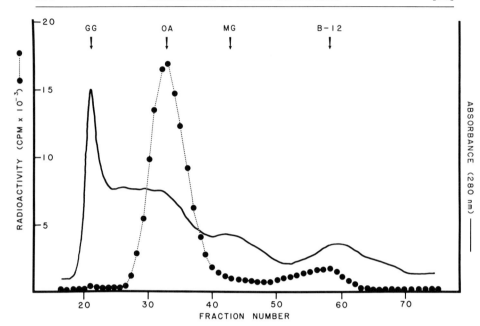

FIG. 7. Sephadex G-100 column (24 × 1.6 cm) chromatography of 1.0 ml of K_c cytosol that was incubated with 1×10^{-6} M [^{3}H]EFDA and followed by photolysis at 254 nm for 8 min. Eluant was TMK at a flow rate of 0.42 ml/min. Fractions (2.1 ml) were analyzed for radioactivity (circles) and absorbance at 280 nm (solid line) at 0.5 AUFS. External standards (Bio-Rad) were γ-globulin (GG), ovalbumin (OA), myoglobin (MG), and vitamin B-12 (B-12). An approximate molecular weight of 49,200 ± 1600 was obtained. EFDA is a photoaffinity analog of JH. Modified from Chang et al.[25]

derived from kinetic data should be within an order of magnitude of that obtained by Scatchard analysis.

When the above kinetic data have been obtained, information may then be collected on the physical parameters of the CJHBP. Molecular weights of the protein can be obtained through the use of chromatography through molecular sieves and centrifugation through sucrose gradients. Gel filtration column chromatography can be performed with Sephadex (Pharmacia) or similar media. Approximate molecular weights can be estimated by comparing the elution profile of macromolecule-associated radiolabeled hormone with those of standard proteins (Fig. 7). Parallel analyses using cytosol incubated with both radiolabeled hormone and excess unlabeled material must be conducted. Only those peaks that are competed away by the unlabeled hormone can be considered as indicative of CJHPBs.

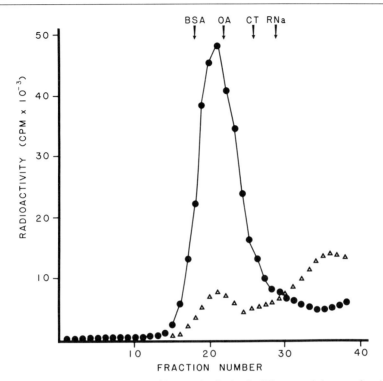

Fig. 8. Sucrose gradients (10–30% in TMK) of 0.2 ml of K_c cytosol that was incubated with 1×10^{-6} M [^{3}H]EFDA. Photolysis was at 254 nm for 8 min. Gradients (4.5 ml) were preformed and centrifuged for 22 hr (190,000 g, Beckman SW 50.1 rotor). The cytosol was incubated with (triangles) or without (circles) a 50-fold excess of unlabeled JH III. External standards (Pharmacia) were bovine serum albumin (BSA), ovalbumin (OA), chymotrypsinogen A (CT), and ribonuclease A (RNa). An approximate molecular weight of 51,000 ± 5800 was obtained. EFDA is a photoaffinity analog of JH. Modified from Chang et al.[25]

Velocity sedimentation through sucrose is conducted using preformed gradients. Gradients of 10–30% sucrose can be formed either in steps by hand, or, preferably, by the use of a gradient maker (Hoeffer) and a peristaltic pump. As with the column chromatography, parallel analyses of cytosol incubated with excess unlabeled hormone must also be conducted (Fig. 8).

Much more accurate determinations of the molecular weights of the CJHBP are obtained by polyacrylamide gel electrophoresis (PAGE) under denaturing conditions.[22,23] Unfortunately, the sodium dodecyl sulfate

[22] K. Weber and M. Osborn, *J. Biol. Chem.* **244**, 4406 (1969).
[23] U. K. Laemmli, *Nature (London)* **227**, 680 (1970).

(or other denaturing reagents) cause disruption of the noncovalent binding between the hormone and the protein. A recent solution to this problem has been the advent of a photoaffinity analog of JH that is also tritiated. The analog, epoxyfarnesyl diazoacetate (EFDA),[24,25] will form covalent bonds with adjacent proteins following irradiation with ultraviolet light. The covalent bond is resistant to high salt or other denaturing conditions. An extensive discussion of EFDA and its use in the elucidation of JH binding proteins is presented by Prestwich et al.[18]

Nuclear Binding of JH

Our laboratory has also demonstrated the presence of macromolecules that bind JH associated with nuclei of K_c cells. At a density of 1×10^8 cells per ml, 10 ml of this cell suspension is preincubated with the esterase inhibitor 1-naphthyl-N-propyl carbamate (generously supplied by Dr. B. D. Hammock) for 30 min at 25°. Cells are then incubated with 10^{-7} M [^{3}H]JH III for 30 additional min. Following pelleting (800 g, 8 min), the cells are disrupted by 20 strokes of a Dounce homogenizer. Two milliliters of TMK buffer is added to the homogenate, it is then centrifuged (800 g, 8 min), the supernatant discarded, and the upper 90% of the pellet transferred to another tube as the nuclear fraction. This fraction is washed once with 2.5 ml of TMET buffer (5 mM Tris/1.5 mM MgCl$_2$/1.5 mM EDTA/0.1% Triton X-100, pH 7.0) and then 4 more times with 2.5 ml TM buffer, with repelleting after each wash. The resulting washed nuclear fraction is suspended with 0.21 ml TM buffer. Ten microliters is diluted for counting in a hemocytometer and 0.25 ml is analyzed in a scintillation spectrometer.

A parallel incubation with [*methoxy*-^{3}H]inulin (New England Nuclear) is used for determination of the contamination of the nuclear fraction by cytosolic [^{3}H]JH III. For this measurement, radiolabeled inulin is added after homogenization and followed by centrifugation at 4500 g for 10 min. The supernatant is then centrifuged at 122,000 g for 90 min to obtain the cytosolic fraction. An aliquot of this cytosolic fraction is counted to determine the amount of inulin present in a known volume of cytosol. Following washing and counting of the nuclear fraction as described above, the amount of cytosol contamination of the nuclei can then be calculated from the amount of inulin radioactivity.

As an alternative method of preparation, nuclei are separated from other cell constituents by means of centrifugation through sucrose. After cell disruption as described above, the homogenate is mixed with 2.0 ml

---

[24] G. D. Prestwich, A. K. Singh, J. F. Carvalho, J. K. Koeppe, G. E. Kovalick, and E. S. Chang, *Tetrahedron* **40**, 529 (1984).

[25] E. S. Chang, M. J. Bruce, and G. D. Prestwich, *Insect Biochem.*, in press (1984).

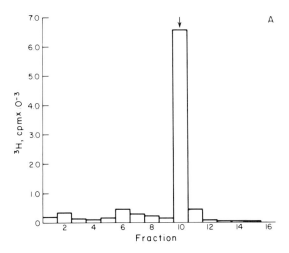
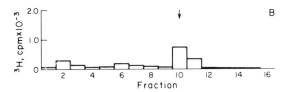
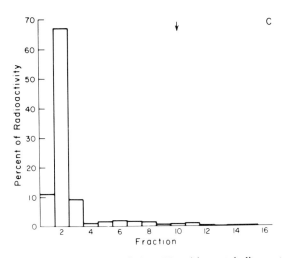

FIG. 9. Thin-layer chromatograms of [^{3}H]JH III and its metabolites extracted from K_c cell nuclei incubated in the absence (A) or presence (B) of a 1000-fold excess of unlabeled JH III. (C) represents the radioactivity found in the cell-free supernatant following pelleting of the cells after incubation with the hormone for 30 min. Samples were extracted with ethyl acetate. Arrows represent the location of authentic external JH III standards.

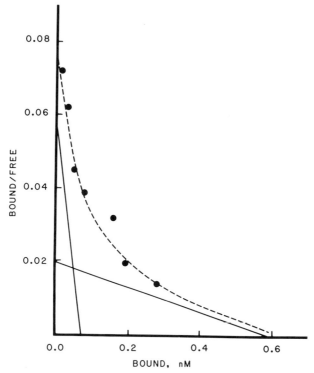

FIG. 10. Scatchard analysis of radiolabeled JH III bound to K_c nuclei as a function of the amount of hormone added to whole cells. [^{3}H]JH III (0.2–20 nM) was incubated with 1×10^8 cells/ml in 10 ml for 30 min. Following correction of the data,[20] K_d values were determined from the negative inverses of the slopes of the resulting lines as determined by least squares analyses.

of TM buffer and layered onto 0.5 M sucrose in TMET buffer prior to centrifugation for 8 min at 4000 g. The pellet is then resuspended in 2.3 ml of 0.5 M sucrose, layered onto 2.5 ml of 1.8 M sucrose in TMK, and then centrifuged (Beckman SW 50.1 rotor) at 122,000 g for 15 min. The pellet is then either fixed for transmission electron microscopy or else resuspended in 0.2 ml of TM buffer for scintillation spectrometry.

Thin sections of the nuclear pellets prepared using either of these methods indicate that they are essentially free of contamination by nonnuclear cellular constituents. The nuclei prepared by multiple washings are somewhat more homogeneous than those obtained by pelleting through sucrose, although the former method is more tedious. These preparation methods minimize contamination by cytosol, as indicated by the negligible amount of radioactivity attributable to inulin.

Scintillation spectrometry of these radiolabeled nuclei indicate that they are able to bind radiolabled JH III with saturability and specificity. Extraction of the bound radioactive ligand with ethyl acetate, followed by thin-layer or high-performance liquid chromatographic analysis,[26] indicates that the nuclear-bound material is native JH III (Fig. 9).

Figure 10 shows a Scatchard analysis of radiolabeled JH associated with nuclei after whole-cell incubation with increasing amounts of [^{3}H]JH III. The binding studies indicate that there are at least two classes of binding sites, one with an approximate K_d of 1.2×10^{-9} M and another one with about half the affinity.

Conclusions

The research described in this chapter only provides a foundation from which further studies of the mode of action of JH may be conducted. Although the data are insufficient to identify conclusively the described CJHBPs as juvenile hormone receptors, all of the accumulated data do support this conclusion. The intriguing experiments that remain, then, concern the translocation of JH from the cytosol to the nucleus, the interaction of JH and chromatin, and ultimately, the way in which JH mediates differential gene expression.

Acknowledgments

I thank M. J. Bruce and A. I. Yudin for valuable assistance and Drs. F. Engelmann and G. D. Prestwich for helpful discussions.

[26] E. S. Chang, *J. Liq. Chromatogr.* **6,** 291 (1983).

[34] Experimental Techniques for Photoaffinity Labeling of Juvenile Hormone Binding Proteins of Insects with Epoxyfarnesyl Diazoacetate

By GLENN D. PRESTWICH, JOHN K. KOEPPE, GAE E. KOVALICK, JOHN J. BROWN, ERNEST S. CHANG, and AMBARISH K. SINGH

Introduction

The use of photoactivatable hormone analogs allows irreversible covalent attachment of radiolabeled substrates to the active sites of hormone binding proteins.[1-4] In the case of juvenile hormone binding proteins (JHBP) of insects, the synthetic JH III analog [10-^{3}H]epoxyfarnesyl diazoacetate ([^{3}H]EFDA) can be used for the selective, irreversible modification of JHBP (Fig. 1).[5] The success of this technique is dependent both on the affinity and specificity of EFDA for the JHBP.

Juvenile hormone III (JH III)

10,11-Epoxy farnesyl diazoacetate (EFDA)

There are four naturally occurring juvenile hormone homologs: JH 0, JH I, JH II, and JH III, each of which may have a different affinity for the binding protein. The presence and proportion of these homologs in an insect is species dependent. The usefulness of EFDA, a JH III analog, will vary considerably from species to species. EFDA has been tested now in several insect systems: hemolymph and ovarian tissue of *Leucophaea maderae*,[6,7] hemolymph from *Manduca sexta*,[7a] cytosol from the *Drosophila* K$_c$ cell line,[8] and hemolymph obtained from nymphal *Locusta migratoria* (unpublished data).

Extreme care must be exercised to avoid overinterpretation of adventitious nonspecific binding. We recommend that all of the criteria for EFDA binding listed below be met to prevent the accumulation of "false positive" JH-binding proteins in the literature. A review of JH cellular and hemolymph binding proteins should be consulted for further background.[9]

[1] H. Bagley and J. R. Knowles, this series, Vol. 46, p. 69.
[2] J. R. Knowles, *Acc. Chem. Res.* **5**, 155 (1972).
[3] V. H. Chowdhry and F. H. Westheimer, *Annu. Rev. Biochem.* **48**, 293 (1979).
[4] J. A. Katzenellenbogen, *Biochem. Actions Horm.* **4**, 1 (1977).
[5] G. D. Prestwich, A. K. Singh, J. F. Carvalho, J. K. Koeppe, G. E. Kovalick, and E. S. Chang, *Tetrahedron* **40**, 529 (1984).
[6] G. D. Prestwich, G. E. Kovalick, and J. K. Koeppe, *Biochem. Biophys. Res. Commun.* **107**, 966 (1982).
[7] J. K. Koeppe, G. E. Kovalick, and G. D. Prestwich, *J. Biol. Chem.* **259**, 3219 (1984).
[7a] J. K. Koeppe, G. D. Prestwich, J. J. Brown, W. G. Goodman, G. E. Kovalick, T. Briers, M. D. Pak, and L. I. Gilbert, *Biochemistry,* in press (1984).
[8] E. S. Chang, M. J. Bruce, and G. D. Prestwich, *Insect Biochem.,* in press (1984).
[9] W. G. Goodman and E. S. Chang, *In* "Comprehensive Insect Physiology, Biochemistry

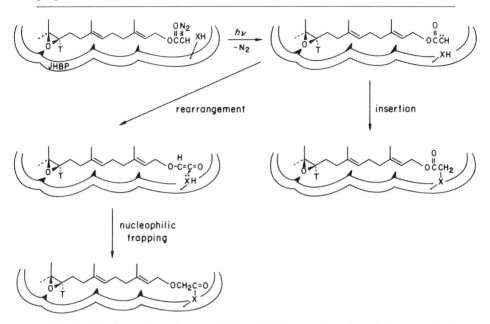

FIG. 1. Photoinduced attachment of tritiated EFDA to an insect juvenile hormone binding protein. Covalent attachment may result from either direct insertion of an acyl carbene or nucleophilic trapping of a rearranged ketene.

Detailed methods for the purification of [^{3}H]EFDA and examples of the experimental protocols we believe to be essential to demonstrate EFDA binding to the juvenile hormone binding site will be described. The chapter will be divided into three sections: (1) the criteria to determine specificity of EFDA binding, (2) properties and handling of EFDA, and (3) the techniques utilized to determine the specificity of binding in three insect systems. The syntheses of both [^{3}H]EFDA[5] and [*carbonyl*-^{14}C]EFDA[10] are described in detail elsewhere and will not be reiterated here.

Criteria for EFDA Binding Specificity

We recommend that the following experiments be completed to determine if this particular analog of JH III will be useful. If [^{3}H]EFDA binds to proteins, the investigator must then demonstrate that [^{3}H]EFDA binds with relatively high affinity and low capacity at the JH binding site on the

---

and Pharmacology" (G. A. Kerkut and L. I. Gilbert, eds.), Vol. 7. Pergamon, Oxford, 1984 (in press).
[10] G. A. Krafft, M. F. Reich, and J. A. Katzenellenbogen, *J. Labelled Comp. Radiopharm.* **9**, 591 (1982).

binding protein. To meet these criteria, and to demonstrate specificity of EFDA binding, known concentrations of binding protein, [^{3}H]EFDA and JH homologs must be utilized. The following descriptions briefly outline the steps and experiments that must be completed when determining if EFDA is a useful probe, and if it binds specifically at the JH binding site.

Displacement of the Natural Hormone from the Binding Protein with EFDA

After identifying a saturable, JH-specific macromolecule in an insect protein preparation, competition studies with JH 0, JH I, JH II, and JH III are needed to establish unambiguously the preferred natural hormone homolog.[11,12] Additionally, the ability of EFDA to displace the natural [^{3}H]JH from the binding protein must be determined under equilibrium conditions. Comparison of these displacement curves will indicate relative binding affinities, and indicate whether or not EFDA is binding at the JH-binding site. If EFDA displays little or no ability to displace the natural hormone, the compound will not be useful.

Photoactivated EFDA Must Irreversibly Inhibit Proteins from Binding JH

Although EFDA might displace JH from the binding protein under equilibrium conditions, the analog may not be properly oriented in the binding site to ensure covalent attachment to the protein(s) when irradiated. In addition, since the diameter of the quartz vessels, the concentrations of the proteins in the extracts, and the sensitivity of the proteins to UV light will be variable, the conditions for photoactivation must be empirically determined. Therefore, the following two series of experiments should be performed. Both the time of UV exposure needed to obtain optimal reduction of the JH binding capacities by EFDA, and the effects of UV exposure on the JH-binding protein in the absence of EFDA should be determined. A third experiment will suggest whether the inactivation takes place at the JH-binding site(s). In this experiment the effect of excess JH on EFDA inactivation of the JH-binding protein will be determined. If the presence of JH prevents loss of binding capacity in protein extracts photolyzed in the presence of EFDA, JH may act by

---

[11] For examples in *Leucophaea* JHBP, see J. K. Koeppe, G. E. Kovalick. and M. C. LaPointe, *in* "Juvenile Hormone Biochemistry" (G. E. Pratt and and G. T. Brooks, eds.), p. 215. Elsevier/North-Holland Biomedical Press, Amsterdam, 1981.

[12] For examples in *Manduca* JHBP, see ref. 11 and W. G. Goodman, W. E. Bollenbacher, H. L. Zvenko, and L. I. Gilbert, *in* "The Juvenile Hormones" (L. I. Gilbert, ed.), p. 75. Plenum, New York, 1976.

filling up the binding sites and preventing EFDA attachment to the protein. If the photoattachment of EFDA has no effect on JH binding capacities in the absence of JH, further studies may prove uninformative.

[³H]EFDA Binding to Protein(s)

If unlabeled EFDA displaces JH from the binding protein and if the photoactivated EFDA can irreversibly reduce the JH-binding capacity of an extract, radiolabeled EFDA should then be used to determine the specificity of EFDA binding. Photoactivation conditions must first be determined for [³H]EFDA as outlined for unlabeled EFDA in the three experiments above. If JH inhibits the binding of [³H]EFDA to proteins, both JH-displaceable and non-JH-displaceable binding can be quantitated. JH-displaceable binding data can be analyzed by the method of Scatchard.[13] These results will provide a relative binding affinity (K_d) of the protein for [³H]EFDA. Such binding may be at a second site on the JH-binding protein and not necessarily at the actual JH-binding site.

Specificity of [³H]EFDA Binding

To determine if [³H]EFDA binds to the JH-binding protein at the JH-binding site, a series of competitive displacement experiments must be undertaken, the underlying assumption being that the binding of the natural hormone will be preferential to one of its homologs. In the first experiments, the ability of the natural unlabeled hormone (i.e., JH III) to displace 50% of the bound natural radioactive hormone (i.e., [³H]JH III) is determined. A comparison of the concentration of unlabeled natural hormone required for displacement of 50% of the bound [³H]JH III with the concentration of an unlabeled homolog (i.e., JH 0, JH I, or JH II) that is required to displace 50% of the bound label, yields a number that is called a displacement ratio. A similar experiment is then conducted with [³H]EFDA as the radioactive ligand. If the ligand (at a concentration equal to or less than its K_D) binds at the JH-binding site, and if the binding protein concentration is similar to that of the above experiment, then the concentrations and the proportions of the two unlabeled homologs needed to displace 50% of the [³H]EFDA should be similar to the values obtained for the displacement of [³H]JH III. If they are different, a second binding site may exist for [³H]EFDA.

To confirm the displacement of [³H]EFDA binding to a JH-specific protein, the labeled proteins are electrophoretically separated and then visualized through the process of fluorography via X-ray film. In these

[13] G. S. Scatchard, *Ann. N.Y. Acad. Sci.* **51,** 660 (1949).

experiments, optimal concentrations of proteins, [³H]EFDA and various concentrations of unlabeled JH are utilized to demonstrate the JH-dependent displacement of [³H]EFDA from specific proteins.

Important Properties, Handling, and Repurification of EFDA

Properties and Handling of EFDA

EFDA (molecular weight 306) is a yellow oil with a UV absorption maximum (in ethanol) at 244 nm ($\varepsilon = 10{,}500$). These two values enable calculations of solution molarities. EFDA, like all JHs, adsorbs to glass. All experiments in aqueous solutions must be carried out with Carbowax 20M (PEG 20,000)-coated glassware. Adsorption to glass is not a problem when ethanol, hexane, or ethyl acetate is used as solvent. All solvents should be distilled or HPLC grade. Being slightly more polar than JH III, EFDA will dissolve in water at $1 \times 10^{-4}\ M$. Solutions in volatile solvents can be evaporated with N_2, handled in ambient light conditions, and held at room temperature during a few hours' experiments without adverse results. Although photolysis is complete in <10 sec at 254 nm, sunlamp irradiation for 1 hr causes <10% chromophore loss. Nonetheless, excessive exposure to light, air, or heat will cause gradual decomposition. Most important, *the diazoacetate group is very acid sensitive* and decomposes rapidly below pH 5. The diazoacetate is also hydrolyzed above pH 10, as expected for most esters.

Photolyses are carried out in quartz vessels, since Pyrex absorbs wavelengths below 280 nm. A Rayonet reactor (Southern New England Ultraviolet) equipped with 4 to 8 Hg-vapor lamps (253.7 nm) was used for irradiations. Carbowax-coated quartz test tubes containing solutions of protein, hormones and EFDA were kept at 0° in a quartz ice bath during the 10 to 120 sec irradiations.

Purification of [³H]EFDA

Currently, [³H]EFDA and the unlabeled material are only available through our laboratories or by total synthesis. We have undertaken to make [³H]EFDA commercially available after 1985. Two methods are available for preparative scale purification or for analytical scale purity determinations. These are described below in sufficient detail to enable their performance in insect biochemistry laboratories. It is important to check purity and to use only radiochemically homogeneous materials. Our experience is that [³H]EFDA at 5 Ci/mmol, stored in sealed ampoules at 4° in the dark in 1:1 ethanol–benzene (~0.3 mg/ml), is 20–70% radiochemically pure after 18 months. As with commercial [^{3}H]JHs, a toluene–hexane mixture is also useful for storing [^{3}H]EFDA solutions. For maximum protection against autoradiolysis, the solvent should *not* freeze during storage. [^{3}H]EFDA is best used within 4–6 months of synthesis. After this time, it should be checked for purity and if necessary repurified as described below.

Normal Adsorption Chromatography on Silica Gel. A sample of [^{3}H]EFDA which had been stored for 18 months as described above was evaporated to dryness with dry N_2. It was transferred in one 250 μl portion of 5% ethyl acetate–hexane (both HPLC grade) to a disposable Pasteur pipet packed with a 5-cm column of 230–400 mesh silica gel G (Merck or Baker, "flash" silica) preequilibrated with the same solvent mixture. Using a pipet bulb or a rubber septum and N_2 line to generate moderate pressure, the EFDA solution was forced onto the silica gel and the ampoule rinsed with two further 250-μl portions of solvent. Twenty 2-ml fractions were collected (the last five were eluted with 10% ethyl acetate–hexane) and a 1-μl aliquot of each was counted (LSC) to monitor elution of label. A second 1-μl aliquot was analyzed by silica gel TLC or by reverse-phase HPLC, both of which are described below. In this experiment the [^{3}H]EFDA was found in fractions 6 to 15, with the major peak of [^{3}H]EFDA as fractions 10 to 14. Removal of solvent (N_2 or rotary evaporation), dilution with 2.00 ml ethanol and determination of UV of an aliquot (diluted 1 : 10) showed 1.51×10^{-4} M for this stock solution.

Thin-Layer Chromatography. The moderate resolution of TLC is sufficient to remove the polar radiolabeled impurities which arise from radiolysis of [^{3}H]EFDA. It also is a quick method to check chemical composition of minicolumn fractions. Using Machery-Nagel Polygram Sil G-UV254 4 × 8-cm plates, we find 25% ethyl acetate–hexane gives R_f = 0.31 for EFDA. Although it is difficult to visualize the small quantities (<1 μg) in the aliquots using I_2, ethanol–vanillin–H_2SO_4 spray, or UV quenching, it is possible to spot unlabeled EFDA and to scrape and count zones at the correct R_f. In this way, the chemical integrity of the chromatographed sample is confirmed.

Reverse-Phase HPLC. A higher resolution and more sensitive method to analyze chromatographic fractions uses a reverse-phase C_8 (octylsilyl) silica microparticle column. Chang[14] reports the use of a radially compressed C_8 column for resolution of JH homologs and analogs, including [^{3}H]EFDA. One can also employ a Whatman C8 PXS 10/25 column, eluting at 2.0 ml/min with 75% methanol–water, and monitoring the eluted

[14] E. S. Chang, *J. Liq. Chromatog.* **6**, 291 (1983).

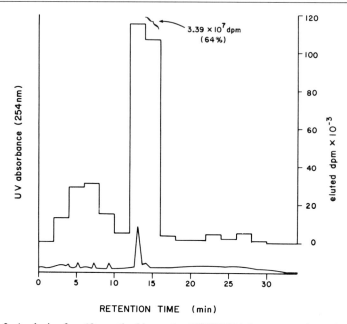

FIG. 2. Analysis of an 18-month-old sample of [^{3}H]EFDA by reverse phase (C$_8$) HPLC. Polar impurities can accumulate due to radiolytic decomposition, but these can be removed by chromatography.

radiolabel by LSC of aliquots and eluted EFDA by absorption at 254 nm. This is the fastest and most accurate means of verifying chemical and radiochemical purity of [^{3}H]EFDA. A sample chromatogram before purification on a silica minicolumn is shown in Fig. 2; after purification, >96% radiochemical purity is achieved.

Photoaffinity Labeling of JH-Binding Proteins in *Leucophaea maderae*

JH-binding proteins have been found in hemolymph,[7,11] ovarian,[7,11] and fat body[15] extracts obtained from reproductively active females of *L. maderae* (Orthoptera: Blaberidae). JH-binding proteins from both the hemolymph and the ovarian extracts have been successfully labeled with [^{3}H]EFDA.[7] In the following discussion we will outline the experimental protocols that were used to determine the specificity of [^{3}H]EFDA binding to hemolymph JH-binding protein(s).

[15] F. Engelmann, *Mol. Cell. Endocrinol.* **24,** 10 (1981).

Extract Preparation

To prepare a hemolymph extract for the JH-binding assay, hemolymph was homogenized in TMK buffer (10 mM Tris–HCl, 5 mM MgCl$_2$, 150 mM KCl, pH 7.4) containing 6 × 10^{-4} M phenylmethylsulfonyl fluoride (PMSF) and 5 mg/ml bovine γ-globulins using a ground glass homogenizer. After centrifugation at 12,000 g for 15 min, the supernatant is then asayed (see following section on JH-binding assays) to determine a dilution that will bind 50% of labeled JH III at a concentration of 1 × 10^{-8} M (the K_D of JHBP for JH III). With this binding capacity for the diluted extract, displacement and competition studies can be performed.

Preparation of JH and EFDA Stock Solutions

Stock solutions of JH were prepared in methanol or hexane (HPLC grade) and the molarity determined by measuring UV absorbance at 217 nm and dividing by the molar absorptivity (ε = 14,770 for JH 0, I, II, or III). For aqueous stock solutions, an aliquot of the organic stock was dryed with N$_2$ in a PEG-treated tube and then resolubilized in ethanol (final concentration not to exceed 1–2%). TMK buffer was then added, and the vortexed solution was allowed to stand 1–3 hr before use.

Handling procedures for EFDA are outlined in the preceding section. It should be stressed that this compound appears to be more "sticky" than JH; constant monitoring of the actual concentration of EFDA in a stock solution is critical.

Binding Assays

Binding Assays for [³H]JH III. In virtually every experiment, regardless of the ligand used, bound ligand must be separated from the unbound ligand. To assay for bound [³H]JH III in extracts from *L. maderae*, three different assays have been tested and all yield similar results. These assays include precipitation of the bound ligand with polyethylene glycol (PEG assay),[12] precipitation of the bound ligand with hydroxylapatite (HAP assay),[13] and the removal of free ligand from solution with dextran-coated charcoal (DCC assay).[15,16] We found that the least reproducible data were obtained with the charcoal assay. This may be due to the rapid rate of dissociation of the [³H]JH III from the complex. In addition, since incubations with charcoal must be completed within 60–90 sec, the number of samples which can be assayed is limited. Hence, we routinely

---

[16] K. J. Kramer, P. E. Dunn, R. C. Peterson, H. L. Seballos, L. L. Sanburg, and J. H. Law, *J. Biol. Chem.* **251**, 4679 (1976).

used the PEG assay, but have confirmed these results with both of the other two assays.

For the PEG assay, 100 μl of an aqueous JH stock solution containing [^{3}H]JH III and a known concentration of unlabeled JH was added to 6 × 50-mm culture tubes previously treated with a 1% solution of polyethylene glycol (MW 20,000). An equal volume of the diluted extract was then added to the hormone solution, and the mixture was vortexed and incubated for 30 min at 23° (equilibrium conditions). To terminate the reaction, 300 μl of 33% PEG 6000 was added to the reaction mixture, vortexed, and then incubated for an additional 30 min at 23°. The samples were then centrifuged at 7000 g for 5 min and the supernatant aspirated off and discarded. The pellet was digested and resuspended in 100 μl of an NCS (Amersham) solution containing 1 part to 3 parts PPO-POPOP–toluene. To this suspension 400 μl of ScintiVerse (Fisher) is added. The tubes were then placed in 3 ml scintillation vials and analyzed by LSC.

Binding Assays for EFDA. To assay for noncovalently bound [^{3}H]EFDA in extracts from *L. maderae* only the PEG assay has been tested, although the addition of γ-globulins increases nonspecific binding to almost 40–50%. To determine the amount of [^{3}H]EFDA which covalently binds to proteins, labeled proteins were precipitated from solution by adding an equal volume of 20% trichloroacetic acid (TCA) to an aliquot of the UV-irradiated solution. After incubating at room temperature for 15 min, unbound [^{3}H]EFDA was washed through a 25-mm-diameter glass fiber filter (Gelman type A/E) with three volumes of 10% TCA and two volumes of 60% ethanol. Proteins remaining on the filter were digested for one hour at 50° with 300 μl of NCS/PPO-POPOP–toluene (1 : 3 v/v). After cooling to 4°, 3 ml of ScintiVerse was added to each vial, and the vials were cooled overnight at 4° to reduce chemiluminescence. To determine the total amount of label present in each reaction mixture, aliquots of the irradiated solutions were transferred directly to 3 ml of ScintiVerse and 300 μl NCS/PPO-POPOP–toluene.

EFDA Binding Specificity

Relative Affinity for the Hemolymph JH-Binding Protein.[6,11] Using a diluted hemolymph extract, the ability of three JH homologs (JH III, JH I, and JH 0) and EFDA to displace bound [^{3}H]JH III from the binding protein was determined. A series of concentrations of each homolog and of EFDA (1×10^{-8} to 1×10^{-5} M) containing 2×10^{-9} M (6000 dpm) [^{3}H]JH III was prepared. The diluted extracts were added, and the samples were vortexed and allowed to reach equilibrium (20 min, 22°). The PEG assay was then employed to separate bound from free hormone/

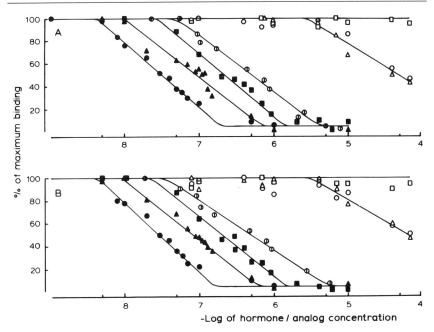

FIG. 3. Displacement of [^{3}H]JH III from the JHBP of *L. maderae* hemolymph (A) and ovaries (B) using increasing concentrations of JH III (filled circles), JH I (filled triangles), JH 0 (filled squares), EFDA (split circles), and methoprene (open figures: circle, racemic; triangle, JH-active enantiomer; squares, inactive enantiomer). The concentration of [^{3}H]JH III in each mixture is 1×10^{-9} M, and the concentration of JHBP is approximately $1-3 \times 10^{-8}$ M. The results are expressed as the percentage of maximum binding found with [^{3}H]JH III in the absence of competitors.

analog. Plots of percentage maximum [^{3}H]JH III bound vs the negative log of the competitor concentration were prepared. The relative binding affinity (RBA) for a given competitor was determined by using the ratio of competitor concentration to the unlabeled JH III concentration required to inhibit 50% of [^{3}H]JH III binding. The results from these studies demonstrate that the photoaffinity label EFDA competes with JH III for the binding sites, although its affinity appears to be almost 50-fold less than that of the natural hormone. Typical competition data are illustrated in Fig. 3.

Binding of Photoactivated EFDA. Protein solutions containing 10^{-5} to 10^{-7} M unlabeled EFDA were photolyzed for 0 to 180 sec as described above. After photolysis, γ-globulins at 5 mg/ml (final concentration) were added to each sample. The proteins were then separated from the unbound analog/hormone by precipitation with 33% PEG 6000 (final concentration 20%), pelleted by centrifugation (7000 g for 5 min), and resuspended in TMK buffer. Duplicate 100-μl aliquots of the resuspended proteins were assayed by adding the protein solution to 100 μl of 1 × 10^{-8} M JH III in TMK buffer. The PEG assay was employed to determine the [^{3}H]JH III binding capacity of each sample. The results from a typical experiment demonstrate no loss of binding at 0 sec of UV exposure, but 40–60% loss within 10–15 sec of exposure.

In order to demonstrate the protective effect of JH III, protein extracts (200 μl) containing 1 × 10^{-5} M JH were photolyzed as above. γ-Globulins were added, followed by addition of 200 μl of a 1:4 dilution of DCC stock solution to remove excess JH III. After standing at 20° for 5 min, the mixture was centrifuged at 7000 g for 10 min, and the [^{3}H]JH III binding capacity of the supernatant was determined by the PEG assay. The results from these studies[6] demonstrated that EFDA, when photoactivated, irreversibly reduced the capacity of an extract to bind [^{3}H]JH III. On the other hand, the addition of a 1000-fold excess of unlabeled JH III prevented this loss of binding activity.

Binding of Photoactivated [^{3}H]EFDA.[7] Hemolymph extracts were diluted to a JHBP concentration of about 1 × 10^{-8} M and incubated for 30 min at 0° with 1 × 10^{-8} M [^{3}H]EFDA and either 1 × 10^{-5} M unlabeled JH III or no JH III. These samples were irradiated for 0, 15, 30, and 60 sec, and the binding was determined by the TCA assay. At these concentrations, 4.4% of the [^{3}H]EFDA became attached to protein in absence of JH III and 0.86% in the presence of 1 × 10^{-5} M JH III (81% specific binding). At 1 × 10^{-7} or 1 × 10^{-6} M [^{3}H]EFDA, nonspecific binding increased to 50%, whereas the percentage of total counts bound remained the same.

Binding Affinity and Specificity of Proteins for [^{3}H]EFDA.[7,11] Labeled EFDA (1 × 10^{-9} to 1 × 10^{-6} M) was incubated (23°, 30 min) with proteins in the presence of excess unlabeled JH III (1 × 10^{-5} M). Binding was measured by PEG precipitation. Nonspecific binding of [^{3}H]EFDA was taken as that amount of binding that occurred in the presence of 1 × 10^{-5} M JH III and was subtracted from total bound ligand. Scatchard analysis of the binding showed a K_D of 1.9 × 10^{-6} M for the hemolymph JHBP. Identical results were obtained when [^{3}H]EFDA was irradiated and covalently linked to the binding protein. To determine the specificity of binding, various concentrations of unlabeled JH III and unlabeled JH I were used to displace [^{3}H]EFDA from the binding proteins. Approximately 4 times more JH I was required relative to JH III for 50% displacement of [^{3}H]EFDA binding; the same ratio was found for 50% displacement of [^{3}H]JH III binding (PEG assay).

Electrophoretic Separations of [^{3}H]EFDA-Labeled Proteins.[7] Samples subjected to electrophoresis were first irradiated for 60 sec in the absence of γ-globulins, precipitated with TCA (final concentration 10%), washed two times with 95% ethanol, and then resuspended in SDS-sample buffer. SDS-polyacrylamide gel electrophoresis (SDS-PAGE) was performed at 20° with 6.25% polyacrylamide in a Bio-Rad slab gel electrophoresis cell. The polypeptides were visualized by staining with Coomassie blue. Molecular weights were estimated using standards. Gels were analyzed by fluorography. After impregnation with EN-3-HANCE (New England Nuclear) for 1 hr, the reactive reagents within the gel were then precipitated in water for 1 hr. The gels were dried on a Bio-Rad Slab Gel Dryer, then placed against a blue-sensitive X-ray film (Kodak X-Omat AR film-XAR-2) and exposed for 1–4 weeks at −70°.

Maximum attachment of [^{3}H]EFDA to proteins occurred when JH was not present at the time of irradiation, while analysis by SDS-PAGE and fluorography confirmed that [^{3}H]EFDA photoattachment to a major protein occurred in each extract. The molecular weight of the hemolymph protein was estimated to be larger than 130,000.

Photoaffinity Labeling of Cytosolic JHBP of *Drosophila melanogaster* (Diptera) K_c Cells[8]

It has been demonstrated that cells from the *Drosophila* K_c line established by Echalier and Ohanessian[17] are useful as a model system for the study of insect endocrinology.[18] The advantages of using an established cell line are (1) endocrine glands and other tissues are absent in cell culture, which permits an unambiguous interpretation of cellular responses to hormone addition; (2) endogenous hormones are not present to dilute the specific activity of any exogenously added radioactive ligands; (3) the cells can be cloned to produce homogeneous populations; and (4) the extensive genetic knowledge of *Drosophila* can be used to assist in the elucidation of problems in insect development.

Previous work indicates that these cells do possess a cytosolic macromolecule that binds radiolabeled JH with high affinity, saturability, and specificity.[19] In addition, these cells do respond to JH at physiological concentrations[20] (L. Cherbas, personal communication). Extensive chemical characterization, however, of the intracellular JHBP has been hampered by the instability of the JH-binding protein complex. The use of

---

[17] G. Echalier and A. Ohanessian, *C.R. Hebd. Seances Acad. Sci.* **268**, 1771 (1969).
[18] J. D. O'Connor and E. S. Chang, *In* "Metamorphosis: A Problem in Developmental Biology" (L. I. Gilbert and E. Frieden, eds.), p. 241. Plenum, New York, 1981.
[19] E. S. Chang, T. A. Coudron, M. J. Bruce, B. A. Sage, J. D. O'Connor, and J. H. Law, *Proc. Natl. Acad. Sci. U.S.A.* **77**, 4657 (1980).
[20] E. S. Chang, A. I. Yudin, and W. H. Clark, Jr., *In Vitro* **18**, 297 (1982).

[³H]EFDA to form a covalent JHBP–ligand complex has enabled the completion of several experiments that were previously not possible.

Cell Culture and Preparation of Cytosol

Cells of the 7-D-11 clone[21] were maintained in D-20 medium[22] as a suspension culture in 3-liter spinner flasks (Bellco Glass). The cells were cultured at 25° at a stirring speed of 60 rpm in an air atmosphere.

Cells were harvested by centrifugation at 800 g for 8 min when the density was $2-5 \times 10^6$ cells/ml. They were then resuspended in TM buffer (10 mM Tris/5 mM MgCl$_2$, pH 6.9) and were disrupted with 20 strokes of a Dounce homogenizer (B pestle). The homogenate was centrifuged at 122,000 g for 1.5 hr to obtain a high-speed supernatant (cytosol). Dilutions of the cytosol were made with TMK buffer (10 mM Tris/5 mM MgCl$_2$/150 mM KCl, pH 7.4).

Competition of EFDA with JH for the Intracellular JHBP

Non-covalently bound hormone was assayed by means of dextran-coated charcoal (DCC) as prepared according to Kramer *et al.*[16] It was diluted to 0.16% final concentration in TMK buffer. The assay was carried out in PEG-coated tubes with radiolabeled hormone as described above. Following the addition of cytosol, the incubation mixtures were kept at 23° for 15 min with frequent gentle mixing. To terminate the incubation, the tubes were placed on ice, 50 μl of an ice-cold 0.3% solution of DCC added, and the tubes held for 15 min at 0°. The solutions were then centrifuged at 14,000 g for 15 min to pellet the charcoal. The amount of DCC used was sufficient to adsorb all of the unbound ligand from the incubation mixture. Proteins of the supernatant were removed for scintillation spectrometry.

Competition experiments similar to those described above for *Leucophaea* were conducted. These consisted of competing unlabeled EFDA with either [³H]JH I or III, labeled EFDA with either JH I or III without photolysis, and competing farnesyl acetate or methoprene against [³H]EFDA. All of the results were consistent with the interpretation that FFDA was competing with JH at the JH binding site.[8] The protection experiments described in previous sections, however, proved to be much more difficult to conduct. This was due to the inability to remove efficiently noncovalently bound ligand from the JHBP without losing most of the binding activity. Dissociation with 0.5 M phosphate or 0.75 M KCl was not successful. Precipitation of the protein with 75% ethanol, 22%

---

[21] Generously provided by Dr. J. D. O'Connor, Department of Biology, UCLA.
[22] G. Echalier and A. Ohanessian, *In Vitro* **5**, 162 (1970).

PEG, 50% ammonium sulfate, or 20% trichloroacetic acid all resulted in a significant loss of binding activity when the protein was resolubilized. In addition, neither concentrated DCC nor hydroxylapatite was effective in adsorption of noncovalently bound hormones. For these reasons, subsequent work was conducted with [^{3}H]EFDA. The following experiments were always conducted in parallel with solution containing a 50- to 500-fold amount of unlabeled JH III and the results consistently demonstrated that the EFDA and JH binding sites were identical.

Gel Permeation Column Chromatography

Gel permeation column chromatography was performed with Sephadex G-100 (Pharmacia, 24 × 1.6 cm bed) using TMK buffer as the eluent. The effluent was monitored at 280 nm with a flowthrough absorbance monitor (Pharmacia). The collected fractions were analyzed by scintillation spectrometry. For three separate analyses, only a single peak of radioactivity eluted from the column. This peak was absent if the cytosol was initially incubated with unlabeled JH III prior to incubation with [^{3}H]EFDA and subsequent ultraviolet irradiation. By means of a standard curve, a molecular weight of 49,200 ± 1600 was obtained.

Sucrose Gradient Centrifugation

Velocity sedimentation was conducted through preformed sucrose gradients (10–30%) in TMK buffer by centrifugation at 190,000 g for 22 hr in a SW 50.1 rotor (Beckman). Fractions were collected and analyzed by scintillation spectrometry. Results indicated that a single peak of radioactivity associated with a macromolecule had entered the gradient (see Fig. 8 in Chang [33]). By means of a standard curve of relative migration into the gradient versus low MW, a molecular weight of 51,000 ± 5800 was obtained for three separate analyses. No [^{3}H]EFDA entered the gradient if cytosol was initially incubated with excess unlabeled JH III.

Polyacrylamide Gel Electrophoresis

Precise determination of the molecular weight of proteins can be obtained with denaturing polyacrylamide gel electrophoresis (PAGE) using sodium dodecyl sulfate (SDS).[23] Since the denaturing conditions also disrupt the hydrophobic and hydrophilic bonds between the JH binding site and the ligand, SDS–PAGE was impossible without the use of covalently bound [^{3}H]EFDA. According to the method of Laemmli,[24] slab gels (0.75

[23] K. Weber and M. Osborn, *J. Biol. Chem.* **244**, 4406 (1969).
[24] U. K. Laemmli, *Nature (London)* **227**, 680 (1970).

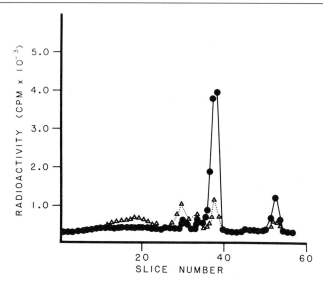

FIG. 4. Distribution of radioactivity in 2 mm slices of an electrophoresed SDS–polyacrylamide gel (12% separating, 5% stacking). A sample of K_c cytosol was photolyzed in the presence of [^{3}H]EFDA (1×10^{-6} M), either with (triangles) or without (circles) excess unlabeled JH III (5×10^{-5} M). Fraction 1 represents the top of the gel.

mm) were prepared with a 12% separating and a 5% stacking gel. Following electrophoresis at 4°, gels were stained for 1 hr in 50% trichloroacetic acid and 0.1% Coomassie blue and destained overnight with 7.5% acetic acid on a reciprocating rocker platform. Gels to be analyzed by scintillation spectrometry were sliced into 2 mm sections and the radioactivity was eluted overnight in 0.4 ml of 1% SDS at 37° in a shaker bath prior to addition of scintillation fluor.

Following electrophoresis, only a single band of radioactivity was observed (Fig. 4). However, the apparent molecular weight of the protein associated with the radiolabeled EFDA was 24,600 ± 1100 for six separate determinations. This is about one-half of the molecular weight obtained from the nondenaturing determinations (column chromatography and sucrose gradient centrifugation). The simplest explanation of this observation is that the JH binding protein consists of two similar subunits, each with a molecular weight of 24,600.

To obtain information about the number of binding sites per subunit of the JHBP, protease digests of the photolyzed cytosol were conducted by the addition of 0.1 volume of 0.01% bacterial protease (Sigma). The protease was added to the cytosol sample just prior to the initiation of electrophoresis. When the dye front had almost entered the separating gel, the current was turned off for 40 min. The current was then turned back on and the gel was electrophoresed as described above. For several determinations, a single radioactive peak was always observed. These results indicate that there is only a single binding site per JHBP subunit.

DNA-Cellulose Column Chromatography

Another characteristic of a hormone binding protein that is acting as a receptor is its ability to bind to DNA.[25] This was demonstrated for the K_c intracellular JHBP by DNA-cellulose (Sigma) column chromatography. The column was 9.5 × 1.6 cm and eluted with a step gradient of 0.0, 0.1, 0.3, 0.5, and 1.0 M KCl in TE buffer (10 mM Tris–HCl/1 mM disodium EDTA, pH 7.4). Effluent was monitored at 280 nm. Both photolyzed (radiolabeled) and unphotolyzed (unlabeled) cytosol were chromatographed. In the former experiments, the eluted fractions were analyzed for radioactivity by scintillation spectrometry. In addition, these fractions were concentrated by ultrafiltration (Amicon) and then analyzed by SDS–PAGE as described above. Following chromatography, it was observed that although only a small amount (less than 0.1%) of the added radioactivity was bound to the column. The majority of the counts that did bind to the column (98%) eluted with the 0.1 M KCl buffer. When this material was subsequently concentrated and electrophoresed on SDS–PAGE, the radioactivity continued to migrate to a position identical to the peak of radioactivity observed in SDS gels of unfractionated photolyzed cytosol.

Alternatively, unlabeled, unphotolyzed cytosol was applied to the DNA-cellulose column and unbound proteins were eluted with several volumes of buffer without KCl. Increasing concentrations of KCl in buffer were then applied to the column. After incubation of these KCl fractions with [^{3}H]EFDA and photolysis, aliquots were removed and assayed for total protein and also analyzed for bound [^{3}H]EFDA by ethanol precipitation. The protein that was not retained by the column (eluted in buffer without KCl) was able to bind 1.1 ng of [^{3}H]EFDA per mg of protein. The protein that was eluted with 0.1 M KCl, however, was able to bind 8.9 ng of labeled EFDA per mg of protein. This indicates that there was significant enrichment of the specific binding activity following affinity chromatography with DNA-cellulose. Both types of experiments described are consistent with the intracellular JHBP of the *Drosophilia* K_c cells acting as a hormone receptor for JH.

[25] W. T. Schrader, this series, Vol. 36, p. 187.

Photoaffinity Labeling of Hemolymph JHBP of *Locusta migratoria*[26]

The final example illustrates alternative assay and purification techniques for a second Orthopteran, the acridid *Locusta migratoria*. Adults of the grasshoppers *L. migratoria* and *Gomphocerus rufus* possess relatively high affinity ($K_D \simeq 10^{-7}$ M), high-molecular-weight (220,000) JH-binding lipoproteins.[27,28] The *Locusta* JHBP can discriminate the (10R)- and (10S)-enantiomers of JH III, but not of JH I.[29] We were thus curious to explore the nature of the JHBP of nymphal *L. migratoria* and to determine their relative binding affinities for EFDA.

Protein Preparation

Fifth-instar nymphal *L. migratoria* were bled into an ice-cooled tube from a puncture at the base of the metathoracic coxa, yielding 200 μl hemolymph per insect. Hemolymph was diluted before use as described below.

[^{3}H]JH III and [^{3}H]EFDA Binding (PEG Assay)

^{3}H-labeled JH III (1 × 10^{-9} M, specific activity 11 Ci/mmol) and varying concentrations of unlabeled EFDA (1 × 10^{-4} to 10^{-9} M) were added in 5 μl of ethanol to PEG 20,000-treated tubes containing hemolymph in 200 μl TMK (pH 7.4) buffer. An additional 100 μl of a 5 mg/ml solution of γ-globulin in TMK buffer was added, and the solutions were incubated 1 hr at 23°. The PEG 6000 assay described above for *Leucophaea* was used to separate bound and free hormone. Binding of [^{3}H]JH III was reduced in the presence of cold EFDA, such that 1 × 10^{-6} and 1 × 10^{-5} M EFDA reduced binding to 50 and 7% (respectively) of the competitor-free binding.

Hemolymph (HL) was serially diluted to find the optimal concentration for specific vs nonspecific binding of [^{3}H]EFDA (1 × 10^{-8} M, specific activity 5 Ci/mmol) under equilibrium (i.e., nonphotolytic) conditions, using 1 × 10^{-5} M unlabeled JH III to compete away the labeled hormone analog. A 1:400 dilution of HL gave 18% total counts bound (unprotected) and 3.8% counts bound in the presence of JH III, or 79% specific binding of [^{3}H]EFDA to an HL-JHBP. Below 1:3200 no binding was detected; above 1:100 nonspecific binding exceeded 50%.

[26] J. J. Brown, G. D. Prestwich, and C. A. D. deKort, unpublished data.
[27] R. Hartmann, *Wilhelm Roux's Arch. Dev. Biol.* **184**, 301 (1970).
[28] H. Emmerich and R. Hartmann, *J. Insect Physiol.* **19**, 1663 (1973).
[29] M. G. Peter, S. Gunawan, G. Gellissen, and H. Emmerich, *Z. Naturforsch.*, **34C**, 588 (1979).

Covalent [³H]EFDA Binding (TCA Filter Assay)

The *Locusta* HL-JHBP was relatively resistant to degradation during photolysis (16 lamps, 254 nm, Rayonet reactor, 0°). [³H]JH III binding capacity was reduced negligibly at 60 sec and 20% after 120 sec irradiation.

The covalent attachment of [³H]EFDA to *Locusta* HL-JHBP was determined by the TCA filter assay as described above for *Leucophaea*. In this case, a 10-sec exposure to 16 254-nm lamps maximized [³H]EFDA bound. Hemolymph samples (1:400) bound 18.5% of the total dpm when incubated with 1×10^{-8} M [³H]EFDA. As controls, γ-globulin solution and HL-JHBP solutions bound less than 2% of the total counts either with or without irradiation. Protection experiments with JH III gave results analogous to those obtained under the equilibrium conditions described above for the PEG assay.

HPLC Separation of JHBP

A 1-μl sample of hemolymph diluted 1:19 with 0.02 M phosphate buffer was injected onto a Spherogel-TSK IEX-545 DEAE (6 × 150 mm) ion exchange column. Delipidation[12] with diethyl ether–ethanol (2:1) at 4° removes free lipids which otherwise cause clogging and irreversible damage to the HPLC column. Isocratic elution at 0.8 ml/min for 10 min with a pH 7.0 phosphate buffer, followed by 30 min gradient from 0 to 70% 0.5 N NaCl allowed separation of the HL proteins (Fig. 6). Total protein was followed by OD_{280} and by the Coomassie blue method.[30] Fractions 34–40 (between 16.5 and 23% of the 0.5 N NaCl gradient), eluted prior to most proteins and showed a peak of [³H]JH III binding activity (Fig. 5). For example, fraction 38 bound 18% of the [³H]JH III compared to 42% for unfractionated hemolymph at the same concentration.

[³H]EFDA Binding to HPLC-Separated Proteins

A 2-μl sample of hemolymph was fractionated by HPLC and eighty 400-μl fractions were assayed for [³H]EFDA binding (2×10^{-9} M) using the PEG assay. This photoaffinity ligand bound to fractions 34 to 41, the same as the [³H]JH III binding fractions (Fig. 5). Experiments in which [³H]EFDA–protein solutions were photolyzed and assayed by the TCA filter method confirmed that irradiation caused covalent attachment to these protein fractions.

---

[30] M. M. Bradford, *Anal. Biochem.* **72**, 248 (1976).

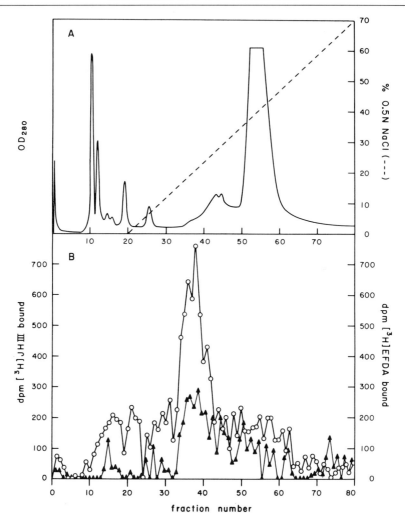

FIG. 5. Ion-exchange HPLC fractionation of *Locusta* hemolymph (see text for conditions): (A) Solid line, proteins (OD_{280}); dashed line, NaCl gradient. (B) dpm [^{3}H]JH III (1×10^{-9} M) bound (circles), and dpm [^{3}H]EFDA (2×10^{-9} M) bound (triangles).

Competitive Binding Studies with HPLC Separated HL-JHBP (Fig. 6)

Fractions 34 to 42 from an HPLC fractionation of 2 μl of *Locusta* HL were divided into four equal 100 μl subfractions. They were incubated with 1×10^{-9} M [^{3}H]JH III and (1) 5 μl ethanol (control), (2) 1×10^{-5} M unlabeled EFDA, (3) 1×10^{-7} M EFDA, or (4) 1×10^{-9} M EFDA. EFDA

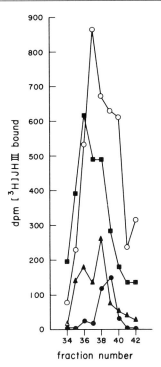

Fig. 6. Competition of EFDA with [^{3}H]JH III for HPLC separated JHBP fractions: dpm 10^{-9} M [^{3}H]JH III bound without EFDA (open circles), with 10^{-5} M EFDA (closed circles), with 10^{-7} M EFDA (triangles), and 10^{-9} M EFDA (squares).

at high concentrations reduced the [^{3}H]JH III binding of the HPLC fractions analogous to its inhibition in assays using whole blood. This further supports the interaction of the photoactivatable ligand with the JH-binding site of the JH-binding protein.

Conclusion

EFDA has clearly demonstrated its potential as a biochemical tool for studying juvenile hormone binding proteins. It competes with natural JHs for the JHBP, and becomes covalently attached to these proteins with good specific binding under mild photolysis conditions. The stage is now set for using EFDA or more highly tritiated analogs as probes for tissue JH receptor proteins. We are now able to ask questions about the amino acid domains at the JH binding site, and to thereby further comprehend the molecular details of JH-JHBP interactions.

Acknowledgments

We thank the Alfred P. Sloan Foundation and the Camille and Henry Dreyfus Foundation for awards to G.D.P. We also acknowledge financial support from NIH Grant GM-30899 to G.D.P. and NSF Grants PCM-8021190 to J.K.K. and CHE-7925081 to G.D.P. We are grateful for a gift of sodium borotritide from Amersham. J.J.B. thanks Dr. C. A. D. deKort for laboratory facilities and intellectual stimulation in Wageningen. E.S.C. acknowledges the technical assistance of M. J. Bruce.

[35] Radiochemical Assay for Juvenile Hormone III Biosynthesis *in Vitro*

By R. FEYEREISEN

The development of a radiochemical assay for juvenile hormone (JH) biosynthesis by Pratt and Tobe[1,2] has greatly facilitated research in various areas of JH biochemistry and physiology, including the identification of secretion products of the corpora allata (CA); the developmental variations in rates of JH synthesis, the endocrine regulation of JH synthesis; the quest for neurohormones stimulating (allatotropins) or inhibiting (allatostatins) JH synthesis; the biochemical pathways of JH synthesis and their substrate specificity and rate limitation; the assay of chemical inhibitors of JH synthesis; the biosynthesis of optically active JH radiolabeled in the methyl ester group and/or in the isoprenoid carbon chain for enzymatic or receptor studies requiring the natural enantiomer rather than a racemic mixture. The radiochemical assay for JH biosynthesis does not share its range of applications with JH radioimmunoassays (RIA)[3] because it is based on a quite different principle.

Pratt and Tobe reported that CA of adult female *Schistocerca gregaria* utilize both [^{3}H]farnesoic acid and [*methyl*-^{14}C]methionine in short-term *in vitro* incubations to produce very large amounts of double-labeled JH III.[1] They showed that the rates of JH III synthesis from these 2 exogenous, labeled precursors were linear over periods of up to 4 hr after removal of the CA from the insect.[4] These experiments also made it possible to measure the stoichiometry of [^{3}H]farnesoic acid and [*methyl*-^{14}C]methionine utilization: a molar incorporation ratio close to unity was obtained *in vitro*, indicating that under those conditions radiolabeled

[1] G. E. Pratt and S. S. Tobe, *Life Sci.* **14**, 575 (1974).
[2] S. S. Tobe and G. E. Pratt, *J. Exp. Biol.* **62**, 611 (1975).
[3] N. A. Granger and W. G. Goodman, *Insect Biochem.* **13**, 333 (1983).
[4] S. S. Tobe and G. E. Pratt, *Biochem. J.* **144**, 107 (1974).

methionine was virtually the exclusive source of the methyl ester group of JH III. The same conclusion was reached by Judy et al. who analyzed by mass spectrometry the ratio of ^{14}C to ^{12}C in JH III biosynthesized by *Manduca sexta* CA.[5]

The conclusion that radioactivity from [*methyl*-^{14}C]methionine is incorporated into the methyl ester moiety of JH without significant dilution has led to the development of a short-term radiochemical assay for the quantitative measurement of the endocrine activity of CA.[6,7]

Principle

The radiochemical assay for CA activity is truly unique in terms of assays for the activity of an endocrine gland. Indeed, its simplicity is that hormone production *in vitro* can be readily measured because *all* the hormone produced is radiolabeled at the methyl ester group. As part of the normal biosynthetic pathway, the S-methyl group of methionine is transferred enzymatically (probably via S-adenosylmethionine[8,9]) to a carboxylic acid precursor of JH. The pool of methionine and S-adenosylmethionine in the CA is extremely small, so that transfer of the glands to an *in vitro* culture medium readily dilutes the nonradioactive methionine from the CA with radioactive methionine from the medium. Consequently, after a short lag period[10] which represents the time necessary to achieve isotopic equilibrium within the CA (i.e., finish dilution) all biosynthesized JH is labeled.

The assay has been used in recent years with corpora allata from locusts, cockroaches, beetles, etc.[7] The procedures described below are used in our laboratory with the viviparous cockroach, *Diploptera punctata*, and integrate a few minor modifications from the original procedure.

Incubation of Corpora Allata

Adult *D. punctata* females are decapitated and the CA are dissected in sterile saline (Yaeger's cockroach saline: 10.93 g NaCl; 1.57 g KCl; 0.85 g

[5] K. J. Judy, D. A. Schooley, M. S. Hall, B. J. Bergot, and J. B. Siddall, *Proc. Natl. Acad. Sci. U.S.A.* **70**, 1509 (1973).
[6] G. E. Pratt and R. J. Weaver, in "Comparative Endocrinology" (P. J. Gaillard and H. H. Boer, eds.), p. 503. Elsevier/North-Holland Biomedical Press, Amsterdam, 1978.
[7] S. S. Tobe and R. Feyereisen, in "Endocrinology of Insects" (R. G. H. Downer and H. Laufer, eds.), p. 161. A. R. Liss, Inc., New York, 1983 (in press).
[8] D. Reibstein and J. H. Law, *Biochem. Biophys. Res. Commun.* **55**, 266 (1973).
[9] R. Feyereisen, G. E. Pratt, and A. F. Hamnett, *Eur. J. Biochem.* **118**, 231 (1981).
[10] G. E. Pratt, S. S. Tobe, R. J. Weaver, and J. R. Finney, *Gen. Comp. Endocrinol.* **26**, 478 (1975).

CaCl$_2$; 0.17 g MgCl$_2$ per liter). Dissection is performed in a laminar flow hood and illumination is provided by fiber optic guides to prevent overheating of the glands during the operation. The CA are freed as much as possible from extraneous tissues such as fat body or nervous tissue. Some corpus cardiacum tissue is left attached to the glands and does in fact keep the glands paired, which helps in the transfer of the CA. In *D. punctata*, the presence or absence of corpora cardiaca during incubation of the CA does not affect the rate of JH synthesis.[11] It can easily be shown, however, that the presence of other tissue will lead to lower apparent rates of JH synthesis. This is caused by the presence of JH esterase which hydrolyzes the labeled methyl ester. Hemolymph JH esterase activity can be removed by careful dissection and preincubation rinses of the glands,[12] but the continued presence of JH esterase in (or around) the CA in certain species (such as *M. sexta*[13]) may invalidate results obtained with the radiochemical assay. After dissection, the CA are transferred with a stainless-steel loop mounted on a drawn out pasteur pipette to a small centrifuge tube with conical bottom (Kimble no. 45150, 3 ml centrifuge tube) containing approximately 100 µl of incubation medium. For *D. punctata* CA the incubation medium of choice is medium TC 199 (Gibco, with L-glutamine, Hanks' salts and 25 mM HEPES buffer, pH 7.2) to which 20 mg/ml Ficoll (type 400, Sigma) is added prior to sterile filtration (Gelman acrodiscs, 0.2 µm pore size). When all glands to be assayed have been dissected, the "holding" incubation medium is removed with a finely drawn pasteur pipette (to prevent suction of the CA; indirect light on a dark background surface helps in keeping visual contact with the single gland pair). It is immediately replaced with 100 µl incubation medium (see above) containing 1.25 µCi of [*methyl*-^{14}C]methionine (Amersham/Searle Corporation). Radiolabeled methionine is prepared as follows. A 250 µCi batch of dry solid [*methyl*-^{14}C]methionine is dissolved in 2 ml of H$_2$O and 100-µl aliquots are dispensed in 1-ml vials and freeze dried, then stored until use at −20°. One vial reconstituted with 1 ml of incubation medium suffices for 10 assays. Final concentration of methionine in the incubation medium is calculated from the radioactivity of 5-µl aliquots of the medium and the specific activity of the radiolabel (50–60 mCi/mmol), taking into account the unlabeled methionine (0.1 mM) present in medium TC 199. The incubation tubes are capped with a piece of parafilm and incubated at 28° in the dark with gentle shaking (Dubnoff incubator) for 3 hr. When assaying glands of a species not previously studied, an optimal culture

[11] B. Stay and S. S. Tobe, *Gen. Comp. Endocrinol.* **33**, 531 (1977).
[12] B. D. Hammock and S. M. Mumby, *Pestic. Biochem. Physiol.* **9**, 39 (1978).
[13] S. J. Kramer and F. Kalish, *J. Insect Physiol.* **30**, 311 (1984).

medium should be found. Parameters such as pH and Na^+/K^+ ratio are important[6] and it is desirable to use chemically defined media, rather than media supplemented with vertebrate or insect serum fractions. Suitability of incubation media is discussed by Weaver et al.[14] and Khan et al.[15] indicate that *Leptinotarsa decemlineata* CA should be dissected in isotonic saline and incubated in medium containing only 10 mg/ml Ficoll. Dose–response curves for the effect of methionine concentrations on rate of JH synthesis have usually revealed a wide plateau with optimal concentrations around 0.3 mM. Recently, Tobe and Clarke[16] have shown that optimal rates of JH III synthesis are obtained at concentrations as low as 40 μM methionine in *D. punctata*.

Assay of Biosynthesized JH III

After incubation of the CA, the biosynthesized JH III can be extracted and quantified by two procedures. The initial procedure developed by Pratt and Tobe[1] involves thin-layer chromatography and allows the quantification of intraglandular and released JH III as well as the intraglandular precursor methyl farnesoate. A more recent partition assay allows only the assay of JH III released into the incubation medium.[17]

TLC Assay. JH III biosynthesis is stopped by the addition of 100 μl 1% EDTA and 200 μl ethanol to the incubation tube. Twenty microliters unlabeled reference carriers/markers is added (a stock solution of markers is prepared by adding 5 μl each of JH III and 2E,6E-methyl farnesoate with a Drummond 5 μl microcap in 5 ml of isooctane; this stock solution is good for 250 assays and is stored at $-20°$). The mixture is then extracted with 1 ml of redistilled chloroform, vortexed, and centrifuged at 2500 g for 5 min to separate phases. The hypophase ($CHCl_3$ phase) is removed with a drawn pasteur pipet and dried over Na_2SO_4 as follows: a pasteur pipet plugged with a minimum of cotton is filled for about 3 cm with Na_2SO_4. This Na_2SO_4 filter is placed in a 3-ml centrifuge tube to collect the chloroform extract (cotton and pasteur pipet are washed with diethyl ether prior to use). The hyperphase is reextracted with 500 μl $CHCl_3$ and the transfer pipet and Na_2SO_4 filter pipet are rinsed with another 500 μl $CHCl_3$. The combined $CHCl_3$ extract (2 ml) is then taken to near dryness under nitrogen in a Pierce Reacti-therm at 30° equipped with a Reacti-vap evaporat-

[14] R. J. Weaver, G. E. Pratt, A. F. Hamnett, and R. C. Jennings, *Insect Biochem.* **10**, 245 (1980).
[15] M. A. Khan, A. Doderer, A. B. Koopmanschap, and C. A. D. de Kort, *J. Insect Physiol.* **28**, 279 (1982).
[16] S. S. Tobe and N. Clarke, *Insect Biochem.*, in press (1984).
[17] R. Feyereisen and S. S. Tobe, *Anal. Biochem.* **111**, 372 (1981).

ing unit. The residue is recovered with diethyl ether and spotted on TLC plates prepared as follows: Plastic-backed 20 × 20-cm plates (Merck silica gel 60 F_{254}, layer thickness 0.2 mm) are washed twice by full development in MeOH (check under UV light). Up to 10 lanes per plate can be easily separated by double tracks scraped off the plate with a spatula (a TLC template, Desaga, is very helpful). The bottom 1 cm of the plate is left unscraped. A spotting area is delineated with two fine pencil lines at 2 and 4 cm from the bottom of the plate. Drummond 100 µl microcaps held by a Clay Adams suction apparatus allow a rapid spotting of the diethyl ether extract. A constant flow of cold air dries the area during spotting. The spotted products are focused by two developments in methanol up to the origin line, i.e., the top of the spotting area, 4 cm from the bottom of the plate. Separation of JH III and methyl farnesoate is achieved by one development in the solvent system hexane/ethyl acetate (3:1 v/v). The position of the unlabeled carriers is visualized under UV light and broad areas around the methyl farnesoate (highest R_f) and JH III (lowest R_f) bands are marked with a pencil and carefully cut out of the plate. The plate sections are assayed for radioactivity in 20-ml scintillation vials with 10 ml of Aquasol (NEN). Recoveries for this extraction and separation procedure are reported to be between 85 and 95% (Fig. 1). More detailed procedures for the localization of JH III and methyl farnesoate on TLC plates and their subsequent identification have been described by the Brighton group.[1,14,18,19] It has been pointed out that isotope effects in chromatography may result in a noncoincidence of the unlabeled markers and the labeled products when using [*methyl*-³H]methionine.[18,19] This problem does not arise, however, with [*methyl*-¹⁴C]methionine. The TLC procedure can be applied to the incubation medium with the CA and will thus measure total JH III and methyl farnesoate synthesis. It can also be applied to the incubation medium and the CA separately, thus measuring JH release on one hand and intraglandular levels of JH III and methyl farnesoate on the other.

Partition Assay.[17] The partition assay was developed as a rapid assay for JH III *released* by the CA, i.e., it is applicable only to the incubation medium. Isooctane (250 µl) (2,2,4-trimethylpentane) is added to the incubation medium and thoroughly mixed. After a 5 min centrifugation at 2500 g, a 200-µl aliquot from the isooctane hyperphase is assayed for radioactivity by liquid scintillation spectrometry (Fig. 2). In a variant of the same procedure, Kramer and Kalish[13] extract the medium with 500 µl isooctane

[18] A. F. Hamnett and G. E. Pratt, *J. Chromatogr.* **158**, 387 (1978).
[19] A. F. Hamnett, G. E. Pratt, K. M. Stott, and R. C. Jennings, *in* "Juvenile Hormone Biochemistry" (G. E. Pratt and G. T. Brooks, eds.), p. 93. Elsevier/North-Holland Biomedical Press, Amsterdam, 1981.

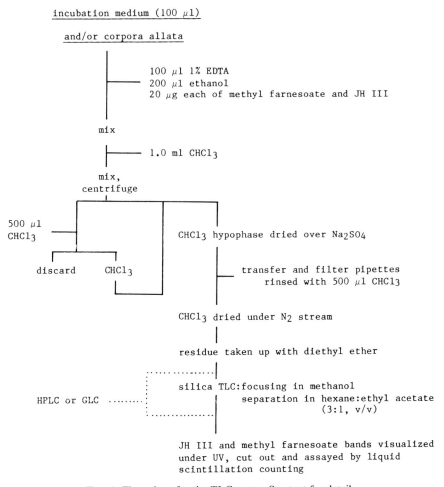

FIG. 1. Flow chart for the TLC assay. See text for details.

and assay a 400-μl aliquot for radioactivity. It is mandatory to run a blank (incubation medium in which no CA were incubated) to measure and subtract the non-JH III radioactivity carried over to the isooctane phase. In a blank series, we have found this contamination to be highly reproducible (401 ± 10 dpm), although the absolute value of the blank may vary slightly between batches of methyl-labeled methionine. Reversed-phase HPLC analysis of the isooctane phase has shown that *D. punctata* CA do not release any labeled non-JH III metabolite of methionine into the incubation medium, and it is thus justified to subtract the radioactivity of a

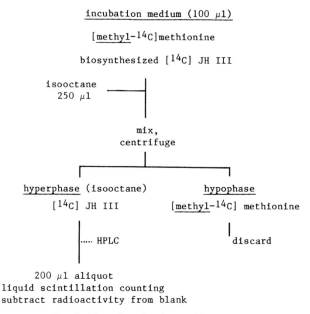

FIG. 2. Flow chart for the partition assay.

blank to obtain the value for the JH III released. Several solvents were compared for the partition assay. Chloroform was eliminated because of its high quenching properties. Ethyl acetate carried over too much methionine (>0.1%). Isooctane was chosen for its high boiling point (i.e., low volatility), its good extraction efficiency and the low contamination by labeled methionine (see the table). Isooctane is effective in extracting JH from incubation media other than TC 199, for instance Grace's medium. However, when small amounts of JH are present in a medium containing high amounts of protein (0.1% BSA), the nonspecific, hydrophobic binding of JH to BSA delays partition of JH into isooctane. Partition becomes time dependent and poor results may be obtained. The partition behavior of methionine is, however, not affected by the presence of massive amounts of protein in the incubation medium. The partition assay is very useful in those species where initial information has been obtained by the TLC assay on the identity of JH produced by the CA, and on the relationship between synthesis and release. When the rate of JH release is extremely low, the value of the blank may become very significant when compared to the amount of radioactivity incorporated into JH III. In this case, a modified TC 199 medium lacking unlabeled methionine can be used. This allows [*methyl*-14C]methionine to be used without dilu-

RECOVERY OF JH III AND CONTAMINATION OF THE ORGANIC
SOLVENT IN THE PARTITION ASSAY

Organic solvent[a]	JH III recovered[b] (%)	Non-JH III radioactivity carried over to the organic solvent[c] (%)
Isooctane	98.1 ± 4.7	0.014
Toluene	97.3 ± 0.6	0.026
Hexane	96.4 ± 3.8	0.013
Xylene	95.5 ± 6.4	0.024
Benzene	93.1 ± 6.9	0.027
Heptane	91.3 ± 1.8	0.011

[a] Incubation medium 199 (100 µl) was extracted with 250 µl of organic solvent.
[b] The percentage of JH III recovered was determined using [10^{3}H]JH III solutions in incubation medium 199.
[c] The percentage contamination was determined using incubation medium 199 containing [*methyl*-^{14}C]methionine. For each procedure 3 to 6 replicates were analyzed. From R. Feyereisen and S. S. Tobe, *Anal. Biochem.* **111**, 372 (1981).

tion. The next step would be to use higher specific activity [*methyl*-^{3}H]methionine.

Validation of the Radiochemical Assay

Although the radiochemical assay for JH synthesis is extremely simple in its principle and procedure, a number of criteria[6] must be met before it is used routinely. These criteria have been reviewed recently[7,20] and are summarized below:

1. The JH synthesized *in vitro* must be identified by unequivocal methods. To this date, only lepidopteran CA have been shown to synthesize JHs other than JH III.

2. Linearity of JH synthesis and/or release after removal of the CA from the insect must be documented for glands of high and low activity. In some insects, e.g., locusts, linearity of synthesis is maintained only for a few hours *in vitro*.

[20] R. Feyereisen, in "Comprehensive Insect Physiology, Biochemistry and Pharmacology" (G. A. Kerkut and L. I. Gilbert, eds.), Vol 7, Chapter 11, p. 391. Pergamon, Oxford, 1985 (in press).

3. Optimal incubation conditions should be found and of course the tissues and incubation medium must be free of any JH esterase activity.

4. A crucial, but more difficult parameter is that radioactivity from the S-methyl group of methionine must be incorporated at a fixed molar ratio, ideally a ratio of 1 : 1 with the carbon skeleton of JH. In other words, the specific activity of radiobiosynthesized JH should be equal to the specific activity of methionine in the incubation medium. As pointed out above, this feature is the basis of the radiochemical assay. It nonetheless needs to be confirmed for each species investigated, because it justifies the calculation of molar quantities of biosynthesized JH. Various techniques have been used to establish the molar incorporation ratio.[1,5,21]

Applications of the Radiochemical Assay

A list of possible applications of the radiochemical assay for JH synthesis *in vitro* is given in the introduction, and most have been reviewed recently[7,20]; only a few additional comments will be presented here.

The sensitivity of the radiochemical assay is a function of the final specific activity of the *methyl*-labeled methionine in the incubation medium. Thus, using normal TC 199 or MEM incubation media which contain 0.1 mM L-methionine, and [*methyl*-^{14}C]methionine at 50–60 mCi/mmol, the sensitivity of the assay is approximately 0.1 pmol per gland pair per hr. Better sensitivities are achieved when using incubation media lacking unlabeled methionine, or when using [*methyl*-^{3}H]methionine.[16] Hamnett and Pratt[18] reported a detection limit in the femtomole per gland pair per hour range when using [*methyl*-^{3}H]methionine at the highest specific activity available (80 Ci/mmol), and discussed some of the applications of a high sensitivity assay coupled with reliable separation techniques. Because the juvenile hormones produced by corpora allata *in vitro* are optically active (10R configuration for JH III, 10R, 11S for the higher homologs) the radiochemical assay can be used to biosynthesize the correct enantiomer for studies where racemic mixtures (i.e., the commercially available labeled JHs) are inadequate, for instance carrier protein and receptor binding studies. Using CA of *D. punctata* which produce massive amounts of JH III at a precise stage of development (adult females aged 4–5 days) and methionine-free TC 199 medium, we have been able to produce up to 0.1 mCi of very high specific activity JH III in a single experiment, from [*methyl*-^{3}H]methionine at 80 Ci/mmol (R. Feyereisen and S. S. Tobe, unpublished). Purification of the high specific activity JH III presents some yet unresolved problems which may be

---

[21] R. Feyereisen, T. Friedel, and S. S. Tobe, *Insect Biochem.* **11**, 401 (1981).

related to the instability of compounds labeled at near maximal isotopic abundance of tritium (H. Emmerich and G. E. Pratt, personal communications). This procedure is different from the Peter *et al.* procedure[22] in which optically pure JH I is prepared enzymatically using $S[methyl-^3H]$-adenosylmethionine, racemic JH I-acid, and an accessory sex gland homogenate of adult male *Hyalophora cecropia* (see G. Weirich, this volume [36]). The corpora allata of *S. gregaria* can be forced to make JH I when presented with bishomofarnesoic acid,[1] but the optical purity of the product has not been determined. It is probably chiral, however, in view of the stereoselectivity of methyl farnesoate epoxidase in another locust.[9] The O-methyltransferase of *L. migratoria* CA is not stereo selective[23] as opposed to the enzyme from *H. cecropia* accessory sex glands, and thus it would be expected that CA of Orthoptera or Dictyoptera presented with racemic JH acid would produce racemic mixtures of JH.

Another interesting application, where very high specific activities are not needed, is the biosynthesis of JH labeled both in the carbon chain and in the methyl ester group. Using [2-^{14}C]acetate and [*methyl*-^{14}C]methionine, double-labeled, optically pure JH III can easily be obtained in reasonable quantities from *D. punctata* CA *in vitro*. This product can be used for metabolic studies *in vitro* or *in vivo*.

Acknowledgments

I wish to thank Dr. G. E. Pratt and Dr. S. S. Tobe for having stimulated and supported my interest in JH biosynthesis.

---

[22] M. G. Peter, S. Gunawan, and H. Emmerich, *Experientia* **35**, 1141 (1979).
[23] G. E. Pratt, K. M. Stott, G. T. Brooks, R. C. Jennings, A. F. Hamnett, and B. A. J. Alexander, in "Juvenile Hormone Biochemistry" (G. E. Pratt and G. T. Brooks, eds.), p. 107. Elsevier/North-Holland Biomedical Press, Amsterdam, 1981.

[36] Epoxyfarnesoic Acid Methyltransferase

By GUNTER F. WEIRICH

JH-I acid	JH-I	$R^1 = CH_3$	$R^2 = CH_3$
JH-II acid	JH-II	$R^1 = CH_3$	$R^2 = H$
JH-III acid	JH-III	$R^1 = H$	$R^2 = H$

Insect juvenile hormones (JHs)[1,2] are a unique group of sesquiterpene esters which are generally produced in the corpora allata, a pair of glands located behind the brain. In the tobacco hornworm, *Manduca sexta*,[3,4] the giant silkmoth *Hyalophora cecropia*,[5-8] and probably in other lepidopteran species, the last step of the biosynthetic pathway for JHs is the methyl transfer from *S*-adenosyl-L-methionine (AdoMet) to the carboxyl group of the precursor acids (JH-I acid, JH-II acid, and/or JH-III acid). This reaction is catalyzed by epoxyfarnesoic acid methyltransferase.[9]

[1] L. I. Gilbert, ed., "The Juvenile Hormones." Plenum, New York, 1976.
[2] G. E. Pratt and G. T. Brooks, eds., "Juvenile Hormone Biochemistry." Elsevier/North-Holland Biomedical Press, Amsterdam, 1981.
[3] D. Reibstein and J. H. Law, *Biochem. Biophys. Res. Commun.* **55**, 266 (1973).
[4] D. Reibstein, J. H. Law, S. B. Bowlus, and J. A. Katzenellenbogen, *in* "The Juvenile Hormones" (L. I. Gilbert, ed.), p. 131. Plenum, New York, 1976.
[5] M. Metzler, D. Meyer, K. H. Dahm, H. Röller, and J. B. Siddall, *Z. Naturforsch., B: Anorg. Chem., Org. Chem., Biochem., Biophys., Biol.* **27**, 321 (1972).
[6] G. F. Weirich and M. G. Culver, *Arch. Biochem. Biophys.* **198**, 175 (1979).
[7] M. G. Peter, P. D. Shirk, K. H. Dahm, and H. Röller, *Z. Naturforsch., C: Biosci.* **36**, 579 (1981).
[8] K. H. Dahm, G. Bhaskaran, M. G. Peter, P. D. Shirk, K. R. Seshan, and H. Röller, *in* "Regulation of Insect Development and Behaviour" (F. Sehnal, A. Zabza, J. J. Menn, and B. Cymborowski, eds.), p. 183. Wroclaw Technical University Press, Poland, 1981.
[9] The enzyme described here reacts mainly with the higher homologs of epoxyfarnesoic acid and might be classified more accurately as (di)homofarnesoic acid methyltransferase. However, for practical reasons the shorter name will be used throughout this article.

Epoxyfarnesoic acid methyltransferase has been demonstrated in corpora allata homogenates obtained from adult *M. sexta*[4] and *H. cecropia*[8] females but is lacking from corpora allata of adult *H. cecropia* males.[8] The corpora allata of males secrete the precursor acid rather than JH into the hemolymph and the acid is sequestered and esterified by the accessory reproductive glands,[8,10] which contain epoxyfarnesoic acid methyltransferase.[6] The methyltransferase of the accessory reproductive glands is a soluble enzyme.[6] Both methyltransferases of *H. cecropia* (in female corpora allata and in male accessory glands) convert the higher precursor homologs (JH-I acid and JH-II acid) more efficiently than the JH-III acid,[7,8] and the accessory gland enzyme is highly enantioselective for the natural $(10R,11S)$-acid.[7,11]

Because of their size, the *H. cecropia* accessory glands are a better source for the methyltransferase than the corpora allata. Enzyme preparations from the accessory glands have been used as a micropreparative tool for the esterification of JH acids.[7,8,11] This method offers the advantages of higher yields, fewer byproducts, and high enantioselectivity over the classical chemical methylation reaction with diazomethane. With high specific activity [*methyl*-^{3}H]AdoMet, [^{3}H]JH of high specific activity can be prepared.[11]

A similar enzyme, farnesoic acid methyltransferase, has been demonstrated in cell-free fractions from corpora allata of *Locusta migratoria*.[12] The name of the latter enzyme is not based on its preference for farnesoic acid over epoxyfarnesoic acid as substrate[13] but on its presumed physiological substrate in *L. migratoria*.

Described below are procedures for extracting epoxyfarnesoic acid methyltransferase from the accessory glands and using the enzyme preparation in reactions with AdoMet and the precursor acids of the insect juvenile hormones.

Enzyme Preparation

Accessory reproductive glands of pharate adult (developing adult inside the pupal cuticle) *H. cecropia* males, during the last days of adult development, or of young adults are used as source for the enzyme. The glands begin to accumulate JH acid and convert it to JH around the time

---

[10] P. D. Shirk, G. Bhaskaran, and H. Röller, *J. Exp. Zool.* **227**, 69 (1983).
[11] M. G. Peter, S. Gunawan, and H. Emmerich, *Experientia* **35**, 1141 (1979).
[12] R. Feyereisen, G. E. Pratt, and A. F. Hamnett, *Eur. J. Biochem.* **118**, 231 (1981).
[13] G. E. Pratt, K. M. Stott, G. T. Brooks, R. C. Jennings, A. F. Hamnett, and B. A. J. Alexander, in "Juvenile Hormone Biochemistry" (G. E. Pratt and G. T. Brooks, eds.), p. 107. Elsevier/North-Holland Biomedical Press, Amsterdam, 1981.

of adult eclosion.[6,8] Therefore, the timing of the animals is crucial if interference from endogenous precursors and/or products of the reaction are to be avoided. These problems can be eliminated by the removal of the corpora allata during the pupal stage from animals later to be used as accessory gland donors.[7,8]

Accessory glands are homogenized in Grace's tissue culture medium or 0.05 M Tris–HCl buffer, pH 7.2, with small all-glass homogenizers, and the homogenate is centrifuged for 60 min at 100,000 g. Both operations are carried out at 0–4°. The supernatant can be used as enzyme source when substrates and cofactors originating from the gland are of no consequence. If precise quantitation of those components is required or dilution of radiolabeled substrate(s) is to be avoided, the 100,000 g supernatant has to be fractionated first on a Sephadex G-25 column.[11] Enzyme activity may vary considerably between animals. To ensure a uniform enzyme supply over a period of time, enzyme is prepared from several glands and stored in suitable aliquots at −80°. At this temperature the methyltransferase is stable for at least 5 months.[11]

Substrates

JH acids are prepared from JHs by enzymatic hydrolysis.[5,7] Synthetic JHs are available from various suppliers as racemic mixtures and will yield the corresponding racemic JH acid mixtures. [^{3}H]JHs are available from New England Nuclear.[14] Hemolymph from insects at a developmental stage rich in JH esterase and low in JH (e.g., wandering fifth-instar larvae of *M. sexta*[15,16]) or, if available, a (partially) purified JH esterase can be used as enzyme preparation. Incubation of 30 μg JH-I in 1 ml of diluted *M. sexta* hemolymph (1/4 dilution in 50 mM Tris–HCl buffer, pH 7.2) will yield approximately 15 μg JH-I acid after 2 hr at 30°. The reaction mixture is extracted with ethyl acetate, and JH acid is purified by thin-layer chromatography (TLC) on silica with ethyl acetate/glacial acetic acid/benzene (15/1/84) as solvent.

AdoMet should be free of *S*-adenosyl-L-homocysteine, a potent inhibitor of methyltransferases. If necessary, AdoMet can be purified by ion exchange chromatography.[17] Radiolabeled AdoMet preparations are supplied in sulfuric acid solutions and have to be neutralized by an adequate amount of buffer (Tris or phosphate) before other components of the reaction mixture are added.

---

[14] Mention of a company or proprietary product does not imply endorsement by the U.S. Department of Agriculture.

[15] G. Weirich, J. Wren, and J. B. Siddall, *Insect Biochem.* **3**, 397 (1973).

[16] R. K. Vince and L. I. Gilbert, *Insect Biochem.* **7**, 115 (1977).

[17] S. K. Shapiro and D. J. Ehninger, *Anal. Biochem.* **15**, 323 (1966).

Reaction Mixture

Epoxyfarnesoic acid methyltransferases have not been characterized completely and no kinetic parameters are available. The accessory gland enzyme works well at 25° in Grace's medium[6] or 50 mM Tris–HCl buffer, pH 7.2,[11] without added cofactors, even after Sephadex G-25 fractionation of the enzyme preparation.

The choice of substrate concentrations depends on the application of the enzyme reaction. If enzyme activity is to be determined, high substrate concentrations will have to be used. If, however, the enzyme is to be employed for a micropreparative application, the concentrations will be determined by the availability of the radiolabeled substrate(s). For the radiochemical assay of the enzyme activity, the radiotracer can be introduced by either one of the substrates (e.g., [^{3}H]JH acid or [*methyl*-^{3}H] AdoMet).

Incubations containing 1 μM JH-I acid (racemic), 100–200 μM AdoMet, and accessory gland homogenate or high-speed supernatant equivalent to 1/4 gland (1/8 gland pair, 100–200 μg protein) have yielded 30–50% methylation in 3-hr incubations[6] (50% methylation is the maximum yield obtainable from a racemic JH acid mixture). A 70-fold increase in the JH-I acid concentration resulted in an approximately 20-fold increase in the yield of JH,[6] indicating that the enzyme is not nearly saturated with JH-I acid at 1 μM.

Accessory glands contain low JH esterase activity,[6] which will cause some loss of JH formed by the methyltransferase. While this loss can be estimated from control experiments with JH and corrected for in determinations of methyltransferase activity, it may be preferable to inhibit the JH esterase with suitable inhibitors[18,19] or to eliminate it by purifying the crude methyltransferase preparation.

Isolation and Analysis of Products

The reaction products are extracted with ethyl acetate and analyzed by silica-TLC and high-performance liquid chromatography (HPLC).[7,20] Silica fractions from the TLC plates and effluent fractions from HPLC are collected and analyzed for radioactivity. In determinations of enzyme activity, TLC separation of the products is adequate if labeled JH acid

---

[18] Phenylmethylsulfonylfluoride, at 1.0 mM, completely inhibits JH esterase of *H. cecropia* hemolymph, but the inhibitor is ineffective in the presence of 35 mM sulfate. G. F. Weirich, unpublished observation (1975).

[19] T. C. Sparks and B. D. Hammock, *Pestic. Biochem. Physiol.* **14**, 290 (1980).

[20] D. A. Schooley, in "Analytical Biochemistry of Insects" (R. B. Turner, ed.), p. 241. Elsevier/North-Holland Biomedical Press, Amsterdam, 1977.

and unlabeled AdoMet are used as substrates and appropriate controls are carried out. If, however, the radiolabel is introduced with [methyl-^{3}H]AdoMet or [methyl-^{14}C]AdoMet, further purification of the products by HPLC is necessary to eliminate labeled methylation products other than JH.

[37] S-Adenosylmethionine : γ-Tocopherol Methyltransferase (*Capsicum* Chromoplasts)

By BILAL CAMARA

During the ripening of *Capsicum* fruits, a massive accumulation of α-tocopherol (α-T) takes place in the chromoplasts.[1] In the chromoplast membranes prepared from semiripened (orange) fruit, γ-tocopherol (γ-T) is also detected and comprises up to 20% of the tocopherol fraction. This chapter summarizes some of the characteristics of the enzymatic conversion of γ-T into α-T.

Principle

The later step in the biosynthesis of α-tocopherol involves the following reaction[1,2]:

$$\text{2,3-dimethylphytylquinol} \xrightarrow{\text{cyclization}} \gamma\text{-tocopherol}$$
$$(1) \qquad\qquad\qquad\qquad (2)$$

and the transfer of the (labeled) methyl group of S-adenosylmethionine to γ-tocopherol (see Fig. 1).

$$\gamma\text{-tocopherol} \xrightarrow{S\text{-adenosyl-L-methionine (SAM)}} \alpha\text{-tocopherol}$$
$$\qquad\qquad\qquad\qquad\qquad\qquad (3)$$

Experimental Procedure

Reagents and Equipment

MgCl$_2$ 1 M
Dithiothreitol (DTT) 1 M

---

[1] B. Camara, F. Bardat, A. Seye, A. d'Harlingue, and R. Moneger, *Plant Physiol.* **70**, 1562 (1982).
[2] J. Soll and G. Schultz, *Phytochemistry* **19**, 215 (1980).

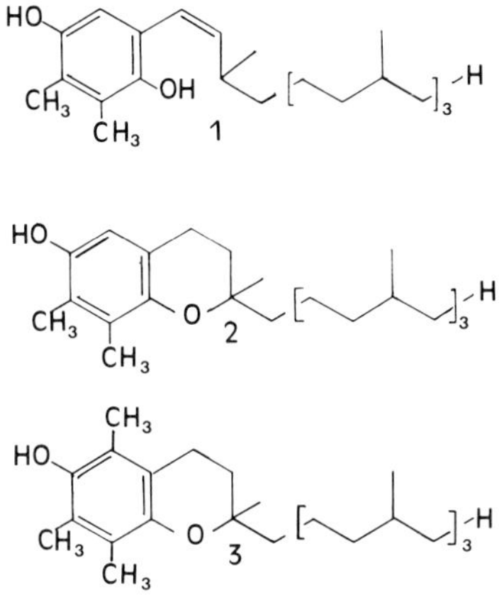

FIG. 1. Substrates and products of tocopherol biosynthesis. **1**, 2,3-Dimethylphytylquinol; **2**, γ-tocopherol; **3**, α-tocopherol.

[*methyl*-^{14}C]SAM, 45 mCi/mmol

Tris–HCl buffer 50 mM (pH 7.6)

Chromoplast membranes prepared as described previously[3] and suspended in 0.4 M of sucrose buffered with Tris–HCl 50 mM (pH 7.6) (approximately 2 mg protein/ml)

Silica gel 60 plates for thin-layer chromatography (TLC) C_{18} μBondapak column for high-performance liquid chromatography (HPLC) (Waters)

Sep-Pak (Silica cartridges, Waters)

Rhodamine 6G, 0.04% in acetone

Dichloromethane

Diethyl ether

Light petroleum 40–60°–diethyl ether (99:1, v/v)

Light petroleum–diethyl ether (9:1, v/v)

Light petroleum–diethyl ether (10:1, v/v)

Methanol–tetrahydrofuran (8:2, v/v) for HPLC

Chloroform–methanol (2:1, v/v)

0.9% sodium chloride

γ-T extracted from ungerminated dry seeds of peas[4]

α-T (Sigma)

Incubation Medium. The enzymatic reaction is carried out for 2 hr at 25° in a medium buffered with 50 mM Tris–HCl (pH 7.6) (1.5 ml final

---

[3] B. Camara, this series, Vol. 110, p 244.
[4] J. Green, *J. Sci. Food Agric.* **9**, 801 (1958).

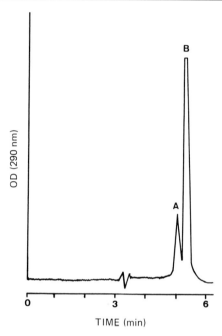

FIG. 2. HPLC separation of γ-tocopherol (A) and α-tocopherol (B) recovered after the incubation procedure described in the text. The sample was analyzed using a Waters system (Pump model 510, variable wavelength detector M450, data module M730). Conditions of analysis (column: C_{18} μBondapak 30 cm; solvent: methanol–tetrahydrofuran (80:20 v/v); flow rate 1 ml/min, detection at 290 nm, AUFS 0.04).

volume) containing 10 mM $MgCl_2$, 1 mM DTT, [*methyl*-^{14}C]SAM (0.5 μCi), and chromoplast membranes (2 mg protein). The reaction is stopped by the addition of 4.5 ml of chloroform–methanol (2:1, v/v).

Extraction and Purification. To the chromoplast membrane suspension previously treated with 4.5 ml of chloroform–methanol (2:1, v/v) is added 4.5 ml of the same solvent mixture containing 50 μg of γ-T and 100 μg of α-T. After shaking for 5 min with a vortex mixer, 2 ml of 0.9% sodium chloride is added. The mixture is homogenized again and the solution is cleared by centrifugation. The chloroform layer is withdrawn and concentrated to dryness *in vacuo* or by a stream of nitrogen. The lipid residue is streaked on a Silica gel 60 chromatoplate and developed with light petroleum–diethyl ether (10:1, v/v).[5] The tocopherol band (R_f 0.3) is revealed by rhodamine 6G and eluted with diethyl ether. A further chromatography on Silica gel 60 developed with dichloromethane allows the

[5] M. Terini and H. K. Lichtenthaler (eds.), *in* "Lipids and Lipid Polymers in Higher Plants," p. 231. Springer Verlag, Berlin and New York, 1977.

separation of α-T (R_f 0.86) and γ-T (R_f 0.65). Alternatively, the chromoplast lipids dissolved in light petroleum–diethyl ether (99:1, v/v) are applied to a Sep-Pak cartridge after which the latter is eluted by 10 ml of light petroleum–diethyl ether (99:1, v/v) and 10 ml of light petroleum–diethyl ether (9:1, v/v). α-T and γ-T are recovered in the latter fraction, which is subjected to HPLC on C_{18} μBondapak column developed isocratically with methanol–tetrahydrofuran (8:2, v/v). The different peaks are detected at 290 nm (Fig. 2).

Results. The endogenous γ-tocopherol present in chromoplast membranes is methylated by exogenous SAM. Up to 0.2 nmol of α-T per hr per mg protein can be formed.

The apparent Michaelis constant (K_m) for SAM is approximately 5 μM.

The reaction is inhibited when the chromoplast membranes are solubilized by Triton X-100.

Acknowledgments

I thank Prof. Moneger for his interest during the work. I appreciate the technical assistance given by C. Agnes, F. Bardat, and A. Nargeot.

[38] Dolichyl Pyrophosphate Phosphatase in Brain

By MALKA G. SCHER and CHARLES J. WAECHTER

Dolichyl Pyrophosphate + H_2O → Dolichyl Monophosphate + P_i

A polyisoprenyl pyrophosphate phosphatase is required to regenerate dolichyl monophosphate(Dol-P)[1] from dolichyl pyrophosphate(Dol-P-P)[1] discharged as precursor glucosylated oligosaccharides are transferred from the glycosyl carrier lipid to asparagine residues in acceptor polypeptides (Fig. 1, for reviews of the lipid intermediate pathway see refs. 2 and 3). Dolichyl pyrophosphate phosphatase activity has been reported for lymphocytes,[4] liver,[5-7] and brain.[8] Procedures for the solubilization and assay of brain Dol-P-P phosphatase activity are described here.

[1] Abbreviations used: Dol-P, dolichyl monophosphate; Dol-P-P, dolichyl pyrophosphate.
[2] S. C. Hubbard and R. J. Ivatt, *Annu. Rev. Biochem.* **50**, 555–583 (1981).
[3] K. A. Presper and E. C. Heath *in* "The Enzymes" (P. Boyer, ed.), Vol. 16, pp. 449–488. Academic Press, New York, 1984.
[4] J. F. Wedgwood and J. L. Strominger, *J. Biol. Chem.* **255**, 1120–1123 (1980).
[5] E. L. Appelkvist, T. Chojnacki, and G. Dallner, *Bioscience Reports* **1**, 619–626 (1981).

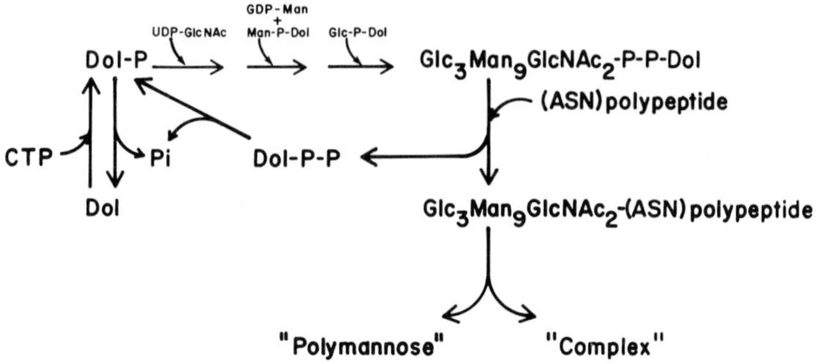

```
                    GDP-Man
                      +
        UDP-GlcNAc  Man-P-Dol  Glc-P-Dol
Dol-P ─┐  ↘    ↘      ↘       → Glc₃Man₉GlcNAc₂-P-P-Dol
       │  ↖                      ↙─ (ASN)polypeptide
 CTP ─┤↓  Pi    Dol-P-P  ←────  │
       │                        ↓
      Dol                Glc₃Man₉GlcNAc₂-(ASN)polypeptide
                                │
                              ↙   ↘
                       "Polymannose"  "Complex"
```

FIG. 1. Role of dolichyl pyrophosphate phosphatase in lipid intermediate pathway for N-glycosylation.

Assay Method

Principle. Dol-P-P phosphatase activity in brain microsomes or detergent extracts can be readily assayed by following the enzymatic release of $^{32}P_i$ from dolichyl [β-^{32}P]pyrophosphate (Dol-P-^{32}P) added exogenously as an aqueous dispersion in Triton X-100. Dol-P-^{32}P provides a convenient substrate for this assay because it obviates the potential complication of $^{32}P_i$ being released from the α position by contaminating Dol-P phosphatase activity in the enzyme preparation. The reaction is terminated by the addition of 20 vol of $CHCl_3$–CH_3OH (2:1), and 1/5 vol of 0.9% NaCl is added to produce a biphasic system. The unreacted substrate remains in the $CHCl_3$-rich (lower) phase, and the radiolabeled product, $^{32}P_i$, partitions into the aqueous–CH_3OH (upper) phase.

Reagents

0.1 M Tris–HCl (pH 7.2)–0.25 M sucrose–1 mM EDTA–10 mM 2-mercaptoethanol
2 mM dolichyl [β-^{32}P]pyrophosphate (300–500 cpm/nmol)
Triton X-100
Enzyme source

Chemical Preparation of Unlabeled Polyisoprenyl Pyrophosphates and Dolichyl [β-^{32}P] Pyrophosphate. Unlabeled polyisoprenyl pyrophosphates can be synthesized by chemically phosphorylating the phosphomonoesters by a modification of the procedure described by Warren and

---

[6] E. Belocopitow and D. Boscoboinik, *Eur. J. Biochem.* **125**, 167–173 (1982).
[7] S. Kato, M. Tsuji, Y. Nakanishi, and S. Suzuki, *Biochem. Biophys. Res. Commun.* **95**, 770–776 (1980).
[8] Malka G. Scher and Charles J. Waechter, *J. Biol. Chem.* **259**, 14580–14585 (1984).

Jeanloz[9] for the phosphorylation of dolichol. Dol-P (300 nmol) and di-triethylammonium phosphate[10] (62 μmol) are dissolved in 1 ml of acetonitrile/toluene (2:3). Acetonitrile is freshly distilled and the toluene is dried over CaH_2. The solution is stirred for 30 min at 24°, and 0.075 ml of freshly distilled trichloroacetonitrile is added. The reaction mixture is kept at 24° for 4 hr, and then 2 ml of $CHCl_3/CH_3OH$ (2:1) is added. The organic mixture is washed with 0.6 ml of 0.9% NaCl to remove water-soluble compounds. The organic (lower) phase is then washed twice with 1.2 ml of $CHCl_3$-CH_3OH-0.9% NaCl (3:48:47). Chromatographic analysis of the polyisoprenyl phosphates indicates that approximately 50% of the starting Dol-P is converted to Dol-P-P. The product has the chromatographic mobility expected for Dol-P-P (R_{Dol-P} = 0.43–0.59) on EDTA-treated[11] SG-81 paper developed with solvent mixtures (A) $CHCl_3$–CH_3OH–H_2O (65:25:4) or (B) $CHCl_3$–CH_3OH–conc. NH_4OH–H_2O (65:35:4:4). The synthetic product can be detected by a phospholipid spray reagent[12] and an anisaldehyde spray reagent that gives a characteristic greenish color for polyprenols.[13] Furthermore, mild acid treatment (0.26 N HCl in $CHCl_3$–CH_3OH–H_2O, 10:10:3, at 80° for 2 hr) releases approximately 50% of the lipid-phosphorus as free phosphate and converts the synthetic product to Dol-P. These chemical and chromatographic properties are consistent with the product formed by this procedure being Dol-P-P. The procedure described here also converts undecaprenyl monophosphate to undecaprenyl pyrophosphate in approximately the same yield.

The procedure used to synthesize Dol-P-^{32}P chemically is essentially the same one used for the preparation of the unlabeled polyisoprenyl pyrophosphates. The principle of this method is to phosphorylate Dol-P with a ^{32}P-labeled phosphorylating reagent. Carrier-free [^{32}P]H_3PO_4 (1–2 mCi) is dried in a 5-ml plastic tube by vacuum centrifugation. Freshly distilled triethylamine (0.5 ml) is then added and evaporated to dryness. Di-triethylammonium phosphate (2 mg) is added and the mixture containing the ^{32}P-labeled phosphorylating agent is suspended in 0.3 ml of acetonitrile. After stirring for 30 min the suspension is transferred to another plastic tube containing 120 nmol of Dol-P. The tube containing the suspension is also rinsed with 0.1 ml of acetonitrile. The remaining reagents are added as outlined above for the preparation of the unlabeled polyiso-

---

[9] C. D. Warren and R. W. Jeanloz, *Biochemistry* **14**, 412–419 (1975).
[10] R. H. Cornforth and G. Popjak, this series, Vol. 15, pp. 359–390.
[11] S. M. Steiner and R. L. Lester, *J. Bacteriol.* **19**, 81–88 (1972).
[12] J. C. Dittmer and R. L. Lester, *J. Lipid Res.* **5**, 126–127 (1964).
[13] P. J. Dunphy, J. D. Kerr, J. F. Pennock, K. J. Whittle, and J. Feeney, *Biochim. Biophys. Acta* **136**, 136–147 (1967).

prenyl pyrophosphate. The radiolabeled product cochromatographs with Dol-P-P, and virtually all of the $^{32}P_i$ is labile under mild acid conditions.

Purification of Unlabeled and β-Labeled Dolichyl Pyrophosphate. The chemically synthesized polyisoprenyl pyrophosphates are purified chromatographically. The mixture of unreacted Dol-P and Dol-P-P is applied as a band on a thin-layer plate of silica gel H. The silica gel H plates are prepared with a slurry of 50 g of silica gel in 125 ml of 2% EDTA adjusted to pH 7.4 with conc. NH$_4$OH. The plates are developed with either solvent mixture A or B. The zones corresponding to Dol-P-P can be detected autoradiographically by adding Dol-P-^{32}P as a radioactive marker. The area containing Dol-P-P is collected by aspiration with a sealing tube and eluted with the following sequence of 10 ml washes: (1) CHCl$_3$, (2) CH$_3$OH, and (3) CHCl$_3$/CH$_3$OH/H$_2$O (10:10:3) containing 50 mM NH$_4$OH. The CHCl$_3$ wash is discarded, and the other two are pooled. The proportion of CHCl$_3$/CH$_3$OH is adjusted to 2:1 and the eluate washed with 1/5 vol of 0.9% NaCl. The lower phase, containing Dol-P-P, is dried under nitrogen and the polyisoprenyl pyrophosphate is stored at $-17°$. Dol-P-^{32}P is similarly purified chromatographically on EDTA-treated SG-81 paper by developing with solvent system B. A major radioactive band ($R_f = 0.3$) is detected by autoradiography. A minor band with a lower mobility than Dol-P-P, possibly corresponding to dolichyl triphosphate, is also observed. The radioactive zone containing Dol-P-^{32}P is cut into small pieces, and the paper strips are soaked in CHCl$_3$–CH$_3$OH–H$_2$O (10:10:3) containing 50 mM NH$_4$OH for 18 hr at 4°. The extracted paper zones are sedimented by centrifugation, and the paper extracted again with the same solvent mixture. The extracts are pooled and the proportion of CHCl$_3$/CH$_3$OH adjusted to 2:1. The extracts are then washed with 1/5 vol of 0.9% NaCl and the lower phase, containing Dol-P-^{32}P, is dried under nitrogen and the labeled substrate is stored at $-17°$.

Solubilization of Dolichyl Pyrophosphate Phosphatase Activity. Dol-P-P phosphatase activity is readily solubilized by extracting calf brain microsomes[14] with Triton X-100. Prior to detergent extraction, the microsomal preparation is sedimented and washed with 0.1 M Tris–HCl (pH 7.2), 0.25 M sucrose, 10 mM 2-mercaptoethanol, and 1 M NaCl. The salt-wash removes substantial amounts of adsorbed nucleic acids, proteoglycans, and extrinsic membrane proteins. Dol-P-P phosphatase is then solubilized by treating the salt-washed microsomes with Triton X-100. The membrane suspension in buffer and the desired concentration of detergent (3 mg detergent/mg membrane protein) is homogenized on ice with

---

[14] C. Sumbilla and C. J. Waechter, this volume [30].

12 strokes of a Kontes Dounce homogenizer (7 ml) pestle B. Extraction is allowed to proceed for 17 hr at 4° with continuous stirring by a magnetic stirring bar. The solubilized fraction is separated from the insoluble membranous material by centrifugation at 100,000 g at 4° for 60 min. The activity recovered in the supernatant fluid is designated as the soluble fraction. Extraction with 3% Triton X-100 solubilizes approximately 65% of the Dol-P-P phosphatase activity and 30% of the membrane protein. The activity in the soluble fraction is stable at 4° overnight, and very little activity is lost when the detergent extracts are stored at $-17°$ for periods as long as 1 month.

Procedure. Typical reaction mixtures consist of solubilized or particulate brain microsomes (0.5 mg protein), 50 mM Tris–HCl (pH 7.2), 124 mM sucrose, 0.5 mM EDTA, 5 mM 2-mercaptoethanol, 1–2% Triton X-100, and 24 μM dolichyl [β-^{32}P]pyrophosphate (300 cpm/nmol) in a total volume of 0.1 ml. Following an incubation at 30° for the desired period of time, the enzymatic reaction is terminated by the addition of 20 vol of CHCl$_3$–CH$_3$OH (2:1) followed by 1/5 vol of 0.9% NaCl. The extraction mixture is cooled to 0–4° in an ice bucket and centrifuged at 1000 g in a benchtop centrifuge to produce a biphasic system. The aqueous (upper) phase, containing the enzymatically released ^{32}P$_i$, is transferred to a scintillation vial. The lower phase, including the delipidated residue at the interface, is washed with 1 ml of CHCl$_3$–CH$_3$OH–0.9% NaCl (3:48:47). The aqueous washes are pooled and dried in a warm water bath under a stream of air. The amount of ^{32}P$_i$ released from Dol-P-^{32}P is measured in a scintillation spectrometer after the addition of 1 ml of 1% SDS and 10 ml of Liquiscint (National Diagnostics).

Properties of Solubilized Dolichyl Pyrophosphate Phosphatase Activity. Calf brain Dol-P-P phosphatase activity does not require a divalent cation, and is substantially stimulated by EDTA. The solubilized activity is optimally active over a pH range of 6.5–7.5 and exhibits an apparent $K_m = 20$ μM for Dol-P-P.

Inhibitors and Specificity. The solubilized phosphatase is markedly inhibited by 20 mM MnCl$_2$ and to a lesser extent by 20 mM CaCl$_2$. Modest inhibition is produced by 20 mM NaF (see the table). Consistent with the dephosphorylation reaction being catalyzed by a polyisoprenyl pyrophosphate phosphatase, the release of ^{32}P$_i$ from Dol-P-^{32}P is reduced by the addition of undecaprenyl pyrophosphate, farnesyl pyrophosphate, or isopentenyl pyrophosphate. A kinetic analysis indicates that undecaprenyl pyrophosphate is a competitive inhibitor (Fig. 2) with an apparent $K_1 = 160$ μM. Relevant to this result, the solubilized enzyme preparations catalyze the conversion of [^{14}C]undecaprenyl pyrophosphate to [^{14}C]undecaprenyl monophosphate. Thiamine pyrophosphate, ADP and UDP, three

SOLUBILIZED DOLICHYL MONO- AND PYROPHOSPHATE
PHOSPHATASE ACTIVITIES IN THE PRESENCE OF
PHOSPHATE AND FLUORIDE IONS[a]

Addition (20 mM)	$^{32}P_i$ released from (% of control)	
	Dol-P-^{32}P	Dol-^{32}P
None	100	100
Sodium orthophosphate	145	10
Sodium pyrophosphate	202	39
Sodium trimetaphosphate	190	69
Sodium fluoride	68	3
Sodium chloride	98	81

[a] From ref. 8.

nonisoprenoid pyrophosphates, do not affect the solubilized Dol-P-P phosphatase activity. The polyisoprenyl pyrophosphate phosphatase activity is virtually abolished by the addition of 0.4 mM bacitracin, presumably by the formation of an unreactive complex[15,16] between the peptide antibiotic and Dol-P-^{32}P.

Properties Distinguishing between Dolichyl Mono- and Pyrophosphate Phosphatase Activities. The Triton X-100 extracts contain both Dol-P and Dol-P-P phosphatase activities. An enzymological comparison of the two activities indicates that these reactions are catalyzed by different enzymes. The two phosphatase activities differ considerably with respect to their thermolability. Dol-P-P phosphatase has a $t_{1/2}$ = 2 min at 50° while the $t_{1/2}$ for Dol-P phosphatase is 8 min. The activities are also distinguished by their responses to phosphate and fluoride ions (see the table). Dol-P phosphatase is extremely sensitive to phosphate and fluoride ions. Dol-P-P phosphatase is actually stimulated by 20 mM sodium phosphate and minimally (32%) inhibited by 20 mM sodium fluoride. Sodium pyro- and trimetaphosphate also inhibit Dol-P phosphatase activity, possibly due to the hydrolytic production of orthophosphate ion. In contrast to this effect, Dol-P-P phosphatase activity is stimulated by pyro- and trimetaphosphate ions (see the table). The two phosphatase reactions are also differentially affected by bacitracin with only the Dol-P-P phosphatase reaction being inhibited by the peptide antibiotic. However, this result is not a valid argument for separate enzymes, because the peptide

[15] K. J. Stone and J. L. Strominger, *Proc. Natl. Acad. Sci. USA* **68**, 3223–3227 (1971).
[16] D. R. Storm and J. L. Strominger, *J. Biol. Chem.* **248**, 3940–3945 (1973).

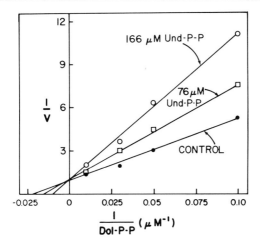

FIG. 2. Double-reciprocal analysis of the inhibitory effect of undecaprenyl pyrophosphate on brain dolichyl pyrophosphate phosphatase. From Scher and Waechter.[8]

discriminates between the substrates, not the enzymes, by selectively forming complexes with polyisoprenyl pyrophosphates.[15,16]

Possible Functions of Polyisoprenyl Pyrophosphate Phosphatase(s). It is quite likely that one function of this type of phosphatase is to re-form Dol-P from Dol-P-P released concomitantly with the transfer of the glucosylated oligosaccharides to acceptor polypeptides (Fig. 1). It is also feasible that a polyisoprenyl pyrophosphate phosphatase plays a role in the *de novo* pathway for Dol-P biosynthesis. If the fully unsaturated polyisoprenyl pyrophosphates are the substrates for the reductase responsible for the saturation of the α-isoprene units, the *de novo* formation of Dol-P would be completed by the removal of the β-phosphate. Saturation of the allylic pyrophosphate group could be a mechanism for terminating the isoprenylation reactions. However, if the reductase acts on the phosphomonoester or the free allylic alcohol, then a polyisoprenyl pyrophosphate phosphatase would be required to convert the pyrophosphate form to the allylic phosphomonoester. The observation that the detergent extracts convert undecaprenyl pyrophosphate to undecaprenyl monophosphate demonstrates that the brain enzyme system has the ability to catalyze this type of dephosphorylation reaction. Determining the topographical orientation of the polyisoprenyl pyrophosphate phosphatase(s) will be relevant to understanding the function of this enzyme(s) in the endoplasmic reticulum.

Author Index

Numbers in parentheses are footnote reference numbers and indicate that an author's work is referred to although the name is not cited in the text.

A

Aarouson, I. R., 340
Abano, D. A., 267, 268(9)
Abrahamson, E. W., 225
Ackman, R. G., 188
Adair, W. L., 201, 202, 203, 204(7), 209(7), 210, 211, 212, 214, 215(36)
Adair, W. L., Jr., 471, 474(6), 475(6), 476
Adams, J. R., 456, 458
Adams, R. P., 172
Adcock, E. W., 92
Addison, C. F., 437
Adler, J., 288, 290(9), 291(9)
Adler, J. H., 8, 48, 334, 343(4)
Adolphson, J. L., 268
Adrian, G. S., 476, 477(13), 480
Ahlberg, J., 105(188), 106, 108(188)
Aitoku, A., 159, 171
Akashi, Y., 105(190), 106, 109(190), 110(190)
Akeroyd, R., 299
Albaiges, J., 186, 188
Albers, J. J., 267, 268(2, 7, 8, 12)
Alberts, B., 467
Alcaide, A., 10, 318
Aldersberg, M., 341
Alexander, B. A. J., 539, 541
Ali, S. S., 91(133), 92, 93(133), 95(133), 96(133)
Allemand, B. H., 220
Allen, C. M., Jr., 471, 474(1), 475
Allen, J. G., 90
Allen, R. R., 167
Almé, B., 66, 81(8), 82(8), 83(8), 84(8, 82), 85(8), 87(8), 88(8), 89(82), 90(82), 94(8), 95(151), 96(82), 97(82), 161, 99(8), 101(8), 102(8), 106(8), 110(166)
Alpers, D. H., 299, 300(18), 302(18)
Altona, C., 3
Amar-Costesec, A., 439, 456
Amos, B., 91, 92, 93(137)

Andersen, N. H., 181
Anderson, I. G., 91(127), 92, 93(137)
Anderson, J. M., 313
Anderson, R. A., 11
Andersson, S., 366, 372(12)
Andersson, S. H. G., 72, 72(47), 79(47), 89(47)
Andrianjafintrimo, M., 388(39), 389, 390(39), 391(39), 393, 400(53), 412, 419(9), 420, 428(9), 444
Angelin, B., 105(186, 187, 188, 193), 106, 108(186–188), 109(193), 111(193)
Appelkvist, E. L., 547
Arebalo, R. E., 299, 302(22)
Arias, I. M., 302
Aringer, L., 80, 83(68), 84(68), 85(68), 86(68), 87(68), 90(68), 91(68), 95(68), 310, 311(31)
Armitt, G. M., 216, 217(5), 218(5), 224, 225(5, 46), 227(46), 232(46), 240(5), 241(5, 46)
Armstrong, D., 214
Arnold, W., 31, 32, 33
Arnon, R., 238
Aron, L., 267, 268(1), 275
Arrendale, R. F., 158
Artland, J., 33
Ashburner, M., 459
Astin, A. A., 339
Astin, A. M., 341
Atsuta, Y., 365
Attallah, N. A., 435
Attramadal, A., 453
Avela, E., 184
Averill, W., 163
Avron, M., 199
Awata, N., 346, 348, 349(13), 350(13), 352(13)
Axelson, M., 89, 90, 100, 101(164), 103(164), 104(164)
Azerad, R., 201
Azzouz, M. A., 151

B

Babashak, J. F., 489
Back, P., 82, 84(72), 96, 97(159, 160), 99, 100(163)
Badenhop, A. A. F., 152
Baerheim Svendsen, M. G., 151, 152, 158, 165(11)
Bagdade, J. D., 267, 268(2)
Bagley, H., 510
Bailey, G. F., 6, 11(5)
Bailey, R. B., 41, 338, 344(60), 345
Baillie, T. A., 82, 83(78), 112
Baisted, D. J., 320
Baker, F. C., 287
Baker, H. N., 293, 294(3), 296(3), 297(3), 298(3), 299(3), 300(3), 302(3)
Balistreri, W. F., 111, 112(198)
Ballantine, J. A., 19, 20(54), 23, 24(54), 25(64)
Ballatore, A. M., 69
Balliano, G., 316, 317(21)
Balmer, J., 275
Baltscheffsky, M., 219
Ban, Y., 66, 72, 73(35), 93(11, 35)
Bang, L., 385
Bankowski, J., 96, 97(158)
Bannai, K., 361
Barbeau, B. K., 484, 485, 487(4)
Barber, M., 68, 72(24)
Barbier, F., 72, 73(49), 99(49)
Barbier, M., 33, 318
Barclay, M., 337
Bard, M., 343, 344(55, 56, 58, 59), 345
Bardat, F., 544
Barghoorn, E. S., 186
Barnes, P. J., 89, 90(115), 92(115)
Barnes, S., 72, 73(48)
Barrett, G. M., 89
Barron, R., 27
Bartha, A., 166
Bartholome, M., 281
Bartholomew, T. C., 105, 106(183), 109(183), 110(183)
Bartle, K. D., 168
Barton, D. H. R., 344
Bastian, E., 166
Batta, A. K., 91(131, 132), 92, 93, 94(143), 95(143)
Baty, J. D., 89, 90(115, 116), 92(115, 116)
Bauer, A. C., 435, 436(38)
Bauernfeind, J. C., 116, 136(10), 148(10)
Baxter, J. D., 451
Bayley, H., 432
Bazyl'chik, V. V., 181
Beach, D. H., 33
Beale, D., 302
Beastall, G. H., 323
Beaudry, J., 199
Beaufay, H., 439, 456
Becker, R. S., 218
Beckers, C., 458
Beckey, H. D., 27, 67
Beckner, C, F., 69
Beekes, H. W., 10
Behlau, H., 166
Behzadan, S., 9, 33, 34(c), 35
Bekaert, A., 33
Beke, R., 72, 73(49), 99(49)
Bell, F. P., 291
Belleau, B., 317
Belobaba, D. T. E., 88, 112(108)
Belosopitow, E., 476(15), 477, 479, 547(6), 548
Ben-Amotz, A., 199
Benecke, I., 184
Bennett, R. D., 320, 321
Benson, M., 8
Bentley, P., 304, 305(5), 306
Benveniste, P., 316, 317(22), 320, 323, 324(22, 43), 331
Beppu, T., 105(106, 185, 191, 194), 108(185, 194), 109(191), 110(191)
Berendes, H. D., 430, 442
Bergamasco, R., 383, 452
Berge, R. K., 288
Berger, B., 249
Berger, H., 217, 225(32), 226(32), 242(32)
Bergot, B. J., 487, 531, 538(5)
Berngruber, O. W., 112, 318
Bernhard, K., 123
Bernhard, R. A., 176
Bernstein, H. J., 216(16), 217, 218(16), 221(38), 232(16), 243(16), 244(16), 245(16)
Beroza, M., 172(123, 124), 173
Berthou, F., 162
Bertsch, W., 161, 162
Besemer, A. C., 317, 318(22)
Beydon, P., 383, 387, 389(22), 390(22),

393(22), 398(22), 399(22), 400(53), 409, 410(80), 444
Beyer, P., 199
Bhaskaran, G., 540, 541(8), 542(8)
Bhat, P. V., 251
Bianchini, J.-P., 185
Bielby, C. R., 391, 394, 428
Bielefeld, B., 18, 19, 20(53), 23, 25(53)
Bieri, F., 306
Bierl-Lieonhardt, B. A., 173
Billheimer, J. T., 286, 287(5), 288(2, 5), 289(5, 8), 290(2, 5, 8, 9), 291(9, 20), 292(20)
Billing, B. H., 84, 105, 106(183), 109(183), 110(183)
Bilton, R. F., 89, 90(115, 116), 92(115, 116)
Bindel, U., 304, 305(5)
Black, D. R., 152
Blackstock, E. J., 105, 106(182), 109(182)
Blais, C., 383, 388(39), 389(22), 390(22, 39), 391(39), 393(22), 398(22), 399(22), 412, 419(9), 420, 424, 428(5, 9), 444
Bligh, E. G., 212, 325
Bligh, E. S., 248
Bloch, C., 58, 59(7), 61(7)
Bloch, K., 20, 205, 296, 301, 346
Blomberg, L., 163, 168, 169
Blondin, G. A., 16, 20(51), 25(51)
Bloom, E., 451
Blumer, M., 187
Blunt, J. W., 33, 36
Body, D. R., 184
Bohm, W., 205
Bollenbacher, W. E., 379, 385(18), 454, 494, 512, 517(12)
Bond, F. T., 42
Boniforte, L., 78
Borbon, J., 188
Borch, G., 216(20), 217, 218(20), 224(20), 225(20, 30, 32), 226(32), 232(30), 242(30, 32), 244(20), 245(20)
Bordier, C., 235, 238
Bordoli, R. S., 68, 72(24)
Borthwick, J. H., 89
Borwitzky, H., 186, 187
Boscoboinik, D., 479, 547(6), 548
Bose, A. K., 67
Boström, H., 365, 367(2), 374(2)
Bottema, C. K., 335, 339(6), 340(6), 344, 345(13, 53)

Botz, F. K., 67
Bouvier, J., 419
Bowen, D. V., 96, 97(160)
Bowlus, S. B., 540, 540(4)
Boyd, G. S., 289, 293(16)
Boyett, R. L., 269, 278
Bradford, M. M., 527
Bragdon, J. H., 277, 282(16)
Brahms, J., 218
Braumann, T., 198, 199
Braverman, D. Z., 111, 112(200)
Brecher, P., 288, 291, 292(26)
Breckenridge, W. C., 202
Brecker, P. I., 291
Breen, M. A., 319
Brehelin, M., 444
Bremmelgaard, A., 66, 72, 73(50), 81(8), 82(8), 83(8, 50), 84(8, 73), 85(8, 50), 87(8), 88(8), 94(8), 99(8), 101(8, 73), 102(8), 106(8), 110(166)
Brent, D. A., 67
Briggs, T., 91(127), 92
Briquet, M., 341
Brissey, G. M., 173
Britten, E. J., 151
Britton, G., 116, 117, 124(11), 129, 132(11, 12), 135(12), 136(11), 148(11), 149(11), 189, 192, 216, 217(5), 218(5), 224, 225(5, 34), 227, 232(34, 46), 240(5, 34), 241(5, 34, 46), 315
Brockmann, H., 127
Brodgen, W. B., Jr., 158
Broich, C. R., 197, 199
Brooks, C., 286, 289(6), 291(6)
Brooks, C. J. W., 11, 24, 83, 84(81), 84(79), 87(81), 88, 89(79, 98), 90, 94, 95(81, 98)
Brooks, G. T., 539, 540, 541
Brown, F. J., 90
Brown, J. J., 500, 506(18), 526
Brown, M. S., 291, 292
Brown, P. R., 41, 189, 198(6), 199
Bruce, M. J., 495, 499(10), 501(10), 502(10), 503(10), 505(25), 506, 510, 521(8), 522(8)
Bruins, A. P., 172
Buchecker, R., 147
Bucher, N. L. R., 349
Büchler, M., 429, 444
Buchwald, M., 216, 218(13), 219(13), 220(13), 221, 222(13), 232(13, 39), 233(13, 39)

Buckle, K. A., 189
Budzikiewicz, H., 64, 80(5, 7), 82(5), 86(5, 7), 87(5), 88(5, 7), 89(5), 90(5, 7), 91(5), 93(5), 147, 172, 403
Bugany, H., 426, 427(17), 428(17), 450
Buijten, J., 163, 169
Burgos, J., 201, 202(2), 206, 208(2)
Burke, C. W., 431
Burkhart, M. F., 291, 292
Burlingame, A. L., 68, 69(21), 71(26, 29), 83(21), 103, 111(26, 173, 176, 177, 178)
Burnett, D. A., 302
Burns, W. F., 169, 180
Burrows, L. S., 344(58), 345
Burstein, S., 353
Burton, W. A., 471, 473(4), 474(4), 476(4, 16), 477, 478(16), 479(16)
Bush, M. T., 307
Buten, B., 216, 217(9), 218(9), 219(9), 220(9)
Butenandt, A., 411
Butterworth, F. M., 430, 442
Butterworth, P. H. W., 480
Buttke, T. M., 20

C

Cabana, V. G., 267, 268(7)
Cafmeyer, N., 203, 471, 474(6), 475(6), 476
Cailla, H. L., 446
Calzolari, A., 358, 364(8)
Camara, B., 544, 545
Campbell, K. P., 280, 282
Canonica, L., 450
Cao, M., 445, 446(24), 448(24), 449(24), 453(24)
Capdevila, J., 304
Caprioli, R. M., 69
Capron, A., 379
Capstick, E., 320
Caputo, O., 316, 317(21)
Caras, I. W., 301
Carey, P. R., 216(16), 217, 218(16), 221(38), 232(16), 243(16), 244(16), 245(16)
Carlson, R. M. K., 327, 329(66)
Carlström, K., 72, 73(41), 82(41), 86(41), 90(41), 91(41)
Carminatti, H., 476(15), 477
Carroll, K. K., 206, 208(20), 210, 212(20), 214, 471, 474(3), 475(3), 476(17), 477, 479(17), 480

Carroll, R. M., 39
Carson, D. D., 213
Carvalho, J. F., 506, 510, 511(5)
Casellato, M., 321
Casey, J., 238
Cattel, L., 316, 317(21), 324
Ceccaldi, H. J., 216, 217(10), 218(10, 14), 219(10), 220(14), 221(10, 14), 222(14), 223(10, 14), 224(10, 14), 225(10), 228(14), 229(14), 230(14), 232(10), 235(10, 14), 237(14), 238(10), 240(10, 75), 243(26, 28), 244(26)
Chacos, N., 304
Chajek, T., 275, 282(6)
Chambaz, E. M., 89
Chambet, J.-C., 428
Chandler, R. F., 185
Chang, E. S., 429, 495, 496, 497, 498, 499(10), 500, 501, 502, 503, 505, 506(18), 509, 510, 511(5), 515, 521(8), 522(8)
Chang, F. C., 72, 73(39), 75(39), 78, 85(63)
Chang, R. C., 162
Chang, T. Y., 286, 289
Chari, V. M., 33
Charlet, M., 377, 378, 385(3), 412
Charlton, J. M., 315
Chataing, B., 275
Chatani, F., 411
Chaudhary, N., 214, 476(17), 477, 479(17)
Cheesman, D. F., 216, 217(10), 218(10), 219(10), 221(10), 223(10), 224(10, 35), 225(10), 231(35), 232(10, 35), 233, 235(10), 238(10), 240(1, 10), 243(1, 26, 28), 244(1, 26)
Chen, C. H., 267, 268(2, 12)
Cheng, F. W., 93, 94(143), 95(143)
Cheng, S. L., 435
Cherbas, L., 429, 459, 467, 496
Cherbas, P., 429, 467
Chichester, C. O., 125
Chihara, C., 459
Child, P., 72, 73(36), 86(36), 87(94), 88(94), 91(94)
Chino, H., 429, 442
Chobanian, A. V., 291
Chojnacki, T., 206, 208(21), 475, 476(10), 547
Chong, K. S., 267, 268(11)
Chong, Y. K., 402
Chorvat, R. J., 291, 292
Chowdhry, V. H., 510

Chu, J.-W., 368
Chung, J., 267, 268(9)
Churacek, J., 255
Clague, A. D. H., 33, 34(a), 35
Claret, J., 387, 388(39), 389, 390(39), 391(39), 412, 419(9), 420, 428(9), 444
Clark, A. J., 346
Clark, B. C., Jr., 159
Clark, C. R., 258
Clark, R. G., 156
Clark, R. J. H., 246
Clark, W. H., Jr., 496, 521
Clarke, N., 533, 538(16)
Clayton, P. T., 82, 85(76), 96(76)
Cleveland, D. W., 238
Coates, J., 151
Cohen, L. H., 461
Cole, W. J., 83, 85(79), 89(79)
Colin, H., 38, 39(6)
Colwell, W. T., 315
Conaway, J. E., 171
Conder, J. R., 164
Connat, J-L., 379, 384(12), 419
Conneely, O. M., 301, 302(24)
Conner, R. L., 6, 9, 10, 11, 26, 34(c), 35
Constantin, E., 389, 390(40, 44), 410(40, 44), 412, 414(21), 415(21), 416(21), 418(20, 21)
Coon, M. J., 365, 372, 376(19)
Cooper, D. Y., 454
Copius-Peereboom, J. W., 10
Cornforth, J. W., 251, 315
Cornforth, R. H., 251, 315, 549
Corrie, J. E. T., 344
Coscia, C. J., 149, 476
Cotter, R. J., 68
Coudron, T. A., 495, 499(10), 501(10), 502(10), 503(10), 521
Courgeon, A. M., 467
Cowen, A. E., 88, 111, 112(108, 198)
Cox, R. E., 188
Cox, T. P. H., 74
Craig, J. C., 310
Cramer, F., 205
Cramers, C., 155
Cremona, T., 421
Crettol-Järvinen, A., 235, 238
Cronholm, T., 72, 73(40), 84(40), 89(40), 94, 95(145), 100, 101(164), 103(40, 164), 104(164, 172), 111(172–178)
Cros, G. S., 446

Croteau, R., 149, 171(1)
Crum, F. C., 214
Crump, D. R., 8, 21(15)
Cuatrecasas, P., 436
Culver, M. G., 540, 541(6), 542(6), 543(6)
Cumingham, F. X., Jr., 200
Curphey, T. J., 315
Curstedt, T., 85, 93(89), 100, 101(164), 103(164), 104(164, 172), 111(172)
Curvers, J., 155
Cytil, F., 299

D

Dabrowa, N., 340
Dahl, S. B., 372, 376(19)
Dahm, K. H., 540, 541(7, 8), 542(5, 7, 8), 543(7)
Dale Poulter, C., 172
Dallner, G., 475, 476(10), 547
Damico, J. N., 27
Danieli, B., 409, 410(80), 450
Danielsson, H., 364, 365(1), 366, 367(13)
Danilov, L. L., 206
Dansette, P. M., 310
Das, P. K., 218
Daum, H. A., 302
Daumas, R., 216, 218(14), 220(14), 221(14), 222(14), 223(14), 224(14), 228(14), 229(14), 230(14), 235(14), 237(14), 238, 240(75)
Davidson, L., 267, 268(11)
Davies, B. H., 117, 124(11, 14), 125(14), 132(11, 14), 136(11, 14), 148(11, 14), 149(11), 186, 192
Davies, D. H., 238
Davies, T. G., 348, 385
Davison, P. M., 268, 274(18)
Day, R. J., 68
Dayal, B., 91(131), 92, 93, 94(143), 95(143)
Dayan, J., 341
Dayton, S., 289
DeAndrade Bruning, I. M. R., 187
Debrauwere, J., 170
de Chaffoy, D., 246
de Chaffoy de Courcelles, D., 232, 246(57), 247
Deedler, R. S., 24
Defago, G., 340, 344(24)
DeGraff, R. A. G., 3

DeGraw, J. I., 315
Dehn, R. L., 315
de Kort, C. A. D., 487, 526, 533
DeLaage, M. A., 394, 431, 443, 446, 450
Delachambre, J., 378
Delbecque, J. P., 378, 388(38), 389, 400(38), 419(8), 420, 428(8)
DeLuca, H. F., 197
de Luca, L. M., 251
Delwicke, C. V., 289
DeMark, B. R., 66, 67(12), 111, 112(200, 201, 203)
Dempsey, M. E., 293, 294(2, 3), 295(2, 5, 8), 296(2, 3, 6), 297, 298, 299(3, 6, 8), 300(3, 4), 301(5), 302(3, 6, 24), 303(6)
DeMoor, P., 438
Dennis, R., 458
DeParscau, L., 279
Depieds, R. C., 446
De Reggi, M. L., 378, 379, 394, 431, 443, 448
DeRosa, M., 322
Desiderio, D. M., 105
Desmond, H., 389, 390(43), 401(43), 404(43), 406(43), 408(43), 410(43)
de Souza, N. J., 10, 11(33), 317
DeVilbiss, E. D., 173
DeVries, G. H., 475, 476, 480
Devys, M., 33, 318
DeWaard, M. A., 340
Dewald, B., 233
De Weerdt, G. A., 72, 73(49), 99(49)
Deykin, D., 276, 288, 289(12), 290(12)
d'Harlingue, A., 544
Dias, J. R., 78, 80(65), 82(64–66), 85(65)
DiBussolo, J. M., 11, 12(44), 13, 14, 15(44), 329
Diehl, P. A., 379, 384(12), 419
Diekman, J., 87
Dietschy, J. M., 313
Dimitriadou-Vafiadou, A., 293, 294(3), 296(3), 297(3), 298(3), 299(3), 300(3), 302(3)
Dinan, L. N., 348, 383, 384, 385, 387(29), 389, 391(45), 393(45), 396(54), 397, 403(63), 408(49), 412, 426, 427, 428(23)
Dingman, C. W., 235
Dittmer, J. C., 549
Djerassi, C., 27, 33, 87, 90, 91, 93, 172, 312, 327, 329(66), 403

Dobryszycka, W. J., 232
Doderer, A., 533
Doi, Y., 267, 268(3)
Doisy, E. A., Jr., 83
Dolphin, D., 319
Donnahey, P. L., 11, 38, 39(10), 329, 393, 396(54), 427, 428(23)
Doolittle, G. M., 286, 289
Döpp, H., 426, 427(17), 428(17), 450
Dorfman, L. F., 25, 26, 27(71)
Dorsey, S., 288, 290(9), 291(9)
Dreano, Y., 162
Dreize, M. A., 291
Drysdale, B., 341
Dubuc, J., 294
Dulaney, J. T., 233
Duncan, J. N., 319
Dunn, P. E., 294, 296(9), 482, 484(1), 485, 486(1), 487(2, 4), 499, 517, 522(16)
Dunphy, P. J., 549
Duque, C., 354, 358, 362, 364(5, 7)
Duriyaprapan, S., 151
Durrant, J. L., 420, 438, 440(2), 441(2)
Durst, F., 454, 457(7)
Dutky, S. R., 402, 419, 420, 437, 440(1)
Duvic, B., 432, 445, 446(24), 448(24), 449(24), 453(24)
Dvornik, D., 294
Dwyer, W., 151
Dyer, W. J., 212, 248, 325
Dygos, J. H., 291, 292

E

Echalier, G., 495, 496(13), 521, 522
Eckers, C., 76
Edenharder, R., 72, 73(43)
Eder, H. A., 277, 282(16)
Edmonds, C. G., 24, 76, 77(56)
Edmonds, V. A., 76
Edwards, R. A., 160
Egan, R. W., 289
Eggermont, E. A., 93
Eggert, H., 33
Eggling, S. B., 154, 156
Eglinton, G., 186
Ehninger, D. J., 542
Ehrhardt, J. D., 321

AUTHOR INDEX

Ehrhardt, M., 187
Einarsson, K., 105(187, 188, 193), 106, 108(187, 188), 109(193), 111(193)
Eisler, W. J., 103, 112(168)
Ekström, T., 475, 476(10)
Elgsaeter, A., 217, 232(33), 240(33)
Ellingboe, J., 11
Ellington, J. J., 158
Elliott, W. H., 11, 51, 52, 54(2), 55(1, 2, 3), 56(4), 57(1), 58(4), 59(1, 4), 61(1, 4), 62(1), 63(1, 3), 64, 68, 71(20), 73(2, 4), 78, 79(2, 4), 82(2), 83(2), 84(2, 4), 85(2), 87(4), 88(88), 89(2), 90(2, 41), 91(2, 4, 133), 92(2), 93(111, 123, 133), 95(133), 96(4, 20, 133), 97(20)
Ellison, B. O., 176
Ellsworth, J. L., 275
Emery, H., 9
Emmerich, H., 426, 428(19), 429, 430, 442, 495, 499(3), 526, 539, 541, 542(11), 543(11)
Enderle, U., 383, 389(22), 390(22), 393(22), 398(22), 399(22), 430, 431(15), 432(14), 433(14), 434(14), 435(14), 453
Endo, K., 319
Eneroth, P., 64, 73(1), 79(1), 80(1), 83(1), 84(1), 85(1), 86(1), 87(1), 88(1), 89(1), 90(1), 91(1), 92(1), 94, 95(147), 310, 311(31), 428
Engel, K.-H., 179
Engelhardt, E. D., 187
Engelmann, F., 495, 499(4, 5, 6), 516, 517(15)
Englert, G., 31, 32, 33, 147, 198, 199, 232
Epstein, W. W., 172
Erdman, J. W., Jr., 197, 199
Erfling, W., 111, 112(200)
Erickson, S. K., 286, 289(6), 291(6)
Eriksson, C.-G., 428
Eriksson, H., 82, 83(78), 85(86), 93(89), 103, 111(175)
Estabrook, R. W., 304
Estienne, J., 33
Ettre, L. S., 175
Eugster, C. H., 131
Evans, R. C., 169
Evans, S., 103, 111 (176, 177)
Everson, G. T., 111, 112(200, 201, 203)
Ewerth, S., 105(193), 106, 109(193), 111(193)
Eyssen, H. J., 93

F

Fahimi, H. D., 306
Fairclough, D. P., 246
Falck, J. R., 304
Falcone, M. S., 181
Fales, H. M., 27, 489
Falk, O., 105(189), 106, 109(189), 110(189)
Farr, A. L., 281, 282(23), 296, 365, 439
Farris, C. L., 38
Farrow, R., 340
Faust, J. R., 291, 292
Favre, M., 233
Feeney, J., 202, 549
Feinstein, A., 436
Fendesack, F., 157
Fenselau, C., 67
Ferguson, K. A., 10, 26
Fernandez, E., 12, 16(49), 17
Ferrari, G., 451
Ferretti, A., 151
Feyereisen, R., 386, 430, 443, 445(8, 9), 446(9), 448(9), 449(8, 9), 454, 457(7), 531, 533, 534(17), 537(7), 538(7, 20), 539(9), 541
Field, F. J., 289, 291(17)
Fielding, C. J., 267, 268(1), 270, 271, 272(28), 273(25), 274(18, 25), 275, 282(6), 283
Fielding, P. E., 267, 268, 270, 271, 272(28), 273(25), 274(18, 25), 275, 279, 283
Fiksdahl, A., 200
Finamore, F. J., 246
Findlay, J. A., 409, 410(80)
Fink, G. R., 339, 342(5), 344(5)
Finney, J. R., 531
Firpo, G., 12, 16(49), 17
Fischer, S. G., 238
Fisher, M. M., 72, 73(33), 96, 97(156, 157)
Fitzpatrick, F. A., 39
Flanagan, V. P., 151
Flath, R. A., 152, 154, 156, 162
Fleischner, G., 302
Fless, G. M., 267, 268(9)
Florini, J. R., 463, 465
Folch, J., 212
Fonteneau, P., 320, 324(43)
Forrey, R. R., 162
Fosket, D. E., 199
Foster, D. O., 185

Fourcans, B., 9
Foury, F., 341
Fox, D. L., 217, 224(31), 225(30), 232(30), 242(30, 31)
Francke, W., 184
Franke, P., 206, 208(21)
Frantz, I. D., 37, 384
Freeman, D. J., 214
Freeman, R. R., 172
Frew, N. M., 24
Friedel, T., 538
Friese, L., 157
Fristrom, J. W., 458, 459, 462(5), 467
Frost, D. J., 33
Fruitwala, N. A., 164
Fryberg, M., 319
Fujii, H., 213
Fujimoto, Y., 33, 346, 347(6), 348(6), 352(6, 11)
Fujino, A., 91
Fujiwara, H., 67
Fukunaga, A., 72, 73(46), 76(46), 79(46)
Fukushima, K., 21, 23, 24(56), 185
Fung, C. H., 292
Fyffe, W., 244

G

Gabriel, O., 233
Galbraith, M. N., 397, 402, 408(62)
Galeazzi, R., 72, 73(32)
Gallagher, T. F., 317
Games, D. E., 68, 76
Games, M. P., 68
Gammack, D. B., 216, 218(15)
Gandhi, C., 471, 473(7)
Gant, J. R., 257
Garber, E. D., 192
Garner, G. V., 68, 72(24)
Gaskell, S. J., 94
Gassiot, M., 12, 16, 17, 188
Gatmaitan, Z., 302
Gaydou, E. M., 185
Gaylor, J. L., 289, 293, 302
Gazdag, M. M., 166
Geckle, M., 72, 73(48)
Gellissen, G., 444
Gerber, L. E., 197, 199

Ghisalberti, E. L., 316, 317(23), 318(23)
Gibbons, G. F., 7, 11(7), 28, 34(*a*), 35, 321
Gibson, J. M., 379, 383, 390(13), 393(13)
Gil-Av, E., 181
Gilbert, J. M., 187
Gilbert, L. I., 379, 385(18), 429, 442, 454, 483, 484, 487, 488, 489(6), 490, 512, 517(12), 540, 542
Gilchrist, B. M., 246, 247(84, 88)
Gildersleeve, N., 475
Gilgan, M. W., 428
Gill, S. S., 303, 304(3), 306, 308, 309(29)
Gillen, D. G., 167
Giss, G. N., 173
Givner, M., 294
Glatz, J. J. C., 299
Glomset, J. A., 275
Glossmann, H., 235
Glover, D. M., 437
Gluck, R., 302
Go, C. J., 293
Goad, L. J., 33, 34(*a*), 35, 311, 312, 315, 316, 317(23), 318(23), 319(32), 320, 321(31), 322, 323, 324(38), 326(41), 329, 330(32, 69), 331(32, 50, 69), 332(38, 59), 404
Godefroot, M., 155, 157, 164
Goerke, J., 276
Gohring, K. E. H., 187
Golan, M., 304, 305(5)
Goldstein, J. L., 291, 292
Gollub, E. G., 341, 342(38)
Goltzené, F., 377, 378(3), 385(3), 412
Gonzales, R. A., 335, 336, 339(6), 340(6)
Gonzales, R. B., 41
Goodman, DeW. S., 227
Goodman, D. S., 251, 276, 288, 289(12), 290(12)
Goodman, W., 379, 385(18), 483, 484, 487, 488, 489(6), 489, 495, 510, 512, 517(12), 530
Goodwin, T. W., 11, 38, 39(10), 116, 117, 132(12), 135(12), 189, 217, 225(34), 232(34), 240(34), 241(34), 244(25), 312, 315, 316, 317(23), 318(23), 319(32), 320, 321(31), 323, 324, 326(41), 329, 330(32), 331(32), 332(59), 348, 378, 379, 383, 385, 390(8, 9), 393(14, 25), 394(14), 395(14), 396(14, 54), 398(14), 399(25), 400(9), 401(9), 402(8, 9, 25), 403(25, 55),

404(8, 9), 405(9), 406(8, 9), 407(9), 408(8, 9, 55), 409(9, 55), 410(8, 9), 412, 413(13), 414(13), 415(12, 13), 416(13), 417(13), 418(12, 13, 19), 427, 428(23)
Gordon, D. B., 68, 72(24)
Gordon, J., 458
Gordon, J. I., 299, 300(18), 302(18)
Gorelik, M., 96, 97(158)
Gorell, T. A., 429, 442
Gorski, J., 458, 462
Gotchel, B. V., 461
Goto, J., 62, 69, 71(28), 110
Goto, R., 364
Goudeau, M., 403
Gould, W., 152
Gower, J. L., 76
Grabowski, G., 93, 302
Granat, M., 172
Grande, A. R., 394
Granger, N. A., 487, 494, 530
Greco, A. V., 78
Gredt, M., 340
Green, J., 545
Greenwood, D. R., 397, 403(63)
Greville, G. D., 436
Griffin, M. J., 304
Grimme, L. H., 198, 199
Grindle, M., 340
Grob, E. C., 124
Grob, G., 161, 169, 170(100), 171, 175
Grob, K., Jr., 161, 162, 167, 169, 170(100), 171, 175
Groeneweg, H., 340
Gronemeyer, H., 432, 434(30), 461
Grossman, D., 164
Gruenke, L. D., 310
Grunberger, D., 69, 71(29)
Grzeszkiewicz, A., 94
Guadagni, D. G., 152
Guadry, R., 294
Guengerich, F. P., 303(4), 304, 306
Guenthner, T. M., 304, 306
Guiltinan, M., 199
Guiochon, G., 38, 39(6)
Gumulka, J., 310
Gunatilaka, A. A. L., 344
Gunawan, S., 539, 541, 542(11), 543(11)
Guo, L. S. S., 276
Gustafsson, J., 365, 366, 367(11)

Gustafsson, J. Å., 82, 83, 85, 86, 88, 89, 91(91), 94, 95(144, 147)
Gut, M., 8, 21(15), 353

H

Haabti, E. O., 25
Haak, R. A., 344(55, 56), 345
Hachey, D. L., 64, 66(3), 67(3), 72, 73(3, 34), 79(3), 88, 91(3, 125), 92, 95(125), 96(125), 103, 112(108, 169)
Hachimi, Z., 201
Hackney, J. F., 302
Hackney, R. J., 402
Hafferl, W., 429, 442
Hager, A., 132, 189
Haken, J. K., 168
Halket, J. M., 24
Hall, H., 316, 317(23), 318(23)
Hall, L. D., 31
Hall, M. S., 531, 538(5)
Hamilton, R. L., 276
Hamilton-Miller, J. M. T., 340
Hammock, B. D., 303, 304(3), 305, 306, 307, 308, 309(29), 487, 488(3, 4), 489(6), 490(3, 4), 532, 543
Hamnett, A. F., 488, 531, 532, 534(14), 538, 539(9), 541
Hann, J. R., 320
Hansbury, E., 11, 38, 39(4, 9), 299, 302(22), 326, 329
Hanson, R. F., 91(125), 92, 93, 95(125), 96(125), 302
Hansson, R., 365, 366
Hansson, V., 453
Hanzlik, R. P., 315, 316(18)
Harada, K., 171
Harano, K., 83, 87(83)
Harano, T., 83, 87(83)
Hardy, P. J., 152
Hardy, R., 156
Harmony, J. A. K., 275
Harris, B., 7, 11(7), 28, 34(a), 35
Harris, W. S., 80
Harrison, A. G., 66
Harry, P., 432, 437, 445, 446(24), 448(24), 449(24), 453(24)
Hart, J. W., 160
Hartkope, A., 257

Hartkopf, A., 164
Hartmann, M. A., 320, 324(43)
Hartmann, R., 526
Harvey, D. J., 88
Hasegawa, M., 110
Hashimoto, S., 289
Haslam, J. M., 339, 341
Haslewood, G. A. D., 82, 91(126, 127), 92, 93(137), 94(122, 126)
Hatta, Y., 72, 73(46), 76(46), 79(46)
Haugen, D. A., 365, 372, 376(19)
Haumann, J. R., 103, 112(168, 169)
Hausberg, P., 437
Hauschka, P. V., 217, 243(27), 244(27), 245(27), 246(27)
Haustein, D., 435
Havel, R. J., 270, 273(25), 274(25), 275, 276, 277, 282(16)
Hawkes, S., 164, 176
Hays, P. R., 42
Hazelby, D., 103, 111(177)
Headon, D. R., 301, 302(24)
Heath, E. C., 547
Hedin, P. A., 158
Hefendehl, F. W., 158, 160
Heftmann, E., 6, 12, 38, 39(7), 320, 321
Heider, J. G., 269, 278, 292, 293
Heinrich, G., 411
Heinrikson, R. L., 484, 485, 487(4)
Heintz, R., 316, 317(22), 318(22), 323, 324(22)
Heip, J., 246, 247(85)
Helgerud, P., 291
Hellström, K., 87
Hemming, F. W., 201, 202(2), 206(2), 208(2), 480
Henderson, W., 83, 84(81), 87(81), 95(81), 160
Hendriks, H., 172
Henneberg, D., 104
Henry, S. A., 343
Hercules, D. M., 68
Herling, J., 181
Herman, G., 96, 97(158)
Herres, W., 175
Herrick, G., 467
Herring, P. J., 217, 224(29), 225(29), 227(29), 231, 232(51), 234(29), 235(29), 238(29), 239(29), 240(29, 51)
Herscovics, A., 202, 213(12)

Herz, J. E., 82, 83(78)
Hesso, A., 75
Hetru, C., 377, 378(3), 383, 385(3), 389, 390(27, 40, 44), 394, 403, 404(27), 407(27), 408(27), 410(40, 44), 411, 412, 414(21), 415(21), 416(21), 418(8, 20, 21), 451
Hewlins, M. J. E., 321
Heyns, W., 438
Hicks, J. B., 339
Higuchi, T., 69, 71(28)
Hikasa, Y., 75
Hilunen, R., 167
Hiremath, S. V., 89, 93(111)
Hirm, M. H., 443
Hirn, M., 378, 394, 412, 431, 448
Hirter, P., 76
Hirth, L., 321, 331
Hjérten, S., 235
Hlavay, J., 166
Hocks, P., 411
Hoffmann, J. A., 377, 378(3), 383, 385(3), 389, 390(27, 40, 41, 44), 394, 399, 403, 404(27, 41), 407(27, 41), 408(27, 41), 410(40, 44), 411, 412, 414(21), 415(21), 416(21), 418(8, 20, 21, 22, 26), 430, 432, 437, 443, 445, 446(24), 448(24), 449(24), 453(24)
Hoffman, N. E., 69
Hoffmeister, H., 411
Hofmann, A. F., 88, 111, 112(108, 198)
Hogge, L. R., 172
Holman, G. M., 420, 438, 440(2), 441(2)
Holmberg, I., 366, 372(12)
Holmbom, B., 184, 185
Holz, G. G., Jr., 33
Holzer, G., 188
Homberg, E., 18, 19, 20(53), 23, 25(53)
Hooper, S. N., 185, 188
Hoppe, W., 411
Horie, Y., 346
Horn, D. H. S., 383, 385, 388(37), 389, 391(37), 395(37), 397(37), 401(37), 402(37), 408(62), 450, 451, 452
Horning, E. C., 24, 83, 87, 89(98), 95(98)
Horning, F. C., 25
Horning, M. G., 83
Hoshita, T., 91(128, 129, 130), 92, 93(129), 94, 95(130, 146, 148–150, 152), 96(128–130)

Hosoi, K., 66
How, P., 160
Howard, D. H., 340
Howard, J. B., 293, 294(3), 296(3), 297(3), 298(3), 299(3), 300(3), 302(3)
Hoyer, G.-A., 426, 427(17), 428(17), 450
Hsiao, C., 412
Hsiao, T. H., 412
Hsiung, H. M., 9
Huang, J. J., 368, 481
Huang, S.-K. S., 87
Hubbard, S. C., 547
Huber, R., 411
Hubert, E. V., 291
Hubert, P., 153, 154
Hudson, C. A., 310
Hug, G., 218
Hughes, L. B., 275
Hummel, H., 411
Hunt, D. F., 67
Hunter, G. L. K., 158
Hunter, I. R., 6, 12, 38, 39(7)
Huntoon, S., 9
Husman, H., 166
Hutchins, R. F. N., 348
Huttash, R. G., 267, 268(11)
Hwang, K.-K., 87
Hyde, P. M., 11

I

Iacobucci, G. A., 159
Iatrides, M. C., 33
Idler, D. R., 11
Idoyaga-Vargas, V., 476(15), 477
Idstein, H., 175
Ihm, J., 275
Iida, T., 33, 34(a), 35, 72, 73(39), 75(39), 78, 85(63)
Ikawa, M., 125, 186, 189
Ikawa, S., 91(126), 92, 94(126)
Ikekawa, N., 33, 185, 346, 347(6), 348(6), 349(13), 350(13), 352(6, 11, 13), 354, 355, 358, 361, 362, 364(5-8), 411
Imai, K., 72, 73(42)
Imai, Y., 365
Ingall, D., 111, 112(197)
Ingraham, D. F., 176
Inman, D. R., 458

Inoue, M., 72, 73(37), 92, 99(37), 106(37)
Ioffe, B. V., 180
Ionova, E. A., 181
Isaac, R. E., 377, 378(4), 383, 387(4), 389, 390(8, 9, 23, 42, 43), 392, 393(25), 398(23), 399(25), 400(9, 42, 52), 401(9, 23, 42, 43), 402(8, 9, 25), 403(25, 55, 74), 404(8, 9, 23, 42, 43, 74), 405(9), 406(8, 9, 23, 42, 43), 407(9, 23, 42), 408(8, 9, 23, 43, 55), 409(9, 55), 410(8, 9, 23, 42, 43), 411, 412, 413(13), 414(13), 415(12, 13), 416(13), 417(13), 418(12, 13, 19), 428
Isaka, Y., 346
Isenberg, J. N., 91(125), 92, 95(125), 96(125)
Isenhour, T. L., 164
Ishibashi, M., 72, 73(37, 45, 46), 76(46), 79(45, 46), 92, 99(37), 106(37, 45, 67), 107(57)
Ishibashi, T., 296
Isidorov, V. A., 180
Isler, O., 116
Isobe, M., 305, 310(12), 311(12)
Itagaki, Y., 69, 71(28)
Itho, M., 76, 108(57)
Ito, M., 92
Itoh, M., 72, 73(37), 99(37), 106(37)
Itoh, T., 10, 21, 23, 24(56), 33, 185
Ivatt, R. J., 547
Izumi, A., 10, 28, 34(b), 35

J

Jackson, A. H., 68
Jackson, R. L., 275
Jacobs, P. B., 123
Jacobs, S., 232
Jacobson, H. I., 458
Jamieson, W. D., 185
Jandera, P., 255
Jankowski, T., 206, 208(21)
Janssen, G. A., 93
Jaouni, T. M., 489
Jarman, T. R., 344
Jasne, S. J., 38, 39(5)
Javitt, N., 72, 73(32)
Jaworska, R., 94
Jeanloz, R., 202, 213(12), 549
Jehle, E., 210
Jelus, B. L., 67

Jencks, W. P., 216, 217(9), 218(9, 13), 219(9, 13), 220(9, 13), 221, 222(13), 224, 232(13, 39), 233(13, 39)
Jennings, R. C., 488, 533, 534(14), 539, 541
Jennings, W., 74, 166, 170, 171, 175, 176
Jennings, W. G., 153, 158, 163, 172(114), 180
Jensen, A., 117, 125(13), 192
Jensen, B. W., 335, 339(6), 340(6)
Jensen, E. V., 458
Jensen, J., 502, 509(20)
Jeon, I. J., 151
Jerina, D. M., 305, 310(13), 311(13)
Joanen, T., 91(131, 132), 92
Johansson, G., 94, 95(145)
John, J. P., 16, 20(51), 25(51)
Johnson, J. H., 152
Johnson, M. R., 203
Johnson, P., 454
Johnson, R. M., Jr., 152
Johnston, A. M., 340
Johnstone, R. A. W., 403, 404(76)
Jollow, D. J., 334
Jolu, E. J. P., 446
Joly, P., 443
Jonas, A., 267, 268(13)
Jones, B. B., 159
Jones, J. H., 187
Jones, R., 216(18), 217, 218(18)
Jones, S., 267, 268(1)
Jordi, H., 39
Joseph, J., 7, 11(7), 28, 34(*a*), 35
Joseph, J. M., 6, 8, 9, 11, 16, 21(12, 13), 23, 33(13), 34(*c*), 35
Judge, J., 160
Judy, K. J., 531, 538(5)
Justice, J. B., 164

K

Kaduce, T. L., 286
Kalasinsky, V. F., 150
Kalin, J. R., 471, 474(1)
Kalish, F., 532, 534
Kallafer, M. E., 289
Kallner, A., 88
Kameyama, K., 233
Kamin, H., 368
Kamio, Y., 267, 268(3)
Kammereck, R., 9
Kanai, M., 305, 310(12), 311(12)
Kanazawa, A., 322
Kandutsch, A. A., 205
Kane, J. P., 270, 273(25), 274(25), 275, 299
Kaplanis, J. N., 346, 385, 402, 419, 420, 437, 440(1), 454, 457(8)
Kappler, C., 385, 403
Karasek, M., 268, 274(18)
Karger, B. L., 257
Karlaganis, G., 72, 73(38), 75(38), 94, 95(151)
Karlaganis, V., 94, 95(151)
Karls, M., 112
Karlsen, J., 160, 171
Karlson, P., 391, 399, 411, 419, 420(1), 421(1, 11), 422(11), 423(1, 11), 424(1, 11), 425(1, 11), 426, 427(1, 11, 17), 428(17), 429(1), 430, 433, 437, 442, 443, 450
Karmat, S. Y., 91, 93(123)
Karst, F., 341, 342
Kasama, T., 105(185), 106, 108(185), 110(185)
Katayama, K., 105(192), 106, 108(192), 111, 112(199)
Katayama, T., 125
Kates, M., 189
Kato, H., 62, 110
Kato, S., 547(7), 548
Katsuki, H., 286
Katz, A., 199
Katzenellenbogen, J. A., 432, 453, 487, 510, 511, 540, 541(4)
Käuser, G., 430, 431(15), 432(14), 433(14), 434(11), 435(11), 437, 453
Kawabe, K., 105(192), 106, 108(192)
Kawata, M., 91
Keeley, L. L., 439, 457
Keenan, R. W., 201, 205, 206(4), 471, 473(7), 476, 477(13), 478, 480
Keim, P. S., 484, 485, 487(4)
Keller, R., 460
Keller, R. K., 201, 202, 203, 204, 209(8), 210, 211, 214, 215(36), 471, 474(6), 475(6)
Kelly, L. A., 292
Kelsey, M. I., 80, 84, 87, 88(88)
Kemple, M. D., 344(55), 345
Kepner, R. E., 154, 172, 173, 176
Kern, F., Jr., 85, 93(89), 111, 112(200, 201, 203)
Kern, H., 340, 344(24)

Kerr, J. D., 549
Kerven, G. L., 151
Ketter, B., 302
Kézdy, F. J., 192, 487(6)
Khachadurian, A. K., 292
Khalil, M. W., 11
Khan, M. A., 533
Khoo, J. C., 481
Khuong-Huu, F., 33
Kibe, A., 94, 95(146)
Kienzle, F., 232
Kihira, K., 91(128, 129), 92, 93(129), 94, 95(148, 149, 150, 152), 96(128, 129)
Kikuchi, H., 95
Kikuchi, N., 11
Kimura, M., 91
Kimura, T., 368
King, D. S., 444, 462(5)
King, R. J. B., 458
King, W., 275
Kinnear, J. F., 436
Kirchner, J. C., 157
Kirk, D. N., 72, 73(41), 82(41), 86(41), 90(41), 91(41)
Kirsch, D. R., 285
Kirschner, M. W., 238
Kisic, A., 10, 28, 34(b), 35
Kitabatake, K., 267, 268(3)
Kizer, D. E., 304
Klages, G., 495, 499(3)
Klause, K. A., 25
Klein, P., 111, 112(201)
Klein, P. D., 37, 64, 66(3), 67(3, 12), 72, 73(3, 34), 79(3), 88, 91(3, 125), 92, 95(125), 96(125), 103, 111, 112(108, 168, 169, 197, 198, 200, 203), 316, 318
Kleinhans, F. W., 344(55, 56, 58), 345
Kleining, H., 199
Klouwen, M. H., 167
Knapp, F. F., 9, 27, 312, 318, 319(32), 330(32), 331(32)
Knight, J. C., 316
Knight, M. E., 76
Knight, S. A., 33
Knights, B. A., 11, 90
Knowles, J. R., 432, 510
Koch, P., 437
Koedam, A., 151, 158, 165(11)
Koeppe, J. K., 495, 499(7, 8), 500, 506(18), 510, 511(5), 512, 516(7, 11), 518(6, 11), 520(6, 7, 11)

Koga, T., 111, 112(202)
Koizumi, N., 33
Kok, E., 72, 73(32)
Kollman, P. A., 451
Kondo, M., 232, 246(57), 247(57, 85)
Kong, J. M., 176
König, M., 437
Konig, W. A., 184
Koolman, J., 382, 383, 386(19), 389(22), 390(22), 391, 393(22), 394, 398(22), 399(22), 405, 408(79), 411, 419(10), 420(1, 4), 421(1, 11), 422(11, 12), 423(1, 11), 424(1, 11), 425(1, 11), 426, 427(1, 4, 11), 428(18, 19), 429(1), 430, 431(15), 432(14, 24), 433(14), 434(14), 435(14), 437, 438, 443, 453
Koopmanschap, A. B., 533
Koppang, N., 214
Koppler, H., 157, 179
Kossa, M., 179
Kovalick, G. E., 495, 499(7, 8), 500, 506(18), 510, 511(5), 512, 516(7, 11), 518(6, 11), 520(6, 7, 11)
Kovats, E. sz., 158, 171
Koyama, T., 213
Koyasako, A., 176
Krafft, G. A., 511
Kramer, K. J., 294, 296(9), 482, 484(1), 486(1), 487(2, 6), 499, 517, 522
Kramer, S. J., 532, 534
Kraml, M., 294
Kratzl, K., 184, 185(157)
Krepinsky, J., 409, 410(80), 451
Kreuz, K., 199
Krevitz, K., 7, 8, 11(7), 21(14), 28, 33(14), 34(a), 35
Krien, P., 393, 400(53)
Krstulovic, A. M., 41, 189, 198(6), 199
Krusczek, M., 201, 206(4)
Kubeczka, K. H., 158, 159
Kubodera, T., 72, 73(37), 99(37), 106(37)
Kugler, E., 158
Kühn, H., 216, 218(11, 12), 219(11), 220(11, 12), 223(12), 231(12), 232(12)
Kuhn, R., 216, 217, 218(11, 12), 219(11), 220(11, 12), 223(12), 231(12), 232(12), 243(23, 24), 244(23, 24)
Kuksis, A., 72, 73(31, 33, 36), 86(36), 87(94), 88(94), 91(94), 96, 97(156)
Kulkarni, B. D., 16, 20(51), 25(51)
Kuntz, I. D., 451

Kuppert, P., 429, 444, 459, 460(9)
Kuramoto, T., 91(128, 129, 130), 92, 93(129), 94, 95(130, 146, 148–150), 96(128–130)
Kuriyama, K., 66, 72, 73(35), 93(11, 35)
Kurosawa, T., 21, 23, 105(195), 106, 108(195)
Kushwaha, S. C., 189
Kutner, A., 94
Kutner, W., 94

L

Laatikainen, T., 75
Lachaise, F., 403
Lacko, A. G., 267, 268(11)
Lacops, H. J. C., 3
Lacroute, F., 341, 342
Laemmli, U. K., 233, 238, 285, 365, 461, 505, 523
Lafont, R., 382, 383, 387, 388(38, 39), 389(22), 390(22, 39), 391(39), 393(22, 38), 394(38), 395(38), 398(22), 399(22), 400(38, 53), 412, 419(8, 9), 420(4), 424, 427(4), 428(5, 8, 9), 444
Laqueux, M., 377, 378(3), 385(5), 389, 390(40, 44), 394, 410(40, 44), 412, 414(21), 415(21), 416(21), 418(20, 21), 430, 437, 443, 445, 448
Lambin, P., 235
Landrey, J. R., 6, 8, 9, 10, 11, 26, 34(c), 35
Langworthy, T. A., 188
Lapointe, M. C., 495, 499(7), 512, 516(11), 518(11), 520(11)
Largeau, C., 318, 321(31)
Larson, P. A., 169
Laskowski, M. B., 476
Lata, G. F., 435
Lau, S. Y. M., 216, 217(5), 218(5), 224, 225(5, 46), 227(46), 232(46), 240(5), 241(5, 46)
Law, J. H., 294, 296(9), 482, 484(1), 485, 486(1), 487(2, 4, 6), 495, 499(10), 501(10), 502(10), 503(10), 517, 521, 522(16), 531, 540, 540(4)
Lawrie, T. D. V., 89
Lawson, A. M., 75, 76(56), 77(56), 80(56), 82, 85(76), 96(76), 105, 106(182, 183), 109(182, 183), 110(183)

Lawson, J. A., 315
Lawson, R., 238
Layne, F., 232
Lazaro, E., 67
Luzdunski, M. J., 480
Leary, J. J., 164
Leavis, P. C., 435
LeBoeuf, R. D., 123
Lee, D. L., 451
Lee, M. L., 168, 169(96)
Lee, T.-C., 125
Lee, T. H., 160
Lee, W. H., 9
Lee, W. L., 216, 218, 225(36), 226(36), 240(1), 243(1), 244(1)
Lees, M., 212
Lees, N. D., 344(55, 56, 58, 59), 345
Lehrer, S. S., 435
Lennarz, W. J., 208, 213, 476(18), 477
Lenton, J. R., 312, 316, 317(23), 318(23), 321
Leroux, P., 340
Lester, R., 92, 111, 112(197)
Lester, R. L., 549
Leuenberger, U., 124
Levenbook, L., 435, 436(38), 432(42)
Levin, W., 304, 305, 310(13), 311(13)
Levine, E., 205
Lewenhaupt, A., 103, 111(179)
Lewin, P. K., 86
Lewis, D., 379, 384(11), 390(11), 393(11)
Lewis, E., 76, 185
Lewis, I. A. S., 404
Lewis, S., 68, 69(21), 83(21)
Lewis, T. A., 342
Li, L. N., 27
Liaaen-Jensen, S., 117, 125(13), 192, 200, 216(20), 217, 218(20), 224(20, 31), 225(20, 30, 32), 226(32), 232(30, 33), 240(33), 242(30–32), 244(20), 245(20)
Lichtenstein, A. H., 288, 291, 292(26)
Lichtenthaler, H. K., 546
Liehr, J. G., 69
Likens, S. T., 154, 155
Lilljequist, A. C., 291
Lin, J., 267, 268(8)
Lin, S.-N., 24
Linday, R. C., 179
Ling, V., 69, 71(29)
Linnane, A. W., 334
Lipsky, S. R., 163

Lisboa, B. P., 7, 24, 82, 83, 89
Liu, I. Y., 202, 213(12)
Liu, K.-P., 341, 342(38)
Liu, P. K., 341
Llewellyn, C. A., 199
Lockley, W. J. S., 348, 385
Lofton, S. L., 344(59), 345
Londershausen, M., 459, 460(9)
Looman, A., 151, 152
Lorbeer, E., 184, 185(157)
Lorsbach, T., 293, 294(3), 296(3), 297(3), 298(3), 299(3), 300(3), 302(3)
Loury, D. N., 308
Lowry, O. H., 281, 282(23), 296, 298, 299(3), 365, 439
Lowry, S. R., 164
Lozano, R., 11
Lu, A. Y. H., 303, 304, 310
Lucas, J. J., 471, 473(4), 474(4), 476(4)
Lugaro, G., 321
Lukacs, G., 33
Lutsky, B. N., 9
Luu, B., 383, 389, 390(27, 40, 41, 44), 394, 404(27, 41), 407(27, 41), 408(27, 41), 410(40, 44), 411, 412, 414(21), 415(21), 416(21), 418(8, 20, 21, 22, 26)
Lysenko, N., 302

M

Maarse, H., 154, 159, 172, 173
McCammon, M. T., 41, 42(19)
MacCarthy, J. J., 320
McClelland, A., 437
McCloskey, J. A., 9, 76
McClure, G. L., 175
McCommas, S. A., 123
McCormick, A. M., 197
McCoy, K. E., 293, 294(3), 296(3), 297(3), 298(3), 299(3), 300(3), 302(3)
McDonald, J. T., Jr., 150
McGarrahan, K., 349
McGarry, J. D., 313
McGee, L. R., 172
McGill, A. S., 156
McGuire, D. M., 294, 296(6), 299(6), 302(6), 303(6)
McGuire, W., 464

McInnes, A. G., 33
McIntyre, H. B., 83, 85(79), 89(79)
McKean, M. C., 312
McKean, M. L., 3, 11, 31
Mackenthun, M. L., 218
Mackenzie, A. S., 188
McKinley, C., 111, 112(200, 201)
MacLachlan, J., 89
McLean-Bowen, C. A., 48, 335, 338, 339(6), 340(6), 344(13), 345(53)
MacLennan, D. H., 280, 281
McNease, L., 91(131, 132), 92
McPherson, A., Jr., 239
McReynolds, W. O., 18, 19
McVittie, L., 275
Madani, C., 89
Maderspack, A., 24
Maestas, P. D., 184
Mahadevan, V., 267, 268(14)
Majumder, M. S. I., 379, 390(13), 393(13)
Makino, I., 72, 73(40), 84(40), 87, 89(40), 103(40), 105(195), 106, 108(195), 111(174)
Mallory, F. B., 10, 26
Malya, P. A. G., 10, 25, 26
Mankowski, T., 206, 208(21)
Manna, S., 304
Manning, J. A., 299, 302
Mantoura, R. F. C., 199
Marai, L., 80, 86, 87(94), 88(94), 91(94), 96, 97(156)
Marcel, Y. L., 267, 268(10)
March, E. W., 163
March, S. C., 436
Markides, K., 163, 169
Maroy, P., 458
Marsh, L. L., 172
Martin, C. E., 340
Martin, J. A., 9
Martin-Somme, G., 428
Martin-Wixtrom, C., 304
Martinez, R. A., 205, 478
Maruyama, S., 347
Masada, Y., 172
Mashige, F., 72, 73(42)
Mason, A. N., 89, 90(115, 116), 92(115, 116)
Massey, I. J., 27
Masters, B. S. S., 368
Matern, S., 103, 111(175)
Mathur, S. N., 286, 289, 291

Matsui, A., 105, 106(184), 109(184)
Matsumoto, N., 91(128), 92, 96(128)
Matsumoto, T., 10, 21, 23, 24(56), 33, 34(*a*), 35, 72, 73(39), 75(39), 78, 85(63), 185
Mattoon, J. R., 341
Matulich, D. T., 451
Matz, C. E., 267, 268(13)
Mauchamp, B., 388(38), 389, 393(38), 394(38), 395(38), 400(38, 53), 419(8), 420, 428(8)
Maume, B. F., 388(38), 389, 393(38), 394(38), 395(38), 400(38), 419(8), 420, 428(8)
Maxwell, J. R., 186, 188
Mayer, H., 123
Mayer, R. T., 420, 427, 438, 440(2), 441(2), 454, 457(5)
Mayr, M., 184, 185
Mead, J. F., 310, 311(32)
Means, G. E., 466
Meijer, J., 306
Meinschein, W. G., 186
Meister, W., 31, 32, 33
Meixun, C., 432
Mejbaum-Fatzenellenbogen, S., 232
Meltzer, P., 459
Mendeloff, A. I., 96, 97(158)
Mendis, A. H. W., 379, 390(13), 393(13, 14), 394(14), 395(14), 396(14), 398(14)
Mentaberry, A., 476(15), 477
Merdes, M., 306
Metzler, M., 540, 542(5)
Meyer, D., 540, 542(5)
Meyer, D. J., 286, 289(6)
Meyer-Bertenrath, T., 189
Michaud, D. P., 305, 310(13), 311(13)
Michelson, A. M., 218
Middleditch, B. S., 88, 105, 353
Middleton, E. J., 402
Mihara, S., 159
Miki, T., 11
Miles, D. H., 159
Miller, J. M., 157
Miller, L., 344
Miller, R., 161
Miller, W. L., 289
Milne, G. W. A., 27
Milner, N. P., 383, 390(23), 392, 393(24), 398(23), 400(52), 401(23), 404(23), 406(23), 407(23), 408(23), 410(23), 428, 438

Minale, L., 322
Mingrone, G., 78
Minyard, J. P., 159
Mishkin, S., 302
Misso, N. L. A., 319, 324(38), 327, 329, 330(69), 331(69), 332(38)
Mitra, A., 254
Mitzner, B., 161, 162(61)
Miwa, G. T., 303
Miwa, H., 111, 112(202)
Miyawaki, H., 322
Miyazaki, H., 72, 73(37, 45, 46), 76(46), 79(45, 46), 92, 99, 105(190), 106(37, 45, 67), 107(57), 109(190), 110(190)
Mizuno, T., 411
Modde, J.-F., 383, 388(39), 389(22), 390(22, 39), 391(39), 393(22), 398(22), 399(22), 411, 412(6), 413(6), 417(6), 418(6)
Modde, J.-L., 419(9), 420, 428(9), 444
Mody, N. V., 159
Moens, L., 246, 247(85)
Mohammad, S. A., 238
Molina, J. E., 87
Molzahn, S. W., 343
Mon, T. R., 154, 156
Moneger, R., 544
Moody, D. E., 308
Moore, T. A., 218
Morgan, E. D., 161, 388(36), 389, 391(36), 394(36), 395(59), 400, 412, 413(24), 417(24), 428, 444
Moriarty, R. M., 86
Morici, M., 379, 384(12), 419
Morioka, Y., 94, 95(152)
Morisaki, M., 346, 347(6), 348(6), 349(13), 350(13), 352(6, 11, 13), 354, 355, 358, 361, 362, 364(5-8)
Morpurgo, G., 334
Morris, J., 451
Morris, N. R., 285
Morris, R. J., 19, 20(54), 24(54), 25(64)
Morrow, C. J., 184
Mortensen, J. T., 200
Mortimer, R. K., 341
Morton, R. A., 201, 202(2), 206(2), 208(2)
Morton, R. E., 275, 281
Mosbach, E. H., 93, 94(143), 95(143)
Moshonas, M. G., 151
Mosora, F., 426, 428(19)
Moss, G. P., 147
Mui, M. M., 84, 88(88)

Mulheirn, L. J., 33, 34(a), 35
Muller, D. P. R., 82, 85(76), 96(76)
Muller, J.-F., 383, 390(27), 404(27), 407(27), 408(27), 411, 418(8)
Muller, R. K., 123
Mullin, C., 305
Mumby, S. M., 303, 304(3), 532
Munn, E. A., 436
Munson, B., 67
Munson, P. J., 502
Murakami, S., 409, 410(80)
Murata, T., 66, 93(11)
Murphy, G. M., 105, 106(182), 109(182)
Muschik, G. M., 66, 67(10)
Musial, B. C., 72, 73(44)
Muth, J. D., 475
Myher, J. J., 72, 73(33, 36), 80, 86(36), 96, 97(156)
Myllek, C., 437

N

Nadeau, R. G., 315, 316(18)
Nair, P. P., 96, 97(158)
Naito, Y., 69, 71(28)
Nakagaki, M., 56
Nakagawa, S., 105(195), 106, 108(195)
Nakai, S., 94, 95(146)
Nakanishi, K., 225, 401, 402(69), 408
Nakanishi, Y., 547(7), 548
Nakashima, T., 66, 72, 73(35), 93(11, 35)
Nakasone, S., 346
Nakata, Y., 21, 23, 105(195), 106, 108(195)
Nakayama, F., 56, 76, 105(190), 106, 107(57), 109(190), 110(190)
Nambara, T., 62, 69, 71(28), 110
Napoli, J. L., 197
Naqvi, S. H. M., 87
Nassim, B., 78, 80(65), 82(64, 65), 85(65)
Nathanson, N., 216
Neal, M. W., 463, 465
Neal, W. D., 344
Nearn, R., 385
Neises, B., 253
Nes, W. D., 7, 8, 11(7), 28, 34(a), 35, 48, 334, 343(4)
Nes, W. R., 3, 6, 7, 8, 9, 10, 11(7, 33), 12(44), 13, 14, 15(44, 45), 20(51), 21(13, 14, 15), 25(51), 26, 28, 31, 33(13, 14), 34(a, c), 35, 36, 48, 286, 287(5), 288(5), 289(5), 290(5), 291(20), 292(20), 312, 317, 320, 321, 329, 334, 343
Neville, D. M., Jr., 235
Nezu, Y., 105(192), 106, 108(192), 111, 112(199)
Ng Ying Kin, N. M. K., 202
Nichols, A. V., 274
Nickerson, G. B., 154, 155
Nicotra, F., 321, 347
Niedmann, D., 281
Nigg, H. N., 420, 437, 440(1)
Nijhout, H. F., 457
Nirde, P., 379
Nishida, T., 111, 112(202), 267, 268(3)
Nishikawa, Y., 79, 106(67)
Nishino, T., 286
Nishioka, A., 33
Noack, K., 147, 232
Noda, K. I., 304
Noland, B. J., 299, 302(22)
Noma, Y., 91(129), 92, 93(129), 94, 95(150), 96(129)
Nooner, D. W., 187
Norby, J. G., 502, 509(20)
Norden, Å., 96, 97(161)
Norden, D. A., 216, 218(14), 220(14), 221(14), 222(14), 223(14), 224(14), 228(14), 229(14), 230(14), 235(14), 237(14)
Nordström, L., 80, 83(68), 84(68), 85(68), 86(68), 87(68), 90(68), 90(68), 91(68), 95(68), 428
Norman, A., 87
Norman, A. W., 335, 342(7), 344(7)
Norrbom, A. L., 11, 15(45)
Norum, K. R., 275, 277, 291
Noteboom, W. D., 458
Nowock, J., 488, 489(6)
Noy, T., 155
Nursten, H. E., 156
Nyström, E., 11

O

Oakley, B. R., 285
Ockner, R. K., 299, 300(18), 302(18)
O'Connor, J. D., 244, 429, 458, 459, 460, 464, 465(22), 495, 499(10), 501(10), 502(10), 503(10), 521, 522
Odani, S., 299

Oelschlager, A. C., 42, 43, 48(24), 318, 340
Oesch, F., 303, 304, 305(5), 306
Ogura, K., 213, 248
Ohanessian, A., 495, 496(13), 499, 521, 522
O'Hern, P. A., 484
Ohnishi, E., 411
Ohnishi, S., 66
Ohtaka, H., 346
Okamoto, R. A., 176
Okubayashi, M., 346
Okuda, K., 83, 365
Olavesen, A. H., 68
Oldham, R. S., 91(127), 92
Olson, C. D., 294, 296(6), 299(6), 301, 302(6, 24), 303(6)
Omura, T., 365
Ong, D. E., 299
Ono, T., 299
Onyiruika, E. C., 173
Orban, E., 24
Oro, J., 187, 188
Osborn, M., 296, 505, 523
Osei-Twum, E. Y., 76
Ostroy, S. E., 225
Osuga, T., 72, 73(42)
Ota, K., 303, 304(3), 308, 309(29)
Otsuka, H., 105(191), 106, 109(191), 110(191)
Ott, P. Ph. L., 317
Ottolenghi, P., 502, 509(20)
Ourisson, G., 321, 331
Ozawa, N., 305, 310(12), 311(12)

P

Paliokas, A. M., 9
Palmer, M. A., 319
Palmer, R. H., 87
Palmisano, G., 409, 410(80)
Pan, M. P., 437
Parcher, J., 164
Parikh, I., 436
Parish, E. J., 9, 26
Parks, L. W., 38, 39(3), 41, 42(3, 19), 48, 50, 333, 335, 336, 337, 338(1, 10), 339(1, 6), 340(6, 19), 341(17), 342(19, 20, 21, 22), 343(19), 344(13, 17, 22, 60, 61), 345(13, 53)
Parmentier, G. G., 93

Pascal, R. A., Jr., 10, 28, 34(*b*), 35, 38
Passi, S., 78
Patel, A. K., 216, 217(5), 218(5), 224, 225(5, 46), 227(46), 232(46), 240(5), 241(5, 46)
Patience, R. L., 188
Patterson, G. W., 7, 11(7), 16, 17, 19(50), 20(50), 21(50), 23, 24, 28, 34(*a*), 35
Paulus, H., 205
Paumgartner, G., 72, 73(38), 75(38), 111, 112(204)
Payne, D. W., 453
Peacock, A. C., 235
Peaden, P. A., 168, 169(96)
Pedersen, J. I., 365
Pekkala, S., 184
Peng, Y.-M., 199
Pennetier, J.-L., 388(39), 389, 390(39), 391(39), 412, 419(9), 420, 428(9), 444
Pennock, J. F., 201, 202(2), 206(2), 208(2), 549
Persinger, H. E., 164
Peter, M. G., 495, 499(3), 526, 539, 540, 541(7, 8), 542(7, 8, 11), 543(7, 11)
Peters, R. H., 315
Peterson, C. E., 179
Peterson, G. L., 232
Peterson, R. C., 294, 296(9), 482, 484(1), 485, 486(1), 487(2, 4), 517, 522(16)
Peterson, R. L., 499
Petrakis, N. L., 310
Pettei, M. J., 225
Pfander, H., 124
Phillips, R. J., 172
Pickens, C. E., 292
Pickett, J. A., 151
Pierce, A. M., 43, 48(24), 340
Pierce, H. D., 340
Pietschy, J. M., 293
Pilkiewicz, F. G., 225
Pillinger, C. T., 186
Pinkson, J. D., 102
Pinto, W. J., 11
Piran, U., 267, 268(3)
Plapinger, R., 96, 97(158)
Plummer, T. L., 253
Polyakova, N. P., 181
Pongs, O., 432, 434(30, 31), 444, 461
Poole, C. F., 388(36), 389, 391(36), 394(36), 395(59)
Poorthuis, B. J. H. M., 299

Popják, G., 251, 315, 549
Popov, S., 327, 327(66)
Porcheron, P., 387
Porter, J. W., 189
Poulter, C. D., 252, 253
Powls, R., 129
Pramanik, B. N., 67
Pratt, G. E., 530, 531, 533(6), 534(1, 14), 537(6), 538(1), 539(9), 540, 541
Presper, K. A., 547
Preston, W. H., 320
Prestwich, G. D., 500, 505(25), 506, 510, 511(5), 516(7), 518(6), 520(6, 7), 521(8), 522(8), 526
Pritchard, D. G., 72, 73(48)
Proudlock, J. W., 334
Puckbridge, J., 240
Puodziukas, J. G., 246
Purris, K., 453
Pyrek, J. S., 92, 310

Q

Quarmby, R., 216, 218(14, 15), 220(14), 221(14), 222(14), 223(14), 224(14), 226(45), 227(45), 228(14), 229(14), 230(14), 235(14), 237(14)
Quilliam, M. A., 76
Quistad, G. B., 305, 487, 488(3), 490(3)

R

Raab, K. H., 317
Radin, D. N., 199
Radin, N. S., 212
Rahier, A., 324
Rahman, F. M., 189
Raisanen, S., 167
Ramaekers, J. J. M., 24
Ramp, J. L., 335, 339(6), 340(6)
Ramp, J. R., 41, 42(18)
Randall, J. M., 153
Randall, R. J., 281, 282(23), 296, 365, 439
Randerath, K., 489
Raper, J. H., 216, 218(15)
Rao, K. R. N., 76
Ratovohery, J. V., 185

Ravi, K., 210
Raz, A., 227
Reaven, R. M., 275
Redant, G. R., 164, 171
Rees, H. H., 11, 38, 39(10), 316, 317(23), 318(23), 323, 329, 348, 377, 378(4, 5), 379(2, 5), 383, 384(2, 5, 11), 385(2, 5), 386(2, 5), 387(4, 29), 389, 390(8, 9, 11, 13, 23, 42, 43), 391(45), 392, 393(11, 13, 14, 24, 25, 45), 394(14), 395(2, 13, 14), 396(14, 54), 397, 398(14, 23), 399(25), 400(9, 42, 52), 401(9, 23, 42, 43), 402(8, 9, 25), 403(25, 55, 63, 74), 404(8, 9, 23, 42, 43, 74), 405(9), 406(8, 9, 23, 42, 43), 407(9, 23, 42), 408(8, 9, 23, 43, 49, 55), 409(9, 55), 410(8, 9, 23, 42, 43), 411, 412, 413(13), 414(13), 415(12, 13), 416(13), 417(13), 418(12, 13, 19), 419, 426, 427, 428(23), 438, 454
Rehnborg, C. S., 274
Reibstein, D., 531, 540, 541(4)
Reich, M. F., 511
Reich, M. R., 487
Reineccius, G. A., 151
Remendal, R., 98, 100(162), 101(164), 103(164), 104(162, 164)
Rendell, N. B., 319
Renner, E., 154
Renstrøm, B., 216(20), 217, 218(20), 224(20), 225(20), 244(20), 245(20)
Resnik, M. A., 341
Reum, L., 391, 394, 428, 430, 431(15), 432(14), 433(14), 434(14), 435(14), 453
Rice, N., 480
Richard-Molard, C., 499
Richards, G., 377, 428, 459
Rickborn, B., 203
Riddiford, L. M., 306, 457, 490
Rigassi, N., 147
Rijks, J., 155
Riley, C. T., 484, 485, 487(4)
Rilling, H. C., 248, 252
Rip, J. W., 210, 214, 471, 474(3), 475(3), 476(17), 477, 479(17)
Ritter, K. S., 11, 15(45), 286, 290(2)
Ritter, M. C., 293, 295
Rittersdorf, W., 205
Ritzen, E. M., 453
Rivetna, M., 52, 54(2)
Robb, J. A., 462

Robbins, W. E., 346, 347, 348, 385, 402, 419, 420, 437, 440(1), 454, 457(8)
Roberts, D. H., 437
Roberts, G. P., 267, 268(8)
Roberts, J. C., 19, 20(54), 23, 24(54)
Roberts, P. E., 495, 499(9)
Robertson, S., 201, 203
Robinson, D. C., 199
Robinson, D. R., 224
Rodbard, D., 502
Rodriguez, D. B., 125
Rodriguez, R. J., 38, 39(3), 42(3), 48, 50, 337, 338(10), 339, 340(19), 342(19, 22), 343(19), 344(22), 345
Roeschlau, C., 270
Rogers, L. B., 171
Röller, H., 540, 541(7, 8), 542(7, 8), 543(7)
Romeo, G., 35
Römer, G., 154
Romers, C., 3
Ronald, R. C., 149, 171(1)
Ronchetti, F., 321, 347
Rønneberg, H., 216(20), 217, 218(20), 224(20, 31), 225(20, 30, 32), 226(32), 232(30), 242(30, 31, 32), 244(20), 245(20)
Ros, A., 89
Rose, M. E., 147, 319, 378, 379, 390(8, 9), 393(14), 394(14), 395(14), 396(14), 398(14), 400(9), 401(9), 402(8, 9), 403(74), 404(8, 9, 74, 76), 405(9), 406(8, 9), 407(9), 408(8, 9), 409(9), 410(8, 9), 412, 413(13), 414(13), 415(12, 13), 416(13), 417(13), 418(12, 13)
Rosebrough, N. J., 281, 282(23), 296, 365, 439
Ross, A. C., 293
Ross, A. L., 192
Rossignol, D. P., 476(18), 477
Rossiter, M., 68
Roth, B. L., 476
Rothblat, G. H., 293
Rottler, G. D., 202, 204(7), 209(7), 471, 474(6), 475(6)
Rowland, S. J., 188
Rubinstein, I., 33, 34(a), 35, 327
Ruddat, M., 125, 149(19), 192
Rudel, L. L., 39
Rupar, A., 476(17), 477, 479(17)
Rupar, C. A., 480
Russell, P. T., 16, 20(51), 25(51)
Russo, G., 321, 347
Ruttimann, A., 123
Ryan, D., 304
Ryback, G., 7, 8(11), 20(11), 21(11), 23, 188
Ryhage, R., 64, 73(1), 79(1), 80(1), 83(1), 84(1), 85(1), 86(1), 87(1), 88(1), 89(1), 90(1), 91(1), 92(1), 251

S

Sabbaiah, P. V., 267, 268(2)
Sack, J., 471, 474(1)
Saeed, T., 171
Sage, B. A., 429, 458, 464, 465(22), 495, 499(10), 501(10), 502(10), 503(10), 521
Sakaguchi, K., 105(192), 106, 108(192), 111, 112(199)
Sakurai, S., 10, 411
Salares, V. R., 216(16), 217, 218(16), 221(38), 232(16), 243(16), 244(16), 245(16)
Salen, G., 91(131, 132), 92, 93, 94(143), 95(143)
Salerno, L. F., 339, 342(20)
Sall, C., 412
Salomon, K., 217, 243(21, 22), 244(22)
Samuel, O., 201
Sanburg, L. L., 294, 296(9), 482, 484(1), 486(1), 487(6), 499, 517, 522(16)
Sanders, H. K., 341
Sanders, J. K. M., 31
Sandra, P., 155, 157, 164, 165, 171, 175, 176
Sanemori, H., 95
Sangare, M., 33
Sanghvi, A. T., 92
Sannasi, A., 399
Sarimento, R., 172(123, 124), 173
Sato, R., 365
Sato, S., 355, 364(6)
Sawaya, T., 91
Sawyer, W. H., 240
Scalia, S., 391, 400, 412, 413(24), 417(24), 428, 444
Scallen, T. J., 11, 38, 39(4, 9), 299, 302(22), 326, 329
Scanlon, J. T., 167
Scanu, A. M., 267, 268(9)
Scatchard, G., 467, 500, 513, 517(13)
Schaltmann, K., 432, 434(31), 444, 461

Scheffer, J. J. C., 151, 152, 158, 165(11)
Scheller, K., 430, 431(15), 436, 437, 453
Schenck, P. A., 187
Schenkel, H., 437
Scher, M. G., 471, 474, 475, 476(16, 18), 477, 478(16), 479(16), 480, 547(8), 548, 552(8)
Schiff, J. A., 200
Schilz, W., 154
Schlager, S. I., 39
Schleyer, H., 454
Schmassmann, H., 304, 305(5)
Schmitz, F. J., 312
Schodder, H., 127
Schoeller, D. A., 103
Schomburg, G., 166, 186, 187
Schooley, D. A., 287, 483, 487, 490, 531, 538(5), 543
Schopf, J. W., 186
Schrader, W. T., 525
Schreier, P., 175
Schroepfer, G. J., Jr., 6, 9, 10, 11(5), 12(5), 26, 27, 28, 33, 34(b), 35, 36, 38, 39(8), 294, 295(8), 299(8), 384
Schroer, J. A., 66, 67(10)
Schubert, R., 111, 112(204)
Schüren, H., 91(127)
Schultz, T. H., 152, 154, 156
Schultz, W. G., 152, 153
Schütz, E., 154
Schwarzenbach, R. P., 72, 73(38), 75(38)
Schwieter, U., 147
Seaton, J. D., 294, 295(8), 299(8)
Seballos, H. L., 294, 296(9), 482, 484(1), 485, 486(1), 487(4), 499, 517, 522(16)
Sedel'nikov, A. I., 181
Sedgwick, R. D., 68, 72(24)
Seidel, D., 281
Sekeris, C. E., 436
Sekula, B. C., 8, 11, 28, 35, 48, 334, 343(4)
Senanayake, U. M., 160
Seo, S., 313
Serlupi-Crescenzi, G., 334
Seshan, K. R., 540, 541(8), 542(8)
Sesták, Z., 189
Setchell, K. D. R., 75, 76(56), 77(56), 80(56), 105, 106(182, 183, 184), 109(182, 183, 184), 110(183), 111(55)
Seto, S., 248
Settine, R. L., 72, 73(48)

Sevanian, A., 310, 311(32)
Severson, R. F., 158
Sexton, S. A., 80
Seyama, Y., 105(185, 191, 194), 106, 108(185, 194), 109(191), 110(185, 191)
Seyer, A., 544
Shabanowitz, J., 67
Shabtai, J., 181
Shackleton, C. H. L., 84
Shank, J. T., 164
Shapiro, S. K., 542
Sharp, H. L., 91(126), 92, 93, 96(125)
Sharpe, F. R., 151
Shaw, P. E., 159
Shaw, R., 51, 52, 54(2), 55(1, 2, 3), 56(4), 57(1), 58(4), 59(1), 61(1), 62(1), 63(1, 3), 68, 71(20), 78, 96(20), 97(20)
Shefer, S., 91(131, 132), 92, 93, 94(143), 95(143)
Sherma, J., 149
Sherman, F., 339
Sherman, M. R., 431
Shibamoto, T., 159, 171, 172(114), 176
Shikita, M., 354, 355, 358, 361, 362, 364(5–7)
Shimizu, N., 33
Shimizu, S., 33
Shimoyama, T., 75
Shingari, M. K., 164
Shino, M., 105(192), 106, 108(192), 111, 112(199)
Shiratori, T., 288, 289(12), 290(12)
Shirk, P. D., 540, 541(7, 8), 542(7, 8), 543(7)
Shoemaker, C. G., 176
Shone, C. C., 216, 217(5), 218(5), 224, 225(5, 34, 46), 227(46), 232(34), 240(5, 34), 241(5, 34, 46)
Shoppee, C. W., 7
Shore, V. G., 267
Showalter, R., 111, 112(200, 201, 203)
Shrewsbury, M. A., 286, 289(6), 291(6)
Shunbo, F., 162
Shuster, L., 233, 296, 298(14)
Siddall, J. B., 429, 442, 531, 538(5), 540, 542(5)
Siegel, M. R., 340
Siggia, S., 39
Signer, R., 124
Simon, P. W., 179
Simpson, E. R., 291, 292

Simpson, K., 189, 198(6), 199
Simpson, K. L., 125
Singer, T. P., 421
Singh, A. K., 500, 506(18), 510, 511(5)
Siouffi, A., 38, 39(6)
Siwon, H., 171
Sjöstrand, U., 27
Sjövall, J., 11, 64, 66, 72, 73(1, 40, 41, 47, 50), 75, 76(56), 77(56), 79(1, 47), 80(1, 56), 81(8), 82(1, 8, 41), 83(1, 8, 50), 84(1, 8, 40, 73, 82), 85(1, 8, 50, 86), 86(1, 41), 87(1, 8), 88(1, 8), 89(1, 40, 47, 82), 90(1, 82), 91(1, 41, 91), 92(1), 93(89), 94(8), 95(144, 151), 96(82), 97(82, 161), 98, 99(8), 100(162, 163), 101(8, 73, 164), 102(8), 103(40), 104(164, 172), 105(162), 106(8), 111(172–178), 112, 364, 365(1)
Sjövall, K., 99, 100(163)
Skallegg, B. A., 438
Slemr, J., 72, 73(43)
Skipski, V. P., 337
Slade, M., 305, 307(17), 487, 488(2)
Sloane-Stanley, G. H., 212
Slover, H. T., 24
Smellie, C. G., 289, 293(16)
Smith, A. G., 24, 83, 85(79), 89(79), 94, 322, 331(50)
Smith, A. R. H., 316, 317(23), 318(23)
Smith, D. F., 437
Smith, D. G., 33
Smith, J. A., 57, 58(4), 59(4), 61(4)
Smith, L., 274
Smith, L. L., 310
Smith, S. L., 176, 454
Smith, W. B., 33
Snook, M. E., 158
Sodano, G., 322
Sofer, S. S., 248
Solel, Z., 340
Soll, J., 544
Soloff, L. A., 267, 268(14)
Somme-Martin, G., 388(38), 389, 393(38), 394(38), 395(38), 400(38, 53), 419(8), 420, 428(8), 444
Song, M-K. H., 293, 295(5), 301(5)
Sörensen, N. A., 217, 243(23, 24), 244(23, 24)
Sparks, T. C., 306, 307, 490, 543
Sparrow, A., 304, 305(5)
Spaziani, E., 444

Spector, A. A., 286, 289
Spielvogel, A. M., 335, 342(7), 344(7)
Spillert, C. R., 67
Spindler, K.-D., 405, 408(79), 426, 428(18, 19), 429, 444, 459, 460(9)
Spiteller, G., 411
Sprinson, D. B., 38, 39(5), 341
Srisukh, S., 217, 244(25)
Stahl, E., 154
Stahl, Y. D. B., 267, 268(7)
Stamoudis, V., 488, 489(6)
Stark, T. J., 169
Starka, L., 85
Starr, P. R., 42, 338
Staubli, W., 306
Stay, B., 532
Steel, G., 83, 84(81), 87(81), 95(81)
Steele, J. A., 7, 26
Steglich, W., 253
Stein, L., 302
Stein, R. A., 310, 311(32)
Steinberg, D., 481
Steiner, S. M., 549
Stellaard, F., 111, 112(204)
Stephenson, E., 91(133), 92, 93(133), 95(133), 96(133)
Stern, K. G., 217, 243(21, 22), 244(22)
Sterzycki, R., 92
Stewart, P. A., 303
Still, W. C., 254
Stokke, K. T., 277
Stone, K. J., 552, 553(15)
Storm, D. R., 552, 553(16)
Stothers, J. B., 33, 36
Stott, K. M., 534, 539, 541
Strandvik, B., 82, 84(71), 85(71), 87
Strange, E. F., 293
Stransky, H., 132
Straub, K. M., 68, 69(21), 83(21)
Strauss, A. W., 299, 300(18), 302(18)
Strauss, J. F., III, 288, 289(8), 290(8)
Strominger, J. L., 202, 476, 478(14), 547, 552, 553(15, 16)
Stroupe, S. D., 435
Stumpf, P. K., 320
Sturzenegger, V., 147
Subbiah, M. T. R., 80
Subbiak, M. T., 25
Suckling, K. E., 293(16)
Suge, S. T., 164

Sumbilla, C., 550
Summerfield, J. A., 84, 105, 106(183), 109(183), 110(183)
Suzue, G., 267, 268(10)
Suzuki, S., 547(7), 548
Svoboda, J. A., 346, 347, 348, 385, 419, 420, 427, 437, 438, 440(1, 2), 441(2), 442, 454, 457(5, 8)
Sweeley, C. C., 102
Sweeney, F. P., 378
Synder, L. R., 159
Szczepanik, P., 92(125), 92, 95(125), 96(125), 111, 112(197)
Szczepanik, P. A., 37, 64, 66(3), 67(3), 72, 73(3, 34), 79(3), 91(3), 111, 112(198)
Szczcepanik-Van Leeuwen, P., 93, 111, 112(200)
Szepesi, G., 166

T

Takagi, S., 233
Takagi, T., 233
Takahashi, I., 248
Takahashi, K., 299
Takahashi, S., 66
Takane, K., 358, 364(7)
Takasu, A., 346, 347
Takeda, K., 233, 313
Takemura, T., 218
Takeoka, G., 166
Takeshita, T., 33
Taketani, S., 286
Takikawa, H., 105(191, 194), 106, 108(194), 109(191), 110(191)
Takino, T., 72, 73(35), 93(35)
Tamkun, J. W., 214, 215(36)
Tamura, T., 10, 21, 23, 24(56), 34(*a*), 35, 72, 73(39), 75(39), 78, 85(63), 185
Tamura, Z., 72, 73(42)
Tanabe, M., 315
Tanaka, N., 261
Tanaka, T., 288, 289(8), 290(8)
Tanaka, Y., 125
Tandon, R., 89
Tani, H., 21, 23, 24(56), 185
Tanida, N., 75, 76(56), 77(56), 80(56)
Taniguchi, S., 365
Tanis, M. A., 464, 465(22)

Tarr, E., 216
Tash, J., 429, 442
Tateyama, T., 105(192), 106, 108(192), 111, 112(199)
Tauber, J. D., 123, 217, 232(33), 240(33)
Tavani, D. M., 286, 287(5), 288(2, 5), 289(5, 8), 290(2, 5, 8), 291(20), 292(20)
Taylor, F. R., 50, 335, 339(6), 340(6), 341(17), 342(21, 22), 344(17, 22)
Taylor, R. F., 125, 186, 189
Taylor, U. F., 10, 28, 34(*b*), 35
Taylor, W., 83, 85(86)
Tchen, T. T., 248, 315
Tecce, G., 334
Teitelbaum, C. L., 159
Temmerman, I., 175
Tenneson, M. E., 89, 90(115), 92(115)
Teranishi, R., 152, 154, 156
ter Heide, R., 167
Teshima, S-I., 322
Tetler, L. W., 68, 72(24)
Thakur, A. K., 502
Thaler, M. M., 68, 69(21), 71(29), 83(21)
Thamer, G., 430, 442
Thomas, E. L., 151
Thomas, G. H., 90
Thomas, P. E., 304, 305, 310(13), 311(13)
Thomas, P. J., 88, 112(108)
Thomassen, P., 66, 81(8), 82(8), 83(8), 84(8), 85(8), 87(8), 88(8), 94(8), 99(8), 101(8), 102(8), 106(8)
Thommen, H., 216
Thompson, E. D., 42, 344(60, 61), 345
Thompson, M. J., 24, 318, 346, 347, 348, 385, 402, 419, 420, 437, 440(1), 442, 454, 457(8)
Thompson, R. H., Jr., 24
Thomson, J. A., 436
Thornton, E. R., 261
Thowsen, J. R., 6, 11(5), 12(5), 38, 39(8)
Tingey, D. T., 169, 180
Tint, G. S., 91(131, 132), 92, 93, 94(143), 95(143)
Tipping, E., 302
Tisse, C., 33
Tobe, S. S., 530, 531, 532, 533, 534(1, 17), 537(7), 538(1, 7, 16)
Tohma, M., 21, 91, 105(195), 106, 108(195)
Tökés, L., 82, 88, 91(126, 127), 92, 93(137), 94(126), 112(108)

Tom, R. D., 218
Toma, L., 347
Toma, M., 23
Tomita, Y., 313
Tomori, E., 24
Toppet, S., 93
Tori, K., 313
Tornabene, T. G., 188
Torpier, G., 379
Toscano, M. A., 35
Touster, O., 233
Towle, H. C., 294, 296(6), 299(6), 302(6), 303(6)
Tressl, R., 157, 179
Trocha, P. J., 38, 39(5), 341
Trockman, R. W., 294, 295(8), 299(8)
Trulzsch, D., 302
Truman, J. W., 457
Trusell, F. C., 186, 187(170)
Trzaskos, J. M., 293, 302
Tsai, L.-S., 310
Tserng, K.-Y., 112
Tsoupras, G., 383, 389, 390(27, 40, 41, 44), 404(27, 41), 407(27, 41), 408(27, 41), 410(40, 44), 411, 412, 413, 414(21), 415(21, 25), 416(21, 25), 417(25), 418(8, 20, 21, 22, 25, 26)
Tsuda, M., 10, 26, 28, 33, 34(*b*), 35, 36
Tsuji, M., 547(7), 548
Tsutsumi, J., 105(192), 106, 108(192), 111, 112(199)
Tun, P., 270, 273(25), 274(25), 275
Turner, R. B., 3
Tyson, B. J., 166

U

Uemura, I., 315
Uetsuki, T., 10, 185
Underhill, E. W., 160
Une, M., 91(128, 129, 130), 92, 93(129), 95(130), 96(128–130)
Ungar, F., 301, 302(24)
Unrau, A. M., 43, 48(24), 318, 340

V

Vagelos, P. R., 481
Valente, F., 334

vanAller, R. T., 16, 20(51), 25(51)
Van Den Berg, J. H. M., 24
Van den Berg, P. M. J., 74
Vanden Heuvel, W. J. A., 25
van der Hoeven, T. A., 365, 372, 376(19)
Van Harreveld, A., 459
Van Holde, K. E., 218
van Lenten, F. J., 171
Van Nisterlrooy, J. G. M., 340
van Os, F. H. L., 159
Van Roelenbosch, M., 164
van Tamelen, E. E., 315
Varaas, T., 453
Varkey, T. E., 8, 21(14, 15), 33(14)
Veares, M. P., 404
Vecchi, M., 123, 198, 199
Venetacei, D., 334
Verdievel, H., 72, 73(49), 99(49)
Verhoeven, G., 438
Verizzo, D., 471, 474(1)
Verstappe, M., 164, 175
Verzele, J., 164
Verzele, M., 155, 157, 164, 165, 170, 171, 176
Vetter, W., 147
Vezina, C., 267, 268(10)
Vigh, Gy., 166
Vigo, C., 212
Vilim, A., 206, 208(20), 212(20)
Vince, R. K., 542
Vinutha, A. R., 167
Viola, F., 316, 317(21)
Vitzthum, O. G., 153, 154
Vlahcevic, Z. R., 103, 104(172), 111(172)
Vocci, M., 96, 97(158)
Vogel-Bindel, U., 306
von Rudloff, E., 149, 158, 160, 167, 171, 172
Vouros, P., 88
Vozkl, A., 306
Vrbanac, J. J., 102
Vuilleme, P., 419

W

Wada, T., 105(194), 106, 108(194)
Wade, A. P., 319
Wadhams, L. J., 161
Waechter, C. J., 471, 473(4), 474(4), 475, 476(4, 16, 18), 477, 478(16), 479(16), 480, 547(8), 548, 550, 552(8)

Waechter, F., 306
Wagniere, G., 147
Wahlgren-Brannström, L., 219
Wakeham, S. G., 24
Wald, G., 216
Walden, M. K., 6, 11(5), 38, 39(7)
Walker, S. L., 217, 243(27), 244(27), 245(27), 246(27)
Wallace, R. A., 217, 243(27), 244(27), 245(27), 246(27)
Walls, F. C., 69
Walter, J. A., 33
Walter, K., 82, 84(72)
Walters, J. D., 175
Wang, P., 306
Wännmam, T., 163
Wannman, T., 169
Ward, J. P., 33
Warner, A. H., 246
Warren, C. D., 202, 213(12), 549
Watabe, T., 305, 310(12), 311(12)
Watkins, J. B., 58, 59(7), 61(7), 111, 112(197)
Watson, R. D., 444
Weaver, R. J., 531, 533(6), 534(14), 537(6)
Weber, K., 296, 505, 523
Wedgwood, J. F., 202, 476, 478(14), 547
Weedon, B. C. L., 116, 147
Weeke, F., 166
Weerasinghe, N. C., 76
Wegmann, A., 327, 329(66)
Weiner, P. H., 257
Weirich, G. F., 419, 420, 438, 440(2), 441(2), 442, 454, 456, 457(5, 8), 458, 540, 541(6), 542(6), 543(6)
Weiss, B. A., 11, 15(45)
Weisz-Kincze, I., 450
Welburn, A. R., 199
Welburn, F. A. M., 199
Wells, M. J. M., 258
Werkhoff, P., 153
Westheimer, F. H., 510
Westphal, U., 429, 431, 432, 435(19), 442
Wetzels, M. L., 24
Wheeldon, L. W., 334
White, B. N., 437
White, R. L., 173
Whitney, J. O., 68, 69(21), 71(26, 29), 83(21), 111(26)
Whittle, K. J., 549
Wickerman, L. J., 335

Widdowson, D. A., 344
Wieland, H., 281
Wielgus, J. J., 454
Wientjens, W. H. J. M., 317
Wiggins, P. L., 253
Wigglesworth, K. P., 379, 384(1), 390(11), 393(11), 419
Wigglesworth, V. B., 495
Wikström, S.-Å, 82, 84(71), 85(71)
Wikvall, K., 365, 366, 367(2, 13), 368, 372(12), 374(2), 375(16)
Wilhelm, S., 444
Wilkins, C. L., 173
Wilkinson, C. F., 305
Will, O. H., 125, 149(19), 192
Williams, C. H., Jr., 368
Williams, C. M., 457, 467
Williams, C. N., 72, 73(44)
Williams, D., 462
Williams, D. H., 172, 403
Williams, G. C., 91(125), 92, 93, 95(125), 96(125), 302
Williams, K., 24, 25(64)
Williams, M. C., 276
Williams, R. F., 205, 478
Williams, R. J. H., 315
Willing, E., 154
Wilson, C. W., III, 159
Wilson, D. M., 103, 111(176, 177, 178)
Wilson, I. D., 391, 394, 412, 428, 444
Wind, M. L., 251
Winkler, H. U., 27
Winterburn, P. J., 68
Wirtz, K. W. A., 299
Withers, M. K., 167
Wixtrom, R. N., 307
Wojciechowski, Z. A., 324, 332(59)
Wold, F., 432
Wolfe, L. S., 202
Wolff, M. E., 451
Wolstromer, R. J., 172
Womack, M., 348
Wong, D., 91(126), 92, 94(126)
Wong, R. G., 335, 342(7), 344(7)
Wong, R. Y., 8
Wong, T. K., 208
Woods, M. C., 206, 208(20), 212(20)
Woods, R. A., 341, 343, 344(56, 59), 345
Wren, J., 542
Wright, B. W., 168, 169(96)
Wright, J. L. C., 33

Wright, J. R., 25
Wright, L. H., 66, 67(10)
Wrolstad, R. E., 158
Wyatt, G. R., 437, 495, 499(9)
Wyllie, S. G., 93

Y

Yamada, H., 105(195), 106, 108(195)
Yamaguchi, K., 159, 171
Yamakawa, T., 105(185, 191), 106, 108(185), 109(191), 110(185, 191)
Yamamoto, M., 111, 112(202)
Yamamoto, R. T., 402
Yamasaki, K., 83, 87(83)
Yamashita, K., 72, 73(45), 74(45), 79, 106(45, 67)
Yamazaki, K., 91
Yanagisawa, J., 76, 108(57)
Yao, T., 111, 112(202)
Yasuda, M., 91(129), 92, 96(129)
Yasuhara, M., 91(128), 92, 93(129), 94, 95(149, 150), 96(128)
Yasukochi, Y., 368
Yon, D. A., 188
Yonge, C. D., 467
Yoshimura, Y., 313
Yoshioka, D., 87
Young, J. S., 87, 89(98), 95(98)
Young, N. L., 249
Young, N. M., 216(16), 217, 218(16), 221(38), 232(16), 243(16), 244(16), 245(16)
Yousef, I. M., 72, 73(33), 86, 96, 97(156, 157)
Yudd, A. P., 225

Yudin, A. I., 496, 497, 498, 521
Yund, M. A., 458, 462(5), 467

Z

Zachary, D., 432, 445, 446(24), 448(24), 449(24), 453(24)
Zafiriou, O. C., 187
Zagalsky, P. F., 216(17–19), 217(10), 218(3, 10, 14, 15, 18, 19), 219(10), 220(14), 221(10, 14), 222(14), 223(10, 14), 224(10, 14, 29), 225(3, 10, 29, 36), 226(3, 36), 227(29), 228(14), 229(14), 230(14), 231, 232(10, 51), 234(29), 235(10, 14, 29), 237(14), 238(10, 19, 29), 239(29), 240(1, 10, 29, 51, 75), 243(1, 3, 26, 28), 244(1, 26), 246(3), 247(80, 84, 88)
Zahler, W. L., 237
Zahra, J. P., 33
Zakaria, M., 189, 198, 199
Zarembo, J. E., 33, 36
Zaretskii, Z. C., 64, 79(6), 80(6), 82(6), 86(6), 87(6), 88(6), 89(6), 90(6), 91(6), 93(6)
Zenkevich, I. G., 180
Zibitt, C. H., 305, 307(17), 487, 488(2)
Ziboh, V. A., 291
Ziffer, H., 25
Ziller, S. A., 83
Zilversmit, D. B., 275, 281
Zimmerman, J., 68
Zinkel, D. F., 185
Zlatkis, A., 162
Zuagg, R. H., 484
Zvenko, H. L., 512, 517(12)
Zweig, G., 149

Subject Index

A

Abies × arnoldiana, volatiles, 151
Acetic acid, labeled, sterol precursor, 313
N-Acetylglucosaminylpyrophosphoryldolichol, biosynthesis, 471
Actinioerythrin, light absorption maxima, 138
Actinioerythrol, chromatography, adsorbents, 125
Acyl-CoA hydrolase, 290
S-Adenosylmethionine, deuterium-labeled, 319
S-Adenosylmethionine-cycloartenol methyltransferase, 324, 325
  reaction products, identification, 322
S-Adenosylmethionine:γ-tocopherol methyltransferase, 544–547
Adonirubin, 241
  light absorption maxima, 138
Adonixanthin, light absorption maxima, 138
Adrenodoxin, in side-chain cleavage of cholesterol, 352, 354
Adrenodoxin reductase, in side-chain cleavage of cholesterol, 352, 354
Aeshna cyanea, 420
Albumin, 429
Aleuriaxanthin
  light absorption maxima, 138
  specific absorbance coefficient, 138
Algae
  carotenes
    HPLC, 199
    separation, 129
  carotenoid, HPLC, 199
  cell-free system, sterol biosynthesis, 323, 324
  chlorophyll, HPLC, 199
  sterol, 312
  sterol synthesis, 324
  uptake of labeled sterol precursors, 320
  xanthophyll, HPLC, 199
24-Alkylsterol
  Δ^{22}, 29
  NMR spectra, 33

Allatostatin, 530
Allatotropin, 530
Allochenodeoxycholic acid, ion current chromatography, 104
Allopora californica. See Coral
Alloporin, 242
Alloxanthin, light absorption maxima, 138
Alumina, 159
  adsorbent, for carotenoids, 125, 126
5-α-Androstane, 8
  ^{1}H NMR, C-18 and C-19 methyl groups, effect of selected groupings on, 32–35
Androsterone, cholesterol acyltransferase inhibitor, 292
Anhydrorhodovibrin
  light absorption maxima, 138
  specific absorbance coefficient, 138
Annona cherimoya. See Cherimoya
Anostracan, carotenoprotein-lipovitellin complexes, 246, 247
Antheraea polyphemus, 429
Antheraxanthin
  chromatography, octyl-columns, 198
  light absorption maxima, 138
Anthracite coal, volatiles, capsule injector, 161
Apiezon L, 170, 171, 187
8′-Apo-β-caroten-8′-al
  light absorption maxima, 138
  specific absorbance coefficients, 138
10′-Apo-β-caroten-10′-al,
  light absorption maxima, 138
  specific absorbance coefficients, 138
12′-Apo-β-caroten-10′-al
  light absorption maxima, 138
  specific absorbance coefficient, 138
8′-Apo-β-caroten-8′-oic acid
  light absorption maxima, 138
  specific absorbance coefficient, 138
Apolipoprotein A-I, 271, 274
Apple essence, volatiles, solvents for, 152
Arapaima gigas, 82
Armyworm, 438
  phytosterol dealkylation, 348
Artemia saline. See Brine shrimp

Arthroderma, sterol mutants, 340
Arthropod
 carotenoprotein, 217
 ecdysteroid carriers, 437
 molting hormone, hemolymph binding protein, 442, 443
Arylphorin, 430
 amino acid composition, 436
 genes, 436
 molecular weight, 436
 properties, 436, 437
 subunits, 436
Arylsulfatase, *Helix,* 378, 389, 398, 417
Aspergillus, sterol mutants, 340
Astacene
 chromatography, adsorbents, 125
 light absorption maxima, 138
 specific absorbance coefficient, 138
Astaxanthin, 216, 241
 chromatography, adsorbents, 125
 column chromatography, 127
 esters, saponification, 123
 light absorption maxima, 138
 molar absorbance coefficient in pyridine, 232
 purification, 134
 spectral properties, 219, 220
 structure, 114, 115
 TLC, 131
Asteriarubin
 properties, 240, 241
 purification, 241, 242
Asterias rubens. See Starfish
 uptake of labeled sterol precursors, 322
Auroxanthin
 light absorption maxima, 138
 specific absorbance coefficient, 138
Azafrin
 light absorption maxima, 138
 specific absorbance coefficient, 138
Azalea indica. See Rhododandron simsii

B

Bacterioruberin
 light absorption maxima, 139
 specific absorbance coefficient, 139
Barley, seedlings, uptake of labeled sterol precursors, 321
Bean, cell-free system, sterol synthesis, 323

Bile acid
 alkyldimethylsilyl ethers, molecular weight calculation, 79
 biosynthesis, cytochrome P-450 in, 364–377
 C_{29}, 92, 93
 conjugated
  fast atom bombardment spectra, quasimolecular ions in, 70
  HPLC, 55
  mass spectra, 96–98
  mass spectrometry, 69
   sample introduction and ionization methods, 65
  reversed-phase liquid chromatography, 55–60
  selected ion monitoring, 109
 deconjugation, 51
 derivatives
  as artifacts of fragmentation, 80, 81
  for GLC and GC-MS, 72, 73
  mass values for calculation of molecular weights, 79
  resolution, 76
 derivatized unconjugated mass spectrometry, 69
  sample introduction and ionization methods, 65
 fragmentation, types, 79
 free, mass spectrometry, 69
  sample introduction and ionization methods, 65
 gas-liquid chromatography, 72
 GC-MS, repetitive scanning, ions useful in screening for common structures, 81
 glucuronides
  mass spectra, 96, 97
  selected ion monitoring, 109, 110
 HPLC, 51–63, 76, 78
  instrumentation, 51, 52
  mobilities, under specific conditions, 52–55
  solvents, 52
 hydroxy, with unsaturated ring system, mass spectra, 86–88
 ion current chromatography, 99–101, 103, 104
  total, 99, 100
 kinetic studies, labeling with stable isotopes, 111–113

mass spectra
  with hydroxyl groups in A-ring, 81–83
  with hydroxyl groups in B-ring, 83–85
  with hydroxyl groups in C-ring, 85
  with hydroxyl groups in D-ring, 85, 86
mass spectrometry, 63–113
  chemical ionization, 66, 67, 69, 96
  conditions, 64–78
  direct chemical ionization, 67
  direct probe, 96
  electron impact, 64–66, 69, 96
  fast atom bombardment, 68–71
  field desorption, 67–69
  ionization methods, 64–71
  laser desorption, 68
  sample introduction, 64–65, 71–78
    direct insertion probe, 71
    gas chromatographic inlet, 71–76
    liquid chromatograpic inlet, 76–78
methyl ester acetates, molecular weight determination, 67
methyl ester silylated derivatives, separation, 76
methyl ester TMS ether derivatives
  chromatographic analysis, 77
  separation on capillary columns, 75
mixtures, resolution, 75–77
molecular weight, 68
monooxygenated, HPLC, 54
normal-phase chromatography, 63
oxo
  mass spectra, 88–91
  quantitation, in blood, 110, 111
  selected ion monitoring, 109
polyoxygenated, HPLC, 54
quantification, 72–74
quantitative GC-MS analysis, 98–111
  capillary columns, 102
  computer methods, 102
  in evaluating radiolabel patterns, 102–104
  repetitive scanning techniques, 98–104
    application, 99–104
    interfering compounds, 102
RPLC, fatty acid analysis column, 56–59

selected ion monitoring, 104–111
  application, 105–111
  standards, 106–110
separation, on capillary columns, 74, 75
side chain, structure, 91–96
solvolysis, 51
structure, stereoselective fragmentations in, 80
sulfates
  FAB spectra, quasimolecular ions in, 70
  mass spectra, 97
  mass spectrometry, 69, 71
  RPLC, 62–63
  selected ion monitoring, 109, 110
TMS ethers, 80
  fragmentation, 78
  fragmentation patterns, 80
  mass spectra, 85–86
unconjugated
  electron impact ionization, 78
  mass spectra, 78–96
    diagnostic fragmentation patterns, 80–96
    general fragmentation patterns, 78–80
    molecular weight, determination, 78, 79
unsaturated, mass spectra, 86
with unsaturated side chain, 93
Bile alcohol, side chain hydroxylated, TMS ethers, side chain fragment ions and losses, 93–96
Bisanhydrobacterioruberin, light absorption maxima, 139
3,4,3',4'-Bisdehydro-β-carotene
  light absorption maxima, 139
  specific absorbance coefficient, 139
Bisdehydrolycopene, light absorption maxima, 139
Bixin
  light absorption maxima, 139
  specific absorbance coefficient, 139
Blattella germanica. See Cockroach
Blowfly, 430
  ecdysone, 411
  ecdysone oxidase, 420–429
Bombyx mori. See Silkworm
Boophilus microplus. See Tick
Brain
  calf

584 SUBJECT INDEX

dolichol esterase, 480–482
dolichol kinase, 471–476
dolichyl phosphate phosphatase, 476–480
dolichyl pyrophosphate phosphatase, 547–553
Bramble, cell-free system, sterol synthesis, 324
Branchipus stagnalis, carotenoid-lipovitellin complex, 246, 247
Brassicasterol, HPLC properties, 43, 45, 47
Brine shrimp, cartenoprotein-lipovitellin complex, 246, 247
Butylated hydroxytoluene, 190
  antioxidant for carotenoids, 118
  reverse-phase chromatography, 195

C

C.p. 450, light absorption maxima, 140
Calliphora vicina. See Blowfly
Calonysterone, 452
  in methanol, ultraviolet absorption, 450
  ratio of equilibrium constants of association, 451
Caloxanthin, light absorption maxima, 139
Δ^7-Campestenol, HPLC properties, 43
Campesterol, 8, 11, 20
  demethylation, 346–348
  HPLC properties, 43, 46
  ^{13}C NMR spectra, 36
Camphene, 180
Candida, sterol mutants, 340
Canthaxanthin
  absorption spectra, shape, 145
  light absorption maxima, 136, 139
  specific absorbances coefficient, 139
Canthaxanthin-lipovitellin, *Artemia*, 246, 247
  isolation, 247
  properties, 246, 247
Capsanthin
  light absorption maxima, 139
  specific absorbance coefficient, 139
Capsicum, chromoplast, tocopherol biosynthesis, 544–547
Capsorubin
  light absorption maxima, 139
  specific absorbance coefficient, 139

Carbowax 20M, 164–166, 170–172
  free fatty acid phase, 167
  temperature limit, 166
Carbowax 400, 183
Carotene, 190
  cis-trans isomers, separation on alumina, 128, 129
  HPLC, stationary and mobile phases, 199
  numbering scheme, 114, 115
  plasma, HPLC, 199
  separation, C_8 columns, 198
  serum, HPLC, 199
  solubility properties, 116
  standards, reverse-phase chromatography, 197
α-Carotene
  absorption spectra, shape, 145
  light absorption maxima, 137, 139
  source, 192
  specific absorbance coefficient, 139
  structure, 190, 191
  TLC, on MgO-kieselguhr, solvent, 132
α- and β-Carotene
  cis isomer, chromatography, on alumina, 129
  TLC, 130, 131
  trans isomer, chromatography, on alumina, 129
β-Carotene, 114
  absorption spectra, shape, 145
  light absorption maxima, 136, 137, 139
  normal-phase chromatography, 200
  precursor for vitamin A, 197
  reverse-phase chromatography, 194–196
  source, 192
  specific absorbance coefficient, 139
  standard, chromatography, 197
  structure, 115, 190, 191
  TLC, on MgO-kieselguhr, solvent, 132
β, γ-Carotene, light absorption maxima, 139
γ-Carotene
  absorption spectra, shape, 145
  light absorption maxima, 136, 137, 139
  normal-phase chromatography, 200
  reverse-phase chromatography, 196
  specific absorbance coefficient, 139
  structure, 190, 191
  TLC, on MgO-kieselguhr, solvent, 132

γ, γ-Carotene, light absorption maxima, 136, 139
δ-Carotene
  light absorption maxima, 137, 139
  specific absorbance coefficient, 139
  TLC, on MgO-kieselguhr, solvent, 132
ε-Carotene, specific absorbance coefficient, 140
ε, ε-Carotene
  absorption spectra, shape, 144
  light absorption maxima, 136, 137, 140
ψ, ψ-Carotene, 114
ζ-Carotene
  cis isomer
    normal-phase chromatography, 200
    reverse-phase chromatography, 196, 197
  15-cis isomer, chromatography, on alumina, 129
  light absorption maxima, 137, 140
  protection against oxygen, 118
  reverse-phase chromatography, 194–196
  specific absorbance coefficient, 140
  structure, 190, 191
  TLC, on MgO-kieselguhr, solvent, 132
  trans isomer
    chromatography, on alumina, 129
    normal-phase chromatography, 200
  unsymmetrical, light absorption maxima, 140
β-Carotene 5,6,5',6'-diepoxide, light absorption maxima, 139
β-Carotene 5,6-epoxide
  light absorption maxima, 139
  specific absorbance coefficient, 139
Carotenoid, 113–149. See also Tetraterpenoid
  5,6-epoxides, test for, 145, 146
  absorption spectra, shape, 144, 145
  artifacts, 117
  avoidance of acid or alkali, 119, 133
  chromatography, 124–132
    absorbents, 119, 120, 125, 126
    on Ca(OH)$_2$, 125, 126
    on MgO, 125, 126, 131
    reagents, 119, 120
    solvent purification, 119, 120
    on ZnCO$_3$, 126, 127
  circular dichroism, 147
  cochromatography, 133
  column chromatography, 124, 125, 126–130
    on alumina, 127, 128
    on silica, 127
    solvents, 128
  cyclization, 113, 114
  elution, 133
  end groups, 114
  extraction, 120–122
  gas chromatography, 185, 186, 189
  HPLC, 124, 125, 149, 186, 189–200
    equipment, 190–192
    operating conditions, 193
    stationary and mobile phases, 199
  hydrocarbon. See Carotene
  instability, 116, 117
  iodine-catalyzed cis-trans isomerization, 146, 147
  isolation, 189
  light-absorption spectral maxima
    effect of increasing length of conjugated double bond chromophore, 135–137
    effect of presence of water, 136–144
    effect of rings, 136, 137
    effect of solvents, 136, 137
  light-absorption spectroscopy, 135–147
    positions of absorption maxima, 135–144
  location, on chromatogram, 133
  mass spectrometry, 147
  microscale spectroscopic tests, 145, 146
    for oxocarotenoids, 146
  NMR spectrometry, 147
  noncovalent association with protein, 216
  oxidation products, 117
  oxygenated. See Xanthophyll
  perhydrogenated, GLC separation, 124, 125
  precautions with, 117
  protection against heat, 118, 119
  protection against light, 118
  protection against oxygen, 117, 118, 132, 133
  purification, 123–132
    for MS, 134
    for NMR, 134
  quantitative determination
    by HPLC, 149
    spectrophotometric method, 147–149

radioactive samples, purification, 134, 135
saponification, 118, 119, 122, 123
semisystematic nomenclature, 114
separation, 123–132
   CN-column, 198
   nonchromatographic procedures, 124
   normal-phase chromatography, 200
   phase, 124
   by reversed-phase chromatography, 193, 194
solubility properties, 116
source, 192
specific absorbance coefficient, 148
specificity of carotenoid attachment, study methods, 225–227
spectroscopic analysis, 135–147
storage, 118, 119
structure, 114, 115, 190, 191
thin-layer chromatography, 124, 125, 126, 130–132, 189
   on MgO-kieselguhr, 131
      solvents, 131, 132
   ready-made plates, 134
   on silica gel, 130, 131
tritium-labeled, purification, 135
type A, apoprotein preparation, 224, 225
zone chromatography, 129, 130
Carotenoid-lipovitellin complexes
   absorption spectra, 247
   of anostracans, 246, 247
   of decapods, 243
Carotenoprotein
   apoproteins, comparison, by peptide mapping, 234, 238, 239
   carotenoid content, determination, 232, 233
   crystallization, 238, 239
   echinoderm, 240–242
   Hydrocorollina, 242, 243
   invertebrate, 216–247
      types, 216
   lobster carapace, 217–240
      extraction, 219–221
   minimum molecular weight, 231, 232
   polyacrylamide gel electrophoresis, 233–238
   protein content, determination, 231, 232
   type A, 216, 217
      general extraction procedures, 239, 240
      properties, 218, 219
   type B, 216, 217, 243–247
      extraction, 244, 245
      properties, 243, 244
      purification, 244, 245
      storage, 245, 246
Chenodeoxycholate
   mass spectrometry, 67
   RPLC, 61, 62
   sulfate, RPLC, 62
Chenodeoxycholic acid
   biosynthesis, 364
   conversion to lithocholic and ursodeoxycholic acids, monitoring, 113
   HPLC, 53–55
   ion current chromatography, 104, 107
   LC-MS, 76
   selected ion monitoring, 108, 109
   in serum, 112
   TMS ether, ion current chromatography, 100
Cherimoya, volatile components, 175
Chiral alcohol, isopropyl urethanes, enantiomer separation, 183, 184
Chlamys opercularis, uptake of labeled sterol precursors, 322
Chlorobactene, light absorption maxima, 140
Chlorophyll, isoprenoids derived from, 186
Chloroplast, carotenoid, HPLC, 199
Chloroxanthin, light absorption maxima, 140
Cholate
   mass spectrometry, 67
   RPLC, 61
   sulfate, RPLC, 62
Cholestane, 6
5α-Cholestane, 7, 8, 19
5β-Cholestane, 7
Cholestane-3,6-diol, TMS ether, 84
Cholestanol, 46, 48
   cholesterol contaminants, 50
   for yeast sterol auxotrophs, 50
5α-Cholestanol, 4, 11
5β-Cholestan-3β-ol, 3, 19
Cholestanone, 6
5α-Cholestan-3-one, 18–19

Cholestatetraenol, HPLC properties, 43
Cholestatrienol, HPLC properties, 43
Δ^4-Cholestenol, HPLC properties, 43–45
Δ^7-Cholestenol, See Lathosterol
$\Delta^{8(14)}$-Cholestenol, HPLC, properties, 43, 44
Cholest-4-en-3-one, ^{13}C NMR spectra, 36
Cholesterol, 6, 8, 11, 16
  analogs
    incubations with cytochrome P-450, 357
    possible cleavage product containing vicinal diol function in side chain, analysis, 358–360
    side-chain cleavage products, analysis, 358
    as substrates for cytochrome P-450, 362–364
  cell membrane, in LCAT reaction, 271–274
  conversion to bile acids, 364
  ecdysteroid precursor, 380, 384, 385
  epoxide
    assay of epoxide hydrolase with 309–311
    hydration, 304
  esterified by lecithin-cholesterol acyltransferase, lipoprotein origin, 270, 271
  HPLC properties, 43–45, 46, 47
  hydroxylated derivatives, 353
    analysis, 355–358
    SIM, 361, 362
    stereoisomers, separation, 358, 359
  hydroxylation
    in bile acid biosynthesis, 364, 365
    assay, 365–367
    C-22, stereochemical course, 361, 362
  mass assay, 268–270
  NMR, 30
  ^{13}C NMR spectra, 36
  precursors, deuterium-labeled, metabolism in insects, 347
  side-chain cleavage, 352–364
    GC-MS assay, 353, 354
      with selected ion monitoring, 354
    intermediates, 353, 355
      identification, 360
    pathway, 353
  synthesis
    from lanosterol, cofactor requirements, 301
    from squalene, 302
  TMS ether, 20
    ion current chromatography, 100
Cholesterol acyltransferase, 286–293, 302
  activation, 293
  assay, 286–289
  inhibitors, 291–293
  properties, 290
  solubilization, 289, 290
  substrate specificity, 290, 291
Cholesteryl, trimethylsilyl derivatives, 19
Cholesteryl ester, 6
  mass transport between HDL and VLDL + LDL in plasma, assay, 278, 279
  indirect method, 280
Cholesteryl ester transfer activity, 274–285
  chromatofocusing, 285
  inhibitor, 281
  in isolated plasma lipoproteins, 277–279
  plasma
    assay, 275–280
    purification, 280–285
  purification, 275
  separation from LCAT, 283–285
  in synthetic liposomes, 275–277
Cholesteryl ester transfer protein, purification, 275
Cholic acid
  biosynthesis, 364
  ion current chromatography, 107
  LC-MS, 76
  molecular weight determination, 67
  selected ion monitoring, 108, 109
  in serum, 112
  TMS ether, ion current chromatography, 100
Choristoneura fumiferana, 420
Chromoplast, *Capsicum*, tocopherol biosynthesis, 544–547
Chrysanthemaxanthin
  light absorption maxima, 140
  specific absorbance coefficient, 140
1,4-Cineole, 151
Cinnamon, volatiles, 160, 161
β-Citraurin, light absorption maxima, 140
Citronellal, 158

Coast redwood, needle, oil, 176
Cockroach
  corpora allata
    endocrine activity, radiochemical assay, 531
    incubation, 531–533
    in vitro biosynthesis of optically pure JH III, 539
  hemolymph extract, for JH binding assay, 517
  JH, preparation, 517
  JH binding proteins, 495
    distribution, 516
    photoaffinity labeling, 510, 516–521
  phytosterol dealkylation, 346
Coelenterate, carotenoprotein, 217
Coprostanol, 101
  ion current chromatography, 107
Coral, carotenoprotein, 242, 243
Corn earworm. *See Heliothis zea*
Corpora allata, 540
  endocrine activity, radiochemical assay, 531
  incubation, 531–533
Crayfish, 429
  hypodermis, ecdysteroid receptor, 459–461
Crocetin
  light absorption maxima, 140
  specific absorbance coefficient, 140
Crustacyanin, 216
  absorption spectra, 219, 220
  α and γ, separation, on Porath column, 223, 224
  apoprotein
    chromatofocusing, 228, 229
    gel filtration, 227, 228
    ion-exchange chromatography, 228
    molecular size, 235–237
    preparation, 224, 225
    subunits
      isolation, 227–230
      separation, 235
  DEAE-cellulose chromatography, gradient elution, 221, 222
  formation, 219
  isomers, 217, 218
  properties, 218, 219
  reversible dissociation, 218
  separation, 221–223, 235
  storage, 219
  structure, 218
α-Crustacyanin
  crystallization, 238, 239
  minimum molecular weight, 232
  preparation, by stepwise DEAE-cellulose chromatography, 223
  reconstitution, 225
β-Crustacyanin
  apoprotein composition, 229
  components, separation, 229–231
Crustaxanthin, light absorption maxima, 140
α-Cryptoxanthin
  light absorption maxima, 140
  specific absorbance coefficient, 140
β-Cryptoxanthin
  light absorption maxima, 136, 140
  specific absorbance coefficient, 140
Cycloartenol, 7, 21
  cholesterol acyltransferase inhibitor, 292
  labeled, 316, 317
  separation from lanosterol, 331
Cycloeucalenol-obtusifoliol isomerase, 323, 324
Cyclolaudenol, identification, 332
Cycloleucalenol, 317
9,19-Cyclosterol, 29
  NMR spectra, 33
p-Cymene, 151
Cytochrome *P*-450
  in bile acid biosynthesis, 364–377
    assay, 365–367
    catalytic properties, 375, 376
    properties, 375–377
  in cholesterol side-chain cleavage, substrate specificity, 362–364
  incubations of cholesterol analogs with, 357
  rabbit liver microsome, with different substrate specifications toward C_{27} steroids
    catalytic properties, 376
    purification, 372–374
  rabbit liver mitochondria, active in 26-hydroxylation of C_{27} steroids
    catalytic properties, 376

purification, 374–376
rat liver microsome
  catalyzing 25-hydroxylation of C_{27} steroids
    catalytic properties, 376
    purification, 370–372
  catalyzing 7α-hydroxylation of cholesterol
    catalytic properties 376
    purification, 368–370
  in side-chain cleavage of cholesteol, 352, 354

D

Daffodil, carotenes, HPLC, 199
Decanal, 151
Decaprenoxanthin, light absorption maxima, 140
3,4-Dehydro-β-carotene
  light absorption maxima, 140
  specific absorbance coefficient, 140
5,7-Dehydrocholesterol, HPLC properties, 47
7-Dehydrocholesterol, 10
  ecdysteroid intermediate, 380
  HPLC properties, 43, 46
Z-17(20)-Dehydrocholesterol, cholesterol acyltransferase inhibitor, 292
3-Dehydroecdysone, hormonal activity, 428
3-Dehydroecdysone reductase, 419, 420
3-Dehydroecdysteroid, 419, 426
  HPLC, 428
  identification, 426–428
  physicochemical properties, 426
  separation, 426–428
3-Dehydro-20-hydroxyecdysone
  hormonal activity, 428
  in methanol, ultraviolet absorption, 450
  ratio of equilibrium constants of association, 451
3,4-Dehydrolycopene
  light absorption maxima, 137, 140
  specific absorbance coefficient, 140
4-Demethylsterol
  HPLC, 329
  labeling, 317–319
  side chain labeling, 318

4-Demethylsterol acetate, TLC migration, 327–329
2-Deoxy-20-hydroxyecdysone, 380, 381, 451
  in methanol, ultraviolet absorption, 450
  ratio of equilibrium constants of association, 451
Deoxycholate
  mass spectrometry, 67
  RPLC, 61, 62
  sulfate, RPLC, 62
Deoxycholic acid
  HPLC, 54, 55
  ion current chromatography, 107
  mass spectrometry, 77
  selected ion monitoring, 108, 109
  TMS ether, ion current chromatography, 100
Deoxycorticosterone, cholesterol acyltransferase inhibitor, 292
Deoxycrustecdysone, 451
2-Deoxyecdysone, 379, 380, 381, 411
Desmosterol, 11, 346, 347
  HPLC properties, 43
Desmosterol 24,28-epoxide, 346–348
Dexsil 300GC, 25
Diadinoxanthin, light absorption maxima, 140
Diapocarotenoid, 116
Diatoxanthin, light absorption maxima, 140
7,8-Didehydroastaxanthin, light absorption maxima, 140
2,22-Dideoxyecdysone, 380, 381
22,25-Dideoxyecdysone, 380, 381
2,22-Dideoxy-20-hydroxyecdysone, 380, 381
cis-Dihydrocarvone, 151
24-Dihydrolanosterol, HPLC properties, 43
5α-Dihydroprogesterone, cholesterol acyltransferase inhibitor, 292
Dihydrotestosterone, cholesterol acyltransferase inhibitor, 292
2,2'-Dihydroxy-β-carotene
  light absorption maxima, 141
  specific absorbance coefficient, 141
3,12-Dihydroxy-7-cholenoic acid, 88
2β,3β-Dihydroxy-5α-cholest-7-en-6-one
  in methanol, ultraviolet absorption, 450
  ratio of equilibrium constants of association, 451

Dimethylallyl pyrophosphate, TLC, 249, 250
4,4-Dimethylcholesterol, 6
5,6-Dimethyl-5-norbornen-*exo*-2-ol, 151
3,7-Dimethyl-1-octyl benzoate, 262
2,3-Dimethylphytylquinol, 545
4,4-Dimethylsterol, 316, 317
    HPLC, 329
4,4-Dimethylsteryl acetate, chromatographic separation, 327, 330
3,7-Dioxocholanoate, 90
Diploptera punctata. See Cockroach
Disodium glycolithocholate-3-*O*-sulfate, FAB spectra, quasimolecular ions in, 70
Disodium taurochenodeoxycholate-3-*O*-sulfate, FAB spectra, quasimolecular ions in, 70
Disodium taurolithocholate-3-*O*-sulfate, FAB spectra, quasimolecular ions in, 70
Distichopora, carotenoprotein, 242
Diterpene resin acids, separation, 184, 185
Dolichol
    assay, by derivitization, 211, 212
    HPLC, 202, 203, 209–211
        solvent removal, 203
        solvents, 203
    isolation, from animal tissue, 207, 208
    isolation and assay, 206–212
    labeled, preparation, 203, 204
    monophosphate esters, chemistry, 202
    TLC, 202
    total, assay, 209–212
Dolichol esterase, 480–482
    assay, 480, 481
    properties, 482
Dolichol kinase, 471–476
    assay, 471–473
    CHAPS-solubilized, properties, 475
    functions, 475, 476
    membrane-bound
        activators, 473
        inhibitors, 473
        pH effect, 473
        properties, 473, 474
        solubilization, 474, 475
        specificity, 473, 474
        subcellular distribution, 475

Dolichyl monophosphate, regeneration from dolichyl pyrophosphate, 547
Dolichyl monophosphate phosphatase, vs. dolichyl pyrophosphate phosphatase, 552, 553
Dolichyl phosphate
    assay, 214, 215
        by derivitization, 214, 215
    dephosphorylation, 476, 477
    glycosylated derivatives, 202
    HPLC, 202, 203, 214
        solvent removal, 203
        solvents, 203
    isolation and assay, 212–215
    ^{3}H-labeled, 205, 206
    ^{32}P-labeled, 204, 205
    TLC, 202
    total
        analysis, 213
        isolation, 213, 214
Dolichyl phosphate phosphatase, 476–480
    activators, 478
    assay, 477, 478
    axolemma-enriched fractions, 479, 480
    inhibitors, 478
    pH effect, 479
    polyisoprenyl phosphatase activity, 477, 479
        functions, 479, 480
    properties, 478, 479
    solubilization, 479
    specificity, 478, 479
    subcellular distribution, 479
Dolichyl phosphomannose, 202
Dolichyl pyrophosphate
    β-labeled
        chemical preparation, 548–550
        purification, 550
    unlabeled, purification, 550
Dolichyl pyrophosphate phosphatase
    assay, 548–550
    brain, 547–553
    inhibitors, 551
    role in lipid intermediate pathway, 547, 548
    solubilization, 550, 551
    solubilized, properties, 551
    specificity, 551, 552
    tissue distribution, 547

vs. dolichyl monophosphate phosphatase, 552, 553
α-Doradecin, light absorption maxima, 141
β-Doradecin, light absorption maxima, 141
α-Doradexanthin, light absorption maxima, 141
Drosophila hydei, 420, 429, 430, 495
Drosophila melanogaster, 420
  ecdysteroid receptors, 458, 461
  K_c cell line. *See* K_c cell line
Durabond Wax, temperature limits, 166

E

Ecdysone, 380, 381
  biosynthesis, 377, 378
  conjugates, 411–419
  conversion to 3-epiecdysone, 437, 438, 442
  hydroxylation, 454
  in methanol, ultraviolet absorption, 450
  quantification, by GLC-MS (SIM), 396, 397
  radiolabeled, purification, 454, 455
  receptor binding, 467
Ecdysone 3-epimerase, 393, 437–442
  assay, 439, 440
  cofactors, 441, 442
  HPLC, 440
  kinetics, 440
  pH optimum, 441
  preparation, 438, 439
  properties, 440, 441
  reaction catalyzed, 437
  specificity, 440
Ecdysone 20-monooxygenase, 454–458
  assay, 455–457
  properties, 458
  tissue distribution, 454
Ecdysone oxidase, 419–429, 438
  characterization, 423–425
  hydrogen peroxide formed by, 429
  inhibitors, 423, 424
  kinetics, 424
  molecular weight, 424
  optical assay, 420–422
  pH optimum, 424
  preparation, 422, 423
  radioassay, 420, 421
  reaction catalyzed, 419, 420, 424
  reaction products, 419, 426–429
  stability, 423
  storage, 423
  substrates, 424–426
  tissue distribution, 425
Ecdysonoic acid, methylation, 398
Ecdysteroid, 377–410
  acetates, 418, 419
  acetonide formation, 397, 398
  acetylation, 397
  acids, 390
  in beam ionization, 403, 404
  biosynthesis
    experimental approaches, 384–386
    pathways, 377–381
    putative intermediates, 385, 386
  catabolism, 411
  characterization, 387, 391
  chemical ionization, 402, 403
  conjugates, 383, 384
    distribution, 411, 412
    enzymatically hydrolyzed, analysis, 398, 399
    enzymatic and chemical hydrolysis, 415, 417, 418
    extraction, 412, 414
    fast atom bombardment, 418
    identification, 413–418
    infrared spectroscopy, 413, 414
    isolated and identified in insects, 390, 418
    mass spectrometry, 416–418
    ^{13}C NMR spectra, 416
    ^{31}P NMR, 410, 416
    NMR spectroscopy, 415, 416
    purification, 412–414
    UV spectroscopy, 414
  cross-linkage, quantitative analysis, 433, 434
  derivatization reactions, 397
  electron impact spectra, 401, 402, 405
  extraction from tissues, 388, 389
  fast atom bombardment, 404, 405
  fractionation, 387–391
  free
    GLC, 394
    GLC-MS, 394, 395
      in SIM mode, 395, 396

HPLC, 392
HPLC-RIA, 393, 394
purification and analysis, 391–398
reversed-phase TLC, 391, 392
TLC on silica gel, 391
free and conjugated
butanol-water partition, 391
^{13}C NMR spectroscopy, 409, 410
^{1}H NMR spectroscopy, 406–409, 415
reversed-phase column chromatography, 390, 391
separation, 389, 390
silicic acid column chromatography, 389, 390
HPLC, 428
inactivation, 379, 380–384, 386, 411
irradiation, method, 432–435
isolation, 387–391
mass spectrometry, 401–405
metabolism
experimental approaches, 386, 387
pathway, 380–384
in methanol, ultraviolet absorption, 450
NMR spectroscopy, 405–410
occurrence, 379
phosphates, 383, 390, 418, 419
HPLC, 399–401
retention volumes on HPLC, 400
poplar fraction
analysis, 398–401
desalting, 401
high voltage paper electrophoresis, 399
silica gel TLC, 399
ratio of equilibrium constants of association (RAC), 450, 451
reversed-phased column chromatography, retention volumes, 392
silylated derivatives
preparation, 394, 395
retention times on GLC-MS (SIM), 396
tissue distribution, 377
3β-Ecdysteroid, 437. *See also* Molting hormone
Ecdysteroid carrier protein, 429–437
assay, 430–432
isolation, 435, 436
by combination of photoaffinity labeling and immunoadsorption, 435, 436
photoaffinity labeling, 432
specificity, 434, 435
properties, 436, 437
Ecdysteroid-26-oic acids, HPLC, 399–401
Ecdysteroid receptor, 458–468
binding determinations, 464, 465
crustacean hypodermis, 459–461
imaginal discs, 462
K_c cell, 459
binding properties, 465–467
binding to DNA cellulose, 466–468
characteristics, 465–468
effect of salt, 465
preparation, 462–465
sedimentation properties, 465, 466
stability, 465
salivary gland, 461
tissue sources, 459–462
Echinenone
light absorption maxima, 136, 141
specific absorbance coefficient, 141
Echinoderm, carotenoprotein, 217, 240–242
Ehrlich ascites tumor, cholesterol acyltransferase, 289
α-Elemene, 151
β-Elemene, 151
Emulphor ON 870, 166
20-Epicholesterol, NMR spectra, 33–36
Epoxide, hydration, 304, 305
in insects, 305
Epoxide hydrase, 303
Epoxide hydratase, 303
Epoxide hydrolase, 303–311
assay, 304
with cholesterol epoxide, 309–311
with juvenile hormone, 305–308
with *trans*- and *cis*-stilbene oxide, 308, 309
cytosolic, 304, 305
preparation, 306
forms, 304
microsomal, 304
reaction catalyzed, 303
5,6-Epoxycarotenoid, TLC, 130, 131
5,8-Epoxycarotenoid, TLC, 130, 131

Epoxyfarnesoic acid methyltransferase, 540–544
    preparation, 541, 542
    reaction mixture, 543
    reaction products
        analysis, 543, 544
        isolation, 543, 544
    sources, 541
    substrates, 542
Epoxyfarnesyl diazoacetate, 506
    binding assay, in cockroach hemolymph extract, 518, 519
    binding specificity, 518–521
        criteria, 510–514
    displacement of natural hormone from binding protein, 512
    future applications, 529
    handling, 514
    molecular weight, 514
    photoactivated
        binding, in cockroach hemolymph extract, 519, 520
        irreversible inhibition of JH binding proteins, 512, 513
        radiolabeled, binding, 520
    photoaffinity labeling of JH binding proteins, 509–530
    photoinduced attachment to JH binding proteins, 510, 511
    properties, 514
    radiolabeled
        binding affinity and specificity of proteins for, 520
        binding to proteins, 513
        normal adsorption chromatography, 515
        proteins labeled by, electrophoretic separation, 520, 521
        purification, 514–516
        reversed-phase HPLC, 515, 516
        specificity of binding, 513, 514
        synthesis, 511
        TLC, 515
    relative affinity for hemolymph JH binding protein, 518, 519
    UV absorption maximum, 514
Ergosta-7,22-dien-3β-ol. *See* Sterol, 3701b-n3
Ergostanol, 46, 48

Ergosterol, 8, 10, 190, 291, 333
    biosynthetic pathway, 48, 343, 344
    cholesterol acyltransferase inhibitor, 292
    conversion to cholesterol, in insects, 346
    HPLC properties, 43, 46–50
    natural auxotrophy in yeast, 334
    reverse-phase chromatography, 195
Eschscholtzxanthin
    light absorption maxima, 141
    specific absorbance coefficient, 141
Essential oils. *See also* Isoprenoid
    analysis, 162, 175, 176
    prefractionation, 157
24α-Ethylcholesterol. *See* Sitosterol
24-Ethylidenesterols, isomers, separation, 10
24-Ethylidinelophenol, labeling, 317
Eucalyptus, terpenoids, 180
Euglena gracilis, carotenoids, separation, 200
Euphol, 10, 33

F

Farnesoic acid methyltransferase, 541
Farnesyl benzoate, 262
Farnesyl monophosphate, ion-pairing chromatography, 251
Farnesyl pyrophosphate, ion-pairing chromatography, 251
Fatty acid binding protein, 296, 298
    amino acid composition, 299
Fatty acyl-CoA ligase, 302
α-Fenchyl alcohol, 151
Ferredoxin
    in bile acid biosynthesis, 365
    purification, 368
Ferredoxin reductase
    in bile acid biosynthesis, 365
    purification, 368
Flavoxanthin
    light absorption maxima, 140
    specific absorbance coefficient, 140
Florisil, 159
Fucosterol, 10, 346
    HPLC properties, 43, 45

Fucoxanthin
  enzymatic hydrolysis, 123
  light absorption maxima, 141
  protection against alkali, 119
Fusarium, sterol mutants, 340

G

Galleria mellonella, ecdysteroid conjugates, 412
Gas chromatography-mass spectrometry, 71–76
  applications, 63, 64
Gas-liquid chromatography, 72. *See also* Sterol, gas-liquid chromatography
  artifacts, 25
  capillary columns, 24, 74, 75
  liquid phases for, 18
  mass transfer, 24
  support coated open tubular columns, 24
  wall coated open tubular columns, 24
Gas-liquid chromatography-mass spectrometry, 24
Geranial, 151
Geranyl acetate, 151
Geranyl benzoate, 261
Geranylgeranyl monophosphate, ion-pairing chromatography, 251
Geranylgeranyl pyrophosphate, ion-pairing chromatography, 251
Geranyl monophosphate, ion-pairing chromatography, 251
Geranyl pyrophosphate, ion-pairing chromatography, 251
Giant silkmoth, 429
  accessory glands, source for epoxyfarnesoic acid methyltransferase, 541
  in vitro JH biosynthesis, 539
  JH biosynthesis, 540, 541
Glucosylphosphoryldolichol, biosynthesis, 471
Glycochenodeoxycholate, RPLC, 56, 59, 60, 61
Glycocholate
  HPLC, 55
  RPLC, 56, 59, 60
Glycocholic acid
  FAB spectra, quasimolecular ions in, 70
  LC-MS, 76
  RPLC, 57
Glycodeoxycholate, RPLC, 56, 59–61
Glycodeoxycholic acid, FAB spectra, quasimolecular ions in, 70
Glycolithocholate
  RPLC, 56, 59
  3-sulfate, RPLC, 63
Glycoursodeoxycholate, RPLC, 56, 59
Gomphocerus rufus, hemolymph binding protein, 526
Gryllus bimaculatus, 420

H

Halostane, 8
Heliothis zea, cholesterol acyltransferase, 286, 288, 290
Helix pomatia. *See* Snail
Hemolymph
  ecdysteroid binding protein, 442, 443
  M. sexta
    acetone powder, preparation, 485
    juvenile hormone carrier protein, 482–487
  preparation, 444
High-pressure liquid chromatography, 76, 78
Homarus sp. *See* Lobster
Homocarotenoid, 116
Hops. *See also* Saaz hops
  essential oils, extraction, 155
  Hersbrucker Spat, 179
    sesquiterpenes, 183
  Northern Brewer, 179
  tricyclic sesquiterpenes, 179
  volatiles, extraction, 154
Hordeum vulgare. *See* Barley
Hyalophora cecropia. *See* Giant silkmoth
Hydrocorallina, 217
  carotenoproteins, 242, 243
2-Hydroxy-β-carotene
  light absorption maxima, 141
  specific absorbance coefficient, 141
7-Hydroxycholanoate, 87
3-Hydroxy-5-cholenoic acid, 87
3β-Hydroxy-5-cholenoic acid, TMS ether, ion current chromatography, 100

3-Hydroxycholestane, isomers, chromatographic mobility, 7
3β-Hydroxy-5α-cholestane, 7
6β-Hydroxycholesterol, 7
7α-Hydroxycholesterol, 6
20-Hydroxyecdysone, 377, 378, 411
  formation, 380
  in methanol, ultraviolet absorption, 450
  quantification, by GLC-MS (SIM), 396, 397
  radiolabeled, biosynthesis, 443, 444
  ratio of equilibrium constants of association, 451
  receptor binding, 467
20-Hydroxyecdysone binding protein, 442–453
  assay, 445, 446
  binding site, 452
  isoelectric point, 448
  kinetics, 448, 449
  molecular weight, 448
  partial purification, 446–448
  photoaffinity labeling, 453
  properties, 448–452
  research, difficulties, 453
  specificity, 449–452
20-Hydroxyecdysonoic acid, methylation, 398
20-Hydroxyecdysteroid, 379
3-Hydroxyechinenone, light absorption maxima, 141
7-Hydroxy-4-en-3-one, 90
7-Hydroxyl-1,4-dien-3-one, 90
3-Hydroxy-3-methylglutaryl-CoA reductase, 313
6α-Hydroxy-3-oxo-5β-cholanoate, 89
11-Hydroxyprogesterone, ^{1}H NMR, 30–31
Hydroxyspheroidene
  light absorption maxima, 141
  TLC, on MgO-kieselguhr, solvent, 132
Hydroxyspheroidenone, light absorption maxima, 141
3α-Hydroxysteroid oxidoreductase, 438
Hyocholic acid, 84
  HPLC, 53, 54
  ion current chromatography, 107
  TMS ether, ion current chromatography, 100
Hyodeoxycholic acid, 84
  HPLC, 54

ion current chromatography, 107
TMS ether, ion current chromatography, 100
6β-Hyodeoxycholic acid, HPLC, 53, 54

I

Ignosterol, HPLC properties, 43
Inokosterone
  in methanol, ultraviolet absorption, 450
  ratio of equilibrium constants of association, 451
Ion monitoring. *See* Selected ion monitoring
Isocaproaldehyde, formation, 353
Isocryptoxanthin, light absorption maxima, 141
22-Isoecdysone
  in methanol, ultraviolet absorption, 450
  ratio of equilibrium constants of association, 451
28-Isofucosterol, 10
Isomenthone, enolization, 167
Isopentenyl pyrophosphate, 248
  sequential 1'-4 condensation, 252
  TLC, 248–250
Isoprenoid. *See also* Essential oils
  benzoates
    capacity factors on reversed-phase column, 255–261
    synthesis, 253
  biosynthesis, 252
  C_5 to C_{20}, reversed-phase HPLC, 252–263
    composition of solvent, 257–259
    correlations of capacity factors, 255–261
    number of isoprene units, 259, 260
    topology and substitutents, 260, 261
  capillary columns for, 150, 161–163, 187
  carbon skeleton analysis, 172, 173
  esters, synthesis, 253, 254
  gas chromatography, 149–188, 170
    stationary phase
      cross-linked, 169
      resilylation, 169, 170
      stability, 168, 169
      viscosity, 168, 169

stationary phase considerations, 163–170
system considerations, 161–163
gas chromatography-Fourier transform infrared spectroscopy, 150, 173–176
GC-MS, 150, 172, 173
gel permeation chromatography, 159
in geological samples, 188
HPLC, 159, 160
hydrocarbons, 149, 150
  extraction, 153
identification, retention index system, 171, 172
identification techniques, 170–176
IR spectroscopy, 170
naphthoates
  capacity factors on reversed-phase column, 256
  synthesis, 253, 254
NMR, 170
preparative GC, 170
sample preparation, 150–160
  direct injection technique, 160, 161
  extraction, 152–154
  prefractionation, 157–160
  simultaneous steam distillation extraction, 154–157
  steam distillation, 150, 151
synthesized by bacteria, 188
ultraviolet spectroscopy, 170
variability, 149
Isoprenoid hydrocarbons, acyclic, 186–188
  preseparation, 187
  stereoisomers, 186
Isorenieratene
  light absorption maxima, 141
  specific absorbance coefficient, 141
Isoterpinolene, 158
Isozeaxanthin
  absorption spectra, shape, 145
  light absorption maxima, 141
  specific absorbance coefficient, 141

J

Juniper, volatiles, 160
Juvenile hormone, 494, 495, 540
  assay, 488
  assay of epoxide hydrolase with, 305–308

binding proteins, photoaffinity labeling, 509-530
biosynthesis, 540
  radiochemical assay, 530
cellular binding protein, 494–509
  assay of JH binding activity, 499, 500
  binding affinities, analysis, 500–502
  cytosolic location, 495
  demonstration of, 500–506
  kinetic data, determination, 502–504
  molecular weight estimation, 504–506
  physical parameters, 504–506
functions, 487
I, optically pure, preparation, 539
III
  biosynthesis, assay, 533–537
  cockroach, radiolabeled, binding assay, 517–519
  in vitro biosynthesis, radiochemical assay, 530–539
  partition assay, 534–537
  radiochemical assay
    applications, 538, 539
    validation, 537, 538
  synthesis, from exogenous, labeled precursors, 530, 531
  TLC assay, 533, 534
labeled in carbon chain and methyl ester group, biosynthesis, 539
metabolite
  analytic method, 490
  HPLC, 490
  identification, 488, 489
  TLC, 488
naturally occurring homologs, 510
nuclear binding, 506–509
nuclear binding protein, 495, 506
photoaffinity analog, 506
radioimmunoassay, 530
radiolabeled, 488
Juvenile hormone carrier protein, 482–487
  assay, 482–484
  binding site, 486, 487
  molecular weight, 486
  properties, 486, 487
  purification, 484–486
  stability, 487
Juvenile hormone esterase, 487–494
  partition assay, 490–493

procedure, 491–493
preparation, 490, 491
substrate preparation, 491
TLC assay, 490, 493, 494

K

Kaladasterone
  in methanol, ultraviolet absorption, 450
  ratio of equilibrium constants of association, 451
K_c cell. See also Ecdysteroid receptor
  clones, 499
  culture, 496–499, 522
  cytosol preparation, 499
  cytosolic JH binding protein
    competition of EFDA and JH for, 522, 523
    DNA-cellulose column chromatography, 525
    molecular weight determination, 523, 524
    number of binding sites, 524, 525
    photoaffinity labeling, 521–525
  cytosol preparation, 522
  JH binding protein, 495, 496
    photoaffinity labeling, 510
  nuclear JH binding protein, 506
  response to JH, 496–498
  source, 496
7-Keto-20-oxacholesterol, cholesterol acyltransferase inhibitor, 292
Kumquat, peel, volatiles, 176–179

L

Lanosterol, 6, 7, 21, 33, 304
  cholesterol acyltransferase inhibitor, 292
  HPLC properties, 43, 46, 47
  labeled, 317
  separation from cycloartenol, 331
Lathosterol
  HPLC properties, 43–46
  ^{13}C NMR spectra, 36
Lecithin-cholesterol acyltransferase, 280
  activity, 268–270
  in cholesterol transport, 268
  inhibition, 277, 278
  isolation, 267, 268
  plasma level, 268

properties, 268
purification, 280, 281
reaction, 267, 274, 275
  utilization of cell membrane cholesterol for, 271–274
  separation from cholesteryl ester transfer activity, 283–285
  specificity, 267
  substrate requirement, 267
Lemon oil, analysis, 163
Lepidoptera, epoxide hydration, 306
Leprotene
  light absorption maxima, 141
  specific absorbance coefficient, 141
Leucophaea maderae. See Cockroach
Leukotriene A, 304
Limonene, 180
  isomerization with silica gel, 158
Linalyl benzoate, 261, 262
Lipoprotein
  cholesteryl ester transfer, 274, 275
  contribution to cholesterol utilized by lecithin-cholesterol acyltransferase, 270–272, 274, 275
  high density, radiolabeling, 277, 278
  isolated, cholesteryl ester transfer between, 277–279
  low-density, unlabeled, preparation, 278
Lipovitellin
  Artemia, 246
  Branchinecta, 247
  Branchipus, 246, 247
  crustacean, 243
  decapod, 243
    purification, 244, 245
  properties, 243, 244
Liquid chromatography-mass spectrometry, 76
Lithocholate
  mass spectrometry, 67
  RPLC, 61, 62
  sulfate, RPLC, 62
  3-sulfate, RPLC, 63
Lithocholic acid
  ion current chromatography, 107
  LC-MS, 76
  mass spectrometry, 77
  selected ion monitoring, 108, 109
  TMS ether, ion current chromatography, 100

Liver
  cholesterol acyltransferase, 286–290
  rat and rabbit
    microsomes, preparation, 367
    mitochondria, preparation, 367, 368
Lobster
  carapace
    carotenoproteins, 217–240
    yellow pigment, 218
      absorption spectrum, 219, 220
      isolation, 221–223
      purification, 222, 223
  ovoverdin, 244
Lophenol, HPLC properties, 43, 45–46
Loroxanthin, light absorption maxima, 141
Lutein
  chromatography, octyl-columns, 918
  light absorption maxima, 136, 141
  source, 192
  specific absorbance coefficient, 141
  standard, chromatography, 197, 198
  structure, 114, 115, 190, 191
  TLC, 130, 131
    on MgO-kieselguhr, solvent, 132
Lutein epoxide, light absorption maxima, 142
Lutein 5,6-epoxide, chromatography, octyl-columns, 198
Lycopene, 113
  absorption spectra, shape, 144, 145
  cis isomer, reverse-phase chromatography, 196
  light absorption maxima, 136, 137, 142
  reverse-phase chromatography, 193, 194, 196
  source, 192
  specific absorbance coefficient, 142
  structure, 115, 190, 191
  TLC, 131
    on MgO-kieselguhr, solvent, 312
  trans isomer, chromatography, on alumina, 129
Lycophyll, light absorption maxima, 142
Lycoxanthin, light absorption maxima, 142

M

Maize, cell-free system, sterol synthesis, 324
Makisterone, A, 379, 451
  in methanol, ultraviolet absorption, 450
  ratio of equilibrium constants of association, 451
Manduca sexta. *See* Tobacco hornworm
Mannosylphosphoryldolichol, biosynthesis, 471
Marthasterias glacialis. *See* Starfish
Mass fragmentography. *See* Selected ion monitoring
Mass spectrometry
  chemical ionization, 66–67
  direct chemical ionization, 67
  electron impact ionization, 64–66
  field desorption, 67–68, 69
p-Menthene, 181
Menthol, 183
Menthone, 183
Methionine, deuterium-labeled, 319
3α-Methoxy-7α-acetoxychol-4-enoate, mass spectrum, 87
Methyl allocholate, TMS ether, 83
3-Methyl-1-butyl benzoate, 262
Methyl cholanoate
  acetate, GLC and GC-MS, 72, 73
  TMS ethers, 67
    GLC and GC-MS, 72, 73
  trifluoroacetates, GLC and GC-MS, 72, 73
4α-Methyl-5α-cholestane, 8
24β-Methyl-5α-cholestane, 8
4α-Methylcholesterol, 6
20-Methylcholesterol, cholesterol acyltransferase inhibitor, 292
24-Methylcholesterol, 29
  HPLC properties, 43, 45
24α-Methylcholesterol. *See* Campesterol
24β-Methylcholesterol
  HPLC properties, 46
  ^{13}C NMR spectra, 36
24-Methylenecholesterol, 9, 10, 347
  epoxide, 347
  HPLC properties, 43, 45
24-Methylenecycloartenol
  identification, 332
  labeled, 316
24-Methylenelophenol, labeling, 317
Methyl 7-hydroxycholanoate, 83
Methyl 8,15-isopiramaradien-18-oate, 185
Methyl levopimarate, 185
Methyl palustrate, 185
Methyl 8(14),15-piramaradien-18-oate, 185
Methyl silicone, 168, 169

4α-Methylsterol
  HPLC, 329
  labeling, 317
Methyl varanate, 95
Mevalonate, TLC, 248, 249
Mevalonate phosphate, TLC, 248–250
Mevalonate pyrophosphate, TLC, 249
Mevalonic acid
  labeled, sterol precursor, 313
  tritium-labeled, 314, 315
Microcell C, 125, 126
Migratory locust, 420
  binding of 20-hydroxyecdysone to hemolymph protein, 452, 453
  corpora allata
    farnesoic acid methyltransferase, 541
    O-methyltransferase, 539
  ecdysteroid biosynthesis, 385
  ecdysteroid catabolism, 411
  ecdysteroid conjugates, 412
  ecdysteroid production, 378
  hemolymph JH binding protein
    covalent radiolabeled EFDA binding, 527
    HPLC-separated
      competitive binding studies, 528, 529
      radiolabeled EFDA binding, 527
    HPLC separation, 527
    photoaffinity labeling, 526–529
    preparation, 526
    radiolabeled JH III and EFDA binding, 526
  hemolymph protein, binding of ecdysteroids, 430
  20-hydoxyecdysone binding protein, 452
  JH binding protein, 495
  photoaffinity labeling, 510
Mint, volatiles, 160
Molting hormone, 437. See also 3β-Ecdysteroid
  hemolymph binding protein, 442, 443
  insect. See Ecdysteroid
Monoterpene, 176
  fractionation, 160
  GC-MS, 167
  hydrocarbons, prevention of isomerization, 158
  hydrogenolysis, 172, 173

Muricholic acid, 84
α-Muricholic acid
  ion current chromatography, 104
  mass spectrometry, 77
β-Muricholic acid, mass spectrometry, 77
Ω-Muricholic acid, mass spectrometry, 77
Muristerone A
  in methanol, ultraviolet absorption, 450
  ratio of equilibrium constants of association, 451
Mutatochrome, light absorption maxima, 142
Mutatoxanthin, light absorption maxima, 142
Myrcene, identification, 171
Mytiloxanthin, light absorption maxima, 142
Myxoxanthophyll
  light absorption maxima, 142
  specific absorbance coefficient, 142

N

NADPH-cytochrome P-450 reductase, 364, 365
  purification, 368
Neochrome
  light absorption maxima, 142
  specific absorbance coefficient, 142
Neoxanthin
  chromatography, octyl-columns, 198
  light absorption maxima, 142
  light absorption spectroscopy, 144
  specific absorbance coefficient, 142
  TLC, on MgO-kieselguhr, solvent, 132
Neral, 151
Neryl acetate, 151
Neryl benzoate, 261
Neurospora, sterol mutants, 340
Neurosporene
  light absorption maxima, 137, 142
  normal-phase chromatography, 200
  reverse-phase chromatography, 194, 195, 196
  specific absorbance coefficient, 142
  structure, 190, 191
  trans isomer, chromatography, on alumina, 129
Nicotiana tobacum, cell-free system, sterol synthesis, 323
Norpristane, 186–188

Nostoxanthin, light absorption maxima, 142
Nystatin, in selection of yeast sterol mutants, 335

O

Ochromonas malhamensis, cell-free system, sterol synthesis, 323
cis-β-Ocimene, 171
trans-β-Ocimene, 171
N-(1-oxo-9-Octadecanyl)-DL-tryptophan(z) ethyl ester, 292
Okenone
   light absorption maxima, 142
   specific absorbance coefficient, 142
Orconectes limosus. See Crayfish
Oscillaxanthin, light absorption maxima, 142
OV-1, 170
OV-101, 170, 172
Ovary, rat, cholesterol acyltransferase, 288, 290
Ovoverdin, 217, 219
   of different lobster species, 244
   properties, 243, 244
2,3-Oxidosqualene, 315, 316
   HPLC, 329
2,3-Oxidosqualene-cycloartenol cyclase, 323
3-Oxo-4-cholenoic acid, 89

P

Pachygrapsus crassipes, 429
Penicillium, sterol mutants, 340
Pepper
   essential oils, extraction, 155
   oil
      constituents, isomerization, 158
      monoterpene isolation, 170
      volatiles, capsule injector, 161
Peppermint, oil, chromatography, 173, 174
Peridinin
   light absorption maxima, 142
   protection against alkali, 119
Phaseolus vulgaris. See Bean
β-Phellandrene, 151, 158
Philosamia cynthia, 429
Phytane, 186–188

Phytoene
   cis-trans isomers, separation, 129
   light-absorption spectral maxima, 136, 137, 143
   location, on chromatogram, 133
   reverse-phase chromatography, 193, 194
   specific absorbance coefficient, 143
   structure, 190, 191
   TLC, on MgO-kieselguhr, solvent, 132
   trans isomer, structure, 114, 115
Phytofluene
   15-*cis* isomer, chromatography on alumina, 129
   light absorption maxima, 136, 137, 143
   location, on chromatogram, 133
   protection against oxygen, 118
   reverse-phase chromatography, 196, 197
   specific absorbance coefficient, 143
   structure, 190, 191
   TLC, on MgO-kieselguhr, solvent, 132
Phytol, isoprenoids derived from, 184
Phytosterol
   conversion to cholesterol, in insects, 346, 347
   dealkylation to insects, 346–352
      incubation, 349–352
      incubation products
         gas-liquid chromatographic assay, 349–352
         tracer assay, 349, 350
      substrate specificity, 351, 352
      substrate synthesis, 348, 349
Pieris brassicae, 419, 420
   ecdysteroid metabolism, 387
Pineapple, terpenes, 162
α-Pinene, 180
Pisum sativum, cell-free system, sterol synthesis, 322
Plant, uptake of labeled sterol precursors, 320–322
Plasma
   cholesterol level, 268
   cholesteryl ester transfer activity, 274, 275
      assay, 275–280
      purification, 280–285
   transport of cholesterol from cells to, 274

Plectaniaxanthin
　light absorption maxima, 143
　specific absorbance coefficient, 143
Pluronics, 166, 167
Polyethylene glycol, temperature limit, 166
Polyethylene glycol 20M, thermal decomposition, 164
Polyhydroxycholanoates, 82
　ion current chromatography, 107
　TMS ethers, 79, 83–84, 84–85
Polyisoprenyl pyrophosphate, unlabeled, chemical preparation, 548–550
Polyisoprenyl pyrophosphate phosphatase, possible functions, 553
Polyphenol ether, 167, 168
Polyphenyl ether, 187
Polyprenol
　assay, by derivitization, 211, 212
　HPLC, 209–211
　isolation, from plant tissue, 208, 209
　isolation and assay, 206–212
　labeled, preparation, 203, 204
　monophosphate esters, chemistry, 202
　total, assay, 209–212
Polyprenol phosphate
　assay, 214, 215
　　by derivitization, 214, 215
　glycosylated derivatives, 202
　HPLC, 214
　isolation and assay, 212, 213
Polyprenyl phosphate, ^{32}P-labeled, 204, 205
POLY-S-176, 25
Pomacea canaliculata. See South American snail
Ponasterone A
　in methanol, ultraviolet absorption, 450
　ratio of equilibrium constants of association, 451
　receptor binding, 462, 467
Poststerone, ratio of equilibrium constants of association, 451
5α-Pregnane, 8
Pregn-5-en-3β-ol, cholesterol acyltransferase inhibitor, 292
Pregnenolone
　cholesterol acyltransferase inhibitor, 292
　determination, by GC-MS in SIM mode, 354–356

　deuterated, preparation, 354
　production, 352, 353
Prenol
　free unesterified, isolation and analysis, 212
　HPLC, 202
　long-chain
　　chemistry, 201
　　direct saponification, 201
　　isolation and characterization, 201
　　oxidation to form prenals, 201
　spectroscopic analysis, 202
Prenyl fatty acid esters, isolation and analysis, 212
Prenyl phosphate
　ion-pair chromatography, separation system, 251, 252
　ion-pairing chromatography, 250–252
　ion-pair reagent, 250
　standard formation, 250, 251
　short chain, HPLC, 248
Presqualene pyrophosphate, conversion to squalene, 322, 323
Pristane, 186–188
Progesterone, cholesterol acyltransferase inhibitor, 292
Prolycopene
　iodine-catalyzed cis-trans isomerization, 146, 147
　light absorption maxima, 143
　specific absorbance coefficient, 143
Prostaglandin, cholesterol acyltransferase inhibitor, 291
Pyrophosphate, allylic, ion-pairing chromatography, 250
Pyrrhocoris apterus, 429
Pythocholate, TMS ether, mass spectrum 86

Q

Quasimolecular ion, 66–68
　in FAB spectra of conjugated bile acids, 70

R

Renieratene, light absorption maxima, 143
Retinoid, HPLC, stationary and mobile phases, 199

Retinol
  source, 192
  standard, chromatography, 197
Retinyl acetate
  source, 192
  standard, chromatography, 197
Rhododandron simsii, essential oils, extraction, 155
Rhodopin, light absorption maxima, 143
Rhodovibrin, light absorption maxima, 143
Rhodoxanthin
  isomers
    HPLC, 199
    separation, 198
  light absorption maxima, 413
  specific absorbance coefficient, 143
Rosmarinus officinalis, 160
Rubixanthin
  light absorption maxima, 143
  specific absorbance coefficient, 143
Rubus fruticosus. See Bramble
Rumen, bovine, liquor, isoprenoid hydrocarbons, 184

S

Saaz hops, essential oil, 176–178, 180, 181
  components, 181
Sabinene
  decomposition, on silicone oil, 167
  isomerization, 158
trans-Sabinene hydrate, 151
Saccharomyces cerevisiae. See Yeast
Santene, 151
Sarcinaxanthin, light absorption maxima, 143
Scenedesmus obliquus. See Alga
Schistocerca gregaria
  corpora allata, *in vitro* JH III biosynthesis, 530
  ecdysteroid biosynthesis, 384, 385
  ecdysteroid catabolism, 411
  ecdysteroid conjugates, 412
  ecdysteroid metabolism, 386, 387
  ecdysteroid production, 378
  ecdysteroids, FAB mass spectra, 405
  in vitro JH biosynthesis, 539
  phytosterol dealkylation, 348
Scots pine needle, essential oils, 167
SCP_2, 302

SE-30 Ultraphase, 25, 170, 171
Selected ion detection. *See* Selected ion monitoring
Selected ion monitoring, bile acids, 104–111
  application, 105–111
  sensitivity, factors affecting, 105
Sequoia sempervirens. See Coast redwood
Sesquiterpene, 176
  fractionation, 160
  hydrogenolysis, 173
  tricyclic, in hops, 179
Silica, adsorbent, for carotenoids, 125, 126
Silica gel, 159
  deactivation, 158
  isomerization of isoprenoids with, 158
Silicon, stationary phase in gas chromatography, 168, 169
Silkworm
  cell-free extracts, preparation, 349
  ecdysteroid metabolism, 411
  phytosterol dealkylation, 347, 348
Siphonaxanthin, light absorption maxima, 143
Sitosterol, 6, 8, 9, 10, 11
  cholesterol acyltransferase inhibitor, 291, 292
  demethylation, 346, 347
  HPLC properties, 43, 45, 46, 47
Snail. *See also* South American snail
  arylsulfatase, 378, 389, 398
  digestive juice, hydrolysis of ecdysteroid conjugates, 417
Soap, perfumed, volatiles, capsule injector, 161
Sodium chenodeoxycholate-3-O-sulfate, FAB spectra, quasimolecular ions in, 70
Sodium chol-5-enoate-3-O-sulfate, FAB spectra, quasimolecular ions in, 70
Sodium glycochenodeoxycholate, FAB spectra, quasimolecular ions in, 70
Sodium lithocholate-3-O-sulfate, FAB spectra, quasimolecular ions in, 70
Sodium taurochenodeoxycholate, FAB spectra, quasimolecular ions in, 70
Sodium taurocholate, FAB spectra, quasimolecular ions in, 70
Sodium taurodeoxycholate, FAB spectra, quasimolecular ions in, 70

SUBJECT INDEX

South American snail, red astaxanthin glycoprotein, 217
Sow thistle, triterpene acetate, 185
Soxhlet extraction, 153
Spheroidene
  light absorption maxima, 143
  specific absorbance coefficient, 143
  TLC, on MgO-kieselguhr, solvent, 132
Spheroidenone
  light absorption maxima, 143
  TLC, on MgO-kieselguhr, solvent, 132
Spices, volatiles, extraction, 154
Spinach
  carotenes, HPLC, 199
  xanthophyll, HPLC, 199
Spirilloxanthin
  light absorption maxima, 143
  specific absorbance coefficient, 143
  TLC, on MgO-kieselguhr, solvent, 312
Spodoptera littoralis. See Armyworm
Spruce, volatiles, 160
Squalane, 187
Squalene, 293, 304
  HPLC, 329
  labeled, 315
  purification, 295–298
Starfish, carotenoproteins, 240
Steroid, cholesterol acyltransferase inhibitor, 291, 292
Sterol
  3701b-n3, HPLC properties, 43, 49, 50
  absorption phenomena, 5
  adsorption chromatography, 5–10
  argentation chromatography, 8–10
  biosynthesis
    algal and plant cell-free preparations for study of, 322–325
    and interconversion, 311–332
  Δ^0, 9
  Δ^5, 9
  Δ^5 ad Δ^7, separation, 7
  Δ^5, 9
  Δ^{22}, 9
  cholesterol acyltransferase inhibitor, 291, 292
  chromatographic mobility
    effect of double bonds, 6, 10–12
    and molecular weight, 4, 16
    and polar groups, 5, 6, 16
    and shape, 8
  and size, 8, 10–13
  and structure, 16
chromatography, 3–4
  rate of movement, 13
  relative retention time, 15
  retention time, 15
    contribution of molecular features to, 14–16
chromic and oxidation, 331
conjugated dienic, separation, 9–10
derivative formation, 331, 332
epoxide formation, for separable derivatives, 331
extraction
  by alkaline saponification, 336
  chloroform-methane method, 336
  DMSO-hexane method, 336, 337
fragmentograms, 27–28
gas-liquid chromatography, 16–25, 338, 339. See also Gas-liquid chromatography
  effect of polarity of liquid phase, 19–20
  relative retention time
    calculation, 21
    contribution of molecular features, 21–23
    and McReynold's constant for stationary phase, 18–19
  retention time, 16, 17
    calculation, 21–22
GC-MS, 338
HPLC, 11–12, 337, 338
identification, 3–37
  methods, 5–37
infrared spectroscopy, 338
isomers, separation, 9–10
marine, 312
mass spectrometry, 27–29
molecular weight, determination, 27
nonconjugated dienols, separation, 10
nuclear magnetic resonance, 3, 338
nuclear magnetic resonance spectra, 29–37
  ^{13}C, 30, 33–37
  ^{1}H, 30–35
selected methyl
  group signals, 34, 35
ordinary phase chromatography, 5–10
  without silver ion, 5–8

with silver ion, 8–10
plant, biosynthetic pathway, 311
polarity, reduced by acetylation, 9
precursors
  deuterium-labeled, 319
  labeled
    choice and preparation, 312–319
    incubation with plant and marine invertebrate tissue, 320–322
purification, from lipid extract, 337, 338
quantitation, 5, 338, 339
radioactive metabolites
  addition of carrier sterols, 325, 326
  chromatography, 326–330
  gas-liquid chromatography, 330, 331
  HPLC, 327–330
  lipid extraction, 325
reversed-phase chromatography, 10–16
reversed-phase high-performance liquid chromatography, 11–12
separation, on basis of double bonds, 10
side chain hydroxylated, TMS ethers, side chain fragment ions and losses, 93–96
σ-value, 12–15
stereochemistry, 3–4
structural elucidation, 27
synthesis
  mammalian, 313
  plant, 313
TLC, 337
tritium-labeled, to investigate interconversion, 314
ultraviolet spectrometry, 25–27
yeast, 333–346
  elution properties, 42–46
  free, detection, 46–48
  HPLC, 37–51
    apparatus, 39, 40
    chromatographic conditions, 40
    columns, 40
    dead volume, 41
    design of system, 38–39
    double bonds vs. alkyl groups on side chain, 46
    effect of alkyl groups, 45–46
    effect of nuclear unsaturations, 43–45
    effect of side-chain unsaturations, 45
    effect of specific structural moieties 42–46
    effect of unsaturations, 42, 43
    materials and methods, 39–41
    resolution, 41
    retention values, calculation, 41
    solvents, 39, 40, 41
    unsaturation effects, mechanism, 45
  sample preparation, 41, 42
  standards, 42
Sterol binding protein
  human liver, amino acid composition, 299
  rat liver, amino acid sequence, 299, 300
  yeast, amino acid composition, 299
Sterol carrier protein, 293–303
  abundance, 300
  assay, 294–296
  functions, 301–303
  occurrence, 300
  properties, 298–303
  purification, 295–298
  rat liver, 293
    diurnal variation, 293, 294, 300–303
  roles, 293
  stability, 298
  structure, 298, 299
  subcellular distribution, 301
  tissue distribution, 300
  turnover, 302
Steryl acetate
  chromic acid oxidation, 331
  epoxidation, 331
Stigmastane-3,6-diol, TMS ether, 84
Stigmasterol, 8, 9, 10, 20
  demethylation, in insects, 346–348
  HPLC properties, 43, 46, 47
Stigmosterol, cholesterol acyltransferase inhibitor, 292
Stilbene oxide, isomers, assay of epoxide hydrolase with, 308, 309
Stylaster, carotenoprotein, 242
Sugar maple, terpenoids, 180
Supernatant protein factor, 301, 302
Superoxes, 164–166

T

Taraxanthin, light absorption maxima, 142
Taurochenodeoxycholate, RPLC, 56, 59, 60, 61
Taurocholate
   HPLC, 55
   RPLC, 56, 59, 60, 61
Taurocholic acid
   LC-MS, 76
   RPLC, 57
Taurodeoxycholate
   RPLC, 56, 59, 60, 61
   3-sulfate, RPLC, 63
Taurolithocholate, RPLC, 56, 59, 60
Tauro-α-muricholate, RPLC, 56, 57
Tauro-β-muricholate, RPLC, 57
Tauroursodeoxycholate, RPLC, 56, 59
Tea, volatiles, extraction, 153, 154
Tenebrio, pupae, ecdysteroid production, 378
Tenebrio molitor, 420
   phytosterol dealkylation, 347
Terpene, 149
   hydrogenation and hydrogenolysis, 172, 173
   identification, 171
   isomers, separation, by GC, 181, 182
   lability, 150, 161–163
   vapor phase sampling, 179, 180
Terpene alcohols
   for analysis and separation, 261
   conversion to benzoate or naphthoate esters, 261, 262
Terpenoid, 176–186
   oxygenated, fractionation, 160
α-Terpinene, 158
γ-Terpinene, 158
Terpinen-4-ol, 151
α-Terpineol, 151
β-Terpineol, 151
Terpinolene, 158
Testosterone, cholesterol acyltransferase inhibitor, 292
Testosterone acetate, RPLC, 56
3,4,3',4'-Tetradehydrolycopene, light absorption maxima, 137, 143
Tetraterpenoid. *See* Carotenoid
Thermal conductivity detector cell, isoprenoid decomposition in, 170, 171
α-Thujene, 151, 158
Thujyl alcohol, isomers, separation, 181–183
Tick, ecdysteroids, 379
   metabolism, 383
Tirucallol, 10
Tobacco
   flue-cured, 159
   volatiles, extraction, 154
Tobacco hornworm
   binding of 20-hydroxyecdysone to hemolymph protein, 453
   corpora allata, JH III biosynthesis, 531
   ecdysone 3-epimerase, 437–442
   ecdysone 20-monooxygenase, 454–458
   ecdysteroid biosynthesis, 385
   juvenile hormone binding proteins, photoaffinity labeling, 510
   juvenile hormone biosynthesis, 540, 541
   juvenile hormone carrier protein, 482–487
   microsomes, isolation, 457, 458
   mitochondria, isolation, 457, 458
   phytosterol dealkylation, 348
Tocopherol, biosynthesis, 544, 545
   products, 544, 545
   substrates, 545
α-Tocopherol
   biosynthesis, 544–547
   HPLC separation from γ-tocopherol, 546
γ-Tocopherol, conversion to α-tocopherol, 544, 545
Tomato
   carotene extract, reverse-phase chromatography, 196, 1979
   carotenes, HPLC, 199
   carotenoid extraction, 192
Torularhodin
   light absorption maxima, 144
   specific absorbance coefficient, 144
Torulene
   light absorption maxima, 144
   specific absorbance coefficient, 144
Transcortin, 429
Trebouxia sp., sterol synthesis, 324
Trichoplusia ni, epoxide hydration, 306

2β,3β,14α-Trihydroxy-5β-cholest-7-en-6-one
  in methanol, ultraviolet absorption, 450
  ratio of equilibrium constants of association, 451
Trioxocholanoate, epimers, mass spectra, 90
Triterpene, 185
Tunaxanthin, light absorption maxima, 144

U

Ucon HB 5100, 166
Urine, bile acid, 101
Ursodeoxycholate, RPLC, 61
Ursodeoxycholic acid
  HPLC, 55
  selected ion monitoring, 108, 109
  TMS ether, ion current chromatography, 100
Ustilago, sterol mutants, 340
Ustilago violacea
  carotene extracts, reversed-phase chromatography, 193–195
  carotenoids, 192

V

Varanic acid, 95
Violaxanthin
  absorption spectra, shape, 144
  chromatography, octyl-columns, 198
  light absorption maxima, 144
  light absorption spectroscopy, 144
  specific absorbance coefficient, 144
  TLC, on MgO-kieselguhr, solvent, 132
Violerythrin
  chromatography, adsorbents, 125
  light absorption maxima, 144
Vitamin A, 190
Vitamin D, 25
Vitellins, 437
Vitellogenin, 437

X

Xanthophyll, 113, 190
  chromatography, octyl-columns, 198
  light-absorption spectral maxima, 136
  reverse-phase chromatrography, 193
  separation, 197, 198, 200
  solubility properties, 116

Y

Yeast
  auxotrophic feeding experiments, 48, 50
  cell extraction, 41, 42
  cholesterol acyltransferase, 286, 288, 290
  effects of altered sterol composition, 344, 345
  ethylmethane sulfonate treatment, 334, 335
  growth conditions, 41
  natural auxotrophs, 334
  sterol auxotrophs, 333
    culture conditions, 339, 340
    effect of aminolevulinic acid, 341
    effect of ergosterol on, 341
    effects of changing sterol structure on cellular physiology, 343
    heme mutations, 341, 342
    isolation, 339
    specificity of sterol uptake in anaerobic conditions, 342
    strains, 339, 340
    for study of specificity of sterols, for satisfying growth requirements, 342, 343
    utilization, 341–343
  sterol metabolism, 48
    and polyene resistance, 343
  sterol mutants, 333
    characterization, 336–339
    isolation, 334–336
    maintenance, 341
    screening, 48–50
    utilization, 341–345
  sterols, 291, 333–346. *See also* Sterol, yeast
    HPLC, 37–51
    sample preparation, 41, 42
    sterol structures preferable to, 344
    strains, 41

Z

Zeacarotene, TLC, on MgO-kieselguhr, solvent, 132
α-Zeacarotene
  light absorption maxima, 144
  specific absorbance coefficient, 144

β-Zeacarotene
  cis isomer, reverse-phase chromatography, 194, 195
  light absorption maxima, 144
  normal-phase chromatography, 200
  reverse-phase chromatography, 194, 195
  source, 192
  specific absorbance coefficient, 144
  structure, 190, 191
  trans isomer, chromatography, on alumina, 129

Zea mays. See Maize
Zeaxanthin
  light absorption maxima, 136–144
  source, 192
  specific absorbance coefficient, 144
  standard, chromatography, 197, 198
  structure, 190, 191
  TLC, 130, 131
    on MgO-kieselguhr, solvent, 132
Zellweger syndrome, 93
Zymosterol, HPLC properties, 43–45, 46, 47

211840